T0327621

Smart Energy for Transportation and Health in a Smart City

IEEE Press
445 Hoes Lane
Piscataway, NJ 08854

Smart Energy for Transportation and Health in a Smart City

Chun Sing Lai
Guangdong University of Technology
China
Brunel University London
UK

Loi Lei Lai
Guangdong University of Technology
China

Qi Hong Lai
University of Oxford
UK

Published by John Wiley & Sons, Inc., Hoboken, New Jersey.
Published simultaneously in Canada.

Library of Congress Cataloging-in-Publication Data applied for:
Hardback ISBN: 9781119790334

Cover Design: Wiley
Cover Images: © girafchik/Shutterstock

Printed in the United States of America.

Set in 9.5/12.5pt STIXTwoText by Straive, Pondicherry, India

Contents

Authors' Biography

Dr. Chun Sing Lai received the B.Eng. (First Class Hons.) in electrical and electronic engineering from Brunel University London, London, UK, in 2013, and the D.Phil. degree in engineering science from the University of Oxford, Oxford, UK, in 2019.

He is currently an Honorary Visiting Fellow of the School of Automation, Guangdong University of Technology, China, and a Lecturer in Circuits & Devices; he is also the Course Director, MSc Electric Vehicle Systems, with the Department of Electronic and Electrical Engineering, Brunel University London, UK. From 2018 to 2020, he was an UK Engineering and Physical Sciences Research Council Research Fellow with the School of Civil Engineering, University of Leeds, Leeds, UK. His current research interests are in power system optimization and data analytics.

Dr. Lai was the Publications Co-Chair for both 2020 and 2021 IEEE International Smart Cities Conferences. He is the Vice-Chair of the IEEE Smart Cities Publications Committee and Associate Editor for IET Energy Conversion and Economics. He is the Working Group Chair for IEEE P2814 Standard; Associate Vice President, Systems Science and Engineering of the IEEE Systems, Man, and Cybernetics Society (IEEE/SMCS); and Chair of the IEEE SMC Intelligent Power and Energy Systems Technical Committee. He received a Best Paper Award from the IEEE International Smart Cities Conference in October 2020. He was awarded the 2022 Meritorious Service Award by the IEEE/SMCS. Award citation is for meritorious and significant service to IEEE SMC Society technical activities and standards development. Dr. Lai has contributed to four journal articles that appeared in Web of Science as Highly Cited Papers, out of which he is the lead author for three of them. He is an IEEE Senior Member, an IET Member, and a Chartered Engineer.

Professor Loi Lei Lai received the B.Sc. (First Class Hons.), Ph.D., and D.Sc. degrees in electrical and electronic engineering from the University of Aston, Birmingham, UK, and City, University of London, London, UK, in 1980, 1984, and 2005, respectively.

Professor Lai is currently a University Distinguished Professor with Guangdong University of Technology, Guangzhou, China. He was a Pao Yue Kong Chair Professor with Zhejiang University, Hangzhou, China, and the Professor and Chair of Electrical Engineering with City, University of London. His current research areas are in smart cities and smart grid. Professor Lai was awarded an IEEE Third Millennium Medal, the IEEE Power and Energy Society (IEEE/PES) UKRI Power Chapter Outstanding Engineer Award in 2000, the IEEE/PES Energy Development and Power Generation Committee Prize Paper in 2006 and 2009, the IEEE/SMCS Outstanding Contribution Award in 2013 and 2014, and the Most Active Technical Committee Award in 2016.

Professor Lai is an Associate Editor of the IEEE Transactions on Systems, Man, and Cybernetics: Systems, Editor-in-Chief of the IEEE Smart Cities Newsletter, a member of the IEEE Smart Cities Steering Committee, and the Chair of the IEEE Systems, Man, and Cybernetics Society (IEEE/SMCS) Standards Committee. He is Conference General Chair of the 12th International Conference on Power and Energy Systems 2022 (ICPES 2022), IEEE. He was a member of the IEEE Smart Grid Steering Committee; the Director of Research and Development Center, State Grid Energy Research Institute, China; a Vice President for Membership and Student Activities with IEEE/SMCS; and a Fellow Committee Evaluator for the IEEE Industrial Electronics Society and IEEE/PES Lifetime Achievement Award Assessment Committee Member. He is an IEEE Life Fellow and IET Fellow.

Ms. Qi Hong Lai studied at Harrow International School Beijing, China, where she was awarded funding under the IEEE Systems, Man, and Cybernetics Society (IEEE/SMC) Pre-College Activities initiative to set up a program on Brain Computer Interface. She went on to gain her Bachelor of Science in Biomedical Science with First Class Honors from King's College London, UK, in 2019.

At present, she is working toward her Doctor of Philosophy in Molecular Cell Biology in Health and Disease at the Sir William Dunn School of Pathology, University of Oxford, UK.

She is the Working Group Secretary of the IEEE P3166 Standard on Smart Cities Terminology. Her current research interests are in transcription, bioinformatics, biotechnology, and smart health. She is an IEEE Student Member.

Foreword

As the world population continues to rise, the optimal management of major cities will play a key role in orchestrating the global responses to challenges posed by rapid urbanization. The notion of smart city is driven by stakeholders' intention to meet increasing societal demands as large city populations grow in all corners of the world. A prosperous smart city would manage a collection of large and critical infrastructures that support socioeconomic initiatives as it celebrates cultural and ethnic diversities. Smart cities manifest a safer, more secure, more economical, and more sustainable environment that promotes optimal resource allocation and utilization, industrial ecology, and energy conservation. However, a smart city is not all about decarbonization and energy sustainability. It also focuses on public safety, clean water utilization and conservation, public waste management, traffic control and congestion management, telemedicine and public health, and cyber-resilient communication for the automation of personal and social services that can improve the quality of life.

Smart cities rely on widely distributed smart devices to monitor and collect the pertinent data in real-time for intelligent decision-making. To accomplish the task, a distributed network of smart sensor nodes and data centers that stores and shares sensor data will make up the multiple levels of hierarchy in smart city infrastructures. Smart cities are operated in affordable and sustainable manners with more sophisticated control and management systems to ensure that social objectives can be attained in a fair and equitable style. The implementation of new technologies is also accelerated in smart cities as decision-makers and city planners seek to improve their effectiveness to manage limited resources in a more resilient fashion.

This book, which is on smart energy management for optimizing the transportation and healthcare infrastructures in a smart city, brings forth the importance of sustaining a secure, clean, and economical energy network in a smart city. In particular, the availability of a reliable, sustainable, and affordable supply of clean energy is critical for the electrification of smart city infrastructures. The respective authors provide a detailed coverage of these forthcoming topics and their roles in building smart cities.

The book is the product of major contributions of well-known experts and technical investigators with the goal of covering all levels of understanding to optimize the delivery of the concept to various interest levels. It explains in depth the compelling reasons for erecting smart cities and touches on analytical models that are deemed critical for analyzing the essence of establishing smart cities. Various practical examples and pertinent technologies are discussed to highlight the nucleus and promote the curtailed subject areas of energy, transportation, and healthcare in smart cities. The book provides various smart city stakeholders including operating managers, planners, practitioners, and research investigators with valuable insights on many levels of practical and academic landscapes as individuals embark on establishing smart cities for better serving their concerned citizens.

Mohammad Shahidehpour
Elected Member, US National Academy of Engineering
Life Fellow, IEEE
Fellow, American Association for the Advancement of Science
Fellow, National Academy of Inventors
University Distinguished Professor, Illinois Institute of Technology, Chicago, United States

Preface

To make city safer, more secure, and environmentally sustainable, environmental governance, public safety, city planning, industrial promotion, resource utilization, energy conservation, traffic control, telemedicine, interpersonal communication, education, social activities, and entertainment are focused upon. Smart cities have been driven by the desire of citizens to meet increasing demands and allow the choice on the basis of price and service provided. The dramatic changes in the organization of city management bring new challenges and opportunities, by a new competitive and marketable framework. This book was written in response to the growing interest in green smart city technology and its deployment on a global scale. People firmly believe that the technology will produce win-win solutions in terms of environmental, social, and economic impacts.

To achieve net-zero emissions by 2050, preserve biodiversity, and mitigate global warming, people are committing to building a better and more sustainable world. Smart energy will play a key role in a carbon-neutral society. Major environmental, economic, and technological challenges such as climate change, economic restructuring, pressure on public finances, digitalization of the retail and entertainment industries, and growth of urban and ageing populations have generated huge interest for cities to be run differently and smartly.

Smart health will enable medical practitioners to manage patient health using digital means in a secure and private environment whenever and wherever care is required. Road traffic accidents are one of the major causes of injury-related deaths. Safety is the highest priority for transportation. The application of the Internet of Things (IoT) is of special interest to support the aim of efficiently transforming cities to acquire more substantial and sustainable development, as well as a higher quality of life by using data for decision-making to control resources and assets more efficiently.

Smart city is now a hot topic, but its definition and specifics remain unclear. This has led to different interpretations of a smart city. A smart city may be described by six basic pillars: (i) smart economy that improves competitiveness, (ii) smart people relating to social and human capital, (iii) smart governance handling social operation decisions, (iv) smart mobility integrating ICT with transportation to minimize fatality and maximize comfortability, (v) smart environment aiming to achieve net-zero emission through the utilization of natural resources, and (vi) smart living seeking to improve quality of life and life expectancy.

This book focuses on delivering a comprehensive and detailed analysis of smart energy, smart transportation, smart infrastructures, and smart health. The purpose is to first inform readers through a more general but comprehensive coverage of the smart city concept, and then go deep into more specific areas, rather than over-specialization, as to avoid only presenting qualitative data and numerical techniques, and where feasible, provide actual case studies and project discussions.

The book is composed of five parts, namely, Part 1 is the Introduction as presented in Chapter 1; Part 2 is related to smart energy for smart cities and is presented in Chapters 2–11 on power systems, battery, PVs economy and cost, planning, demand response, network microgrids, home energy management, virtual energy storage, reliability modeling of CPS, and vehicle-to-grid; Part 3 is related to smart transportation to related to fast-charging station, electric vehicles, and parking vehicles based on machine learning and wireless communication and is presented in Chapters 12–16; Part 4 is related to smart health and is presented in Chapters 17–19; and Part 5 is the concluding remarks and proposed future directions for smart cities and this is given in Chapter 20. The details for each chapter are given as follows.

The first chapter discusses the definition of a smart city and explains its functions, characteristics, and domains. It will go through some case studies and the established standards of smart cities worldwide.

Chapter 2 introduces a state-of-the-art financial model that has achieved novel and meaningful financial and economic results when applied to lithium-ion (Li-ion) electrical energy storage (EES). Real solar irradiance and load and retail electricity price data from Kenya were used to develop a set of case studies. EES is combined with photovoltaic and anaerobic digestion biogas power plants.

Due to the diurnal and intermittent nature of solar irradiance, photovoltaic (PV) power plants will introduce power generation and load power imbalance issues. Anaerobic digestion (AD) biogas power plants also have a partial load operation constraint that needs to be met. To overcome these limitations, EES is needed to provide power generation flexibility. Chapter 3 reports on the optimal operating mechanism designed for the PV-AD-EES hybrid system, followed by the study of the levelized cost of electricity (LCOE). The degradation cost per kilowatt-hour and the degradation cost per cycle of EES are considered. The study used the 22 years (1994–2015) irradiance data of Kenya's Tkwell Canyon Dam (1.90 °N, 35.34 °E) and Kenya's national load.

With demand-side management (DSM), several electricity prices have emerged, and residential customers are faced with the challenge of choosing a plan that meets their individual needs. The Electricity Plan Recommender System (EPRS) can alleviate this problem. Chapter 4 proposes a new EPRS model integrated with electrical instruction-based recovery (EPRS-EI) to restore electrical appliance usage and set the recovered data as features that represent the customer's life pattern. With these functions, a personal electricity plan is recommended.

Chapter 5 proposes a new classifier network construction method: non-intrusive load monitoring (NILM) and semi-intrusive load monitoring (SILM). This method is not to create a classifier for NILM or SILM but to help decision-makers choose different types of classifiers and optimize the location of the classifiers. In this method, the economy of each classifier is considered to ensure that the cost of decision-makers is reduced. A combinatorial optimization problem is established on the tree-type model for the optimized classifier network. Numerical studies on public data sets and industrial operation data have demonstrated the benefits obtained.

Demand response (DR) is one of the typical methods to optimize the load characteristics of the power system. Chapter 6 introduces the boundary model framework for the construction and transformation of consumer behavior of household appliances. Electricity tariffs are analyzed by this model for their load variation potentials.

Case studies are also included to reflect the implementation potential of the model framework in terms of pricing and smart meter deployment.

Chapter 7 proposes a novel two-stage game-theoretic residential PV panels planning framework for distribution grids with potential PV prosumers. A residential PV panels location-allocation model is integrated with the energy sharing mechanism to increase economic benefits to PV prosumers and meanwhile facilitate the reasonable installation of residential PV panels. Simulations on IEEE 33-node and 123-node test systems prove the effectiveness of the proposed method.

Chapter 8 proposes a two-stage energy management strategy for networked microgrids in the presence of a large number of renewable resources. It decomposes the microgrids energy management into two stages to offset the intra-day stochastic variations of renewable energy resources, electricity load, and electricity prices. According to the simulation results, the proposed method can identify optimal scheduling results, reduce operation costs of risk-aversion, and mitigate the impact of uncertainties.

Chapter 9 proposes a novel framework for home energy management (HEM) based on reinforcement learning in achieving efficient household DR. The Extreme Learning Machine (ELM) processes real data on electricity prices and solar PV power generation promptly in a rolling time window to make uncertain predictions. The simulation was performed at the residential level, which included multiple household appliances, an electric car, and multiple PV panels. The test results prove the effectiveness of the proposed data-driven HEM.

Chapter 10 proposes a two-level consensus-driven distributed control strategy to coordinate virtual energy storage systems (VESSs), i.e. residential households with air conditioners, to avoid the violation of voltage and loading, which are regarded as part of the main power quality issues in future distribution networks. Changes in dynamic communication network topology are studied to prove their impacts on system performance. Simulation results based on an actual system in NSW, Australia, are used to demonstrate the proposed control scheme that can effectively manage voltage and loading and is scalable and robust.

Chapter 11 proposes a reliability modeling and evaluation method for the power information system, i.e. cyberspace in power system. The proposed composite Markov model will couple physical characteristics and information flow performances in a two-layer model. The proposed reliability method combines sequential Monte Carlo simulation with a linear programming model to obtain the maximum flow that can meet the power demand.

Chapter 12 introduces the co-simulation integration of the direct-execution simulator, which provides special support for distributed smart grid software. A case study of agent-based smart grid restoration using this new type of co-simulation platform is conducted. The results show that the proposed direct-execution simulation framework can promote the understanding, evaluation, and debugging of distributed smart grid software. A case study on vehicle-to-grid voltage support application is given.

Chapter 13 reports on the development of Advanced Metering Infrastructure (AMI), which is an effective tool to reshape the electric vehicle (EV) charging load curve by adopting appropriate DSM strategies. An overall solution for an electric vehicle charging service platform (EVAMI) based on power line and Internet communication is proposed. EV owners understand their energy usage, so they can effectively carry out energy-saving activities.

Since plug-in hybrid electric vehicles (PHEVs) are expected to be widely used in the near future, in Chapter 14 a mathematical model is developed based on the traditional security-constrained unit commitment (SCUC) formulation to address the power system dispatching problem with PHEVs taken into account. A real system in China is used to study the impact of PHEV charging on the distribution system. It is proved that charging brings peak load to the grid, and control is essential to reduce the risk of instability.

As more and more electric buses (EBs) are put into use, the reasonable location of charging stations plays an important role in the process of bus electrification. Chapter 15 proposes a location planning model for EB fast-charging stations that considers the bus operation network and the distribution network. The goal of the model is to minimize the sum of the construction cost of charging stations, the operation and maintenance costs, the cost to go to charging stations, and the distribution network losses. The model is applied to simulate and analyze the bus public transportation of a coastal city in South China. The case study shows that the model can effectively optimize the layout of a city's bus charging stations.

Infrastructure and applications based on the IoT are essential for smart cities deployment. The low power wide area network (LPWAN) plays a key role in IoT techniques due to its wide coverage and low power consumption. However, it is hard to decide which one of the LPWAN techniques to be implemented in a specific application to obtain best practice. Therefore in Chapter 16, the main characteristics of the three popular LPWAN technologies, namely, LoRaWAN, NB-IoT, and Sigfox are discussed, and LP-INDEX is proposed to weigh performance factors according to application requirements. To further distinguish the differences, a comparative test based on parking detection sensors using the three different technologies was carried out as a case study.

It is foreseen that the trends for the next decade in healthcare will include more patients requiring care, increased use of technology, the need for greater information storage capacity, development of new healthcare delivery models, error reduction, more emphasis on preventative healthcare, and faster disease diagnosis and innovation-driven by competition. It is important not only to improve patient care processes but also to decrease costs while maintaining quality. Chapter 17 reviews the benefits and challenges of innovations in healthcare, with emphasis on the IoT and smart devices. In Chapter 18, an electrocardiogram (ECG) scheme based on multiple criteria decision-making approach and analytic hierarchy process is proposed to detect drunk status for drivers. Chapter 19 explains the use of bioinformatics and telemedicine for healthcare. Some models, designs, and frameworks for potential applications will be illustrated.

In addition to the smart energy, transportation, and health mentioned in the first19 chapters, there are more elements in a green smart city, such as water and waste; biology, food, and agriculture; education, safety and well-being; government engagement with society and citizens; social entrepreneurship, digital finance, and legal and economic development; sustainable flexible buildings and infrastructure; and open data, privacy, and security for research. In the final Chapter 20, the authors formulate the roadmap and the interrelationships between certain elements. Based on current work and existing information, some suggestions are made, and an overall view of the development and deployment of green smart cities in the next ten years or so has been put forward, and the progress of smart energy, health, transportation, and construction has also been critically evaluated.

This book addresses the latest problems and solutions of smart cities in a coherent manner. It is the product of the contributions of world-class experts, educators, and students, so it covers all levels of understanding to optimize its delivery. Therefore, we believe it will provide decision-makers, engineers, doctors, educators, system operators, managers, planners, practitioners, and researchers with valuable insights on all levels of professional and academic progress.

Chun Sing Lai
Loi Lei Lai
Qi Hong Lai
From Guangzhou, China, and London and Oxford, UK

Acknowledgments

First of all, the authors wish to thank the late Professor Mohamed E. El-Hawary, a great professor and real gentleman, in inviting them to write the so-called the most authoritative and definitive book on Smart Cities. At the time, Professor El-Hawary was Editor of the series of books entitled Advances in Electric Power and Energy, which is part of the Power Engineering Series of Wiley/IEEE Press books. The authors would also like to thank Mary Hatcher, Editor for the Wiley-IEEE Press book program, and Victoria Bradshaw, Senior Editorial Assistant for Electrical Engineering in supporting the project management.

The authors wish also to thank friends, colleagues, and students, without their support this book could not have been completed. In particular, the authors thank Dr. Kim Fung Tsang, Dr. Xiaomei Wu, Professor Alfredo Vaccaro, Dr. Dongxiao Wang, and Zhanlian Li. The permission to reproduce copyright materials by the IEEE and Elsevier for a number of papers mentioned in some of the chapters is most helpful.

Last but not least, we thank Teresa Netzler, Senior Managing Editor, Academic and Professional Learning; Ms. Priyadharshini Arumugam, Permissions Specialist and Jeevaghan Devapal, Content Refinement Specialist, Wiley for supporting the production of this book and for the extremely pleasant co-operation. Special help from Li Rong Li and Qi Ling Lai in selecting the book cover is very much appreciated as well.

1

What Is Smart City?

1.1 Introduction

One of the reasons behind the lack of unified definitions of a smart city is because of the various entities involved and the functions the smart city provides. Hence, existing definitions can vary greatly. There are several definitions for a smart city which are defined by various organizations and stakeholders.

The most common consensus is that the smart city employs various kinds of digital and electronic technologies to transform the living environments with Information and Communications Technologies (ICTs) [1, 2]. Deakin [3] labeled the smart city as a city that employs ICT to meet the market (the citizens') needs. There is a need for larger community involvement to achieve a smart city. A smart city does not simply contain ICT technology but has also developed the technology to achieve positive impacts to the local community. Some definitions for a smart city from major professional organizations and government agencies are given as follows:

Association of Southeast Asian Nations [4]: "A smart city in ASEAN harnesses technological and digital solutions as well as innovative non-technological means to address urban challenges, continuously improving people's lives and creating new opportunities. A smart city is also equivalent to a 'smart sustainable city,' promoting economic and social development alongside environmental protection through effective mechanisms to meet the current and future challenges of its people, while leaving no one behind. As a city's nature remains an important foundation of its economic development and competitive advantage, smart city development should also be designed in accordance with its natural characteristics and potentials."

British Standard Institution [5]: A smart city is an "effective integration of physical, digital, and human systems in the built environment to deliver a sustainable, prosperous, and inclusive future for its citizens."

Department for Business, Innovation, and Skills, UK [6]: "A Smart City should enable every citizen to engage with all the services on offer, public as well as private, in a way best suited to his or her needs. It brings together hard infrastructure, social capital including local skills and community institutions, and (digital) technologies to fuel sustainable economic development and provide an attractive environment for all."

European Commission [7]: "A smart city is a place where traditional networks and services are made more efficient with the use of digital and telecommunication technologies for the benefit of its inhabitants and business. A smart city goes beyond the use of ICT for better resource use and less emissions. It means smarter urban transport networks, upgraded water supply and waste disposal

Smart Energy for Transportation and Health in a Smart City, First Edition. Chun Sing Lai, Loi Lei Lai and Qi Hong Lai.

facilities and more efficient ways to light and heat buildings. It also means a more interactive and responsive city administration, safer public spaces and meeting the needs of an ageing population."

Innovation and Technology Bureau, Hong Kong [8]: "Embrace innovation and technology to build a world-famed Smart Hong Kong characterized by a strong economy and high quality of living."

Institute of Electrical and Electronics Engineers Smart Cities Community [9]: A smart city gathers government, technology, and society to achieve a minimum of the following factors: smart mobility, a smart economy, a smart environment, smart cities, smart governance, smart people, and smart living.

International Electrotechnical Commission [10]: "A smart city is one where the individual city systems are managed in a more integrated and coherent way, through the use of new technologies and specifically through the increasing availability of data and the way that this can provide solid evidence for good decision making."

Japan Smart Community Alliance [11]: The expression "Smart Community" is more widespread than "Smart City" in Japan [7]. "A smart community is a community where various next-generation technologies and advanced social systems are effectively integrated and utilized, including the efficient use of energy, utilization of heat and unused energy sources, improvement of local transportation systems and transformation of the everyday lives of citizens."

Ministry of Housing and Urban Affairs, India [12]: "The conceptualization of Smart City, therefore, varies from city to city and country to country, depending on the level of development, willingness to change and reform, resources and aspirations of the city residents. A smart city would have a different connotation in India than, say, Europe. Even in India, there is no one way of defining a smart city."

According to the above, the similarity and differences in smart city definitions can be summarized as follows:

- Similarities:
  - Enhancement of living standards by making informed decisions with advanced technologies to collect, process, and evaluate data.
  - Systems are integrated to exchange information.
  - Citizens are better informed about their surroundings.
  - Sustainability and environmental conservation should be maximized.
- Differences:
  - Smart city domains or elements, e.g. transport, energy, and health (explained in the following section), can be different due to regional interests.

From the above summary, it is shown that for a city to become smart, multiple sources of data from a range of urban activities and domains must be connected to reveal opportunities to bring innovation to today's connected citizens. Deloitte [13] stated that a smart city is driven by the innovation success of six key domains including:

1) Energy and environment: sustainable growth is created by technology and cities make better use of resources from electronic sensors that monitor leakages, as well as gamification and behavioral economics to support citizens to conduct considerate decisions on resource utilization [14]. Renewable energy including solar and wind will be important sources of energy generation [15–17]. Data analytics will be used to enhance energy and power system operation [18].

2) Economy: the economy will be affected by digitization and disruptive technologies, which will change the needs of several types of jobs. Smart cities need to create strategies to adopt future jobs that will power Industry 4.0 and beyond [19].
3) Safety and security: as criminals will make use of technology to commit advanced crimes, public safety and security authorities will also use technology for crime prevention by assessing multiple streams of social and crowdsourced information, including super-resolution images [20] and image fusion [21].
4) Health and living: the lives of citizens are enhanced with technology and connectivity. Connected communities are achieved with smart buildings. Enhanced social programs and innovated health care sector are data-driven [22].
5) Mobility: the integrated mobility systems include autonomous vehicles and shared mobility services achieved with the Internet of Things (IoT). The concept of IoT occurs when devices are communicating with other devices on behalf of people and will dominate the future of Internet communications [23]. Advanced analytics allow citizens and goods to travel in ways that are safer, cheaper, cleaner, and faster [24].
6) Education and government: technological advancement will aid government procedures and give a seamless experience to businesses. Smart cities use analytics to assist authorities to create insight-driven policies, monitor performance and outcomes, allow constituent engagement, and enhance government efficiency. Data and analytics will also assist next-generation teachers to familiarize their counseling and teaching for greater student achievement. More creative and personalized education plans can be created such as virtual learning environments [25].

Similarly, Giffinger et al. [26] described the smart city as having six domains, including:

1) Smart economy: consists of features surrounding economic competitiveness including entrepreneurship, innovation, flexibility, the productivity of the labor market, trademarks, and participation in the global market.
2) Smart people: concerns not only the level of qualification or education received by citizens but also additional social interactions and perceptions of public life.
3) Smart governance: concerns political involvement, citizen services, and administration functions.
4) Smart mobility: includes local and global accessibility with the presence of ICTs and sustainable and relevant transport systems.
5) Smart environment: concerns attractive natural conditions including green space, less extreme climate, reduced pollution, resource management, and working to achieve environmental protections.
6) Smart living: includes many features of quality of life composed of health, housing, culture, tourism, and safety.

It is worth noting that there are other domains apart from smart energy, smart transportation and smart health (to be discussed further later), including:

Smart water [27, 28]: Smart water systems employ IoT-enabled sensors to collate real-time data. With precise and reliable data, smart water systems can drive great transformations in water sector transparency and accountability. There will be governance improvements, risk reductions, water quality control, and eventually novel business cases for water sector investment [28]. The data allows water facilities optimization by detecting leaks or observing how water is distributed in the water network. The optimization model empowers citizens to make better decisions about water management. Smart sensors can detect water pipe leaks and quickly inform engineers to take action and resolve the issue. Smart water is critical as an estimated 3.3 billion liters of water is wasted daily in Wales and England due to leaks in water networks [27].

Smart waste [29, 30]: Interreg Europe [30] described smart waste as being used "to improve public policy instruments supporting innovation within waste management procedures. The final result? Smarter, more effective, sustainable, and cost-efficient waste management, benefiting all territorial stakeholders." In the United Kingdom, illegal waste activity including fly-tipping costs the UK economy approximately £600 M annually [29]. The present systems for monitoring commercial and household waste are out-of-date and mainly paper based. Smart waste employs technology including blockchain [31], electronic chips, and sensors for monitoring waste, waste containers, and waste vehicles. Smart waste is an element of smart living and smart environments.

In summary, a smart city is an ambitious and crucial transformation of many cities worldwide. Benefits including improved living conditions are reaped from several sectors/domains. However, a smart city consists of the development and application of novel technologies. There is a need for standardized uniform engineering or technical criteria, methods, processes, and practices. The next section examines how international standards help to build a smart city.

1.2 Characteristics, Functions, and Applications

1.2.1 Sensors and Intelligent Electronic Devices

A sensor is a device aiming to detect events or changes in its environment and send the information to other electronic modules. It responds to a stimulus, such as heat, light, or pressure, and generates a signal that can be measured.

A good sensor must have the features sensitive to the measured property, but insensitive to any other property likely to be encountered in its application, and it will not influence the measured property.

Sensors are used in everyday objects such as touch-sensitive lamps which dim or brighten by touching. With the advancement in industry, the use of sensors has expanded beyond temperature, pressure, or flow measurement. Applications include vehicles, manufacturing, machinery, airplanes, medicine, and many other areas of our day-to-day life. There are a wide range of other sensors, measuring chemical and physical properties of materials. Some examples consist of optical sensors for refractive index measurement, vibrational sensors for fluid viscosity measurement and electro-chemical sensor for monitoring pH of fluids.

A sensor's sensitivity indicates how much the sensor's output changes when the input quantity being measured changes. For instance, if the mercury in a thermometer moves 1 cm when the temperature changes by 1 °C, the sensitivity is 1 cm/°C by assuming there is a linear relationship. Some sensors can also be affected what they measure, namely, a room temperature thermometer inserted into a hot cup of liquid cools the liquid while the liquid heats the thermometer. Sensors are usually designed to have a small effect on what is measured; therefore by making the sensor smaller can often improve accuracy and introduce other advantages, for example, convenience in installation.

Technological progress allows more and more sensors to be manufactured such as microsensors which can have a significantly faster measurement time and higher sensitivity compared with macroscopic approaches. Due to the increasing demand for rapid, affordable, and reliable information, low-cost and easy-to-use devices for short-term monitoring or single-shot measurements have gained growing importance. Using this class of disposal sensors, critical analytical information can be obtained by anyone, anywhere, and at any time, without the need for recalibration and worrying about contamination.

Monitoring sensors are used for house monitoring, office, and agriculture monitoring. Traffic monitoring includes car speed, traffic jams, and traffic accidents; weather monitoring is for rain, wind, lightning, and storms; defense monitoring is for monitoring temperature, humidity, air pollution, fire, health, security, and lighting. Some sensors are used to detect carbon monoxide, sulfur dioxide, hydrogen sulfide, ammonia, and other gas substances. Sensors include intelligent sensors and wireless sensor network (WSN) technology.

The intelligent sensor can give a digital signal, communicate the signal, and execute logical instructions. To qualify as an intelligent sensor, the sensor must be part of the same physical unit. A sensor whose only function is to detect and send an unprocessed signal to an external system which performs some action is not considered intelligent.

Common types of intelligent electronic devices (IEDs) include protective relaying devices, circuit breaker controllers, capacitor bank switches, and recloser controllers. A typical IED can contain several protection functions, control functions controlling separate devices, an auto-reclose function, self-monitoring function, and communication functions. Some recent IEDs are designed to support the IEC61850 standard for substation automation, which provides interoperability and advanced communications capabilities.

1.2.2 Information Technology, Communication Networks, and Cyber Security

Information technology is the development, maintenance, or use of systems, especially computer systems, software, and networks for storing, retrieving, and sending information. Computer networking is the process of electronically linking computing devices to exchange information through data connections.

Communication Networks can be of the following five types: local area network (LAN), metropolitan area network (MAN), wide area network (WAN), wireless, and inter network (Internet).

LAN is designed for small physical areas such as an office, group of buildings, or a factory. LANs are used widely as it is easy to design. Personal computers and workstations are connected to each other through LANs. Different types of topologies such as star, ring, bus, and tree could be used. LAN can be a simple network like connecting two computers, to share files, resources such as printers, shared hard-drive, and network among each other while it can also be as complex as interconnecting an entire building.

MAN was developed in the 1980s. It is basically a bigger version of LAN. It is designed to extend over the entire city. It is mainly operated by a single private or public company. MAN generally covers towns and cities of 50 km. The communication medium used for MAN are in general optical fibers and cables.

WAN can be private or it can be public leased network. It is used for the network that covers large distance such as cover states of a country. It is not easy to design and maintain. WAN generally covers large distances between states, countries, or continents. Communication medium used are satellite and public telephone networks which are connected by routers.

Wireless networks can be divided into three main categories, namely, System interconnection, wireless LANs, and wireless WANs.

System interconnection is all about interconnecting the components of a computer using short-range radio. Some companies design a short-range wireless network called Bluetooth to connect various components such as monitor, keyboard, mouse, digital cameras, headsets, scanners, and printer, to the main unit, without wires.

Inter network or Internet is a combination of two or more networks. Inter network can be formed by joining two or more individual networks by means of various devices such as routers, gateways, and bridges.

Cyber security is the practice of defending computers, servers, mobile devices, electronic systems, networks, and data from malicious attacks. It is also known as information technology security.

1.2.3 Systems Integration

System integration is the process of integrating all the physical and virtual components of an organization's system. In engineering, it is the process of bringing together the component subsystems into one system. The physical components consist of the various machine systems and computer hardware. The virtual components consist of data stored in databases, software, and applications. An aggregation of subsystems cooperates so that the system is able to deliver the overall functionality.

In general, there are six steps in systems integration process that is, requirements gathering, analysis, architecture design, systems integration design, implementation, and maintenance. The main reason for businesses to use system integration is the growing need to improve productivity and quality of day-to-day operations. The goal is to get organization and business IT systems to communicate with each other through integration. This accelerates information outflow and reduces operational costs and improves efficiency.

Before integrating systems, usually, the organizations need to rekey the same information into multiple systems. This makes the flow of information within the organization very slow. One of the main benefits of system integration is after automatically integrating systems, the automated exchange of information will make the information consistent between disparate systems. Critical information is more quickly available throughout the organizations. This allows for faster and better decision-making.

1.2.4 Intelligence and Data Analytics

Data analytics help the business users to analyze historical and present data, thereby predicting future trends and change the proposed business model. This helps businesses optimize their performances. Implementing it into the business model means organizations can help reduce costs by identifying more efficient ways of doing business, making decision, and by storing large amounts of data.

Therefore, data analytic techniques enable the users to take raw data and uncover patterns to extract valuable insights from it. Today, many data analytics techniques use specialized systems and software that integrate artificial intelligent algorithms and automation.

When strategizing for something as comprehensive as data analytics, including solutions across different facets is necessary. Depending on the stage of the workflow and the requirement of data analysis, there are four main kinds of analytics, namely, descriptive, diagnostic, predictive, and prescriptive.

Turning to big data analytics, this helps businesses to get insights from huge data resources. People, organizations, and devices now produce massive amounts of data. Social media, cloud applications, and machine sensor data are some examples.

1.2.5 Management and Control Platforms

A management control system (MCS) gathers and uses information to evaluate the performance of different organizational resources such as human, physical, financial, and also the organization as a whole in respect to the organizational strategies followed.

One MCS characteristic is the organizing and planning of the relationship between these different structures and centers of responsibility. The other characteristic is about the processes or set of activities the organization takes in order to achieve its aims.

1.3 Smart Energy

Smart energy is the process of adopting intelligent devices such as smart sensors for increasing energy efficiency. It focuses on large-scale sustainable renewable energy sources that promote greater eco-friendliness while reducing costs and increasing reliability. To accommodate for ever-increasing data, the application of smart devices to human lifestyles and services, secure computer systems that meet the needs of smart cities are essential. This includes new architecture, concepts, algorithms in machine learning, and artificial intelligence. Smart energy networks require fast and intelligent decisions, which will only be possible with the help of intelligent and complex computer systems. Urban energy networks are becoming increasingly linked and integrated. This is crucial for cities aiming to achieve energy efficiency and environmental sustainability.

There are ongoing advances in renewables and energy storage systems, along with innovative information, communication and control technologies, computational intelligence, as well as power electronics. Thus, there are opportunities and challenges emerging in the design, planning, and operation of more distributed energy system architectures with a significant amount of local energy consumptions [32]. The enabling technologies and methodologies aimed at addressing complex challenges include decentralized computing, self-organizing sensor networks, proactive control, and holistic computing frameworks [18].

The development of smart grids has given various challenges facing on power system operation and planning due to increased penetration of many new technologies of diversified properties. On one hand, system operators and many other participants have to deal with increased uncertainties and risks involved in daily operation and planning activities. On the other hand, applications of many new metering and measurement devices, capable of closely monitoring and sensing grid operation in real time, result in overwhelming amount of measurement data of high precision and resolution. By far, how to make the best use of the massive data remains quite a challenging task facing power system researchers and practitioners. The availability of the high-quality data could potentially facilitate risk hedging and decision-making in system operation and planning, of which the prerequisite calls for innovative informatics approaches that are intelligent, data-driven, and capable of handling various complex problems. Zhao et al. [33] reported some recent work done in some areas related to smart energy. For instance, the application of advanced data mining techniques has enabled better prognosis of renewable generations that are highly uncertain and intermittent. Specially, a revolutionary change of forecasting paradigm moving from the point-based forecasting approaches toward the probabilistic interval-based ones by utilizing advanced data analytic techniques such as extreme learning machine to achieve better prognosis and representation of the uncertainties and intermittencies involved in renewable energies. The wide area measurement systems data measured by widely deployed phasor measurement units has been exploited to develop advanced online power system security surveillance and visualization tools by many power utilities. There are also industrial applications of intelligent and data-oriented approaches dedicated for resolving complex control and operation problems in transmission or distribution grids with high renewable energy penetration. In terms of electricity market operation, applications of advanced data analytic techniques have been popular to construct optimal bidding strategies for generators or electric vehicle aggregators, as well as short load forecasting and electricity market price forecasting. There are also many applications of risk-based techniques to counterbalance the uncertainties facing construction/expansion planning, as well as operation planning in context of deregulated electricity markets.

Since solar resource is intermittent and noncontrollable which depends on the location, time of day, day of year, and weather pattern, the difficulty of short-term solar irradiance forecasting lies in challenges to predict variations of weather patterns and quantify their impacts. Approaches including numerical weather prediction models, image-based models, statistical models, machine learning models, and hybrid models for solar irradiance forecasting have been widely discussed [15].

Huang et al. [34] reported the study of short-term forecasting of solar irradiance at a targeted site with consideration of its time series and measurements at neighboring sites. A data-driven framework for forecasting solar irradiance based on fusing spatial and temporal information was proposed. In the framework, data-driven approaches including boosted regression trees (BRTs), artificial neural network (NN), support vector machine, and least absolute shrinkage and selection operator are applied to model spatial dependence among solar irradiance time series thereby to generate forecasts of solar irradiance at the targeted site. Bench marking models including scaled persistence model, auto regressive model, and auto regressive exogenous model are employed to further validate the effectiveness of data-driven forecasting models. Computational results of multiple steps ahead forecasting demonstrate that the BRT model gives the best performance with the lowest normalized root mean squared error for forecasting horizons up to two hours.

Regarding day-ahead power market (DAM), load serving entities (LSEs) are required to submit their future load schedule to market operator. As a more accurate load forecasting model may not lead to a lower cost for LSEs, accurate pursuing load forecast model may not target a solution with optimal benefit. Facing this issue, Xu et al. [35] initiated a beneficial correlated regularization (BCR) for neural network (NN) load prediction. The training target of NN contains both accuracy and power cost consideration. The result shows that NN with BCR can reduce power cost with acceptable accuracy level.

There are two major ways to make energy more sustainable, namely, using renewable energy such as solar, wind, geothermal, and improving energy utilization efficiency. The generation of renewable energy is still not very stable and scalable in comparison to traditional energy generation methods, e.g. coal and nuclear power plants. With the fact that 40% of global energy is consumed by residential and commercial buildings, effective management of the energy usage is essential to improve energy efficiency. Load monitoring has great potential in many useful applications, for examples, energy awareness and energy conservation, controllable load quantitative evaluation, human behavior, and load prediction. Load monitoring helps to understand the energy consumption of specific appliances in a house and make a more energy efficient plan. If the electricity customers are aware of the average consumption of a type of appliance, more personalized and specific energy saving models of appliances can be recommended to those who are using inefficient devices.

Current methods for nonintrusive load monitoring (NILM) problems assume that the number of appliances in the target location is known which may be unrealistic. In the real-world, the initial settings of the site can be known, but new appliances can be added by the user after a period of time, especially in a household. Zhang et al. [36] proposed an appliance detection and a training algorithm for multi-label classification in NILM. Then the Stochastic Sensitivity Measure-based Noise

Filtering and Oversampling (SSMNFOS) is applied to train base classifier for an appliance to form the multi-class ensemble of NN for the multi-label classification. Experimental results show that the SSMNFOS yields a better performance than the widely used SMOTE (Synthetic Minority Oversampling Technique); the SSMNFOS is more robust to noisy samples. This is important to the NILM problem because the load measured by the smart meter may have noise interference from wires and the environment.

Wu et al. [37] presented a probabilistic prediction interval (PI) model based on variational mode decomposition (VMD) and a kernel extreme learning machine using the firefly algorithm (FA-KELM) to tackle the problem of photovoltaic (PV) power for intra-day-ahead prediction. First, considering the nonstationary and nonlinear characteristics of a PV power output sequence, the decomposition of the original PV power output series is carried out using VMD. Second, to further improve the prediction accuracy, KELM is established for each decomposed component and the firefly algorithm is introduced to optimize the penalty factor and kernel parameter. Finally, the point predicted value is obtained through the summation of predicted results of each component and then using the nonlinear kernel density estimation to fit it. The cubic spline interpolation algorithm is applied to obtain the shortest confidence interval. Results from practical cases show that this probabilistic prediction interval could achieve higher accuracy as compared with other prediction models.

As load forecasting is a complex nonlinear problem with high volatility and uncertainty, Lai et al. [38] reported a novel load forecasting method known as deep neural network and historical data augmentation (DNN–HDA). The method utilizes HDA to enhance regression by DNN for monthly load forecasting, considering that the historical data to have a high correlation with the corresponding predicted data. To make the best use of the historical data, one year's historical data is combined with the basic features to construct the input vector for a predicted load. In this way, if there is C years' historical data, one predicted load can have C input vectors to create the same number of samples. DNN–HDA increases the number of training samples and enhances the generalization of the model to reduce the forecasting error. The method is tested on daily peak loads from 2006 to 2015 of Austria, Czech, and Italy. Comparisons are made between DNN-HDA and several state-of-the-art models. DNN–HDA outperforms DNN by 60%.

Accurate load forecasting is essential to the operation and planning of power systems and electricity markets. Lai et al. [39] proposed an ensemble of radial basis function neural networks (RBFNNs) for short-term and mid-term load forecasting. Exogenous features and features extracted from load series with long short-term memory networks and multi-resolution wavelet transform in various timescales are used to train the ensemble of RBFNNs. Multiple RBFNNs are fused as an ensemble model with high generalization capability using a proposed weighted fusion method based on the localized generalization error model. Experimental results on three practical datasets show that compared with other forecasting methods, the proposed method reduces the mean absolute percentage error (MAPE), mean squared error (MSE), and mean absolute error (MAE) by up to 0.30%.

Wang et al. [40] also reported a short-term electric load forecasting model based on deep autoencoder with localized stochastic sensitivity (D-LiSSA). D-LiSSA can learn informative hidden representations from unseen samples by minimizing the perturbed error (including the training error and stochastic sensitivity) from historical load data. Specifically, this general deep autoencoder network as a deep learning model improves prediction accuracy and reliability. Moreover, a nonlinear fully connected feedforward NN as a regression layer is applied to forecast the short-term load, with the generalization capability of the proposed model using hidden representations learned by D-LiSSA. The performance of D-LiSSA is evaluated using real-world public electric load markets of France (FR), Germany (GR), Romania (RO), and Spain (ES) from ENTSO-E. Extensive experimental results and comparisons with the classical and state-of-the-art models show that D-LiSSA yields accurate load forecasting results and achieves desired reliable capability. For instance, with the French case, D-LiSSA yields the lowest mean absolute error, mean absolute percentage error, and root mean squared error; providing up to 63% forecasting accuracy improvements as compared to the benchmark model for forecasting hourly horizon.

Solar radiation forecasting is a key technology to improve the control and scheduling performance of photovoltaic power plants. Lai et al. [41] proposed a deep learning-based hybrid method for one-hour ahead Global Horizontal Irradiance (GHI) forecasting. Specifically, a deep learning-based clustering method, deep time-series clustering, is adopted to group the GHI time series data into multiple clusters to better identify its irregular patterns and thus providing a better clustering performance. Then, the feature attention deep forecasting (FADF) DNN is built for each cluster to generate the GHI forecasts. The developed FADF dynamically allocates different importance to different features and utilizes the weighted features to forecast the next hour GHI. The solar forecasting performance of the proposed method is evaluated with the National Solar Radiation Database. Simulation results show that the proposed method yields the most accurate solar forecasting among the smart persistence and state-of-the-art models. The proposed method reduces the root mean square error as compared to the smart persistence by up to 13%.

Large-scale deployment of wind energy is constrained by the inherent intermittence and volatility. Therefore, an accurate prediction of wind power is needed. Yin et al. [42] reported a hybrid forecasting model for the short-term wind power prediction by using a secondary hybrid decomposition approach to decompose the original time series into several intrinsic mode functions. The proposed approach has great advantage over other previous hybrid models in terms of prediction accuracy.

With the increasing penetration of wind power in renewable energy systems, it is important to improve the accuracy of wind speed prediction. However, wind power generation has great uncertainties which make high-quality interval prediction a challenge. Existing multi-objective optimization interval prediction methods do not consider the robustness of the model. Thus, trained models for wind speed interval prediction may not be optimal for future predictions. Chen et al. [43] proposed the prediction interval coverage probability, the prediction interval average width, and the robustness of the model used as three objective functions for determining the optimal model of short-term wind speed interval prediction using the multi-objective optimization. Furthermore, a new Stochastic Sensitivity for Prediction Intervals (SS_PIs) was proposed to measure the stability and robustness of the model for interval prediction. Using wind farm data from countries on two different continents as case studies, experimental results show that the proposed method yields better prediction intervals than other methods under different confidence levels. In terms of the SS_PIs.

The satisfactory operating condition and performance of transformers are essential for a reliable power grid operation. The condition diagnosis and monitoring of oil/paper systems in the transformer are thus of great interest to engineers and researchers. The dissolved gas analysis (DGA) is generally considered to be the most convenient and common approach for reconditioning the energized transformers. Since the diagnostic accuracy of the traditional DGA can no longer meet the requirements of modern power grids, the artificial intelligence (AI) diagnostic algorithm based on DGA has become a prevailing method for the fault diagnosis of the oil/paper system in power transformers [44].

Moisture is one of the critical factors to determine the service life of transformers. The moisture inside the transformer oil-immersed insulation could be quantified with feature parameters.

Liu et al. [45] proposed and developed a genetic algorithm support vector machine (GA-SVM) model to carry out the moisture diagnosis. The finding revealed that these feature parameters can be obtained by using frequency-domain spectroscopy. Therefore, a novel model for predicting the frequency-domain spectroscopy curves is first reported based on a small number of samples, which could be utilized to obtain the feature parameters data base to develop GA-SVM. The model is under exploration as an intelligent based moisture diagnosis tool for power transformers.

Some of the benefits derived from smart energy technology such as a more diverse power supply, a reduction in air pollution and emissions that cause global warming and a growing independence from traditional energy sources that harm the environment. In brief, the move toward smart energy is one that will create a brighter, eco-friendlier future. With global warming and other climate issues currently a major policy concern for governments globally, the switch to smart energy is becoming more important than ever before.

1.4 Smart Transportation

Smart transportation system is the application of sensing, analysis, control, and communications technologies to transportation to improve safety, mobility, and efficiency. Smart transportation system includes a wide range of applications that process and share information to ease congestion, improve traffic management, and minimize environmental impact.

Smart transportation includes the use of several technologies, from basic management systems such as car navigation, traffic signal control systems, container management systems, automatic number plate recognition, or speed cameras to monitor applications, such as security CCTV systems, and to more advanced applications. The use of IoT-enabled technology is changing the way that the transportation sector operates. A smart transportation network is clean and efficient. As suggested, reduced traffic congestion results in cleaner air, less wasted time, and reduced energy consumption. And cities that are working to accommodate electric and, eventually, autonomous connected vehicles can expect to realize even greater environmental benefits. Smart transportation system gives regular information to the daily commuters about public buses, timings, seat availability, the current location of the bus, time taken to reach a particular destination, next location of the bus, and the density of passengers inside the bus. Therefore, advantages of smart transportation system include travel time improvement, capacity management, and incident management. There is a range of information and communications technologies that enables the development of smart transportation. For instance, fiber optics, GPS, laser sensors, digital map databases, and display technologies. Enabling technologies can be divided into the following several classes [46].

Data Acquisition:

To monitor traffics using several means such as traffic sensor. Examples of traffic sensors are ultrasonic and radar, video image detector (VID), and visual images from closed circuit television (CCTV) which provide live images to help the traffic center operator to monitor complicated traffic situations and make decisions.

1.4.1 Data Processing

Information collected at data management center required to be processed, verified, and consolidated into a format that is useful for the operators. This can be done using data fusion process. Automatic incident detection (AID) may also be used for data processing. Global positioning system (GPS) can be used on vehicles to process data.

Information Usage:

This involves controlling the flow of vehicles merging onto an expressway, and coordination of traffic control within large urban areas taking place at the traffic management center. In addition to dynamic route guidance which permits the user to make strategic decisions on minute interval, and adaptive cruise control which allows the driver to change vehicle speed to maximize safety.

1.5 Smart Health

The European Commission [47] described smart health permits healthcare providers to reduce illnesses occurrence, to care for patients more efficiently, and to cure illnesses more effectively. Smart health also reduces healthcare expenditure in the growing aging population. Smart health solutions consist of technological developments in portable and mobile devices, sensor technology, application development, mobile data connectivity, cloud computing, and big data analytics, with new ideas on patient comanagement, health tracking of remote neighborhoods, and minimizing unhealthy lifestyles. Deloitte [48] stated that smart health consists of five features, including to: (i) empower proactive health and well-being management to make choices that can proactively improve health, well-being, and quality of life to reduce adverse health outcomes in the future; (ii) foster a sense of community and well-being with virtual and in-person community meetings; (iii) enable digital technology and behavioral science with mobile applications for users to enter and track data and seek information, e.g. fitness tips and recipes, and deploy the use of coaching and guide to support adherence and uptake of behaviors associated with healthy, active living; (iv) meaningfully use data to improve outcomes and allow users to track their progress. Consent would be requested from users to share and use data, to enhance the program and for it to make improved recommendations; (v) enable new and innovative ecosystems to consist of the collaboration of businesses with all kinds of organizations, e.g. government agencies and academia to align on health outcome measures and coordinate on investments in communities.

1.6 Impact of COVID-19 Pandemic

Mobility behavior was impacted severely by the COVID-19 crisis. Crisis managers need access to credible and timely data; need to look at the potential of traffic management data for crisis management, list the different categories and types of traffic data sources and provide an overview of how policymakers, research institutions, and private companies can repurpose their data to monitor the effect of the crisis and the accompanying lockdown measures on mobility behavior [49].

Over the past couple of years, information gathered from connected travelers through smartphones and GNSS (Global Navigation Satellite System) trackers has become increasingly important for commercial purposes, research studies, and policy making. Aggregated mobility tracking data is an excellent source of information on people's transportation habits. While this data is already beneficial for general traffic management, these aggregated traffic flows are also invaluable to examine the impact of crisis-specific policies on population mobility patterns. Several commercial entities active in this domain have already published resources on how their datasets could be used to assess the impact of crises, such as the ongoing pandemic on urban mobility.

Connected vehicle technology enables cars, buses, trucks, and other infrastructure to sense their environment and communicate with each other. Data from connected vehicles can be useful in aggregated form. A large fraction of commercial cars and car-sharing companies have built-in location and activity trackers for fleet management. This data is useful to monitor real-time link speeds and to analyze changes in traffic intensity in certain areas during a crisis.

Data from Wi-Fi and Bluetooth scanners, which have been used previously for long-term monitoring of visitors to shops and touristic hot spots, can also be repurposed to monitor real-time business in shopping districts. Automated systems can use this information to send alerts to city officials, so that they can temporarily restrict access to certain areas to avoid overcrowding in times where social distancing is required.

Time-series information from ticketing machines in public parking garages and on-street parking zones can be a new source of information in crisis management. Unlike data from static vehicle counters, which merely convey information on traffic intensity, parking garage and on-street parking occupancy data can also inform crisis managers about trip destinations and trip purposes. The ticketing machines and parking garages typically have geographical coordinates attached to them, and these can be used as the actual destination of a car trip within the city.

The increasing popularity of shared cars, bicycles, and steps in European urban areas provides an additional opportunity for monitoring changes in mobility behavior in case of crises. These vehicles typically accommodate some forms of real-time location trackers. These location trackers, along with information on ticketing and reservations, are typically only used by the sharing companies for billing and fleet management. However, crisis managers can quickly repurpose it toward monitoring mobility, as the start and end locations, along with vehicle reservation times, can be used to build time-dependent transport modes. These can offer insight into when and where people are traveling.

As global communities respond to COVID-19, public health officials said that aggregated, anonymized products such as Google Maps could be helpful to make critical decisions.

The Community Mobility Report [50] aims to provide insights into what has changed in response to policies at combating COVID-19. The reports chart movement trends over time by geography, across different categories of places such as retail and recreation, groceries and pharmacies, parks, transit stations, workplaces, and residential. This can show how the community is moving around differently under COVID-19 [51].

In the reports, no personally identifiable information, like an individual's location, contacts, or movement, is made available at any point. Insights in these reports are created with aggregated, anonymized sets of data from users who have turned on the location history setting. These reports are powered by the world-class anonymization technologies that keep your activity data private and secure. These reports use differential privacy, which adds artificial noise to the datasets enabling high-quality results without identifying any individual person.

To monitor social distancing interventions, rather than showing individual travel or behavior patterns, information from multiple devices is aggregated in space and time, so that the data reflect an approximation of population-level mobility [52]. The estimates of aggregate flows of people are incredibly valuable. A map that examines the impact of social distancing messaging or policies on population mobility patterns will help county officials understand what kinds of messaging or policies are most effective. Comparing the public response to interventions, in terms of the rate of movement over an entire county from one day to the next.

1.7 Standards

1.7.1 International Standards for Smart City

The International Organization for Standardization (ISO) has described standards as "the first step towards the holy grail of an interoperable, plug-and-play world where cities can mix and match solutions from different vendors without fear of lock-in or obsolescence or dead-end initiatives" [53]. International standards are best practices created by global experts. Standards can be used to benchmark functional and technical performances. Standards make sure that technologies deployed in cities are efficient, safe, and well-integrated.

The largest and most well-established international standards organizations include ISO [54], the International Electrotechnical Commission (IEC) [55], and the International Telecommunication Union (ITU) [56] which were founded between 50 and 150 years ago. The description of these organizations are as follows:

- ISO is a nongovernmental and independent global organization with 164 national standards bodies as members. The standards body for each country (e.g. Bureau de Normalization (NBN) in Belgium and Ghana Standards Authority (GSA) in Ghana) works directly with ISO and aims to minimize diversity in technical definitions. ISO standards are applied in various fields including quality management, environmental management, IT security, energy management, health and safety, and food safety [54].
- IEC is the world's forefront organization for the groundwork and publication of international standards for electronic, electrical, and relevant technologies, i.e. "electrotechnology" [55]. IEC described technical and international standards as reflecting "agreements on the technical description of the characteristics to be fulfilled by the product, system, service or object in question. They are widely adopted at the regional or national level and are applied by manufacturers, trade organizations, purchasers, consumers, testing laboratories, governments, regulators and other interested parties." Standards help researchers, industry, regulators, and consumers globally to achieve an optimal experience and meet mutual needs for various countries. Standards establish one of the vital bases for the elimination of technical obstacles to trade.
- ITU is the United Nations bespoke agency for ICTs and enables global connectivity of communications networks [56]. ITU manages international satellite orbits and radio spectrum, creates the international standards that allow technologies and networks to be continuously interconnected, and aims to enhance ICT access for global communities.

The above organizations have developed standards to specify and establish definitions and methodologies for a set of smart cities indicators. For example, ISO 37122:2019 (Sustainable Cities and Communities – Indicators for Smart Cities) [54] intends to give a holistic set of indicators to evaluate advancement in developing a smart city. The standard includes multiple domains including education, energy, economy, environment and climate change, finance, governance, health, housing, population and social conditions, recreation, safety, solid waste, sport and culture, telecommunication, transportation, urban/local agriculture and food safety, urban planning, wastewater, and water. The World Council on City Data is a prominent initiative in using standardized city data to create smart cities [57]. The initiative hosts a network of innovative cities dedicated to refining quality of life and services with open city data and delivers a reliable and holistic platform for standardized urban metrics. The World Council on City Data is an international hub for international organizations, education partnerships across cities, corporate partners, and academia to expand innovation, envisage alternative futures, and construct enhanced cities. The initiative developed the first city data standards, namely, ISO 37120 (Sustainable development of communities: Indicators for city services and quality of life).

The IEC has identified over 1800 standards that already impact smart cities [58]. The SyC Smart cities promote the coordination of standards efforts of several IEC committees and other organizations, including ISO, to promote the development of standards to achieve integration, interoperability, and effectiveness of city systems. SyC Smart City is presently developing IEC 63152 as the best practice tool for city planners. Considering the higher frequency of natural disasters and destruction in some urban areas, IEC 63152 proposes guidelines to sustain several city services after a disruption occurs. IEC 63152 provides the fundamental concepts of how several city services can cooperate to uphold the supply of electricity.

ITU established Study Group 20 and United for Smart Sustainable Cities to develop standard activities in supporting the utilization of ICTs in a smart city [59]. These standards focus on terminologies for the IoT and smart cities, high-performing ICT infrastructures requirements, and the interoperability between various ICT or IoT networks. The ICT standard consists of four layers, namely the "application and support layer," "data layer," "communication layer," and "sensing layer."

In addition to the above three organizations, the Institute of Electrical and Electronics Engineers (IEEE) also develops international standards for smart cities. One of the most well-known IEEE standards is the IEEE 802 family, which was established in the early 1980s and covers LANs and MANs [60]. In recent years, IEEE has established the IEEE Smart Cities Community, which "brings together IEEE's broad array of technical societies and organizations to advance the state of the art for smart city technologies for the benefit of society and to set the global standard in this regard by serving as a neutral broker of information amongst industry, academic, and government stakeholders" [61].

In 2017, the IEEE P2784 (Smart City Planning Guide) [62] was proposed to develop a framework that mentions the processes and technologies for planning the smart city transformation. A smart city requires a unified process planning framework to use IoT to guarantee agile, interoperable, and scalable solutions that can be used and supported sustainably. The framework is a method for technology and cities integrators to plan for technology and innovative solutions for smart cities. Some of the most recent and first-of-a-kind standard initiatives from IEEE are presented in Tables 1.1–1.5.

In summary, this section has presented the need for international standards in smart city research and development. Some of the emerging standard projects are presented for various smart city domains. The next section examines the different smart cities projects worldwide and focuses on the standards examined and adopted.

Table 1.1 Recent Institute of Electrical and Electronics Engineers (IEEE) Standards in development for a smart grid and smart energy.

Year	Title	Scope
2016	1889–2018 (Evaluating and Testing the Electrical Performance of Energy Saving Devices) [63]	• Tests and evaluates energy saving devices' electrical performance • Measurement methods for observing the power generated or used by the observed load or generator: (i) without the energy saving devices connected to the circuit, (ii) with the energy saving devices connected and powered in the circuit

(Continued)

Table 1.1 (Continued)

Year	Title	Scope
		• Detailed protocols for testing circuits, the accuracy and details of evaluation instrumentation, and the sequence of the test measurements • Bespoke details for possible sources of measurement errors including from (i) wrong instrumentation connection, (ii) inadequate instrumentation, or (iii) wrong results interpretation • Can be used for all kinds of electrically connected energy saving devices that control electrical power given by a source and powering an electrical load
2016	P1922.1 (A Method for Calculating Anticipated Emissions caused by Virtual Machine Migration and Placement) [64]	• Methods to compute anticipated emissions created by virtual machine migration and allocation in distributed locations created by various electricity sources • Identify the anticipated electric grid's marginal emissions due to the change in power generation capacity, reflected by the additional power demand from server accepting the virtual machine and the network supporting virtual machine migration • Creates a technique to study anticipated gaseous (also greenhouse gases) and particle emissions created by virtual machine migration and allocation in distributed servers situated in various regions
2019	P2814 (Techno-economic Metrics Standard for Hybrid Energy and Storage Systems) [65]	• Techno-economic metrics for operation, construction, and development of electrical energy storage systems and renewable energy systems
2020	P2852 (Intelligent Assessment of Safety Risk for Overhead Transmission Lines Under Multiple Operating Conditions) [66]	• Artificial intelligence methods for a 3-D model of overhead transmission line and locational surroundings, to achieve a precise perception of overhead conductors to the ground and near buildings and trees • Identifies safety risk information including distance for overhead transmission lines in various operating conditions and provides an intelligent security assessment technique • Useful for intelligent evaluation and control overhead transmission lines safety risk during a typhoon, hot weather, icing, and alternative operating conditions

Source: From Lai et al. [A]/MDPI/CC BY 4.0.

Table 1.2 Recent IEEE Standards in development for smart health.

Year	Title	Scope
2014	1708–2014 (Wearable Cuffless Blood Pressure Measuring Devices) [67]	• Creates objective performance and normative definition for assessment of wearable cuffless blood pressure measuring devices • Works for all forms of the device or the vehicle where the device is attached or where it is embedded • Works for all kinds of wearable blood pressure measurement devices, such as epidermal and unconstrained blood pressure devices that have various operation modes • Limited to the assessment of devices that do not use a cuff while measuring • No assessment of all sphygmomanometers operated with an inflatable or occluding cuff for the nonintrusive assessment of blood pressure on the upper wrist or arm • Manufacturers guidelines to certify and confirm their products, possible purchasers or users to examine and choose potential products, and health care professionals to perceive the manufacturing practices on wearable blood pressure devices
2017	P3333.2.5 (Bio-CAD File Format for Medical Three-Dimensional (3D) Printing) [68]	• Establishes the Bio-Computer Aided Design format for 3-D printing from sectional scan image data comprising of volumetric and surface information • Related to medical 3-D printing services including pathologic services, anatomic models, and medical instrument printing with 2-D images, 3-D medical data, and alternative medical data
2017	P1752 (Mobile Health Data) [69]	• States requirements for mobile health data applications programming interface and standardized representations for mobile metadata and health data • Mobile health data consists of personal health data collated from mobile applications and sensors
2019	1847–2019 (Common Framework of Location Services for Healthcare) [70]	• The framework consists of location services for healthcare conceptual information model and location services for healthcare common terminology
2020	P2621.1 (Wireless Diabetes Device Security Assurance: Product Security Evaluation Program) [71]	• A connected electronic product security evaluation program framework consists of: 1) A method to use the ISO/IEC 15408 security evaluation framework in a security evaluation program 2) Allowing independent testing labs for security evaluation program

(Continued)

Table 1.2 (Continued)

Year	Title	Scope
		3) Confirming results from authorized labs 4) Defining and certifying novel security requirements and adjustments to security requirements, from protection profiles and security targets for security evaluation program 5) Assurance post-certification maintenance

Source: From Lai et al. [A]/MDPI/CC BY 4.0.

Table 1.3 Recent IEEE Standards in development for smart mobility and transportations.

Year	Title	Scope
2013	P1884 (Stray Current/Corrosion Mitigation for DC Rail Transit Systems) [72]	• Principles, methods, and data for engineering design, commissioning, installation, observing and evaluating; including mitigation and control techniques for stray currents in direct current rail transit systems
2013	P1883 (Electrical and Electro-Mechanical Bench Test Equipment (BTE) for Transit Rail Projects) [73]	• Design factors, documentation, construction materials, and the satisfactory requirement for bench test equipment for novel and current electrical and electromechanical equipment use in transit rail systems
2014	P2406 (Design and Construction of Non-Load Break Disconnect Switches for Direct Current Applications on Transit Systems) [74]	• Design, usage, and application of direct current nonload or no-load break disconnect switches for isolating direct current power distribution circuits in transit applications
2016	P2020 (Automotive System Image Quality) [75]	• Deals with key elements of image and quality for applications in automotive advanced driver assistance systems applications, includes recognizing current metrics and alternative meaningful information associated with these elements • Formulates objective and subjective evaluation methods for measuring automotive camera image quality • Presents tools and evaluation techniques to provide standards-based communication and contrast amid original equipment manufacturer and Tier 1 system integrators, and component vendors concerning automotive advanced driver assistance systems image quality
2017	P2685 (Energy Storage for Stationary Engine-Starting Systems) [76]	• Selection, installation design, sizing, installation, maintenance, and evaluating techniques for optimizing the performance and life of energy storage devices and associated systems for starting stationary engines

Table 1.3 (Continued)

Year	Title	Scope
		• Identify when energy storage devices need replacing. Energy storage devices and related systems including (i) nickel-cadmium and lead-acid batteries, (ii) supercapacitors and electric double-layer capacitors, (iii) air-start systems, (iv) start/control battery chargers, (v) parallel battery blocking diode systems, and (vi) monitoring systems

Source: From Lai et al. [A]/MDPI/CC BY 4.0.

Table 1.4 Recent IEEE Standards in development for smart education.

Year	Title	Scope
2018	P7919.1 (eReaders to Support Learning Applications) [77]	• Arranges and explains the eReaders abilities for working as a platform for education, training, learning, and using other approaches for developing these abilities • Methodology comprises of industry standards applications and may cover open-source reference code
2019	P2834 (Secure and Trusted Learning Systems) [78]	• Details technical specifications for privacy protection and student data management in learning online services and systems
2019	1876 (Networked Smart Learning Objects for Online Laboratories) [79]	• Techniques for saving, retrieving, and using online laboratories as interactive and smart learning objects • Defines techniques for combining online laboratories as smart learning objects in learning object repositories and learning environments
2020	1589–2020 (Augmented Reality Learning Experience Model) [80]	• Develop an overarching integrated theoretical model to identify interactivities across the real world, digital information, and the user and the conditions for augmented reality-assisted learning of the environment • Defines two data models and interface to Extensible Markup Language and JavaScript Object Notation for depicting learning activities and the learning environment as the tasks are executed

Source: From Lai et al. [A]/MDPI/CC BY 4.0.

1.7.2 Smart City Pilot Projects

Having examined the importance of international standards and the emerging ones, the following section presents some of the smart cities pilot projects from various countries. The focus is on the application of international standards. Following the alphabetical order of the continents, the smart cities pilot projects in Africa and Asia are presented in Tables 1.6 and 1.7, respectively.

In Australia, the Australian Government established the Smart Cities Plan in 2016 [97]. The Plan highlights the Government's vision for productive and habitable cities that boost innovation,

Table 1.5 Recent IEEE Standards in development for smart governance.

Year	Title	Scope
2017	P7005 (Transparent Employer Data Governance) [81]	• Methods to assist employers to validate to access, collect, share, utilize, store, and destroy employee data • Specific metrics and conformance needs on (i) when handling data from trusted global partners and (ii) how vendors and employers can react to handling data
2017	P7004 (Child and Student Data Governance) [82]	• Methodologies for stakeholders in certifiable and responsible student and child data governance
2020	P2145 (Framework and Definitions for Blockchain Governance) [83]	• A generic framework and nomenclature for explaining blockchain governance for all kinds of contexts and use cases, comprising of private, public, permissionless, hybrid, and with permission • The standard is normative concerning terminology and non-normative considering the particular blockchain systems and protocols design
2020	P2863 (Organizational Governance of Artificial Intelligence) [84]	• States the governance basis including safety, responsibility, accountability, transparency, and reducing bias, and procedures for performance auditing, useful implementation, training, and compliance in the formulation or deployment of artificial intelligence in organizations

Source: From Lai et al. [A]/MDPI/CC BY 4.0.

Table 1.6 Smart cities pilot projects in Africa.

Name	Description
Kenya	Konza will be a smart city with a connected urban ICT network that provides urban services connections and efficient management of those services on a great scale [85]. The city will connect the following four key services: infrastructure services including transportation, utilities, public safety and environment; citizen services including access and participation; city services including city information, planning, and development; business services including support services for local commerce. There is no information regarding the standards adopted in the Konza project
South Africa	Slavova and Okwechime [86] examined the broader transformative processes taking place in Africa and developed a vision of the future African cities. The authors showed the alignment of critical aspects of the smart city concept with the African Union's Agenda 2063. Several factors could impact on the transformative process including: (i) balancing the power dimension of smart city projects in Africa, (ii) the dichotomy dividing rural regions from urban spaces needs to be reduced, and (iii) the rapid adoption of technologies to implement a smart city. The relevant standards were not discussed by the authors. The Stellenbosch Smart Mobility Lab helps with developing Stellenbosch to be the first transportation leaning "Smart City" in South Africa [87]. The lab assists with the planning and implementation of mobility applications of the Smart City model. The applications include parking for enhanced vehicle distribution, real-time traffic and transport operations control systems, bicycle-sharing schemes, information services for commuters, and traffic planning platforms. Transport engineers employ big data gained from probe vehicles and human movement patterns as input into several of the aforementioned applications. There is no information regarding the standards adopted in the Smart City model

Source: From Lai et al. [A]/MDPI/CC BY 4.0.

Table 1.7 Smart cities pilot projects in Asia.

Name	Description
China	Alibaba's cloud project, City Brain, uses data collected from video feeds at traffic lights to relieve traffic congestion and gridlock in Hangzhou, China. The traffic management is 92% precise in recognizing traffic violations, aids emergency vehicles to reach their destinations 50% faster than before, and has permitted traffic speeds to grow by 15% [88]. City government leaders and planners can also utilize City Brain to overcome other pressing issues such alleviating a reduced water supply. Alibaba cloud had adopted numerous international standards to meet security compliance, including ISO 27001 and ISO 20000 [89]
Dubai	The Smart Dubai initiative improves the living standards of Dubai citizens [90]. There are more than 130 initiatives with the joint effort from government and private sector entities. Examples of initiatives include the Dubai Data Initiative, the Dubai Blockchain Strategy, the Happiness Agenda, the Dubai AI Roadmap, and the Dubai Paperless Strategy. Khan et al. [91] identified the best practices linked to Dubai's smart city and smart tourism. The city had a large amount of data that were unorganized, unstructured, and had very poor relationships. The Dubai Data initiative reinforces the Smart Dubai strategy and its applicable major components that enable the efficient exchange of data and information and modernizes the continuous connectivity for the private and public sectors. The innovative data-sharing initiative will be managed by international standards and best practices for safe, seamless, and fair exchanges of data [92]. However, there is no detail regarding the standards that will be adopted
Hong Kong	Hong Kong's smart city domains include smart mobility, smart living, smart environment, smart people, smart government, and smart economy [8]. The Next Generation GovCloud and Big Data Analytics Platform will modernize the current government cloud infrastructures and implement a new application architecture. Bureaux and departments can accelerate the development and delivery of digital government services, comprising big data analytics and artificial intelligence applications [93]. There is no information regarding the standards adopted in the Next Generation GovCloud and Big Data Analytics Platform. Ma and Lam [94] explored the interrelationships between several obstacles to open data adoption and suggested practical recommendations to improve open data development for smart cities. The study concluded that the lack of open data policy should be confronted as a matter of urgency in Hong Kong. An open mindset and IT literacy in the government organizations continue to be developed
Japan	Japan Smart Community Alliance is the leading authority to promote smart communities in Japan [11]. As in Feb. 2019, there are 259 members within the Japan Smart Community Alliance consisting of businesses from the manufacturing, electricity, gas, heat supply and water, information, and communications sectors. The alliance has also developed four working groups as follows [11]: • The International Strategy Working Group: Determines policy technological, and market developments concerning smart communities and shares information with alliance members and international organizations. The group creates strategies to assist the contributions of Japanese businesses in smart community development across the globe • The International Standardization Working Group: Expedites efforts in different fields in cooperation with Japan's Ministry of Economy, Trade, and Industry to attain international smart grid standardization. The group examines worldwide developments in smart grid standardization and encourages compliance in international standardization development. • Roadmap Working Group: Formulates smart community technology development roadmaps. The group supports technology development by creating scenarios for next-generation societies in which smart grid technologies have been implemented. The findings would create a synergistic effect between technology development and usage.

(Continued)

Table 1.7 (Continued)

Name	Description
	• Smart House and Building Working Group: Conducts activities to promote the dissemination of smart houses and smart buildings by creating work schedules for individual task, comprising the identification and maintenance of underlying critical devices, evaluating the development of individual task, and stimulating appropriate activities. Regarding smart cities projects, Woven City is a completely connected ecosystem run by hydrogen fuel cells to be constructed from early 2021 at the bottom of Mount Fuji [95, 96]. This "living laboratory," a 175-acre urban development in Higashi-Fuji will comprise full-time residents and researchers who will research and develop technologies including robotics, autonomy, smart homes, and personal mobility in a real-world setting. Buildings will be mainly built using carbon-neutral wood and adopt a mixture of robotic production methods and traditional Japanese joinery methods. Rooftops will be roofed in photovoltaic panels to generate solar power. Electricity will also be produced by hydrogen fuel cells. All homes will have state-of-the-art human assistance technologies, from maintaining basic needs and improving daily life to sensor-based artificial intelligence for monitoring personal's health. Woven City is an opportunity to utilize connected technology with security and integrity. The original plan is for 2000 people living in Woven City. The dwellers include Toyota employees and their families, industrial partners, retailers, retired couples, and visiting scientists. More residents will be invited as the project progresses. There is no information regarding the standards adopted in the Woven City.

Source: From Lai et al. [A]/MDPI/CC BY 4.0.

upkeep growth, and generate jobs. The Plan embodies a framework for a cities policy at the federal level. City Deals are the important drivers for executing the Smart Cities Plan. They are partnerships between the three levels of government and the community to strive for a shared vision for livable and productive cities. Standards Australia is the country's leading independent, nongovernmental, not-for-profit standards organization [98]. The organization is actively participating in national and international discussions on smart cities, including being involved in the ISO Technical Management Board United Nation Sustainable Development Goals Taskforce. The task force will: revisit the mapping of ISO standards to the Sustainable Development Goals; identify the importance of Sustainable Development Goals for ISO, leading to the design of a database that can be used by businesses and organizations to determine the useful standards in promoting Sustainable Development Goals; create guidance for committees on how to proactively identify the right partnerships including the United Nations and other international organizations; offer recommendations for which organizations including ISO should work in standards promotion to support Sustainable Development Goals.

The smart cities pilot projects in Europe, North America, and South America are presented in Tables 1.8 and 1.9, respectively.

In summary, the review on some smart cities developments from various countries and international standards shows:

1) There is a lack of discussions on the use of international standards in implementing smart cities. It is important to acknowledge the currently available standards in development when structuring and developing a smart city.

Table 1.8 Smart cities pilot projects in Europe.

Name	Description
Barcelona	Bakici et al. [99] examined the city of Barcelona and analyzed its Smart City initiative, including its urban policy implications. This article analyzes Barcelona's transformation in the areas of Smart City management, drivers, bottlenecks, conditions, and assets. The authors described the Barcelona Smart City model and examined the key factors of the Smart City strategy while considering living labs, Open Data, e-Services, smart districts, initiatives, and infrastructures. The Barcelona Smart City model consists of four domains including smart governance, smart economy, smart living, and smart people. The 22@ Barcelona region is a central point for innovation and economic development, as small-medium enterprises use the region as a test-bed to trial novel technologies. There is no information regarding the standards adopted in the 22@ Barcelona region
Romania	Pop and Proştean [100] studied the implementation approaches for different smart cities in Romania. The considered cities include Craiova, Napoca, Sibiu, and Timisoara. All Romanian smart cities offer mobile applications to citizens to notify them in real time of the timing for each individual public transport station. In Cluj-Napoca, Craiova, and Timisoara, traffic management lefts are used to monitor traffic lights. In addition, parking systems are available to offer citizens to pay for parking with short message service and identify the number of accessible parking places in real-time. International standards were not discussed on how the services were managed and achieved The Romanian Association for Smart City [101] is the leading authority of the Smart City Industry in Romania, which consists of professionals and experts from various industries. The association is also supported by over 200 national and international partners. The association aims to create creative-intelligent communities in Romania and achieve this by developing activities related to the Smart City ecosystem. The association has introduced 8 ISO international standards in legislation; however, there are no details regarding them Alba Iulia is located in the west-central part of Romania. The pilot project Alba Iulia Smart City has many distinctive features. The project promotes collaborative partnerships across governmental organizations, research institutions, local administration, companies, universities, citizens, and associations. The partnerships are not driven by commercial interests. The solutions are developed and examined by partner companies, with the local administration providing the required support and infrastructure. It is worth mentioning that there is no technical standard discussed for the Alba Iulia Smart City project [102]
Sweden	Stockholm aims to achieve environmental goals and efficient cooperation between various stakeholders, including the private and public sectors [103]. Kista Science City is an important venue for ICT research and development. The prominent ICT businesses including IBM and Ericsson settled in Kista during the 1970s and more than 1000 other ICT companies have joined. The city hosts one of the world's leading ICT clusters and largest urban business districts. Robèrt [104] examined a local travel planning network in Kista Science City where the travel demand is probable to surpass the capacity of the transport system in the future. There is no information regarding the standards adopted in Kista Science City
United Kingdom	Caird and Hallett [105] described that the British Standards Institution (BSI) collaborating with ISO has established a substantial body of work on smart city standards and urban performance metrics. BSI is the leading organization in developing smart city standards and urban performance metrics in the United Kingdom. Publicly Available Specification (PAS) 180 details industry-concurred understanding of smart city definitions and terms in the United Kingdom, to help develop a robust foundation for imminent standardization and good practices [106]. PAS 180 also helps enhancing smart cities understanding by setting a common language for designers, developers, clients, and manufacturers. The

(*Continued*)

Table 1.8 (Continued)

Name	Description
	standard will support industry to work more effectively and efficiently and reduce the probabilities of confusion in the supply chain. PAS 180 defines terms for smart cities encompassing smart city concepts for various infrastructure and systems' components. It covers processes, materials, applications, and methodologies PAS 181 is a smart city framework for city leaders to create, concur, and provide smart city strategies that can assist transform their city's capability to encounter its impending challenges and deliver its potential aspirations [107]. The smart city framework is based on current good practices and is a set of dependable and repeatable tasks that city leaders can use to support create and execute their smart city plans. The framework does not expect to describe a one-size-fits-all model for the UK cities. PAS 181 emphases on the enabling processes for the pioneering usage of data and technology, composed with organizational modification, can assist deliver the varied visions for potential UK cities in increased efficient, sustainable, and effective habits Manchester CityVerve [108] uses IoT technologies to transform the city. The program focuses on four aspects of transformation including "culture and public realm," "energy and environment," "health and social care," and "travel and transport." Milton Keynes is a fast-growing city in the United Kingdom. MK:Smart [109] is a large collaborative initiative to develop innovative solutions to support economic growth in Milton Keynes. The state-of-the-art "MK Data Hub" plays a critical role in the project which facilitates the acquisition and management of big data of city systems from numerous data sources. The data concerns with energy and water consumption, transport data, data from satellite technology, economic and social datasets, and crowdsourced data from social media or specialized apps. Caird and Hallett [105] examined both projects and concluded that the city authorities were unfamiliar with the smart city indicator frameworks [105]

Source: From Lai et al. [A]/MDPI/CC BY 4.0.

Table 1.9 Smart cities pilot projects in North America and South America.

Name	Description
Brazil	Macke et al. [110] described that the city of Curitiba, Southern Brazil to be a green, inclusive, and livable city. It is the top ten smartest cities globally speaking. Curitiba has several well-known sustainability programs. Effective leadership and devotion to intelligent transportation planning aided Curitiba to turn into a sustainable city and the standard for effective urban planning. The city's achievements are considered in six factors namely integrated urban planning, pedestrian priority zones, environmental awareness, waste management system, effective public transport system, and social justice. Smart living can be attained by delivering the four factors namely, community integration socio-structural relations, material well-being, and environmental well-being. International standards are not discussed in the work. Afonso et al. [111] studied Brazilian capital indicators and developed a maturity model called Brazilian Smart City Maturity Model to allow transform public databases into useful indicators to assist city managers in planning. The authors mentioned that the ISO 37120 standard provides 100 different performance indicators for cities. The standard consists of 17 themes, 46 core indicators, and 54 indicators that can help define public policies based on different domains. The model is an ongoing work
Canada	The Smart Cities Challenge is a national competition open to all municipalities, local or regional governments, and indigenous communities [112]. The Challenge promotes communities to adopt a smart cities approach to enhance the living

Table 1.9 (Continued)

Name	Description
	standards of citizens through data, innovation, and connected technology. The Challenge aims to address four areas including (i) to realize outcomes for residents, (ii) empower communities to innovate, (iii) forge new partnerships and networks, and (iv) spread the benefit to all Canadians Edmonton is experiencing a resident-led digital transformation supported by the city's council. The city developed the Business Technology Strategy, the first-of-its-kind in Canada to guide data usages, different technologies, and business solutions to enhance citizen's life [113]. The Edmonton's Smart City Strategy is an innovation ecosystem of academia, government, residents, and industry that abides by ISO 37106:2018. This standard is guidance for leaders in smart cities and communities across the private, public, and voluntary sectors concerning how to create a collaborative, open, digitally enabled, and citizen-centric operating model for their city that drives a sustainable future. The standard focuses on creating cities that (i) makes present and future citizen needs as the driver behind investment decision-making, planning and delivery of entire city spaces and systems, (ii) combine physical and digital planning, (iii) determine, foresee and react to emerging challenges in an agile, sustainable, and systematic manner, and (iv) develop changes in the capacity for joined-up delivery and innovation within organizational boundaries for the city [114] Saskatoon aims "to be the city that breaks the cycle of Indigenous youth incarceration by creating a new cycle focused on building purpose, belonging, security and identity" [115]. The ConnectYXE initiative is based on three pillars (i) to empower indigenous youth and their families by giving real-time information and choices for how to use services across the city, (ii) to work with partners by developing a data repository for all relevant programs and services accessible, and (iii) to exploit innovative technology by connecting systems, distributing data, and using artificial intelligence. The collective data will give a city-wide image of what is accessible and the needs of those supports at all times. This enables service providers and decision-makers to frequently study and recognize gaps, changes, and better approaches to respond to the needs. Presently, technical standards were not discussed in the proposal for ConnectYXE
United States of America	New York aims to become an equitable and smart city to improve government services and citizens' living standards [116]. The transformation contains multiple programs including "New York City Connected Communities," where the government develops computer lefts in the places with highly concentrated poverty rates. Over 100 lefts have been developed, which have improved the level of digital literacy and enhanced the quality of life by developing employment opportunities. The digital lefts are in parks, computer resource lefts, New York City Housing Authority Centers, recreation lefts, libraries, and senior citizen lefts. Another initiative is "LinkNYC" developed in 2014. The purpose was to develop a free ultra-high speed WiFi network to connect the whole city with free high-speed internet service. The city has installed over 7500 communication junctions with free WiFi network, domestic phone calls, and cell phone charging facility. Kansas City, Missouri is one of the smartest cities due to its successful technology utilization [117]. Along the two-mile track of the Kansas City Streetcar, a 15 million USD public–private partnership has facilitated the placement of 328 Wi-Fi access points, 178 smart streetlights that can monitor traffic patterns and available parking spaces, 25 video kiosks, pavement sensors, and video cameras. They are all connected by the city's fiber-optic data network. It was determined that the three smart city projects including New York City Connected Communities, LinkNYC, and the Kansas City Streetcar have not discussed technical standards

Source: From Lai et al. [A]/MDPI/CC BY 4.0.

2) With standards and technologies swiftly evolving, many cities need to avoid getting locked into one vendor's integrated solutions, which makes it more difficult for the city to share data with citizens, developers, and other cities.
3) International standards should be developed to address some of the most pressing challenges in a smart city, including potential solutions to a pandemic such as COVID-19 [118]. In combating the COVID-19, ISO has made some standards freely available to the public, including ISO 13688:2013 (Protective clothing – General requirements) and ISO 19223:2019 (Lung ventilators and related equipment – Vocabulary and semantics) [119]. Simultaneously, IEC also decided to make some standards and most relevant normative references for critical care ventilators free of charge to industries who are creating products or converting their existing assembly lines to ventilator production [120]. In the current pandemic, many organizations and governments are sharing or publishing data. For example, government health agencies are publishing data concerning regional cases and deaths; symptom trackers are distributing data with researchers and making data public; technology companies are obtaining mobility data which can help us to understand the impact of the coronavirus on our lives. Standards need to support data interoperability, the ability of services and systems that create, exchange, and use data to have clear, shared expectations for the contents, context, and meaning of that data [121, 122].

1.8 Challenges and Opportunities

By 2050, it is expected that 66% of the global population will dwell in urban regions [123]. The challenge will be to supply these populations with essential resources including sufficient energy, clean water, and safe food while simultaneously warranting complete economic, social, and environmental sustainability.

Several cities today have aspirations of transforming into the smart cities of tomorrow. However, the challenges to be overcome to accomplish this include the planning of a complicated plan that comprises public and private participants, product vendors, and information technology infrastructure providers. A smart city needs the foundation of standards-based information technology infrastructure that fulfills and supports a wide range of requirements and can adapt to novel technologies, such as advanced sensors, measurement and analytics tools, and solutions driven by machine learning and artificial intelligence. Smart city development requires support from public organizations, citizens, state and local government, and private enterprises. The benefits of a smart city include the creation of major prospects for sustainability, disaster prevention, business, public safety, and quality of life enhancements. However, there are key challenges that need to be addressed for a smart city including:

Commodification: As discussed by Gandy and Nemorin [124], a major concern regarding the smart cities' development is the motivation to support this worldwide initiative, from the pursuit of new markets by transnational corporations. Corporate organizations are keen to mine personal data, such as biometric data [125]. Data brokers could create consumer profiles including biometric information, and identities can be located and tracked as citizens move in a smart city. These profiles can also intensify commodification by mining the freshly available sources of data, with the ubiquity of sensors as dynamic data collection points.

Social and digital exclusion: In designing smart cities solutions, it is important to use suitable means to engage and empower population groups which are hard to reach, such as citizens living in poverty and/or social exclusion, migrants, younger and older people, or people with disabilities

[126]. Smart city technologies should be made affordable and able to be accessed by all groups of consumers. A smart city should be an age-friendly environment. The World Health Organization defines age-friendly environments as ones which "foster health and well-being and the participation of people as they age" [127]. These environments are accessible, equitable, inclusive, safe and secure, and supportive. Senior citizens may experience negative attitudes and discrimination based on their age. Creating age-friendly environments acknowledges diversity, fights ageism, and ensures that everyone has the opportunity to fully participate.

Privacy and surveillance: Privacy becomes a major concern when the data collected could lead to linking or identifying an individual, especially when gathered from numerous information sources. Data storage by governments is generally nontransparent. The likelihood for cross-sharing data within government services could lead to third parties to have access to the data, where the provider has no intention for it to happen. Zoonen [128] constructed a four-quadrant privacy framework to theorize if and how smart city technologies and urban big data produce privacy concerns among the people in these cities. The framework is developed according to two recurring dimensions in research toward people's concerns about privacy: one dimension signifies that people see specific data as more personal and sensitive than others, the other dimension signifies that people's privacy concerns vary according to the purpose for which data is gathered, in contrast to the surveillance and service purposes which are the most dominant. The work concludes that the smart technology options and the use of specific data and analytic tools are important factors to comprehend people's privacy concerns in smart cities, as well as to their awareness of what type of data to use to serve a purpose. A smart city should address (i) an applied need to substantiate the empirical relation between purpose and technologies, and (ii) to produce a theoretical and situated comprehension of people's privacy anxieties in smart cities.

In addition, building a smart city is a gigantic task as there are several working parts and components involved, namely the smart cities domains. Many smart cities are not constructed from scratch or all in a single attempt. Smart city development is a gradually evolving process that witnesses the city becoming smarter, bit by bit. As time progresses, the individual regions of smartness develop together and interconnect, but on the condition for them using the same consistent technical rules that are stipulated by technical standards.

Several researchers have reviewed smart city projects from different perspectives. Camero and Alba [129] explored the computer science and information technology used for a smart city. There is no agreement on a smart city definition and in fact, several definitions are being developed. One explanation is for the iterative process where cities become smarter as time progresses. There are very few studies on the inclusion of policy and urban planning recommendations in information technology and computer science literature.

Caird and Hallett [105] examined the creation of appropriate, valid, credible, and valuable approaches to smart city evaluation by studying conceptual, measurement, and evaluation challenges for five UK smart city projects. Caird and Hallett [105] identified that a critical challenge for evaluation design is in creating standardized smart city development and performance indicators that give useful citizen and city-centered evaluations. There is a significant amount of work on standardization and smart urban metrics driven by international standards organizations. Specifically, the Smart and Sustainable Cities and Communities Coordination Group advises on European interests and requirements concerning standardization on Smart and Sustainable cities and communities. The International Organization for Standardization (ISO) has concurred on standards for "Smart Community Infrastructures" performance metrics. ISO Technical Report 37150:2014 (Smart community infrastructures – Review of existing activities relevant to metrics) [130] reports community infrastructures including water, energy, waste transportation, and

Information and Communications Technology (ICT). The standard concentrates on the technical features of current activities which are available. Political, societal, or economic aspects are not studied in this standard. ISO Technical Report 37151:2015 (Smart community infrastructures – Principles and requirements for performance metrics) [130] details the principles and stipulates requirements for the definition, identification, optimization, and harmonization of community infrastructure performance metrics, and gives recommendations for analysis, including smartness, interoperability, synergy, resilience, safety, and security of community infrastructures. Funded by the European Union HORIZON 2020 program, the CITYkeys project [131] is an important European Commission EUROCITIES initiative that aims to create acceptable city performance measurement frameworks: key performance indicators. The initiative creates standardized data collection processes to increase the adoption rate of smart city solutions. It is anticipated that comparable, scalable, and replicable smart city solutions can be achieved across cities. The authors concluded that standardized smart urban metrics and indicators are not widely adopted by cities while the development of standards is at the early stages.

Hasija [88] examined the current global advancements in smart city initiatives. The study was categorized into three themes, namely data access and collection, end-user utility, and economic feasibility of different solutions. The economic viability is crucial to the success of a smart city initiative. The potential ideas to enhance city operations could not be delivered if they are economically unsustainable. For business strategies, prudent analysis is required to examine the trade-offs that determine the efficacy of such initiatives. A bike-sharing scheme is an affordable and convenient mode of transportation in China. However, not all bike-sharing companies are successful. Some of the issues contribute to the failure of bike-sharing initiatives include (i) no regulation: bikes could end up in different places and be dumped along city streets, (ii) lack of operational sustainability: many bike-sharing platforms do not need a security deposit, and (iii) no optimization: lack of consideration for how and where the bikes should be located to maximize utilization and to avoid bikes piling up on streets.

Anthopoulos [132] examined 20 smart cities projects of various scales in different countries and continents. Furthermore, the review documented the challenges that the cities meet as they work toward being a smart city. The review examines smart cities in relation to climate change, sustainability, natural disasters, and community resiliency. A smart city project is complex and expensive. Anthopoulous [132] first examined the project management guidelines and frameworks for agile and complex projects, including a smart city. ISO 21500:2012 Guidance on project management is an international standard that can be used by private or public organizations for all kinds of projects. The aim is to provide a guide to project managers on how to apply project management disciplines into a business environment to increase the possibilities for enhanced business results and project success. An important aspect is the use of the common language and processes by all project stakeholders, which enhances communication and cooperation. ISO 21500 gives a high-level description of concepts and processes to create good practices in project management. The cities reviewed focuses on the project management perspective including scope, organization, time, cost, quality, risk, and procurement. The smart city projects are well documented with great detail in the project development. However, there is a lack of discussion on technical standards of the smart city projects apart from project management.

Van Winden and Van den Buuse [133] analyzed the procedures of upscaling, concentrating on smart city pilot projects where numerous stakeholders with dissimilar missions, agendas, and incentives work together. If technical standards can be smoothly adapted to fit with the geospatial context, then the solution becomes more attractive to many cities. Numerous works on smart cities have been conducted and review literature for smart cities exists. However, most recent literature lacks discussions on an important topic of international standards for smart cities. International standards are technical standards developed by international organizations. International standards can greatly assist tailor-made solutions development for bespoke conditions of a city. Standards stipulate the anticipated level of performance and technologies compatibility. Standards are generic metrics that allow solutions to be benchmarked and compared.

1.9 Conclusions

Smart cities are intelligent and sustainable cities. It is well known that a smart city requires the use of novel technologies, including robust ICT infrastructures and sensor devices. First, this paper has revisited and identified some of the new smart city definitions. The definition of a smart city is continuing to evolve, and one must accept that different terminologies exist due to the different scope considered, e.g. the region and community involved. This paper then examines the smart city from the view of international standards. It is identified that numerous international standards are currently in development to develop a smart city and the old standards are being revised to become relevant to address current society needs. Six smart city domains were identified including smart energy, smart health, smart education, smart mobility, smart economy, and smart governance. There is a need for researchers and city developers to acknowledge the different kinds of standards currently available and in development, in order to build a city that is functional and sustainable. The review identified that international standards are by no means as yet pervasive: there is a need for smart city projects to present details on the international standards adoption, and its implications for a smart city. Well-defined standards allow meaningful comparisons among smart cities implementation. With the presence of many standard bodies, challenges exist if international standards are not agreed on by standards developers and users. This paper serves as a guide on international standards for smart city researchers and developers.

Acknowledgements

The permission given in using the materials in the following paper is very much appreciated.

[A] Lai, C.S., Jia, Y., Dong, Z. et al. (2020). Review of technical standards for smart cities. *Clean Technologies*. MDPI 2 (3): 290–310.

References

1 Deakin, M. and Al Waer, H. (2011). From intelligent to smart cities. *Intell. Build. Int.* 3: 140–152.
2 Chamran, M.K., Yau, K.-L.A., Noor, R., and Wong, R. (2020). A distributed testbed for 5G scenarios: an experimental study. *Sensors* 20: 18.
3 Deakin, M. (2013). *Smart Cities: Governing, Modelling and Analysing the Transition*. Oxford, UK: Routledge.
4 ASEAN(2019). Smart Cities Framework. https://asean.org/storage/2019/02/ASCN-ASEAN-Smart-Cities-Framework.pdf (Accessed 1 May 2021).
5 Smart Cities Overview – Guide (2015). BSI Standards Publication. http://shop.bsigroup.com/upload/Shop/Download/PAS/30313208-PD8100-2015.pdf (accessed 22 April 2021).

6 Department for Business Innovation & Skills (2013). Smart Cities: Background Paper. https://assets.publishing.service.gov.uk/government/uploads/system/uploads/attachment_data/file/246019/bis-13-1209-smart-cities-background-paper-digital.pdf (accessed 7 February 2021).
7 European Commission Smart Cities (n.d.). https://ec.europa.eu/info/eu-regional-and-urban-development/topics/cities-and-urban-development/city-initiatives/smart-cities_en (accessed 21 October 2020).
8 Office of the Government Chief Information Vision & Mission. https://www.smartcity.gov.hk/vision/ (accessed 7 August 2020).
9 IEEE Smart Cities. https://smartcities.ieee.org/images/files/pdf/IEEE_Smart_Cities-_Flyer_Nov_2017.pdf (accessed 7 May 2021).
10 Strategic Business Plan (SBP) SMB/6817/R (2019). International Electrotechnical Commission. https://www.iec.ch/public/miscfiles/sbp/SYCSMARTCITIES.pdf (accessed 1 May 2021).
11 Japan Smart Community Alliance (n.d.). Smart Community Development. https://www.smart-japan.org/english/ (accessed 22 April 2021).
12 Ministry of Housing and Urban Affairs (n.d.). Government of India Smart Cities Mission. http://smartcities.gov.in/content/innerpage/what-is-smart-city.php (accessed 7 August 2020).
13 Deloitte Consulting (n.d.). Deloitte Define Your Smart City Strategy. https://www2.deloitte.com/us/en/pages/consulting/solutions/smart-cities-strategies.html (accessed 7 May 2021).
14 Fedorova, E., Caló, A., and Pongrácz, E. (2019). Balancing socio-efficiency and resilience of energy provisioning on a regional level, case Oulun Energia in Finland. *Clean Technolies* 1: 273–293.
15 Lai, C.S., Jia, Y., Lai, L.L. et al. (2017). A comprehensive review on large-scale photovoltaic system with applications of electrical energy storage. *Renewable and Sustainable Energy Reviews* 78: 439–451.
16 Xu, X., Jia, Y., Xu, Y. et al. (2020). A multi-agent reinforcement learning based data-driven method for home energy management. *IEEE Transactions on Smart Grid* 11 (4): 3201–3211.
17 Wang, D., Wu, R., Li, X. et al. (2019). Two-stage optimal scheduling of air conditioning resources with high photovoltaic penetrations. *Journal of Cleaner Production* 241: https://doi.org/10.1016/j.jclepro.2019.118407.
18 Vaccaro, A., Pisica, I., Lai, L.L., and Zobaa, A.F. (2019). A review of enabling methodologies for information processing in smart grids. *International Journal of Electrical Power & Energy Systems* 107: 516–522.
19 Lasi, H., Fettke, P., Kemper, H.-G. et al. (2014). Industry 4.0. *Business and Information Systems Engineering* 6: 239–242.
20 Dong, Z., Lai, C.S., He, Y. et al. (2019). Hybrid dual-complementary metal–oxide–semiconductor/memristor synapse-based neural network with its applications in image super-resolution. *IET Circuits, Devices and Systems* 13: 1241–1248.
21 Dong, Z., Lai, C.S., Qi, D. et al. (2018). A general memristor-based pulse coupled neural network with variable linking coefficient for multi-focus image fusion. *Neurocomputing* 308: 172–183.
22 Taha, A., Wu, R., Emeakaroha, A., and Krabicka, J. (2018). Reduction of electricity Costs in Medway NHS by inducing pro-environmental behaviour using persuasive technology. *Future Cities Environ.* 4: 1–10.
23 Guerrero-Ibanez, J.A., Zeadally, S., and Contreras-Castillo, J. (2015). Integration challenges of intelligent transportation systems with connected vehicle, cloud computing, and internet of things technologies. *IEEE Wireless Communications* 22: 122–128.
24 Yin, J., Su, S., Xun, J. et al. (2019). Data-driven approaches for modeling train control models: comparison and case studies. *ISA Transactions* https://doi.org/10.1016/j.isatra.2019.08.024.

25 Hoel, T. and Mason, J. (2018). Standards for smart education–towards a development framework. *Smart Learning Environments* 5: 3.

26 Giffinger, R., Christian, F., Hans, K. et al. Smart Cities: Ranking of European Mid-Sized Cities. https://ec.europa.eu/digital-agenda/en/smart-cities (accessed 7 May 2021).

27 Hitachi (n.d.). Smart Water in Smart Cities. https://www.hitachi.eu/en-gb/social-innovation-stories/communities/smart-water-smart-cities (accessed 7 May 2021).

28 Hope, R., Foster, T., Money, A. et al. (2011). Smart Water Systems. Project report to UK DFID, April 2011. Oxford University, Oxford. https://assets.publishing.service.gov.uk/media/57a08ab9e5274a31e000073c/SmartWaterSystems_FinalReport-Main_Reduced__April2011.pdf (accessed 7 May 2021).

29 GOV.UK (2019). £1 Million Boost for UK Smart Waste Tracking. https://www.gov.uk/government/news/1-million-boost-for-uk-smart-waste-tracking (accessed 7 May 2021).

30 Interreg Europe (n.d.) Smart Waste Interreg Europe Innovation in Waste Management Policies. https://www.interregeurope.eu/smartwaste/ (accessed 7 May 2021).

31 Swan, M. (2015). *Blockchain: Blueprint for A New Economy*. Sebastopol, USA: O'Reilly Media, Inc.

32 Kumar, R., Strasser, T., Deconinck, G. et al. (April 2021). Guest editorial special issue on recent advances for intelligence in power and energy systems. *IEEE Transactions on Systems, Man and Cybernetics: Systems* 51 (4): 2036–2040.

33 Zhao, X., Lai, L.L., Wong, K.P. et al. (October 2017). Guest editorial special section on emerging informatics for risk hedging and decision making in smart grids. *IEEE Transactions on Industrial Informatics* 13 (5): 2507–2510.

34 Huang, C., Wang, L., and Lai, L.L. (2019). Data-driven short-term solar irradiation forecasting based on information of neighboring sites. *IEEE Transactions on Industrial Electronics* 66 (12): 9918–9927. https://doi.org/10.1109/TIE.2018.2856199.

35 Xu, F.Y., Cun, X., Yan, M. et al. (November 2018). Power market load forecasting on neural network with beneficial correlated regularization. *IEEE Transactions on Industrial Informatics* 14 (11): 5050–5059.

36 Zhang, J., Chen, X., Ng, W.W.Y. et al. (August 2019). New appliance detection for non-intrusive load monitoring. *IEEE Transactions on Industrial Informatics* 15 (8): 4819–4829.

37 Wu, X., Lai, C.S., Bai, C. et al. (2020). Optimal Kernel ELM and variational mode decomposition for probabilistic PV power prediction. *Energies* 13 (14): 3592.

38 Lai, C.S., Mo, Z., Wang, T. et al. (2020). Load forecasting based on deep neural network and historical data augmentation. *IET Generation Transmission and Distribution* https://doi.org/10.1049/iet-gtd.2020.0842.

39 Lai, C.S., Yang, Y., Pan, K. et al. (2020). Multi-view neural network ensemble for short and mid-term load forecasting. *IEEE Transactions on Power Systems* https://doi.org/10.1109/TPWRS.2020.3042389.

40 Wang, T., Lai, C.S., Ng, W.W.Y. et al. (2021). Deep autoencoder with localized stochastic sensitivity for short-term load forecasting. *International Journal of Electrical Power & Energy Systems* 130: 106954.

41 Lai, C.S., Zhong, C., Pan, K. et al. (September 2021). A deep learning based hybrid method for hourly solar radiation forecasting. *Expert Systems with Applications* 177 (1): 114941.

42 Yin, H., Dong, Z., Chen, Y. et al. (2017). An effective secondary decomposition approach for wind power forecasting using extreme learning machine trained by crisscross optimization. *Energy Conversion and Management* 150: 108–121.

43 Chen, X., Lai, C.S., Ng, W.W.Y. et al. (2021). A stochastic sensitivity-based multi-objective optimization method for short-term wind speed interval prediction. *International Journal of Machine Learning and Cybernetics* 12: 2579–2590.

44 Su, Q., Mi, C., Lai, L.L., and Austin, P. (May 2000). A fuzzy dissolved gas analysis method for the diagnosis of multiple incipient faults in a transformer. *IEEE Transactions on Power Systems* 15 (2): 593–598.

45 Liu, J., Fan, X., Zhang, C. et al. (July 2021). Moisture diagnosis of transformer oil-immersed insulation with intelligent technique and frequency-domain spectroscopy. *IEEE Transactions on Industrial Informatics* 17 (7): 4624–4634.

46 What is intelligent transportation system? Its working and advantages. https://theconstructor.org/transportation/intelligent-transportation-system/1120/ (accessed 7 May2021)

47 European Commission (n.d.). Case Study Summary: Smart Health. https://ec.europa.eu/growth/content/smart-health_en (accessed 7 May 2021).

48 Deloitte Development (2019). Smart Health Communities and the Future of Health. https://www2.deloitte.com/content/dam/Deloitte/lu/Documents/life-sciences-health-care/lu-smart-health-communities.pdf (accessed 7 May 2021).

49 Van Gheluwe, C., Lopez, A.J., Semanjski, I., and Gautama, S. (n.d.) Repurposing existing traffic data sources for COVID-19 crisis management. In *2020 IEEE International Smart Cities Conference*, Piscataway, NJ, USA (28 September – 1 October 2020).

50 Google (2020). COVID-19 Community Mobility Report. https://www.gstatic.com/covid19/mobility/2020-10-13_BE_Mobility_Report_en.pdf (accessed 7 May 2021).

51 Google (n.d.). See how your community is moving around differently due to COVID-19. https://www.google.com/covid19/mobility/ (accessed 7 May 2021).

52 Buckee, C.O., Balsari, S., Chan, J. et al. (10 April 2020). Aggregated mobility data could help fight COVID-19. *Science* 368 (6487): 145–146.

53 ISO (n.d.). What Are Smart Cities? https://www.iso.org/sites/worldsmartcity/ (accessed 7 May 2021).

54 ISO (n.d.). Standards. https://www.iso.org/standards.html (accessed 7 May 2021).

55 About the IEC. https://www.iec.ch/about/ (accessed 7 May 2021).

56 International Telecommunication Union (n.d.). About International Telecommunication Union (ITU). https://www.itu.int/en/about/Pages/default.aspx (accessed 7 May 2021).

57 World Council on City Data (n.d.). ISO 37120 World Council on City Data. https://www.dataforcities.org (accessed 7 May 2021).

58 IEC (n.d.). Editorial Team Standards Support Smart Cities. https://blog.iec.ch/2019/07/standards-support-smart-cities/ (accessed 7 May 2021).

59 Sang, Z. and Li, K. (2019). ITU-T standardisation activities on smart sustainable cities. *IET Smart Cities* 1: 3–9.

60 IEEE Standards Association. (n.d.). IEEE 802® – Keeping the World Connected. https://standards.ieee.org/featured/802/index.html (accessed 7 May 2021).

61 IEEE Smart Cities (n.d.). About IEEE Smart Cities. https://smartcities.ieee.org/about (accessed 7 May 2021).

62 IEEE Standards Association (n.d.). P2784 – Guide for the Technology and Process Framework for Planning a Smart City. https://standards.ieee.org/project/2784.html (accessed 7 May 2021).

63 IEEE Standards Association (2019). 1889–2018 – IEEE Guide for Evaluating and Testing the Electrical Performance of Energy Saving Devices. https://standards.ieee.org/content/ieee-standards/en/standard/1889-2018.html (accessed 7 May 2021).

64 IEEE Standards Association (n.d.). P1922.1 – Standard for a Method for Calculating Anticipated Emissions Caused by Virtual Machine Migration and Placement. https://standards.ieee.org/project/1922_1.html (accessed 7 May 2021).

65 IEEE Standards Association (n.d.). P2814 – Techno-economics Metrics Standard for Hybrid Energy and Storage Systems. https://standards.ieee.org/project/2814.html (accessed 7 May 2021).
66 IEEE Standards Association (n.d.). P2852 – Intelligent Assessment of Safety Risk for Overhead Transmission Lines Under Multiple Operating Conditions. https://standards.ieee.org/project/2852.html (accessed 7 May 2021).
67 IEEE Standards Association (2014). 1708–2014 – IEEE Standard for Wearable Cuffless Blood Pressure Measuring Devices. https://standards.ieee.org/standard/1708-2014.html (accessed 7 May 2021).
68 IEEE Standards Association (n.d.). P3333.2.5 – Standard for Bio-CAD File Format for Medical Three-Dimensional (3D) Printing. https://standards.ieee.org/project/3333_2_5.html (accessed 7 May 2021).
69 IEEE Standards Association (n.d.). P1752 – Standard for Mobile Health Data. https://sagroups.ieee.org/1752/ (accessed 7 May 2021).
70 IEEE Standards Association (2020). 1847–2019 – IEEE Recommended Practice for Common Framework of Location Services for Healthcare. https://standards.ieee.org/standard/1847-2019.html (accessed 7 May 2021).
71 IEEE Standards Association (2022). P2621.1 – Standard for Wireless Diabetes Device Security Assurance: Product Security Evaluation Program. https://standards.ieee.org/project/2621_1.html (accessed 7 May 2021).
72 IEEE Standards Association (n.d.). P1884 – Guide for Stray Current /Corrosion Mitigation for DC Rail Transit Systems. https://standards.ieee.org/project/1884.html (accessed 7 May 2021).
73 IEEE Standards Association (n.d.). P1883 – Recommended Practice for Electrical and Electro-Mechanical Bench Test Equipment (BTE) for Transit Rail Projects. https://standards.ieee.org/project/1883.html (accessed 7 May 2021).
74 IEEE Standards Association (n.d.). P2406 – IEEE Draft Standard for Design and Construction of Non-Load Break Disconnect Switches for Direct Current Applications on Transit Systems. https://standards.ieee.org/project/2406.html (accessed 7 May 2021).
75 IEEE Standards Association (n.d.). P2020 – Standard for Automotive System Image Quality. https://standards.ieee.org/project/2020.html (accessed 7 May 2021).
76 IEEE Standards Association (n.d.). P2685 – Recommended Practice for Energy Storage for Stationary Engine-Starting Systems. https://standards.ieee.org/project/2685.html (accessed 7 May 2021).
77 IEEE Standards Association (n.d.). P7919.1 – Requirements for eReaders to Support Learning Applications. https://standards.ieee.org/project/7919_1.html (accessed 7 May 2021).
78 IEEE Standards Association (n.d.). P2834 – Standard for Secure and Trusted Learning Systems. https://standards.ieee.org/project/2834.html (accessed 7 May 2021).
79 IEEE Standards Association (2019). 1876–2019 – IEEE Standard for Networked Smart Learning Objects for Online Laboratories. https://standards.ieee.org/standard/1876-2019.html (accessed 7 May 2021).
80 IEEE Standards Association (2020). 1589–2020 – IEEE Standard for Augmented Reality Learning Experience Model. https://standards.ieee.org/standard/1589-2020.html (accessed 7 May 2021).
81 IEEE Standards Association (2021). P7005 – Standard for Transparent Employer Data Governance. https://standards.ieee.org/project/7005.html (accessed 7 May 2021).
82 IEEE Standards Association (n.d.). P7004 – Standard for Child and Student Data Governance. https://standards.ieee.org/project/7004.html (accessed 7 May 2021).
83 IEEE Standards Association (n.d.). P2145 – Standard for Framework and Definitions for Blockchain Governance. https://standards.ieee.org/project/2145.html (accessed 7 May 2021).

84 IEEE Standards Association (n.d.). P2863 – Recommended Practice for Organizational Governance of Artificial Intelligence. https://standards.ieee.org/project/2863.html (accessed 7 May 2021).

85 Konza Technopolis Development Authority (n.d.). SmartCity. https://www.konza.go.ke/smart-city/ (accessed 7 May 2021).

86 Slavova, M. and Okwechime, E. (2016). African smart cities strategies for Agenda 2063. *Africa Journal of Managment* 2: 210–229.

87 Stellenbosch University (2013). Smart City. http://www.sun.ac.za/english/faculty/eng/ssml/research-focus/smart-city (accessed 7 May 2021).

88 Hasija, S., Shen, Z.-J.M., and Teo, C.-P. (2020). Smart city operations: modeling challenges and opportunities. *Manufacturing and Service Operations Management* 22: 203–213.

89 Alibaba Cloud (n.d.). Security & Compliance Center. https://www.alibabacloud.com/trust-center (accessed 7 May 2021).

90 Digital Dubai Authority (n.d.). Our vision is to make Dubai the happiest city on earth. https://www.smartdubai.ae (accessed 7 May 2021).

91 Khan, M.S., Woo, M., Nam, K., and Chathoth, P.K. (2017). Smart city and smart tourism: a case of Dubai. *Sustainability* 9: 2279.

92 Digital Dubai Authority (2019). Smart Dubai and Dubai Economy Launch New Data Initiative for Retail Sector. https://www.smartdubai.ae/newsroom/news/smart-dubai-and-dubai-economy-launch-new-data-initiative-for-retail-sector (accessed 7 May 2021).

93 Government Chief Information Officer (2022). Office of the Government Chief Information Officer The Government of the Hong Kong Special Administrative Region. https://www.ogcio.gov.hk/en/about_us/facts/doc/Fact_Sheet-OGCIO-EN.pdf (accessed 7 May 2021).

94 Ma, R. and Lam, P.T.I. (2019). Investigating the barriers faced by stakeholders in open data development: a study on Hong Kong as a "smart city". *Cities* 92: 36–46.

95 Toyota (2021). Woven City. https://www.woven-city.global (accessed 7 May 2021).

96 Clifford, J. (2020).Toyota to build a hydrogen-powered city of the future. https://blog.toyota.co.uk/toyota-woven-city-hydrogen-power (accessed 7 May 2021).

97 The Department of Infrastructure, Transport, Regional Development, Communications and the Arts (n.d.). Cities. https://www.infrastructure.gov.au/cities/ (accessed 7 May 2021).

98 Standards Australia (2020). Smart Cities Standards, August 2020. https://www.standards.org.au/getmedia/bfe42f98-011e-4798-8fa5-5b70c8a2a6bd/SA_Smart_Cities_Roadmap.pdf.aspx (accessed 18 August 2022).

99 Bakıcı, T., Almirall, E., and Wareham, J. (2013). A smart city initiative: the case of Barcelona. *Journal of the Knowledge Economy* 4: 135–148.

100 Pop, M.-D. and Proștean, O. (2018). A comparison between smart city approaches in road traffic management. *Procedia - Social and Behavioral Sciences* 238: 29–36.

101 Smart Cities in Romania. https://businessforsmartcities.com/load/118/presentation/2_alexa_dimitrscu_1_28a81.pdf (accessed 7 May 2021).

102 World Bank Group (n.d.). The City of Alba Iulia Alba Iulia Project Prioritization for 2014–2020. http://documents1.worldbank.org/curated/en/527401468190739988/pdf/Alba-Iulia-project-prioritization-for-2014-2020.pdf (accessed 7 May 2021).

103 City of Stockholm (2021). The Smart City. https://international.stockholm.se/city-development/the-smart-city/ (accessed 7 May 2021).

104 Robèrt, M. (2017). Engaging private actors in transport planning to achieve future emission targets – upscaling the Climate and Economic Research in Organisations (CERO) process to regional perspectives. *Journal of Cleaner Production* 140: 324–332.

105 Caird, S.P. and Hallett, S.H. (2019). Towards evaluation design for smart city development. *Journal of Urban Design* 24: 188–209.

106 BSI (n.d.). PAS 180 Smart Cities Vocabulary. https://www.bsigroup.com/en-GB/smart-cities/Smart-Cities-Standards-and-Publication/PAS-180-smart-cities-terminology/ (accessed 7 May 2021).

107 BSI (n.d.). PAS 181 Smart City Framework. https://www.bsigroup.com/en-GB/smart-cities/Smart-Cities-Standards-and-Publication/PAS-181-smart-cities-framework/ (accessed 7 May 2021).

108 O'Rourke, H. (2017). CityVerve and what makes Manchester a 'Smart City'. https://www.manchestereveningnews.co.uk/business/business-news/cityverve-smart-city-manchester-oxford--13048666 (accessed 7 May 2021).

109 Willetts, D. (n.d.). About MKSmart. http://www.mksmart.org/about/ (accessed 7 May 2021).

110 Macke, J., Casagrande, R.M., Sarate, J.A.R., and Silva, K.A. (2018). Smart city and quality of life: Citizens' perception in a Brazilian case study. *Journal of Cleaner Production* 182: 717–726.

111 Afonso, R.A., dos Santos Brito, K., do Nascimento, C.H. et al. (2015).Brazilian smart cities: using a maturity model to measure and compare inequality in cities. In *Proceedings of the 16th Annual International Conference on Digital Government Research*, Phoenix, AZ, USA (27–30 May 2015), pp. 230–238. New York, NY: Association for Computing Machinery

112 Infrastructure Canada (n.d.). Smart Cities Challenge. https://www.infrastructure.gc.ca/cities-villes/index-eng.html (accessed 7 May 2021).

113 City of Edmonton (2019). Smart City Challenge Edmonton Final Proposal. https://www.edmonton.ca/city_government/documents/CityofEdmontonSmartCitiesProposal_21MB.pdf (accessed 7 May 2021).

114 ISO 37106:2018(en) Sustainable cities and communities – Guidance on establishing smart city operating models for sustainable communities. https://www.iso.org/obp/ui/#iso:std:iso:37106:ed-1:v1:en (accessed 7 May 2021).

115 Connectyxe Smart Cities Challenge. https://www.saskatoon.ca/sites/default/files/documents/corporate-performance/communications/Engagement/connectyxe_saskatoon_march_5_2019.pdf (accessed 7 May 2021).

116 smartcity (2017). The Equitable City – A New Name for New York. https://www.smartcity.press/new-yorks-smart-city-initiatives/ (accessed 7 May 2021).

117 Smart Cities and the Journey to the "Cloud". https://www2.deloitte.com/content/dam/Deloitte/us/Documents/about-deloitte/us-about-deloitte-smart-cities-journey-cloud.pdf (accessed 7 May 2021).

118 Andersen, K.G., Rambaut, A., Lipkin, W.I. et al. (2020). The proximal origin of SARS-CoV-2. *Nature Medicine* 26: 450–452.

119 ISO (2021). COVID-19 response: Freely available ISO standards. https://www.iso.org/covid19 (accessed 7 May 2021).

120 IEC (n.d.). Access to key standards for critical care ventilators. https://webstore.iec.ch/webstore/webstore.nsf/xpFAQ.xsp?OpenXPage&id=GFOT-BNAEXA (accessed 7 May 2021).

121 Thereaux, O. Data and Covid-19: Why standards matter. https://theodi.org/article/data-and-covid-19-why-standards-matter/ (accessed 7 May 2021).

122 Mehta, S. (2021). What is data interoperability? https://datainteroperability.org (accessed 7 May 2021).

123 IEC (2020). Smart Cities. https://www.iec.ch/smartcities/introduction.htm (accessed 7 August 2020).

124 Gandy, O.H. Jr. and Nemorin, S. (2019). Toward a political economy of nudge: smart city variations. *Information, Communication & Society* 22: 2112–2126.

125 Sadowski, J. and Pasquale, F.A. (2015). The spectrum of control: A social theory of the smart city. *First Monday*, p. 20.

126 Smart Cities Marketplace (2016). Inclusive Smart Cities: A European Manifesto on Citizen Engagement. https://smart-cities-marketplace.ec.europa.eu/sites/default/files/Conference%20Inclusive%20Smart%20Cities%20-%20Brussels%2C%20November%2023rd%202016_0.pdf (accessed 7 August 2020).

127 Ageing and Life-Course. https://www.who.int/ageing/projects/age-friendly-environments/en (accessed 7 August 2020).

128 Van Zoonen, L. (2016). Privacy concerns in smart cities. *Government Information Quarterly* 33: 472–480.

129 Camero, A. and Alba, E. (2019). Smart City and information technology: a review. *Cities* 93: 84–94. https://doi.org/10.1016/j.cities.2019.04.014.

130 ISO/TC 268/SC 1 Smart Community Infrastructures (2015). Principles and Requirements for Performance Metrics. https://www.iso.org/standard/61057.html (accessed 30 March 2021).

131 Technical Research Centre of Finland (n.d.). Citykeys. http://citykeys-project.eu (accessed 7 August 2020).

132 Anthopoulos, L. (2019). *Smart City Emergence: Cases From Around the World.* Amsterdam, The Netherlands: Elsevier.

133 van Winden, W. and van den Buuse, D. (2017). Smart city pilot projects: exploring the dimensions and conditions of scaling up. *Journal of Urban Technology* 24: 51–72.

2

Lithium-Ion Storage Financial Model

2.1 Introduction

For the future decarbonized society, cheaply, and compactly storing the energy delivered by sustainable but variable renewable energy sources (RESs), mainly solar and wind will be important to preserve our living standards. In the present COVID-19 pandemic, one of its good impacts on energy industry may be on the smooth supply of fuels to generate electricity. However, during the lockdown, it is clear that this is a big challenge to transport fuels to the generation site. The use of renewable such as PV could remove this requirement of fuel delivery and minimize the risk imposed on security of power supply. However, the high penetration of noncontrollable and partly unpredictable renewables could lead to major concerns for the operation of the power system. These include keeping the system synchronized in the presence of a generation mix with much lower mechanical inertia, which needs to be able to withstand sudden and steeper up or down ramps, as well as many more start up or down of peak generators. Various energy storage (ES) technologies are expected to provide a wide range of advanced services such as frequency restoration, restoration reserves, ramping support, and energy balance.

As mentioned, to achieve the goal of decarbonizing the energy sector, more and more energy systems are heavily reliant on non-dispatchable intermittent renewables, such as solar photovoltaics (PV) and wind energy. Electrical energy storage (EES) can store the surplus generation produced by renewables until a time when it is needed, thus smoothing the energy system operation by acting as an additional generator or load, and reducing curtailment of renewables [1].

EES comprises a wide range of technologies, covering mechanical, electrical, electrochemical, chemical and thermal storage systems [2]. By discharging or charging, EES systems can release or absorb electricity to/from a power system. Electrochemical storage with rapid response times in the order of milliseconds can avoid short-term abnormal phenomena such as voltage and frequency deviations. State of charge (SOC) and depth of discharge (DOD) are parameters considered for hybrid energy system planning and operation [3, 4]. While there is an abundance of studies about the economics of EES, financial studies are remarkably rare. This chapter focuses on financial analysis of EES, with a case study on graphite/$LiCoO_2$ batteries.

Accurate financial analysis of EES must deal with uncertainties surrounding technical and economic performance. As such, various models have been developed to examine EES economics which will be reviewed in Section 2.2.2, with many studies emphasizing the levelized cost of electricity (LCOE) in particular [5–10]. However, these works have focused on the economic aspects of

Smart Energy for Transportation and Health in a Smart City, First Edition. Chun Sing Lai, Loi Lei Lai and Qi Hong Lai.

EES such as costing. A key to successfully deploying EES is the need to have an in-depth understanding of the financial aspects, i.e. the quantification of assets and liabilities and their allocation over the entire project lifetime.

The financing of EES was studied in [11, 12] and a review of the literature was given in [13]. However, these works do not provide all the key indicators, e.g. net present value (NPV), internal rate of return (IRR), and LCOE. In addition, the modeling of EES in many techno-economic studies (such as [11, 14–16]) does not consider EES degradation. This work aims to address the research gap in EES by building and presenting a more realistic financial model. The novelty of the work is described as follows:

- A comprehensive cash flow model is developed for Li-ion EES. The model includes detailed technical (e.g. degradation), financing (e.g. cost of debt), and economic (e.g. capital cost) parameters.
- The detailed techno-economic and financial study is conducted using a two-stage simulation. The technical aspect includes the hybrid energy system operation, with the results used as input for the cash flow model.
- A case study for a hybrid energy system composing of PV/biogas generator/Li-ion EES in Gorge Dam, Kenya is conducted with real-life solar, load, and electricity price data.

In particular, the work aims to:

- Identify the research gaps in techno-economic analysis of PV/biogas generator/Li-ion EES hybrid energy systems, with a focus on EES.
- Examine the financial performance of EES.
- Describe how EES economics relate to key financial parameters e.g. weighted average cost of capital (WACC).
- Provide a research agenda for financial and economic analysis of low-carbon energy storage.

The rest of this chapter is organized as follows: Section 2.2 provides a review of the literature on the techno-economic analysis and financing of EES and biogas/PV/EES hybrid energy systems. Section 2.3 presents the energy system context, and a case study on the LCOE of EES is given in Section 2.4. To examine the financing of EES, Sections 2.5 and 2.6 present the cash flow model and case studies respectively. A sensitivity analysis of the effect on NPV of various technical and economic parameters is provided in Section 2.7. Finally, discussion and conclusions are given in Sections 2.8 and 2.9, respectively.

2.2 Literature Review

This work is concerned with the financing and economics of hybrid energy systems under a range of EES capital costs and operating conditions. EES degradation is also considered, which can affect system lifetime. The first part of this literature review covers the techno-economic analysis of biogas, PV, and EES hybrid energy systems. After this, previous work on EES degradation is presented. Finally, financial studies of EES and renewable energy systems are evaluated.

2.2.1 Techno-economic Studies of Biogas, PV, and EES Hybrid Energy Systems

Das et al. [17] presented a techno-economic analysis of an off-grid PV/biogas generator/pumped hydro energy storage/battery hybrid renewable energy system for a radio transmitter station, using metaheuristic optimization approaches. Metaheuristic algorithms can outperform genetic algorithms in techno-economic optimization. The total net present cost and LCOE are examined. The LCOE from the hybrid energy system is found to be 0.4864 $/kWh, but the effect of storage on LCOE is not discussed.

Biomass for electricity is gaining popularity in rural areas of Pakistan. Ahmad et al. [18] used HOMER to conduct techno-economic analysis of a wind/PV/biogas generator hybrid energy system for rural electrification. The cost of energy, net present cost, and LCOE are examined. For a 50 MW system, the LCOE of the hybrid energy system is 0.058 $/kWh. A grid-connected PV/biomass/wind system can have a lower LCOE than a grid-connected PV/biomass system, showing wind to be an important component of hybrid energy systems in Pakistan. However, this work has not examined EES. Similarly, Shahzad et al. [19] used HOMER to perform techno-economic analysis of a PV/biomass off-grid system for rural areas in Pakistan. The economic indicators examined are cost of energy, net present cost and payback period. For the hybrid system, the capital cost has the highest share of net present cost, followed by the replacement, and finally the operating cost. Batteries are included in this work but there is no information on the LCOE for storage or the consideration of battery degradation. Das et al. [20] presented a techno-economic study for an off-grid biogas generator/PV/diesel/battery EES hybrid renewable energy system for application in remote areas of Bangladesh, where cow dung is a commonly-accessible resource for biogas production [20]. The cost of energy, net present cost, payback period and emissions are studied. The optimal system has PV, diesel generation, and biogas generation with capacity shares of 49, 36, and 15%, respectively.

In summary, a large number of studies uses the HOMER software, showing it to be a comprehensive and powerful tool for microgrid planning and techno-economic analysis of energy systems. However, it is a black-box model and not open source, and it is difficult to modify the optimization algorithm and cost calculation methodologies [5].

2.2.2 EES Degradation

Electrochemical EES has been used extensively in many electronic and electrical applications, such as mobile phones, laptops, and uninterruptible power supplies [21, 22]. In recent decades, EES has been extended to grid applications and applications requiring high energy and power densities, such as electric vehicles (EVs).

There are numerous parameters that may affect the state of health of an EES system, and for an electrochemical system the most prominent of these are temperature, charge/discharge rate (C-rate), and change in SOC. Therefore development of a comprehensive model that quantifies the capacity and power fade is complicated and challenging [23]. Reference [23] provides a technical discussion of the mechanisms that cause Li-ion cell degradation, affecting the electrolyte, electrodes, separator, and current collectors.

Both power systems and EVs require high EES energy (kWh) and power (kW) capacities to meet the energy demand. Since the electrochemical EES technologies used for EVs and power systems are the same, the issue of degradation is present in both research areas. In the area of EVs, one of the main barriers to their wide-scale adoption is the degradation of battery packs [24], leading to reductions in range and power output. EES degradation is affected by the dynamic battery temperature, ambient temperature, heat generated from chemical reactions in battery cycling, electrical resistance, and wear of mechanical components. The work in reference [24] provided a methodology to quantify EV battery degradation with different vehicle-to-grid services. The trade-off for the vehicle to provide grid services with maximum value with minimal impact on vehicle battery life was identified.

Degradation has a significant impact on the performance of electrochemical storage systems. It affects storage and power capacities, and hence the ability of the storage to meet electrical demands [23]. Li-ion cells degrade due to operation and environmental conditions. The degradation can be classified as cycling-induced degradation and calendar aging.

Regarding cycling-induced degradation, this is caused by operation of the EES system, C-rate, temperature, and energy throughput. The degradation is caused by mechanical strains in the lithium plating or electrodes' active materials, and is promoted by deep discharge, high C-rate, temperature, and energy throughput. As such, $LiFePO_4$ storage can potentially achieve 3200 cycles at 20% DOD (depth of discharge) or 760 cycles at 80% DOD [10].

To determine the rated cycle-life of a Li-ion EES under cycling from different SOC levels, Saxena et al. [25] determined that the capacity loss is affected by the mean SOC, change in SOC (ΔSOC), and C-rate. A power law model for the capacity loss was also developed based on the experimental results. The mean SOC is calculated with Eq. (2.1) during a discharge event.

$$\mathrm{SOC}_{\mathrm{Mean}} = \frac{\mathrm{SOC}_{\mathrm{Upper}} + \mathrm{SOC}_{\mathrm{Lower}}}{2} \tag{2.1}$$

$\mathrm{SOC}_{\mathrm{Lower}}$ is the SOC when the charging starts for each partial cycle and $\mathrm{SOC}_{\mathrm{Upper}}$ is the SOC when the discharge starts. The change in SOC is given in Eq. (2.2) as follows:

$$\Delta SOC = SOC_{Upper} - SOC_{Lower} \tag{2.2}$$

Subsequently, the rated cycle-life is calculated with Eqs. (2.3) and (2.4) [25] where NDC is the normalized discharge capacity with a depth of discharge at 80%.

$$Ratedcycle_{SOC_{Upper},SOC_{Lower}} = e^{\frac{\ln\left(\frac{100^{0.453} * (100 - NDC)}{a_{EES}}\right)}{0.453}} \tag{2.3}$$

$$a_{EES} = 3.25 * SOC_{Mean} * \left(1 + 3.25 * \Delta SOC - 2.25 * \Delta SOC^2\right) \tag{2.4}$$

For year n and discharge cycle k, the cost of EES degradation in each cycle ($) is calculated with Eq. (2.5) [10]. $C_{Cap_{EES}}$ is the EES capital cost ($/kWh) and $E_{EES_{Rated}}$ is the rated energy capacity (kWh).

$$C_{EES_{Degcycle}}(n, k) = \frac{C_{Cap_{EES}} E_{EES_{Rated}}}{Ratedcycle_{SOC_{Upper},SOC_{Lower}}(n, k)} \tag{2.5}$$

The EES system will reach end-of-life when the accumulated cost of degradation reaches the EES capital cost [10, 26], i.e. when W_{EES} equals to one, as provided in Eq. (2.6) as follows:

$$W_{EES} = \frac{\sum_{n}^{N} \sum_{k}^{K} C_{EES_{Degcycle}}(n, k)}{C_{Cap_{EES}} E_{EES_{Rated}}} \tag{2.6}$$

Turning to calendar aging, this class of degradation is independent of charge-discharge cycling. Calendar aging is mainly caused by time and temperature exposure due to the change in passivation layers at the electrode–electrolyte interfaces.

2.2.3 Techno-Economic Analysis for EES

Techno-economic analysis examines research, development, and deployment areas with a focus on benefits, costs, risks, time frames, and uncertainties [27]. LCOE is widely used to compare generation cost for an asset or energy system [28, 29]. An energy system typically operates over a long lifetime; a PV system, for example, may last for 25 years [30]. As such, LCOE includes a discount rate that converts future cash flows into their present value. A classical formulation of LCOE is given in Eq. (2.7) below [5]:

$$LCOE = \frac{\sum_{n=0}^{N} C_{cap_n} + \frac{C_{O\&M_n}}{(1+d)^n}}{\sum_{n=0}^{N} \frac{E_n}{(1+d)^n}} \tag{2.7}$$

where C_{cap} is the capital cost ($), $C_{O\&M}$ is the operation and maintenance (O&M) cost ($), E is the energy output (kWh), N is the system lifetime in years, and d is the discount rate. One of the key challenges in estimating the LCOE is to identify the costs (fixed, variable, direct, and indirect) and energy produced (accounting for round-trip efficiency).

In techno-economic analysis of EES, the costs can be separated into two types, namely direct and indirect costs [31]. Direct costs can be traced in an economically viable manner, whereas indirect costs cannot. The costs and revenues can be broken down into four categories as detailed below [32]:

- Monetary savings and profits: Revenues or savings are accumulated based on power, energy, or reliability related applications.
- Investment cost: Direct storage cost such as a battery, casing, and electrolyte. In addition, there is the grid coupling cost such as the transformers and power electronics.
- Operation and maintenance cost: Indirect cost such as conversion losses due to component efficiencies, auxiliary consumptions such as thermal management systems, and direct operating costs such as labor and insurance.
- Degradation and replacement cost: Battery performance degradation due to increased resistance and capacity fade, and fatigued materials replacement cost for battery and power electronics. Replacement cost needs to be considered if the unit of analysis is the hybrid system. Many studies consider degradation as an indirect cost [16, 17, 21].

The cost of EES can be evaluated via the levelized cost of storage (LCOS). The LCOS metric is derived from LCOE. The LCOS is given in Eq. (2.8) as follows [5, 8]:

$$LCOS = \frac{\sum_{n=0}^{N} C_{capEES_n} + \frac{C_{O\&MEES_n}}{(1+d)^n}}{\sum_{n=0}^{N} \frac{E_{out}}{(1+d)^n}} \tag{2.8}$$

C_{capEES} and $C_{O\&MEES}$ are the capital and O&M costs of EES respectively. E_{out} is the *EES* energy discharge.

Having summarized the components such as cost and revenue of EES economics, the following section provides a review on the recent works in EES techno-economics.

Obi et al. [9] proposed a methodology to calculate the LCOE for utility-scale storage systems. The purpose is to provide financiers, policy makers, and engineers a means by which to evaluate different EES systems with a common economic metric. Zakeri and Syri [21] examined life cycle costs and LCOS, using the Monte Carlo method to consider uncertainties. The study presents the economy of different EES for three main applications, i.e. frequency regulation, transmission and distribution support services, and bulk energy storage. Jülch [8] examined the LCOS for electrochemical EES, pumped hydro storage, and compressed air energy storage. The LCOS depends on the cost data, plant design, and annual operation hours. Belderbos et al. [33] proposed three different LCOS metrics and their application to EES for electricity price arbitrage. These metrics are known as "required average price spread," "required average discharge price," and "required average operational profit." Lai and McCulloch [5] studied the LCOS for vanadium redox flow batteries and Li-ion batteries for a PV system. The lifetime, costs, and efficiency can affect the LCOS. The works reviewed in this paragraph have not considered storage degradation.

Having reviewed the LCOS metric, the rest of this review covers general techno-economic analysis of EES. Shaw-Williams et al. [34] conducted a techno-economic analysis to evaluate the economic impacts on distribution networks of PV and EES investments. PV-only installations achieve the largest return, and the economic viability of a combined EES and PV system is marginal based on the current EES capital cost.

Kaldellis et al. [35] presented a mathematical model to maximize the contribution of a PV generator and to minimize the life-cycle electricity generation cost for remote island networks containing one or more PV generators and an EES system. It is determined that for islands with abundant solar resources, it is more cost effective to use a PV-EES system than thermal power stations. Xia et al. [16] proposed a stochastic cost-benefit analysis model. The energy system consists of wind generation and conventional generators. The model considers both the generation fuel cost expectation and the EES's amortized daily capital cost. Based on the cost-benefit analyses, it is indicated that the EES charging/discharging efficiency, capital cost, and lifetime are critical factors for optimizing the EES size, whilst it is not always economically viable to use EES in power systems. However, the degradation cost is not examined in neither work. Bordin et al. [26] presented linear programming models for the optimal management of off-grid systems. Battery degradation is included in the optimization model and the terms "cost per cycle" and "cost per kWh" for batteries were presented.

All in all, there is an increasing importance of and interest in studying the EES economics. As such, LCOS is a widely examined metric due to the simplicity of calculation and the ability to compare different EES costs "at a glance." Due to the complexities of battery degradation and its effect on cycle-life, as discussed in Section 2.2.1, battery degradation is only considered at a primitive level or not considered at all.

2.2.4 Financing for Renewable Energy Systems and EES

Due to high capital cost and uncertainties, financing is a key aspect for renewables power plants [36, 37]. Financing decentralized renewable energy infrastructures is a complicated task. Private investors are commonly reluctant to invest due to risk-return concerns and high transaction costs [38–41]. For many renewable energy projects, start-ups rely on their own capital, government support (grants and seed funds), or private funding sources such as angel investor and venture capital [39, 42].

The merit of a specific investment in a renewable energy technology can be examined by calculating indicators such as PP, NPV, and IRR [43]. The selection of financing structures, e.g. corporate financing, sales before construction, and leveraged lease for renewable energy projects, depends on technical maturity, financial viability of renewable energy technologies, and the availability of natural resources, in addition to the supported regulatory environment and government policies [39]. In brief, projects can be financed through debt and equity [37]. The financing cost is a crucial input for the calculations, since it changes the rate by which both electricity output and costs are discounted [44]. The WACC is used to determine a realistic discount rate to be used in a financial project appraisal, and can be calculated using Eq. (2.9) as follows [44–46]:

$$WACC = D.K_d.(1 - t) + E.K_e \tag{2.9}$$

D and E are percentage of debt (%) and percentage of equity (%) respectively, and sum to 100%. K_d and K_e are cost of debt (%) and cost of equity (%) respectively. t is the corporate tax rate (%).

There are several categories of technology related risk which need to be scrutinized for an investment decision. These risks are major determinants of the financing cost and structure, and can be broken down into six categories, namely Construction, Technological, O&M, Supply, Market, and Political [36, 46].

For new technologies, many of the above risks are often judged to be high, and this is reflected in a higher cost of financing, i.e. K_d and K_e increase with the perceived investment risk. For instance, loans can be obtained from banks and usually guarantees are required; these guarantees and the cost of the loan increase with the risk of the project [39].

Cucchiella et al. [47] used a discounted cash flow (DCF) model to examine the financial feasibility and NPV of PV integrated lead acid battery systems. It is found that subsidies are needed for the energy system to be profitable. Avendano-Mora and Camm [15] used the DCF model to examine the benefit-cost ratio, NPV, IRR, and PP of battery storage systems, for market-based frequency regulation service in a regional transmission organization. It shows that systems greater than 5 MW with minimal battery replacements are expected to have the best financial performance. Jones et al. [48] combined life cycle assessment and DCF analysis to find the carbon dioxide and financial impact of adding battery storage to a PV system. Battery costs need to be reduced rapidly, or extra revenue from delivering electricity system services is required to make batteries financially attractive in areas with reduced insolation. Financial studies of EES considering EES degradation are not examined.

Krupa and Harvey [46] examined the current and future financing of renewable electricity options. Over the past 10 years, private equity has contributed to the growth of the U.S. renewable electricity industry. Part of the capital came from commercial banks [49] and large investment banks, which exercised private equity funds to create public companies. Venture capital and private equity funds are pooled investment vehicles that raise money from large investors (such as pension funds) and wealthy individuals for targeted investments.

As reported by Yildiz [38], financial citizen participation is a financing approach that is increasingly popular in Germany, where private individuals can invest in renewable energy projects. The two main equity-based financial citizen participation business models are "The energy cooperative" and "Closed-end funds." Karltorp [36] studied the challenges of financing the development of offshore wind power and biomass gasification in Europe. Renewable energy tends to have high risks and low return. Therefore, it needs support from public and private finance. Energy bonds can be used to promote energy system investment.

To summarize, Table 2.1 presents an overview of the recent works in the techno-economic and financing studies of EES. It can be observed that many financial and economic indicators have been examined for different EES technologies. However, EES degradation is seldom taken into account, and the consideration of both financial and economic metrics for EES is missing.

Table 2.1 Comparison of recent techno-economic and financing studies for EES.

	This work	Berrada et al. [11]	Shaw-Williams et al. [34]	Locatelli et al. [12]	Guinot et al. [50]	Xia et al. [16]	Avendano-Mora and Camm [15]	Cucchiella et al. [47]	Jones et al. [48]
Country considered	Kenya	Spain	Australia	U.K.	Ilorin, Nigeria	Unspecified	U.S.A.	Italy	U.K.
Research context	Presenting a financial model for EES coupled with PV and AD biogas power plant. A DCF model for the Li-ion storage is introduced	A cost-benefit analysis is performed to determine the economic viability of energy storage used in residential and large-scale applications	Evaluating the scope for promoting distributed generation and storage from within existing network spending	Examining the value of real options valuation on the development of the ESS project	The techno-economic feasibility of a hybrid solar energy system, including lithium batteries and a hydrogen chain for an off-grid application	Using a stochastic model to size the ESS for power grid planning with intermittent wind generation.	Examining the expected financial performance of battery storage systems by providing market-based frequency regulation service	A DCF analysis is conducted to evaluate the financial feasibility of photovoltaic-integrated lead acid battery systems	Combining a life cycle assessment approach and DCF analysis to assess the carbon dioxide and financial impact, of adding battery storage to a PV system.
Financial and economic indicators examined	NPV, IRR, Debt duration, LCOE, and LCOS	NPV	NPV, IRR, LCOE, value of deferred augmentation, value of customer reliability	NPV	LCOE	Present value	NPV, PP, IRR, benefit-cost ratio, simple payback	NPV	NPV
Types of storage	Li-ion	Gravity storage	Batteries	Compressed air energy storage and pumped hydro storage	Li-ion	Lead acid, superconducting magnet, zinc bromine, and sodium-sulphur	Li-ion	Lead-acid	Li-ion
Storage degradation	Yes (cycle degradation considers the change in SOC)	No	Yes (as a percentage per year)	No	Yes, (cycle degradation and calendar degradation) DOD only	No	No	No	No

Findings	The existing market is unprofitable to use Li-ion when the capital cost is at 1500 $/kWh, unless participating in grid services with high payments	Storage is unprofitable for residential applications except if it is used as a stand-alone system	There are power network benefits with a more rapid adoption of distributed generation and residential battery storage	Real option analysis increases the economic performance of ESS	Not considering battery degradation leads to significant difference in estimating system size and LCOE values	The ESS charging/ discharging efficiency, amortized daily capital cost, and lifetime are crucial to affect the system cost-benefits	The finance can be affected by the system size, regulation market capability clearing prices, and EES replacements	The profitability of PV-integrated battery system is affected by the EES energy self-consumption and the presence of subsidies	The battery storage costs would have to drop significantly to contribute positively to the financial performance of PV systems in current U.K. market conditions

Source: From Lai et al. [A]/Elsevier/CC BY 4.0.

Another emerging financing method for renewables projects is crowdfunding [39, 51–54]. It is the practice of project funding by securing small amounts of cash from many people, typically via the Internet. Compared with traditional financing, crowdfunding has the advantage of low search and transaction costs, and savings can be passed on to investors [53]. It is possible to obtain project feedback via comment features on the crowdfunding page. However, due to the viral nature of this financing method, the project is prone to public failure if the funding campaign's goals are not achieved. Cybersecurity is also an issue as the funding is conducted via the internet. At present, there is no literature on financing EES with crowdfunding.

In summary, the deployment of EES in renewable energy systems is limited more by economics and financing than by the technology itself [13]. These include high capital costs and a lack of financing incentives and options. Similar to renewable energy, regulations, and market rules can impact strongly on whether EES is economically viable. As such, Miller and Carriveau [13] evaluated the factors and mechanisms of renewable energy financing that could be adapted for the EES industry. Compared to renewable energy, EES financing is more difficult to comprehend due to multifunctional capabilities and services.

2.3 Research Background: Hybrid Energy System in Kenya

The government in Kenya aims to provide energy access for all by 2020 [55]. Rural electrification in remote areas faces multiple challenges including the inability to extend the national grid to provide electricity in rural areas.

The Nationally Determined Contribution in Kenya has pledged to reduce its greenhouse gas emissions by 30% in 2030 from the emissions level in 2016 [56]. This is achievable with a timely deployment of renewable technology, and strict climate change policies in the transport and residential sectors [56].

Nowadays, businesses can take advantage of the opportunities presented by the changed regulatory environment and the abundant natural solar resource in Kenya [10, 57, 58]. Solar PV is becoming increasingly attractive as a grid electricity source. As commented by Ondraczek [57], previous studies suggest that solar PV in developing countries should "forever" only be used in off-grid applications, due to its high LCOE. Nevertheless, solar PV can be less expensive than the most expensive conventional generation technologies, e.g. emergency power plants running on heavy fuel oil [57].

In addition to the abundant solar resources, electricity generation can be achieved via biogas power plants with anaerobic digestion (AD). The large agricultural industry in Kenya produces significant animal and crop waste [10, 30].

Having discussed energy in Kenya in a wider context, the following section provides the hybrid system sizing and operating methodologies, to examine the EES finance.

2.3.1 Hybrid System Sizing and Operation

Figure 2.1 presents the layout of a hybrid PV-AD-EES energy system. The EES can be charged from the PV and from the AD biogas power plant. The AD plant has a minimum and maximum output power, denoted as $P_{AD_{Min}}$ and $P_{AD_{Max}}$, respectively. Load is met by the electricity output from the PV, biogas power plant, and EES. The solar charge controller is used to constrain the rate at which electric current is drawn from or added to the EES. The bi-directional inverter converts DC electricity to AC electricity and vice versa, to charge the EES from the AD. According to the sizing methodology

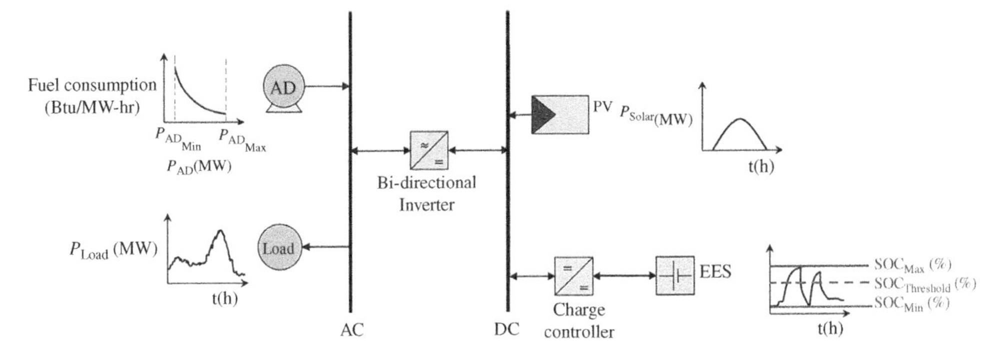

Figure 2.1 Diagram of the hybrid energy system. *Source:* Based on C. S. Lai [59], Chun Sing Lai et al. [A].

proposed in [30], the EES is sized with a power capacity and energy capacity of 2 MW and 5 MWh, respectively.

The optimal generator/EES dispatch or scheduling is challenging for the hybrid renewable energy system, due to the intermittent nature of solar power and the unpredictability of demand. Hence, this chapter adopts the operating regime proposed by Lai et al. [10], a deterministic rule-based approach. Since the biogas generator and EES are the dispatchable sources in the system, during times when PV is unavailable, the load can be met by biogas power or electricity stored in the EES. Hence in this regime, a SOC threshold, $SOC_{Threshold}$, has been defined for the EES to discharge its energy content to meet the demand before operating the biogas generator. A lower $SOC_{Threshold}$ reduces the solar curtailment (storing surplus energy) by cycling the EES more frequently.

Having presented the system operating method and optimal sizing, the following section presents the real-life solar and electricity price data to conduct the research.

2.3.2 Solar and Retail Electricity Price Data

This model considers an isolated community. There is no "opportunity cost" in the financial analysis because the assumption is that the community already receives the electricity it needs. The EES stores the surplus of electricity when is produced and not needed.

This research employs real-life data to examine the effectiveness of the proposed cash flow model for Kenya. These are solar irradiance data and national retail electricity market data, discussed as follows:

Solar irradiance data: Solar irradiance data is crucial for solar energy studies [30]. In this study the location for the solar irradiance is Turkwel Gorge Dam, Kenya, with longitude 35.34°, latitude 1.90°, and elevation at 1170 m above sea level. The solar irradiance data is obtained from SOLARGIS [60]. The sampling period is from 1 January 2012 to 31 December 2012 with a sampling interval of one sample every 15 minutes. Figure 2.2 displays the solar irradiance intensity for the location in 2012. The sunrise and sunset hours are consistent throughout the year and as expected the peak irradiance is at noon. The intermittency of solar irradiance can be seen in the figure during the daytime, shaded in blue.

Retail electricity market data: In Kenya as of January 2018, power distribution was maintained by a monopoly, Kenya Power and Lighting Company (Kenya Power). However, the government has recently introduced new companies for electricity retail [61]. Figure 2.3 shows the CI3

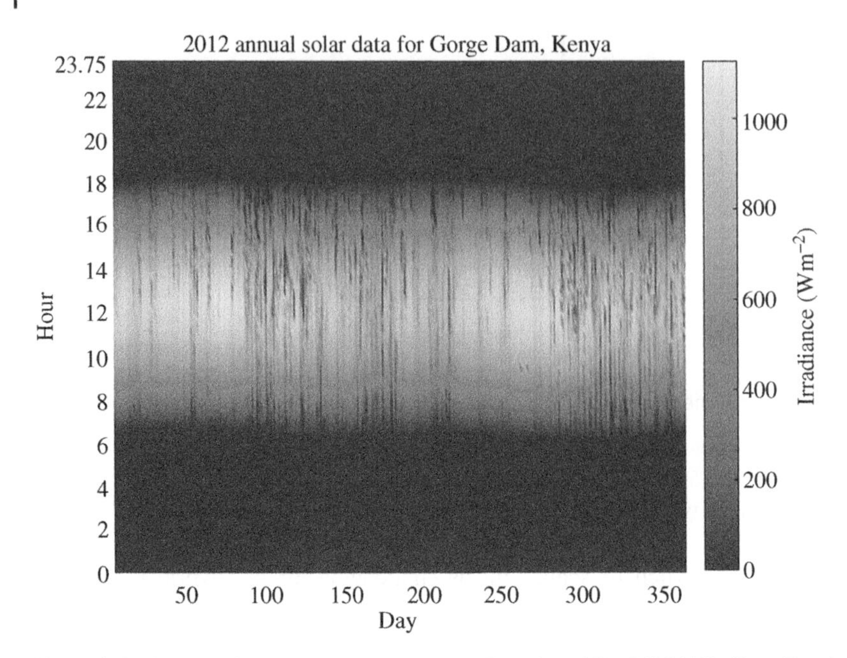

Figure 2.2 Kenya solar irradiance data. *Source:* Based on SOLARGIS [60], Chun Sing Lai et al. [A].

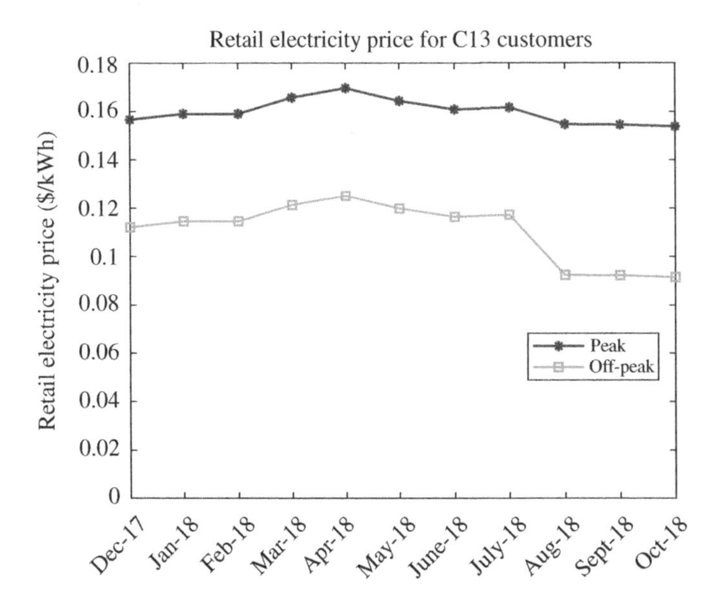

Figure 2.3 Retail electricity price in Kenya for CI3 customers [62]. *Source:* Based on Shah [62], Chun Sing Lai et al. [A].

customer's peak and off-peak retail electricity prices for Kenya, set by Kenya Power. Currently, there are seven core tariffs. These are known as DC (Domestic, 240 V), SC (Small Commercial, 240 V), CI1 (Commercial, 415 V), CI2 (Commercial, 11 kV), CI3 (Commercial, 33 kV), CI4 (Commercial, 66 kV), and CI5 (Commercial, 132 kV) [62]. The CI3 tariff is adopted based on the size of the hybrid energy system under consideration. The Government of Kenya has announced special off-peak rates for commercial customers with effect from December 2017. For CI3 customers, the average peak and off-peak electricity prices from December 2017 to October 2018 were 0.1632 and 0.1129 $/kWh, respectively [62].

Having described the research background in Kenya, i.e. hybrid energy system and data, the following section begins to examine the economics of the hybrid energy system with EES.

2.4 A Case Study on the Degradation Effect on LCOE

As degradation is an important aspect for EES cost-benefit analysis, this section examines how the degradation cost affects the LCOE of the hybrid energy system (by including and excluding the degradation cost in the LCOE calculation).

The key assumptions are:

- Operating life: 22 years [10];
- Discount rate: 6% [5, 30];
- PV capital cost: 0.36 $/W [63];
- AD rated capacity: 2.4 MW [10];
- Kenya load curve at 2 MW peak [10]; and
- Fixed 'operational cost' to EES energy discharge: 0.42 $/MWh [10].

The cost and technical parameters for the system can be found in [10].

The LCOE for the hybrid energy system can be calculated by Eq. (2.10) as follows [10]:

$$LCOE_{System} = \frac{C_{PV} + C_{Con} + C_{EES} + C_{AD} + C_{Inv}}{E_{AD_{Direct}} + E_{EES} + E_{PV_{Direct}}} \tag{2.10}$$

C_{PV}, C_{Con}, C_{EES}, C_{AD}, and C_{Inv} are the lifetime PV panels, solar charge controller, EES, AD biogas plant, and inverter costs (in NPV), respectively. E_{EES} is the lifetime energy output (discounted) from EES. $E_{AD_{Direct}}$ and $E_{PV_{Direct}}$ are the lifetime energy outputs that are used to meet the load directly (i.e. no storage) from AD and PV, respectively. The details for calculating the cost and energy output for the system consisting of each generation type and storage can be found in [10].

The SOC constraints are enforced by the operating regime and the power balance between generation and demand is achieved. For the case with no degradation cost, $C_{EESDegkWh}$ is not included in the LCOE. The degradation cost equation obtained from a capacity fade model can be found in Section 2.2.1. In this work, system LCOE refers to the LCOE for the hybrid system which considers the lifetime system, i.e. PV, AD, EES, inverters and solar charge controller costs and energy productions that meet the energy demand. The details of the mathematical modeling for the cost and energy calculations can be found in [10].

2.4.1 Sensitivity Analysis on the $SOC_{Threshold}$

This case study analyses how the dispatch priority for EES will affect the LCOE with respect to the degradation cost. The PV rated capacity is at 5 MW and the EES energy capacity is at 5 MWh [10]. Figure 2.4 presents the results for the sensitivity analysis with various values of $SOC_{Threshold}$. The diamond and circle symbols denote the maximum and minimum LCOE, respectively.

Without degradation cost, the least LCOE is achieved when storage is regularly discharged, i.e. $SOC_{Threshold}$ at 25% and the highest LCOE happens when storage is at minimal use. With

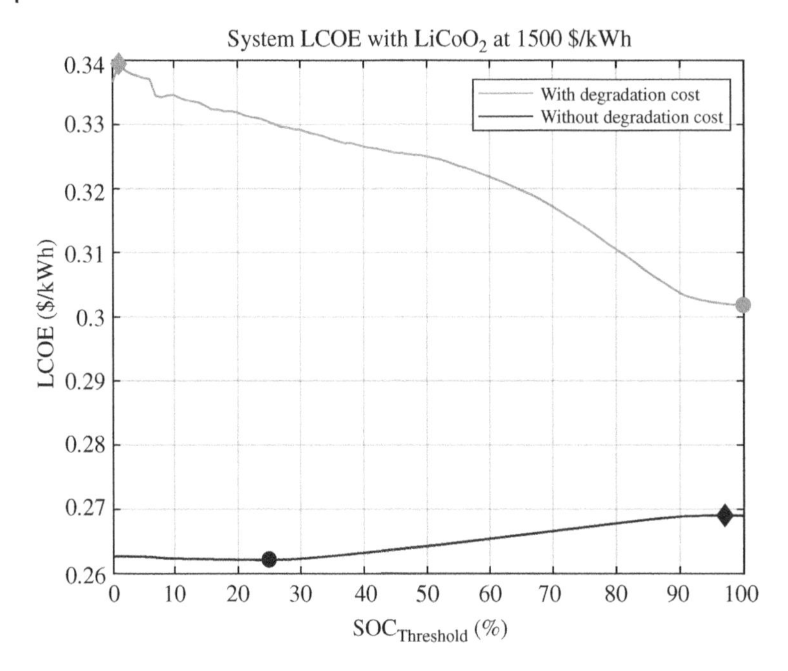

Figure 2.4 System LCOE studies with various $SOC_{Threshold}$. *Source:* From Lai et al. [A]/Elsevier/CC BY 4.0.

degradation cost, the least LCOE is achieved when storage is at minimal use, i.e. $SOC_{Threshold}$ at 100% and the highest LCOE occurs when storage is used as much as possible. This can be explained due to the degradation cost is included in the cycle-life degradation, the cost for each cycle can contribute to the loss in capital value and life expectancy of storage. When degradation is not considered, the frequent use of storage is ideal since it maximizes the use of the asset and the "fuel cost" for storage is minimal, the marginal cost for PV is approximately zero.

The EES degradation can affect the energy system's LCOE. By excluding degradation costs at scenarios with high EES capital costs, it is learned that the lowest LCOE can be achieved when EES is given dispatch priority over AD. This appears to be the opposite when degradation cost is included. Hence, degradation is an important aspect in EES techno-economic studies.

As the rated capacities of PV system and EES can affect the LCOE for the energy system, the next section examines the LCOE based on different EES and PV farm capacities, with and without EES degradation.

2.4.2 Sensitivity Analysis on PV and EES Rated Capacities

This case study investigates the energy system's LCOE at different energy storage capacity (MWh), and PV rated capacity (MW) when degradation cost is studied with EES at 1500 $/kWh energy capital costs [22]. Different results with the EES at 200 $/kWh energy capital costs were reported in [64]. Here, a $SOC_{Threshold}$ of 30% is used to frequently cycle the EES. Figures 2.5 and 2.6 depict the results for the system LCOE when degradation cost is considered and not considered respectively.

The energy system's LCOE increases proportionally to the degradation cost. The minimal LCOE is achieved when no EES is installed and has a 1.5–2.5 MW of PV rated capacity. The reduced capital

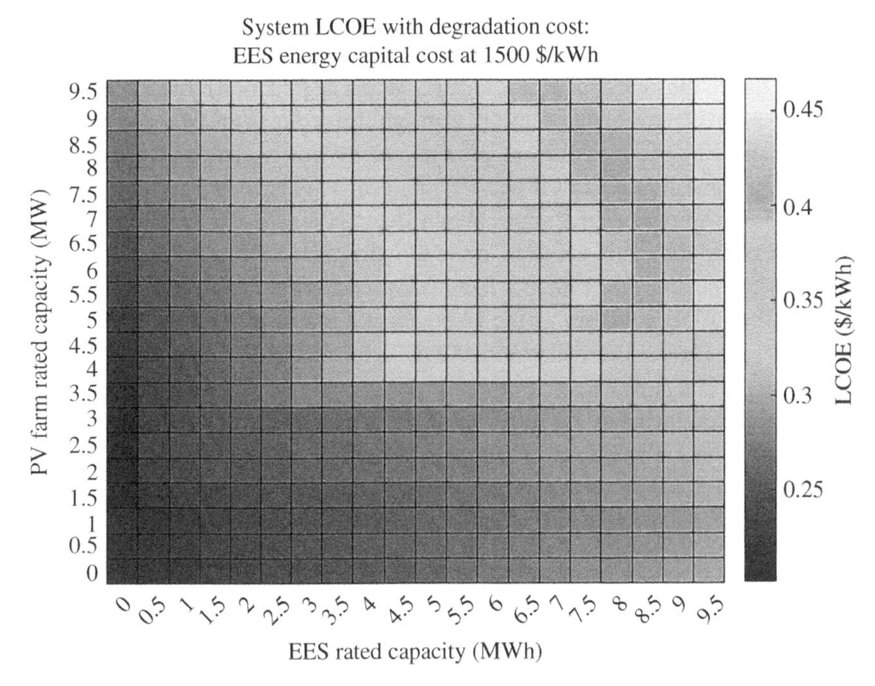

Figure 2.5 System LCOE with the degradation cost considered. *Source:* From Lai et al. [A]/Elsevier/CC BY 4.0.

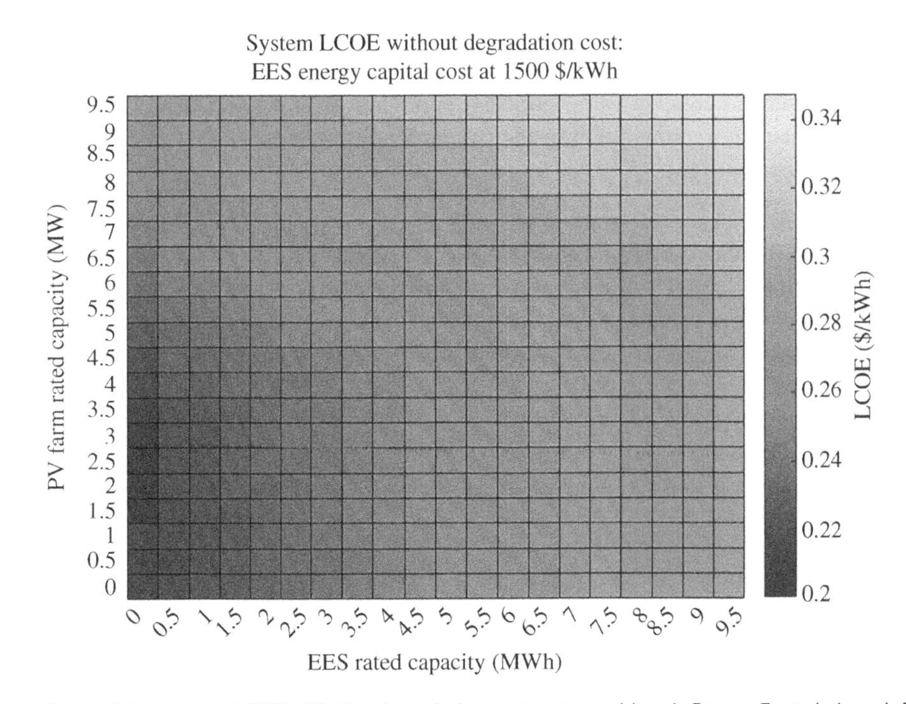

Figure 2.6 System LCOE with the degradation cost not considered. *Source:* From Lai et al. [A]/Elsevier/CC BY 4.0.

cost and negligible marginal cost for PV can produce less expensive electricity than the biogas. In this case with no import electricity, storing the surplus energy produced by PV for later use is not the most economic choice, due to the high capital cost for EES. When degradation cost is considered with the PV capacity below 3.5 MW, the change in LCOE is insignificant due to the battery cycling is reduced. This can be explained by the insignificant presence of PV power. The LCOE escalates when EES rated capacity is larger than 6.5 MWh and PV rated capacity is above 4.5 MW. This is the contribution of significant storage degradation. When degradation cost is not considered, the nonlinear mathematical relationship between cycle-life degradation (cycles) and cost is excluded in the techno-economic analysis. Due to a fixed O&M cost is applied to EES discharge, it could be observed that the LCOE increases as the EES capacity increases. Similar to the case where degradation cost is considered, the maximum LCOE is located at PV is at 9.5 MW and EES is at 9.5 MWh.

This section concludes that the inclusion of an EES cannot be justified only by the economic merit, at least with the actual market in Kenya. However, as discussed in the Introduction, there are political reasons to support the deployment of EES and renewable. Hence, the next section will examine EES from the financial perspective.

2.5 Financial Modeling for EES

The NPV is an important concept for economic and financial studies alike. The NPV is the summation of the present value of a series of present and future cash flows in both outbound and inbound generated by an investment in this case the EES with discount [12]. Figure 2.7 summarized how different types of cash flow are considered in the financial model. In the construction of an infrastructure, the capital can be provided in several ways with the most relevant are debt and equity.

Ideally, the investment has to create a value sufficient enough to gain support from the debt holders and to provide adequate remuneration to the equity. Realistic financial models consider three types of NPV as follows [45, 65]:

- **NPV for economic studies**: This is the "traditional" NPV used in economic studies as it does not consider taxes or how the finance is divided between equity and debt. The cash flow is discounted at WACC calculated with Eq. (2.9). The finance and economic data are presented in Section 2.5.2. The WACC is at 3.55%, which is viable as examined by Sidhu et al. [66]. This indicator only considers the cost aspect or the outbound cash flow. This NPV is useful for engineers and policy

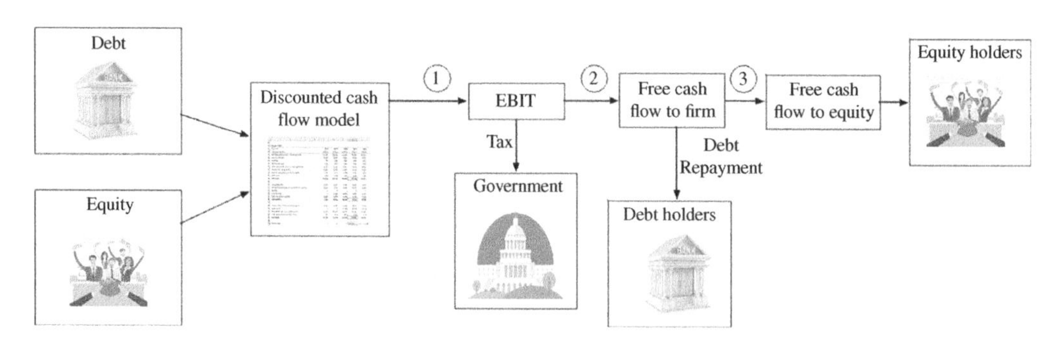

Figure 2.7 Exemplification of the financial model for EES. *Source:* From Lai et al. [A]/Elsevier/CC BY 4.0.

makers to calculate the LCOE and electricity price to break-even, since the LCOE is the average minimum price at which the electricity must be sold (at lifetime) for the project to break-even. The point of cash flow used for examining the LCOE is ① in Figure 2.7.

- **NPV to the firm:** This is the "free cash flow to the firm" or the sum of the unlevered cash flows discounted with the WACC. Debt holders may consider financing the EES project if the NPV is larger than zero. This signifies the investment generates enough value to pay off the debt. With respect to the "NPV for economic studies," this NPV considers the taxes and the financial structure of the investment. The point of cash flow used for examining the free cash flow to the firm (FCFF) is displayed as ② in Figure 2.7.
- **NPV to equity holders:** This is the sum of the levered cash flows. Specifically, the equity can be determined by discounting the "free cash flow to the equity" at equity cost. It is the NPV from the perspective of the equity holders, e.g. consider the payment of taxes and the repayment of debt to debt holders. The equity holders receive a remuneration equal to the cost of equity if the NPV is zero. The point of cash flow used for examining the free cash flow to equity (FCFE) is displayed as ③ in Figure 2.7.

2.5.1 Model Description

The inputs for the EES financial model can be separated into three major categories, namely, technical, economic, and financial as detailed in Figure 2.8. Figure 2.8 presents the financial modeling process for the EES with the model inputs and outputs. Remarkably the EES variable O&M cost also includes the biogas labor and fuel cost, when the biogas energy is stored in EES.

The model calculates the NPVs, debt durations, and IRRs with FCFF and FCFE. The LCOE is calculated with the EBIT cash flow. Equations (2.11) and (2.12) present the NPV and IRR to the firm. This is similar for "to the equity" calculations by changing the cash flow to the equity. The equity NPV needs to be discounted at the cost of equity and not the WACC.

$$NPV\ to\ the\ firm = \sum_{t=0}^{n} \frac{Cash\ flow\ to\ the\ firm_t}{(1 + WACC)^t} \tag{2.11}$$

$$\sum_{t=0}^{n} \frac{Cash\ flow\ to\ the\ firm_t}{\left(1 + IRR_{firm}\right)^t} = 0 \tag{2.12}$$

The debt duration is the amount of time for the project to repay the debt to the debt holders. This is calculated as $\sum_t^n Debt_{Duration}(t)$. Let D_{Cum_t} and E_{Cum_t} be the debt cumulated in million dollars (M$) and equity cumulated (M$) respectively, the debt duration is calculated with Eqs. (2.13) and (2.14) as follows:

$$Debt_{Duration}(t) = \begin{cases} 1, & if\ \ Debt_{Percentage}(t) > 0 \\ 0, & otherwise \end{cases} \tag{2.13}$$

$$Debt_{Percentage}(t) = \begin{cases} \dfrac{D_{Cum_t}}{D_{Cum_t} + E_{Cum_t}}, & if\ \ D_{Cum_t} > 0 \text{ and } E_{Cum_t} > 0 \\ 1, & if\ \ E_{Cum_t} \leq 0 \\ 0, & otherwise \end{cases} \tag{2.14}$$

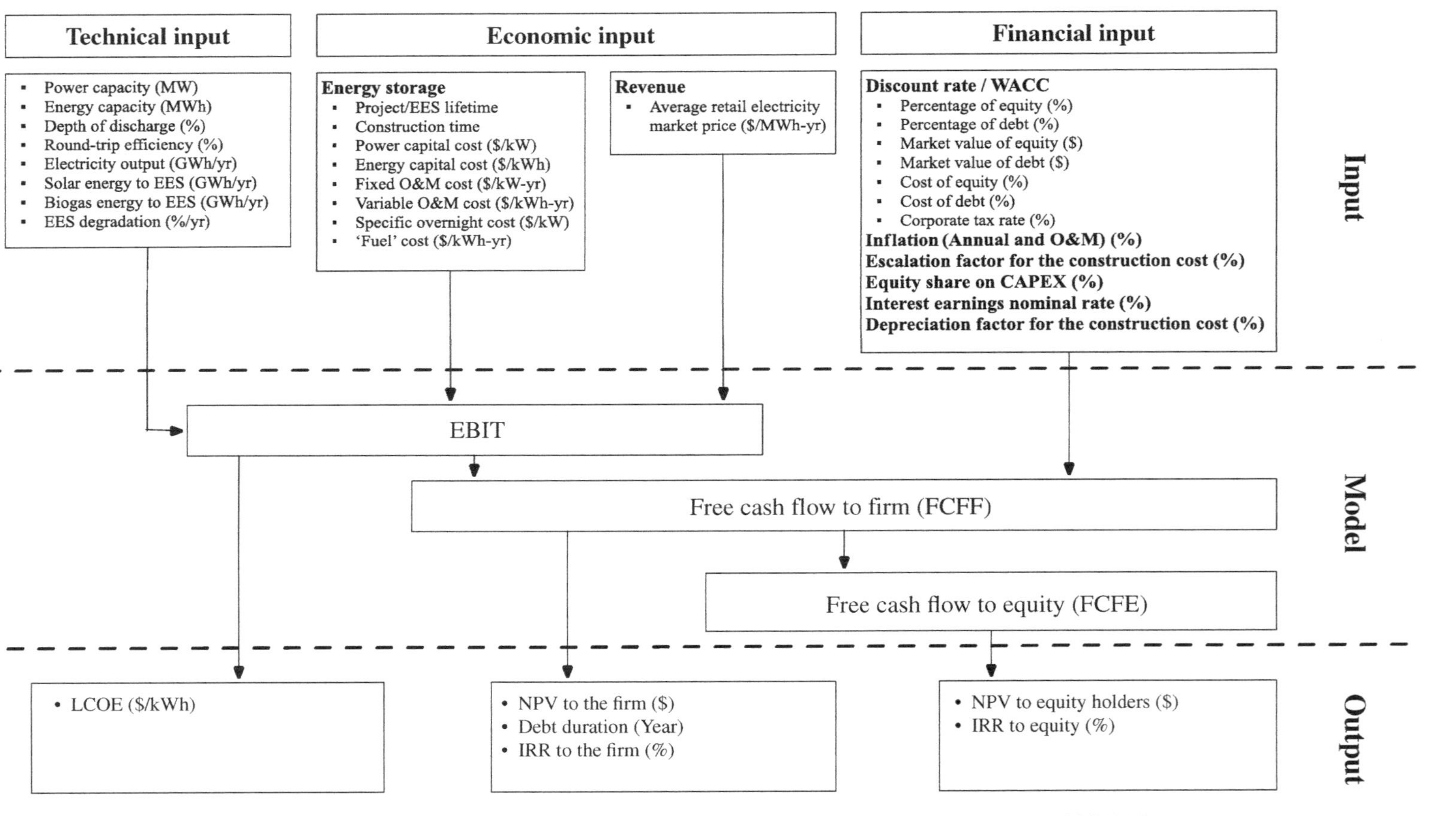

Figure 2.8 Technical, financial, and economic inputs for EES financial assessments. *Source:* From Lai et al. [A]/Elsevier/CC BY 4.0.

2.5.2 Case Studies Context

Technical, economic, and financing factors affect the finance and economic viability of EES. In this study, the focus will be on the financing studies based on the change in EES capital costs and operating conditions. The $SOC_{Threshold}$ affects how the electricity demand is met by the generators and storage.

This work considers six scenarios based on three operating methods and two EES capital costs as presented in Table 2.2. The names "High-PV," "Balanced," and "High-AD" represent different $SOC_{Threshold}$ values at 20, 50, and 100%, respectively. Table 2.2 presents the technical data for

Table 2.2 Technical and economic specifications.

Index	Input	Scenario					
		1	2	3	4	5	6
A1	$SOC_{Threshold}$ (%)	20	50	100	20	50	100
A2	EES capital cost ($/kWh)	1500			200		
A3	Energy capacity $E_{EES_{Rated}}$ (MWh)	5 [30]					
A4	Power capacity $P_{EES_{Rated}}$ (MW)	2 [30]					
A5	Construction time (yr)	1					
A6	EES operating lifetime (yr)	8	9	18	8	9	18
A7	EES energy output during first year of operation (GWh)	1.56	1.14	0.61	1.56	1.14	0.61
A8	Equivalent EES degradation cost (M$/yr)	0.86	0.75	0.40	0.114	0.10	0.05
A9	Round-trip efficiency (%)	95 [22]					
A10	EES fixed O&M costs ($/kW-yr)	2.12 [10]					
A11	Specific overnight cost ($/kWh)	1500			200		
A12	Total overnight cost (M$)	7.5			1		
A13	Biogas energy to storage (GWh/yr)	0.12	0.014	0.005	0.12	0.014	0.005
A14	Solar PV energy to EES (GWh/yr)	1.53	1.18	0.64	1.53	1.18	0.64
A15	Biogas labor and fuel cost for EES (M$/yr, fuel at 6.97 $/mcf + labor at 0.05 $/kWh [10])	0.014	0.0018	0.0006	0.014	0.0018	0.0006
A16	EES's NDC losses E_{Deg} (%/yr)	2.29	2.01	1.08	2.29	2.01	1.08
A17	Average retail electricity price ($/MWh)	138.10					

Source: From Lai et al. [A]/Elsevier/CC BY 4.0.

the financial modeling. The energy input and output of storage are calculated according to [10], relevant inputs are:

- Construction time: One year
- Annual inflation (%): 5.50 (2005–2018 average) [67]
- O&M inflation (%): 5.50 (2005–2018 average) [67]
- Escalation factor for the construction cost (%): 5.50 (2005–2018 average) [67]
- Depreciation factor for the construction cost (%): 10.00
- Equity share on capital expenditure (CAPEX) (%): 50.00
- k_d (%): 3.00
- k_e (%): 5.00
- Tax rate (%): 30.00 [68]
- Interest earnings nominal rate (%): 5.00 [69]
- Average retail electricity price ($/MWh): 138.10

The O&M inflation rate and the escalation factor for construction costs are assumed to be the same as the annual inflation rate. The average retail electricity price is calculated by taking the average of the peak and off-peak electricity prices, as discussed in Section 2.3.2. The EES will be "grid parity" if the project breaks even or makes a profit from the revenue based on the average retail electricity price.

The unit of analysis is the single EES project. Hence, the EES operating lifetime (years) and the NDC degradation per year need to be determined with the equivalent degradation cost as discussed in Section 2.2.1, as follows:

EES operating lifetime: The number of years for the EES to operate can be calculated with Eq. (2.6). The lifetime N needs to be an integer with W_{EES} to be less than and close to one.

NDC loss: The EES will reach the end-of-life when the NDC reaches 80% [10, 25]. The equivalent EES degradation cost ($) for the year can be calculated with $\sum_k^K C_{EES_{Degcycle}}(k)$ for k cycles. Based on the assumption that the perfect EES (i.e. NDC at 100%) is equal to the EES capital cost ($), Eq. (2.15) calculates the NDC degradation per year.

$$E_{Deg}(\%/\text{yr}) = \frac{(100 - NDC)\sum_{k}^{K} C_{EES_{Degcycle}}(k)}{C_{Cap_{EES}} E_{EES_{Rated}}} \tag{2.15}$$

The equivalent EES degradation cost per year, NDC losses per year, and the operating lifetime is presented in Table 2.2.

A Li-ion loss of NDC is approx. 2% per year [5]. The operating lifetime for the scenarios can be different. The "High-AD" scenario has a longer EES lifetime due to the reduced cycling and less energy input and output of EES. The NDC losses increase as more cycling occurs.

The biogas energy and costs for EES reduces as the $SOC_{Threshold}$ increases as seen in Table 2.2. "High-AD" refers to the operating strategy for the hybrid energy system to utilize more biogas. Hence, the total biogas energy output increases with a higher $SOC_{Threshold}$, as presented in [10]. However, the amount of biogas energy stored in EES may not necessarily increase with a higher $SOC_{Threshold}$. Biogas is a form of stored energy (i.e. in an anaerobic digester) and is inefficient to store the energy in EES in electricity form, due to conversion cost and energy losses. A low $SOC_{Threshold}$ causes the EES to be at a low SOC more often. To avoid energy deficits in the energy system, the biogas power plant needs to charge the EES to the $SOC_{Threshold}$ [10]. This section has presented the model description and the financial, economic and technical input data to conduct the study. The case studies to be examined are described. The next section presents the case study results for the EES financial feasibility in Kenya.

2.6 Case Studies on Financing EES in Kenya

This section presents a financial appraisal of Li-ion EES using the Kenyan scenario.

2.6.1 Influence of WACC on Equity NPV and LCOS

WACC is a key input for financial model, especially for capital intensive infrastructure. Hence, this section performs a sensitivity analysis on the WACC and to examine the effects to the equity NPV and LCOS of the EES. The WACC is considered as an overall combined effect of the cost of debt, cost of equity, share of CAPEX, and the corporate tax rate.

International Renewable Energy Agency (IRENA) [70] mentioned that the WACC for six low carbon generation technologies (i.e. wind, PV, concentrating solar power, hydro, biomass, and geothermal) is 10% for countries excluding OECD and China. A 10% WACC is also used by the Institute of Development Studies, U.K. for renewable energy projects in Kenya [71].

The WACC values are between 8 and 32% for 46 African countries and nine power generation technologies (i.e. concentrating solar power, PV, onshore wind, small hydro, geothermal, large hydro, coal, natural gas, and diesel) [72]. In African context, Sweerts et al. [72] reported that PV electricity needs a WACC of 6% to be cost competitive with natural gas. Similarly, Rose et al. [58] examined the prospects for grid-connected solar PV systems in Kenya with a 5% WACC. Grant Thornton [73] suggested that the WACC for ground mount solar PV for Kenya is between 11.8 and 16.25%.

Considering the literature, this chapter analyses a WACC between 0 and 20%.

Due to the importance and uncertainty in WACC, Figures 2.9 and 2.10 show the LCOS and equity NPV for the EES, respectively, under different WACCs. The EES capital cost is at 200 $/kWh. In Figure 2.9, compared to the other two scenarios, the LCOS has a higher rate of change with respect to the WACC for "High-AD." "High-AD" contributes to a higher life-cycle cost (i.e. LCOS) due to lower annual energy output. The increased rate of change can be explained by the nonlinearity of LCOS in Eq. (2.8). The LCOS is reasonable by considering the order of magnitude, compared to the Lazard's LCOS analysis [74] and the works in [5, 8–10]. In addition, Jülch [8] claimed that battery technologies can achieve 0.22 $/kWh in the future. From this study, the LCOS can be reduced if the system operates in "High-PV" scenario. The range of WACC required for the LCOS to be greater than the retail electricity price is between 2% (High-AD) and 10% (High-PV). This is an important indicator as the EES can be economic if the cost of electricity per kWh (LCOS) is less than the revenue generated per kWh.

The following examines the value of EES (NPV to equity) with respect to the WACC.

Rose et al. [58] claimed that a one percent increase in the discount rate results in a 6% decrease in the value of solar ($/W). For the case of EES, the relationship between NPV to equity and WACC is quadratic as depicted in Figure 2.10. Compared to other scenarios, "High-AD" can have a higher rate of change in the NPV at low WACCs because of the higher LCOS (Figure 2.9). It is profitable to invest in EES (i.e. $NPV > 0$) when the WACC is larger than 5.5, 7, and 9.8% for "High-AD," "Balanced," and "High-PV" operating scenarios, respectively. With reference to the WACC from IRENA, EES can be profitable in "High-PV" scenario. ESS can be profitable for all scenarios with the WACC used by Rose et al. [58] for PV plants. However, the NPV for the "High-AD" scenario is below zero with the WACC discussed by Sweerts et al. [72]. This analysis concludes that the WACC to make the EES profitable depends on the energy system operating strategy, and has not been examined and discussed in other studies.

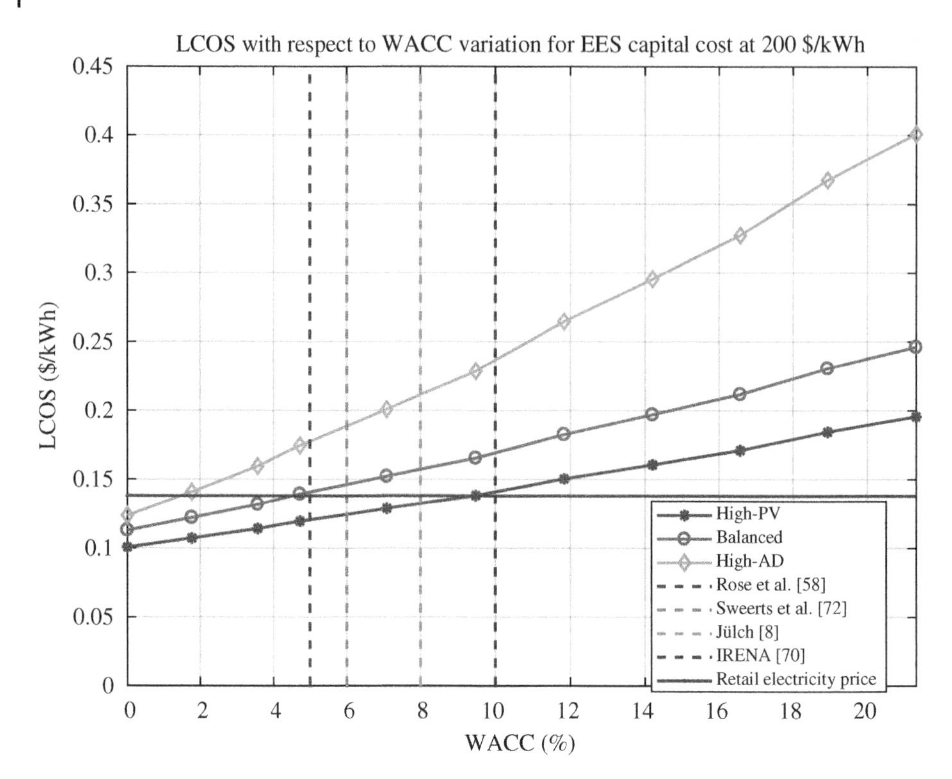

Figure 2.9 LCOS with respect to various WACC. The vertical dashed lines are the references from the literature, while the horizontal continuous line is the retail price of electricity. *Source:* From Lai et al. [A]/Elsevier/CC BY 4.0.

2.6.2 Equity and Firm Cash Flows

2.6.2.1 Cash Flows for EES Capital Cost at 1500 $/kWh

Figure 2.11 displays the non-discounted annual cash flows in firm and equity for the three operating scenarios based on an EES capital cost at 1500 $/kWh. It can be observed that none of the scenarios is profitable with the EES, as the cash flow at the end of project life is negative. Usually for a project, the cash flow repays the debt first as depicted in Figure 2.7. Hence, the cash flow to the equity is less than the cash flow to the firm. The FCFE for "High-PV" and "Balanced" scenarios are constant with no profit to the equity. However, the project is making a profit in overall and the debt gradually decreases. For "High-AD" scenario, the FCFF is decreasing, this means that the equity is losing money as the EES revenue does not even cover operating cost and debt repayment. Consequently, it is better to use the EES more frequently with a reduced lifetime to avoid extra capital loss.

2.6.2.2 Cash Flows for EES Capital Cost at 200 $/kWh

Figure 2.12 displays the non-discounted annual cash flows in both firm and equity for the three operating scenarios based on an EES capital cost at 200 $/kWh. The EES is financially viable under the three operating scenarios. The debt duration for the firm for "High-PV," "Balanced," and "High-AD" are approximately 3, 4, and 7 years, respectively.

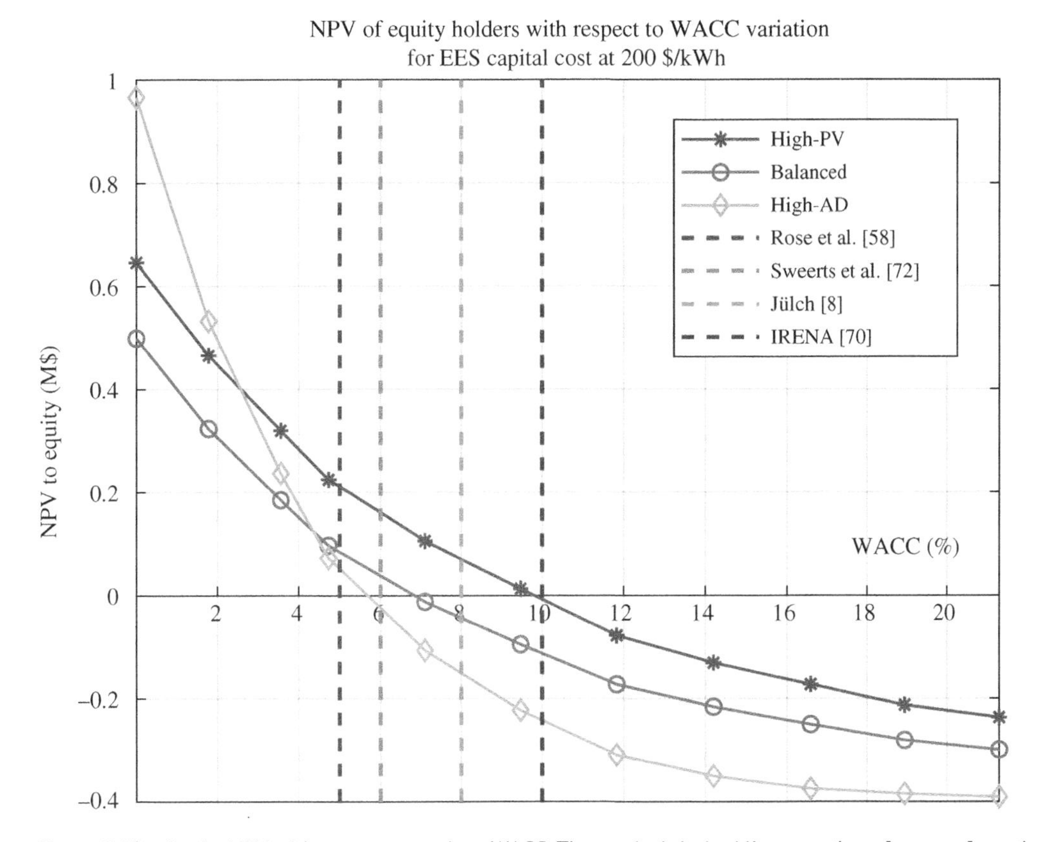

Figure 2.10 Equity NPV with respect to various WACC. The vertical dashed lines are the references from the literature. *Source:* From Lai et al. [A]/Elsevier/CC BY 4.0.

Although the degradation is higher due to more energy input and output of the EES and resulting in a shorter lifetime, this operating approach will be more financially attractive than "Balanced" or "High-AD" operating methods. To explain, the DD for the EES is shorter compared to the other scenarios with the highest cumulated cash flow by the end of the project life. Table 2.3 presents the financial modeling results for the six scenarios. The LCOS is the lowest amongst all scenarios when the EES discharges electricity more frequently with a capital cost at 200 $/kWh as mentioned in Scenario 4, with the following observed phenomena:

- The equity NPV and the IRR to the equity holders are the highest.
- The maximum and total exposition for firm and equity are the smallest.
- The retail electricity price in Kenya is greater than the LCOE, hence there will be revenue generated.

Having examined the cash flows for the three operating scenarios at two different EES's capital costs, it can be concluded that the EES operation method can significantly affect the economic and finance of the EES. At 1500 $/kWh, Li-ion proves to be too expensive for investment. When the cost drops to 200 $/kWh, it is financially viable by using the EES more frequently although the EES lifetime will be shorter.

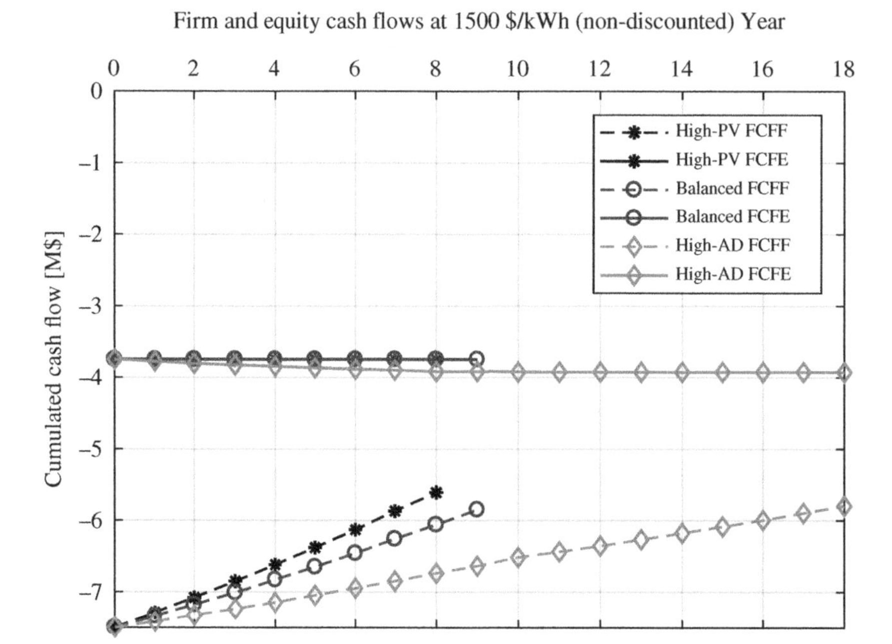

Figure 2.11 Cumulated cash flow to the firm and cumulated cash to the equity for three operating scenarios with EES capital cost at 1500 $/kWh. *Source:* From Lai et al. [A]/Elsevier/CC BY 4.0.

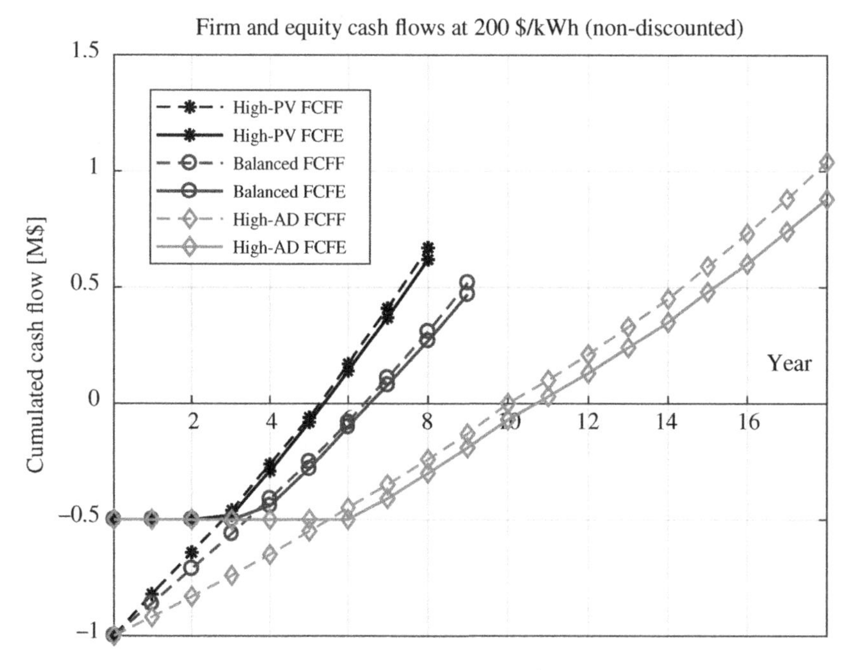

Figure 2.12 Cumulated cash flow to the firm and cumulated cash to equity for three operating scenarios with EES capital cost at 200 $/kWh. *Source:* From Lai et al. [A]/Elsevier/CC BY 4.0.

Table 2.3 Financial modeling results for the six scenarios.

	Scenarios					
	1	2	3	4	5	6
Electricity price – LCOE ($/kWh)	−0.62	−0.80	−0.96	0.02	0.01	−0.02
LCOS ($/kWh)	0.76	0.94	1.10	0.11	0.13	0.16
NPV (cash flow for LCOE) (M$)	−6.01	−6.21	−6.38	0.27	0.06	−0.10
IRR (cash flow for LCOE) (%)	0.00	0.00	0.00	9.95	4.99	2.13
NPV to the firm (M$)	−5.68	−5.91	−6.05	0.41	0.26	0.43
IRR to the firm (%)	0.00	0.00	0.00	12.11	8.66	7.89
Equity NPV (M$)	−5.20	−5.35	−5.17	0.32	0.18	0.24
Equity IRR (%)	0.00	0.00	0.00	14.61	10.25	8.57
Debt duration (year)	9.00	10.00	19.00	3.00	4.00	7.00
Max. exposition firm (M$)	−7.50	−7.50	−7.50	−1.00	−1.00	−1.00
Total exposition firm (M$)	−59.34	−67.13	−126.44	−3.25	−3.87	−5.86
Max. exposition equity (M$)	−3.75	−3.75	−3.93	−0.50	−0.50	−0.50
Total exposition equity (M$)	−33.75	−37.50	−73.90	−2.34	−2.82	−4.46

Source: From Lai et al. [A]/Elsevier/CC BY 4.0.

2.6.3 LCOS and Project Lifecycle Cost Composition

Since cost is an important aspect in finance, this section examines the LCOS and the lifecycle cost composition.

Figure 2.13 presents the breakdown of the EES costs in variable O&M, fixed O&M, and capital calculated from the cash flow for LCOE for the six scenarios. Capital is a major part of the cost, with the variable and fixed costs, e.g. servicing and import energy cost, constitute a small portion of the lifecycle cost. With reference to [19], this observation is reasonable as EES is a capital-intensive technology, similar to the cost of a hybrid energy system. At 200 $/kWh, it can be seen that the percentage of capital cost reduces with the "High-PV" scenario at approx. 85% due to the project being less capital-intensive. The "Balanced" scenario has the highest percentage of capital cost. This is due to the similar percentage of variable O&M cost to "High-AD" due to a comparable biogas consumption and a similar percentage of fixed O&M cost to "High-PV" due to a comparable lifetime, where both of them are low.

Since EES capital cost is uncertain with the economy of scale expansion and technology breakthrough, it is necessary to examine the LCOS based on different capital costs. Figure 2.14 presents the LCOS for the three operating scenarios under different EES capital costs. The relationship is linear for the three operating scenarios. The LCOS reduces as the EES cycles more frequently (with the "High-PV" scenario as most frequent), although the lifetime is also reduced. According to Eq. (2.8), this means that the energy output from EES is greater than the accumulated lifecycle cost.

This section shows the dependency of the capital cost, fixed O&M, and variable O&M cost on the lifecycle cost. Having examined the cost composition and the LCOS under different EES capital costs, the next section will study on the revenue aspect in the financing.

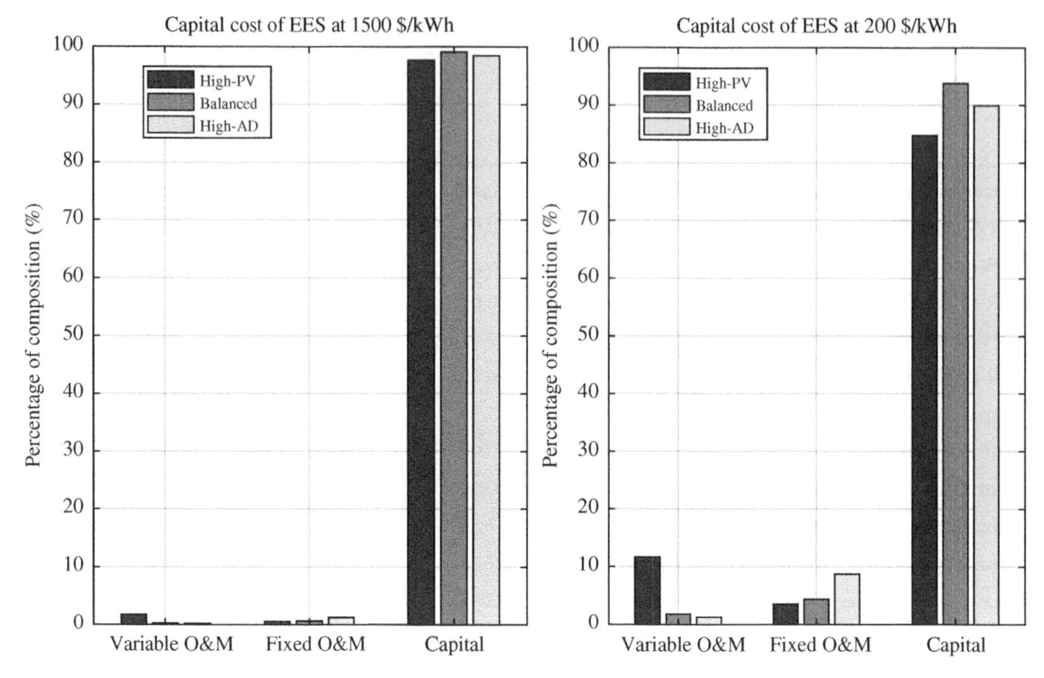

Figure 2.13 Percentage of the costs for three operating scenarios with EES capital cost at 1500 and 200 $/kWh. *Source:* From Lai et al. [A]/Elsevier/CC BY 4.0.

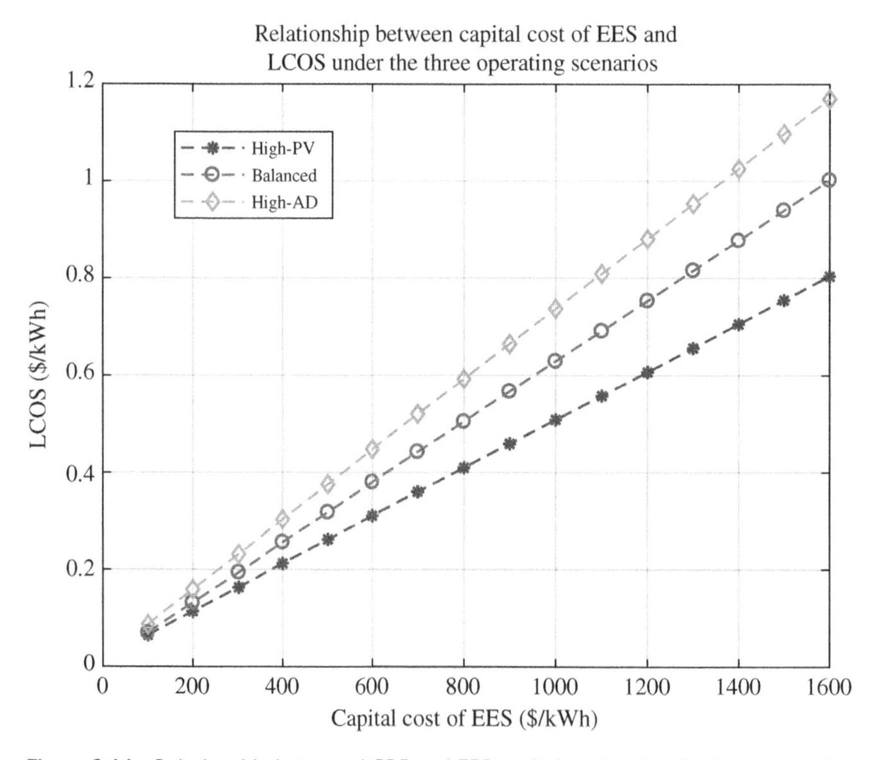

Figure 2.14 Relationship between LCOS and EES capital cost under the three operating scenarios.

2.6.4 EES Finance Under Different Electricity Prices

The electricity price affects the EES revenue. Due to intermittent generation, the electricity price can change in a low carbon electricity system [75]. EES has the ability to provide many grid services such as operating reserve and power quality improvement [22]. Hence, the electricity price for the EES energy is uncertain. As an example in the United Kingdom, the electricity price to perform short-term operating reserve (STOR) is approximately three times the electricity market price [12]. This section examines the EES finance under different retail electricity prices.

2.6.4.1 Study on the Retail Electricity Price

This section examines the equity cash flow under different retail electricity price, for the EES with a capital cost at 1500 $/kWh. The "High-PV" scenario is adopted as it is the one to have the best chance to gain the highest profit, as shown in Section 2.6.1. Figure 2.15 shows the cash flow for the EES under different retail electricity prices. It can be seen that a low price makes the project unprofitable. The equity will begin to have a positive cumulated cash flow when the electricity price is at 600 $/kWh. The cumulated cash flow to the equity increases with a higher electricity price and will take reduced number of years to break-even.

To examine the impact of different electricity prices on the EES financing in a more general context, Table 2.4 presents the equity NPV for the three operating scenarios according to different retail electricity prices. Achieving a positive cumulated cash flow to the equity does not guarantee to give a positive equity NPV. The NPV will begin to be positive when the electricity price is at 700 $/MWh for the "High-PV" scenario and EES capital cost at 1500 $/kWh. With the capital cost at 200 $/kWh, the EES can make a profit when the retail electricity price is at 100 $/MWh or above.

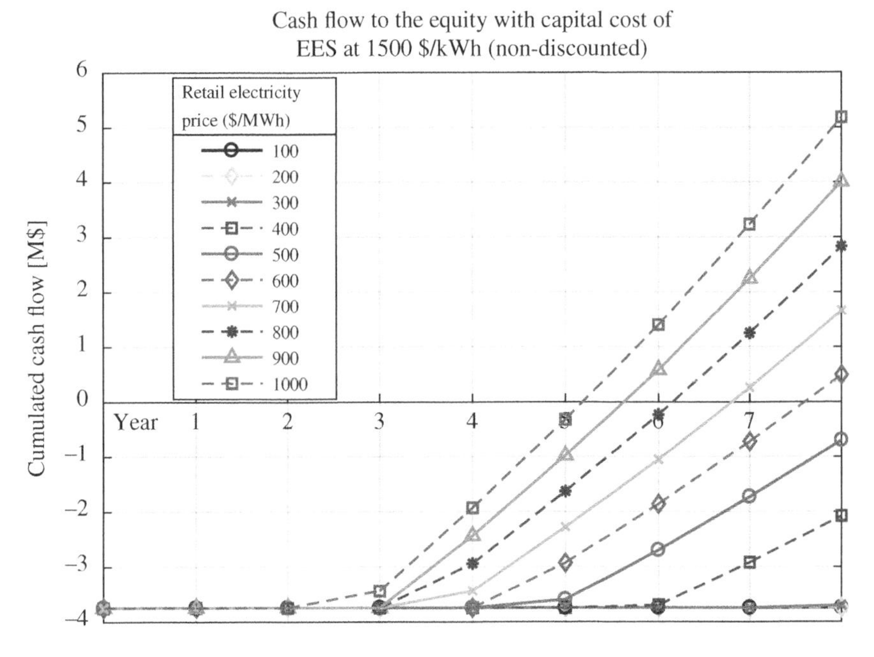

Figure 2.15 The cumulated cash flow to the equity under different retail electricity prices for EES. *Source:* From Lai et al. [A]/Elsevier/CC BY 4.0.

Table 2.4 Equity NPV (M$) under different EES's capital costs and retail electricity prices.

		EES capital cost ($/kWh)					
		1500			200		
		High-PV	Balanced	High-AD	High-PV	Balanced	High-AD
Retail electricity price ($/MWh)	100	−5.59	−5.66	−5.84	−0.01	−0.03	−0.06
	200	−4.57	−4.85	−4.65	0.86	0.63	0.76
	300	−3.55	−4.03	−4.04	1.75	1.35	1.57
	400	−2.47	−3.20	−3.42	2.64	2.08	2.39
	500	−1.50	−2.29	−2.67	3.53	2.80	3.21
	600	−0.65	−1.45	−1.80	4.43	3.53	4.03
	700	0.20	−0.70	−0.86	5.33	4.27	4.86
	800	1.06	0.00	0.11	6.22	5.00	5.68
	900	1.91	0.69	1.07	7.12	5.74	6.51
	1000	2.79	1.39	2.00	8.02	6.47	7.33

Source: From Lai et al. [A]/Elsevier/CC BY 4.0.
Positive number, Profitable
Negative number, Unprofitable

In summary, this section examined the cash flow to equity under different retail electricity prices. The equity NPV was also studied. It is identified that the EES profitability is affected by the retail electricity price and the EES operating conditions.

2.7 Sensitivity Analysis of Technical and Economic Parameters

Consider that there are many technical and economic uncertainties as presented in Table 2.2, this section presents a sensitivity analysis on the key parameters and to examine its effect on the NPV and LCOS.

Figures 2.16–2.18 present the sensitivity analysis results for "High-AD," "Balanced," and "High-PV" scenarios. The description for the index at the x-axis can be referred to Table 2.2. For the sake of this analysis, the parameters are varied +/− 10% independently from the others. The EES capital cost is set here at 200 $/kWh.

The EES lifetime has the largest impact on the NPV for "High-PV" and "Balanced," since the lifetime is relatively short, e.g. 8 or 9 years of operation. This is not the case for "High-AD." Energy related parameters, such as PV energy to EES and round-trip efficiency, have a stronger influence on the NPV than economic parameters for example fixed and operating costs which are negligible on the NPV since EES is a capital-intensive project. The only economic parameter that affects the NPV is the average wholesale electricity price as it affects the revenue. There is a larger "swing" in the NPV with the change of parameters value for "High-PV" as compared to others, as the operating lifetime is shorter and the annual energy output is higher.

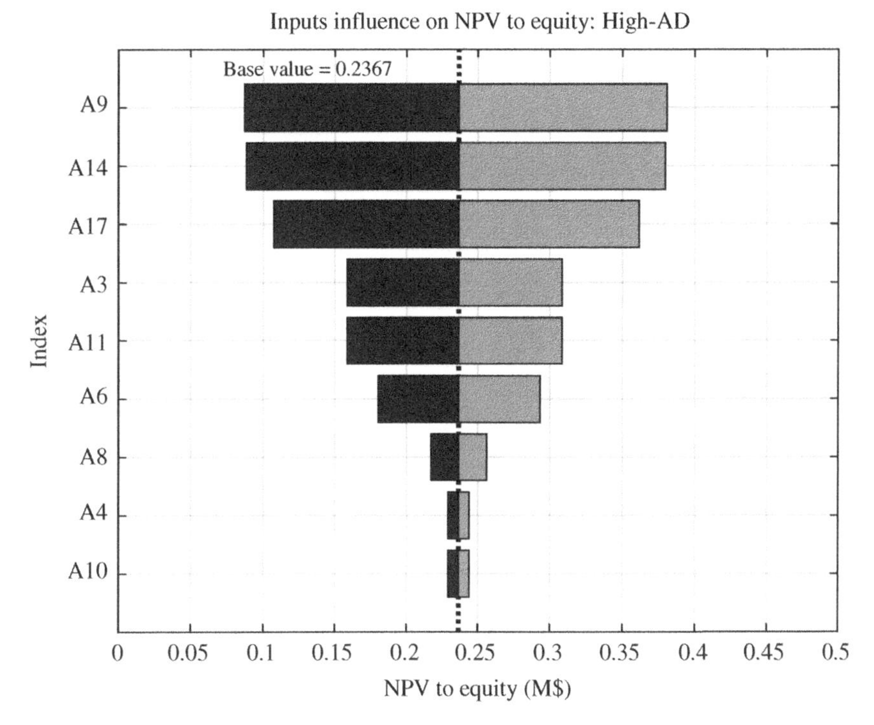

Figure 2.16 Sensitivity analysis on parameters for "High-AD" scenario. *Source:* From Lai et al. [A]/Elsevier/CC BY 4.0.

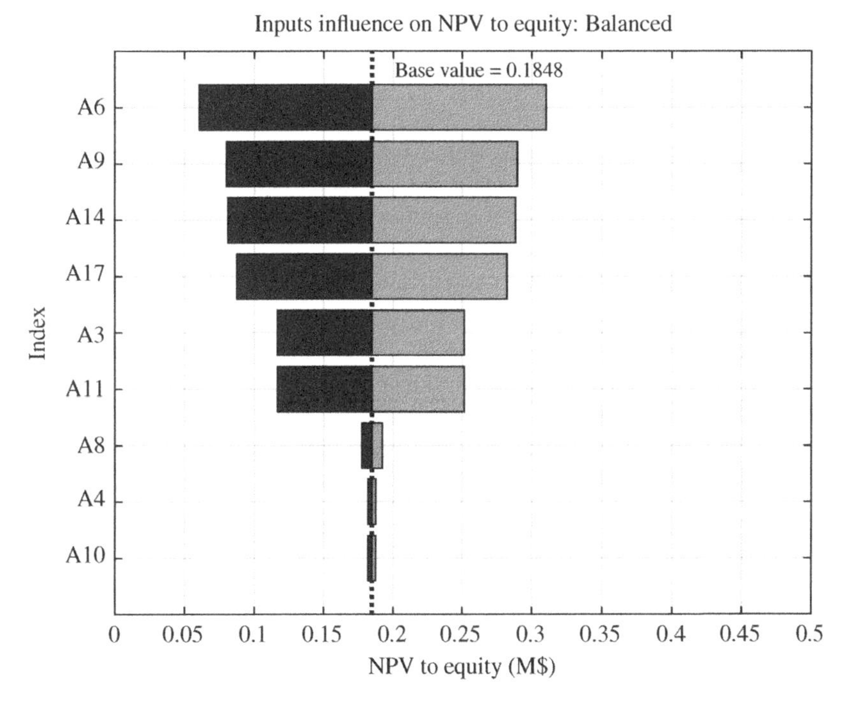

Figure 2.17 Sensitivity analysis on parameters for "Balanced" scenario. *Source:* From Lai et al. [A]/Elsevier/CC BY 4.0.

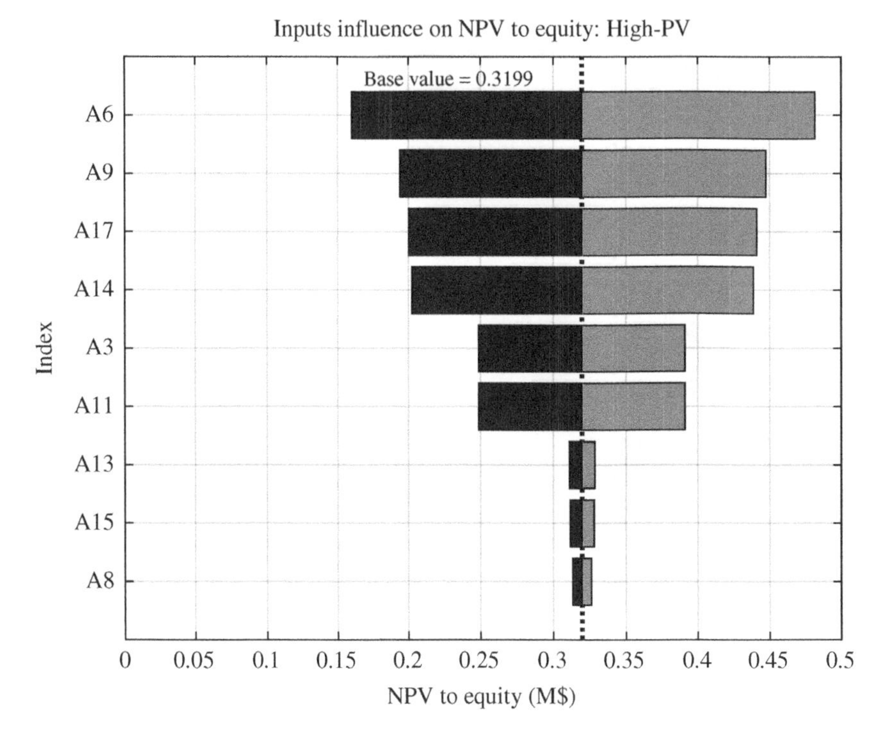

Figure 2.18 Sensitivity analysis on parameters for "High-PV" scenario. *Source:* From Lai et al. [A]/Elsevier/CC BY 4.0.

2.8 Discussion and Future Work

EES technologies such as Li-ion batteries are an increasingly important asset to support the rising penetrations of intermittent renewables and provide grid support such as energy balancing. With respect to policy implications, this work has examined the impact of WACC on the LCOS and equity NPV. Cost of debt and equity are reducing with respect to time due to lower risk and reduced borrowing costs promoted by governments [70]. For EES and low-carbon technology investments, the LCOE can be effectively reduced by reducing the WACC [70]. In Section 2.6.1, the LCOS has been studied and compared to values determined in previous studies. Apart from cost of debt and equity, the share of CAPEX and corporate tax rate also influences the financing of EES.

From the investment perspective, financing the EES can be challenging due to the technical difference with a generator, i.e. EES is not an electrical generator that transforms primary energy to electricity. Most of the current works as presented in Section 2.2 examines the economic feasibility of EES via techno-economic studies. However, the financing aspect is usually disregarded. It is important to identify the flow of cash whether inbound or outbound and the stakeholders whether with debt or equity clearly for a more in-depth cost-benefit study. In this study, it shows that the EES project financial feasibility and profitability depend on the retail electricity price, especially at high capital cost scenarios.

This work firstly examines the LCOE for a PV, AD biogas power plant, and EES hybrid energy system. Kenya is used for the case study. The research then focuses on the EES and a financial model is built to examine its financing and economics. The following phenomena are discovered in this research:

- The economics for EES with and without the degradation can be very different.
- The EES capital cost plays an important role in the financing. It constitutes most of the lifecycle cost. The EES is unprofitable under most operating situations when the capital cost is at 1500 $/kWh. When the capital cost drops to 200 $/kWh, the EES becomes profitable when it operates more frequently with a reduced lifetime.
- A higher retail electricity price is needed to make the EES profitable at 1500 $/kWh. For the three operating scenarios namely "High-PV," "Balanced," "High-AD," the retail electricity prices required to make EES profitable are 700, 900, and 800 $/MWh, respectively. Hence, the EES should participate in high revenue generating activities such as STOR.
- With the EES capital cost at 200 $/kWh, the EES can be used in regular discharging activities and will be profitable. The retail electricity price needs to be 200 $/MWh and above.

The following describes how the aims of this work are addressed:

- Section 2.2 has identified the research gaps in techno-economic analysis for PV/biogas generator/Li-ion EES hybrid energy system and EES. Subsequently, the EES degradation and financing are reviewed. It is identified that there is little work done in examining the finance of EES, with a focus on storage degradation.
- Section 2.5 has presented an EES cash flow model that consists of storage degradation. The technical results such as energy output and percentage of EES degradation per year are used as inputs for the cash flow model. Section 2.6 presented the financial and economic analysis. In particular the model has been verified, with the LCOS results compared to other works as presented in Section 2.6.1.
- Section 2.6 studied the economics and financing of EES in detail. Specifically, Section 2.6.1 examined how the WACC affects the NPV and LCOS of the EES. Section 2.6.2 investigated the cash flows for the three operating scenarios. Section 2.6.3 presented the breakdown of the cost of the EES by comparing how different operating strategies can affect the EES cost. Section 2.7 presents a sensitivity analysis on the economic and technical parameters and identified the key parameters that affects the equity NPV.

The present research has opened many future works. Future research directions on the financial and economic analysis for low-carbon energy storage are as follows:

- This work focuses the development of a financial model for the EES. Future work will develop and study the financial model for the hybrid energy system.
- As reviewed in Section 2.2, there are other types of financing methods for renewable and EES (such as crowdfunding). Models based on other financing theories will be useful for EES financing.
- The degradation aspect could be enhanced by including other models (e.g. calendar aging) and additional data (e.g. operating and ambient temperatures) as these affect the cycle-life and lifetime of the EES. This will impact the financing results.
- In this work, the revenue for EES is based on the solar surplus energy and the biogas energy that are used to charge the EES to the predefined $SOC_{Threshold}$. An energy arbitrage algorithm can be included to maximize the revenue. This is particularly important for grid-connected systems.
- Real options valuation will be useful in the financial model to take the technical, economic, and financing uncertainties into account [12].

- The current financial model can be expanded by including additional EES technical details. It is particularly useful to examine generation integrated energy storage systems by taking the exergy and transmission efficiency into account [76].

2.9 Conclusions

The EES profitability is difficult to establish due to uncertainties related to both key technical aspects such as the EES degradation, the operating lifetime and economic aspects such as EES capital cost and retail electricity price. Moreover, there is a need to integrate advanced EES technical models, e.g. accounting for EES degradation to the cash flow model for the EES project financial appraisal. The literature in the energy storage field is mostly grounded in simplistic economic models. The key contribution of this chapter is to detail a state-of-the-art financial model and apply it to the novel case of Li-ion EES. Three EES operating scenarios namely "High-PV," "Balanced," and "High-AD" were examined. It is identified that the project is unprofitable with the EES capital cost at 1500 $/kWh under current economic settings. The EES needs to participate in high revenue generating activities, e.g. short-term operating reserve for the EES to be profitable. For the case with the EES capital cost at 200 $/kWh, the project can be profitable when the EES is operated frequently, i.e. more energy discharge and higher number of EES cycling, even if the lifetime is reduced due to the increase in EES degradation. The chapter also shows the key importance of the weighted average cost of capital with respect to different system operating scenarios.

Acknowledgments

The permission given in using the materials from the following paper is very much appreciated.

[A] Lai, C.S., Locatelli, G., Pimm, A. et al. (2019). A financial model for lithium-ion storage in a photovoltaic and biogas energy system. *Applied Energy* 251: 113179.

References

1 Locatelli, G., Palerma, E., and Mancini, M. (2015). Assessing the economics of large Energy Storage Plants with an optimisation methodology. *Energy* 83: 15–28.

2 Luo, X., Wang, J., Dooner, M., and Clarke, J. (2015). Overview of current development in electrical energy storage technologies and the application potential in power system operation. *Applied Energy* 137: 511–536.

3 Hamedi, A.-S. and Rajabi-Ghahnavieh, A. (2016). Explicit degradation modelling in optimal lead–acid battery use for photovoltaic systems. *IET Generation, Transmission & Distribution* 10 (4): 1098–1106.

4 Zhang, Z., Wang, J., and Wang, X. (2015). An improved charging/discharging strategy of lithium batteries considering depreciation cost in day-ahead microgrid scheduling. *Energy Conversion and Management* 105: 675–684.

5 Lai, C.S. and McCulloch, M.D. (2017). Levelized cost of electricity for solar photovoltaic and electrical energy storage. *Applied Energy* 190: 191–203.

6 Mundada, A.S., Shah, K.K., and Pearce, J. (2016). Levelized cost of electricity for solar photovoltaic, battery and cogen hybrid systems. *Renewable and Sustainable Energy Reviews* 57: 692–703.

7 Singh, N. and McFarland, E.W. (2015). Levelized cost of energy and sensitivity analysis for the hydrogen–bromine flow battery. *Journal of Power Sources* 288: 187–198.

8 Jülch, V. (2016). Comparison of electricity storage options using levelized cost of storage (LCOS) method. *Applied Energy* 183: 1594–1606.

9 Obi, M., Jensen, S., Ferris, J.B., and Bass, R.B. (2017). Calculation of levelized costs of electricity for various electrical energy storage systems. *Renewable and Sustainable Energy Reviews* 67: 908–920.

10 Lai, C.S., Jia, Y., Xu, Z. et al. (2017). Levelized cost of electricity for photovoltaic/biogas power plant hybrid system with electrical energy storage degradation costs. *Energy Conversion and Management* 153: 34–47.

11 Berrada, A., Loudiyi, K., and Zorkani, I. (2017). Profitability, risk, and financial modeling of energy storage in residential and large scale applications. *Energy* 119: 94–109.

12 Locatelli, G., Invernizzi, D.C., and Mancini, M. (2016). Investment and risk appraisal in energy storage systems: a real options approach. *Energy* 104: 114–131.

13 Miller, L. and Carriveau, R. (2018). A review of energy storage financing – learning from and partnering with the renewable energy industry. *Journal of Energy Storage* 19: 311–319.

14 Dufo-López, R. and Bernal-Agustín, J.L. (2015). Techno-economic analysis of grid-connected battery storage. *Energy Conversion and Management* 91: 394–404.

15 Avendano-Mora, M. and Camm, E.H. (2015). Financial assessment of battery energy storage systems for frequency regulation service. In *Power & Energy Society General Meeting, 2015 IEEE*, Denver, CO, USA (26–30 July 2015), pp. 1–5: IEEE.

16 Xia, S., Chan, K., Luo, X. et al. (2018). Optimal sizing of energy storage system and its cost-benefit analysis for power grid planning with intermittent wind generation. *Renewable Energy* 122: 472–486.

17 Das, M., Singh, M.A.K., and Biswas, A. (2019). Techno-economic optimization of an off-grid hybrid renewable energy system using metaheuristic optimization approaches–Case of a radio transmitter station in India. *Energy Conversion and Management* 185: 339–352.

18 Ahmad, J., Imran, M., Khalid, A. et al. (2018). Techno economic analysis of a wind-photovoltaic-biomass hybrid renewable energy system for rural electrification: a case study of Kallar Kahar. *Energy* 148: 208–234.

19 Shahzad, M.K., Zahid, A., ur Rashid, T. et al. (2017). Techno-economic feasibility analysis of a solar-biomass off grid system for the electrification of remote rural areas in Pakistan using HOMER software. *Renewable Energy* 106: 264–273.

20 Das, B.K., Hoque, N., Mandal, S. et al. (2017). A techno-economic feasibility of a stand-alone hybrid power generation for remote area application in Bangladesh. *Energy* 134: 775–788.

21 Zakeri, B. and Syri, S. (2015). Electrical energy storage systems: a comparative life cycle cost analysis. *Renewable and Sustainable Energy Reviews* 42: 569–596.

22 Lai, C.S., Jia, Y., Lai, L.L. et al. (2017). A comprehensive review on large-scale photovoltaic system with applications of electrical energy storage. *Renewable and Sustainable Energy Reviews* 78: 439–451.

23 Birkl, C.R., Roberts, M.R., McTurk, E. et al. (2017). Degradation diagnostics for lithium ion cells. *Journal of Power Sources* 341: 373–386.

24 Wang, D., Coignard, J., Zeng, T. et al. (2016). Quantifying electric vehicle battery degradation from driving vs. vehicle-to-grid services. *Journal of Power Sources* 332: 193–203.

25 Saxena, S., Hendricks, C., and Pecht, M. (2016). Cycle life testing and modeling of graphite/$LiCoO_2$ cells under different state of charge ranges. *Journal of Power Sources* 327: 394–400.

26 Bordin, C., Anuta, H.O., Crossland, A. et al. (2017). A linear programming approach for battery degradation analysis and optimization in offgrid power systems with solar energy integration. *Renewable Energy* 101: 417–430.

27 NREL (n.d.). Techno-economic analysis. NREL, [Online]. https://www.nrel.gov/analysis/techno-economic.html (accessed 28 July 2022).
28 Darling, S.B., You, F., Veselka, T., and Velosa, A. (2011). Assumptions and the levelized cost of energy for photovoltaics. *Energy & Environmental Science* 4 (9): 3133–3139.
29 Kang, M.H. and Rohatgi, A. (2016). Quantitative analysis of the levelized cost of electricity of commercial scale photovoltaics systems in the US. *Solar Energy Materials and Solar Cells* 154: 71–77.
30 Lai, C.S. and McCulloch, M.D. (2017). Sizing of stand-alone solar PV and storage system with anaerobic digestion biogas power plants. *IEEE Transactions on Industrial Electronics* 64 (3): 2112–2121.
31 Horngren, C.T., Maguire, H.W., Datar, S.M. et al. (2017). *Horngren's Cost Accounting: A Managerial Emphasis*. Pearson Education Australia.
32 Hesse, H.C., Schimpe, M., Kucevic, D., and Jossen, A. (2017). Lithium-ion battery storage for the grid – a review of stationary battery storage system design tailored for applications in modern power grids. *Energies* 10 (12): 2107.
33 Belderbos, A., Delarue, E., Kessels, K., and D'haeseleer, W. (2017). Levelized cost of storage – introducing novel metrics. *Energy Economics* 67: 287–299.
34 Shaw-Williams, D., Susilawati, C., and Walker, G. (2018). Value of residential investment in photovoltaics and batteries in networks: a techno-economic analysis. *Energies* 11 (4): 1022.
35 Kaldellis, J., Zafirakis, D., Kaldelli, E., and Kavadias, K. (2009). Cost benefit analysis of a photovoltaic-energy storage electrification solution for remote islands. *Renewable Energy* 34 (5): 1299–1311.
36 Karltorp, K. (2016). Challenges in mobilising financial resources for renewable energy – the cases of biomass gasification and offshore wind power. *Environmental Innovation and Societal Transitions* 19: 96–110.
37 Steffen, B. (2018). The importance of project finance for renewable energy projects. *Energy Economics* 69: 280–294.
38 Yildiz, Ö. (2014). Financing renewable energy infrastructures via financial citizen participation–the case of Germany. *Renewable Energy* 68: 677–685.
39 Lam, P.T. and Law, A.O. (2018). Financing for renewable energy projects: a decision guide by developmental stages with case studies. *Renewable and Sustainable Energy Reviews* 90: 937–944.
40 Kayser, D. (2016). Solar photovoltaic projects in China: high investment risks and the need for institutional response. *Applied Energy* 174: 144–152.
41 Wang, X., Lu, M., Mao, W. et al. (2015). Improving benefit-cost analysis to overcome financing difficulties in promoting energy-efficient renovation of existing residential buildings in China. *Applied Energy* 141: 119–130.
42 Owen, R., Brennan, G., and Lyon, F. (2018). Enabling investment for the transition to a low carbon economy: government policy to finance early stage green innovation. *Current Opinion in Environmental Sustainability* 31: 137–145.
43 Yang, S., Zhu, X., and Guo, W. (2018). Cost-benefit analysis for the concentrated solar power in China. *Journal of Electrical and Computer Engineering* 2018: 4063691.
44 Ondraczek, J., Komendantova, N., and Patt, A. (2015). WACC the dog: the effect of financing costs on the levelized cost of solar PV power. *Renewable Energy* 75: 888–898.
45 Locatelli, G. and Mancini, M. (2010). Small–medium sized nuclear coal and gas power plant: a probabilistic analysis of their financial performances and influence of CO_2 cost. *Energy Policy* 38 (10): 6360–6374.
46 Krupa, J. and Harvey, L.D. (2017). Renewable electricity finance in the United States: a state-of-the-art review. *Energy* 135: 913–929.

47 Cucchiella, F., D'Adamo, I., and Gastaldi, M. (2017). The economic feasibility of residential energy storage combined with PV panels: The role of subsidies in Italy. *Energies* 10 (9): 1434.

48 Jones, C., Peshev, V., Gilbert, P., and Mander, S. (2017). Battery storage for post-incentive PV uptake? A financial and life cycle carbon assessment of a non-domestic building. *Journal of Cleaner Production* 167: 447–458.

49 Steffen, B. (2018). The importance of project finance for renewable energy projects. *Energy Economics* 69: 280–294.

50 Guinot, B., Champel, B., Montignac, F. et al. (2015). Techno-economic study of a PV-hydrogen-battery hybrid system for off-grid power supply: impact of performances' ageing on optimal system sizing and competitiveness. *International Journal of Hydrogen Energy* 40 (1): 623–632.

51 Lam, P.T. and Law, A.O. (2016). Crowdfunding for renewable and sustainable energy projects: an exploratory case study approach. *Renewable and Sustainable Energy Reviews* 60: 11–20.

52 Chen, J., Chen, L., Chen, J., and Xie, K. (2018). Mechanism and policy combination of technical sustainable entrepreneurship crowdfunding in China: a system dynamics analysis. *Journal of Cleaner Production* 177: 610–620.

53 Miller, L., Carriveau, R., and Harper, S. (2018). Innovative financing for renewable energy project development–recent case studies in North America. *International Journal of Environmental Studies* 75 (1): 121–134.

54 Zhu, L., Zhang, Q., Lu, H. et al. (2017). Study on crowdfunding's promoting effect on the expansion of electric vehicle charging piles based on game theory analysis. *Applied Energy* 196: 238–248.

55 Micangeli, A., Del Citto, R., Kiva, I.N. et al. (2017). Energy production analysis and optimization of mini-grid in remote areas: the case study of Habaswein, Kenya. *Energies* 10 (12): 2041.

56 Dalla Longa, F. and van der Zwaan, B. (2017). Do Kenya's climate change mitigation ambitions necessitate large-scale renewable energy deployment and dedicated low-carbon energy policy? *Renewable Energy* 113: 1559–1568.

57 Ondraczek, J. (2014). Are we there yet? Improving solar PV economics and power planning in developing countries: the case of Kenya. *Renewable and Sustainable Energy Reviews* 30: 604–615.

58 Rose, A., Stoner, R., and Pérez-Arriaga, I. (2016). Prospects for grid-connected solar PV in Kenya: a systems approach. *Applied Energy* 161: 583–590.

59 Lai, C.S. (2017). Hybridisation of high penetration photovoltaic, anaerobic digestion biogas power plant and electrical energy storage. D.Phil. thesis, Department of Engineering Science, University of Oxford. [Online]. https://ora.ox.ac.uk/objects/uuid:d2935f9e-d560-4b01-b040-8bd3f1beb121 (accessed 28 July 2022).

60 SOLARGIS (n.d.). SOLARGIS. [Online]. http://solargis.info/doc/free-solar-radiation-maps-GHI (accessed 28 July 2022).

61 Herbling, D. (2018). Kenya power may lose distribution monopoly with new law. Bloomberg.com, [Online]. https://www.bloomberg.com/news/articles/2018-07-04/kenya-power-may-lose-distribution-monopoly-with-new-law (accessed 28 July 2022).

62 Shah, S. (2019). Electricity cost in Kenya. https://www.stimatracker.com/ (accessed 28 July 2022).

63 Swanson, R.M. (2006). A vision for crystalline silicon photovoltaics. *Progress in Photovoltaics: Research and Applications* 14 (5): 443–453.

64 Lai, C.S., Locatelli, G., Pimm, A. et al. (2018). Levelized cost of electricity with storage degradation. Presented at the Offshore Energy and Storage Summit, Ningbo, China.

65 Arnaboldi, M., Azzone, G., and Giorgino, M. (2014). *Performance Measurement and Management for Engineers*. Academic Press.

66 Sidhu, A.S., Pollitt, M.G., and Anaya, K.L. (2018). A social cost benefit analysis of grid-scale electrical energy storage projects: a case study. *Applied Energy* 212: 881–894.

67 (n.d.). Kenya inflation rate. Trading Economics. https://tradingeconomics.com/kenya/inflation-cpi (Visited on 8 November 2020).

68 (n.d.). Kenya personal income tax rate. Trading Economics. https://tradingeconomics.com/kenya/personal-income-tax-rate (Visited on 8 November 2020).

69 (n.d.). Kenya interest rate. Trading Economics. https://tradingeconomics.com/kenya/interest-rate (Visited on 8 November 2020).

70 (2018). Renewable power generation costs in 2017. International Renewable Energy Agency. [Online]. https://www.irena.org/-/media/Files/IRENA/Agency/Publication/2018/Jan/IRENA_2017_Power_Costs_2018.pdf (Visited on 8 November 2020).

71 Pueyo, A., Bawakyillenuo, S., and Osiolo, H. (2016). Cost and returns of renewable energy in sub-Saharan Africa: A comparison of Kenya and Ghana. Institute of Development Studies. [Online]. https://opendocs.ids.ac.uk/opendocs/bitstream/handle/123456789/11297/ER190_CostandReturnsofRenewableEnergyinSubSaharanAfricaAComparisonofKenyaandGhana.pdf;jsessionid=CF0A1824599DB72C938387E8DD0BC00B?sequence=1 (Visited on 8 November 2020).

72 Sweerts, B., Dalla Longa, F., van der Zwaan, B.J.R., and Reviews, S.E. (2019). Financial de-risking to unlock Africa's renewable energy potential. *Renewable and Sustainable Energy Reviews* 102: 75–82.

73 (2018). Africa renewable energy discount rate survey – 2018. Grant Thornton. [Online]. https://www.grantthornton.co.uk/globalassets/1.-member-firms/united-kingdom/pdf/documents/africa-renewable-energy-discount-rate-survey-2018.pdf (Visited on 8 November 2020).

74 (2018). Lazard's levelized cost of storage analysis – version 4.0. Lazard. [Online]. https://www.lazard.com/media/450774/lazards-levelized-cost-of-storage-version-40-vfinal.pdf (Visited on 8 November 2020).

75 Vijay, A., Fouquet, N., Staffell, I., and Hawkes, A. (2017). The value of electricity and reserve services in low carbon electricity systems. *Applied Energy* 201: 111–123.

76 Garvey, S.D., Eames, P.C., Wang, J.H. et al. (2015). On generation-integrated energy storage. *Energy Policy* 86: 544–551.

3

Levelized Cost of Electricity for Photovoltaic with Energy Storage

Nomenclature

ΔSOC	Change in state of charge (%)
Δt	Hour interval
η_{EES}	EES round-trip efficiency (%)
η_{PV}	PV array efficiency (%)
σ	Degradation rate for PV (%)
ε	Solar irradiance (Wm^{-2})
a, b and c	Quadratic fuel cost function constants for biogas generator
AD	Anaerobic digestion
$C_{AD_{Fuel}}$	Fuel cost for biogas generator ($)
$C_{AD_{Labor}}$	Labor cost for operating the biogas power plant ($0.05/kWh)
C_{Asset}	Net present value of asset, i.e. PV, AD, EES, controller, or inverter ($)
$C_{Asset_{O\&M_{Fixed}}}$	Fixed operation and maintenance cost for asset, i.e. AD, EES or PV ($/kW)
$C_{Asset_{O\&M_{Total}}}$	Total O&M cost for asset, i.e. AD, EES, or PV ($)
$C_{Asset_{O\&M_{Var}}}$	Variable operation and maintenance cost for asset, i.e. AD, EES, or PV ($/kWh)
$C_{Asset_{Storetotal}}$	Net present value of electricity production from asset, i.e. AD or PV to be stored in EES ($)
$C_{Cap_{Asset}}$	Capital cost for asset, i.e. AD, controller, EES, inverter, or PV (Unit is asset dependent)
C_{EES}	Net present value of electrical energy storage ($)
$C_{EES_{Replacement}}$	Replacement cost per discharge cycle ($)
$C_{EES_{DegkWh}}$	EES degradation cost due to energy discharge ($)
C_{Gas}	AD gas cost (6.97 $/mcf)
$C_{Inst_{Asset}}$	Installation cost for asset, i.e. AD, controller, EES, inverter or PV (Unit is asset dependent)
$C_{O\&M_{Asset}}$	Operation and maintenance cost for asset, i.e. AD, controller, EES, inverter, or PV (Unit is asset dependent)
$C_{O\&M_{PVint}}$	Operation and maintenance cost for PV per hour ($)
CF	Capacity factor (%)
d	Discount rate (%)
d_{ees}, e_{ees} and f_{ees}	Three-parameter equation constants for EES rated cycle life at deep discharges

Smart Energy for Transportation and Health in a Smart City, First Edition. Chun Sing Lai, Loi Lei Lai and Qi Hong Lai.

DOD	Depth of discharge (%)
$E_{Asset_{Directtotal}}$	Net present value of electricity produced by asset, i.e. AD or PV for direct consumption (kWh)
E_{AD}	Net present value of electricity production of biogas generator (kWh)
E_{EES}	Net present value for EES electricity output (kWh)
$E_{EES_{Store}}$	Electricity to be stored in EES (kWh)
$E_{EES_{Rated}}$	Rated energy capacity of EES (kWh)
$E_{EES-S(X)}$	Electricity discharge by EES at stage X, $X \in \{1, 2\}$ (MW)
E_{PV}	Net present value of electricity generated by PV farm (kWh)
$E_{Surplus}$	Surplus electricity generated by PV system (kWh)
EES	Electrical energy storage
EMS	Energy management system
FiT	Feed-in tariff
LCOD	Levelized cost of delivery ($/kWh)
LCOE	Levelized cost of electricity ($/kWh)
$LCOE_{Asset}$	Levelized cost of electricity for generation asset, i.e. PV or AD ($/kWh)
$LCOE_{System}$	Levelized cost of electricity for system ($/kWh)
LCOS	Levelized cost of storage ($/kWh)
$LiCoO_2$	Lithium cobalt oxide
LHV	Lower heating value (905 btu/ft^3)
m	Number of EES cycles (integer)
n	System lifetime (years)
N_{Con}	Number of controllers (integer)
N_{EES}	Number of EES replacements (integer)
N_{Inv}	Number of inverters (integer)
N_{PV}	Total number of PV panels (integer)
N_{Store}	Number of PV panels for generating electricity for storage (integer)
NDC	Normalized discharge capacity (%)
NaS	Sodium-sulfur
NiMH	Nickel-metal hydride
O&M	Operation and maintenance
P_{AD}	Output power of biogas power plant (MW)
$P_{AD_{Max}}$	Rated power capacity of biogas power plant (MW)
$P_{AD_{Min}}$	Minimum output power of biogas power plant (MW)
$P_{Asset_{Direct}}$	Power generated with asset, i.e. AD or PV for direct consumption (MW)
$P_{Asset_{Store}}$	Power generated with asset, i.e. AD or PV for storage (MW)
P_{Con}	Rated power of controller (kW)
$P_{EES_{S(X)}}$	EES power discharge at stage X, $X \in \{1, 2\}$ (MW)
$P_{Generation}$	Power generation (MW)
P_{Load}	Power demand (MW)
P_{Inv}	Rated power of inverter (kW)
P_{PV}	Output power of PV plant (kW)
$P_{PV_{Rated}}$	Rated capacity of PV plant (kW)
$P_{Surplus}$	Surplus power generated by PV farm (MW)
PV	Photovoltaic
$Ratedcycle$	Rated cycle life of EES (integer)

SOC	State of charge (%)
SOC_{Lower}	The minimum SOC value of a cycle (%)
SOC_{Max}	Maximum state of charge (%)
SOC_{Mean}	Mean state of charge (%)
SOC_{Min}	Minimum state of charge (%)
$SOC_{Threshold}$	SOC threshold (%)
SOC_{Upper}	The maximum SOC value of a cycle (%)
SSR	Self-sufficiency ratio (unitless)
t	Time (hour)

3.1 Introduction

Electrical energy storage (EES) plays an increasingly important role in electrical power systems, in particular for energy balancing in off-grid systems. With the escalation of energy demand and the pressure to reduce environmental pollution, renewable energy source such as solar photovoltaic (PV) needs to be adopted [1, 2]. For countries located at the equator, e.g. Kenya in Africa, there is an abundant amount of solar insolation throughout the year. In addition, the waste product generated from the large agricultural industry in Kenya makes electrical power generation from biogas power plant via anaerobic digestion (AD) a desirable option [1]. Therefore, an optimal hybrid energy system for a rural community in Kenya should consist of solar PV and AD biogas power plant. In this chapter, the term AD represents the combination of the anaerobic digester and the biogas power plant.

In general, off-grid hybrid renewable energy systems perform better with multiple energy sources as compared to a single one [3]. This can be explained by the fact that different energy sources have different technical constraints and may be used to complement each other so to maximize the security of supply. The generation costs could also be potentially reduced. However, the control, design, and optimization of such systems is a complicated matter. In general, many of these systems were designed with the aim to minimize the total generation cost such as the levelized cost of electricity (LCOE) [3, 4].

The operation strategy for a system with an EES and PV generator is relatively simple. That is, surplus energy is stored in EES and discharges if the load is greater than generation. However, interesting questions may arise for systems with multiple energy sources. For the case where a dispatchable source such as AD is included, it is required to determine how the EES is charged and which dispatchable source (AD or EES) is to be used when the load demand is greater than the generation. As mentioned in [4], there are three basic control strategies for a PV-diesel-EES system, namely, the zero-charge strategy, full cycle-charge strategy and predictive control strategy. The EES is never charged with the diesel generator in the zero-charge strategy. Diesel generator is used to charge the EES to 100% state of charge (SOC) when the generator is on for the full cycle-charge strategy. Predictive control strategy requires the forecast of renewable generation and load demand to charge the EES. The advantage of this strategy is that energy wastage in surplus energy production from renewables is reduced. But it is essential to determine the optimal point for the SOC, between 0 and 100% to be charged with AD in order to provide a minimum operational cost [4]. In other words, the strategy will be less of an extreme and is between zero-charge and full cycle-charge.

Scheduling regimes such as rule-based strategies [5] have the advantages in avoiding the need of renewable and load forecasting for optimal operation. Additionally, complexity is further reduced when online optimization is not required. The work did not mention the degradation and costs of EES and have highlighted them as a future work.

There are numerous amounts of research works in cycle life studies and the costs due to EES degradation for hybrid renewable energy systems [6–9]. However, most of them do not consider partial charge–discharge cycles and the use of depth of discharge (DOD), i.e. only accurate for initial SOC at 100% for EES cycle life calculations. Electrical energy delivered is also used to consider the DOD in some literatures such as [6] and the actual values of the two SOCs may be neglected. Theoretically as an example, the electrical energy output from EES at SOCs of 100 to 80% may be the same as a situation for 40 to 20%. Recent literatures [10, 11] have confirmed that partial charge–discharge cycles at different SOC states have a profound effect to the State of Health, i.e. discharge capacity of the EES, and as a result affects the total available cycle life.

Due to irregular load demands and the PV power fluctuations induced from stochastic solar irradiance, the hybrid power system is highly affected by irregular SOCs and charge and discharge cycles. It is of paramount importance that the degradation costs and replacement costs are accurately accessed for EES in hybrid power system analysis and optimization. This work aims to provide a study on the economic projection of the hybrid system with the battery degradation costs included by considering the SOCs of each partial discharge cycle. To achieve this, an operating regime is proposed for the hybrid system that provides the optimal dispatch of PV, AD, and EES. Having identified the power output and SOCs for the system with respect to the system lifetime, LCOE is calculated for each asset and for the system. The cost and energy production will be critically analyzed.

Section 3.2 provides a literature review on renewable hybrid system and EES operations. Section 3.3 presents the data analysis and operating regime for the system. Case studies will be conducted and discussed. The cost analysis for the system will be presented in Section 3.4. The models for degradation per cycle and degradation per kWh will be derived. Finally, Section 3.5 draws the conclusion and future work will be discussed.

3.2 Literature Review

In recent years, there is an upsurge in research related to using EES for power systems, such as in charging regimes for electric vehicles [6, 7, 12], profit maximization with energy tariffs [13, 14], and off-grid system operations [2, 4]. Many of these literatures have included the degradation effect of EES to an extent and have made attempts to consider the storage costs. However, the effects of irregular cycling and partial cycles of EES are not properly addressed. Some works have considered to use reference DOD to compute the rated cycle life for the EES.

Reference [15] presented the grid tie electrical systems with distributed generation and energy storage for providing the electricity demand in Sri Lanka. The SOC for the EES has been considered for the control of the dispatchable energy source. However, the cost of storage and degradation effect has not been included.

The economic operation for a diesel-PV-EES island microgrid in rural areas with a two-stage model predictive control strategy was proposed in [16]. The system's economic is assessed with the LCOE metric. Replacement and operational costs for EES are included in the study, which is a function of DOD and cycle life. However, it is unclear how the DOD and cycle life are considered.

Levelized cost of delivery (LCOD) was proposed in [17] to compute the levelized cost of EES in PV system. The highlight of the work is that it is necessary to include the electrical energy production costs, i.e. generators cost into the storage costs. The work could be improved by also including the cost aspect of EES degradation.

EES degradation was considered with a fixed degradation cost in [18]. However, the cost competitiveness of the EES with respect to conventional generation is not provided.

Comparison of different energy storage technologies in a variety of realistic microgrid settings has been developed in [19]. The Energy System Model aims to provide similar functions to the well-known microgrid software HOMER, but with improvements in battery modeling. It is not known if and how the costs in partial cycles and irregular discharges were considered.

The economic feasibility of EES for electric bill management applications had been provided in [20]. A range of EES has been included, these are Li-ion, zinc battery, an advanced lead-acid, a sodium-sulfur (NaS) and a flow battery with no degradation consideration. The techno-economic assessment of distributed EES for load shifting application was provided in [21]. A Li-ion, NaS, and a redox battery were included in the study but again no degradation is included.

The work in [22] aims to develop degradation cost functions for optimization models in off-grid power system. The battery replacement cost has been provided for different ratios for battery degradation cost to diesel cost. Levelized cost is not calculated for the system.

An exhaustive study in determining the optimal combination of renewable electricity sources and EES connected to a 72 GW grid system was provided in [23]. The aim is to find the least-cost combination of renewable generation and EES to match the load demand. It concluded that excess renewable production can result in lower costs for the system. The argument is that the electrical load will be met with less storage, lowering the total system cost. The LCOE has not been studied and it is unknown how storage degradation costs are considered.

Reference [7] proposed an optimal generation scheduling model for a system with thermal power plant, PV plant, wind farm and EES whilst considering the degradation cost of ESS. Lead-acid and nickel metal hydride (NiMH) batteries were considered. This work aims to include the temperature and DOD for the battery degradation cost. The battery degradation cost is modeled and fitted with a piecewise linear function. However, cycle life effects with irregular cycling were not considered and the degradation model is only applicable for the cycles with initial SOC at 100%.

As reported in [13], a Feed-in tariff (FiT) scheme has already been established in Japan to promote the use of PV generators with electric batteries. Also mentioned in [24], FiT schemes could benefit the deployment of EES in the current and upcoming PV systems. This has already happened in other countries such as Germany, where a FiT scheme that maximizes self-consumption by using EES has been established. There are many researchers studied the operation of grid-connected PV-EES system, however, hardly any attention has been paid to the impact of the EES degradation characteristics. The work in [13] aims to evaluate the energy and cost savings for a grid-connected PV-EES system for a household. Operational optimization model was constructed for considering the degradation characteristics of EES. The model has two objective functions, known as the operating costs and energy savings. The study concluded that a grid-connected PV-EES system was more useful in energy saving operation. The cycle degradation is defined by the number of cycles and the cycle degradation rate. DOD degradation is considered to have occurred when the DOD is more than 70%.

The stochastic problem for microgrid energy scheduling with distributed generator and controllable loads such as plug-in electric vehicles was presented in [6]. EES degradation costs and cycle life were calculated with DOD which is defined as the ratio of the energy delivered from the EES to the rated energy capacity. Partial cycling is not included or discussed.

HOMER had been used for the design of a PV-biomass-diesel hybrid systems for off-grid and grid-connected scenarios in [25]. The system's configuration is optimized for different load profiles.

Surrtte 6CS25P flooded battery is used for the off-grid case. Details or studies for the EES have not been provided.

Reference [26] proposed a method to simultaneously optimize the rule-based operation strategy and the EES capacity for a grid-connected PV-EES system. Given the present cost of storage, it is claimed that by using EES with the conventional operation strategy, it is not profitable but increases the self-sufficiency ratio (SSR); the system uses more renewables with a higher SSR.

A microgrid charging and discharging strategy was given in [8]. A three-parameter function is fitted to estimate the battery cycle life under different DODs. With this function, the depreciation cost model of Li-ion battery was developed. It was acknowledged that the battery charge and discharge cycles under working conditions are composed of several micro-cycles with different DOD. To overcome this problem, the rain-flow counting method was employed to decompose the complex cycles to micro cycles of different DOD. Charge/discharge cycles were counted with corresponding to each range of the DOD for the year. Partial cycling was not included, and this method may not be appropriate if the initial SOC is not at 100%. A sizing methodology with loss of load probability for an off-grid PV system with lead-acid battery was presented in [27]. This work also adopts the rain-flow counting method for battery lifetime estimation. It is unclear if the rain-flow counting method can accurately represent the EES degradation, and a comparison with capacity fade model should be made.

A model for the operation of distribution companies in regulating price or locational marginal price mechanisms was given in [9]. Two types of EES were considered namely lead-acid and lithium-ion. It could be observed that there were multiple partial charging and discharging, at different SOCs throughout the time interval. It needs to take into account the partial cycles for a more accurate optimization problem formulation.

Reference [5] presented the rule-based strategies for operating an EES connected to a self-consumed PV system to reduce payments to the grid. As stated by the authors, presently the work does not consider the degradation cost of using the EES.

As suggested in the literature review, at present there is little to no work in considering the costs of power system operation for EES energy discharge at different SOC ranges. This research aims to provide a method in incorporating partial cycles at different SOC ranges for the techno-economic assessment of a PV-AD-EES hybrid system with a proposed operating regime.

3.3 Data Analysis and Operating Regime

3.3.1 Solar and Load Data Analysis

Real-life solar irradiance data were obtained from Solargis for Turkwel Gorge Dam, Kenya with coordinates 1.90 °N, 35.34 °E and elevation at 1170 m. The dataset contains 22 years (1994–2015) of data with sampling rate at 1 sample/15 minutes. Since the PV output power is influenced by the solar irradiance, it is important to know how season variations could impact the amount of solar irradiance received by the panels. The monthly solar insolation is computed for the dataset and for comparison purposes, the 22-year monthly solar insolation data from NASA were obtained in [28]. As solar PV is a non-dispatchable source, capacity factor (CF) is used to represent its annual energy output. CF is the ratio of the actual electrical energy output over the year to the maximum possible electrical energy output over the same time period. The CF is calculated for a 5 MW PV farm with inclusion of solar panel efficiency losses at 15% [1]. Figure 3.1 presents the monthly solar insolation and the annual capacity factor for the PV farm.

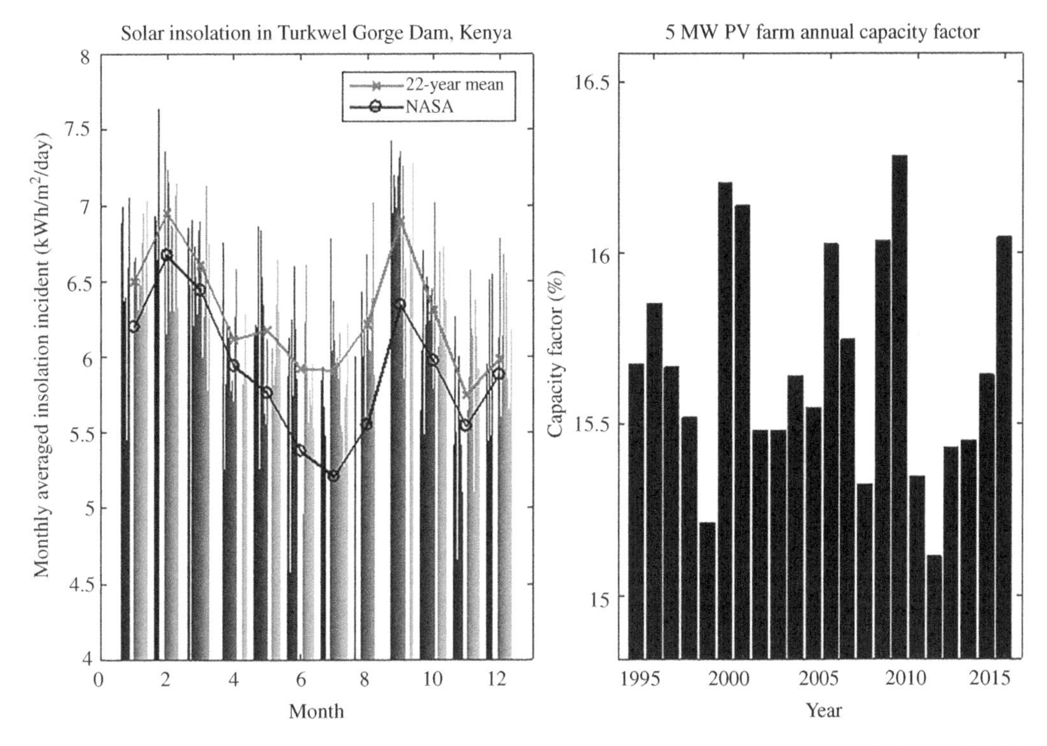

Figure 3.1 Solar insolation and capacity factor for solar farm *Source:* With permission from Lai et al. [A]/with permission of Elsevier.

As shown in Figure 3.1, the differences in annual CF could be significant. The CFs are 15.2 and 16.3% for 2011 and 2009, respectively, which is more than 1% difference. This shows that it may be inaccurate to assume CF to be the same annually throughout the system's lifetime. As reported in [29], a method to overcome this problem is to perform sensitivity analysis to CF to calculate the LCOE for hybrid system. However, CF only produces a single value by taking into account the energy generation, hence the information regarding the fluctuations and stochastic information will be missed. This should be taken into account for a more accurate LCOE analysis by using actual generation data per interval. The insolation results in Figure 3.1 show a notable trend and are similar to the results from NASA. The least insolation is received during Spring (March to May) and Summer (June to August), with the most is received in Autumn (September to October) and Winter (December to February), with the exception in November due to repeated rain.

The Kenya national load data is similar to the studies used in [1, 17]. Table 3.1 presents the normalized load data for Kenya. The normalized load is multiplied with a 2 MW demand to represent the electrical load demand for a community in a rural area.

3.3.2 Problem Context

The system aims to provide the required electrical power, with PV, AD, and EES to meet the load demand at all times. This means that the generation and load power should be balanced. The equality constraints for power balance are given in Eqs. (3.1) and (3.2) below:

$$P_{Generation}(t) = P_{Load}(t) \tag{3.1}$$

Table 3.1 Kenya normalized national load demand [1]

Hour	Normalized load	Hour	Normalized load
0	0.61	12	0.81
1	0.57	13	0.80
2	0.55	14	0.78
3	0.55	15	0.79
4	0.55	16	0.81
5	0.62	17	0.81
6	0.72	18	0.81
7	0.82	19	0.86
8	0.82	20	1.00
9	0.86	21	0.96
10	0.84	22	0.86
11	0.82	23	0.71

Source: With permission from Lai et al. [A]/with permission of Elsevier.

$$P_{Generation}(t) = P_{PV_{Direct}}(t) + P_{AD_{Direct}}(t) + P_{EES_{S1}}(t) + \eta_{EES} P_{EES_{S2}}(t) \tag{3.2}$$

$P_{PV_{Direct}}$ and $P_{AD_{Direct}}$ are the PV and AD power used to meet the load directly. $P_{EES_{S1}}$ and $P_{EES_{S2}}$ are the Stage 1 and Stage 2 power output from EES. In Stage 1, the EES decides to output power to meet the demand when the energy stored in EES is above a SOC threshold. While in Stage 2, the EES outputs power to meet the energy deficit in the system. η_{EES} is the round-trip efficiency for the storage system. This will be further elaborated with the introduction of the operating regime in Section 3.3.3.

The SOC limits are defined in Eq. (3.3) to prevent EES over charging/discharging. The power output of biogas generator should not exceed its rated capacity as shown in Eq. (3.4). Due to the minimum loading constraint for the biogas power plant, it is required to turn off when the power output is below 40% of its rated power output [1], given in Eq. (3.5).

$$SOC_{Min} \leq SOC(t) \leq SOC_{Max} \tag{3.3}$$

$$P_{AD_{Min}} \leq P_{AD_{Direct}}(t) + P_{AD_{Store}}(t) \leq P_{AD_{Max}} \tag{3.4}$$

$$P_{AD_{Min}} = 0.4 * P_{AD_{Max}} \tag{3.5}$$

$P_{AD_{Store}}$ is the power output from AD that is stored in EES. $P_{AD_{Min}}$ and $P_{AD_{Max}}$ are the maximum and minimum power output from AD respectively. SOC_{Min} and SOC_{Max} are the maximum and minimum SOCs, respectively. The instantaneous power output of PV plant is calculated as a function of panel area N_{PV}, panel efficiency η_{PV} and instantaneous irradiance $\varepsilon(t)$ as given in Eq. (3.6).

$$P_{PV}(t) = \varepsilon(t) \cdot N_{PV} \cdot \eta_{PV} \tag{3.6}$$

The schematic for the hybrid system is presented in Figure 3.2. It shows that the bi-directional inverter is used to provide DC power to AC load, and is also used to charge the EES from AD. The cost and size of components are given in Table 3.2.

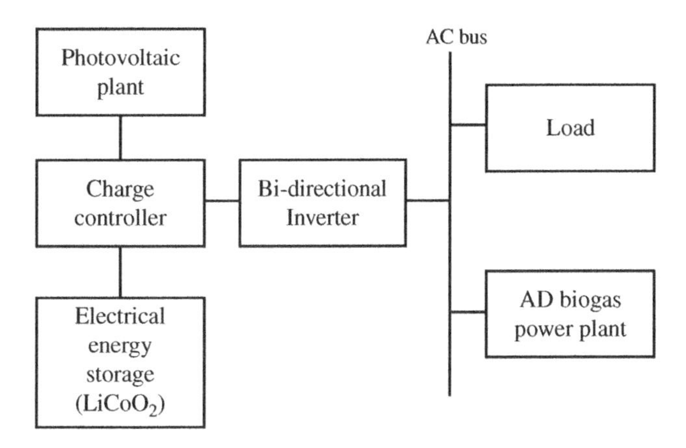

Figure 3.2 Schematic diagram of the hybrid system. *Source:* With permission from Lai et al. [A]/with permission of Elsevier.

Table 3.2 Cost and components size.

	PV (Sharp ND-250QCS)	AD (Caterpillar G3512E Genset)	EES (graphite/ $LiCoO_2$)	Controller (Outback FM 80)	Inverter (Schneider Electric XW6048)
Size	250 W	1.2 MW [30]	2 MW, 5 MWh	2 kW [1]	6 kW [31]
Unit	20 000 [1]	2 [1]	1 [1]	2500	834
$C_{Cap_{Asset}}$	120 ($/unit) [1]	$7.5M [1]	1500 ($/kWh) [17]	335 ($/unit) [1]	1518 ($/unit) [31]
$C_{InstAsset}$	108 ($/unit) [1]	N/A	N/A	6.7 ($/unit) [1]	30.38 ($/unit) [31]
$C_{Asset_{O\&M_{Var}}}$	N/A	$C_{AD_{Fuel}} + C_{AD_{Labour}}$	0.42 ($/MWh) [17] + $C_{EES_{DegkWh}}$	N/A	N/A
$C_{Asset_{O\&M_{Fixed}}}$	6 ($/unit/ year) [1]	300 ($/kW-year) [32]	2.12 ($/kW-year) [17]	1.005 ($/unit/ year) [1]	4.6 ($/unit/year) [31]

Source: With permission from Lai et al. [A]/with permission of Elsevier.

3.3.3 Operating Regime

Energy management system (EMS) is employed by grid operators to monitor, control, and optimize the performance of the microgrid. The main aims are to reduce electricity production costs and to continuously fulfill the load requirement. In explicit terms, the definition of energy management given in [33] is the procedure to collect all the systematic procedures to control and minimize the quantity and the cost of energy used to provide a certain application with its requirements. Generally, the control architecture of energy management can be classified into two types, that is, centralized control and decentralized or distributed control. Hybrid control featuring both centralized and decentralized is also an emerging architecture. As reviewed in [33, 34], the choice of the control architecture depends on the scale, configuration, and assets ownership of the microgrid.

The work in [35] presented a multi-agent decentralized EMS for autonomous polygeneration microgrids. In a decentralized architecture, each system component features a local controller, as opposed to the centralized architecture where a single controller executes the energy management for the system. Hence, decentralized architecture can increase the system reliability and reduce the chance of a complete system failure. On the other hand, decentralized architecture requires complex communication systems for local controllers to communicate and the achievement of global optimization can be challenging [33].

Traditionally, the optimization strategy employed by EMS for hybrid renewable energy systems are commonly achieved by artificial intelligent techniques or linear programming [33]. Artificial intelligent techniques that have been adopted are particle swarm optimization, genetic algorithm, fuzzy logic, and neural network. These techniques have improved the optimization and performance of energy systems over the years. However, real-time and robust energy management techniques are still a major focus in future research, due to the drawbacks in determining the global optimal and the computational costs. This work employs rule-based approach to avoid the drawbacks of artificial intelligence methods and to achieve optimal dispatch.

In smart energy, microgrids can be categorized into five categories, where each category is designed to meet specific goals. These are the commercial/industrial, community/utility, campus/institutional, military, and remote [36]. Taking this into consideration, this work focuses on developing an EMS for a remote community with generation assets located closely together, simple data acquisition and inexpensive architecture; a centralized architecture is adopted for this system.

The off-grid system will operate at the most cost-effective manner with the proposed regime. Basically, the operating regime will operate with the following characteristics:

- As the fuel cost for PV is zero, PV will be used first to meet the load demand before the deployment of AD or EES.
- Surplus PV energy will be stored in EES while fulfilling the SOC constraints.
- AD will be used to meet the load demand when PV power is not available.
- EES will be used to meet the deficit energy, i.e. the situation when no PV power is available, and AD is switched off.
- When there is not enough PV power, AD is switched off and EES has insufficient stored energy to meet the load demand; AD will then operate at minimum power output to meet the load demand. Any surplus energy generated from AD will be stored in EES.

This leads to a question to be explored, that is, "how does the EES dispatch method will influence the cost of the system and its components?" To answer this, a SOC threshold is defined as $SOC_{Threshold}$ to classify the dispatch priority for EES and AD. The operating regime can be broken down into three phases as shown in Figure 3.3 as follows:

Phase 1 (Blue): In this phase, the SOC at time t will be compared with the predefined $SOC_{Threshold}$. If the SOC is greater than or equal to the threshold, then EES will be discharged first to meet the demand, given that the load is greater than the PV output. The power to be discharged is $P_{EES_{S1}}$. It is expected that a higher $SOC_{Threshold}$ will reduce energy spillage and surplus energy from PV will be better utilized. Less biogas will be used. However, more EES cycles will occur throughout the system lifetime due to excessive discharging.

Phase 2 (Brown): The regime will determine if there is deficit power. If so, EES will be discharged to meet the load while satisfying the constraint. The power to be discharged is $P_{EES_{S2}}$. AD will be turned on to minimum output to meet the load if EES does not have enough stored energy. During this time, any surplus power from AD will be stored in EES.

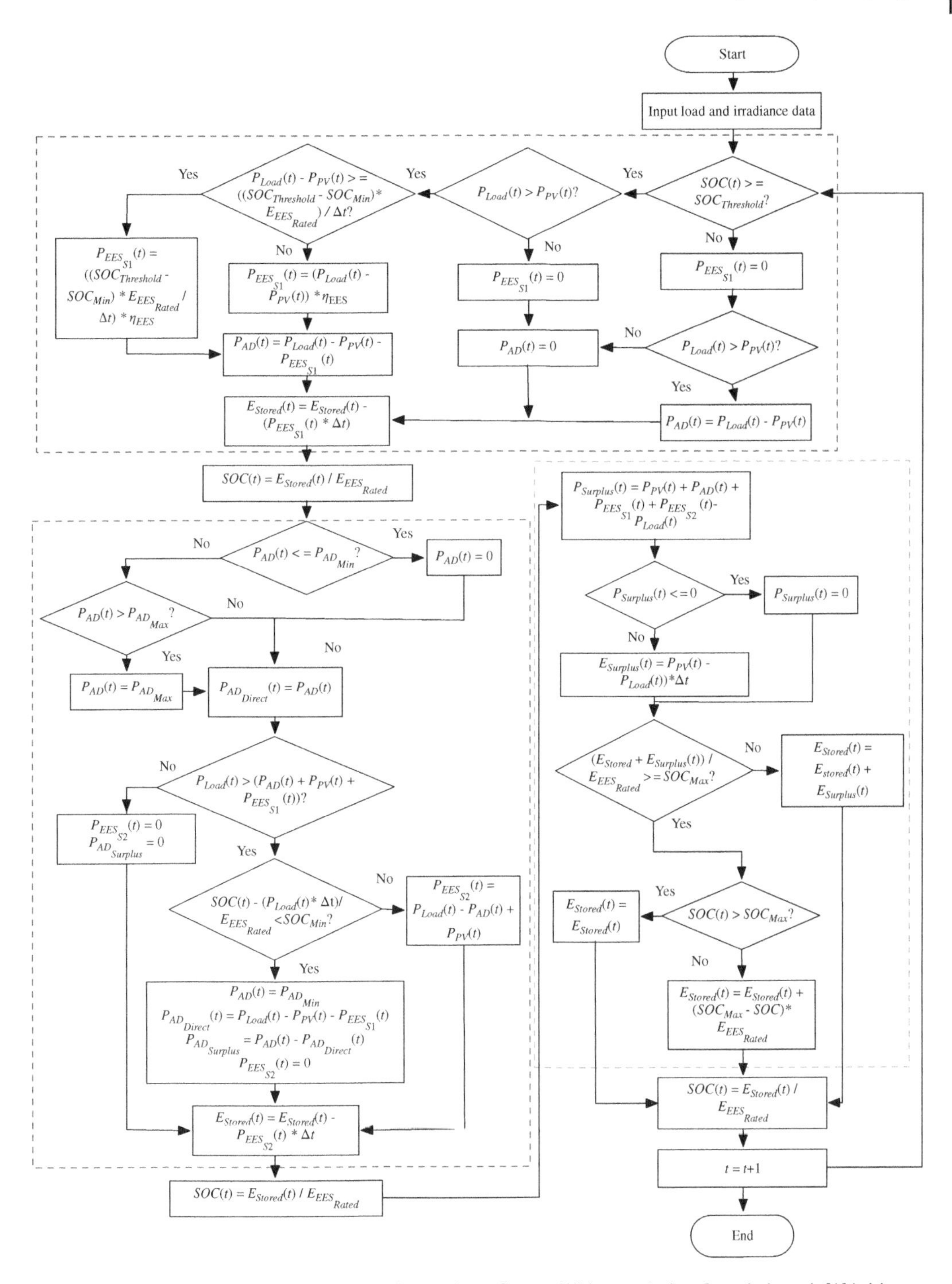

Figure 3.3 Proposed hybrid system operating regime. *Source:* With permission from Lai et al. [A]/with permission of Elsevier.

Phase 3 (Red): In this phase, the energy to be stored from the surplus power from AD and PV will be calculated.

3.3.4 Case Study

Case studies are conducted with MATLAB to verify the dispatch methods and operating regime. $SOC_{Threshold}$ is at 100%, i.e. the EES will not have the priority to discharge over AD. Figure 3.4 presents the results for five days of system operation in January. It can be seen that power curtailment has been applied to PV and the discarded surplus energy produced by PV is shown in green. At this point, the EES is fully charged. The EES is discharged to meet the power deficit, when PV output is increasing during sunrise and decreasing during sunset. The AD part loading constraint is triggered at these instances.

Figure 3.5 shows the simulation results for five days of system operation in November. As mentioned before, November is the month with most repeated rain in the year. This will introduce severe fluctuations to the PV power output. All the PV power is used either to meet the demand directly or stored in EES. Due to the power deficit, there are instances where both AD and EES are used to meet the demand. It can be confirmed that the operation of the hybrid system is achieved with the proposed regime.

The SOC for the system with $SOC_{Threshold}$ at two extremes is presented in Figure 3.6. It can be seen that deep discharges occur in Jan when $SOC_{Threshold}$ is at 0%. The solar power is better utilized, i.e. less energy spillage due to the storage is fully used at the beginning of the regime. Considering that when $SOC_{Threshold}$ is set at 100% as shown in Figure 3.6, for January case the battery is rarely discharged and is kept at high SOC. Surplus power will be wasted, i.e. it will not be stored since EES is full, most of the time.

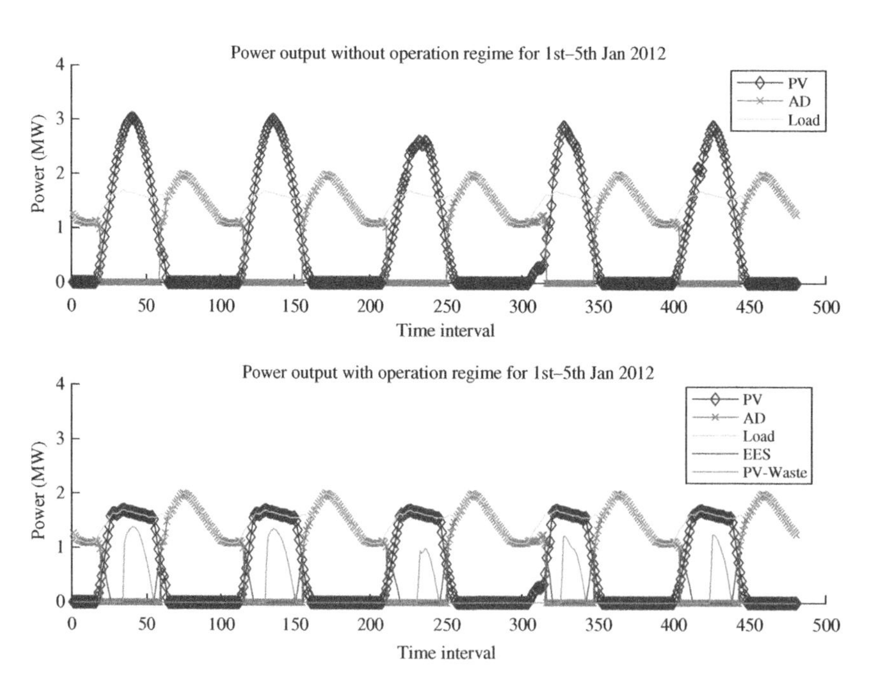

Figure 3.4 System's power output for summer case. *Source:* With permission from Lai et al. [A]/with permission of Elsevier.

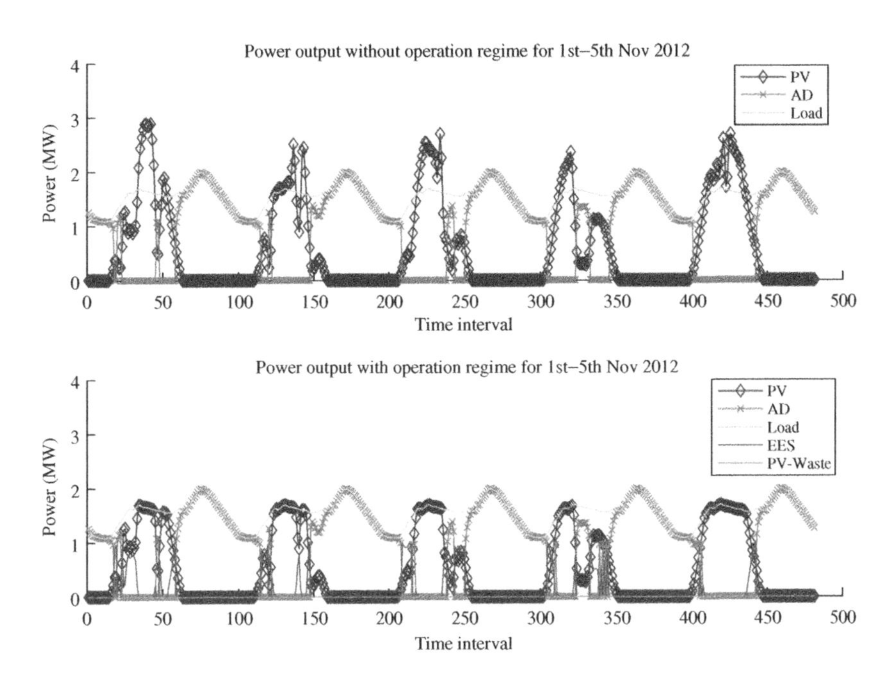

Figure 3.5 System's power output for spring case. *Source:* With permission from Lai et al. [A]/with permission of Elsevier.

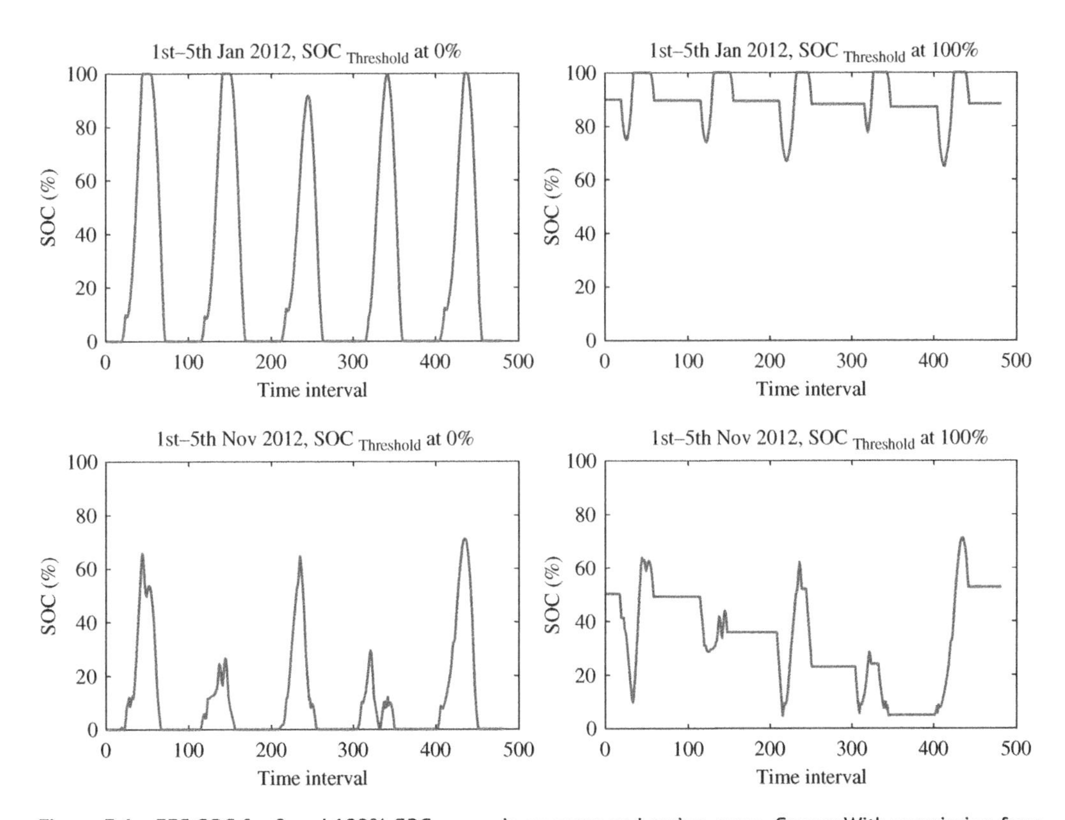

Figure 3.6 EES SOC for 0 and 100% $SOC_{Threshold}$ in summer and spring cases. *Source:* With permission from Lai et al. [A]/with permission of Elsevier.

For the case in November, the surplus power is not a problem as the EES is never fully charged. When $SOC_{Threshold}$ is set at 0%, the SOC is mostly at a minimum of 0%. This can be explained by the fact that the EES is used with priority, similar to the case in Jan with $SOC_{Threshold}$ at 0%. The EES is less susceptible to full discharges when $SOC_{Threshold}$ is at 100%. In general, the EES of a practical hybrid system will experience numerous partial cycles with different SOC ranges to fulfill the system energy requirement.

A sensitivity analysis for $SOC_{Threshold}$ with steps at 0.01, from 0 to 1 is carried out for the system's lifetime operation. The power output and SOC are stored for cost analysis and will be presented in Section 3.4.

3.4 Economic Analysis

Apart from capital costs, there are operational costs associated with EES and AD. These can be separated into fixed and variable costs. For AD, fixed operational costs include labor and routine generator maintenance. Fuel cost to produce electrical power is the variable operational costs for AD. The fixed operational cost for EES is the operation and maintenance (O&M) costs, such as routine battery services. As stated in [37], identifying the operating cost for EES is challenging since it does not consume fuel unlike the AD source. The energy cost can be assigned to the EES to represent the cost for charging and act as the "fuel" cost. In addition to this, another variable operational cost should be included to model the degradation of EES due to charging and discharging. This is known as degradation cost and can be broken down into two types, namely, the degradation cost per kWh and degradation cost per cycle.

3.4.1 AD Operational Cost Model

The Caterpillar G3512E Genset [30] is adopted to model the biogas power plant, with a rated capacity of 1.2 MW. The cost for AD power generation is given in Eq. (3.7) below:

$$C_{AD_{Fuel}}(t) = C_{Gas} \cdot \frac{P_{AD}(t) \cdot [a \cdot P_{AD}{}^2(t) + b \cdot P_{AD}(t) + c]}{LHV} \tag{3.7}$$

C_{Gas} is the AD gas cost at 6.97 \$/mcf [38], LHV is the lower heating value at 905 btu/ft^3 [30]. The fuel consumption is a quadratic function with constants a, b, and c at 0.0016, −3.935, and 10 641, respectively. The fuel consumption curve is displayed in Figure 3.7. The consumption is at the lowest when the power output is at the maximum due to the generator achieving the highest operation efficiency at rated capacity. The generator set should be avoided to meet partial load demand, hence the constraint.

3.4.2 $LiCoO_2$ Degradation Cost Model and Number of Replacements

In contrast to many works which use DOD or energy discharge content in kWh to calculate the EES degradation, this chapter uses a capacity fade model to quantify the degradation costs. In recent years, there are published materials on the study in Li-ion capacity fading. The factors that are studied are mainly on DOD, discharge C-rate and operating temperature. In [39], the cycle life of a Graphite/$LiFePO_4$ battery was researched and subsequently, the cycle life models were built. Three parameters were studied, with six temperatures from −30 to 60 °C, five DODs from 10 to 90% and four C-rates ranging from C/2 to 10C. A power law relationship was established between the capacity fade and the charge throughput, which represents the amount of charge delivered by EES during

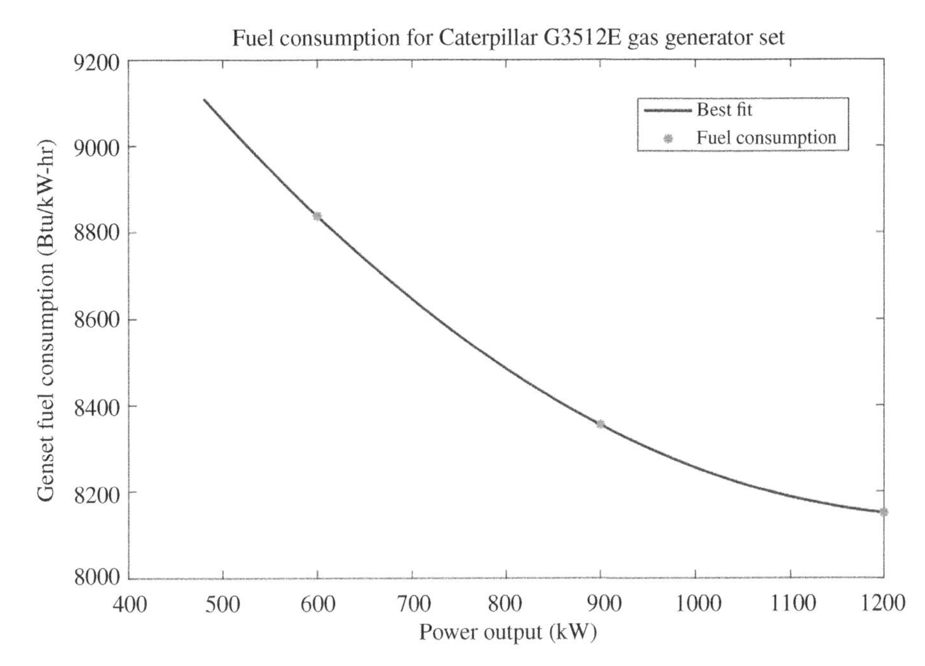

Figure 3.7 Fuel consumption curve for AD [30]. *Source:* With permission from Lai et al. [A]/with permission of Elsevier.

cycling. The highlight is that the capacity loss was mainly affected by temperature and time, while the effect of DOD was less of a concern with a C-rate of C/2.

The capacity loss rate of a Li-ion cell with nickel cobalt aluminum cathode under a restricted range (30–90% SOC) and full discharge (0–100% SOC) for two temperatures (25 and 60 °C) were studied in [40]. It was found that capacity fading can be reduced with a reduced SOC range. These can be explained by the formation of new resistance layer and the lack of contact between the primary particles with the micro-crack generation.

Further analysis in cycle life study with restricted range for Li-ion cell were conducted in [11] with Graphite/$LiCoO_2$ cells. It is found that the capacity loss can be affected by the change in SOC (ΔSOC), mean SOC, and C-rate. A power law model is developed for capacity fade of graphite/$LiCoO_2$ cells under different SOC ranges. It was discussed that the ΔSOC and mean SOC during operation should be minimized in order to decrease the long-term capacity fade rate, which results in higher number of equivalent cycles or increased cumulative discharge capacity throughout the EES lifetime. Also reported in [10], increased cycle life can be a result of cycling in lower charged states for Li-ion cells.

The degree of EES degradation is a function of number of cycles, change in SOC and amount of energy that flows throughout it [22]. The normalized discharge capacity model in [11] is adopted to study the degradation cost for EES. However, there are several limitations with the model. First, it is not able to capture the unexpected capacity loss behavior for low SOC ranges, such as 0–60% in the presented case study. However, the results show that the proposed model gives the pessimistic results for the low SOC ranges. The experiment results for normalized discharge capacity with respect to the equivalent full cycle are above the fitted curve. It is expected that there will be more cycles than the computed result from the model, therefore the degradation should be less for low SOC ranges.

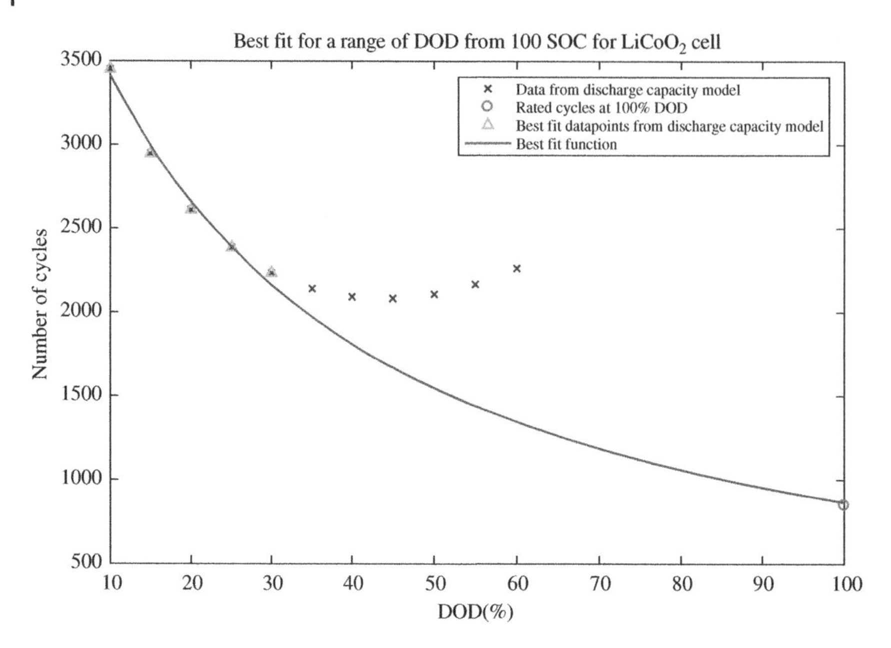

Figure 3.8 $LiCoO_2$ cycle life function for discharge from 100% SOC. *Source:* With permission from Lai et al. [A]/with permission of Elsevier.

The second limitation is that in the study for complete discharge range i.e. 0–100% SOC, there is a sudden increase in capacity loss rate after 500 equivalent full cycles. This suggests that a new degradation mechanism is triggered, and it is in a form of linear capacity fade model. To overcome this problem, a function is fitted with the data points obtained from the normalized discharge capacity model with the number of equivalent full cycles for 100% DOD between 0–100% SOC. With inspection, this number is approximately 850 full cycles. The fitted function is shown in Figure 3.8 and the data points generated from normalized discharge capacity model after 30% DOD are neglected as it prevents the best fit to the 100% DOD. Also after 45% DOD, it shows an increase in cycle life which is unreasonable.

During a discharge cycle, the mean SOC is calculated with Eq. (3.8) below:

$$SOC_{Mean} = \frac{SOC_{Upper} + SOC_{Lower}}{2} = \frac{SOC_{t_1} + SOC_{t_2}}{2} \tag{3.8}$$

SOC_{Upper} is the SOC when discharge begins and SOC_{Lower} is the SOC when charging begins for each partial cycle. The time t_1 is used to locate the time when the SOC is at the maximum of a cycle. The change in SOC is given in Eq. (3.9) below:

$$\Delta SOC = SOC_{Upper} - SOC_{Lower} \tag{3.9}$$

The rated cycle life and a_{ees} are calculated with Eqs. (3.10) and (3.11) [11] respectively.

$$Ratedcycle_{SOC_{Upper}, SOC_{Lower}} = e^{\frac{\ln\left(\frac{100^{0.453} * (100 - NDC)}{a_{ees}}\right)}{0.453}} \quad when\ SOC_{Upper} \neq 100\% \tag{3.10}$$

$$a_{ees} = 3.25 * SOC_{Mean} * \left(1 + 3.25 * \Delta SOC - 2.25 * \Delta SOC^2\right) \tag{3.11}$$

where, NDC is the normalized discharge capacity at 80%. The three-parameter equation in Eq. (3.12) [8, 26] is used to fit the rated cycle for deep discharge situations and SOC_{Upper} is at 100% during a discharge cycle.

$$Ratedcycle_{100\%,SOC_{Lower}} = \frac{d_{ees}}{(DoD - e_{ees})^{f_{ees}}} \quad if\ SOC_{Upper} = 1\&SOC_{Lower} < 0.55 \tag{3.12}$$

where d_{ees}, e_{ees}, and f_{ees} are constants with values 1278, −0.36, and 1.265, respectively.

The purpose of calculating the degradation cost per cycle is to determine the number of EES lifetime replacements. The value for the whole EES is calculated and then the cost of every discharge cycle is recorded. The EES needs to be replaced once the max capital cost is reached [22]. The replacement cost per discharge cycle for year i and cycle k is calculated with Eq. (3.13) and the number of lifetime replacements is calculated with Eq. (3.14) as follows:

$$C_{EES_{Replacement}}(i,k) = \frac{C_{Cap_{EES}} \cdot E_{EES_{Rated}}}{Ratedcycle_{SOC_{Upper},SOC_{Lower}}(i,k)} \tag{3.13}$$

$$N_{EES} = \frac{\sum_{i}^{n}\sum_{k}^{m} C_{EES_{Replacement}}(i,k)}{C_{Cap_{EES}} \cdot E_{EES_{Rated}}} \tag{3.14}$$

$C_{Cap_{EES}}$ and $E_{EES_{Rated}}$ are the capital cost and rated energy capacity of EES. $Ratedcycle_{SOC_{Upper},SOC_{Lower}}$ is the rated cycle life for the EES given a discharge cycle from SOC_{Upper} to SOC_{Lower}. Figure 3.9

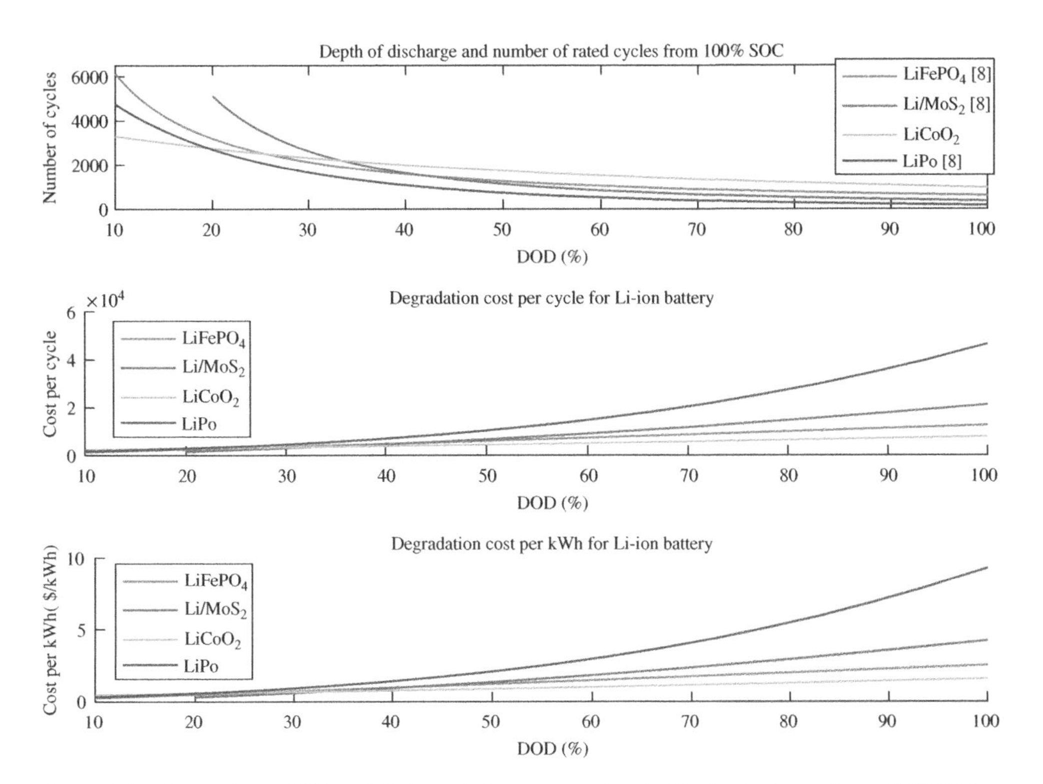

Figure 3.9 Comparison of Li-ion EES degradation costs and cycle life for discharge from 100% SOC. *Source:* With permission from Lai et al. [A]/with permission of Elsevier.

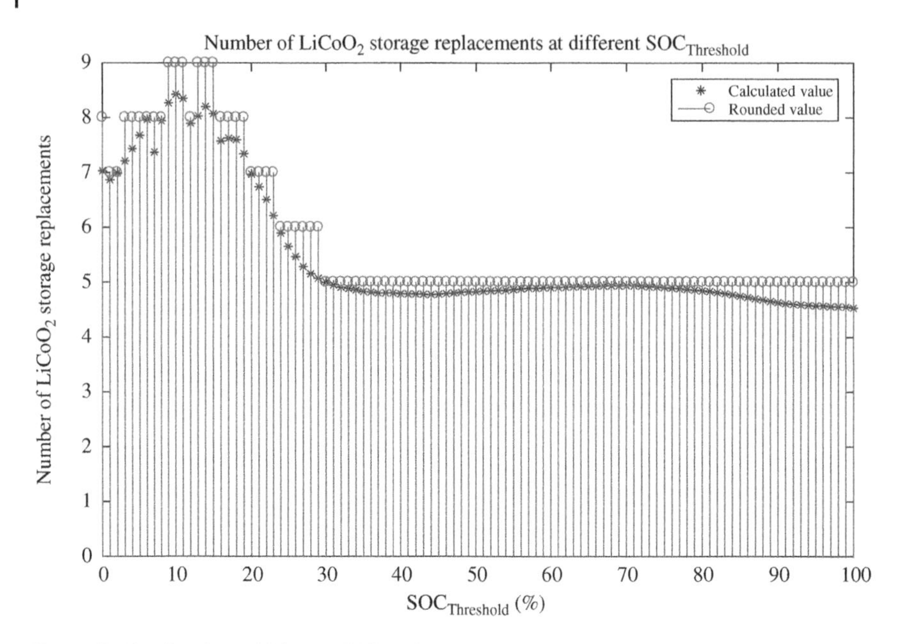

Figure 3.10 Number of lifetime EES replacements for different $SOC_{Threshold}$. *Source:* With permission from Lai et al. [A]/with permission of Elsevier.

shows a comparison of Li-ion cells with different chemistries for the cycle life under a range of DOD from 100% SOC. $LiCoO_2$ cells have a shorter life span as compared to other chemistries and $LiFePO_4$ can double the rated cycle life at low DODs [41]. A lower degradation per cycle and degradation per kWh can be achieved with a higher rated cycle life.

A sensitivity analysis with a range of $SOC_{Threshold}$ has been performed on the system to study the required number of EES lifetime replacements. Figure 3.10 displays the simulation results. Since the number of EES cannot be a fraction, the obtained result is rounded up to the nearest integer. The number of replacements increases as the $SOC_{Threshold}$ is reduced. To explain these phenomena, Figure 3.11 provides the histogram for changes in SOC and SOC occurrences for the lifetime operation with $SOC_{Threshold}$ at 0 and 100%. With $SOC_{Threshold}$ at 100%, the SOCs are mostly situated in the higher region and results in a higher SOC_{Mean}. However, the total number of discharges, which is given by the integral of the number of occurrences for ΔSOC, is less than for the case when $SOC_{Threshold}$ is at 0. There are more deep discharges for the case with $SOC_{Threshold}$ at 0% as given by ΔSOC. In summary, a $SOC_{Threshold}$ at 0% will generally on average result in a lower SOC_{Mean}, but the increase in total number of discharges and the deep discharge cycles have a larger impact to the EES degradation and result in an increased number of EES replacements.

3.4.3 Levelized Cost of Electricity Derivation

The LCOE for individual components and the system are studied with the established operational cost models for AD and EES. It is of crucial importance to understand the cost implications of how each generation asset and EES will react to the operating regime. The cost is largely affected by the

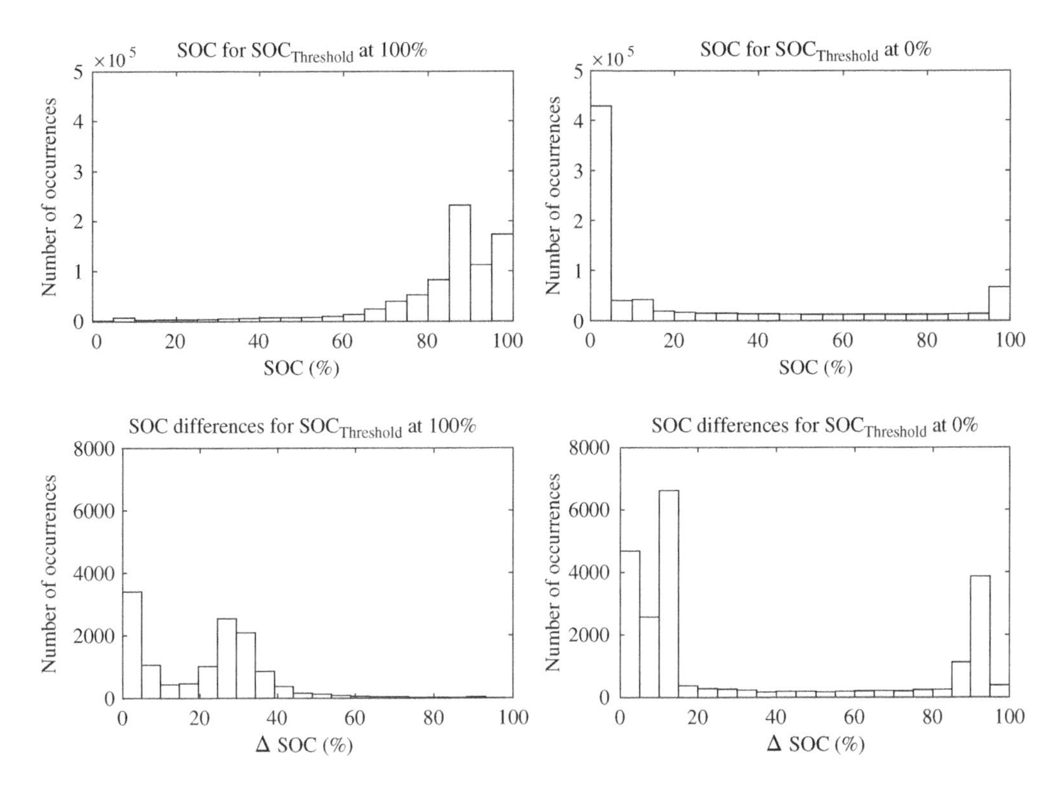

Figure 3.11 System's lifetime SOCs and SOC differences for $SOC_{Threshold}$ at 0 and 100%. *Source:* With permission from Lai et al. [A]/with permission of Elsevier.

asset lifetime electricity production, as well as the operational costs or degradation costs for the case with EES. The original definition of LCOE is given in Eq. (3.15) [1, 17] below:

$$\text{LCOE} = \frac{\textit{lifecycle cost}\ (\$)}{\textit{lifetime energy production}\ (\text{kWh})} = \frac{C_{Cap} + \sum_{i=0}^{n} C_{O\&M}(i) \cdot (1+d)^{-i}}{\sum_{i=0}^{n} E(i) \cdot (1+d)^{-i}} \tag{3.15}$$

C_{Cap} is the fixed capital cost of the asset and may include installation costs. $C_{O\&M}$ is the total operation and maintenance costs and E is the total energy production, it is often given per year i for asset lifetime n, and a discount rate d needs to be included to account for the depreciation in value for costs and energy. The discount rate used for the studies is 8% [17].

3.4.3.1 LCOE for PV

The LCOE for PV can be calculated with Eqs. (3.16)–(3.18) below:

$$\text{LCOE}_{\text{PV}} = \frac{C_{PV}}{E_{PV}} \tag{3.16}$$

$$C_{PV} = N_{PV} \cdot \left(C_{Cap_{PV}} + C_{Inst_{PV}}\right) + \sum_{i=0}^{n} \frac{N_{PV} \cdot C_{PV_{O\&M_{Fixed}}}}{(1+d)^{i}} \tag{3.17}$$

$$E_{PV} = \sum_{i=0}^{n} \frac{(1-\sigma)^i \cdot \sum_{j=1}^{t} P_{PV_{Direct}}(i,j) + P_{PV_{Store}}(i,j)}{(1+d)^i} \tag{3.18}$$

$C_{Cap_{PV}}$, $C_{Inst_{PV}}$, and $C_{PV_{O\&M_{Fixed}}}$ are the capital, installation, and operation and maintenance costs respectively. σ is the annual degradation constant for the PV panel at 0.5% [17]. $P_{PV_{Direct}}$ and $P_{PV_{Store}}$ are the power output from PV that is used to meet the demand and to be stored in EES, respectively, at hour j and year i. Since the system is an off-grid system, the wasted energy from PV generator, i.e. not used directly to meet the demand or stored in EES is not included in the lifetime energy production as the generated electricity will not be sold and the cost cannot be recovered, resulting in a higher levelized cost.

3.4.3.2 LCOE for AD

The LCOE for AD can be calculated with Eqs. (3.19)–(3.22) below:

$$\text{LCOE}_{\text{AD}} = \frac{C_{AD}}{E_{AD}} \tag{3.19}$$

$$C_{AD} = C_{Cap_{AD}} + \sum_{i=0}^{n} \frac{C_{AD_{O\&M_{Fixed}}} \cdot P_{AD_{Max}} + \sum_{j=1}^{t} C_{AD_{O\&M_{Var}}}(i,j)}{(1+d)^i} \tag{3.20}$$

$C_{Cap_{AD}}$ and $C_{AD_{O\&M_{Fixed}}}$ are the capital and fixed annual O&M costs for AD. The variable operation and maintenance cost for AD, $C_{AD_{O\&M_{Var}}}$, is calculated with Eq. (3.21) below:

$$C_{AD_{O\&M_{Var}}}(i,j) = C_{AD_{Fuel}}(i,j) + (C_{AD_{Labour}}/2) \cdot P_{AD}\Delta t(i,j) \tag{3.21}$$

$C_{AD_{Labour}}$ is the labor cost for operating the AD at \$0.05/kWh [32]. The AD lifetime energy production at present value is calculated with Eq. (3.22) below:

$$E_{AD} = \sum_{i=0}^{n} \frac{\sum_{j=1}^{t} P_{AD_{Direct}}(i,j) + P_{AD_{Store}}(i,j)}{(1+d)^i} \tag{3.22}$$

$P_{AD_{Direct}}$ is the AD power output that is used to meet the demand directly and $P_{AD_{Store}}$ is the surplus power from AD that is produced while meeting the deficit demand.

3.4.3.3 Levelized Cost of Storage (LCOS)

Recently, the term LCOS has received attention in both the industry and academia [42, 43]. Since EES can provide a diverse range of services for power systems operation, e.g. frequency support, renewable integration, uninterruptible power supply, it is of significant interest in studying the cost effectiveness of deploying a specific type of EES, such as lead-acid, Li-ion, etc. to perform certain tasks. LCOS has similar attributes to LCOE and is given by Eq. (3.23) below:

$$\text{LCOS} = \frac{C_{EES}}{E_{EES}} \tag{3.23}$$

$$C_{EES} = C_{Cap_{EES}} \cdot E_{EES_{Rated}} + \sum_{i=0}^{n} \frac{C_{EES_{O\&M_{Total}}}}{(1+d)^i} \tag{3.24}$$

$$C_{EES_{O\&M_{Total}}} = C_{EES_{O\&M_{Fixed}}} \cdot E_{EES_{Rated}} + \sum_{k=1}^{m} C_{EES_{DegkWh}}(i,k) + \sum_{j=1}^{t} C_{EES_{kWh}}(i,j) \tag{3.25}$$

$C_{EES_{O\&M_{Fixed}}}$ is the fixed annual $EES_{O\&M}$ cost. The variable operational costs for EES include the degradation per kWh cost and energy cost are calculated with Eqs. (3.26) and (3.27), respectively. m is the number of discharge cycles per year.

$$C_{EES_{DegkWh}}(i,k) = \frac{C_{Cap_{EES}}}{Ratedcycle_{SoC_{Upper},SoC_{Lower}}(i,k)} \cdot \left[SOC_{Upper}(i,\ k) - SOC_{Lower}(i,\ k)\right] \cdot E_{EES_{Rated}} \tag{3.26}$$

$$C_{EES_{kWh}}(i,j) = C_{EES_{O\&M_{Var}}} \cdot E_{EES_{Store}}(i,j) \tag{3.27}$$

$C_{EES_{O\&M_{Var}}}$ is the variable O&M cost for the energy $E_{EES_{Store}}$ to be stored in EES. The lifetime electricity production is calculated with Eq. (3.28) below:

$$E_{EES} = \eta_{EES} \sum_{i=0}^{n} \frac{\sum_{j=1}^{t} \left[E_{EES-S1}(i,\ j) + E_{EES-S2}(i,\ j)\right]}{(1+d)^i} \tag{3.28}$$

η_{EES} is the round-trip efficiency of EES at 90% for Li-ion.

3.4.3.4 Levelized Cost of Delivery (LCOD)

Since EES is a storage device and is not a conventional electrical energy generating source, i.e. the asset does not generate electricity from primary energy source, it is necessary to take account of the cost for the energy conversion process from the primary form, i.e. solar irradiance or biomass into electricity. This is studied in [17] and the term LCOD was proposed. Eq. (3.29) gives the LCOD for the system.

$$\text{LCOD} = \frac{C_{EES} + C_{PV_{Storetotal}} + C_{Con} + C_{AD_{Storetotal}}}{E_{EES}} \tag{3.29}$$

The charge controller cost is given in Eq. (3.30) below:

$$C_{Con} = N_{Con} \cdot \left(C_{Cap_{Con}} + C_{InstCon}\right) + \sum_{i=0}^{n} \frac{C_{O\&M_{Con}} \cdot N_{Con}}{(1+d)^i} \tag{3.30}$$

$$N_{Con} = ceil\left(\frac{P_{PV_{Rated}}}{P_{Con}}\right) \cdot N_{Con-rep} \tag{3.31}$$

N_{Con} is the number of controllers; $C_{Cap_{Con}}$, $C_{InstCon}$, and $C_{O\&M_{Con}}$ are the capital, installation and O&M costs for the controller. The ceil function converts the number of controllers into the nearest integer above. $N_{Con-rep}$ is 2 and is the number of controller replacements during the system's lifetime [1]. $P_{PV_{Rated}}$ and P_{Con} are the rated power output for the PV farm and the controller respectively. The cost of producing the electricity from AD to be stored in EES is given in Eq. (3.32) below:

$$C_{AD_{Storetotal}} = C_{Cap_{AD}} + \sum_{i=0}^{n} \frac{C_{AD_{O\&M_{Fixed}}} \cdot P_{AD_{Max}} + \sum_{j=1}^{t} C_{AD_{O\&M_{Store}}}(i,j)}{(1+d)^i} \tag{3.32}$$

$$C_{AD_{O\&M_{Store}}}(i,j) = C_{AD_{Fuel}}(i,j) + C_{AD_{Labour}} \cdot P_{AD_{Store}}(i,j) \tag{3.33}$$

$P_{AD_{Store}}$ is the power generated from AD to be stored in EES as energy, the cost for producing the electricity from PV to be stored in EES is given in Eq. (3.34) below:

$$C_{PV_{Storetotal}} = N_{PV} \cdot \left(C_{Cap_{PV}} + C_{Inst_{PV}}\right) + \sum_{i=0}^{n} \frac{\sum_{j=1}^{t} N_{Store}(i,j) \cdot C_{O\&M_{PVint}}(i,j)}{(1+d)^{i}} \tag{3.34}$$

$$N_{Store}(i,j) = \frac{P_{PV_{Store}}(i,j)}{\varepsilon(i,j) \cdot \eta_{PV}} \tag{3.35}$$

N_{Store} is the PV panel area that is used to produce the surplus electricity for storage. The maintenance cost per hour for PV is calculated in Eq. (3.36) below:

$$C_{O\&M_{PVint}}(i,j) = \frac{C_{O\&M_{PV}}(i,j)}{365 * 24} \tag{3.36}$$

3.4.3.5 LCOE for System

The system's LCOE is derived in Eq. (3.37). It takes into account the total costs for operating the system and the energy output to meet the load demand.

$$\text{LCOE}_{\text{system}} = \frac{C_{PV} + C_{EES} + C_{Con} + C_{Inv} + C_{AD}}{E_{EES} + E_{PV_{Directtotal}} + E_{AD_{Directtotal}}} \tag{3.37}$$

$$C_{Inv} = N_{Inv} \cdot \left(C_{Cap_{Inv}} + C_{Inst\,Inv}\right) + \sum_{i=0}^{n} \frac{C_{O\&M_{Inv}}(i) \cdot N_{Inv}}{(1+d)^{i}} \tag{3.38}$$

C_{CapInv}, $C_{InstInv}$, and $C_{O\&M_{Inv}}$ are the capital, installation and O&M costs for the bi-directional inverter. The number of inverters required is calculated with Eq. (3.39) below:

$$N_{inv} = ceil\left(\frac{P_{PV_{Rated}}}{P_{Inv}}\right) \cdot N_{Inv-rep} \tag{3.39}$$

P_{Inv} is the rated capacity of the inverter. $N_{Inv-rep}$ is 2 and is the number of inverter replacements during the system's lifetime [1]. The total energy produced by PV and AD during the operation of the system that is used to meet the load demand directly are given in Eqs. (3.40) and (3.41), respectively as shown below:

$$E_{PV_{Directtotal}} = \sum_{i=0}^{n} \frac{(1-\sigma)^{i} \cdot \sum_{j=1}^{t} P_{PV_{Direct}}(i,j)}{(1+d)^{i}} \tag{3.40}$$

$$E_{AD_{Directtotal}} = \sum_{i=0}^{n} \frac{\sum_{j=1}^{t} P_{AD_{Direct}}(i,j)}{(1+d)^{i}} \tag{3.41}$$

$P_{PV_{Direct}}$ and $P_{AD_{Direct}}$ are the power generated from PV and AD, respectively for direct consumption.

3.4.4 LCOE Analyses and Discussion

The LCOE results with $SOC_{Threshold}$ at different values are displayed in Figure 3.12. The cost for AD is lower when $SOC_{Threshold}$ is at a higher value due to AD having a higher dispatch priority than EES and more electricity is generated from AD. The cost for EES will increase as a result of a decreased

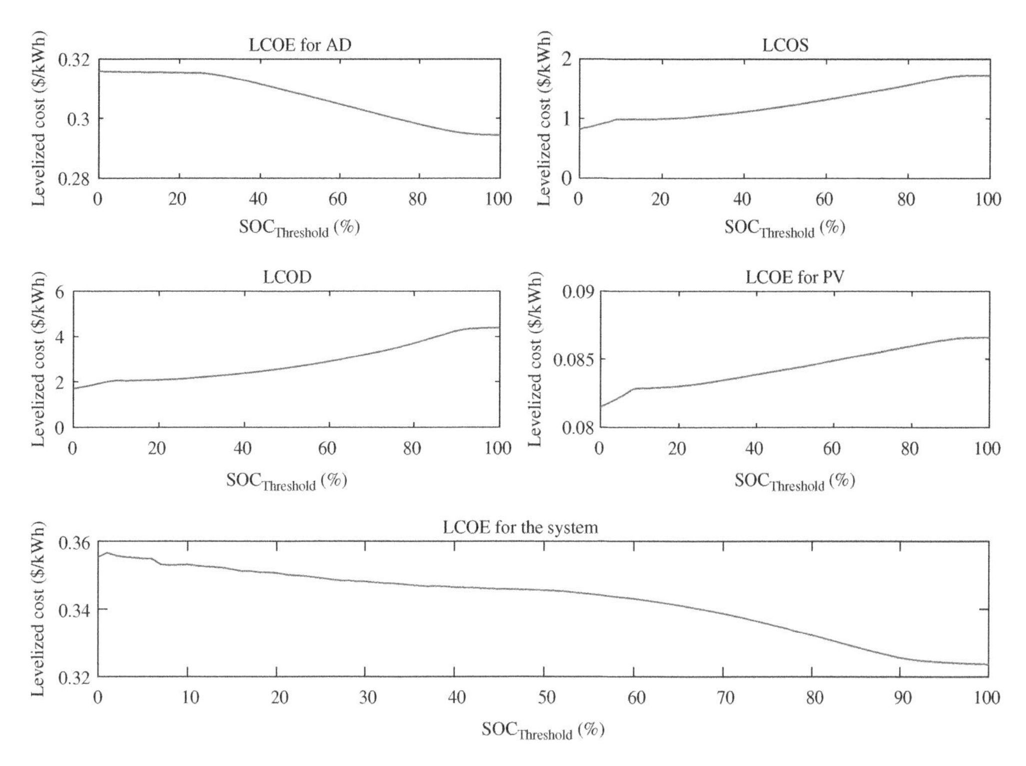

Figure 3.12 LCOE for AD, EES and system at different $SOC_{Threshold}$ values. *Source:* With permission from Lai et al. [A]/with permission of Elsevier.

energy output. LCOD has a similar curve to LCOS, but with a larger cost. As explained earlier, LCOD takes into account the energy conversion costs from primary source to electricity for storage. The capital costs for AD and PV have a large contribution to the actual energy storage costs. The cost for PV increases as $SOC_{Threshold}$ increases, since less energy delivered by EES and produced from PV is used to meet the load demand. By taking all the costs and electricity generation into consideration, it is more economical to have the $SOC_{Threshold}$ at maximum value and run the system at minimum costs.

Due to the excellent round-trip efficiency, high energy density and high power density, it is expected that Li-ion batteries will be widely used in renewable hybrid systems [44]. However, the capital costs such as manufacturing and the cost of cobalt in cathodes are the major factors that prevent the wide-scale adoption at present. As discussed in [45], it is possible for Li-ion batteries to have a capital cost of 200 $/kWh by 2020. Therefore, Figure 3.13 presents the results for the system LCOE with a sensitivity analysis on the EES capital cost.

At 1000 $/kWh, the system cost is reduced as compared to the original study at 1500 $/kWh. As the capital cost decreases, it can be observed that the differences in levelized cost are minimized with $SOC_{Threshold}$ variation. This signifies that the cost of EES degradation has less influence on the system's cost. At 200 $/kWh, the largest difference in system's levelized cost is 0.0028 $/kWh. The lowest levelized cost for the system is achieved with a $SOC_{Threshold}$ at 33%. It is more cost effective to cycle the EES more often and utilized the surplus PV energy than meeting the load demand with biogas. In future work, the carbon emission cost can be included in the cost generation for the analysis. Although biogas is seen as a renewable energy source, it still generates little amount of carbon emission due to the combustion process.

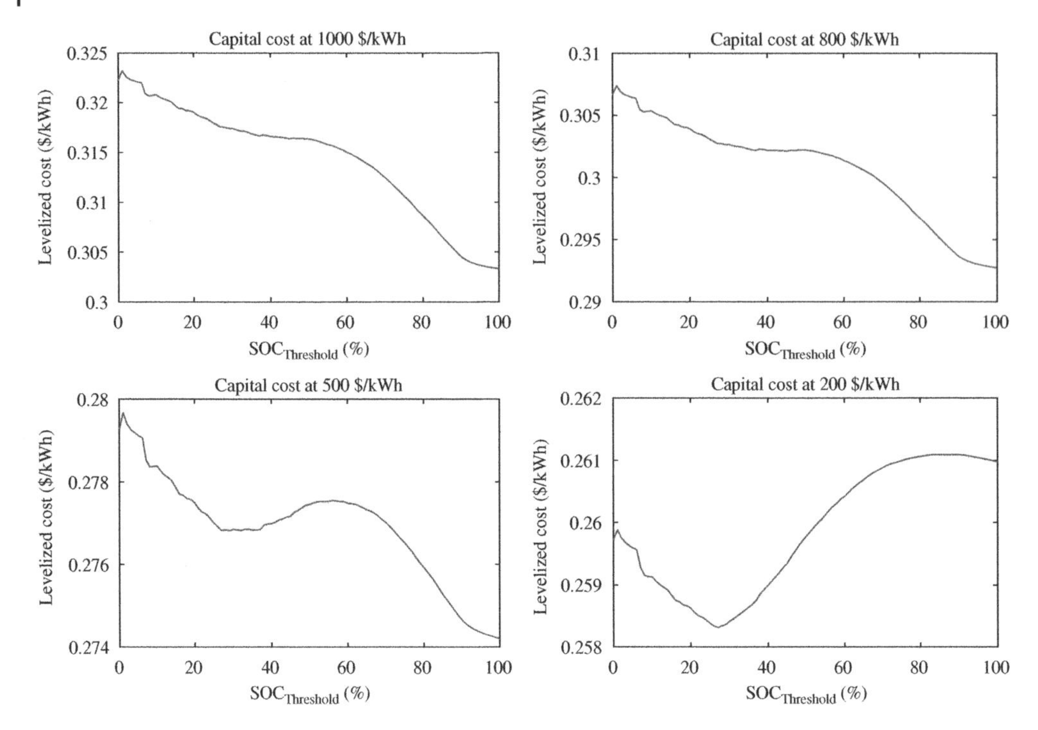

Figure 3.13 System LCOE with different EES capital costs. *Source:* With permission from Lai et al. [A]/with permission of Elsevier.

3.5 Conclusions

Naturally no/low carbon energy can substantially reduce greenhouse gas emissions, but in order to reach the 2050 Paris targets, additional reductions will be needed. This work provides a techno-economic analysis of an off-grid photovoltaic, anaerobic digestion biogas power plant (AD) renewable energy system with Graphite/$LiCoO_2$ storage. The highlight of this work is that the accuracy of degradation costs for EES is enhanced by utilizing a capacity fade model, by obtaining the cycle life on partial SOC range cycling. This work opens a range of opportunities and future work to be conducted, listed as follows:

1) At present, most EES degradation effects focus on the discharging phase, i.e. depth of discharge and energy delivered by EES. The degradation effect due to charging phase should also be given.
2) The optimization problem formulations for power system with EES should be modified to incorporate capacity fade model.
3) Temperature and C-rate should be included in the capacity fade model to provide a more realistic analysis.
4) For illustration purposes, the proposed regime is of static nature. The variable to control the dispatch priority of EES and AD, $SOC_{Threshold}$, can be of dynamic nature to minimize the system lifetime cost. This may need generation and load forecasting.

5) More studies in partial SOC ranges is required in order to build cell degradation model for all possible SOC ranges.
6) Capacity fade model for dynamic SOC cycling conditions, i.e. the cycle life due to a combination of different SOC ranges is needed.

Based on the results obtained in this study, it shows that the Graphite/$LiCoO_2$ is cost competitive compared to the dispatchable generator, i.e. AD biogas generator to provide electricity as the capital cost reduces to 200 $/kWh.

Acknowledgment

The permission given in using materials from the following paper is very much appreciated by the authors.

[A] Lai, C.S., Jia, Y., Xu, Z. et al. (2017). Levelized cost of electricity for photovoltaic/biogas power plant hybrid system with electrical energy storage degradation costs. *Energy Conversion and Management* 153: 34–47.

References

1 Lai, C.S. and McCulloch, M.D. (2017). Sizing of stand-alone solar PV and storage system with anaerobic digestion biogas power plants. *IEEE Transactions on Industrial Electronics* 64: 2112–2121.

2 Akikur, R.K., Saidur, R., Ping, H.W., and Ullah, K.R. (2013). Comparative study of stand-alone and hybrid solar energy systems suitable for off-grid rural electrification: a review. *Renewable and Sustainable Energy Reviews* 27: 738–752.

3 Tezer, T., Yaman, R., and Yaman, G. (2017). Evaluation of approaches used for optimization of stand-alone hybrid renewable energy systems. *Renewable and Sustainable Energy Reviews* 73: 840–853.

4 Bernal-Agustín, J.L. and Dufo-López, R. (2009). Simulation and optimization of stand-alone hybrid renewable energy systems. *Renewable and Sustainable Energy Reviews* 13: 2111–2118.

5 Kazhamiaka, F., Rosenberg, C., and Keshav, S. (2016). Practical strategies for storage operation in energy systems: design and evaluation. *IEEE Transactions on Sustainable Energy* 7: 1602–1610.

6 Su, W., Wang, J., and Roh, J. (2014). Stochastic energy scheduling in microgrids with intermittent renewable energy resources. *IEEE Transactions on Smart Grid* 5: 1876–1883.

7 Zhou, B., Liu, X., Cao, Y. et al. (2016). Optimal scheduling of virtual power plant with battery degradation cost. *IET Generation, Transmission & Distribution* 10: 712–725.

8 Zhang, Z., Wang, J., and Wang, X. (2015). An improved charging/discharging strategy of lithium batteries considering depreciation cost in day-ahead microgrid scheduling. *Energy Conversion and Management* 105: 675–684.

9 Zheng, Y., Dong, Z.Y., Luo, F.J. et al. (2014). Optimal allocation of energy storage system for risk mitigation of DISCOs with high renewable penetrations. *IEEE Transactions on Power Systems* 29: 212–220.

10 de Vries, H., Nguyen, T.T., and het Veld, B.O. (2015). Increasing the cycle life of lithium ion cells by partial state of charge cycling. *Microelectronics Reliability* 55: 2247–2253.

11 Saxena, S., Hendricks, C., and Pecht, M. (2016). Cycle life testing and modeling of graphite/$LiCoO_2$ cells under different state of charge ranges. *Journal of Power Sources* 327: 394–400.
12 Hoke, A., Brissette, A., Smith, K. et al. (2014). Accounting for lithium-ion battery degradation in electric vehicle charging optimization. *IEEE Journal of Emerging and Selected Topics in Power Electronics* 2: 691–700.
13 Yoshida, A., Sato, T., Amano, Y., and Ito, K. (2016). Impact of electric battery degradation on cost-and energy-saving characteristics of a residential photovoltaic system. *Energy and Buildings* 124: 265–272.
14 Ippolito, M.G., Di Silvestre, M.L., Sanseverino, E.R. et al. (2014). Multi-objective optimized management of electrical energy storage systems in an islanded network with renewable energy sources under different design scenarios. *Energy* 64: 648–662.
15 Perera, A.T.D., Nik, V.M., Mauree, D., and Scartezzini, J.-L. (2017). Electrical hubs: an effective way to integrate non-dispatchable renewable energy sources with minimum impact to the grid. *Applied Energy* 190: 232–248.
16 Sachs, J. and Sawodny, O. (2016). A two-stage model predictive control strategy for economic diesel-pv-battery island microgrid operation in rural areas. *IEEE Transactions on Sustainable Energy* 7: 903–913.
17 Lai, C.S. and McCulloch, M.D. (2017). Levelized cost of electricity for solar photovoltaic and electrical energy storage. *Applied Energy* 190: 191–203.
18 Hamedi, A.-S. and Rajabi-Ghahnavieh, A. (2016). Explicit degradation modelling in optimal lead–acid battery use for photovoltaic systems. *IET Generation, Transmission & Distribution* 10: 1098–1106.
19 Hittinger, E., Wiley, T., Kluza, J., and Whitacre, J. (2015). Evaluating the value of batteries in microgrid electricity systems using an improved Energy Systems Model. *Energy Conversion and Management* 89: 458–472.
20 Telaretti, E., Graditi, G., Ippolito, M., and Zizzo, G. (2016). Economic feasibility of stationary electrochemical storages for electric bill management applications: the Italian scenario. *Energy Policy* 94: 126–137.
21 Graditi, G., Ippolito, M., Telaretti, E., and Zizzo, G. (2016). Technical and economical assessment of distributed electrochemical storages for load shifting applications: an Italian case study. *Renewable and Sustainable Energy Reviews* 57: 515–523.
22 Bordin, C., Anuta, H.O., Crossland, A. et al. (2017). A linear programming approach for battery degradation analysis and optimization in offgrid power systems with solar energy integration. *Renewable Energy* 101: 417–430.
23 Budischak, C., Sewell, D., Thomson, H. et al. (2013). Cost-minimized combinations of wind power, solar power and electrochemical storage, powering the grid up to 99.9% of the time. *Journal of Power Sources* 225: 60–74.
24 Hassan, A.S., Cipcigan, L., and Jenkins, N. (2017). Optimal battery storage operation for PV systems with tariff incentives. *Applied Energy* 203: 422–441.
25 Rajbongshi, R., Borgohain, D., and Mahapatra, S. (2017). Optimization of PV-biomass-diesel and grid base hybrid energy systems for rural electrification by using HOMER. *Energy* 126: 461–474.
26 Zhang, Y., Lundblad, A., Campana, P.E. et al. (2017). Battery sizing and rule-based operation of grid-connected photovoltaic-battery system: a case study in Sweden. *Energy Conversion and Management* 133: 249–263.
27 Mandelli, S., Brivio, C., Colombo, E., and Merlo, M. (2016). A sizing methodology based on Levelized Cost of Supplied and Lost Energy for off-grid rural electrification systems. *Renewable Energy* 89: 475–488.
28 Stackhouse, P.W. (2013). NASA Surface meteorology and Solar Energy - Available Tables. https://ntrs.nasa.gov/citations/20080012141 (accessed 29 October 2020).

29 Mundada, A.S., Shah, K.K., and Pearce, J. (2016). Levelized cost of electricity for solar photovoltaic, battery and cogen hybrid systems. *Renewable and Sustainable Energy Reviews* 57: 692–703.

30 Caterpillar Inc. (2013). G3512E Caterpillar Gas engine technical data, Ref. Data Set DM8811-06-001. Industrial motor power corporation. http://attachments.impcorporation.com/21373414/ Performance Data G3512E.pdf (accessed 29 October 2020).

31 Hassan, A., Saadawi, M., Kandil, M., and Saeed, M. (2015). Modified particle swarm optimisation technique for optimal design of small renewable energy system supplying a specific load at Mansoura University. *IET Renewable Power Generation* 9: 474–483.

32 Gifford, J.S. and Grace, R.C. (2013). CREST Cost of Renewable Energy Spreadsheet Tool: A Model for Developing Cost-Based Incentives in the United States. User Manual Version 4, Subcontract Report, NREL/SR-6A20-50374.

33 Olatomiwa, L., Mekhilef, S., Ismail, M., and Moghavvemi, M. (2016). Energy management strategies in hybrid renewable energy systems: a review. *Renewable and Sustainable Energy Reviews* 62: 821–835.

34 Meng, L., Sanseverino, E.R., Luna, A. et al. (2016). Microgrid supervisory controllers and energy management systems: a literature review. *Renewable and Sustainable Energy Reviews* 60: 1263–1273.

35 Karavas, C.-S., Kyriakarakos, G., Arvanitis, K.G., and Papadakis, G. (2015). A multi-agent decentralized energy management system based on distributed intelligence for the design and control of autonomous polygeneration microgrids. *Energy Conversion and Management* 103: 166–179.

36 Maitra, A., Pratt, A., Hubert, T. et al. (2017). Microgrid controllers: expanding their role and evaluating their performance. *IEEE Power and Energy Magazine* 15: 41–49.

37 Nguyen, T.A. and Crow, M. (2016). Stochastic optimization of renewable-based microgrid operation incorporating battery operating cost. *IEEE Transactions on Power Systems* 31: 2289–2296.

38 Beddoes, J.C., Bracmort, K.S., Burns, R.T., and Lazarus, W.F. (2007). An analysis of energy production costs from anaerobic digestion systems on US livestock production facilities. *USDA NRCS Technical Note.*

39 Wang, J., Liu, P., Hicks-Garner, J. et al. (2011). Cycle-life model for graphite-$LiFePO_4$ cells. *Journal of Power Sources* 196: 3942–3948.

40 Watanabe, S., Kinoshita, M., Hosokawa, T. et al. (2014). Capacity fading of $LiAl_yNi_{1-x-y}Co_xO_2$ cathode for lithium-ion batteries during accelerated calendar and cycle life tests (effect of depth of discharge in charge–discharge cycling on the suppression of the micro-crack generation of $LiAl_yNi_{1-x-y}Co_xO_2$ particle). *Journal of Power Sources* 260: 50–56.

41 Hanusa, T.P. (2015). *The Lightest Metals: Science and Technology from Lithium to Calcium*. John Wiley & Sons.

42 Jülch, V. (2016). Comparison of electricity storage options using levelized cost of storage (LCOS) method. *Applied Energy* 183: 1594–1606.

43 Lazard Ltd (2015). Lazard's Levelized Cost of Storage Analysis V1.0. Lazard. https://www.lazard.com/media/2391/lazards-levelized-cost-of-storage-analysis-10.pdf (accessed 29 October 2020).

44 Nitta, N., Wu, F., Lee, J.T., and Yushin, G. (2015). Li-ion battery materials: present and future. *Materials Today* 18: 252–264.

45 Nykvist, B. and Nilsson, M. (2015). Rapidly falling costs of battery packs for electric vehicles. *Nature Climate Change* 5: 329–332.

4

Electricity Plan Recommender System

Nomenclature

n	The number of samples
d	The number of features
X	The sample matrix (kWh) with the size $d \times n$
X^*	The recovered sample matrix (kWh)
E	The noise of the sample matrix (kWh)
Z	The representation matrix of the sample matrix with the size $n \times n$
S	The similarity matrix among customers with the size $n \times n$
Y	Total appliance electricity usages (kWh) with the size $n \times 1$
R	The rating matrix with the size $n \times c$
U_m	The mth customer
U_m^k	The K Nearest Neighbors of the mth customer
Λ	The unknown appliance usage set
Ω	The known appliance usage set
$X_{i,j}$	The ith row and j^{th} column element of matrix X
Y_i	The ith element of vector Y
$X_{i,*}$	The ith row vector of matrix X
$X_{*,i}$	The ith column vector of matrix X
X^t	The tth iteration of X
r	The rank of sample matrix
μ	The penalty parameter
μ_{max}	The maximum penalty parameter
σ_i	Singular value of sample matrix
I	Identity matrix
W	The projection matrix with the size $1 \times d$
J	The auxiliary computation matrix with the size $d \times n$
Y_1, Y_2	Lagrange multipliers matrices

Smart Energy for Transportation and Health in a Smart City, First Edition. Chun Sing Lai, Loi Lei Lai and Qi Hong Lai.

4.1 Introduction

An increasing number of factors including intermittent renewable power generation and load consumption have posed a threat to the stability of the power system. These factors cause the fluctuation of the power system and the growing peak value of electricity demand. To deal with the problems, demand side management (DSM) [1–5] is used to regulate the demand of energy consumers. In DSM, with the purpose of shaving peak and filling valley, pricing-based demand response (PBDR) [6–9] is proposed to provide residential customers with various electricity plans, indirectly influencing their energy consumption patterns. For example, if a customer selects a plan with a lower charge in the morning, he may shift the use of some appliances from evening to morning.

In a matured electricity market, thousands of electricity plans are listed in the electricity plan interface, which brings challenges to residential customers for making choices among the great number of electricity plans. If a customer chooses an improper electricity plan, to compromise the electricity cost, he may have to change the living pattern and sacrifice the living comfort. Faced with this problem, a new technique named Electricity Plan Recommender System (EPRS) is introduced to help residential customers to choose proper electricity plans. In a project named Smart Grid Smart City (SGSC) [10], 200 residential customers are selected to make a comparison between choosing plans with and without EPRS. It shows that aggregated daily load profiles in the two scenarios are similar in shape but slightly different in the lowest and highest values. This project inspires some electricity market platforms, foster them to provide electricity plan recommender service, such as Energy Made Easy, iSelect, and Power to Choose [11–13].

The current EPRS methods can be classified into the direct method and indirect method. The direct method is relatively easy to be realized, and the above EPRS models [10–14] belong to this class. These methods directly calculate the residential customers' electricity charges through multiplying their total usages by the unit charge of electricity plans and recommend the electricity plans to make less charges to the residential customers. The main drawback of direct methods is that they lack consideration of the personal needs of customers, because two customers having the same electricity usages may have different living patterns.

In the last decade, the electricity meter can only count for the total appliance electricity usages of customers, so direct methods are the mainstream EPRS methods. Fortunately, with the development of the smart meter, the monitoring of household appliances has become an increasingly attractive research field. Unlike the traditional meters, smart meters and intelligent home devices [14–23] can be utilized to monitor the living patterns of various residential customers, which give the possibility to extract key factors affecting personal living patterns. Based on this technology, indirect methods are introduced to recommend electricity plans based on such factors. Indirect methods are a dual-stage model, consisting of feature formulation stage and recommender stage. In the feature formulation stage, primary data and certain features are set as input and output respectively, and the outputted features are the key factors to represent the living patterns. In the recommender stage, the similarity [24–26] of customers is calculated and the testing of personalized electricity plans can be obtained through the calculation based on similarity and training personal electricity plans. The dual-stage framework of EPRS is shown in Figure 4.1.

Similar to the dual-stage of indirect methods, there are two stages as well in these methods. In the feature formulation stage, more explicit features will be obtained. In the recommender stage, personalization will be achieved. The creation will be firstly made in the recommender stage. Reference [27] proposed cluster-based recommender system (CB-RS), which set daily electricity usages of different hours as the features and used them to cluster customers. In CB-RS, customers in the same cluster shared the same series of recommended electricity plans. If a new testing feature is inputted

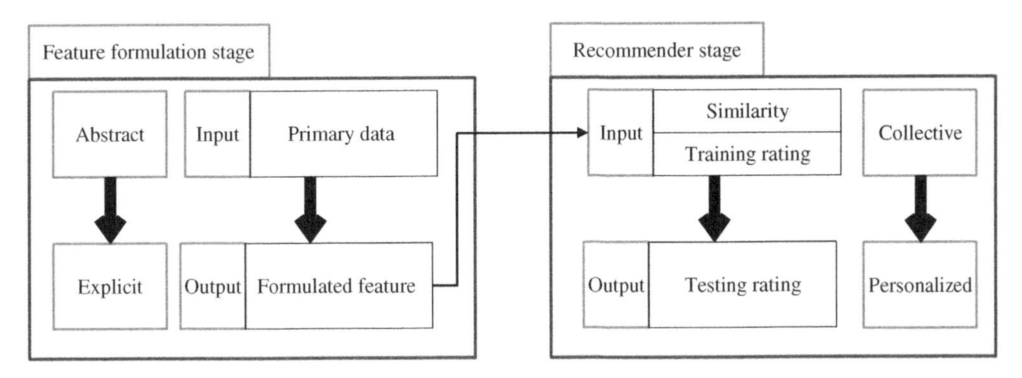

Figure 4.1 The dual-stage framework of EPRS. *Source:* From [A]/with permission of Elsevier.

into CB-RS, the recommended plans can be computed once the cluster is known. Compared to direct methods, CB-RS proposed clustering methods to recommend personal electricity plans to a new customer based on the cluster, but the disadvantage is that customers in the same cluster are allocated to the same range of electricity plans. The method social filtering EPRS (SF-EPRS) in [28] solved this problem by introducing collaborative filtering recommender system (CFRS) [29] into the recommender stage. In SF-EPRS, the feature formulation stage was similar to that of CB-RS, but in recommender stage, a weighting function was used to compute the recommended electricity plans of a testing customer. With this weighting function, the recommended plans of all testing customers are possible to be different from each other, to satisfy the need for personalization in EPRS. The recommender methods of following papers [30, 31] are all based on CFRS.

When the recommender stage is matured, researchers tend to utilize more explicit features in the feature formulation stage. In [30], collaborative filtering-based EPRS (CF-EPRS) was proposed. In this model, electricity usages of several selected appliances are transferred into operation duration in the feature formulation stage, and the operation duration is set as features used to compute similarity. Compared with the feature used in CF-EPRS [30], for example, the operation time of washing machine, the feature in CB-RS [27] and SF-EPRS [28], such as the average evening usage or average summer usage, is abstract and implicit. Further progress can be seen in Bayesian hybrid collaborative filtering-based EPRS (BHCF-EPRS) [31]. To avoid the data incompletion, BHCF-EPRS additionally utilizes probabilistic matrix factorization (BPMF) [32] to recover the extracted operation time, which alleviates the sparse problem. Besides, BHCF-EPRS introduces a classification machine to compute the similarity, which makes customers with similar total electricity usages more possible to be set as nearest neighbors. To give the difference of typical indirect methods, Table 4.1 presents the two stages of these methods.

Table 4.1 Some typical indirect EPRS methods.

Method	Feature extraction stage	Recommender stage
CB-RS [27]	Manual extraction	Cluster
SF-EPRS [28]	Manual extraction	CFRS
CF-EPRS [30]	Turn-on threshold and probability density function	CFRS
BHCF-EPRS [31]	BPMF	User classification, CFRS

Source: From [A]/with permission of Elsevier.

However, for BHCF-EPRS, although BPMF is an applicable way to recovery appliance usages, progress is possible to be made in improving the recovery precision, so we can extract more explicit features. According to [33], a sample matrix is an instinctively low-rank space, which means variables of a sample matrix only depend on a comparably smaller number of factors. The core of matrix recovery is to extract these factors and use them to reconstruct the corrupted data. For example, when predicting the rating of a movie, it is reasonable to assume that the rating may only be determined by a few preferences [34].

In matrix recovery, to exploit low-rank space, two methods are proposed, namely, matrix factorization (MF) [34] and nuclear-norm minimization (NNM) [35–37]. The difference between these two methods is the treatment of the rank of the sample matrix. For MF, the rank of the sample matrix and the probability distribution of parameters are set before learning, for example, parameters in BPMF are set to follow Gaussian–Wishart distribution. Instead, NNM [35–37] does not set any prior information into low-rank space and they apply nuclear regularization [38–40] to regulate the sample matrix. A theorem in [35] shows that NNM can achieve global optimum if the sample matrix is relatively complete, while MF does not have similar convergence property.

In this chapter, to get more explicit features, we introduce novel NNM methods to recovery appliance usages. According to the difference of recovery principle, we reformulate two classic NNM methods into novel ones, and they are Robust principal component analysis (RPCA) [35] and low-rank representation (LRR) [36], which apply nuclear regularization to learn low-rank data matrix and LRR matrix, respectively. Different from the prototypes, the novel ones are combined with electrical recovery instructions, which makes novel methods specialize in recovering appliance usages. Therefore, the new methods are named RPCA with electrical instructions (PRCA-EI) and LRR with electrical instructions (LRR-EI), respectively, and the new EPRS model is named EPRS with electrical instruction-based recovery (EPRS-EI).

The contributions of this chapter are as follows:

1) We propose EPRS-EI, which is a dual-stage model consisting of feature formulation stage and recommender stage. In feature formulation stage, appliance usages are recovered by PRCA-EI or LRR-EI and set as features, while in recommender stage, CFRS with K-Nearest Neighbors and adjusted similarity is applied to recommend personal electricity plans for customers.
2) Different from the classical matrix recovery models, electrical recovery instructions are applied in PRCA-EI and LRR-EI, which makes matrix recovery models specialize in recovering appliance usages. The recovery instructions we use are appliance classification and total electricity usage. The appliance classification is utilized to keep the known appliance data unchanged, while the total electricity usage is used to make recovered data constrained.
3) We also utilize a novel adjusted similarity evaluation in CRFS to computing the testing electricity plans. The total electricity usages are introduced into similarity evaluation for computing living pattern similarity among residential customers. In this case, residential customers with similar total electricity usages are more possible to be set as nearest neighbors.
4) We provide algorithms to solve our proposed methods, together with the convergence behavior and computational complexity analysis. Finally, the results in a recovery simulation and application simulation of EPRS confirm the effectiveness of our proposed methods in comparison to the state-of-the-art methods.

The rest of the chapter is organized as follows. Section 4.2 describes the proposed matrix recovery methods. In Section 4.3, the framework of the proposed EPRS is proposed. In Section 4.4, simulation results are conducted and discussed. Finally, the conclusion and future work are presented in Section 4.5.

4.2 Proposed Matrix Recovery Methods

4.2.1 Previous Matrix Recovery Methods

To show the difference between our methods and other methods, Table 4.2 provides information on the matrix recovery methods.

From Table 4.2, it can be seen that methods are classified into two parts, and they are MF and NNM. The objective of matrix recovery methods is to learn the low-rank sample matrix and find the intrinsic information in the sample matrix [35]. For MF methods, such as BPMF, the rank and the probability distribution of parameters are set before machine learning, while for NNM methods, for example, RPCA, the rank is adjusted during machine learning. NNM methods are thought as better methods because they perform better in convergence.

In addition to this, NNM methods can be classified into single subspace recovery (like RPCA), and multiple subspaces recovery (like LRR), and the difference is the unit used to recover the sample matrix. For RPCA, the unit is a certain recovered sample, while for LRR, the unit is a series of samples.

To improve performance, additional information is set as recovery instructions. NSHLRR [37] is the method that introduces graph construction and sparse regularization into LRR. Our proposed models also use appliance classification and total electricity usages as recovery instructions, and these instructions are introduced into both RPCA and LRR.

Partridge et al. [35] proposed RPCA to learn a low-rank matrix with the same dimensions of the sample matrix. The objective function of RPCA is:

$$
\begin{aligned}
&\min_{Z,E} \|Z\|_* + \lambda\|E\|_1 \\
&s.t. \quad X = Z + E
\end{aligned}
\tag{4.1}
$$

where $Z \in \mathbb{R}^{d\times n}$ denotes the low-rank sample matrix, $E \in \mathbb{R}^{d\times n}$ denotes the noise matrix. λ is the trade-off parameter of two terms. In Model (4.1), the first and second terms are nuclear-norm regularization of recovered matrix and L1-norm regularization of the noise matrix, respectively. The constraint is the formulation of the sample matrix. The L1-norm regularization of matrix E is computed by $\|E\|_1 = \sum_{i=1}^{d}\sum_{j=1}^{n}|E_{i,j}|$, and the nuclear-norm regularization of matrix Z is computed by $\|Z\|_* = \sum_{j=1}^{r}|\sigma_i|$ with r denoting the rank of matrix Z and σ_i denoting the ith singular value of matrix Z. Through the optimization of Model (4.1), the low-rank sample matrix Z can be learned and viewed as the recovered data of sample matrix X.

Table 4.2 Some typical matrix recovery methods.

Method	Classification	Recovery base	Additional instructions
BPMF [32]	MF	Gaussian–Wishart distribution	
RPCA [35]		Single subspace	
LRR [36]			
NSHLRR [37]	NNM	Multiple subspaces	Graph construction and sparse regularization

Source: From [A]/with permission of Elsevier.

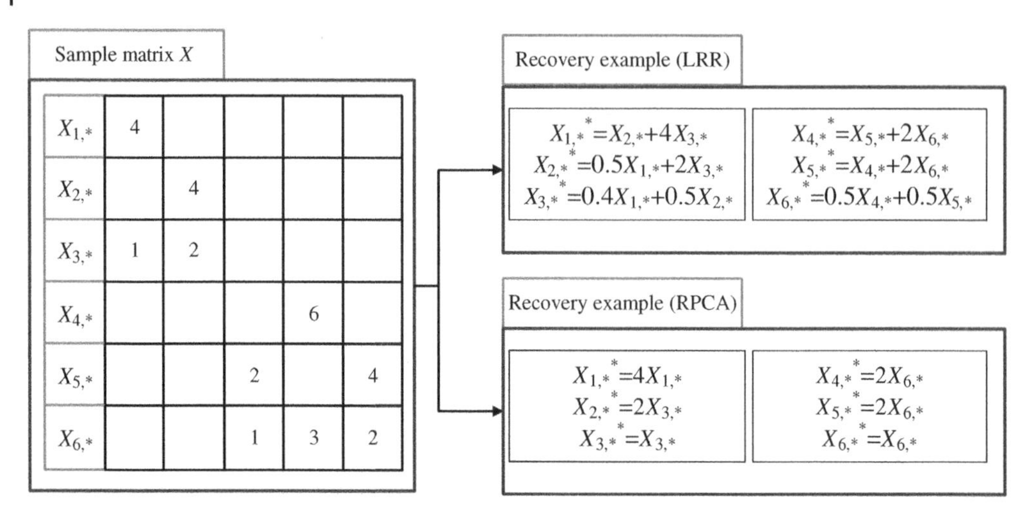

Figure 4.2 The recovery example of LRR and RPCA. *Source:* From [A]/with permission of Elsevier.

The principle of RPCA is that nuclear-norm regularization is applied to extract low-rank bases of the sample matrix, utilizing the bases to reconstruct the recovered samples. Details are shown in Figure 4.2. In Figure 4.2, X_3 and X_6 are selected as a base to recovery of the sample matrix.

Liu et al. [36] proposed LRR to seek the lowest rank representation among all the candidates that can represent the data samples as linear combinations of the bases from primary data, where the objective function is formulated as:

$$\begin{aligned}&\min_{Z,E} \|Z\|_* + \lambda\|E\|_1 \\ &s.t.\ \ X = XZ + E\end{aligned} \tag{4.2}$$

where $Z \in \mathbb{R}^{n\times n}$ denotes the LRR matrix of the sample matrix. λ is a trade-off parameter of two terms. In Model (4.2), the first term is the nuclear-norm regularization of the representation matrix and the second term is the L1-norm regularization of the noise matrix. The constraint is the formulation of a sample matrix. After obtaining an optimal solution (Z, E), the original data can be recovered by using XZ (or $X - E$). Since rank$(XZ) \leq$ rank(X), XZ is also a low-rank recovery to the corrupted data X.

Different from RPCA, LRR introduces low-rank regularization into the representation matrix, instead of the sample matrix. In this case, the data matrix is the first one to split into various subspaces, and the recovered sample is linearly reconstructed by the samples belonging to the subspace, same as the recovered sample. Details are shown in Figure 4.2. In Figure 4.2, X_1, X_2, and X_3 are allocated to one subspace, while X_4, X_5, and X_6 are allocated into other subspace.

Theoretically, this difference of RPCA and LRR lies on the computation flexibility. In [36], it is assumed that the data matrix consists of p low-rank subspaces $\{S_i\}_{1<i<p}$, RPCA can only exploit the sum of these subspaces $\sum_{i=1}^{m} S_i$, while LRR can exploit the union of the subspaces $\bigcup_{i=1}^{m} S_i$. Therefore, LRR can exploit more reconstruction units, having more computation flexibility.

4.2.2 Matrix Recovery Methods with Electrical Instructions

Novel methods are proposed by introducing electrical recovery instructions into RPCA and LRR according to the specifics of the electricity usage, and the novel ones are named RPCA with electrical instructions (PRCA-EI) and LRR with electrical instruction (LRR-EI).

These electrical recovery instructions are appliance classification and total electricity usage. Firstly, for the appliance classification, there are two appliance categories, that is, the known ones and the unknown ones. The goal is to recover unknown data based on the known appliance data, and the known data is kept unchanged. Secondly, the incompletion can be computed by the difference between total appliance electricity usage and the sum of appliance electricity usages. By introducing total electricity usage into the model, the recovered data can get close to complete total electricity usage.

The objective function of LRR-EI is formulated as:

$$\begin{aligned} \min_{Z,E} \ & \|Z\|_* + \lambda_1 \|P_\Omega(E)\|_1 + \lambda_2 \|Y - WXZ\|_F^2 \\ s.t. \ & X = XZ + E, \ \ P_\Lambda(E) = 0 \end{aligned} \tag{4.3}$$

where the third term is the Frobenius-norm regularization of data incompletion error, the second constraint is used to make the noise of measurable appliances zero. λ_1 and λ_2 are trade-off parameters. The projection matrix $W \in \mathbb{R}^{1\times d}$ is a vector that only contains "1", which is used to compute the sum of recovered appliance electricity usage. P_Λ and P_Ω are sets of measurable and measurable appliances, respectively. The Frobenius-norm regularization of matrix X is computed by $\|X\|_F = \sqrt{\sum_{i=1}^{d}\sum_{j=1}^{n} X_{i,j}^2}$.

Since the appliances only contain known and unknown parts, $\Lambda = \bar{\Omega}$ and Model (4.3) can be transformed into:

$$\begin{aligned} \min_{Z,E} \ & \|Z\|_* + \lambda_1 \|P_\Omega(E)\|_1 + \lambda_2 \|Y - WXZ\|_F^2 \\ s.t. \ & X = XZ + E, \ \ P_{\bar{\Omega}}(E) = 0 \end{aligned} \tag{4.4}$$

When $\lambda_2 = 0$ and the appliance sets are cancelled, LRR-EI degenerates into the primary LRR. Similarly, the objective function of PRCA-EI is formulated as:

$$\begin{aligned} \min_{Z,E} \ & \|Z\|_* + \lambda_1 \|P_\Omega(E)\|_1 + \lambda_2 \|Y - WZ\|_F^2 \\ s.t. \ & X = Z + E, \ \ P_{\bar{\Omega}}(E) = 0 \end{aligned} \tag{4.5}$$

When $\lambda_2 = 0$ and the appliance sets are cancelled, PRCA-EI degenerates into the primary RPCA.

4.2.3 Solution

(1) LRR-EI

In this part, the optimization algorithm of LRR-EI is developed.

To make the model unconstrained, the model is reformulated by augmented Lagrange method (ALM) [41].

Theorem 4.1 ALM [41]:
With the following objective function:

$$\begin{aligned} & \min \ g(X) \\ s.t. \ & h(\mathrm{X}) = 0 \end{aligned} \tag{4.6}$$

where $g : \mathbb{R}^n \rightarrow \mathbb{R}$, $h : \mathbb{R}^n \rightarrow \mathbb{R}^m$.

Model (4.6) can be reformulated to the following Model (4.7):

$$\begin{aligned} &\min L(X, Y1, \mu) \\ &s.t.\ \ L(X, Y1, \mu) = g(X) + tr\left((Y1)^{\mathrm{T}} h(X)\right) + \frac{\mu}{2}\|h(X)\|_F^2,\ \ \mu > 0 \end{aligned} \tag{4.7}$$

where $Y1$ is the Lagrange multiplier and μ is the penalty parameter. The optimum of Model (4.7) is the same as that of Model (4.6)

Firstly, by adding an auxiliary matrix J and reformulate Model (4.3) into:

$$\begin{aligned} &\min_{Z,E,J} \|J\|_* + \lambda_1\|P_\Omega(E)\|_1 + \lambda_2\|Y - WXZ\|_F^2 \\ &s.t.\ \ X = XZ + E,\ \ P_{\bar{\Omega}}(E) = 0,\ \ Z = J \end{aligned} \tag{4.8}$$

Secondly, through ALM, Model (4.8) can be reformulated into the following augmented Lagrangian function:

$$\begin{aligned} &\min_{Z,E,J,Y1,Y2} \|J\|_* + \lambda_1\|P_\Omega(E)\|_1 + \lambda_2\|Y - WXZ\|_F^2 \\ &tr\left((Y1)^{\mathrm{T}}(X - XZ - E)\right) + tr\left((Y2)^{\mathrm{T}}(Z - J)\right) \\ &+ \frac{\mu}{2}\left(\|X - XZ - E\|_F^2 + \|Z - J\|_F^2\right) \\ &s.t.\ \ P_{\bar{\Omega}}(E) = 0 \end{aligned} \tag{4.9}$$

where $Y1$, $Y2$ are Lagrange multipliers, and μ is a positive penalty parameter.

Model (4.9) is convex and can be optimized by updating Z, E, J, $Y1$ and $Y2$ alternatively. To optimize Model (4.9), the following lemma should be provided.

Lemma 4.1 ([42]): Given the following model:

$$\min_W \frac{1}{\alpha}\|W\|_* + \frac{1}{2}\|W - S\|_F^2 \tag{4.10}$$

W can be optimized by:

$$W = Udiag\left\{\max\left(\left(\sigma_i - \frac{1}{\alpha}\right), 0\right)\right\}_{1 \le i \le r} V^T \tag{4.11}$$

where r is the rank of S, σ_i, U and V are obtained by the Singular Value Decomposition (SVD) of S, i.e. $S = Udiag\{\sigma_i\}_{1 \le i \le r} V^T$.

Then Z, E, J, $Y1$, and $Y2$ can be subsequently updated. To compute Z when other variables are fixed, Model (4.9) can be reformulated into:

$$\begin{aligned} f_Z = &\lambda_2\|Y - WXZ\|_F^2 + tr\left((Y1)^{\mathrm{T}}(X - XZ - E)\right) + tr\left((Y2)^{\mathrm{T}}(Z - J)\right) \\ &+ \frac{\mu}{2}\left(\|X - XZ - E\|_F^2 + \|Z - J\|_F^2\right) \end{aligned} \tag{4.12}$$

With complete square formula, Eq. (4.12) can be reformulated into:

$$f_Z = \lambda_2\|Y - WXZ\|_F^2 + \frac{\mu}{2}\left(\left\|X - XZ - E + \frac{Y1}{\mu}\right\|_F^2 + \left\|Z - J + \frac{Y2}{\mu}\right\|_F^2\right) \tag{4.13}$$

By taking the derivative of f_Z with respect to Z and setting it to zero, we have:

$$\begin{aligned}
&\frac{\partial f_Z}{\partial Z} = 0\\
&\lambda_2(WX)^T(WX)Z + \mu X^T XZ - \mu X^T\left(X - E + \frac{Y1}{\mu}\right)\\
&\quad + \lambda_2(WX)^T Y + \mu Z - \mu\left(J - \frac{Y2}{\mu}\right) = 0\\
&Z = \left(\lambda_2(WX)^T(WX) + \mu X^T X + \mu I\right)^{-1}\\
&\quad\left(\mu X^T\left(X - E + \frac{Y1}{\mu}\right) + \mu\left(J - \frac{Y2}{\mu}\right) + \lambda_2(WX)^T Y\right)
\end{aligned} \tag{4.14}$$

To compute J when other variables are fixed, Model (4.9) can be reformulated into:

$$f_J = \|J\|_* + tr\left((Y2)^{\mathrm{T}}(Z - J)\right) + \frac{\mu}{2}\|Z - J\|_F^2 \tag{4.15}$$

With complete square formula, Eq. (4.15) can be reformulated into:

$$f_J = \frac{1}{\mu}\|J\|_* + \frac{1}{2}\left\|J - \left(Z + \frac{Y2}{\mu}\right)\right\|_F^2 \tag{4.16}$$

J in Eq. (4.16) can be solved by Lemma 4.1.

To compute E when other variables are fixed, Model (4.9) can be reformulated into:

$$\begin{aligned}
f_E = \lambda_1\|P_\Omega(E)\|_1 + tr\left((Y1)^{\mathrm{T}}(X - XZ - E)\right) + \frac{\mu}{2}\|X - XZ - E\|_F^2\\
s.t.\ \ P_{\bar{\Omega}}(E) = 0
\end{aligned} \tag{4.17}$$

With complete square formula, Eq. (4.17) can be reformulated into:

$$\begin{aligned}
f_E = \frac{\lambda_1}{\mu}\|P_\Omega(E)\|_1 + \frac{1}{2}\left\|X - XZ - E + \frac{Y1}{\mu}\right\|_F^2\\
s.t.\ \ P_{\bar{\Omega}}(E) = 0
\end{aligned} \tag{4.18}$$

According to [36], Eq. (4.18) can be solved by $E = P_\Omega\left(S_{\frac{\lambda}{\mu}}\left(X - XZ - E + \frac{Y1}{\mu}\right)\right)$, where $S_\tau(x)$ is a shrinkage operator denoted as $S_\tau(x) = \mathrm{sgn}(x)\max(|x| - \tau, 0)$.

Finally, since $Y1$ and $Y2$ are Lagrange multipliers, and μ is a positive penalty parameter, these matrices can be updated by the inexact ALM in [42].

From the above analysis, the overall algorithm for optimizing LRR-EI is described in Algorithm 4.1 as follows:

Algorithm 4.1 LRR-EI

1) Input: X, Y, and trade-off parameter λ.
2) Initialization: Initialize $J^0 = 0$, $E^0 = 0$, $Y_1 = Y_2 = 0$, $\mu^0 = 10^{-6}$, $\mu_{max} = 10^6$, $\rho = 1.05$, and $t = 0$.
3) While not converged do
   - Given other variables, update Z^{t+1} via Eq. (4.14).
   - Given other variables, update J^{t+1} via Eq. (4.16).
   - Given other variables, update E^{t+1} via Eq. (4.18).
   - Update the multipliers $Y1$ and $Y2$ via
   - $Y1^{t+1} = Y1^t + \mu^t(X - XZ^t)$.
   - $Y2^{t+1} = Y2^t + \mu^t(Z - J^t)$.
   - Update the parameter μ^{t+1} by $\mu^{t+1} = \min\{\rho\mu^t, \mu_{max}\}$.
   - $t = t + 1$.
4) End while
5) Output: The recovered data $X^* = P_\Omega(XZ)$.

---

(2) RPCA-EI

Same as LRR-EI, RPCA-EI can be made unconstrained by ALM. In this case, Model (4.4) can be reformulated into the following augmented Lagrangian function:

$$\begin{aligned} &\min_{Z,E,J,Y1,Y2} \|J\|_* + \lambda_1\|P_\Omega(E)\|_1 + \lambda_2\|Y - WZ\|_F^2 \\ &tr\left((Y1)^{\mathrm{T}}(X - Z - E)\right) + tr\left((Y2)^{\mathrm{T}}(Z - J)\right) \\ &+ \frac{\mu}{2}\left(\left\|X - Z - E + \frac{Y1}{\mu}\right\|_F^2 + \left\|Z - J + \frac{Y2}{\mu}\right\|_F^2\right) \\ &\qquad s.t.\ \ P_{\bar{\Omega}}(E) = 0 \end{aligned} \tag{4.19}$$

where $Y1$, $Y2$ are Lagrange multipliers, and μ is a positive penalty parameter.

Then we can subsequently update Z, E, J, $Y1$, and $Y2$. To compute Z when other variables are fixed, Model (4.19) can be reformulated into:

$$f_Z = \lambda_2\|Y - WZ\|_F^2 + \frac{\mu}{2}\left(\left\|X - Z - E + \frac{Y1}{\mu}\right\|_F^2 + \left\|Z - J + \frac{Y2}{\mu}\right\|_F^2\right) \tag{4.20}$$

By taking the derivative of f_Z with respect to Z and set it to zero, we have:

$$\begin{aligned} Z &= \left(\lambda_2 W^T W + 2\mu I\right)^{-1} \\ &\left(\mu\left(X - E + \frac{Y1}{\mu} + J - \frac{Y2}{\mu}\right) + \lambda_2 W^T Y\right) \end{aligned} \tag{4.21}$$

To compute J when other variables are fixed, Model (4.19) can be reformulated as

$$f_J = \frac{1}{\mu}\|J\|_* + \frac{1}{2}\left\|J - \left(Z + \frac{Y2}{\mu}\right)\right\|_F^2 \tag{4.22}$$

J in Eq. (4.22) can be solved with Lemma 4.1.

To compute E when other variables are fixed, Model (4.19) can be reformulated as

$$f_E = \frac{\lambda}{\mu}\|P_\Omega(E)\|_1 + \frac{1}{2}\left\|X - Z - E + \frac{Y1}{\mu}\right\|_F^2 \quad s.t.\ P_{\bar{\Omega}}(E) = 0 \tag{4.23}$$

it can be solved by $E = P_\Omega\left(S_{\frac{\lambda}{\mu}}\left(X - Z - E + \frac{Y1}{\mu}\right)\right)$.

From the above analysis, the overall algorithm for optimizing RPCA-EI is described in Algorithm 4.2 as follows:

Algorithm 4.2 RPCA-EI

---

1) Input: X, Y, and trade-off parameter λ.
2) Initialization: Initialize $J^0 = 0$, $E^0 = 0$, $Y_1 = Y_2 = 0$, $\mu^0 = 10^{-6}$, $\mu_{max} = 10^6$, $\rho = 1.05$, and $t = 0$.
3) While not converged do
   - Given other variables, update Z^{t+1} via Eq. (4.21).
   - Given other variables, update J^{t+1} via Eq. (4.22).
   - Given other variables, update E^{t+1} via Eq. (4.23).
   - Update the multipliers $Y1$ and $Y2$ via
   - $Y1^{t+1} = Y1^t + \mu^t(X - Z^t)$.
   - $Y2^{t+1} = Y2^t + \mu^t(Z - J^t)$.
   - Update the parameter μ^{t+1} by $\mu^{t+1} = \min\{\rho\mu^t, \mu_{max}\}$.
   - $t = t + 1$.
4) End while
5) Output: The recovered data $X^* = P_\Omega(Z)$.

4.2.4 Convergence Analysis and Complexity Analysis

First, the convergence of the alternating optimization algorithm is discussed. For both RPCA-EI and LRR-EI, the whole model is solved by inexact ALM [36], while Z, J and E are optimized by derivation, singular value thresholding (SVT) [38] and sparse function [39]. These algorithms converge and the proof is demonstrated in [36, 38, 39].

Then, we analyze the computational complexity of LRR-EI (Algorithm 4.1) as well as RPCA-EI (Algorithm 4.2), and the notation O is used to represent the time complexity.

As for LRR-EI, according to Eqs. (4.14), (4.16), and (4.18), the time complexities of calculating Z, J, and E are $O(2n^2d + n^3)$, $O(nr^2 + n^2r + n^3)$, and $O(nd)$, respectively, where r is the rank of Z. Let t be the iteration number of the overall algorithm, we have $O(t(2n^2d + 2n^3 + nr^2 + n^2r))$ as the overall time complexity of Algorithm 4.1. In the real application, we have $n \gg d$, and therefore, the time complexity of Algorithm 4.1 can approximately be taken as $O(2tn^3)$.

For RPCA-EI, according to Eqs. (4.21)–(4.23), the time complexities of calculating Z, J and E are $O(d^3)$, $O(dr^2 + ndr + n^3)$, and $O(nd)$ respectively, where r is the rank of Z. Let t be the iteration number of the overall algorithm. Thus, we have $O(t(dr^2 + ndr + n^3 + d^3))$ as the overall time complexity of Algorithm 4.2. In the real application, we have $n \gg d$, and therefore, the time complexity of Algorithm 4.2 can approximately be taken as $O(tn^3)$.

4.3 Proposed Electricity Plan Recommender System

A detailed analysis of EPRS with electrical instruction-based recovery (EPRS-EI) is given. The recommender system used is neighborhood-based collaborative filtering in [29], and a total electricity usage is introduced into similarity, to propose a novel adjusted similarity. The process of neighborhood-based collaborative filtering can be viewed as a utility function: $U \times T \rightarrow R$ which can generate a mapping from a set user U and set item T to set rating R. The set user U stands for the customers, the set item T stands for the users to choose the electricity plans, and the set rating R stands for the preference of customers to different electricity plans. The recommender process consists of two stages, and they are feature formulation stage and recommender stage. The detail is shown as follows:

4.3.1 Feature Formulation Stage

In this stage, we need to get features which represent the customers' living patterns. Appliance usages are set as feature data, but it is impossible to utilize the raw appliance usages because the loss of data is inevitable. The incompletion can be found out when the user's total appliance electricity usage fails to match with the sum of the appliance electricity usages.

To recover the corrupted data, both RPCA-EI and LRR-EI are applied to recover the data, and then the recovered data is inputted into the recommender stage.

4.3.2 Recommender Stage

In this stage, feature data is transformed into a similarity matrix by similarity evaluation and testing rating is calculated by a weighting function based on training rating and similarity matrix.

(1) Similarity Evaluation

In neighborhood-based collaborative filtering, the neighbors of the testing customers need to be searched. Firstly, KNN is utilized to search for each testing customer's neighbors. Through KNN, we can avoid computing the similarity of pair customers, making it efficient to detect training customers having a similar living pattern to that of the testing customers.

Radial basis function (RBF) is then used to evaluate the similarity of the neighbors selected by KNN, which is shown as follows:

$$S_{i,j} = e^{-\frac{\left\|X_{*,i}-X_{*,j}\right\|_2^2}{2\rho^2}} \tag{4.24}$$

where $S \in \mathbb{R}^{n \times n}$ is the similarity matrix indicating the similarity between pair customers, $X \in \mathbb{R}^{d \times n}$ denotes the feature data of customers which represents customers' electricity living patterns, and $S_{i,j}$ indicates the similarity of the customers U_i and U_j. The larger the element is, the more similar the pair customers are. The L2-norm regularization in Eq. (4.24) is computed by $\|Y\|_2 = \sqrt{\sum_{i=1}^{n} Y_i^2}$.

To improve the performance, we introduce total electricity usage into similarity evaluation and call the new similarity adjusted similarity. The computation is:

$$S_{i,j} = \frac{1}{1 + \|Y_i - WX_{*,i}\|_2} \times \frac{1}{1 + \|Y_j - WX_{*j}\|_2} \times e^{-\frac{\|X_{*,i} - X_{*,j}\|_2^2}{2\rho^2}} \tag{4.25}$$

where $\frac{1}{1 + \|Y_i - WX_{*,i}\|_2}$ and $\frac{1}{1 + \|Y_j - WX_{*j}\|_2}$ are the penalty factors computed by total electricity usages. If the error between customer's total appliance electricity usage and the sum of the appliance electricity usages enlarges, the penalty factors will make the similarity small, and vice versa.

(2) Item Recommender

For the testing customer U_m, let U_m^k be the K-Nearest Neighbors of U_m, the rating of U_m can be calculated by the weighting function as shown below:

$$R_{m,i} = \frac{\sum\limits_{n \in U_m^k} S_{m,n} R_{n,i}}{\sum\limits_{n \in U_m^k} S_{m,n}} \tag{4.26}$$

If we set the yearly charge of electricity plans as the rating, a lower rating indicates more possibilities that customers choose the electricity plan.

The metric of EPRS is the precision for a top N recommendation, and it is calculated by:

$$\text{Precision} = \frac{\left|\tilde{I}_m^N \cap I_m^N\right|}{N} \tag{4.27}$$

where $\tilde{I}_m^N$ and I_m^N are the set of top N possible predictive and real electricity plans customer U_m would like to choose. In our experiment, N is set as 6.

4.3.3 Algorithm and Complexity Analysis

Overall, we have Figure 4.3 to show the framework, and the Algorithm 4.3 of EPRS-EI is shown as follows:

Algorithm 4.3 EPRS-EI

1) Input: Appliance electricity usage of training customers X_{tr} and testing customers X_{te}, the yearly charge of electricity plans of training customer is (training ratings) R_{tr}.
2) Recovering X_{tr} and X_{te} by RPCA-EI (or LRR-EI)
3) Detecting neighbors of testing customers via KNN and Eq. (4.25).
4) Predicting testing customers via Eq. (4.26).
5) Output: Testing ratings R_{te}.

For EPRS-EI (Algorithm 4.3), let n_{te} be the number of testing samples and k be the number of their nearest neighbors. We have $O(2tn^3)$ (or $O(tn^3)$), $O(n^2d)$, and $O(n_{te}k)$ as the time complexity of Steps 2–4 for EPRS-EI (LRR-EI) and EPRS-EI (RPCA-EI), and the overall time complexity of EPRS-EI (LRR-EI) and EPRS-EI (RPCA-EI) are $O(2tn^3)$ and $O(tn^3)$ respectively.

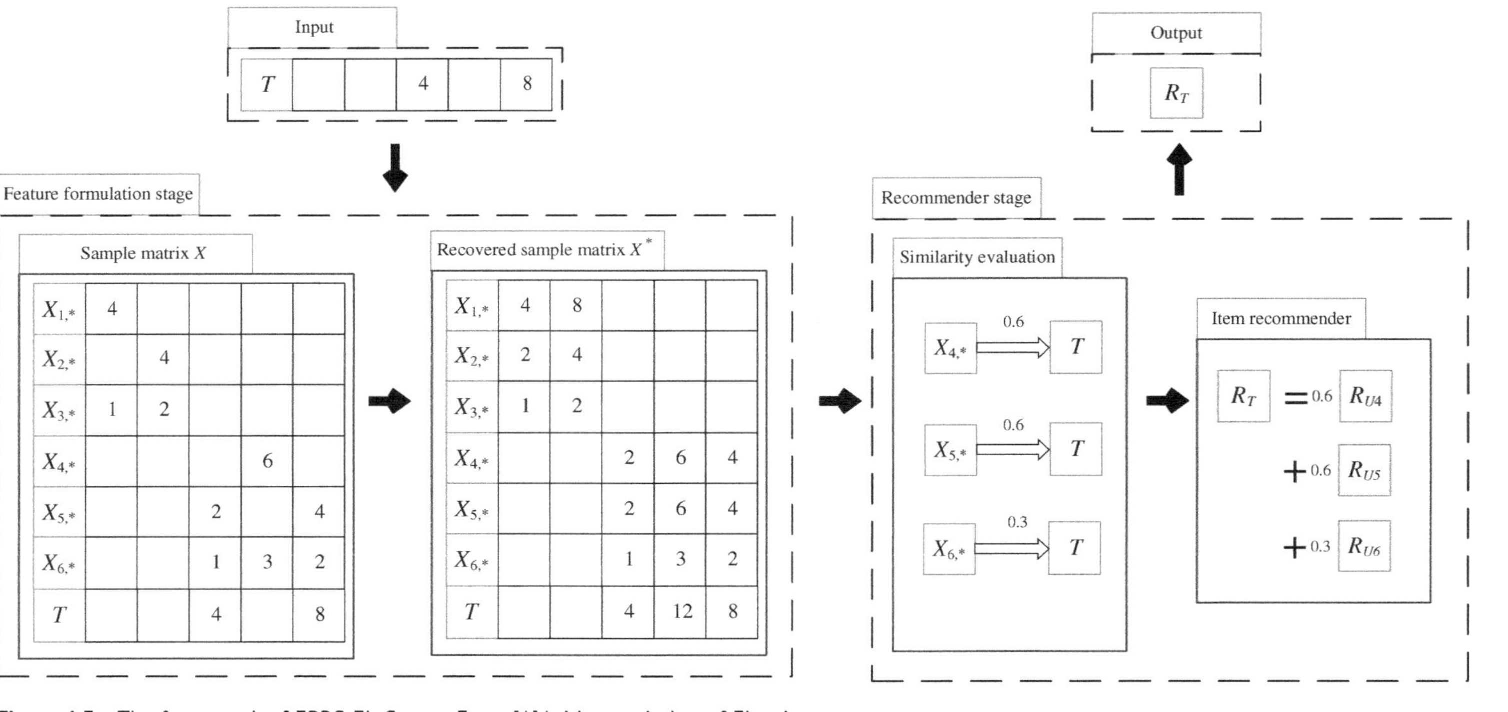

Figure 4.3 The framework of EPRS-EI. *Source:* From [A]/with permission of Elsevier.

4.4 Simulations and Discussions

In this part, two types of simulations are conducted. The first type is for recovery, which is utilized to test the recovery capability of the models, while the second simulation is for application, which is utilized to test these models' application potential in EPRS.

4.4.1 Recovery Simulation

In this simulation, we extract yearly load data from 2014 to 2018 of the three cities (Austin, Boulder, and San Diego), in Dataport [43]. In Dataport, the total appliance electricity usage is the sum of 67 appliance electricity usages. For simplifying the computation, regular electricity usage including wall outlets are cancelled, and only 13 appliances are taken into consideration. Besides, because for a customer, the appliance usage in the weekday and the weekend is different. Therefore, the number of features is set at 26, and the first 13 features represent the usage in the weekdays while the other 13 features represent the usage at the weekends. The selected appliances are shown in Table 4.3. The data size for 2014–2018 data is 622×26, 577×26, 415×26, 364×26, 310×26, respectively.

To conduct the recovery, we select a certain number of appliance electricity usage from known appliance electricity usage and allocate those data into unknown data. In this case, the known data is allocated into training data and the unknown one is allocated into testing data. The testing data is recovered based on the remaining data (training data) by matrix recovery.

To avoid bias, 10 runs are conducted in each database by selecting 30, 40, 50, and 60% known appliance usage as training data randomly in each yearly data, and 10 cross-validations are adopted to select parameters. After recovering, root mean square error (RMSE), and RRMSE are computed by:

$$\text{RMSE} = \sqrt{\sum_{x_i \in D_{test}} (\widetilde{x}_i - x_i)^2} \tag{4.28}$$

where $\widetilde{X}$ denotes the mean of the testing data. The lower RMSE indicates better performance.

To test and verify the effectiveness of our proposed method, several competing methods are considered including Bayesian probabilistic matrix factorization (BPMF) [32], RPCA [35], LRR [36], NSHLRR [37], RPCA-EI and LRR-EI. All the regulation parameters in the models are chosen from the set $\{10^{-4}, 10^{-3}, \ldots, 10^{1}\}$, the rank parameter in BPMF is set to be five times smaller than the number of appliances. The simulation results of RMSE performance are shown in Tables 4.4–4.8, and t-test result between proposed methods and other methods are shown in Tables 4.9–4.13. In t-test, "W" means the proposed method performs better. "M" means other approaches perform

Table 4.3 The selected appliances.

Air-conditional	Electric vehicle	Clothes washer	Dishwasher
Dryer	Furnace	Heater	House fan
Microwave	Oven	Range hood	Water heater
Refrigerator			

Source: From [A]/with permission of Elsevier.

Table 4.4 The RMSE performance for 2014 data.

Method \ Training/Total data set	30%	40%	50%	60%	Average
BPMF	13.89 ± 0.22	11.72 ± 0.18	9.83 ± 0.16	8.03 ± 0.09	10.87
RPCA	5.89 ± 0.07	5.37 ± 0.09	4.82 ± 0.09	4.07 ± 0.12	5.03
LRR	5.75 ± 0.07	5.20 ± 0.09	4.61 ± 0.10	3.82 ± 0.13	4.85
NSHLRR	4.47 ± 0.07	5.47 ± 0.08	4.47 ± 0.02	3.17 ± 0.12	4.39
RPCA-EI	5.10 ± 0.07	4.67 ± 0.02	4.21 ± 0.08	3.54 ± 0.13	4.38
LRR-EI	**3.76 ± 0.08**	**3.54 ± 0.08**	**3.29 ± 0.09**	**2.74 ± 0.14**	**3.33**

Source: From [A]/with permission of Elsevier.

Table 4.5 The RMSE performance for 2015 data.

Method \ Training/Total data set	30%	40%	50%	60%	Average
BPMF	12.66 ± 0.12	10.70 ± 0.13	9.01 ± 0.16	7.39 ± 0.13	9.94
RPCA	5.68 ± 0.11	5.17 ± 0.12	4.61 ± 0.12	3.97 ± 0.11	4.86
LRR	5.55 ± 0.11	5.00 ± 0.12	4.40 ± 0.13	3.72 ± 0.12	4.67
NSHLRR	5.87 ± 0.01	5.17 ± 0.12	4.61 ± 0.12	3.47 ± 0.01	4.78
RPCA-EI	4.90 ± 0.11	4.47 ± 0.12	3.99 ± 0.13	3.44 ± 0.12	4.20
LRR-EI	**3.53 ± 0.16**	**3.28 ± 0.12**	**2.99 ± 0.15**	**2.60 ± 0.14**	**3.10**

Source: From [A]/with permission of Elsevier.

Table 4.6 The RMSE performance for 2016 data.

Method \ Training/Total data set	30%	40%	50%	60%	Average
BPMF	12.72 ± 0.21	10.69 ± 0.16	8.98 ± 0.13	7.29 ± 0.16	9.92
RPCA	4.56 ± 0.07	4.16 ± 0.09	3.71 ± 0.11	3.22 ± 0.09	3.91
LRR	4.44 ± 0.07	4.01 ± 0.10	3.52 ± 0.11	3.00 ± 0.09	3.74
NSHLRR	4.56 ± 0.07	3.16 ± 0.09	3.31 ± 0.11	3.12 ± 0.09	3.54
RPCA-EI	3.89 ± 0.06	3.56 ± 0.08	3.19 ± 0.10	2.77 ± 0.08	3.35
LRR-EI	**2.49 ± 0.06**	**2.32 ± 0.05**	**2.11 ± 0.06**	**1.83 ± 0.07**	**2.19**

Source: From [A]/with permission of Elsevier.

Table 4.7 The RMSE performance for 2017 data.

Method \ Training/Total data set	30%	40%	50%	60%	Average
BPMF	11.74 ± 0.30	13.87 ± 0.20	11.70 ± 0.23	13.88 ± 0.21	12.80
RPCA	5.28 ± 0.06	5.86 ± 0.11	5.31 ± 0.08	5.81 ± 0.09	5.57
LRR	5.11 ± 0.07	5.73 ± 0.12	5.14 ± 0.08	5.68 ± 0.09	5.41
NSHLRR	4.47 ± 0.09	5.47 ± 0.10	4.47 ± 0.18	5.47 ± 0.19	4.97
RPCA-EIR	4.58 ± 0.07	5.07 ± 0.11	4.60 ± 0.08	5.03 ± 0.09	4.82
LRR-EI	**3.46 ± 0.07**	**3.72 ± 0.11**	**3.48 ± 0.07**	**3.71 ± 0.06**	**3.59**

Source: From [A]/with permission of Elsevier.

Table 4.8 The RMSE performance for 2018 data.

Method \ Training/Total data set	30%	40%	50%	60%	Average
BPMF	13.87 ± 0.20	11.74 ± 0.30	9.82 ± 0.15	8.04 ± 0.09	10.87
RPCA	5.86 ± 0.11	5.28 ± 0.06	4.75 ± 0.09	4.10 ± 0.12	5.00
LRR	5.73 ± 0.12	5.11 ± 0.07	4.54 ± 0.09	3.86 ± 0.13	4.81
NSHLRR	5.47 ± 0.01	5.60 ± 0.01	4.47 ± 0.09	3.77 ± 0.01	4.83
RPCA-EIR	5.07 ± 0.11	4.58 ± 0.07	4.13 ± 0.09	3.57 ± 0.13	4.33
LRR-EI	**3.72 ± 0.16**	**3.46 ± 0.07**	**3.17 ± 0.11**	**2.77 ± 0.14**	**3.28**

Source: From [A]/with permission of Elsevier.

Table 4.9 *t*-test result between proposed models and other methods for 2014 data.

Method \ Training/Total data set	RPCA-EI				Method \ Training/Total data set	LRR-EI			
	30%	40%	50%	60%		30%	40%	50%	60%
BPMF	W (0.00)	W (0.00)	W (0.00)	W (0.00)	BPMF	W (0.00)	W (0.00)	W (0.00)	W (0.00)
RPCA	W (0.00)	W (0.00)	W (0.00)	W (0.00)	RPCA	W (0.00)	W (0.00)	W (0.00)	W (0.00)
LRR	W (0.00)	W (0.00)	W (0.00)	W (0.00)	LRR	W (0.00)	W (0.00)	W (0.00)	W (0.00)
NSHLRR	W (0.00)	F (0.04)	W (0.01)	W (0.02)	NSHLRR	W (0.00)	W (0.03)	W (0.00)	W (0.00)
LRR-EI	M (0.00)	M (0.00)	M (0.01)	M (0.00)	RPCA-EI	W (0.02)	W (0.00)	W (0.01)	W (0.00)

Source: From [A]/with permission of Elsevier.

Table 4.10 *t*-test result between proposed models and other methods for 2015 data.

Method \ Training/Total data set	RPCA-EI 30%	RPCA-EI 40%	RPCA-EI 50%	RPCA-EI 60%	Method \ Training/Total data set	LRR-EI 30%	LRR-EI 40%	LRR-EI 50%	LRR-EI 60%
BPMF	W (0.00)	W (0.00)	W (0.00)	W (0.00)	BPMF	W (0.00)	W (0.00)	W (0.00)	W (0.00)
RPCA	W (0.00)	W (0.00)	W (0.00)	W (0.00)	RPCA	W (0.00)	W (0.00)	W (0.00)	W (0.00)
LRR	W (0.02)	W (0.00)	W (0.00)	W (0.00)	LRR	W (0.00)	W (0.00)	W (0.00)	W (0.00)
NSHLRR	W (0.00)	W (0.00)	W (0.01)	W (0.00)	NSHLRR	W (0.00)	W (0.00)	W (0.08)	W (0.00)
LRR-EI	M (0.01)	M (0.00)	M (0.00)	M (0.04)	RPCA-EI	W (0.00)	W (0.01)	W (0.00)	W (0.00)

Source: From [A]/with permission of Elsevier.

Table 4.11 *t*-test result between proposed models and other methods for 2016 data.

Method \ Training/Total data set	RPCA-EI 30%	RPCA-EI 40%	RPCA-EI 50%	RPCA-EI 60%	Method \ Training/Total data set	LRR-EI 30%	LRR-EI 40%	LRR-EI 50%	LRR-EI 60%
BPMF	W (0.00)	W (0.00)	W (0.00)	W (0.00)	BPMF	W (0.00)	W (0.00)	W (0.00)	W (0.00)
RPCA	W (0.00)	W (0.00)	W (0.00)	W (0.00)	RPCA	W (0.00)	W (0.00)	W (0.00)	W (0.00)
LRR	W (0.00)	W (0.00)	W (0.00)	W (0.00)	LRR	W (0.00)	W (0.00)	W (0.00)	W (0.00)
NSHLRR	W (0.00)	F (0.09)	W (0.00)	W (0.00)	NSHLRR	W (0.00)	W (0.00)	W (0.00)	W (0.00)
LRR-EI	M (0.00)	M (0.00)	M (0.02)	M (0.00)	RPCA-EI	W (0.00)	W (0.00)	W (0.00)	W (0.02)

Source: From [A]/with permission of Elsevier.

Table 4.12 *t*-test result between proposed models and other methods for 2017 data.

Method \ Training/Total data set	RPCA-EI 30%	RPCA-EI 40%	RPCA-EI 50%	RPCA-EI 60%	Method \ Training/Total data set	LRR-EI 30%	LRR-EI 40%	LRR-EI 50%	LRR-EI 60%
BPMF	W (0.00)	W (0.00)	W (0.00)	W (0.00)	BPMF	W (0.00)	W (0.00)	W (0.00)	W (0.00)
RPCA	W (0.00)	W (0.00)	W (0.00)	W (0.00)	RPCA	W (0.00)	W (0.00)	W (0.00)	W (0.00)
LRR	W (0.00)	W (0.00)	W (0.00)	W (0.00)	LRR	W (0.00)	W (0.01)	W (0.00)	W (0.00)
NSHLRR	M (0.03)	W (0.00)	M (0.02)	W (0.00)	NSHLRR	W (0.01)	W (0.00)	W (0.00)	W (0.00)
LRR-EI	M (0.00)	M (0.00)	M (0.00)	M (0.01)	RPCA-EI	W (0.00)	W (0.00)	W (0.00)	W (0.03)

Source: From [A]/with permission of Elsevier.

Table 4.13 *t*-test result between proposed models and other methods for 2018 data.

Method \ Training/Total data set	RPCA-EI 30%	RPCA-EI 40%	RPCA-EI 50%	RPCA-EI 60%	Method \ Training/Total data set	LRR-EI 30%	LRR-EI 40%	LRR-EI 50%	LRR-EI 60%
BPMF	W (0.00)	W (0.00)	W (0.00)	W (0.00)	BPMF	W (0.00)	W (0.00)	W (0.00)	W (0.00)
RPCA	W (0.01)	W (0.00)	W (0.00)	W (0.00)	RPCA	W (0.00)	W (0.00)	W (0.00)	W (0.00)
LRR	W (0.05)	W (0.00)	W (0.00)	W (0.00)	LRR	W (0.00)	W (0.00)	W (0.00)	W (0.00)
NSHLRR	W (0.01)	W (0.00)	W (0.00)	W (0.00)	NSHLRR	W (0.03)	W (0.00)	W (0.00)	W (0.00)
LRR-EI	M (0.00)	M (0.00)	M (0.00)	M (0.01)	RPCA-EI	W (0.00)	W (0.00)	W (0.00)	W (0.02)

Source: From [A]/with permission of Elsevier.

better. Value in the bracket is the associate p-value. Statistical significance of *t*-test is 5%. The smaller p-value means the higher assurance of the conclusion.

4.4.2 Recovery Result Discussions

From Figure 4.4, it can be seen that as the trade-off parameters which have the closest relationship with the objective low-rank matrix, this may lead to greater change. Comparing with λ_2, as λ_1 changes, the amplitude of RMSE changes dramatically. This is because λ_1 in both RPCA-RI and LRR-EI, is the parameter which compromises between nuclear-regularization term and other errors.

For Tables 4.4–4.13, some discussions are given as follows:

- It is known that with the increase of the training samples, RMSE decreases for all methods. This is because as the known information increases, it is easier to recover the appliance electricity usage.
- The performance of BPMF cannot exceed other methods. This is because BPMF belongs to MF method, we need to input the matrix's rank as a parameter. If this rank fails to match well with the data, then it cannot perform better.
- Conversely, other NNM methods can adjust the rank of data matrix automatically in the learning process, and it can be drawn that all methods with instructions, i.e. NSHLRR, RPCA-EI, and LRR-EI, outperform their prototypes (RPCA and LRR). The result implies that additional instructions can improve recovery performance, but the additional instructions may cause incorrect recovery results. That is why NSHLRR cannot outperform the prototypes in some databases.
- Fortunately, the performances of RPCA-EI and LRR-EI are better than that of other methods, which indicates that our specific recovery instructions, appliance classification and total electricity usages, are effective in the electronic application.
- To present the effectiveness of proposed methods, $\frac{\text{RMSE}_{pre} - \text{RMSE}_{pro}}{\text{RMSE}_{pre}} \times 100\%$ is proposed to compute the comparisons, where RMSE_{pre} denotes RMSE of the previous methods and RMSE_{pro} denotes that of the proposed methods. Specifically, in Table 4.4, comparing to NSHLRR ($\text{RMSE}_{\text{NSHLRR}} = 4.39$), our method LRR-EI ($\text{RMSE}_{\text{LRR-EI}} = 3.33$) gives up to $\frac{4.39 - 3.33}{4.39} \times 100\% = 24.14\%$ improvement. Similarly, in Table 4.6, we have $\text{RMSE}_{\text{NSHLRR}} = 3.54$, $\text{RMSE}_{\text{LRR-EI}} = 2.19$,

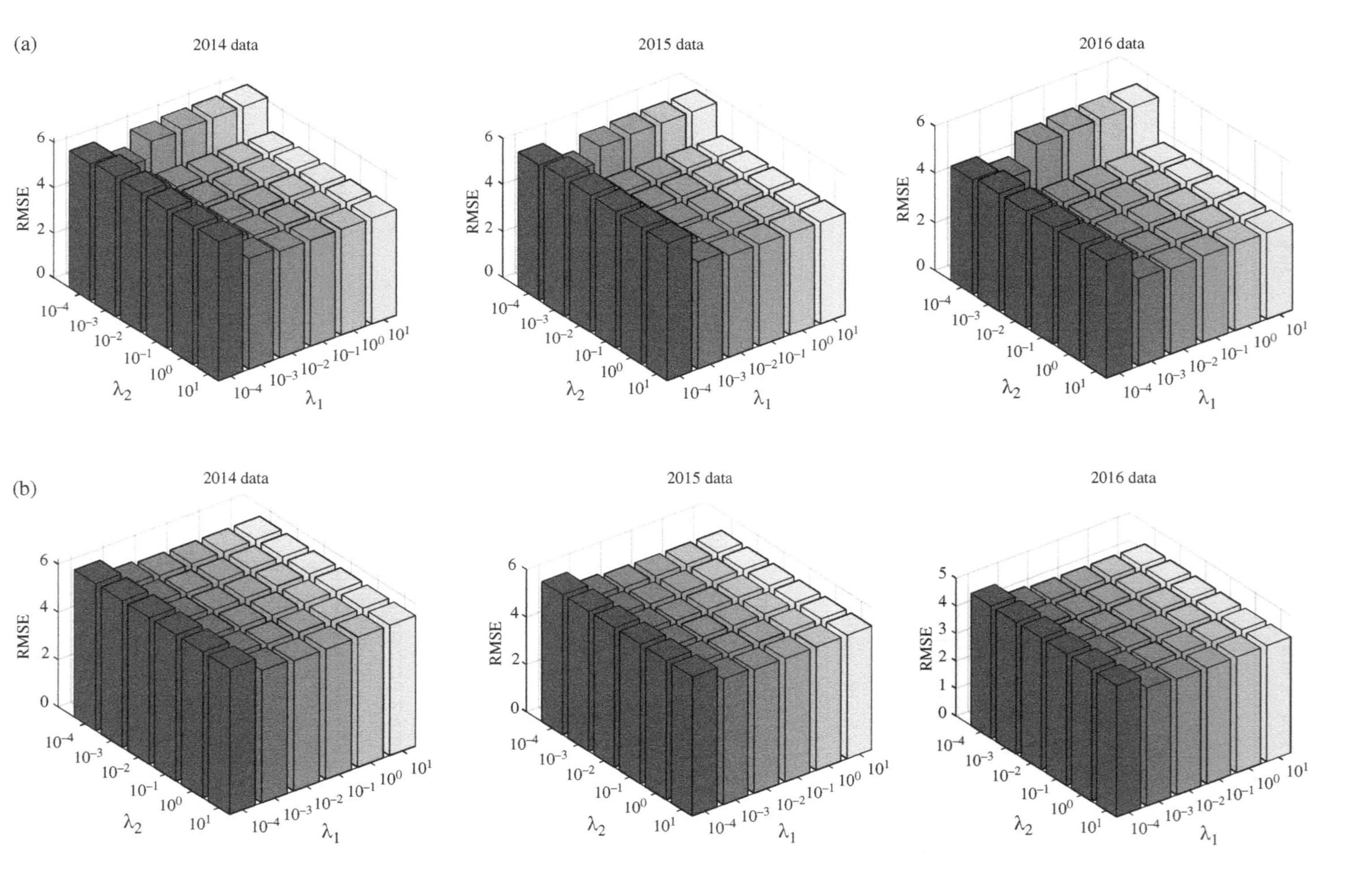

Figure 4.4 RMSE for different databases with different parameters. (only show 2014–2016's 60% testing data cases) (a) RPCA-EI. (b) LRR-EI. *Source:* From [A]/with permission of Elsevier.

so LRR-EI gives up to $\frac{3.54-2.19}{3.54}\times 100\%=38.15\%$ improvement. Improvements can also be seen in other datasets with the range from 24.14 to 38.15%.

- In Table 4.4, comparing to LRR-EI's prototype LRR ($\text{RMSE}_{\text{LRR}}=4.85$), our method LRR-EI ($\text{RMSE}_{\text{LRR}-\text{EI}}=3.33$) gives up to $\frac{4.85-3.33}{4.85}\times 100\%=31.22\%$ improvement, while in Table 4.6, we have $\text{RMSE}_{\text{LRR}}=3.74$, $\text{RMSE}_{\text{LRR}-\text{EI}}=2.19$, so LRR-EI gives $\frac{3.74-2.19}{3.74}\times 100\%=41.50\%$ improvement. Improvements can also be seen in other datasets with the range from 31.22 to 41.50%.
- In Table 4.4, comparing to RPCA-EI's prototype RPCA ($\text{RMSE}_{\text{RPCA}}=5.03$), RPCA-EI ($\text{RMSE}_{\text{RPCA}-\text{EI}}=4.38$) gives up to $\frac{5.03-4.38}{5.03}\times 100\%=12.99\%$ improvement, while in Table 4.6, we have $\text{RMSE}_{\text{RPCA}}=3.91$, $\text{RMSE}_{\text{RPCA}-\text{EI}}=3.35$, so LRR-EI gives $\frac{3.91-3.35}{3.91}\times 100\%=14.32\%$ improvement. Improvements can also be seen in other datasets with the range from 12.99 to 14.32%.

4.4.3 Application Study

In this study, customers' yearly appliance electricity usage is set as feature data. The rating is the yearly charge ranking of different electricity plans. The electricity plans are collected from Energy Made Easy [10], including seven time-of-use tariffs and seven single-rate tariffs.

The rating is calculated from 2015, 2016, 2017, 2018 data, and the feature data used is the year-ahead rating. The data size is 343×26, 314×26, 242×26, 153×26. The number of nearest neighbors is chosen from 2 to 5, N in Eq. (4.27) is set as 6, 10 runs are conducted in each data by selecting 30, 40, 50, and 60% of all the users as training users randomly.

To test and verify the effectiveness of our proposed method, several competing methods are considered including DWR (data without recovery), SF-EPRS, BHCF-EPRS, EPRS-EI (RPCA-EI), and EPRS-EI (LRR-EI), where DWR is the EPRS model by setting corrupted data as features. All the regulation parameters in the models are chosen from the set $\{10^{-4}, 10^{-3}, \ldots, 10^{1}\}$. Figure 4.5 shows the learning curve of EPRS-EI, while Figure 4.6 shows precision for these different databases with

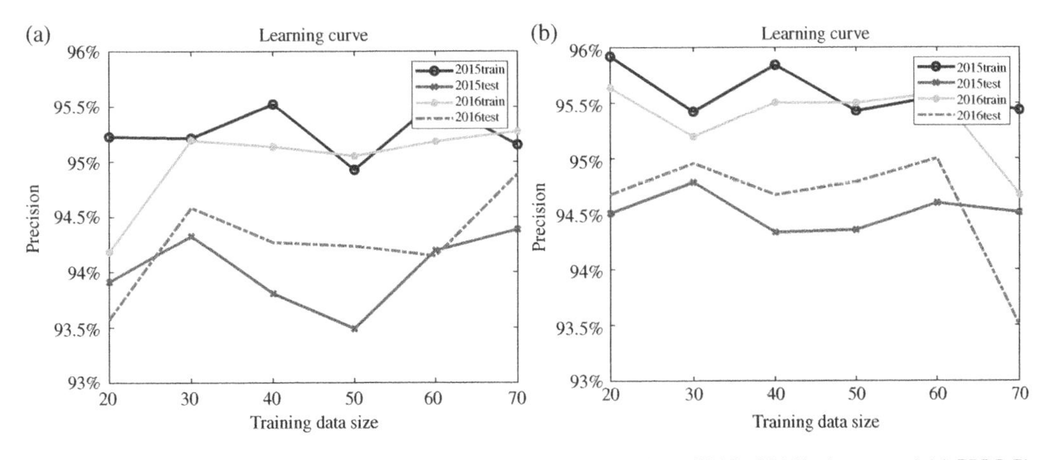

Figure 4.5 Learning curve with different training data size (only show 2015–2016's data cases) (a) EPRS-EI (RPCA-EI). (b) EPRS-EI (LRR-EI). *Source:* From [A]/with permission of Elsevier.

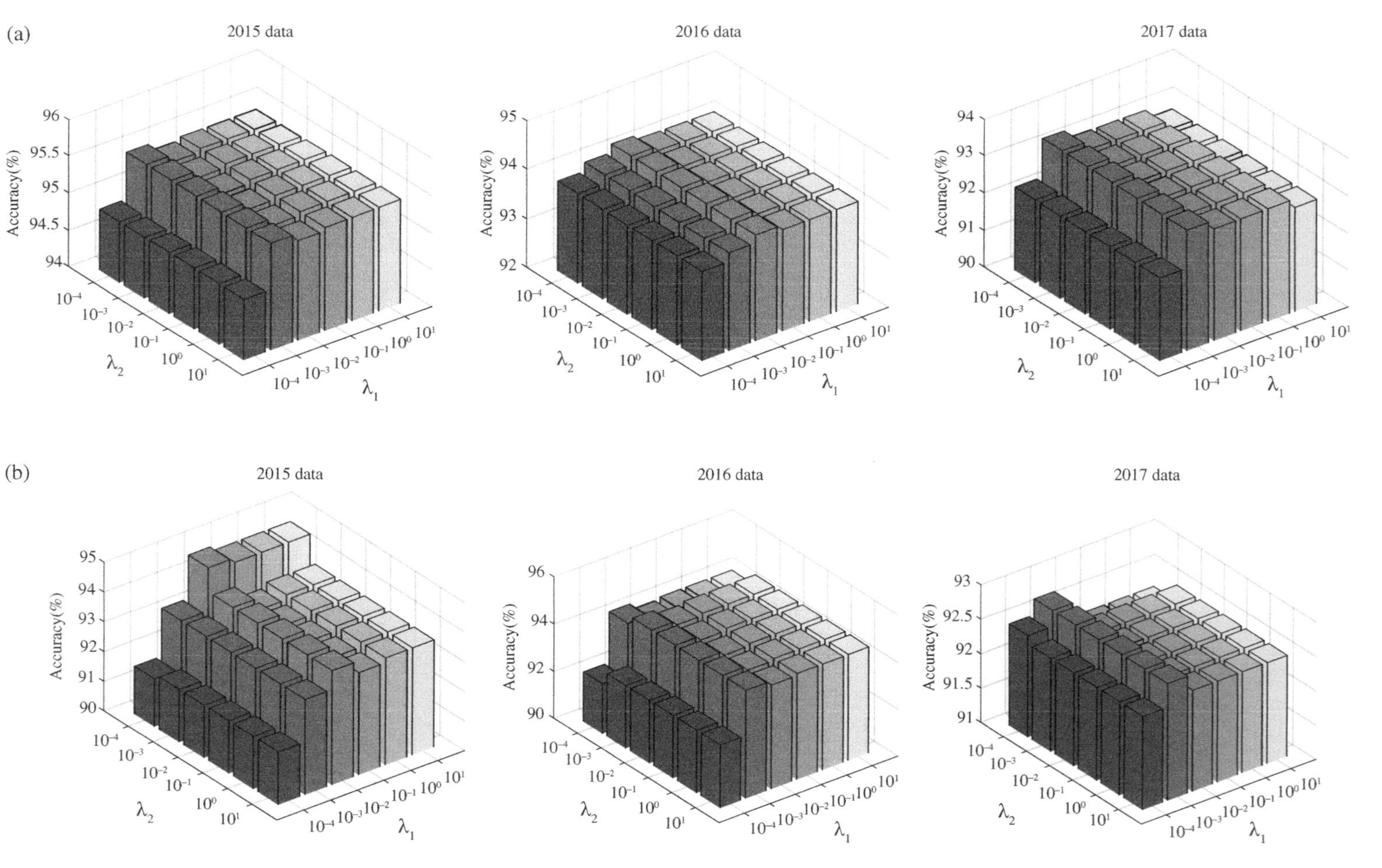

Figure 4.6 Precision for different databases with different parameters (only show 2015–2017s 60% testing data cases) (a) EPRS-EI (RPCA-EI). (b) EPRS-EI (LRR-EI).
Source: From [A]/with permission of Elsevier.

different parameters. The simulation results of precision performance are shown in Tables 4.14–4.17, and *t*-test result between EPRS-EI and other methods are shown in Tables 4.18–4.21. In *t*-test, as mentioned before, "W" means that EPRS-EI performs better. "M" means that other approaches performs better. Also "B" means that EPRS-EI and other approaches cannot outperform each other. Value in the bracket is the associate p-value. Statistical significance of *t*-test is 5% and we have not reported the p-value when the mark is "B". The smaller p-value means the higher assurance of the conclusion.

Table 4.14 The precision performance for 2015 data.

Method \ Training/Total data set	30%	40%	50%	60%	Average
DWR	92.86 ± 1.19%	92.69 ± 1.19%	92.03 ± 0.57%	92.43 ± 0.69%	92.50%
CF-EPRS	93.24 ± 0.86%	93.43 ± 0.60%	93.75 ± 0.57%	92.80 ± 0.82%	93.30%
BHCF-EPRS	92.58 ± 0.99%	93.15 ± 0.87%	93.71 ± 0.66%	93.93 ± 0.89%	93.09%
ERPS-EI(RPCA-EI)	94.78 ± 0.33%	94.67 ± 0.44%	**94.36 ± 0.20%**	94.60 ± 0.62%	**94.60%**
ERPS-EI(LRR-EI)	**94.83 ± 0.35%**	**94.77 ± 0.51%**	94.00 ± 0.46%	**94.70 ± 0.56%**	94.57%

Source: From [A]/with permission of Elsevier.

Table 4.15 The precision performance for 2016 data.

Method \ Training/Total data set	30%	40%	50%	60%	Average
DWR	91.99 ± 1.09%	92.67 ± 1.19%	92.26 ± 1.22%	92.45 ± 1.55%	92.34%
CF-EPRS	93.07 ± 0.73%	93.22 ± 0.74%	93.09 ± 0.69%	92.88 ± 0.78%	93.06%
BHCF-EPRS	93.54 ± 0.50%	93.39 ± 0.69%	93.38 ± 0.72%	93.20 ± 0.82%	93.38%
ERPS-EI(RPCA-EI)	94.95 ± 0.48%	94.67 ± 0.48%	**94.79 ± 0.69%**	**95.00 ± 0.72%**	**94.85%**
ERPS-EI(LRR-EI)	**95.08 ± 0.61%**	**94.77 ± 0.54%**	94.73 ± 0.57%	94.64 ± 0.77%	94.81%

Source: From [A]/with permission of Elsevier.

Table 4.16 The precision performance for 2017 data.

Method \ Training/Total data set	30%	40%	50%	60%	Average
DWR	92.66 ± 0.50%	92.36 ± 0.81%	93.06 ± 1.00%	92.99 ± 1.39%	92.77%
CF-EPRS	92.90 ± 0.99%	92.39 ± 0.85%	92.56 ± 0.69%	92.32 ± 0.80%	92.54%
BHCF-EPRS	93.06 ± 0.64%	92.98 ± 0.58%	93.13 ± 0.68%	92.49 ± 0.79%	92.91%
ERPS-EI(RPCA-EI)	93.33 ± 0.54%	**93.86 ± 0.51%**	93.62 ± 0.57%	93.42 ± 0.84%	93.56%
ERPS-EI(LRR-EI)	**93.47 ± 0.28%**	93.77 ± 0.96%	**94.22 ± 0.50%**	**94.69 ± 0.95%**	**93.79%**

Source: From [A]/with permission of Elsevier.

Table 4.17 The precision performance for 2018 data.

Method \ Training/Total data set	30%	40%	50%	60%	Average
DWR	92.84 ± 1.14%	91.59 ± 1.40%	91.82 ± 1.09%	92.96 ± 0.73%	92.30%
CF-EPRS	93.41 ± 1.04%	93.28 ± 1.15%	92.59 ± 1.05%	92.90 ± 1.50%	93.04%
BHCF-EPRS	93.54 ± 0.64%	92.92 ± 0.45%	92.87 ± 0.85%	93.11 ± 1.08%	93.11%
ERPS-EI(RPCA-EI)	**93.80 ± 1.24%**	93.73 ± 0.30%	**94.06 ± 0.53%**	**94.04 ± 0.62%**	**93.91%**
ERPS-EI(LRR-EI)	93.74 ± 1.11%	**93.77 ± 0.71%**	94.01 ± 0.59%	94.28 ± 1.02%	93.70%

Source: From [A]/with permission of Elsevier.

Table 4.18 *t*-test result between proposed models and other methods for 2015 data.

Training/Total data set	ERPS-EI (RPCA-EI)				Training/Total data set	ERPS-EI (LRR-EI)			
Method	30%	40%	50%	60%	Method	30%	40%	50%	60%
DWR	W (0.00)	W (0.03)	W (0.00)	W (0.00)	DWR	W (0.00)	W (0.05)	W (0.00)	W (0.00)
CF-EPRS	W (0.00)	W (0.00)	W (0.01)	W (0.00)	CF-EPRS	W (0.00)	W (0.00)	B(−)	W (0.00)
BHCF-EPRS	W (0.00)	W (0.00)	W (0.00)	W (0.00)	BHCF-EPRS	W (0.00)	W (0.00)	W (0.00)	W (0.00)
ERPS-EI (LRR-EI)	B(−)	B(−)	W (0.03)	B(−)	ERPS-EI (RPCA -EI)	B(−)	B(−)	M (0.03)	B(−)

Source: From [A]/with permission of Elsevier.

Table 4.19 *t*-test result between proposed models and other methods for 2016 data.

Training/Total data set	ERPS-EI (RPCA-EI)				Training/Total data set	ERPS-EI (LRR-EI)			
Method	30%	40%	50%	60%	Method	30%	40%	50%	60%
DWR	W (0.04)	W (0.00)	W (0.00)	W (0.03)	DWR	W (0.02)	W (0.00)	W (0.03)	W (0.21)
CF-EPRS	W (0.00)	W (0.00)	W (0.00)	W (0.00)	CF-EPRS	W (0.00)	W (0.00)	W (0.00)	W (0.00)
BHCF-EPRS	W (0.00)	W (0.00)	W (0.00)	W (0.00)	BHCF-EPRS	W (0.00)	W (0.00)	W (0.00)	W (0.00)
ERPS-EI (LRR-EI)	B(−)	B(−)	B(−)	B(−)	ERPS-EI(RPCA-EI)	B(−)	B(−)	B(−)	B(−)

Source: From [A]/with permission of Elsevier.

Table 4.20 *t*-test result between proposed models and other methods for 2017 data.

Method \ Training/Total data set	ERPS-EI (RPCA-EI) 30%	40%	50%	60%	Method \ Training/Total data set	ERPS-EI (LRR-EI) 30%	40%	50%	60%
DWR	W (0.17)	W (0.09)	W (0.05)	W (0.29)	DWR	W (0.12)	W (0.15)	W (0.07)	W (0.14)
CF-EPRS	B(−)	W (0.01)	B(−)	W (0.02)	CF-EPRS	B(−)	B(−)	W (0.05)	W (0.04)
BHCF-EPRS	W (0.02)	W (0.09)	B(−)	W (0.01)	BHCF-EPRS	W (0.00)	W (0.00)	W (0.00)	W (0.02)
ERPS-EI(LRR-EI)	B(−)	M (0.22)	M (0.13)	M (0.39)	ERPS-EI(RPCA-EI)	B(−)	W (0.22)	W (0.13)	W (0.39)

Source: From [A]/with permission of Elsevier.

Table 4.21 *t*-test result between proposed models and other methods for 2018 data.

Method \ Training/Total data set	ERPS-EI (RPCA-EI) 30%	40%	50%	60%	Method \ Training/Total data set	ERPS-EI (LRR-EI) 30%	40%	50%	60%
DWR	W (0.00)	W (0.02)	W (0.10)	W (0.06)	DWR	W (0.08)	W (0.01)	B(−)	W (0.00)
CF-EPRS	B(−)	B(−)	W (0.00)	W (0.08)	CF-EPRS	B(−)	B(−)	W (0.00)	W (0.04)
BHCF-EPRS	B(−)	W (0.05)	W (0.00)	W (0.08)	BHCF-EPRS	B(−)	W (0.05)	W (0.00)	W (0.03)
ERPS-EI(LRR-EI)	B(−)	B(−)	W (0.19)	B(−)	ERPS-EI(RPCA-EI)	B(−)	B(−)	M (0.19)	B(−)

Source: From [A]/with permission of Elsevier.

4.4.4 Application Result Discussions

From Figure 4.5, we can see that the simulations in both training and testing data. The horizontal coordinate is training data size, and "20" (to "70") in the coordinate represent the training data takes up 20% (to 70%) of total data and testing data takes up 80% (to 30%) of total data. When the training data size is enlarged, performance in the testing data cannot exceed performance in the training data, which indicates that our models can avoid overfitting.

From Figure 4.6, it can be seen that same as Figure 4.4, the trade-off parameter λ_1, which has the closest relationship with the objective low-rank matrix, is the leading factor to affect the EPRS simulations.

For Tables 4.14–4.21, some discussions are given as follows:

- It is known that in 2017–2018 data (Tables 4.16 and 4.17), with the increase of the training samples, the precision of all methods often rises. This is because as the data size increases, it is much easier to include the training customers having the same living pattern with the testing customer.

However, as data size increases in 2015–2016 data (Tables 4.14 and–4.15), we can observe fluctuations. This is because the whole data size of 2015–2016 data is larger than that of other data, and ever in the simulations, with the increase in training data, the training customers have no more changes in the diversity of living patterns. Besides, the training customers which influences our result is only 2–5 customers, which are set as nearest neighbors, and the number of neighbors is also too small to influence the whole simulations, especially in the large data such as 2015–2016 data.

- In most cases, the performance of DWR fails to have better influence than other methods, and DWR only exceeds other methods when the data size is small, which indicates the effectiveness of the utilization of explicit features and it proves that researchers may be hard to get explicit features when data size is small.
- Also, it can be seen that BHCF-EPRS normally has better performance than CF-EPRS, which proves that BHCF-EPRS can get more explicit features because of the utilization of matrix recovery and user classification. With the user classification technique, even the diversity of customers is scarce in the small data and the nearest neighbors are hard to be detected by similarity evaluation, BHCF-EPRS can also detect nearest neighbors, which strengthen the ability to compute personalized plans.
- Our proposed EPRS-EI still has the best performance, which indicates the effectiveness of the specific recovery instructions, i.e. appliance classification and total electricity usage. All the numbers in Tables 4.14–4.17 are computed by Eq. (4.27). In Table 4.14, EPRS-EI (RPCA-EI) and EPRS-EI (LRR-EI) have 94.60 and 94.57% of customers correctly recommended respectively. Comparing with BHCF-EPRS (93.09%), their improvements are 94.60 − 93.09%=1.62% and 94.57 − 93.09%=1.59%, respectively. In Table 4.16, EPRS-EI (RPCA-EI) and EPRS-EI (LRR-EI) do not show such superiority, with 93.56% and 93.79% of customers being correctly recommended, respectively. Comparing with BHCF-EPRS (92.91%), their improvements are 93.56 − 92.91 = 0.70% and 93.79 − 92.91 = 0.95%, respectively. Improvements can also be seen in other datasets with the range from 0.70 to 1.62%.
- Overall, EPRS-EI has the worst performance in Table 4.16, with 93.56% are correctly recommended, while it has the best performance in Table 4.15, with 94.85% of customers are correctly recommended.
- It can also be seen that compared to the simulation in recovery simulation, the gap between two proposed methods, i.e. EPRS-EI (RPCA-EI) and EPRS-EI (LRR-EI) is reduced. It indicates that though the data can be well recovered and explicit features can be obtained, the number of nearest neighbors is only 2–5 customers, which imposes a limit on the effectiveness of recommender stage. To improve performance, additional techniques such as user classification in BHCF-EPRS can be utilized in EPRS-EI. To invoke user classification, total usages need to be reformulated into regressed targets, which is discrete value, but the total usages in EPRS-EI are set as a continuous value to compute the incompletion between total usages and the sum of appliance usages. Additional preprocessing techniques should be set before user classification is introduced into EPRS-EI.

4.5 Conclusion and Future Work

Among several electricity tariffs, residential customers can choose a tariff to facilitate DSM. To mitigate the negative effect of corrupted data and recommend personalized electricity plans according to the living patterns, electrical instructions, i.e. appliance classification and total electricity usage, are introduced into improving the performance of EPRS.

Firstly, we propose a novel EPRS model EPRS with Electric Instruction-based Recovery (EPRS-EI) which is a dual-stage model consisting of both feature formulation stage and recommender stage. In the feature formulation stage, novel matrix recovery models with recovery instructions, namely RPCA-EI and LRR-EI are used to obtain more explicit features. These explicit features represent the living patterns of customers. In the recommender stage, KNN and adjusted similarity are utilized to detect nearest neighbors of testing customers, and with the help of the nearest neighbors, testing electricity plans can be computed.

Algorithms are developed to solve the proposed methods. Simulation results on recovery and recommendation confirm the effectiveness of our proposed methods. According to Section 4.4, LRR-EI recovers up to 24.14–38.15% of more data compared to NSHLRR, while comparing with BHCF-EPRS, more than 0.70–1.62% of customers are correctly recommended in EPRS-EI. Besides, there are two workable improvements for EPRS-EI. The first one is extensive applications to recommend other items. EPRS-EI is a recommender system based on collaborative filtering whose input data consists of feature data and rating. Therefore, if the rating is replaced by other items, this model can be utilized to recommend these items. Two possible recommended items are energy-saving electrical appliances and demand response schedules, and what needs to be done is setting customers' preferences of those appliances or schedules as the rating. The second one is the potential improvement in recommender stage. Since the number of nearest neighbors is limited, if we cannot select the nearest neighbors correctly, the improvement is also limited. The possible selection method is user classification in BHCF-EPRS. However, to invoke user classification, total usages need to be reformulated into regressed targets, additional preprocessing techniques should be set before user classification is introduced into EPRS-EI. In our future work, we also look for the detailed information on electricity plans, usages and charges to accurately compute the saving gained by customers with the help of EPRS.

Acknowledgments

The permission in using materials from the following paper is very much appreciated.

[A] J. Zheng, C. S. Lai, H. Yuan, Z. Y. Dong, K. Meng and L. L. Lai. Electricity plan recommender system with electrical instruction-based recovery. *Energy* 203 (2020) 117775.

References

1 Liu, X., Gao, B., Wu, C., and Tang, Y. (2018). Demand-side management with household plug-in electric vehicles: a Bayesian game-theoretic approach. *IEEE Systems Journal* 12 (3): 2894–2904. https://doi.org/10.1109/JSYST.2017.2741719.

2 Su, H., Zio, E., Zhang, J. et al. (2019). A systematic data-driven demand side management method for smart natural gas supply systems. *Energy Conversion and Management* 185: 368–383. https://doi.org/10.1016/j.enconman.2019.01.114.

3 Hayes, B., Melatti, I., Mancini, T. et al. (2017). Residential demand management using individualized demand aware price policies. *IEEE Transactions on Smart Grid* 8 (3): 1284–1294. https://doi.org/10.1109/TSG.2016.2596790.

4 Liu, N., Yu, X., Wang, C. et al. (2017). Energy-sharing model with price-based demand response for microgrids of peer-to-peer prosumers. *IEEE Transactions on Power Systems* 32 (5): 3569–3583. https://doi.org/10.1109/TPWRS.2017.2649558.
5 Stavrakas, V. and Flamos, A. (2020). A modular high-resolution demand-side management model to quantify benefits of demand-flexibility in the residential sector. *Energy Conversion and Management* 205: 112339. https://doi.org/10.1016/j.enconman.2019.112339.
6 Yilmaz, S., Chambers, J., and Patel, M.K. (2019). Comparison of clustering approaches for domestic electricity load profile characterisation – implications for demand side management. *Energy* 180: 665–677. https://doi.org/10.1016/j.energy.2019.05.124.
7 McKenna, K. and Keane, A. (2016). Residential load modeling of price-based demand response for network impact studies. *IEEE Transactions on Smart Grid* 7 (5): 2285–2294. https://doi.org/10.1109/TSG.2015.2437451.
8 Lai, C.S., Xu, F., McCulloch, M., and Lai, L.L. (2016). Application of distributed intelligence to industrial demand response. In: *Smart Grid Handbook* (ed. C.-C. Liu, S. McArthur and S.-J. Lee). Piscataway: *IEEE Press & Wiley*.
9 Xu, F.Y., Zhang, T., Lai, L.L., and Zhou, H. (2015). Shifting boundary for price-based residential demand response and applications. *Applied Energy* 146: 353–370. https://doi.org/10.1016/j.apenergy.2015.02.001.
10 Smart Grid (2014). Smart City is a project providing aggregated daily load profile of 200 residential customers with and without EPRS. http://www.environment.gov.au/energy/programs/smartgridsmartcity (accessed 4 February 2016).
11 Australian Energy Regulator (2019). About energy made easy. https://www.energymadeeasy.gov.au/article/about-us (accessed 4 January 2019).
12 iSelect (2019). iSelect is Australia's electricity plan website for comparing plans and saving money. http://www.iselect.com.au/energy/ (accessed 4 January 2019).
13 Power to Choose (2019). Power to Choose is the official, unbiased, electric choice website of the Public Utility Commission of Texas. http://www.powertochoose.org (accessed 4 January 2019).
14 Wang, S., Chen, H., Wu, L., and Wang, J. (2020). A novel smart meter data compression method via stacked convolutional sparse auto-encoder. *International Journal of Electrical Power & Energy Systems* 118: 105761. https://doi.org/10.1016/j.ijepes.2019.105761.
15 Wen, L., Zhou, K., Yang, S., and Li, L. (2018). Compression of smart meter big data: a survey. *Renewable and Sustainable Energy Reviews* 91: 59–69. https://doi.org/10.1016/j.rser.2018.03.088.
16 Kong, W., Dong, Z.Y., Hill, D.J. et al. (2018). A hierarchical hidden Markov model framework for home appliance modeling. *IEEE Transactions on Smart Grid* 9 (4): 3079–3090. https://doi.org/10.1109/TSG.2016.2626389.
17 Kong, W., Dong, Z.Y., Hill, D.J. et al. (2017). Improving nonintrusive load monitoring efficiency via a hybrid programming method. *IEEE Transactions on Industrial Informatics* 12 (6): 2148–2157. https://doi.org/10.1109/TII.2016.2590359.
18 Kim, K.-B., Cho, J.Y., Jabbar, H. et al. (2018). Optimized composite piezoelectric energy harvesting floor tile for smart home energy management. *Energy Conversion and Management* 171: 31–37. https://doi.org/10.1016/j.enconman.2018.05.031.
19 Kabalci, E. (2015). A smart monitoring infrastructure design for distributed renewable energy systems. *Energy Conversion and Management* 90: 336–346. https://doi.org/10.1016/j.enconman.2014.10.062.
20 Eissa, M.M. (2019). Developing incentive demand response with commercial energy management system (CEMS) based on diffusion model, smart meters and new communication protocol. *Applied Energy* 236: 273–292. https://doi.org/10.1016/j.apenergy.2018.11.083.

21 Yildiz, B., Bilbao, J.I., Dore, J., and Sproul, A.B. (2017). Recent advances in the analysis of residential electricity consumption and applications of smart meter data. *Applied Energy* 208: 402–427. https://doi.org/10.1016/j.apenergy.2017.10.014.

22 Xu, F., Huang, B., Cun, X. et al. (2018). Classifier economics of semi-intrusive load monitoring. *International Journal on Electrical Power and Energy Systems* 103: 224–232. https://doi.org/10.1016/j.ijepes.2018.05.010.

23 Zhang, J., Chen, X., Ng, W.W.Y. et al. (2019). New appliance detection for non-intrusive load monitoring. *IEEE Transactions on Industrial Informatics* 15 (8): 4819–4849. https://doi.org/10.1109/TII.2019.2916213.

24 Lu, L., Medo, M., Chi, H.Y. et al. (2012). Recommender systems. *Physics Reports* 519 (1): 1–49. https://doi.org/10.1016/j.physrep.2012.02.006.

25 He, X., He, Z., Song, J. et al. (2018). NAIS: neural attentive item similarity model for recommendation. *IEEE Transactions on Knowledge and Data Engineering* 30 (12): 2354–2366. https://doi.org/10.1109/TKDE.2018.2831682.

26 Kwon, H. and Hong, K. (2011). Personalized smart tv program recommender based on collaborative filtering and a novel similarity method. *IEEE Transactions on Consumer Electronics* 57 (3): 1416–1423. https://doi.org/10.1109/TCE.2011.6018902.

27 Zhang, Y., Meng, K., Xu, D. et al. (2016). Recommending electricity plans: a data-driven method. *2016 IEEE International Conference on Smart Grid Communications*, Sydney, NSW, Australia (12 December 2016), 668–673. https://doi.org/10.1109/SmartGridComm.2016.7778838.

28 Luo, F., Ranzi, G., Wang, X., and Dong, Z.Y. (2019). Social information filtering based electricity retail plan recommender system for smart grid end users. *IEEE Transactions on Smart Grid* 10 (1): 95–104. https://doi.org/10.1109/TSG.2017.2732346.

29 Ramlatchan, A., Yang, M., Liu, Q. et al. (2018). A survey of matrix completion methods for recommendation systems. *Big Data Mining and Analytics* 1 (4): 308–323. https://doi.org/10.26599/BDMA.2018.9020008.

30 Zhang, Y., Meng, K., Kong, W., and Dong, Z.Y.G. (2019). Collaborative filtering-based electricity plan recommender system. *IEEE Transactions on Industrial Informatics* 15 (3): 1393–1404. https://doi.org/10.1109/tii.2018.2856842.

31 Zhang, Y., Meng, K., Kong, W. et al. (2019). Bayesian hybrid collaborative filtering-based residential electricity plan recommender system. *IEEE Transactions on Industrial Informatics* 15 (8): 4731–4741. https://doi.org/10.1007/978-3-540-24685-5_37.

32 Salakhutdinov, R. and Mnih, A. (2008). Bayesian probabilistic matrix factorization using Markov chain Monte Carlo. *Proceedings of the 25th International Conference on Machine Learning*, Helsinki, Finland (5–9 July 2008), 25: 880–887. https://doi.org/10.1145/1390156.1390267.

33 Candes, E.J. and Plan, Y. (2010). Matrix completion with noise. *Proceedings of the IEEE* 98 (6): 925–936. https://doi.org/10.1109/jproc.2009.2035722.

34 Mnih, A. and Salakhutdinov, R.R. (2007). Probabilistic matrix factorization. *Advances in Neural Information Processing Systems* 20 (2): 1257–1264. https://doi.org/10.3233/IFS-141462.

35 Candès, E.J., Li, X., Ma, Y., and Wright, J. (2009). Robust principal component analysis? *Journal of the ACM* https://doi.org/10.1145/1970392.1970395.

36 Liu, G., Lin, Z., Yan, S. et al. (2013). Robust recovery of subspace structures by low-rank representation. *IEEE Transactions on Pattern Analysis and Machine Intelligence* 35 (1): 171–184. https://doi.org/10.1109/TPAMI.2012.88.

37 Yin, M., Gao, J., and Lin, Z. (2016). Laplacian regularized low-rank representation and its applications. *IEEE Transactions on Pattern Analysis and Machine Intelligence* 38 (3): 504–517. https://doi.org/10.1109/TPAMI.2015.2462360.

38 Li, H., Chen, N., and Li, L. (2012). Error analysis for matrix elastic-net regularization algorithms. *IEEE Transactions on Neural Networks and Learning Systems* 23 (5): 737–748. https://doi.org/10.1109/tnnls.2012.2188906.

39 Tibshirani, R. (1996). Regression shrinkage and selection via the lasso. *Journal of the Royal Statistical Society* 58 (1): 267–288. https://doi.org/10.1111/j.2517-6161.1996.tb02080.x.

40 Recht, B., Fazel, M., and Parrilo, P.A. (2010). Guaranteed minimum-rank solutions of linear matrix equations via nuclear norm minimization. *SIAM Review* 52 (3): 471–501. https://doi.org/10.1137/070697835.

41 Lin, Z., Chen, M., and Ma, Y. (2013). The augmented Lagrange multiplier method for exact recovery of corrupted low-rank matrices. *Journal of Structural Biology* 181 (2): 116–127. https://doi.org/10.1016/j.jsb.2012.10.010.

42 Zhang, Y. (2010). Recent advances in alternating direction methods: practice and theory. http://lsec.cc.ac.cn/~sjom/SJOMInvitedTalks/ProfYZhang.pdf (accessed 8 November 2020).

43 Pecan Street Dataport (2019). Dataport hosts all the data collected via Pecan Street's water and electricity research. https://dataport.cloud/data (accessed 4 January 2019).

5

Classifier Economics of Semi-intrusive Load Monitoring

5.1 Introduction

5.1.1 Technical Background

Information and Communication Technologies (ICT) and Intelligent Data Analytical Technologies (IDAT) become a new trend for various industries' development. On one hand, Internet of Things (IoT) provides high degree linkage among multiple devices, which can talk to each other [1–3]. On the other hand, the fast developing artificial intelligent techniques have improved the capabilities of devices' coordination [4–6]. Following this trend, more and more new implementations with ICT and IDAT appear in multiple industries.

Nonintrusive load monitoring (NILM) is one of the ICT and IDAT implementing cases in power system. Other than recognizing devices' operation status by device-based specified metering data traditionally, NILM receives information of detail power profiles at the aggregated point and find out devices operation status by a machine learning-based classifier [7, 8]. Facing requirement on devices operation monitoring, NILM only contains a local meter and a local data analyzing unit, instead of constructing a data transmission network and more than one meter in traditional monitoring system. So intuitively, NILM can decrease the construction cost of load monitoring system [9–11]. Moreover, installing one meter at the aggregated point is said to have less privacy violation than distributing a specified meter for each device inside consumers [12–14]. Relevant researches keep placing effort on NILM development. Wang created an algorithm based on the features of V–I trajectory for NILM with 1% minimal error with Adaboost framework [15]. Chang presented a novel NIFM algorithm to improve the associated recognition accuracy and the average test accuracy of using the proposed method is higher than the 98.13% in DLGF [16]. Kong introduced a Hybrid Programing Method to improve the NILM efficiency and push the NILM being used in the daily life [17].

NILM is still a developing technology [18, 19]. Facing different devices combination, NILM may not always achieve a good enough accuracy. For example, the Type I devices are classified with a high accuracy but the other type devices like Dishwasher are with a lower accuracy [20]. Some NILM schemes only perform well on distinguishing special devices [21]. Generally speaking, performance of NILM will decrease when more devices are integrated together. Because devices with similar power profile may confuse classifier's recognition. Thus, Tang introduced a method named semi-intrusive load monitoring (SILM) [22]. SILM does not accumulate all devices' recognition into one aggregated point. Instead, by considering performances of classifiers, SILM selects several aggregated points with classifiers. Each classifier only classifies the operating status of a section of devices. Thus, each classifier will not suffer overburdening task and the total recognition accuracy will increase.

Smart Energy for Transportation and Health in a Smart City, First Edition. Chun Sing Lai, Loi Lei Lai and Qi Hong Lai.

Saving construction cost of load monitoring system is one of the most attracting advantages from NILM and SILM. Various investigations on NILM or SILM are trying to decrease the cost by decreasing numbers of metering points [23, 24]. Some studies had quantized the cost reduction [11, 25]. But most researches assume that the cost of meter is same/similar with each other. In fact, meter cost with different metering functions fluctuates significantly in the market. If a meter with advanced power profile metering costs too much, NILM system with high cost metering function may cost even higher than traditional load monitoring system. Furthermore, accuracy of traditional load monitoring system is nearly 100% with only active power metering. So NILM and SILM may appear to be even worse than traditional load monitoring method on both accuracy and construction cost. Decision makers of load monitoring require new methods with consideration on both accuracy and construction cost to support their solution design.

5.1.2 Original Contribution

Facing the issue above, this chapter initiates a new classifier network construction method for SILM. Instead of creating a classifier for NILM or SILM, this method helps decision maker to select different types of classifiers and optimally allocates the classifiers' positions. In this method, economics of each type of classifier is considered to ensure decision maker's cost reduction, including cost of feature metering and classifier hardware constructions. A combinatorial optimization problem is established on a tree-type model to the optimized classifier network. In this model, NILM becomes a special SILM case with one classifier only. Numerical studies on a public dataset REDD and an industrial operational data are implemented to support the feasibility of the method.

The content of this chapter is summarized as follows. Section 5.2 provides a technical introduction for NILM and SILM. The classifier network construction is introduced in Section 5.3. Based on this network, the tree-type combinatorial optimization problem construction is shown in Section 5.4. Two numerical case studies are selected to verify the effect of proposed method in Section 5.5.

5.2 Typical Feature Space of SILM

NILM or SILM analyzes devices' operation status from the power information at the aggregated point.

From Figure 5.1, all power operational information downstream devices are accumulated at the aggregated point with classifiers. Theoretically, voltage and current obey Thevenin Theorem. Classifiers should fully recognized devices' operational differences on power profile information to differentiate operational status. At the initial stage of NILM, Massachusetts Institute of Technology (MIT) has summarized four types of appliances with different operational patterns [26], including ON/OFF appliances (Electric Kettle, Toaster), Finite State Machines (Electric Fan, Air Conditioning, Washing Machine), Continuously Variable Consumer Device (Speaker Box, Dimmer Light), and Permanent Consumer Device.

Facing different types of appliances, various features are abstracted to portray each downstream devices. The most popular features are introduced below.

- **Active Power**. Although the focusing point of NILM researches are different, active power appears to be a common feature under their feature space [11, 27]. This feature represents the energy consuming speed and performs well when number of downstream devices is small. The meter price of active power metering only is the lowest in the market.

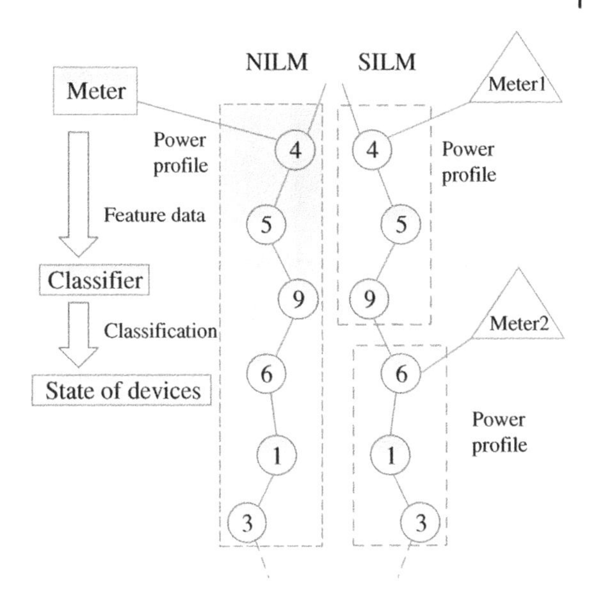

Figure 5.1 SILM/NILM monitoring structure. *Source:* From Xu et al. [A]/with permission of Elsevier.

- **Reactive Power**. Reactive power is preferred by several NILM researches together with active power [28]. Meters with reactive power measuring is higher than those with active power only.
- **Time and Frequency Domain Characteristics**. Features under this title includes Vrms, Irms, Ipeak, and harmonics of voltage and current. It improves the accuracy from models with only reactive power and active power [29]. These features are abstracted by high sampling frequency and thus the metering function cost is much higher than that with active power and reactive power only.
- **Transient Related Features**. Typical features under this title include transient power, start-up current transients and high-frequency sampling of voltage noise [30]. Transient monitoring requires not only high sampling frequency but also a suitable capture triggering scheme for waveforms. The processed data is large and cost for the metering function is usually the most expensive.

Classifiers in NILM or SILM with different feature spaces require different cost of metering functions. Table 5.1 shows a sample of prices for meters shown in Figure 5.2.

Table 5.1 Price sample of meters in Figure 5.2.

Meter	Product no.	Market price ($)
Figure 5.2a	DDS1531	200
Figure 5.2b	DTS3533	300
Figure 5.2c	PAC4200	1000
Figure 5.2d	ION7650	2500

Source: From Xu et al. [A]/with permission of Elsevier.

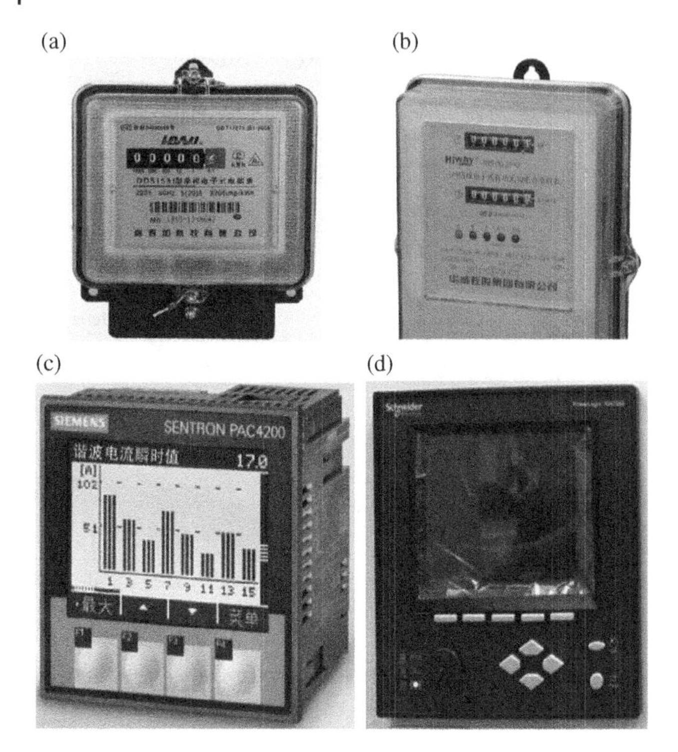

Figure 5.2 Meters. (a) For active power only. (b) For both active power and reactive power. (c) With harmonics monitoring. (d) With waveform capture function. *Source:* From Xu et al. [A]/Elsevier.

5.3 Modeling of SILM Classifier Network

5.3.1 Problem Definition

There are many types of implementations for SILM. Owners of SILM or NILM aim to construct monitoring system with less construction cost and sufficient accuracy. So construction cost and accuracy are the two main factors for SILM system installation consideration.

In terms of accuracy, nearly all types of classifiers will suffer accuracy reduction with large scale of devices. Because large scale of devices will have a high probability on containing similar performing devices (or even several same type of devices). It is difficult for classifier to separate devices with similar features. A preliminary experiment is selected to reveal this phenomenon in Figure 5.3.

Random forest is selected as classifiers in this experiment. A public dataset REDD is selected for analysis [31]. Two feature spaces are selected for preliminary experiments. The feature Space1 has the P and Q. The feature Space2 contains the fundamental to the eighth harmonic component and feature Space1.

From Figure 5.3, the accuracy decreases when number of devices increases. Classifier cannot find out exactly which device is switched on/off as devices' feature contribution is similar. Implementation cases with exact device recognition will suffer the decrease in accuracy, which is the case selected in this chapter. But this decrease may not be significant if decision maker only prefers the total power of same type devices. Also, one better classifier may achieve less accuracy when using multiple classifiers on each section of devices set.

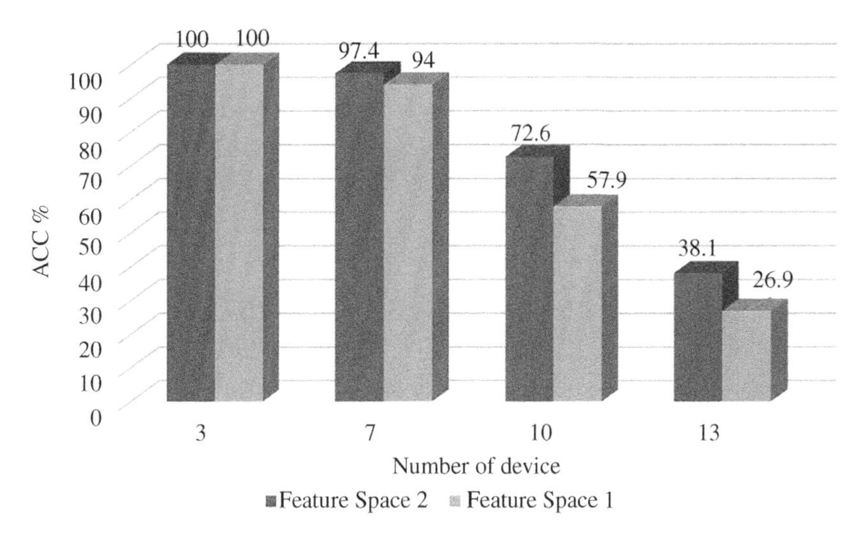

Figure 5.3 Preliminary experiment on classifier performances by increasing device numbers. *Source:* From Xu et al. [A]/with permission of Elsevier.

Regarding cost, previous researches tend to equate it to the number of meters. But price of meters is one of the important sections in the cost as well. For example, one consumer has 10 different devices. Traditional load monitoring system requires a specified meter for each device. As feature space for one device recognition is simple in most situations (active power is enough), so each meter with data transmission function may cost 150$ (about 1000CNY). A new NILM solution requires one meter at the aggregated point only. But this classifier requires high sampling frequency and 20th harmonics from FFT. Although this NILM requires one meter, this meter may cost 3000$ (about 20 000CNY). So the total construction cost of NILM is still higher than traditional system. This example indicates that a less meter solution may not be the best choice for decision maker when considering the price difference on feature space construction (price of meter).

Considering the two aspects above, decision maker requires a method to consider the device scale and meters' price into their SILM system construction.

5.3.2 SILM Classifier Network Construction

Figure 5.4 introduces a typical topology of grid-connected consumers. Each consumer connects to a downstream position of feeder. Details of connection may be changed by consumer types:

- Each residential consumer may connect to only one phase within the 3-phase circuit. The aggregated point may be a single phase meter at the grid connecting point.
- A commercial consumer may require a 3-phase connection to support larger power supply. The aggregated point may be a 3-phase meter at the grid connecting point.
- An industrial consumer may seize more than one feeder or even more than one bus of transformer. The smallest number of meters will be the same as the number of feeders.

In Figure 5.4, each triangle represents a consumer. Inside each consumer, all devices are connected to the inner grid structure, which contains a connecting node to external power grid.

From Figure 5.4, the solution of SILM with the smallest number of meters will use only one classifier with one meter at the grid connecting node. But this solution will not guarantee the minimum

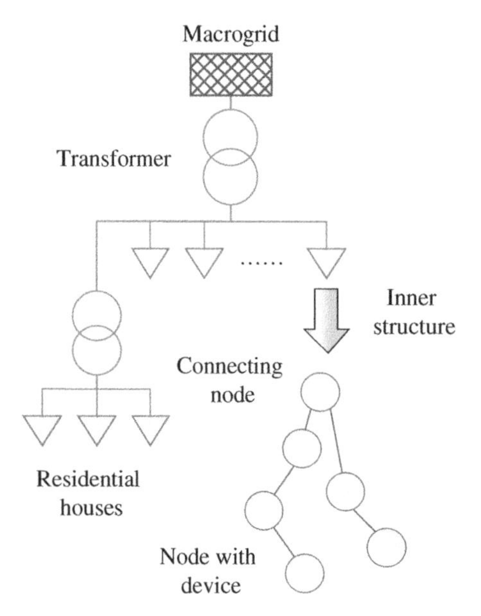

Figure 5.4 Grid connection of each consumer. *Source:* From Xu et al. [A]/with permission of Elsevier.

cost for classifier construction and sufficient accuracy. So following the idea of SILM, classifiers can be laid at any node with a device in Figure 5.4. To fit for the optimization of classifiers' node locating, each node in Figure 5.4 is modelled with Eqs. (5.1)–(5.6).

$$Node(i) \cdot Type \in \{Consumer, Device\} \tag{5.1}$$

Equation (5.1) represents that there are two types of nodes. Device node means that the node is associated with a device connection. Consumer node means that the node is the highest aggregated point inside the consumer.

$$\begin{cases} Node(i) \cdot Dev \in Device\_Base \\ Device\_Bace = \{Empty,\ Dev_1,\ Dev_2,...,\ Dev_m\} \end{cases} \tag{5.2}$$

Equation (5.2) enables the model to place a specified device at any specific node or leave it empty. Node_Dev means

$$\begin{cases} Node(i).Father \in Node\_Set \\ Node(i).Children \subset Node\_Set \end{cases} \tag{5.3}$$

Equation (5.3) enables the model to construct an entire connection tree by setting father node and children nodes. Each node can only have one father node. Node_Set represents the set of all nodes.

$$Const\,1: \begin{cases} Node(i).FS \in Fea\_Space\_Set \\ Fea\_Space\_Set = \{Empty,\ FS_1,\ FS_2,\ ...,\ FS_l\} \end{cases} \tag{5.4}$$

$$\begin{cases} Node(i).CM \in Classfier\_Set \\ Classfier\_Set = \{FFNN,\ RF,\ ...\} \end{cases} \tag{5.5}$$

Equations (5.4) and (5.5) enable the model to select different feature space and classifying models. Model user can preset different set of feature space in Fea_Space_Set (FS represents a candidate feature space). If FS equals to empty, it means that there is no classifier at the specific node. Model user can also preset different classifying model in Classifier_Set. In Eq. (5.5), FFNN means feed-forward neural network. RF means random forest. As feature space selection is completed within optimization process, so Eq. (5.4) is also the constraint of optimization.

$$Node(i).Sep \subset Node\_Set \tag{5.6}$$

Equation (5.6) indicates the node information of devices that classifier in node(i) is responsible for. In other words, classifier in node(i) need to recognize the operation status of devices at nodes inside Node(i).Sep.

With Eqs. (5.1)–(5.6), decision makers can locate classifier at any possible position in consumer's microgrid to construct any possible SILM network. If only one classifier is laid at the highest aggregated point, the solution becomes a NILM solution. If each node is placed with a classifier, the solution becomes a traditional load monitoring solution.

5.4 Classifier Locating Optimization with Ensuring on Accuracy and Classifier Economics

5.4.1 Objective of SILM Construction

The aim of SILM implementation is to find out the lowest SILM construction cost with sufficient accuracy. So the construction cost of SILM is the optimization objective. When locating a classifier at a node, it should be associated with a meter and a data transmission module which are capable of supporting data sampling on specified feature space. So the optimization objective should follow the expression in Eq. (5.7).

$$Min : Object = \sum_{i=1}^{I}(Cost(Node(i) \cdot FS)) \tag{5.7}$$

where, Cost(·) is a function which receives a feature space and outputs the cost of meter supporting feature sampling in inputted feature space. The detail value of cost depends on the market price of the required meter. So the value of this function is a boundary of optimization. The total number of nodes is i.

5.4.2 Constraint of Devices Covering Completeness and Over Covering

SILM should cover all important devices that need to be recognized. In other words, any device should be recognized and only recognized by one classifier. As devices can only be recognized by classifiers located at upstream nodes, the constraint of device covering is given in Eqs. (5.8) and (5.9).

$$Const\,2 : \begin{cases} \forall i,\ Node(i),\ Dev \neq Empty, \\ then \exists n \in NNI \Rightarrow A \text{ is true} \\ A \Leftrightarrow Fa\left(Fa\left(...\overset{nth\ iteration}{Fa\ (Node(i))}\right)\right) \cdot FS \neq Empty \end{cases} \tag{5.8}$$

$$Const\,3 : \begin{cases} \forall n_1,\ n_2 \in B \Rightarrow n_1 = n_2 \\ B = \{n \in NNI | A \text{ is true}\} \end{cases} \tag{5.9}$$

Equation (5.8) ensures that each device will be covered by at least one node with classifier. NNI means the set of nonnegative integers. Function Fa(·) returns the father node of inputted node. Equation (5.9) ensures that each device can be only covered by one node with classifier.

5.4.3 Constraint of Bottom Accuracy and Accuracy Measurement

To monitor device operational status accurately is the basic requirement of all load monitoring systems. Solution from SILM must satisfy the lowest accuracy requirement from decision maker. Equation (5.10) reveals the constraints of accuracy.

$$Const\,4 : Acc\_tot \geq MA \tag{5.10}$$

where Acc_tot is the entire accuracy. MA is the minimum requirement of entire accuracy given by decision makers. Practically, there are multiple selections on accuracy measurement expression. Two typical measurement expressions are introduced by Eqs. (5.11) and (5.12). Equation (5.11) sets the minimum requirement for the device with lowest accuracy. Equation (5.12) sets the minimum requirement for average performance of all devices. Acc_i represents the recognition accuracy of device i.

$$Acc\_tot = min\,(ACC_i) \tag{5.11}$$

$$Acc\_tot = mean(Acc_i) \tag{5.12}$$

5.4.4 Constraint of Sampling Computation Requirements

During optimization process, assume that Figure 5.5 is a section of candidate SILM networks.

In Figure 5.5, classifier 1 is responsible for devices 1, 2, and 3. Classifier 2 is responsible for devices 4, 5, and 6. Practically, the meter at node of classifier 2 can only collect feature information of operation combination among devices 1, 2, 3, 4, 5, and 6. So separating operational information of devices 4, 5, and 6 from devices 1–6 should use metering data of classifier 2 to subtract data of classifier 1. But if the sampling frequency of meter associated with classifier 2 is higher than classifier 1, it is difficult for the subtraction operation as some of data in classifier 2 cannot find data from classifier 1 at corresponding time points. From this point of view, sampling frequency of classifiers' meters should be equal to or lower than downstream classifiers' meters. Equation (5.13) reveals this constraint.

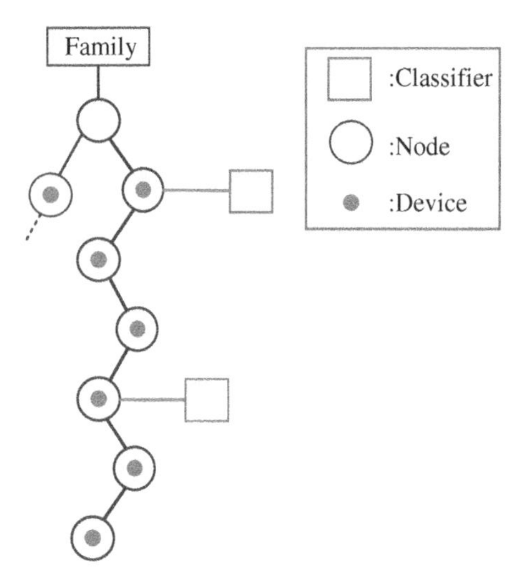

Figure 5.5 Sample section of candidate SILM network. *Source:* From Xu et al. [A]/with permission of Elsevier.

$$Const\ 5 \begin{cases} \forall i, Node(i).FS \neq Empty, \\ if\ B \neq \emptyset, \forall n \in B \\ \Rightarrow \begin{cases} SF\left(Fa\left(Fa\left(...Fa \overset{nth\ iteration}{(Node(i))}\right)\right) \cdot FS\right) \\ \leq SF(Node(i) \cdot FS) \end{cases} \end{cases} \tag{5.13}$$

In Eq. (5.13), function SF (·) receives the feature space and outputs the minimum sampling frequency which supports the inputted feature space. This sampling frequency depends on the meter selections.

5.4.5 Optimization Algorithm

A modified genetic algorithm is selected for the optimization process above. Each candidate feature space distribution is represented as one individual. After heredity, crossover, and mutation in each iteration, each feature space will form its own training dataset and testing dataset for its corresponding classifier. The classifier will be trained and tested and then the accuracy of each feature space will be formed. After that, the constraints fitness will be verified. Individuals satisfying all constraints will be left for next iteration computation. Details of optimization are shown in Figure 5.6.

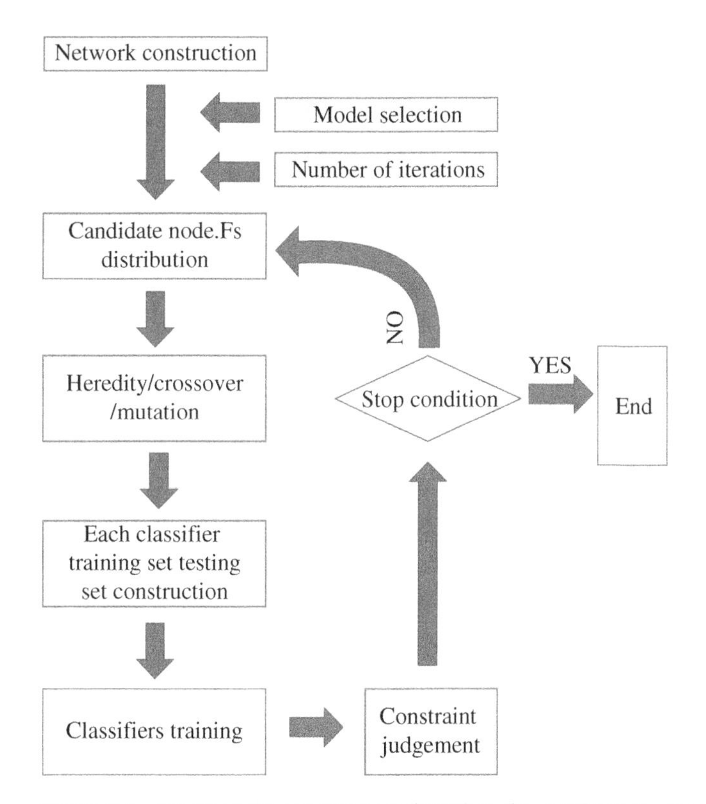

Figure 5.6 Modified GA optimization algorithm. *Source:* From Xu et al. [A]/with permission of Elsevier.

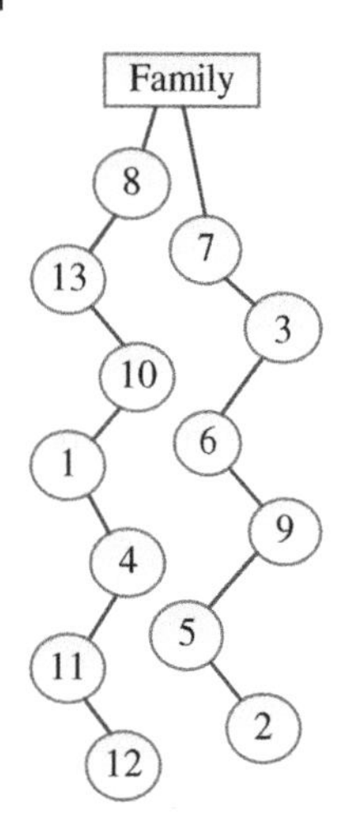

Figure 5.7 Typical residential network topology for numerical study. *Source:* From Xu et al. [A]/with permission of Elsevier.

5.5 Numerical Study

5.5.1 Devices Operational Datasets for Numerical Study

To verify the model feasibility, two devices operational datasets are selected for numerical study. The first dataset is a public residential dataset named 'The Reference Energy Disaggregation Data Set' (REDD) created by Kolter and Johnson [31], which is widely used for numerical study in NILM and SILM [32–33]. As REDD only provides operational data without network topology, a typical residential network topology is assumed with REDD and details are given in Figure 5.7 and Table 5.2.

The second dataset is obtained from a practical factory on optical measuring equipment manufacturing. The original monitoring system in this factory is to distribute a smart meter for each device. Now this factory is planning to update its device monitoring system and prefer to obtain a cost-effective monitoring solution. The microgrid network topology of this factory is shown in Figure 5.8 and Table 5.3.

5.5.2 Feature Space Set for Numerical Study

Considering traditional scheme, offering an independent meter for each equipment with high recognition accuracy only requires active power measurement in most cases. So the feature space for machine learning model in Eq. (5.5) is the active power only. Any classifier selecting this feature space only needs to pay for the price of an active power measuring meter.

For dataset REDD, the numerical study considers two feature spaces for classifiers, including a feature space with active power only. The second feature space for REDD contains single-phase active power, reactive power, order 2–8 harmonics, which can be measured by typical residential smart meter. Any classifier selecting this feature space needs to pay for the price of a residential smart meter.

For the factory dataset, not all smart meters in old monitoring system are damaged. So the second feature space follows the capability of the in-use smart meters, which are 3-phase total active power, 3-phase total reactive power, total power factor, total harmonic active power, total harmonic

Table 5.2 Device selections in Figure 5.7.

Dev seq no	Dev name	Dev seq no	Dev name
1	Microwave	7	Subpanel 2
2	Light 1	8	Electric heat 1
3	Light 1	9	Electric heat 2
4	Dishwasher	10	Light 2
5	Washer dryer	11	Bathroom GFI
6	Subpanel 1	12	Light 3
13	Furnace		

Source: From Xu et al. [A]/with permission of Elsevier.

reactive power. Any classifier selecting this feature space needs to pay for the price of a 3-phase industrial power meter.

No matter which dataset, the meter for second feature space is named advanced meter. The price of advanced meter and pure active power meter is different.

5.5.3 Numerical Study 1: Classifier Economics via Different Meter Price and Different Accuracy Constraints

In this study, price of an active power measuring meter is assumed to be 100$. Random Forest is selected as classifier. The optimization is computed with Matlab. Parameters of classifier model and optimization are introduced in Table 5.4. Result of optimization is shown in Table 5.5.

In Table 5.5, optimization is implemented on both REDD dataset and factory dataset. For each data set, each row represents the results under the same accuracy constraint in Eq. (5.10). Each column represents results under the same price of the second feature space (the price of the more advanced meter). In each cell of table, there are three elements. The left element represents the total cost of the optimization solution. The middle element represents the number of classifiers with a pure active power measuring meter for the first

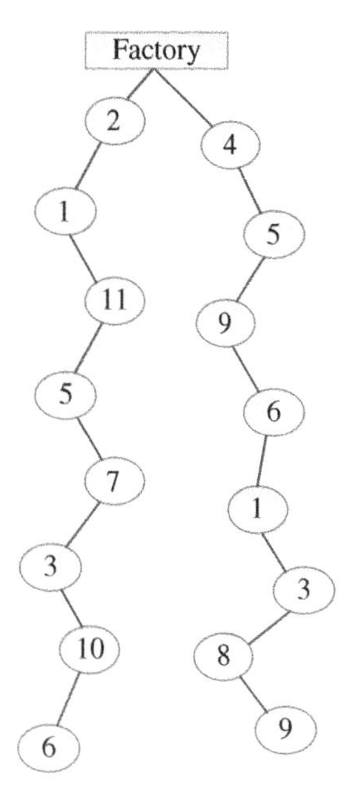

Figure 5.8 Factory topology for numerical study. *Source:* From Xu et al. [A]/with permission of Elsevier.

Table 5.3 Device selections in Figure 5.8.

Dev seq no	Dev name	Dev seq no	Dev name
1	Light	7	Solidification Stove 4
2	Injection molding machine	8	Solidification Stove 5
3	Water Purifier	9	Air conditioner
4	Solidification Stove 1	10	Mould Cleaner 1
5	Solidification Stove 2	11	Mould Cleaner 2
6	Solidification Stove 3		

Source: From Xu et al. [A]/with permission of Elsevier.

Table 5.4 Parameters of random forest and genetic algorithm.

Random forest		Genetic algorithm	
Min parent size	10	Population size	1000
Min leaf size	1	Mutation Fraction	0.01
Max num splits	Training sample size – 1	Crossover Fraction	0.8

Source: From Xu et al. [A]/with permission of Elsevier.

Table 5.5 Classifier economics optimization result on REDD dataset and factory dataset.

					REDD dataset				
	100$	**130$**	**150$**	**180$**	**200$**	**230$**	**250$**	**280$**	**300$**
25%	[100,1,0]	[100,1,0]	[100,1,0]	[100,1,0]	[100,1,0]	[100,1,0]	[100,1,0]	[100,1,0]	[100,1,0]
30%	[100,0,1]	[130,0,1]	[150,0,1]	[180,0,1]	[200,0,1]	[200,2,0]	[200,2,0]	[200,2,0]	[200,2,0]
40%	[200,2,0]	[200,2,0]	[200,2,0]	[200,2,0]	[200,2,0]	[200,2,0]	[200,2,0]	[200,2,0]	[200,2,0]
50%	[200,2,0]	[200,2,0]	[200,2,0]	[200,2,0]	[200,2,0]	[200,2,0]	[200,2,0]	[200,2,0]	[200,2,0]
60%	[200,2,0]	[200,2,0]	[200,2,0]	[200,2,0]	[200,2,0]	[200,2,0]	[200,2,0]	[200,2,0]	[200,2,0]
70%	[200,2,0]	[200,2,0]	[200,2,0]	[200,2,0]	[200,2,0]	[200,2,0]	[200,2,0]	[200,2,0]	[200,2,0]
80%	[200,2,0]	[200,2,0]	[200,2,0]	[200,2,0]	[200,2,0]	[200,2,0]	[200,2,0]	[200,2,0]	[200,2,0]
90%	[200,2,0]	[200,2,0]	[200,2,0]	[200,2,0]	[200,2,0]	[200,2,0]	[200,2,0]	[200,2,0]	[200,2,0]
95%	[200,1,1]	[230,1,1]	[250,1,1]	[280,1,1]	[300,1,1]	[300,3,0]	[300,3,0]	[300,3,0]	[300,3,0]
96%	[200,1,1]	[230,1,1]	[250,1,1]	[280,1,1]	**[300,1,1]**	[300,3,0]	[300,3,0]	[300,3,0]	[300,3,0]
97%	[200,1,1]	[230,1,1]	[250,1,1]	[280,1,1]	[300,1,1]	[300,3,0]	[300,3,0]	[300,3,0]	[300,3,0]
98%	[200,0,2]	[260,0,2]	[300,0,2]	[300,3,0]	[300,3,0]	[300,3,0]	[300,3,0]	[300,3,0]	[300,3,0]
99%	[200,0,2]	[260,0,2]	[300,0,2]	[360,0,2]	[400,0,2]	[400,4,0]	[400,4,0]	[400,4,0]	[400,4,0]
100%	[300,1,2]	[360,1,2]	[400,1,2]	[400,4,0]	[400,4,0]	[400,4,0]	[400,4,0]	[400,4,0]	[400,4,0]
					Factory dataset				
25%	[200,0,2]	[260,0,2]	[300,0,2]	[300,3,0]	[300,3,0]	[300,3,0]	[300,3,0]	[300,3,0]	[300,3,0]
30%	[300,3,0]	[300,3,0]	[300,3,0]	[300,3,0]	[300,3,0]	[300,3,0]	[300,3,0]	[300,3,0]	[300,3,0]
40%	[300,3,0]	[300,3,0]	[300,3,0]	[300,3,0]	[300,3,0]	[300,3,0]	[300,3,0]	[300,3,0]	[300,3,0]
50%	[400,4,0]	[400,4,0]	[400,4,0]	[400,4,0]	[400,4,0]	[400,4,0]	[400,4,0]	[400,4,0]	[400,4,0]
60%	[400,2,2]	[460,2,2]	[500,2,2]	[500,5,0]	[500,5,0]	[500,5,0]	[500,5,0]	[500,5,0]	[500,5,0]
70%	[400,0,4]	[520,0,4]	[600,0,4]	[660,3,2]	[700,3,2]	[700,7,0]	[700,7,0]	[700,7,0]	[700,7,0]
80%	[600,3,3]	[690,3,3]	[700,7,0]	[700,7,0]	[700,7,0]	[700,7,0]	[700,7,0]	[700,7,0]	[700,7,0]
90%	[600,3,3]	[690,3,3]	[700,7,0]	[700,7,0]	[700,7,0]	[700,7,0]	[700,7,0]	[700,7,0]	[700,7,0]
95%	[600,2,4]	[720,2,4]	[800,2,4]	[800,8,0]	[800,8,0]	[800,8,0]	[800,8,0]	[800,8,0]	[800,8,0]
96%	[600,2,4]	[720,2,4]	[800,2,4]	[800,8,0]	[800,8,0]	[800,8,0]	[800,8,0]	[800,8,0]	[800,8,0]
97%	[600,2,4]	[720,2,4]	[800,2,4]	[800,8,0]	[800,8,0]	[800,8,0]	[800,8,0]	[800,8,0]	[800,8,0]
98%	[600,1,5]	[750,1,5]	[800,8,0]	[800,8,0]	[800,8,0]	[800,8,0]	[800,8,0]	[800,8,0]	[800,8,0]
99%	[600,1,5]	[750,1,5]	[800,8,0]	[800,8,0]	[800,8,0]	[800,8,0]	[800,8,0]	[800,8,0]	[800,8,0]
100%	[600,1,5]	[750,1,5]	[800,8,0]	[800,8,0]	[800,8,0]	[800,8,0]	[800,8,0]	[800,8,0]	[800,8,0]

Source: From Xu et al. [A]/with permission of Elsevier.

feature space. The right element represents the number of classifiers with an advanced meter for the second feature space. For example, if the price of the advanced meter is 200$ and the accuracy constraint is 96%, the optimized solution is constructed by one classifier with a pure active power meter and one classifier with an advanced meter. The total cost of the solution is 300$.

5.5.3.1 Result Analysis via Row Variation in Table 5.5

Figure 5.9 introduces the average meter number of each row in Table 5.5. It is obvious that the total number of meters increases while accuracy requirement increases. This is because that one classifier model with a meter cannot support unlimited number of devices with sufficient accuracy. If accuracy requirement increases, more meters should be added to share the devices so that device number under each classifier decreases. Then higher accuracy can be achieved.

Let 97 and 98% with 150$ under REDD dataset as an example. Figure 5.10 shows the details of classifier location changes.

5.5.3.2 Result Analysis via Column Variation in Table 5.5

Figure 5.11 introduces the average number of advanced meter of each column in Table 5.5.

From Figure 5.11, the number of classifiers with advanced meter decreases while the price of advanced meter increases. The reason is that although classifier with advanced meter and the second feature space can achieve a better accuracy for a set of devices, similar accuracy can be achieved by a classifier with pure active power meter for less devices. Then similar total accuracy of one classifier with advanced meter is equivalent to multiple classifiers with pure active power meter that each classifier is responsible for a section of devices only. In this case, if the price of one advanced meter is higher than the total cost of multiple pure active power meters, the solution of multiple pure active power meter will be selected. So if the price of advanced meter increases, the advanced meter will have less chance to be implemented in SILM solution. Figure 5.12 shows a detailed comparison on advanced meter price increase from 130 to 150$ with 90% accuracy constraint.

From Figure 5.12, the total cost of SILM construction when price of advanced meter is 130$ is $3^*100 + 3^*130 = 690\$$. When price of advanced meter is raised to 150$, the total cost of a same SILM solution becomes $3^*100 + 3^*150 = 750\$$, which is higher than that of the optimal solution (700$) in Table 5.5. So the optimization model has selected a better solution.

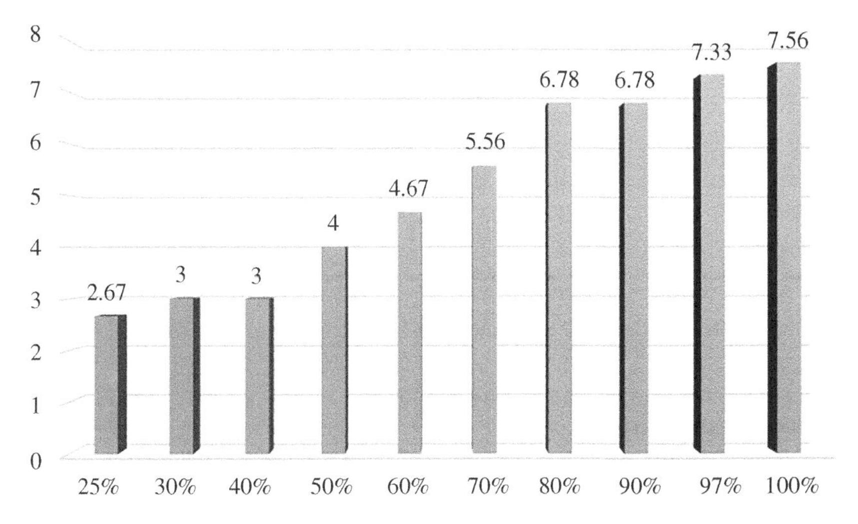

Figure 5.9 Average meter number of each row in Table 5.5. *Source:* From Xu et al. [A]/with permission of Elsevier.

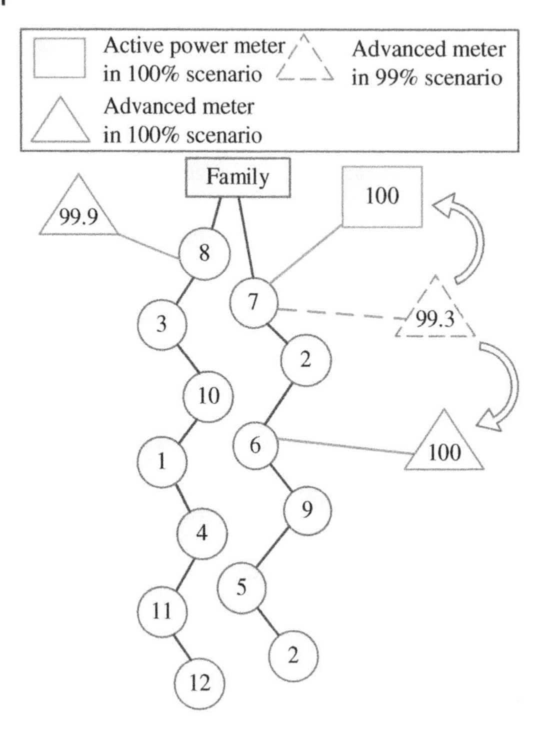

Figure 5.10 Average meter number of each row in Table 5.5. *Source:* From Xu et al. [A]/with permission of Elsevier.

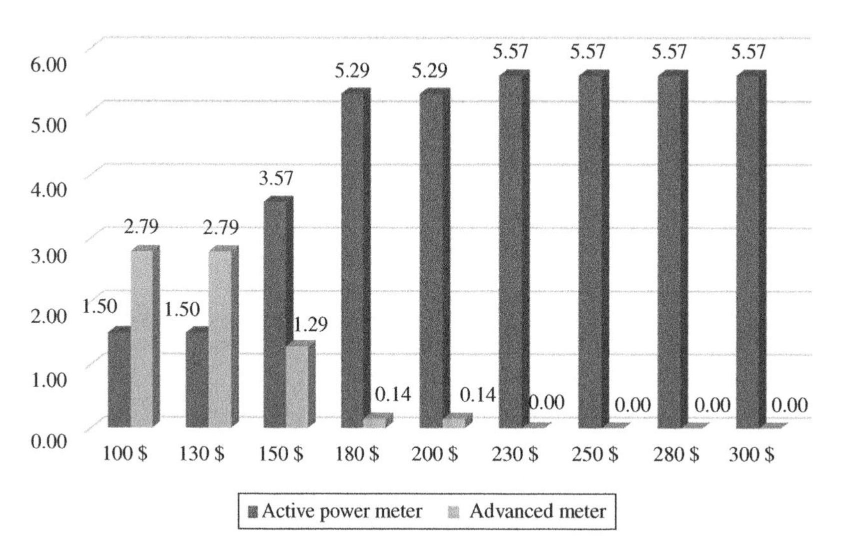

Figure 5.11 Average number of advanced meter of each column in Table 5.5. *Source:* From Xu et al. [A]/with permission of Elsevier.

5.5.3.3 Result Converging via Price Variation

Figure 5.13 shows the entire variation via accuracy constraints and price of second feature space with advanced meter.

It is obvious that no matter what the value of accuracy constraint is, the optimal solution will converge while the price increases over a threshold value. In the converging state, there is no

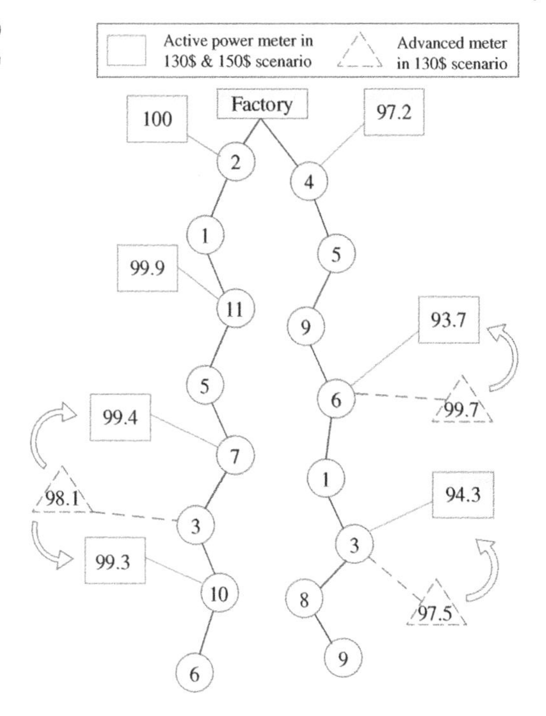

Figure 5.12 SILM solution comparison between 130 and 150$ with 90% accuracy constraint. *Source:* From Xu et al. [A]/with permission of Elsevier.

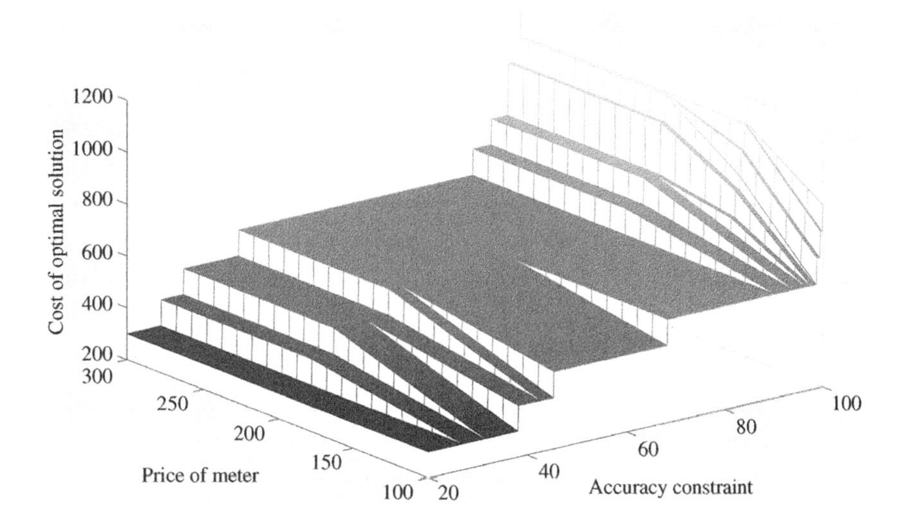

Figure 5.13 Cost variation of SILM solution via price of advanced meter and accuracy constraint. *Source:* From Xu et al. [A]/with permission of Elsevier.

classifier with advanced meters. The reason is that classifier with pure active power meter can achieve similar accuracy to classifier with advanced meter with less devices. It means that a classifier with advanced meter on a set of devices is equivalent to having same accuracy to multiple classifiers with pure active power meter (each classifier covers a small section of devices). When the price of advanced meter becomes much higher than pure active power meter case, the solution of multiple classifiers with pure active power meter will have less cost than one classifier with advanced meter. So the converging state of each accuracy constraint does not have a classifier with advanced meters.

5.5.4 Numerical Study 2: Classifier Economics via different Classifiers Models

In this study, price of an active power measuring meter is still 100$. Different classifiers are selected to perform the simulation. Result is shown in Table 5.6.

In Table 5.6, four types of typical classification models are tested. Results show that a model with better classifying capability can cover more devices within accuracy constraints. So a good model may achieve a solution with less classifiers and meters. But the optimal solution will still change when price of advanced meter increases. The reason is that no matter how good the classifier is, the classifier performance will decrease while more and more devices are responsible from the classifier, especially for devices with similar electric profile. So, devices set will be separated into several subset with independent classifier with simple meter while the price of advanced meter becomes expensive. Figure 5.14 shows a comparison on performance among different classifier models.

Table 5.6 Classifier economics optimization result on REDD dataset with different classifier model.

	Classifier on liner regression			**Classifier on naive bayes**		
	100$	**200$**	**300$**	**100$**	**200$**	**300$**
80%	[700,0,7]	[1400,0,7]	[1500,15,0]	[200,0,2]	[200,2,0]	[200,2,0]
85%	[700,0,7]	[1400,0,7]	[1500,15,0]	[200,0,2]	[200,2,0]	[200,2,0]
90%	[800,0,8]	[1500,15,0]	[1500,15,0]	[200,0,2]	[200,2,0]	[200,2,0]
95%	[1200,0,12]	[1500,15,0]	[1500,15,0]	[200,0,2]	[300,3,0]	[300,3,0]
97%	[1200,0,12]	[1500,15,0]	[1500,15,0]	[300,0,3]	[300,3,0]	[300,3,0]
98%	[1200,0,12]	[1500,15,0]	[1500,15,0]	[400,4,0]	[400,4,0]	[400,4,0]
99%	[1200,0,12]	[1500,15,0]	[1500,15,0]	[400,4,0]	[400,4,0]	[400,4,0]
100%	[1200,0,12]	[1500,15,0]	[1500,15,0]	[400,4,0]	[400,4,0]	[400,4,0]
	Classifier on random forest			**Classifier on KNN**		
80%	[200,2,0]	[200,2,0]	[200,2,0]	[200,2,0]	[200,2,0]	[200,2,0]
85%	[200,2,0]	[200,2,0]	[200,2,0]	[200,0,2]	[300,3,0]	[300,3,0]
90%	[200,2,0]	[200,2,0]	[200,2,0]	[200,0,2]	[300,3,0]	[300,3,0]
95%	[200,1,1]	[300,1,1]	[300,3,0]	[200,0,2]	[400,4,0]	[400,4,0]
97%	[200,0,2]	[300,3,0]	[300,3,0]	[300,2,1]	[400,4,0]	[400,4,0]
98%	[200,0,2]	[400,0,2]	[400,4,0]	[300,2,1]	[400,4,0]	[400,4,0]
99%	[300,1,2]	[400,4,0]	[400,4,0]	[400,3,1]	[500,5,0]	[500,5,0]
100%	[300,1,2]	[400,4,0]	[400,4,0]	[400,3,1]	[500,5,0]	[500,5,0]

Source: From Xu et al. [A]/with permission of Elsevier.

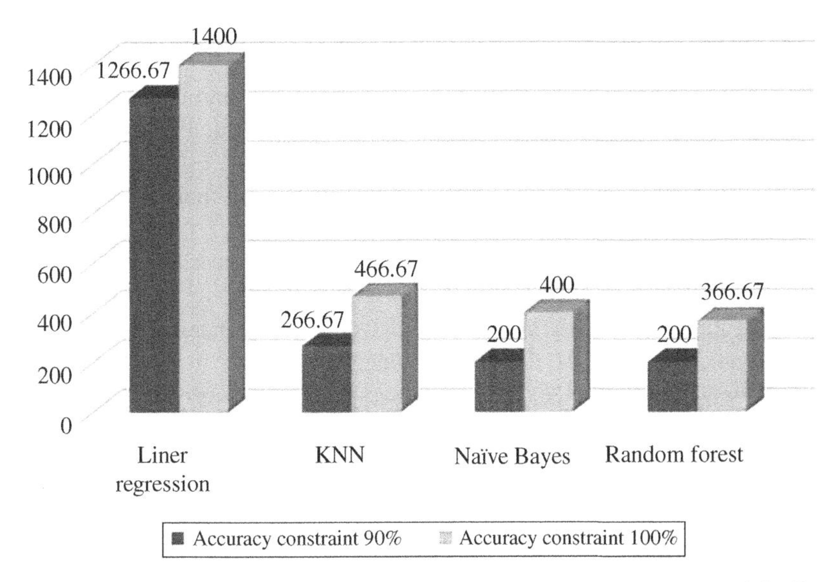

Figure 5.14 Typical performance comparison among different classifier models. *Source:* From Xu et al. [A]/ with permission of Elsevier.

5.6 Conclusion

This chapter introduces a new classifier network construction method for SILM or NILM. The economics of each type of classifier is introduced into this method. Thus, decision maker can obtain a solution of optimal construction economic within accuracy constraints instead of optimal classifiers quantity. A combinatorial optimization problem is established on a tree-type model to the optimized classifier network. Numerical Studies are deployed to support the method feasibility and further analysis. Although only two types of meters are selected in numerical study, this method operates well in cases with situations of more meter selections.

This method indicates that a more accurate model in NILM or SILM with more expensive metering cost may not win the competition from a multiple classifiers network on simpler meters.

Acknowledgements

[A] Xu, F., Huang, B., Cun, X. et al. (2018). Classifier economics of semi-intrusive load monitoring. *International Journal on Electrical Power and Energy Systems*, Elsevier 103: 224–232.

References

1 Tonyali, S., Akkaya, K., Saputro, N. et al. (2017). Privacy-preserving protocols for secure and reliable data aggregation in IoT-enabled smart metering systems. *Future Generation Computer Systems* 78 (2): 547–557.

2 Pocero, L., Amaxilatis, D., Mylonas, G. et al. (2017). Open source IoT meter devices for smart and energy-efficient school buildings. *Hardwarex* 1: 54–67.

3 Spanò, E., Niccolini, L., Pascoli, S.D. et al. (2014). Last-meter smart grid embedded in an Internet-of-Things platform. *IEEE Transactions on Smart Grid* 6 (1): 468–476.
4 Hossain, M., Mekhilef, S., Olatomiwa, L. et al. (2017). Application of extreme learning machine for short term output power forecasting of three grid-connected PV systems. *Journal of Cleaner Production* 167 (20): 395–405.
5 Makridakis, S. (2017). The forthcoming artificial intelligence (AI) revolution: its impact on society and firms. *Futures* 90: 46–60.
6 Mata, J., Miguel, I.D., Durán, R.J. et al. (2018). Artificial intelligence (AI) methods in optical networks: a comprehensive survey. *Optical Switching & Networking* 28: 43–57.
7 Sadeghianpourhamami, N., Ruyssinck, J., Deschrijver, D. et al. (2017). Comprehensive feature selection for appliance classification in NILM. *Energy & Buildings* 151: 98–106.
8 Baets, L.D., Ruyssinck, J., Develder, C. et al. (2017). On the Bayesian optimization and robustness of event detection methods in NILM. *Energy & Buildings* 145 (15): 57–66.
9 Nardello, M., Rossi, M., and Brunelli, D. (2017). An innovative cost-effective smart meter with embedded non intrusive load monitoring. *2017 IEEE PES Innovative Smart Grid Technologies Conference*, Europe (ISGT-Europe), Ghaziabad, India (22–23 February 2013). 8260242
10 Das, S., Srikrishna, Shukla, A., et al. (2013). A low-cost non-intrusive appliance load monitoring system. *Advance Computing Conference. IEEE*, 1641–1644.
11 Biansoongnern, S. and Plungklang, B. (2016). Non-intrusive appliances load monitoring (NILM) for energy conservation in household with low sampling rate. *Procedia Computer Science* 86: 172–175.
12 Ducange, P., Marcelloni, F., and Antonelli, M. (2014). A novel approach based on finite-state machines with fuzzy transitions for nonintrusive home appliance monitoring. *IEEE Transactions on Industrial Informatics* 10 (2): 1185–1197.
13 Paris, J., Donnal, J.S., and Leeb, S.B. (2014). NilmDB: the non-intrusive load monitor database. *IEEE Transactions on Smart Grid* 5 (5): 2459–2467.
14 Tabatabaei, S.M., Dick, S., and Xu, W. (2017). Towards non-intrusive load monitoring via multi-label classification. *IEEE Transactions on Smart Grid* 8 (1): 26–40.
15 Wang, A.L. and Chen, B.X. (2018). Non-intrusive load monitoring algorithm based on features of V–I trajectory. *Electric Power Systems Research* 157: 134–144.
16 Chang, H.H. (2016). Non-intrusive fault identification of power distribution systems in intelligent buildings based on power-spectrum-based wavelet transform. *Energy & Buildings* 127: 930–941.
17 Kong, W., Dong, Z.Y., Hill, D.J. et al. (2016). Improving nonintrusive load monitoring efficiency via a hybrid programing method. *IEEE Transactions on Industrial Informatics* 12 (6): 2148–2157.
18 Huang, T.D., Wang, W.S., and Lian, K.L. (2017). A new power signature for nonintrusive appliance load monitoring. *IEEE Transactions on Smart Grid* 6 (4): 1994–1995.
19 Zhang, J., Chen, X., Ng, W.W.Y. et al. (August 2019). New appliance detection for non-intrusive load monitoring. *IEEE Transactions on Industrial Informatics* 15 (8): 4819–4829.
20 Henao, N., Agbossou, K., Kelouwani, S. et al. (2017). Approach in nonintrusive type I load monitoring using subtractive clustering. *IEEE Transactions on Smart Grid* 8 (2): 812–821.
21 Zhang, P., Zhou, C., Stewart, B.G. et al. (2011). An improved non-intrusive load monitoring method for recognition of electric vehicle battery charging load. *Energy Procedia* 12 (39): 104–112.
22 Tang, G., Wu, K., and Lei, J. (2016). A distributed and scalable approach to semi-intrusive load monitoring. *IEEE Transactions on Parallel & Distributed Systems* 27 (6): 1553–1565.
23 Cominola, A., Giuliani, M., Piga, D. et al. (2017). A hybrid signature-based iterative disaggregation algorithm for non-intrusive load monitoring. *Applied Energy* 185: 331–344.
24 Aladesanmi, E.J. and Folly, K.A. (2015). Overview of non-intrusive load monitoring and identification techniques. *IFAC-PapersOnLine* 48 (30): 415–420.

25 Rafsanjani, H.N. and Ahn, C. (2016). Linking building energy-load variations with occupants' energy-use behaviors in commercial buildings: non-intrusive occupant load monitoring (NIOLM). *Procedia Engineering* 145: 532–539.

26 Lee, D.K. (2003). Electric load information system based on non-intrusive power monitoring. Ph.D. dissertation. Department of Mechanical Engineering, MIT.

27 Figueiredo, M., Almeida, A.D., and Ribeiro, B. (2012). Home electrical signal disaggregation for non-intrusive load monitoring (NILM) systems. *Neurocomputing* 96 (3): 66–73.

28 Esa, N.F., Abdullah, M.P., and Hassan, M.Y. (2016). A review disaggregation method in non-intrusive appliance load monitoring. *Renewable & Sustainable Energy Reviews* 66: 163–173.

29 Basu, K., Debusschere, V., Douzal-Chouakria, A. et al. (2015). Time series distance-based methods for non-intrusive load monitoring in residential buildings. *Energy & Buildings* 96: 109–117.

30 Liu, B., Luan, W., and Yu, Y. (2017). Dynamic time warping based non-intrusive load transient identification. *Applied Energy* 195: 634–645.

31 Kolter, J.Z. and Johnson, M.J. (2011). REDD: a public data set for energy disaggregation research. *Proceedings of the SustKDD 2011*, San Diego, USA (August 2011).

32 Giri, S. and Bergés, M. (2015). An energy estimation framework for event-based methods in Non-intrusive load monitoring. *Energy Conversion & Management* 90: 488–498.

33 Aiad, M. and Peng, H.L. (2016). Non-intrusive load disaggregation with adaptive estimations of devices main power effects and two-way interactions. *Energy & Buildings* 130: 131–139.

6

Residential Demand Response Shifting Boundary

6.1 Introduction

Demand response (DR) in power is a concept defined as "changes in electric usage by end-use customers from their normal consumption patterns in response to changes in the price of electricity over time, or to incentive payments designed to induce lower electricity use at times of high wholesale market prices when system reliability is jeopardized" [1]. DR entails "all intentional modifications to consumption patterns of electricity of end-use customers that are intended to alter the timing, level of instantaneous demand, or the total electricity consumption" [2]. DR is a very important concept in smart grid research, as a core benefit of smart grid is changing electric users' behavior by enabling effective demand-side management (DSM) tactics. DR is an effective method to improve the reliability and flexibility of the power distribution system. Based on the behavior-driven mechanism, different DR programs have been divided into two types: incentive-based programs (IBP) and price-based programs (PBP) [2]. IBPs rely on the operation mechanism in which program sponsors (i.e. electricity companies) pay participating electric users to reduce their electricity loads at requested times, usually triggered by either grid reliability issues or high electricity prices [3]. IBPs are usually adopted by electricity companies to change the electricity consumption patterns of large electric users (e.g. industrial and commercial organization users). IBPs, as suggested by [2], include classical IBP (e.g. direct control) and market-based IBP (e.g. emergency demand response programs). PBPs rely on the operation mechanism in which electric users economically tend to use less electricity at times when electricity prices are high. The time-varying rates can reflect the value and cost of electricity at different times given by electricity companies via different tariffs [3]. PBPs are usually adopted by electricity companies to change the behavior of small and price-sensitive electric users (e.g. small commercial and domestic electric users). PBPs include time of use (TOU), critical peak pricing (CPP), extreme day CPP (ED-CPP), extreme day pricing (EDP), and real-time pricing (RTP). References [2, 3] provide a comprehensive account of the operations these IBPs and PBPs

The importance of DR in power system can be reflected from the broad range of benefits it brings around the concept of smart grid. From a power system economics point of view, the core benefit of DR is "improved resource-efficiency of electricity production due to closer alignment between customers' electricity prices and the value they place on electricity" [2]. Broadly speaking, the variety of benefits of DR can fall into four groups [2, 3] as follows:

1) Participant financial benefits: by changing electricity consumption in response to time-varying electricity rates or incentives offered by electricity companies, electric users can gain electricity bill savings and incentive payments.

Smart Energy for Transportation and Health in a Smart City, First Edition. Chun Sing Lai, Loi Lei Lai and Qi Hong Lai.

2) Market-wide financial benefits: DR programs can enable electric users to reduce electricity consumption at peak times and the need to use the most costly-to-run power plants during periods of high demand. Therefore sustained DR can lower the wholesale market prices and the aggregated power system capacity requirements, which, from a long term point of view, can enable electricity generation and supply companies to purchase or build less capacity. These marketing savings can eventually be passed onto electric users as bill savings.
3) Reliability benefits: effective DR can reduce the probability and consequences of forced outages, which if happened can impose substantial financial costs and inconvenience on electric users. Therefore, DR can improve power system's operational security and reliability. System resiliency could be further improved through the integration of automated switches and sensors.
4) Market performance benefits: with DR programs (especially market-based programs and dynamic pricing programs) customers can have more choices in the market (even when there is no retail competition). This can alleviate electricity companies' ability to use market power by raising electricity prices significantly above production costs.

DR attracts researchers as an important means to promote smart energy. Effect of DR is analyzed statistically by various data, like researches reported in [4]. The impact from DR to power system operation level has been studied [5]. But more researches are focusing on DR modeling for quantifying DR's effect [6, 7]. A popular method is price-elasticity-style modeling, which emphasizes on establishing a price-load elasticity matrix as the bridge between tariff and load of customer group [8–10]. The elasticity matrices in these models are summarized by large set of historical data and contain advantages on practical cases whose load pattern is similar to the selected historical data set. But for other cases, such as impact analysis of an unimplemented tariff format in tariff planning, this method will lose effectiveness. Unlike elasticity, more attention is paid into customer consumer behaviors on appliance level. Load is decoupled by different appliances and the behavior toward each appliance is modeled. The logical sequence is behavior incentive -> behavior change on appliances -> load change on appliances -> total load change. This method targets and develops models on behavior variation incentives and has more stable result than elasticity matrix as the primary cause of load variation related to behavior incentives. Various works are implemented in relevant load model or DR analysis. Ref. [11] proposed a residential load model. This model decouples the load into four types of appliances, namely space cooling/heating, water heating, clothes dryer, and electric vehicle. But for DR research on large group of residential customers, this model lacks of appliances' types and appliances penetration, as well as the behavior variation. In [12], a load model was introduced with emphasis on operational level of appliance, which is more suitable for customers' power quality analysis than DR. Electric Power Research Institute (EPRI) in the United States has also revealed a DR model framework in [13]. The load modeling in this framework introduces not only the appliances analysis but also general residential lifestyle behaviors. But this framework is used as a guide for further model research and it lacks the details on model construction. It points out that coordinated multidisciplinary research to study how feedback information influences household electricity usage. Models proposed in [14–16] appear to contain more suitability than the previous cases. In these works, appliances usage is considered with home activities and penetration level. Their time steps are also less than one hour. But these models are established for individuals so that they do not consider behavioral tropism and uncertainty from probability. Also, the behavior variation is not considered. References [17, 18] provide a DR model with consideration of load decoupling and behavior pattern variation. But these two models do not cover the constraint of behavior variation, such as appliance switch-on limitation from leaving home or time limitation for dining.

In this chapter, a model for residential DR analysis is introduced. By considering different utilizations of typical home appliances at each hour in a day, power load of residential customer group is decoupled. Also, a new concept, "Shifting Boundary," is introduced to measure the largest load shifting potential of several price-based demand response tariff by analyzing residential behavior variations. With Shifting Boundary, applications of TOU price planning and smart meter installation scale planning are implemented as case studies, revealing application potential of this model.

In this chapter, a model framework of individual consumer behavior and group power consumption is introduced. Section 6.2 gives the residential customer behavior modeling. Section 6.3 explains residential customer shifting boundary concept. Section 6.4 provides a case study on the whole model. Sections 6.5 and 6.6 provide another two case studies to indicate that different PBPs or different smart meter installation scale can be evaluated by using Shifting Boundary. Conclusions and Future Work are given in Section 6.7.

6.2 Residential Customer Behavior Modeling

6.2.1 Multi-Agent System Modeling

Agent-based model or multi-agent system is a simulation system for reproducing the interaction between environment and a group of individuals. In this system, agent is an individual unit with intelligence and independent ability for action and decision-making. Agents receive information from environment and then generate their own actions toward the environment. On the other hand, environment changes its status by actions from agents and then generate new information to agents [19].

Price-based demand response possesses a loop interaction structure as shown in Figure 6.1. Price generator creates price information by power consumption information collected from consumers. In many cases, price generator will be a power utility. Sometimes it will be an independent governmental organization in some areas. Power consumers receive price information through smart meters. With this information, power consumers make or update their consuming plan so as to increase their benefits through economic incentives.

Smart meter provides a platform for bidirectional information flow. It passes the price information, summary of consumption information or power utilization advices to various types of power consumers. Reversely, it collects information of consumers for price generator, including power consumption, real-time power load and power quality. In this chapter, multi-agent system is used to construct the loop interaction structure as shown in Figure 6.1.

6.2.2 Multi-agent System Structure for PBP Demand Response

Agent is an individual unit with independent ability for action and decision-making. In loop interaction structure, price generator and power consumer are recognized as agents with independent decision-making function, as shown in Figure 6.2.

Agent of price generator is responsible for price making. Under different tariffs, the price signal is generated with various schemes. Some typical tariffs are given below:

Static price is the most traditional power tariff. All electricity is sold under a same price value. It contains a simple pricing scheme in Eq. (6.1).

$$pri_i = const, const \in R^+ \tag{6.1}$$

where *prii* is the price value at the *i*th hour in a day.

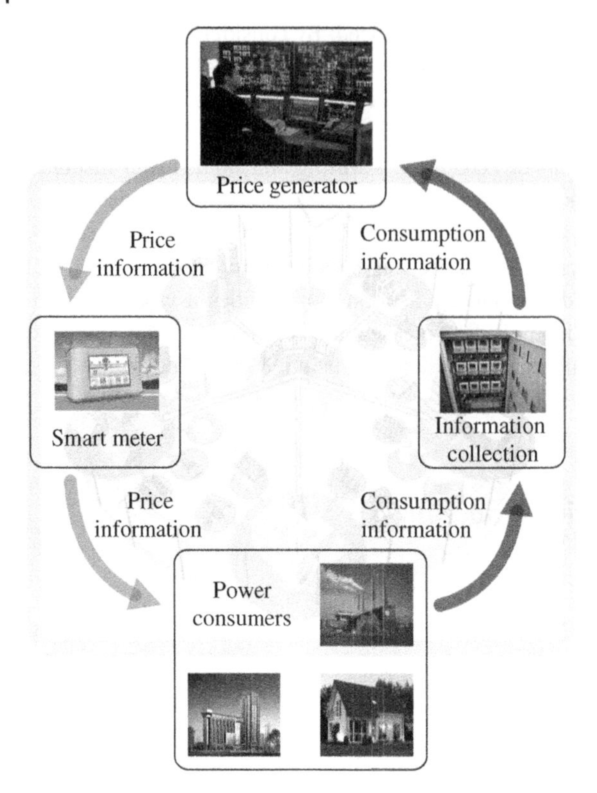

Figure 6.1 Loop interaction structure of PBPs. *Source:* From Xu et al. [A]/Elsevier.

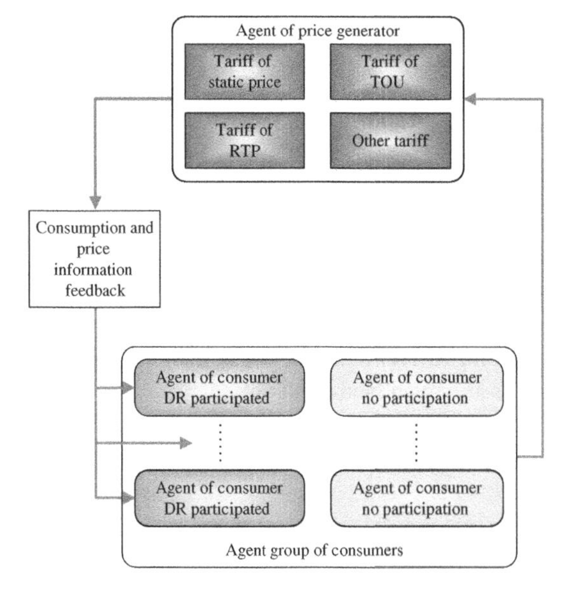

Figure 6.2 Multi-agent system model for loop interaction structure. *Source:* From Xu et al. [A]/with permission of Elsevier.

TOU pricing tariff separates a day into several periods (usually less than five periods). It incentivizes customers to permanently alter their energy consumption by using static price rate that are different during peak and off-peak periods [20]. Under this tariff, a detailed tariff is updated in advance and is kept unchanged for some periods, such as half a year. To a certain extent, TOU reflects the relationship between power demand and power supply. Eq. (6.2) introduces the pricing scheme of TOU tariff.

$$pri_i = \begin{cases} const_1,\ i \in (h_{a_1},\ h_{a_2}) \\ const_2,\ i \in (h_{a_2},\ h_{a_3}) \\ \vdots \\ const_k,\ i \in (h_{a_{k-1}},\ h_{a_k}) \end{cases}$$

$$\sum_{j=1}^{k-1}\left(h_{a_{j+1}} - h_{a_j}\right) = 24 \tag{6.2}$$

In Eq. (6.2), a day (24 hours) is separated into k periods. Each period is provided a specific price level known as *constk*.

Similar to TOU tariff, RTP tariff separates a day into several smaller time periods. But each time period could be less than one hour, and represents a much higher changing frequency of time-varying price. This price tariff reflects the utility's cost of generating and/or purchasing electricity at the wholesale level. In general operation, RTP and real-time power load have a positive correlation. A typical relationship is revealed in Eq. (6.3) below:

$$pri_i = \alpha \times Area\_Load_i \tag{6.3}$$

where *Area_Loadi* is the total load generated from a consumer group. α is the rate between price and load, which is usually set by utilities but limited by policies and regulations. Practically, the positive correlation mapping is much more complex than Eq. (6.3). Whole sale price, real-time status of operation, line losses, emergent events are all the influencing factors toward RTP scheme. But electricity price is also limited by government policies and other organizations, e.g. ISO. For example, electricity price for urban residential customers in China are constrained under 1 CNY/kWh by China National Development and Reform Commission.

6.2.3 Agent of Residential Consumer

Residential consumer is one of the main components for domestic electricity consumption. Load from residential consumers are mainly generated from home appliances. People make behavioral decisions for power consumption based on their lifestyle.

In the loop interaction structure, each agent of consumer represents an individual family with a set of home appliances. Agent makes decision to control all home appliances and generates power load. Some influencing factors of residential behavior makings could be summarized as below:

Home appliances are used for satisfaction of residential home area requirements. Basic requirements cover food, health, comfort, and entertainment. Toward these requirements, typical appliances include computer, cooking devices, fridge, air-condition, lighting, TV, electric-shower, and washing machine. Utilization of home appliances is deeply dependent on lifestyle. For example, TV will not be switched on when people have left home or fallen asleep. So "At Home" status and "Awake" status are recognized as enabling condition for controlling certain appliances, such as computer, cooking devices, air-condition, lighting, TV, and electric shower. Washing machine and dishwasher have timing function and so are not limited by these situations.

Under DR scheme, information availability is an important foundation of behavior modification. Potential of DR occurs within people who can access necessary information feedback, including consumption history and price. Other consumers who cannot access information feedback will lack support for their decision-making and so cannot take part in DR program. The consumers' response will deeply be influenced by smart meter installation scale. In DR loop interaction, people response to PBP is due to financial incentives. People will more actively respond to price variation if they are short of financial support. Information availability of financial situation can be collected directly by surveys of DR statistics such as fuel poverty rate. Considering the above impact factors, the structure of consumer agent is represented in Figure 6.3.

Figure 6.3 reveals agent structure of consumers who prefer to take part in PBPs. Definitions of the symbols are given in Table 6.1. Each appliance in the agent is controlled by a specified time-varying switch-on probability. Also, each agent contains an independent decision-making process. Life routine, financial situation, and information availability are collected for decision-making. The final decision will be a rescheduled switch-on probability for each appliance. For those agents without DR participation, there is a lack of input of "Consumption & Price Information Feedback" as shown in Figure 6.3. Equation (6.4) reveals the decision-making process in Figure 6.3. As revealed in Eq. (6.4), people will not change their behavior without information feedback or financial support.

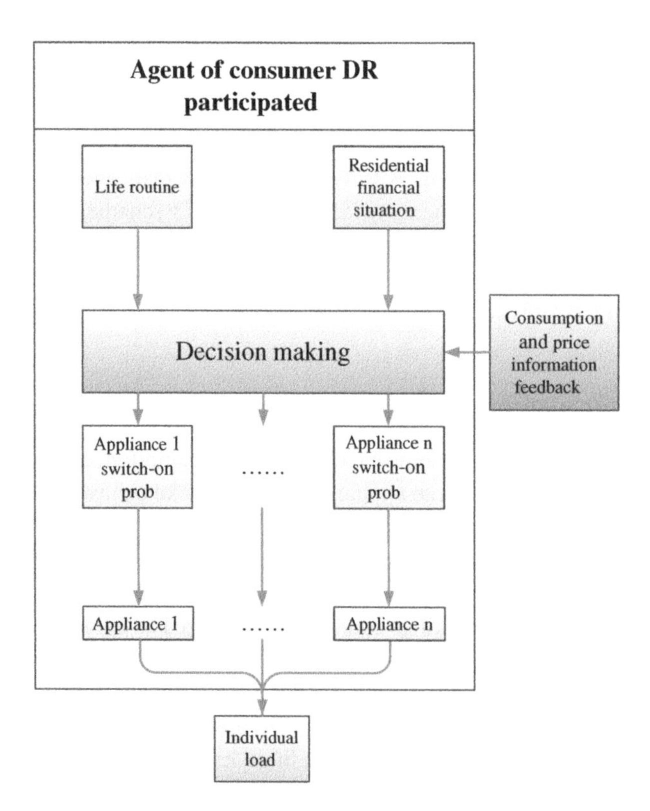

Figure 6.3 Agent structure of consumers DR participated. *Source:* From Xu et al. [A]/with permission of Elsevier.

Table 6.1 Parameters for Eq. (6.4).

Parameters	**Description**
LR_i	Life routine at hour i
$Sw_i(on/LR_i)$	Switch-on probability vector of all appliances given life routine at hour i.
sw_{in}	Switch-on probability of appliance n at hour i.
pri_i	Electricity price at hour i
IFA	Logical parameter. "1" means Information Feedback available
FS	Logical parameter. "1" means Financial Support available

Source: From Xu et al. [A]/with permission of Elsevier.

Only those people who have information feedback and do not have strong financial support will have potential of PBPs participation.

$$\begin{aligned} Sw_i(on/LR_i) &= (sw_{i1}, sw_{i2}, \cdots, sw_{in}) \\ &= f(LR_i, pri_i, IFA, FS) \\ &= \begin{cases} const, IFA\cdot(1-FS) = 0 \\ g(LR_i,\ pri_i), IFA\cdot(1-FS) = 0 \end{cases} \end{aligned} \tag{6.4}$$

6.3 Residential Customer Shifting Boundary

6.3.1 Consumer Behavior Decision-Making

By considering the information received and its own situation that is life routine and financial situation, residential consumers alter their behavior decision to reduce their bill. The inner structure of "decision-making" in Figure 6.3 is revealed in Figure 6.4.

Initially when receiving the information feedback, residential customers will consider its financial situation. If a family belongs to fuel poverty, it will start its behavior alternation for bill reduction. Block "Behavior Optimization" helps this agent to reschedule behaviors. Otherwise, the bill does not take a notable section of family's income and cannot provide a strong-enough incentive for behavior changing. In this case, this family will keep its usual behavioral habit.

6.3.2 Shifting Boundary

Behavioral alternation is the main influencing factor for load variation in DR. Usage reduction and usage transformation are two primary behavior rescheduling methods.

Reducing utilization of appliances is a straight consideration of bill reduction. People reduce its consumption with high electricity price and recover its consumption when price decreases. Manually, consumption reduction is simple and easy to promote. But it significantly decreases the bill as well as the comfort and satisfaction of residential customers. Price-consumption elasticity is the usual index for usage reduction quantization.

Usage transformation is another method for behavior rescheduling. People can also transform their consumptions from time period with high price to the time period with lower price. Usage transformation does not decrease utilization of appliances and so its reduction of comfort and

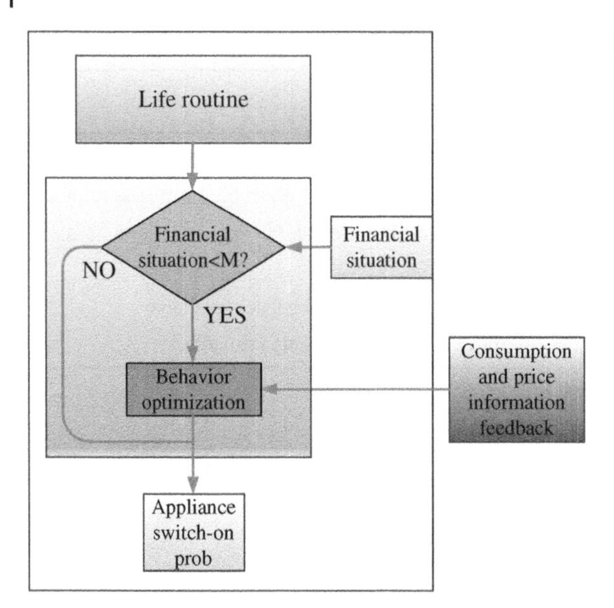

Figure 6.4 Inner structure of decision-making. *Source:* From Xu et al. [A]/with permission of Elsevier.

satisfaction is less than usage reduction. But the promotion and education to consumers are more complex. Different educational levels or different working background will also influence the transformation. Difficulty also appears in quantifying usage transformation.

To find out the quantization of usage transformation, shifting boundary is introduced. That is, under a certain PBPs participation rate and a certain consumers' financial situation distribution, shifting boundary is the maximum usage transformation that a certain consumer group can achieve without usage reduction.

Based on this concept of shifting boundary, PBPs participation rate and consumers' financial situation distribution are derived. Consumers outside this DR participation will not reschedule their behaviors and consume electricity in their usual way. Consumers within the participation are all assumed to take the most efficient usage transformation for bill reduction. Practically, not all consumers can achieve the most efficient usage transformation and so the term of boundary is given. A suitable promotion pattern and a successful consumer education will approximate the practical usage transformation to the shifting boundary. Shifting boundary indicates the maximum load-shift for a certain group of consumers.

6.3.3 Target Function and Constraints

Consumers transform their usage for bill reduction and the target of shifting boundary is to make a behavioral rescheduling for bill minimization. Equation (6.5) indicates the target function in shifting boundary. Definitions of the symbols are given in Table 6.2.

$$\begin{cases} On\_sta_{ijm} = h\left(Sw_{ijm}(on/LR_i)\right) \\ Load_{ij} = \sum\limits_{m=1}^{M} Power_{ijm} \times On\_sta_{ijm} \\ \min : Bill_j = \sum\limits_{i=1}^{24} Load_{ij} \times pri_i \times 3600 \end{cases} \tag{6.5}$$

Table 6.2 Parameters for Eq. (6.5).

Parameters	Description
$h(Sw_{ijm}(on/LR_i))$	Appliance working status determinations function. At hour i, this function determines the working status of the m_{th} appliance in the j_{th} agent. Output of this function is a Boolean number with 1 representing on-status for the appliance and 0 representing off-status
On_sta_{ijm}	Appliance working status of the m_{th} appliance in the j_{th} agent at hour i
$Power_{ijm}$	Working power of the m_{th} appliance in the j_{th} agent at hour i
$Load_{ij}$	Power load generated from the j_{th} agent at hour i
$Bill_j$	Total daily bill for the j_{th} agent

Source: From Xu et al. [A]/with permission of Elsevier.

Further to the concept of shifting boundary, utilization of each appliance is a constant. This represents that all daily requirements are satisfied as usual. This limit is set as a constraint in Eq. (6.6) for bill optimization. The total daily power consumption for any appliance under any agent should be kept constant.

$$Constr : const_j = \sum_{i=1}^{24} Load_{ij} \times 3600 \tag{6.6}$$

Behavioral rescheduling will be influenced by lifestyle, for example, At home status and awake status. Figure 6.5 provides typical time varying at home rate and awake rate for residential

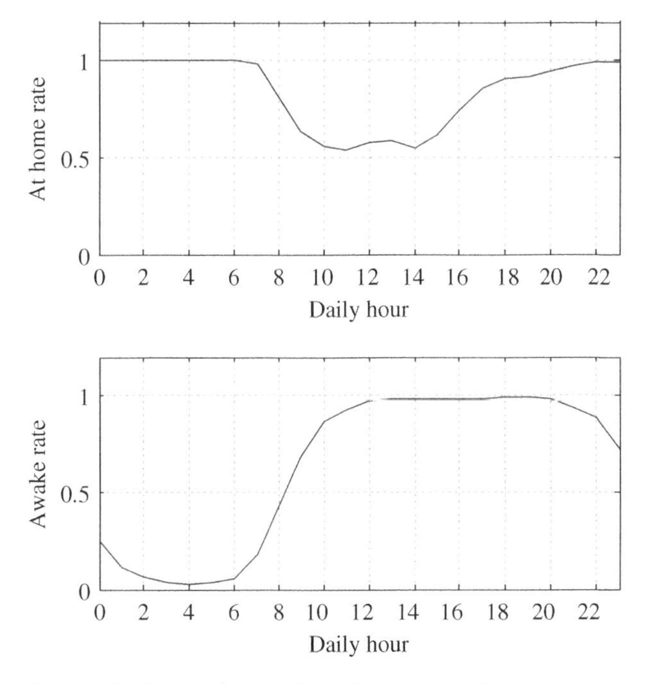

Figure 6.5 Typical time varying at home rate and awake rate for residential consumers. *Source:* From Xu et al. [A]/with permission of Elsevier.

Table 6.3 Parameters for Eq. (6.7).

Parameters	Description
At_h_{ij}	"At Home" rate of the j_{th} agent at hour i
Awk_{ij}	"Awake" rate of the j_{th} agent at hour i

Source: From Xu et al. [A]/with permission of Elsevier.

Table 6.4 Typical transformation range of daily meals.

Subperiod	Meals
4:00–9:59	Breakfast
10:00–14:59	Lunch
15:00–21:59	Dinner
22:00–3:00	Supper

Source: From Xu et al. [A]/with permission of Elsevier.

consumers. This typical rate is abstracted from a survey of 4-people family and so the at home rate is generally over 0.5.

Most of home appliances can only be controlled when people are at home and awake. As such when new behavior is rescheduled, this constraint is still in use. Equation (6.7) indicates the detail of this constraint. Definitions of the symbols are given in Table 6.3. Some of the home appliances namely, washing machine and dishwasher can be controlled even people are in sleeping status.

$$Constr: sw_{ijm} \leq 1 \times At\_h_{ij} \times Awk_{ij} \tag{6.7}$$

Features of appliances may limit the usage transformation. For example, breakfast will not be transformed to 21:00 in rescheduling process. The transformation of breakfast has to be limited within a daily subperiod. Table 6.4 reveals a typical transformation range of daily meals.

In Table 6.4, each meal can only be shifted within the subperiod for food requirement. In other words, the consumption of cooking appliances within each subperiod is the same. Equation (6.8) indicates this constraint which is similar to Eq. (6.6). But Eq. (6.6) is the daily consumption constraint while Eq. (6.8) is subperiod consumption constraint.

$$Constr: const_{jn} = \sum_{i=n1}^{nk} Load_{ij} \times 3600 \tag{6.8}$$

6.4 Case Study

6.4.1 Case Study Description

A typical Chinese resident group of 1000 people in severe winter is selected as target consumers for case study. Each family is represented by an independent agent. Altered pricing schemes are implemented on target consumers to analyze their behavior changes and load variation in weekdays.

All target consumers are equipped with home appliances from a typical appliance set. All appliances are independently distributed on the target group with certain owning rate from a survey on a typical Chinese city residents' group. Considering the utilizing feature, each appliance is with appropriate constraints from Section 6.4. Table 6.5 reveals the details of appliance set.

In Table 6.5, Fridge is usually switched on for 24 hours for food freshness detainment. It is not controllable and its behavior will not change with different pricing scheme. Behavior changes on electric heater without energy storage, and lighting are considered to cause serious problems to residential consumers. Their behaviors do not change with different pricing scheme as well. All other appliances in Table 6.5 are behavioral transformable and will be the main force of load variation.

Behavior constraints are another influencing factor for appliance control. In the appliance set such as electric heater, computer, cooking appliances, lighting, electric shower, and TV are only probable to be switched on when people at home. Moreover, computer, cooking appliances, lighting, electric shower, and TV are controlled only by awake people. Therefore, behaviors toward these appliances are limited by lifestyle, which is shown with the constraint in Eq. (6.7). Washing machine is an appliance with timing control, they may operate any time during a day and will not be limited by constraint of lifestyle.

Most of the appliances in the case study are limited by constraint of the same power as given in Eq. (6.6) for shifting boundary analysis. But this constraint is not sufficient for cooking appliances and TV as they have a "stronger" limit on usage transformation. For example, breakfast can only be shifted during morning period. So, for TV and cooking appliances, same power consumption should be kept during each subperiod within a day, which is the constraint given in Eq. (6.8). Lifestyle is a critical constraint for behavior variation. All agents are operating under their lifestyle rate. From our survey, agents' time-varying at home status and sleeping status are shown in Figure 6.6:

In Figure 6.6, size of families is classified with the membership number from 1 to 6. Time varying at home status and sleeping status are sampled per hour in a day. In each sub-figure for at home status, curve "0p" represents 0 people. Curve "2p" represents that there are two persons in the family. Other curves are with similar meaning. For the target consumer group, distribution of family membership is given in Table 6.6.

Table 6.5 Details of home appliances in case study.

Appliance	Owning rate (%)	Power (W)	Behavior change	Constraints equation enabled
Electric Heater	5	3000	NO	(6.7)
Computer	77	290	YES	(6.6) (6.7)
Washing Machine	78	2500	YES	(6.6)
Cooker	100	1000	YES	(6.7) (6.8)
Fridge	100	80	NO	NO
Lighting	100	300	NO	(6.7)
Electric Shower	5	3500	YES	(6.6) (6.7)
TV	97	250	YES	(6.7) (6.8)

Source: From Xu et al. [A]/with permission of Elsevier.

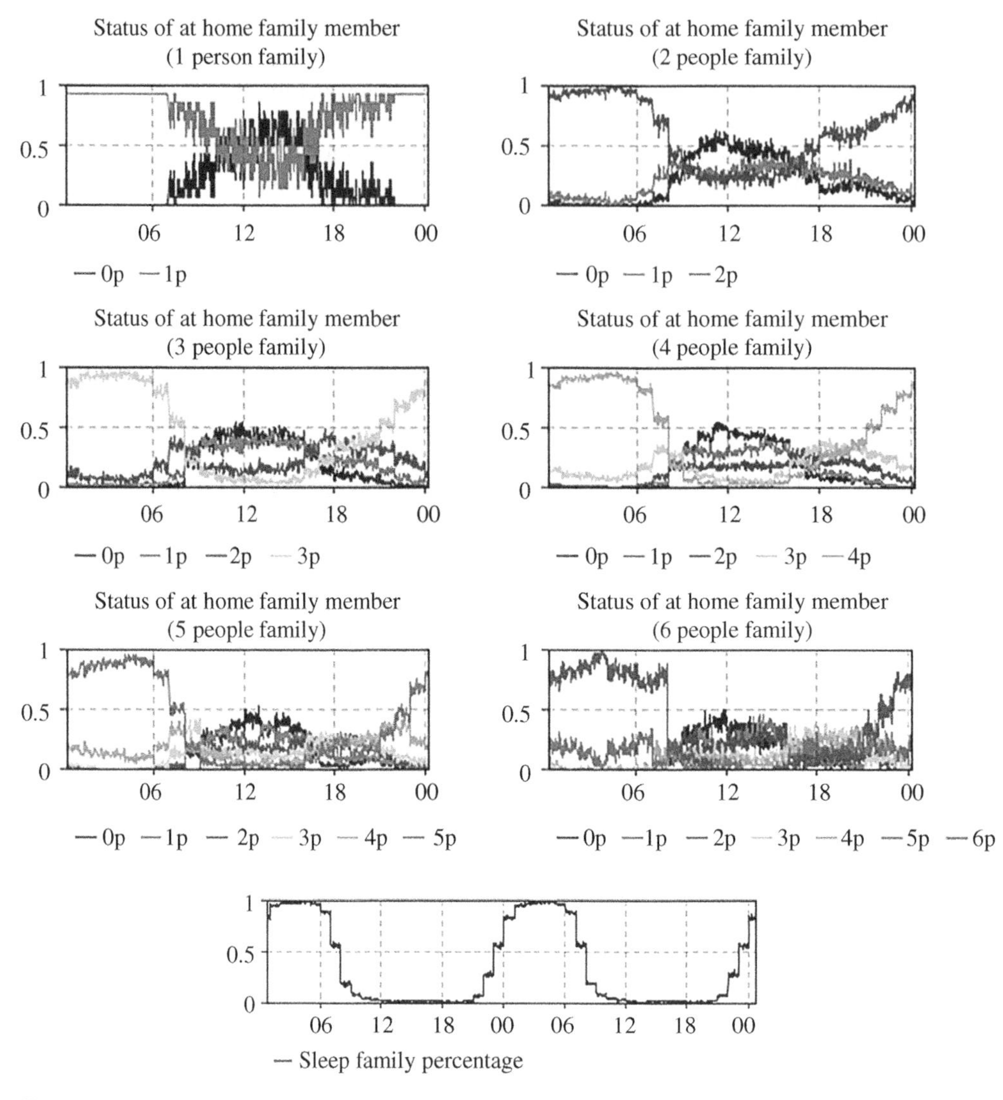

Figure 6.6 Agents' time-varying at home status and sleeping status. *Source:* From Xu et al. [A]/with permission of Elsevier.

Table 6.6 Distribution of family membership.

Number of people in family	Percentage in target group (%)
1	1
2	15
3	22
4	37
5	18
6	7

Source: From Xu et al. [A]/with permission of Elsevier.

Behavior is a basic and critical factor of each agent. Agents output their behavior when responding to price change. In this case study, initial behavior of each appliance is abstracted from residential time-varying activities from our survey in Figure 6.7, as conditional switch-on probability. These initial behaviors are set as the initial point for behavior optimization. Also, it is recognized as the consumer behavior under static price, to represent the general requirements without impact of daily time-varying price. Behavior for each appliance is summarized by appliance switch-on probability.

Regarding simulation environment, Anylogic Professional 6.8.0 is used as simulation platform for this research. It provides a platform for agent-based modeling and multi-agent system simulation.

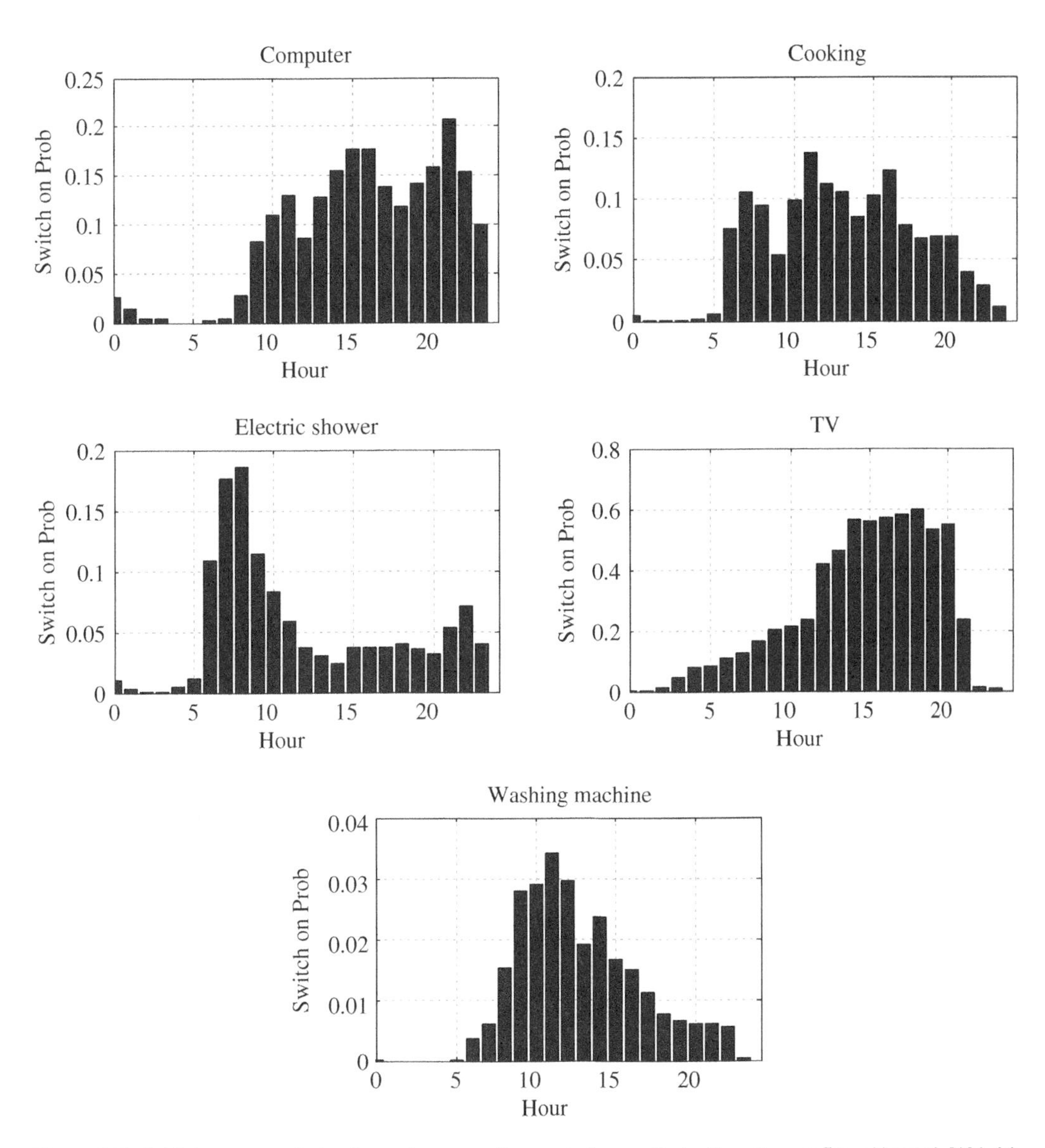

Figure 6.7 Initial consumer behavior on home appliances before optimization. *Source:* From Xu et al. [A]/with permission of Elsevier.

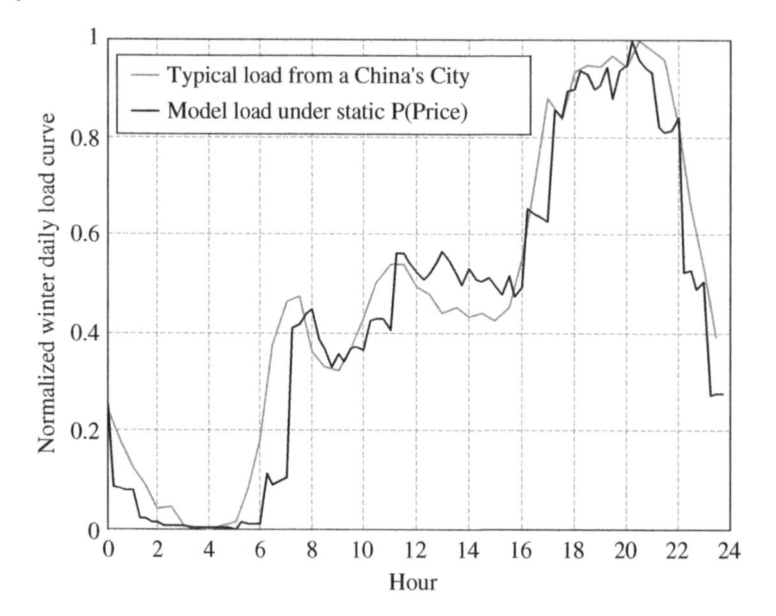

Figure 6.8 Normalized load comparison between the load constructed from survey and load of a China's City. *Source:* From Xu et al. [A]/with permission of Elsevier.

MATLAB is used for all simulation data pretreatment and output analysis. To verify the quality of our survey on consumer behavior, Figure 6.8 reveals a comparison between the normalized load constructed from our survey and the normalized load from a China's City.

From Figure 6.8, the general variation trends of both curves are the same. Practical load from a China's city is smoother as it contains more samples.

6.4.2 Residential Shifting Boundary Simulation under TOU

TOU tariff is typical non-static pricing scheme that can reshape the daily load curve by changing consumer behavior. A typical TOU tariff in China as shown in Figure 6.9 is selected for this case study.

In target consumer group, 80% of families have installed Smart Meters to receive information feedback. Considering the financial situation, 30% of families have inclinations to change their behaviors for lower bills with the same daily requirements of all home appliances. So, it results with 24% participation rate of target group, daily load curve of shifting boundary is shown in Figure 6.10.

From Figure 6.10, load of TOU peak period (8:00–21:00) is lower than load under static price as the result of consumption transformation. But TOU valley period (21:00–8:00) increases the load. TOU load in 7:00–8:00 and 21:00–22:00 are obviously increased by concentrated consumption. Concentrated consumption represents that lots of families tend to consume electricity at a certain period.

In Table 6.5, consumer behavior toward electric heater, lighting, and fridge will not change under different pricing scheme. A same price variation appears under TOU and static price.

From Figure 6.11, daily load variation under static price and TOU are almost the same. Small differences still exist as the result of randomness from behavior generation on switch-on probability. Electric heater will be switched on once home is not empty. Electric load of heating decreases in the morning and increases in the evening according to the lifestyle of the people. About 90% of the

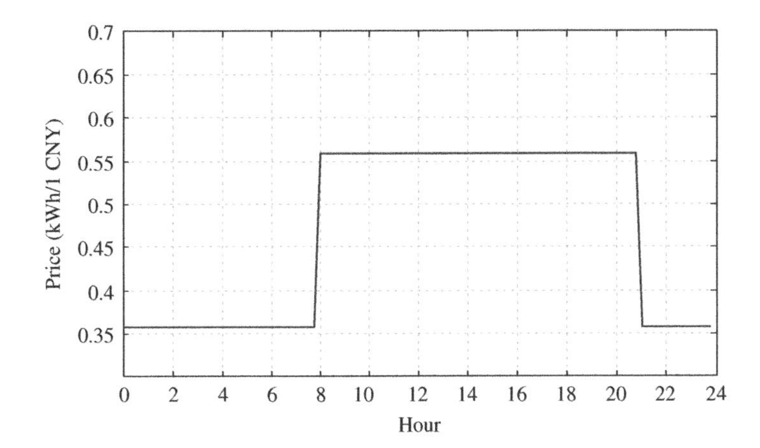

Figure 6.9 A typical 2-period TOU tariff for Chinese Residents. *Source:* From Xu et al. [A]/with permission of Elsevier.

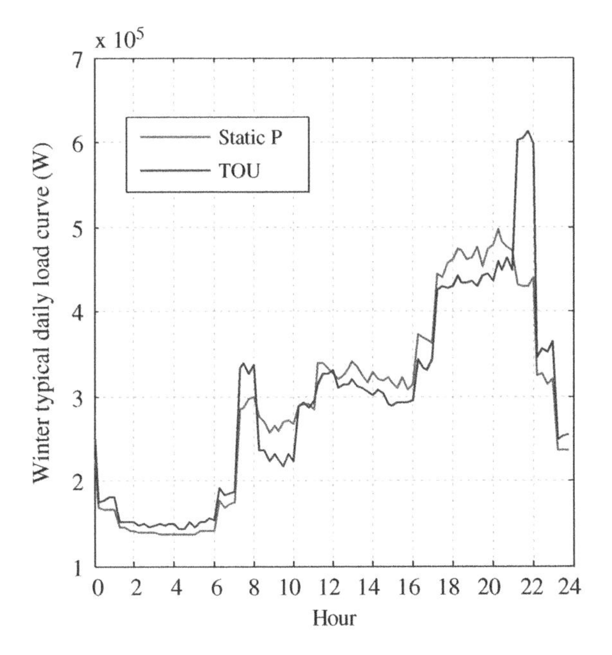

Figure 6.10 Daily load curve comparison between static price and TOU shifting boundary. *Source:* From Xu et al. [A]/with permission of Elsevier.

lights will be applied when people are awake at home and the sky is dark so there is the load of light in the morning and in the evening. Fridge will be switched on for 24 hours, so no distinct daily variation appears in its load curve.

When computer, dishwasher, electric shower, and washing machine are considered. Their daily consumptions are kept the same while behavior is changing. Figure 6.12 reveals the load curve variation of these four appliances.

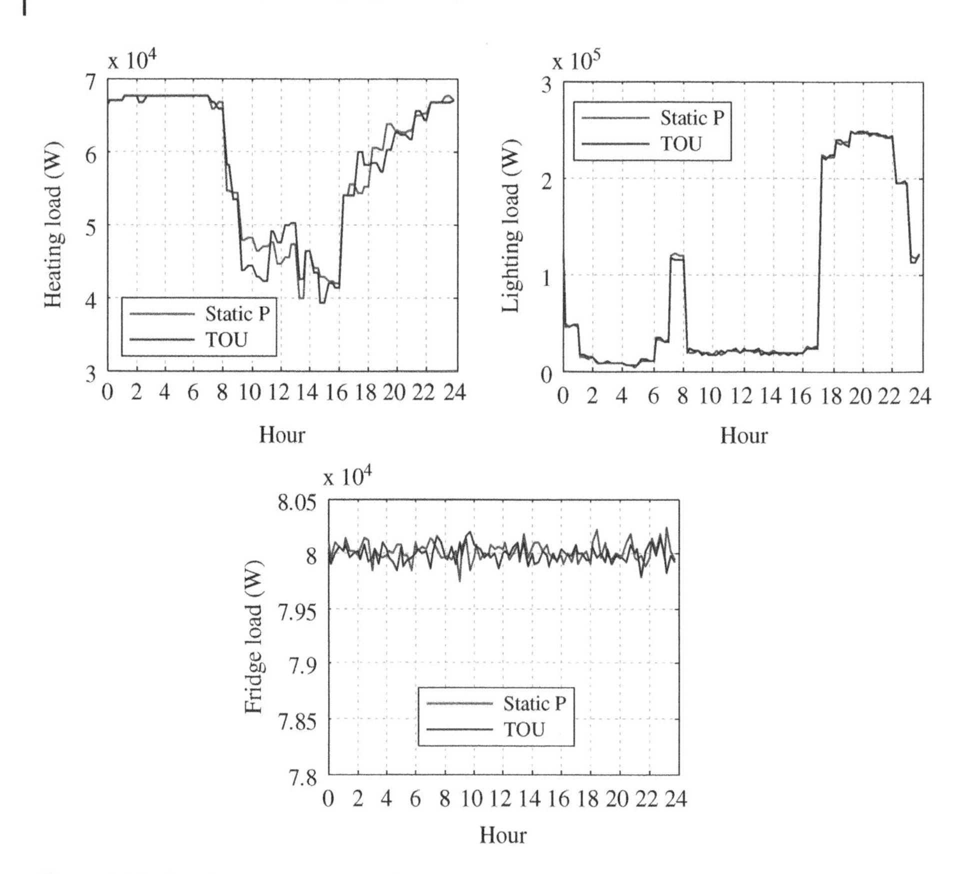

Figure 6.11 Load curve comparison between static price and TOU shifting boundary for heater, lighting, and fridge. *Source:* From Xu et al. [A]/with permission of Elsevier.

From Table 6.5, washing machine is only constrained by its daily total consumption without condition of sleep or at home. People's lifestyle does not have any effect on the control of these two appliances for their timing function. As such in Figure 6.11, load of price-peak period can be shifted to any time of price-valley period, thus load between 21:00 and 8:00 of the next morning increases. Computer and electric shower are not only covered by constraints of same daily consumption but also the constraint of lifestyle. Therefore, for these two appliances, load of price-peak period are mainly shifted to 21:00–22:00, 22:00–23:00 and 7:00–8:00, because these three hours have the highest at home rate and lowest sleeping rate of price-valley period. Load is seldom shifted to period between 23:00 and 7:00 as most of people are sleeping. In other words, load computer and electric shower have formed a morning peak and a new evening peak under TOU shifting boundary, and this represents a risk of concentrated consumption.

Turning to cooker and TV, not only their daily consumptions, but also consumptions of subperiod in Table 6.4 should be kept the same while behavior is changing. Figure 6.13 reveals the load curve variation of cooking and TV.

For cooking, the first subperiod is from 4:00 to 9:59, in which 4:00–7:59 belongs to price-valley period; cooking could be changed from period of 8:00–9:59 to period of 4:00–7:59. Also, cooking appliances can only be switched on when people are available. So 7:00–7:59 has most load and there is a new morning peak of cooking load at 7:00–7:59. The second subperiod is from 10:00 to 14:59. The whole subperiod belongs to price-peak period. Electricity price during this subperiod is kept the

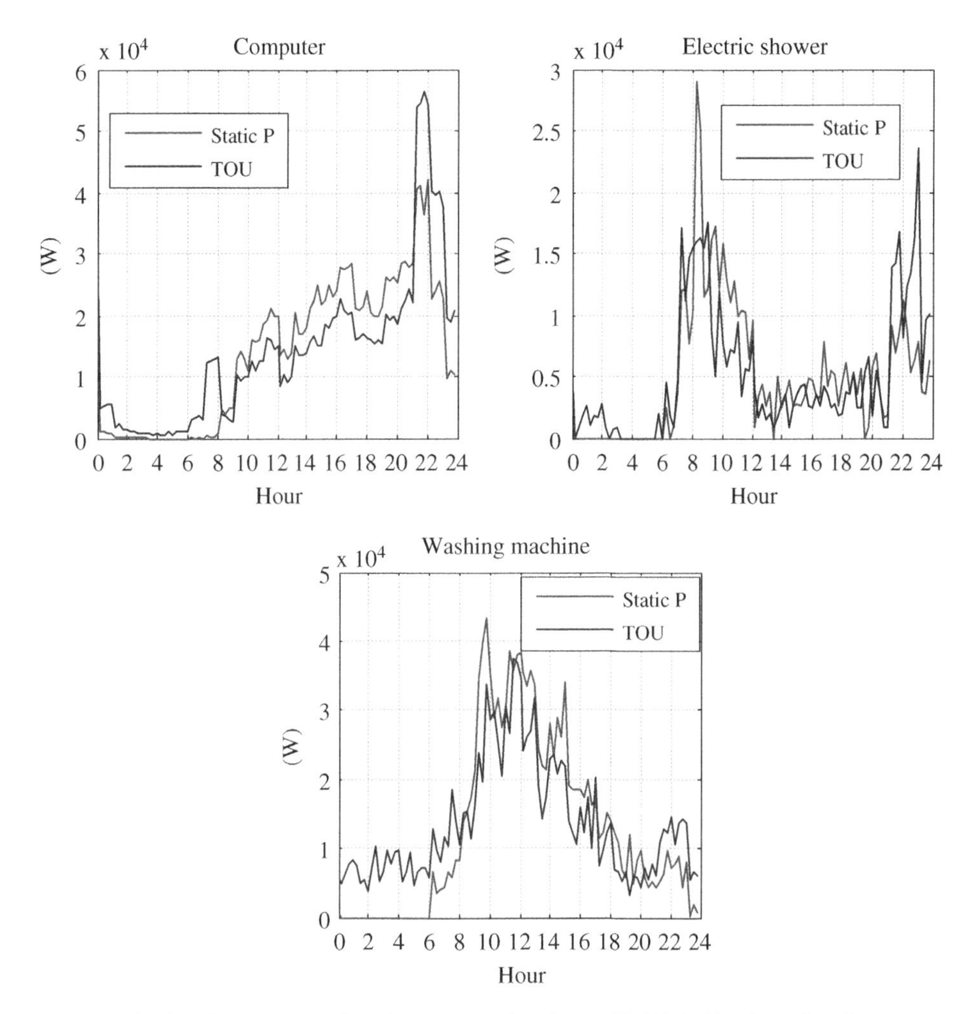

Figure 6.12 Load curve comparison between static price and TOU shifting boundary for computer, electric shower, and washing machine. *Source:* From Xu et al. [A]/with permission of Elsevier.

same. There are no incentives for load shifting. The third subperiod is from 15:00 to 21:59. Within this subperiod, only one hour (21:00–21:59) belongs to price-valley period. Therefore, other available will be shifted to this hour, forming a concentrated cooking consumption. The last subperiod is from 22:00 to 3:59. The whole subperiod belongs to price-valley period. Electricity price is kept the same and no incentives are given for load shifting. TV has a similar subperiod as cooking.

Load shifting of target consumer group is a comprehensive combination of all appliances' utilization shifting. Residents shift their daily using time of appliances to reduce consumption in peak-price period so as to reduce the bill. Table 6.7 reveals feature variation of appliances' load curve between static price and TOU. In Table 6.7, "Ori Peak" represents the average load variation between TOU and static price in period of 18:00–20:00; "Ori Valley" represents the average load variation in period of 3:00–5:00; "Evening Concentration" represents the average load variation in period of 21:00–21:59; "Morning Concentration" represents the average load variation in period of 7:00–7:59. Negative value in Table 6.7 means load of static price is lower than TOU and vice versa.

In Table 6.7, there is 30.3 kW potential reduction which is about 6.1% of original daily peak. This potential is mainly contributed by consumption shifting of cooking, TV, and computer. On the

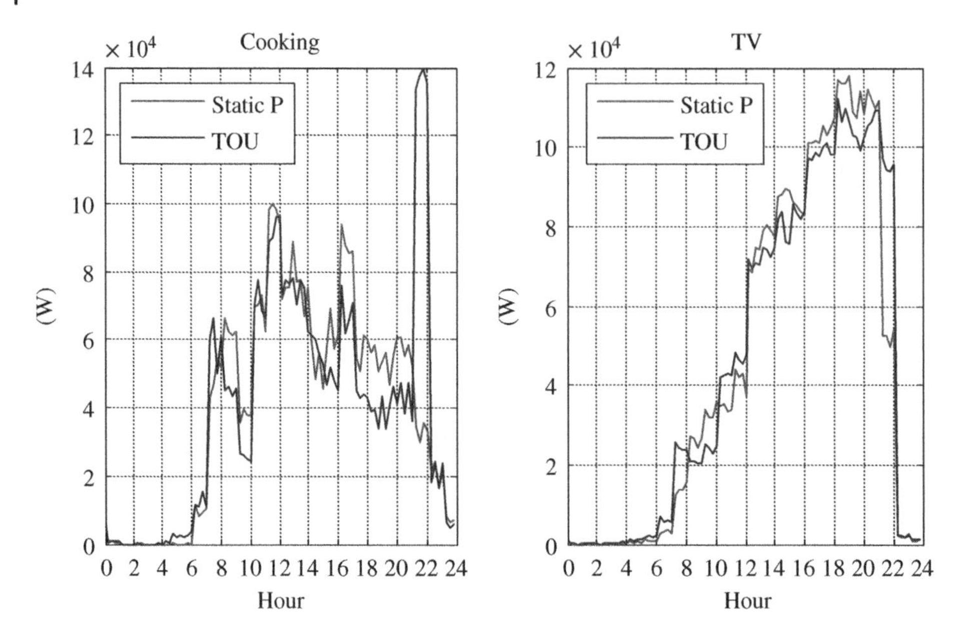

Figure 6.13 Load curve comparison between static price and TOU shifting boundary for cooking and TV. *Source:* From Xu et al. [A]/with permission of Elsevier.

Table 6.7 Load variation between TOU and static price.

Appliance	Ori peak (kW)	Ori valley (kW)	Evening concentration (kW)	Morning concentration (kW)
Computer	5.6	−0.6	−10.1	−9.9
Cooker	15.1	−0.5	−73.6	−8.4
Fridge	0	0	−0.1	0
Heater	0.3	0	0	0.5
Electric shower	0.3	−0.1	−4.8	−3.0
Light	−0.3	−0.1	−1.3	1.3
TV	7.4	−0.3	−32.0	−9.2
Washing machine	1.9	−7.7	−4.1	−7.4
Total	30.3	−9.2	−126.0	−36.1

Source: From Xu et al. [A]/with permission of Elsevier.

other hand, valley of daily load curve has an increasing potential of 9.2 kW which is about 1.8% of original daily peak. This potential is mainly contributed by dish washer and washing machine as they can be switched on without constraints from lifestyle. Due to cooking and TV watching having extra constraints on their consumption shifting, they cause risk of concentrated consumption in period of 21:00–21:59 and period of 7:00–7:59. This risk may generate a new daily peak.

By analyzing the load boundary, trend and risk of variation in daily load curve are revealed. Practically, effect of load shifting will not exceed this boundary because not all of consumer group can find the best transformation methods and some consumers may have their own limitations. Other than load characteristics, TOU in this case study can save residents' bill from 3.44 CNY/Day to 3.27 CNY/Day, which is 4.9% reduction. In shifting boundary analysis, daily power consumption under static price is 6.98 kWh/Day, which is the same as that under TOU.

6.4.3 Residential Shifting Boundary Simulation Under RTP

RTP is a dynamic pricing scheme that reflects the real-time pressure suffered on power systems. In this study, hourly price is proportional to the real-time power load of target group of consumers, which is shown in Eq. (6.3). The proportional rate between group load and price is 11.5/8000000. Same as study of TOU, 80% of families have installed Smart Meters to receive information feedback, and 30% of families have inclinations to change their behaviors for lower bills with the same daily requirements of all home appliances. So, considering 24% participation rate of target group, daily load curve of shifting boundary under RTP is shown in Figure 6.14:

From Figure 6.14, RTP has more potential for a smooth daily load curve than TOU and static price scheme. A better distribution of load shifting and less concentrated consumption can be achieved.

Power consumption can only be moved from period of higher price level to period of lower price level. In previous TOU analysis, large segment of lower price level is in the sleeping time. For those appliances with constraints of lifestyle, time zones that can accommodate their shifting consumption are small, thus concentrated consumption occurs. In the previous analysis, cooking appliances, computer, and TV are used intensively under TOU, which generate new peak in load curve. However, RTP can provide a much more dynamic price curve which can generate more prices for each hour. Therefore, consumers obtain more zones to accommodate their shifting consumption. The concentrated consumptions are decreased. Figure 6.15 reveals the daily load curve of appliances under RTP.

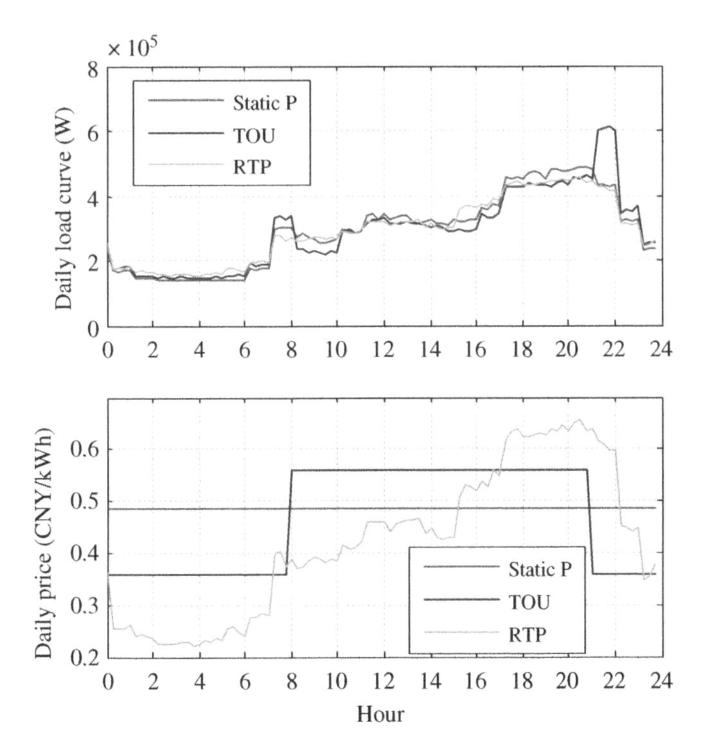

Figure 6.14 Daily load curve and price comparison among 3 PBPs. *Source:* From Xu et al. [A]/with permission of Elsevier.

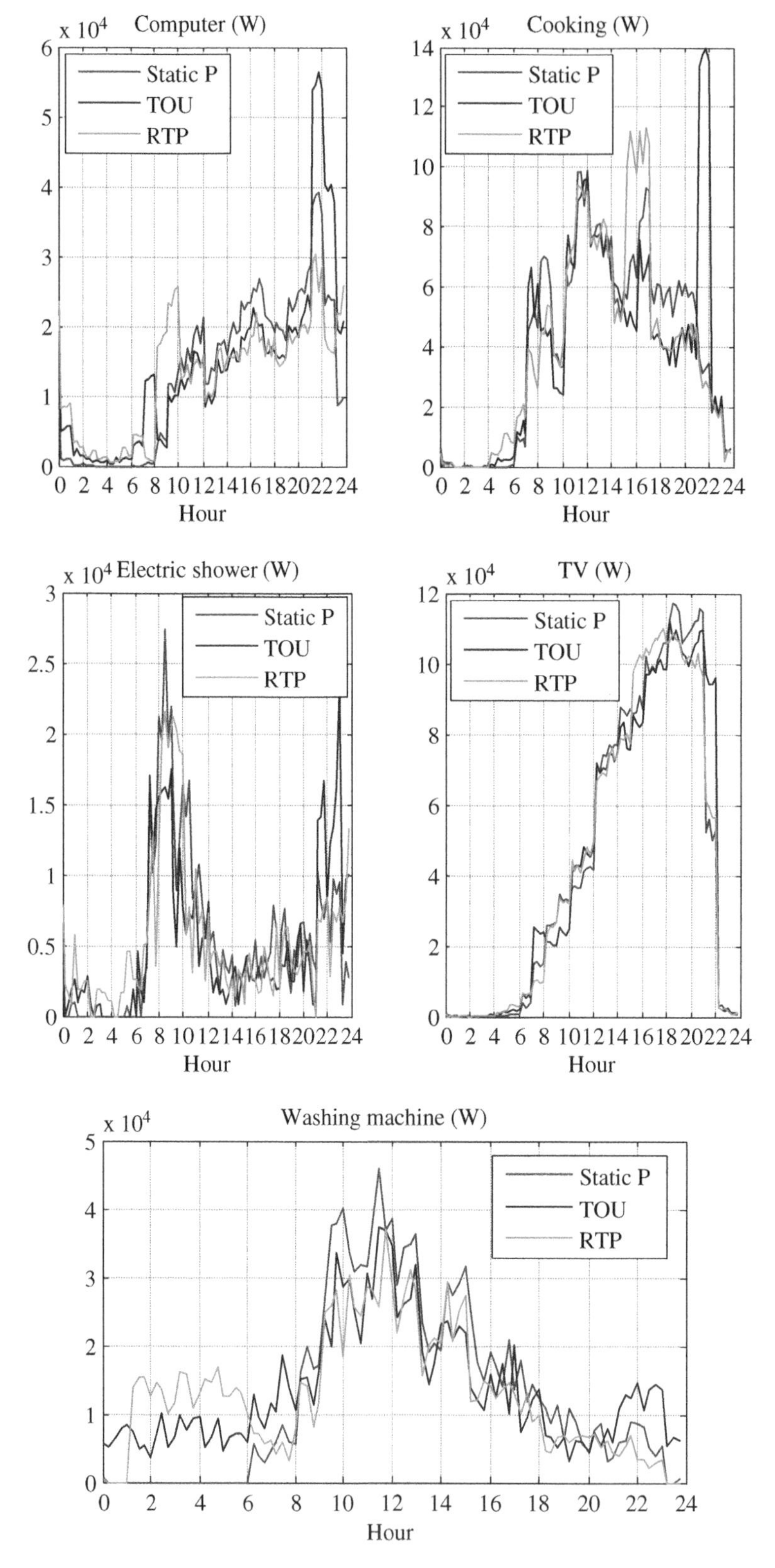

Figure 6.15 Daily load curve of appliances under different PBPs. *Source:* From Xu et al. [A]/with permission of Elsevier.

Under TOU tariff, obvious peak of concentrated consumption appears in load curves of computer, electric shower, and cooking. RTP provides more time zones for low price level so that consumers have more selections for shifting. It is clear that less concentrated consumptions occur and load curve of RTP boundary becomes much smoother than TOU case, this implies that load curve under RTP has less risk of generating a new peak.

The price from RTP program aims to reflect the real-time load pressure of power utility. Practically load is a time-varying variable with certain varying tendency and uncertainty. Price varying tendency reflects the actual tendency of general power requirement and is predictable. Price varying uncertainty reflects the small change of requirement by random factors and is unpredictable. In this model, varying tendency is represented by switched on probability of appliances. Appliances states are decided by random number generated on the probability. Varying uncertainty is generated during random number generation. Mapping to RTP program, varying tendency and uncertainty is passed to daily price curve. Figure 6.16 provides a 5-day price curve variation of RTP from simulation.

Unlike the stable change in TOU tariff, every daily price curve under RTP tariff has a similar varying tendency but with slightly uncertain differences. If standard deviation (STD) is used to analyze the daily differences on hourly price, the average STD of RTP price curve is 0.023 CNY/kWh, that is, 6% of the price uncertainty.

Price uncertainty decreases the satisfaction of consumers and increases the complexity of behavior changing. This impact reduces initiative of DR participation so that implementation of RTP may require more education for promotion. Also, functions to support automatic consumption scheduling will contribute to the result.

TOU and RTP are both dynamic pricing schemes that intend to optimize consumer behaviors so as to optimize daily load curve. Figure 6.17 reveals the statistic result of boundary load curve under TOU and RTP.

In Figure 6.17, daily peak of load curve, daily valley of load curve, peak-valley differences, and load between 18:00 and 20:00 are selected for comparison.

Under static price, daily peak of load curve appears between 18:00 and 20:00. In the simulation of TOU and RTP, a higher price level is given to this time period to promote consumer to shift consumption out. If load between 18:00 and 20:00 is recognized as 100%, the load will reduce to 93.5% under TOU or 94.0% under RTP. But TOU has an effect of concentrated consumption, so a new daily peak is generated between 21:00 and 21:59. This peak is 123.4% of daily peak under static price. In other words, daily peak under TOU increases instead of an obvious reduction. RTP does not have serious concentrated consumption so its daily peak can be reduced by about 6%.

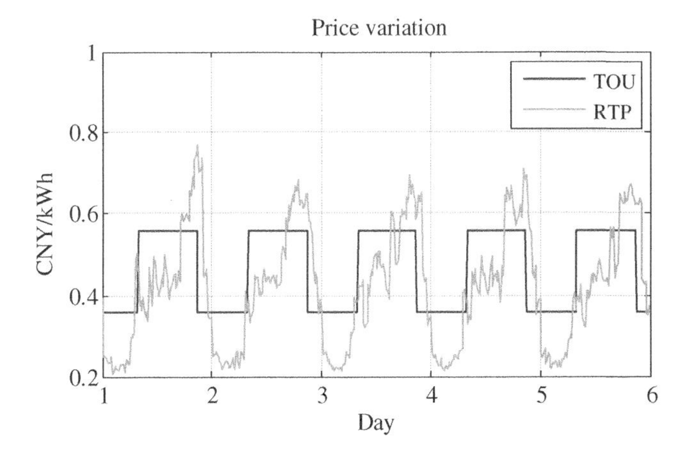

Figure 6.16 Five-day (weekdays) price curve variation comparison between RTP and TOU. *Source:* From Xu et al. [A]/with permission of Elsevier.

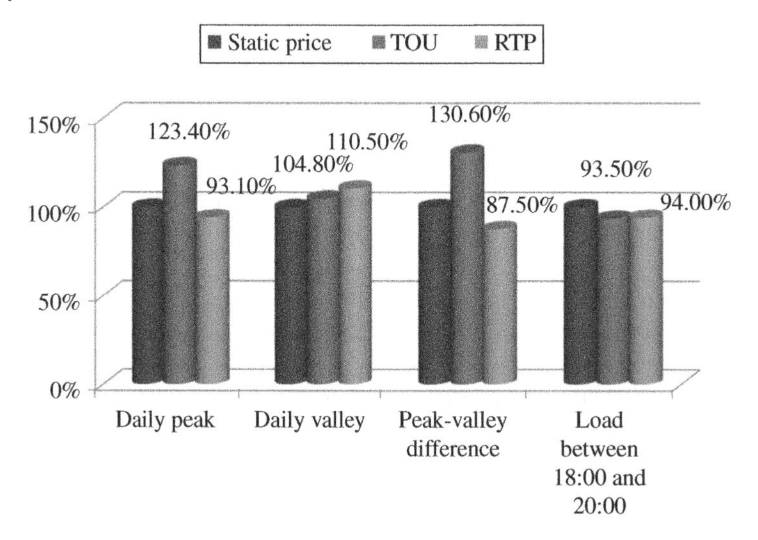

Figure 6.17 Boundary comparison between static price, TOU, and RTP. *Source:* From Xu et al. [A]/with permission of Elsevier.

Daily valley is another feature of power load. Price level at time zone of Daily Valley is set to lower level so that consumers can shift their consumption into this period. All daily valleys of three prices occur during 2:00–4:00 in the morning. If daily valley of static price is recognized as 100%, it will increase to 104.8% under TOU or 110.5% under RTP from simulation.

From Figure 6.17, TOU and RTP can alter daily peak and increase daily valley. If peak-valley difference of static price is recognized as 100%, it will increase to 130.6% under TOU or 87.5% under RTP. Reduction of daily peak and peak-valley difference may decrease the requirement of generation capacity and spinning reserve, which increase the energy efficiency of power system operations.

Dynamic pricing may change power bills from consumers. Figure 6.18 reveals variation of daily bill from those DR-participated consumers.

Daily bill of target consumer group in this case study under static price is 3.44 CNY/Day. By shifting consumer behaviors, residents can achieve 3.27 CNY/Day under TOU boundary with the same

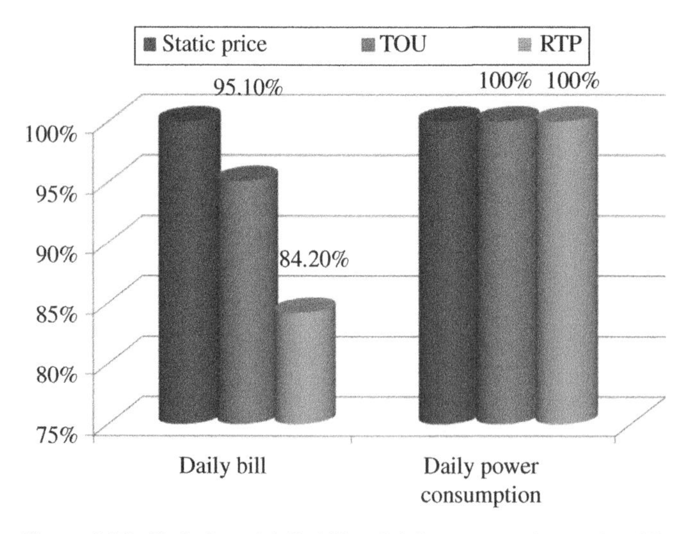

Figure 6.18 Variation of daily bill and daily consumption under different pricing schemes. *Source:* From Xu et al. [A]/with permission of Elsevier.

power consumption. Under RTP boundary, residents can achieve 2.89 CNY/Day with the same power consumption. From Figure 6.18, TOU and RTP in this case study can both reduce the daily bills and RTP has a stronger reduction effect.

6.5 Case Study on Residential Customer TOU Time Zone Planning

6.5.1 Case Study Description

From previous case study, shifting boundary of a TOU tariff can reveal the load shifting trend by generating a typical daily load curve of target consumer group as well as typical daily load curves of all appliances. In other words, TOU tariff can be evaluated by its shifting boundary. This case study intends to use shifting boundary to evaluate different TOU tariffs for supporting tariff planning.

TOU distributes different price levels into different time zone in a day. For example, TOU in Figure 6.8 contains two time zones. The one of higher price is from 8:00 to 21:00 in the day time, called price peak period. The other with lower price is from 21:00 to 8:00 in the night time, called price valley period. Time points for period substitution are 8:00 and 21:00. In fact, time zone of TOU may have other selections. The substitution time point in the morning can be 6:00, 7:00 or 9:00. Some selections are shown in Table 6.8:

Shifting Boundary is computed from each selection in Table 6.8 for a time zone planning with the best daily load curve. Target consumer group is the same as that in previous case study. Peak period starting time is varying from 6:00 to 10:00 and peak period ending time is varying from 19:00 to 23:00.

6.5.2 Result and Analysis

Simulations are implemented for each selection. Table 6.9 provides the indices of load curves under different TOU time zone planning. In Table 6.9, "Load between 7:00 and 8:00" measures the consumption level between 7:00 and 8:00, where morning peak occurs within this period. "Load between 18:00 and 20:00" measure power load between 18:00 and 20:00, where daily peak under static price occurs within this period. "Load between 21:00 and 22:00" measures the new peak of load generated by concentrated consumption.

Table 6.8 some TOU tariffs selections of time zone planning.

TOU time zone selection	Peak period starting time	Peak period ending time
1	6:00	20:00
2	6:00	21:00
3	6:00	22:00
4	7:00	20:00
5	7:00	21:00
6	7:00	22:00
7	8:00	20:00
8	8:00	21:00
9	8:00	22:00

Source: From Xu et al. [A]/with permission of Elsevier.

Table 6.9 Morning peak variation.

Peak period end time	19:00				20:00				
Peak period start time	Load between 7:00 and 8:00 (kW)	Load between 18:00 and 19:00 (kW)	Load between 21:00 and 22:00 (kW)	New evening peak appear hour	Peak period start time	Load between 7:00 and 8:00 (kW)	Load Between 18:00 and 19:00 (kW)	Load between 21:00 and 22:00 (kW)	New evening peak appear hour
6:00	260.7	466.7	534.2	19:00–21:00	6:00	274.3	445.9	582.7	20:00–21:00
7:00	290.1	483.9	553.4	19:00–21:00	7:00	266.5	436.1	568.3	20:00–21:00
8:00	300.8	466.7	532.7	19:00–21:00	8:00	292.1	429.0	539.4	20:00–21:00
9:00	286.3	463,7	533.5	19:00–21:00	9:00	288.4	438.3	542.4	20:00–21:00
10:00	280.4	471.0	529.4	19:00–21:00	10:00	269.9	428.4	534.7	20:00–21:00
Peak period end time	21:00				22:00				
6:00	262.9	450.9	580.9	21:00–22:00	6:00	267.0	468.1	465.1	19:00–21:00
7:00	266.3	444.5	571.0	21:00–22:00	7:00	270.8	462.0	459.9	19:00–21:00
8:00	296.3	436.0	566.9	21:00–22:00	8:00	303.6	457.7	451.3	19:00–21:00
9:00	278.5	436.1	552.2	21:00–22:00	9:00	298.9	475.3	473.9	19:00–21:00
10:00	270.7	440.3	549.4	21:00–22:00	10:00	286.0	469.6	468.4	19:00–21:00

Peak period end time	23:00								
6:00	253.4	466.1	462.1	19:00–21:00					
7:00	272.1	464.5	458.3	19:00–21:00					
8:00	313.6	454.8	448.9	19:00–21:00					
9:00	293.7	460.0	456.8	19:00–21:00					
10:00	283.1	463.2	462.5	19:00–21:00					

Source: From Xu et al. [A]/with permission of Elsevier.

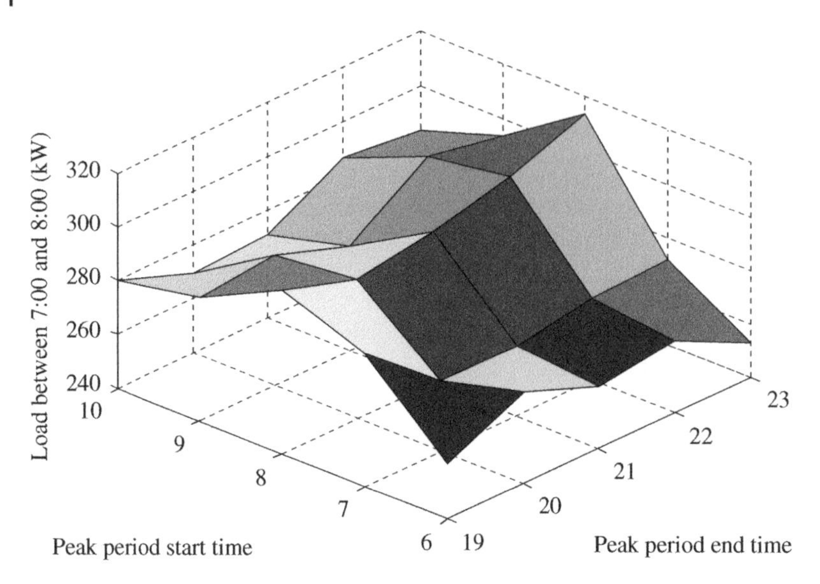

Figure 6.19 Daily morning peak variation. *Source:* From Xu et al. [A]/with permission of Elsevier.

Under static price, morning load peak appears between 7:00 and 8:00. Figure 6.19 reveals morning peak variation from Table 6.9.

From Figure 6.19, daily morning peak is mainly influenced by peak period start time. When peak period start time changes from 6:00 to 8:00, the morning peak will increase. If it changes from 8:00 to 10:00, the morning peak decreases.

When peak period start time is set at 6:00 or 7:00, period between 7:00 and 8:00 is with high price level. Consumption is shifted out from this time period then the morning peak is comparatively low.

When peak period start time is set at 8:00, period between 7:00 and 8:00 has an hour with low price level. Power from other periods with high price is shifted into this period. For those appliances, cooking, and TV, constrained by Eq. (6.8), the hours with low price level in morning subperiod are 4:00–5:00, 5:00–6:00, 6:00–7:00, and 7:00–8:00. Except 7:00–8:00, all other hours in the morning subperiod are with the high sleeping rate. As cooking and TV should be switched on while awake, 7:00–8:00 become the only time to suffer most of the shifting load from morning subperiod, thus a concentrated consumption occurs in this hour, and increases the morning peak.

When peak period start time is set at 9:00 or 10:00, more hours in morning subperiod are with low price level while people awake. More time space to suffer the shifting load and consumption can be distributed into more hours. Therefore, the morning power peak is not as high as 8:00. Figure 6.19 provides a daily morning peak comparison among different peak period start time.

Under static price, daily evening peak appears between 18:00 and 19:00. By analyzing previous case study, concentrated consumption may occur and generate new peak load. Figure 6.20 reveals the variation of original daily evening peak and new evening peak generated by concentration consumption.

From Figure 6.21, peak period end time mainly influences daily evening peak and load between 21:00 and 22:00. Comparing to use 19:00 as peak period end time, using 20:00 or 21:00 may achieve a lower daily peak. This is because the whole 18:00–20:00 will be fully covered with high electricity price. Also, appliances constrained by Eq. (6.8) still have time space with low price in evening subperiod to suffer the load shift. The consumption is shifted out from 18:00 to 20:00 when use 20:00 or

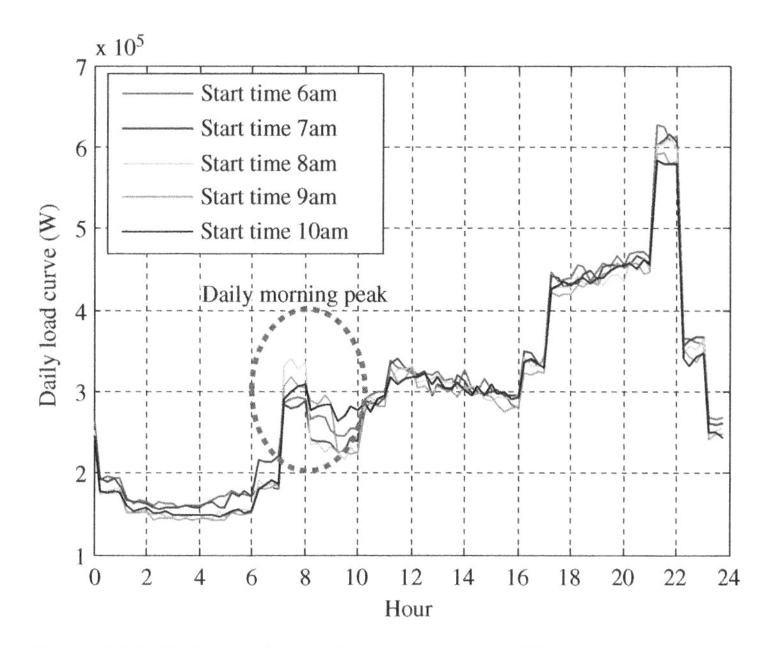

Figure 6.20 Daily morning peak compare among different peak period start time (peak period end time is 21:00). *Source:* From Xu et al. [A]/with permission of Elsevier.

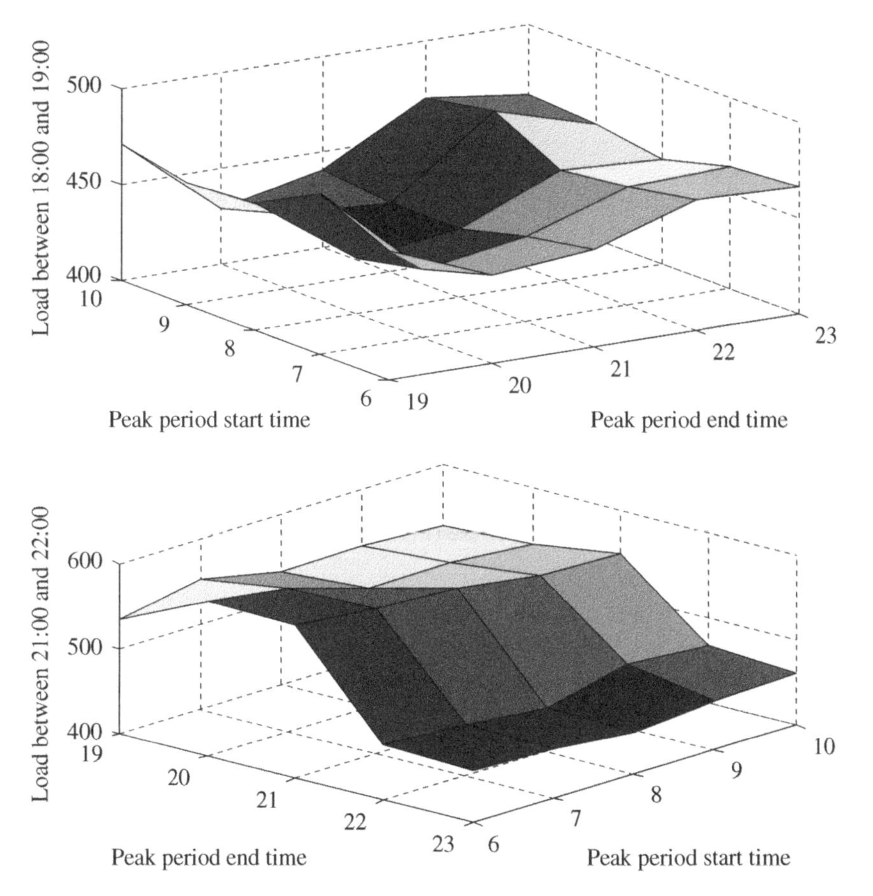

Figure 6.21 Variation of original daily evening peak and new peak generated by concentrated consumption (peak period start time is 8:00). *Source:* From Xu et al. [A]/with permission of Elsevier.

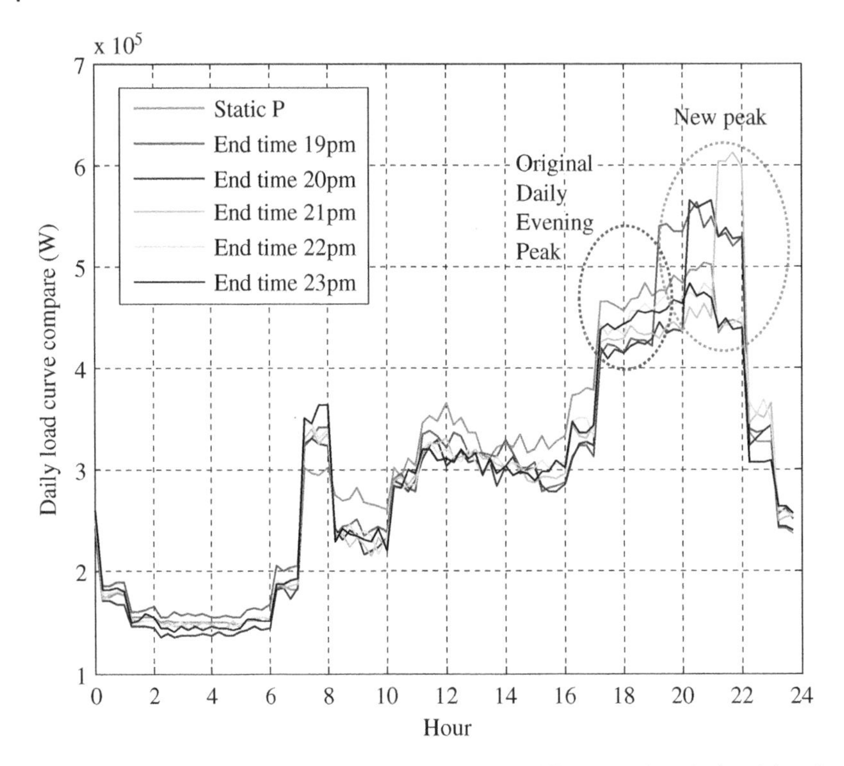

Figure 6.22 Daily morning peak comparison among different peak period end time (peak period start time is 8:00). *Source:* From Xu et al. [A]/with permission of Elsevier.

21:00 as peak period end time. The shifted consumption generates a new peak at different hour. The shifting load under TOU will lead to risk of concentration consumption.

When using 22:00 or 23:00 as peak period end time, the whole evening subperiod is covered by high price level. Time zones after 23:00 are associated with high sleeping rate and cannot suffer shifted load. Thus, load at daily evening peak has limited transformation to other time zone and the daily peak appears at the same period as original peak. Figure 6.22 provides a daily evening peak comparison among different peak period start time.

6.6 Case Study on Smart Meter Installation Scale Analysis

6.6.1 Case Study Description

In the case study, participation rate of a tariff is constructed by smart meter installation scale and consumer scale which have an interest for lower bills. The smart meter installation scale is the main impact factor of PBP consumer participation. This case study intends to use shifting boundary to evaluate the impact from smart meter installation scale. Assuming 30% consumers have the interest for lower bills in the target group. Table 6.10 provides relations between scales of smart meter installation and consumer participation rate.

Table 6.10 Relation between smart meter scale and consumer participation rate.

Scenario	Smart meter scale (%)	Consumer participation willingness (%)	Consumer participation rate (%)
1	10	30	3
2	20	30	6
3	30	30	9
4	40	30	12
5	50	30	15
6	60	30	18
7	70	30	21
8	80	30	24
9	90	30	27
10	100	30	30

Source: From Xu et al. [A]/with permission of Elsevier.

6.6.2 Analysis on Multiple Smart Meter Installation Scale under TOU and RTP

Daily load curve is influenced by smart meter installation scale under various PBP. Only those whose pricing information is easily available can take part in behavior changing. To evaluate the load curve variation trend, shifting boundary is computed under each scenario of Table 6.10. Target consumer group is the same as before. TOU tariff in Figure 6.9 is selected for simulation. Figures 6.23 and 6.24 reveal shifting boundary of various scenarios.

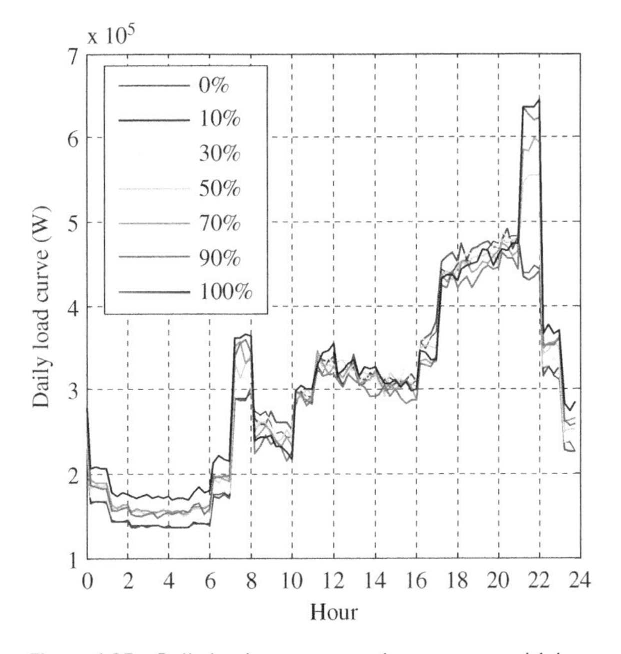

Figure 6.23 Daily load curve comparison among multiple smart meter installation scale under TOU. *Source:* From Xu et al. [A]/with permission of Elsevier.

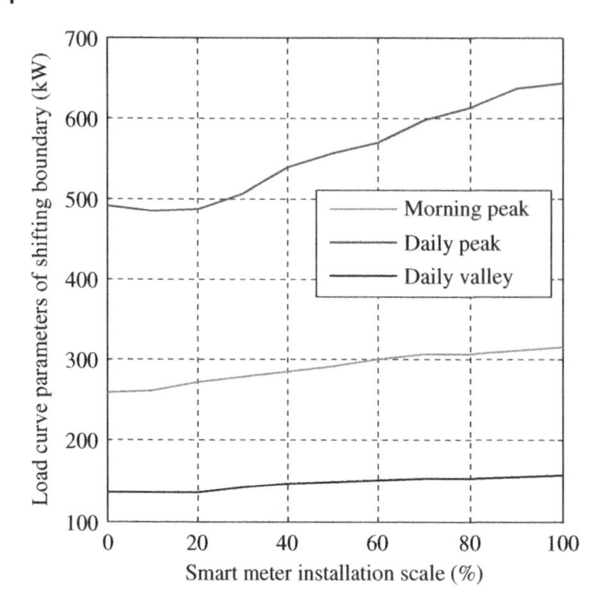

Figure 6.24 Comparison between morning peak, daily peak, and daily valley among multiple smart meter installation scale under TOU. *Source:* From Xu et al. [A]/with permission of Elsevier.

From Figures 6.23 and 6.24, more consumers decide to change their behaviors when smart meter installation scale increases. The load shifting within daily load curve tends to be more obvious. The concentrated consumption becomes much clearer as well. But limited by the willingness to participate financially, when all consumers have installed smart meter, only 30% will join TOU in this case study. In other words, the ultimate effect of smart meter under TOU is shown in the black curve in Figure 6.23. Similar results for RTP simulation are given in Figures 6.25 and 6.26.

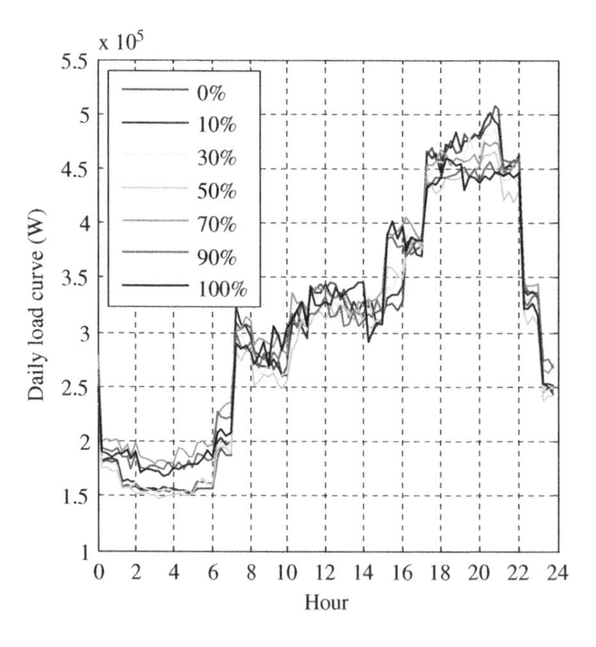

Figure 6.25 Daily load curve comparison among multiple smart meter installation scale under RTP. *Source:* From Xu et al. [A]/with permission of Elsevier.

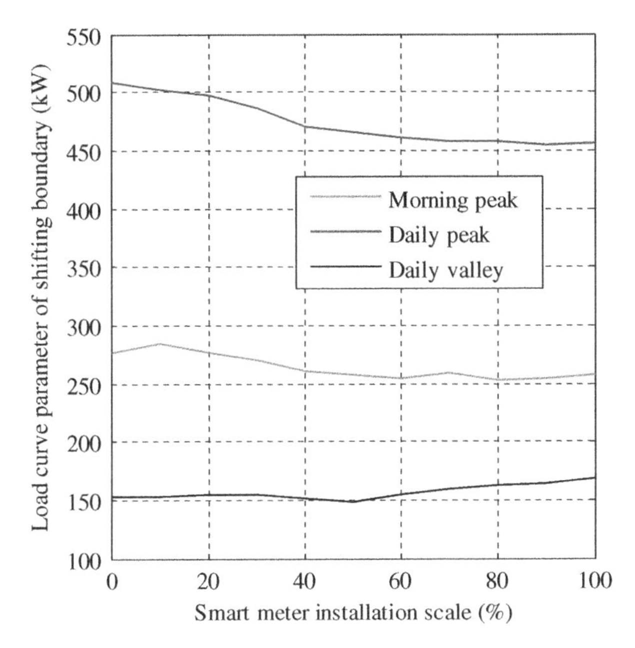

Figure 6.26 Comparison between morning peak, daily peak, and daily valley among multiple smart meter installation scale under RTP. *Source:* From Xu et al. [A]/with permission of Elsevier.

6.7 Conclusions and Future Work

This chapter introduces a model framework for consumer behavior PBPs analysis. This framework reveals the closed-loop relationship between individual consumer behaviors and group daily load. Based on this framework, this chapter also provides a quantification method of usage transformation other than usage reduction. Three case studies are implemented to indicate the model utilization in load analysis, PBPs evaluation and optimization, and evaluation of the effect from smart meter installation scale.

However, further work is anticipated. This chapter only reports a quantification method of usage transformation. Practically, DR effect is the combination between usage reduction and usage transformation. Therefore, the combination effect will be studied in the future. Behavior of industrial and commercial consumers should have a different pattern. An extension of the model framework to cover industrial and commercial consumers would be useful. With the framework extension, the impact of PBPs toward a complete consumer group could be better demonstrated.

Acknowledgements

[A] Xu, F.Y., Zhang, T., Lai, L.L., and Zhou, H. (2015). Shifting boundary for price-based residential demand response and applications. *Applied Energy*, Elsevier 146: 353–370.

References

1 US Department of Energy (2006). Benefits of Demand Response in Electricity Markets and Recommendations for Achieving Them. A report to the United States Congress Pursuant to Section 1252 of the Energy Policy Act of 2005. https://www.energy.gov/sites/prod/files/oeprod/DocumentsandMedia/DOE_Benefits_of_Demand_Response_in_Electricity_Markets_and_Recommendations_for_Achieving_Them_Report_to_Congress.pdf (accessed 9 November 2020).
2 Albadi, M.H. and El-Saadany, E.F. (2007). Demand response in electricity markets: an overview. *Proceedings of IEEE Power Engineering Society General Meeting*, Tampa, Florida, USA (24–28 June 2007).
3 Strbac, G. (2008). Demand side management: benefits and challenges. *Energy Policy* 36: 4419–4426.
4 Ueno, T., Sano, F., Saeki, O., and Tsuji, K. (2006). Effectiveness of an energy-consumption information system on energy savings in residential houses based on monitored data. *Applied Energy* 83: 166–183.
5 Zakariazadeh, A., Homaee, O., Jadid, S., and Siano, P. (2014). A new approach for real time voltage control using demand response in an automated distribution system. *Applied Energy* 117: 157–166.
6 Lai, C.S., Xu, F., McCulloch, M., and Lai, L.L. (2016). Application of distributed intelligence to industrial demand response. In: *Smart Grid Handbook* (ed. C.-C. Liu, S. McArthur and S.-J. Lee). IEEE Press & Wiley.
7 Xu, F.Y., Wang, X., Lai, L.L., and Lai, C.S. (2014). Agent-based modeling and neural network for residential customer demand response. *Proceedings of the International Conference on Systems, Man and Cybernetics*, IEEE, UK, Manchester, UK (13–16 October 2013).
8 Parsa Moghaddam, M., Abdollahi, A., and Rashidinejad, M. (2011). Flexible demand response programs modeling in competitive electricity markets. *Applied Energy* 88: 3257–3269.
9 Wang, F., Xu, H., Xu, T. et al. (2017). The values of market-based demand response on improving power system reliability under extreme circumstances. *Applied Energy* 193: 220–231.
10 Venkatesan, N., Solanki, J., and Solanki, S.K. (2012). Residential demand response model and impact on voltage profile and losses of an electric distribution network. *Applied Energy* 96: 84–91.
11 Shao, S., Pipattanasomporn, M., and Rahman, S. (2013). Development of physical-based demand response-enabled residential load models. *IEEE Transactions on Power System* 28: 607–614.
12 Ghorbani, M.J., Rad, M.S., Mokhtari, H., and Honarmand, M.E. (2011). Residential load modeling by norton equivalent model of household loads. *2011 Asia-Pacific Power and Energy Engineering Conference*, Wuhan, China (25–28 March 2011).
13 US Electric Power Research Institute (2009). Residential Electricity Use Feedback: A Research Synthesis and Economic Framework. Report, 126 pages.
14 Sharifi, R., Fathi, S.H., and Vahidinasab, V. (2016). Customer baseline load models for residential sector in a smart-grid environment. *Energy Reports* 2: 74–81.
15 Grandjean, A., Adnot, J., and Binet, G. (2012). A review and an analysis of the residential electric load curve models. *Renewable and Sustainable Energy Reviews* 16 (9): 6539–6565.
16 Collin, A.J., Tsagarakis, G., Kiprakis, A.E., and McLaughlin, S. (2014). Development of low-voltage load models for the residential load sector. *IEEE Transactions on Power Systems* 29 (5): 2180–2188.
17 Meng, F.-L., Zeng, X.-J., and Ma, Q. Learning customer behaviour under real-time pricing in the smart grid. *2013 IEEE International Conference on Systems, Man, and Cybernetics*, October 2013, Manchester, UK, 3186–3191.
18 Mohsenian-Rad, A.-H., Wong, V.W.S., Jatskevich, J. et al. (2010). Autonomous demand-side management based on game-theoretic energy consumption scheduling for the future smart grid. *IEEE Transactions on Smart Grid* 1: 320–331.

19 Lai, L.L., Motshegwa, T., Subasinghe, H. et al. (2002). Feasibility study with agent on energy trading. *Proceedings of the Fifth International Conference on Advances in Power System Control, Operation and Management*, Vol. 2, IEE, Publication Number CP478 (30 October to 1 November 2000), 505–510.

20 Richard, J.C. (2018).The smart grid: status and outlook. Congressional Research Service Report. https://fas.org/sgp/crs/misc/R45156.pdf (accessed 9 November 2020).

7

Residential PV Panels Planning-Based Game-Theoretic Method

Nomenclature

Indices and Sets

j	Index of nodes/lines/agents
k/i	Index of prosumer agents/consumer agents
t	Index of time periods
w	Index of scenarios
$\mathcal{N}/\mathcal{L}/\mathcal{T}/\mathcal{W}$	Set of nodes/lines/time periods/scenarios
$\mathcal{C}_j$	Set of child nodes of node j
$\mathcal{N}^{PV}/\mathcal{N}^{C}$	Set of prosumer agents/consumer agents

Variables

P_{jtw}/Q_{jtw}	Active/reactive power flow on distribution line j at time t in scenario w
v_{jtw}	Squared voltage magnitude on node j at t, w
l_{jtw}	Squared line current magnitude on line j at t, w
p_{jtw}/q_{jtw}	Active/reactive power injection on node j at j, t, w
P_{itw}^{PVc}	PV electricity sold to the consumer agent i at t, w
P_{ktw}^{PVg}	PV electricity sold by the prosumer agent k at t, w
$P_{ktw}^{Curtail}$	PV energy curtailment on prosumer agent node k at t, w
P_{jtw}^{G}	Electricity purchased from the utility grid by agent j at t, w
c_{tw}^{U}	Uniform price applied to all consumer agents at t, w
u_{k}^{PV}	Number of PV panels purchased by prosumer agent k
ϕ^{PV}	Total installation number of PV panels
λ_{itw}^{L}	Dual variable of supply-demand balance (6g)
$\mu_{itw}^{PVc}/\mu_{itw}^{G}$	Dual variable of inequality constraint (6h)/(6i)
U_{itw}^{C}	The utility function of consumer agent i at t, w
Rev^{C}	Revenue of the coalition formed by all prosumer agents

Smart Energy for Transportation and Health in a Smart City, First Edition. Chun Sing Lai, Loi Lei Lai and Qi Hong Lai.

Parameters

c^{G}_{tw}/c^{MCP}_{tw}	Electricity price /wholesale market clearing price at t, w
$c^{PV}_{inv}/c^{PV}_{o\&m}$	Investment cost/operation and maintenance cost of each PV panel
c^{E}	Greenhouse gas emission cost
r^{C}	The proportion of the incomes of sharing PV electricity to consumer agents via the energy sharing platform
w^{E}_{i}	Individual weight factor for greenhouse gas emissions reduction of consumer agent i
c^{WTP}_{itw}	Willingness-to-pay of consumer agent i at t, w
p_{w}	Probability of uncertainty scenario w
$\gamma^{PV}_{tw}/\gamma^{L}_{tw}$	PV energy output factor/load factor at t, w
P^{L}_{jtw}/Q^{L}_{jtw}	Active/reactive load on node j at t, w
E^{PV}_{size}	Size of each PV panel
E^{PVmax}	PV hosting capacity of the distribution grid
u^{PVmax}_{k}	Maximum allowable installation number of PV panels on node prosumer agent node k
v^{min}/v^{max}	Squared lower/upper bound of voltage magnitude
l^{max}_{j}	Squared upper bound of current magnitude on distribution line j
η^{PV}	Efficiency of PV panels output
r_j/x_j	Resistance/reactance of distribution line j

7.1 Introduction

RESIDENTIAL photovoltaic (PV) panels-based distributed renewable generation becomes a promising alternative power generation technology to reduce greenhouse gas emissions and promote a low-carbon lifestyle. Energy policies introduced by many countries encourage end-users to install self-consumed PV systems on their rooftops [1]. As a consequence, PV prosumers have emerged in large numbers [2], which are end-use consumers who can also act as energy providers if they install residential PV panels. In practice, PV prosumers can behave as sellers or buyers according to their net power profiles as well as electricity price when they participate in the energy sharing process. However, the proliferation of PV generators leads to the oversupply of PV energy in some areas, thus, PV curtailment may be needed when all consumers are fully supplied with local PV energy and meanwhile the wholesale market clearing prices are negative. For PV prosumers, they try to maximize the revenue from PV energy sharing to offset the investment, operation, and maintenance cost of PV panels. As for the normal consumers without renewable sources, they are interested in reducing their electricity bills and some of them also concern the potential valuation of environmental benefits. Therefore, the energy sharing between PV prosumers and their nearby consumers becomes an effective approach to improve the local PV generation consumption and even reduce negative impacts caused by PV energy integration into the upstream electrical network. In this regard, from the perspective of the residential prosumers, this chapter aims to optimally plan residential PV panels for maximizing the revenue of PV prosumers in a distribution grid via energy sharing.

The idea of energy sharing has drawn attention recently. Wang et al. [3] reports a bi-level transactive energy trading (TEM) framework is proposed to improve the energy scheduling and operation efficiency for multi-carrier energy systems which are modeled as energy hubs. TEM introduces new challenges for distribution system operators (DSOs) since the unconstrained TE trading between energy hubs (EHs) connected with distribution system could change the status of power systems, increase power losses, and lead to congestion problems. As microgrids (MGs) interact with power distribution systems, the transactive energy (TE) trading among MGs poses new challenges on operation and trading functions of DSOs. Wang et al. [4] proposed a bi-level programming framework. The framework enables distributed TE trading decisions at the multi-MG level and effective market clearing determination by coordinating MGs with DSO. Cui et al. [5] proposes a two-stage peer-to-peer energy sharing model for an energy building cluster. Zhao et al. [6] propose a double-layer framework of energy transactions based on blockchain in multi-microgrids to provide decentralized trading, information transparency and mutual trust system of each node in the trading market. In Mediwaththe et al. [7], a novel energy sharing strategy is presented for demand-side management in a neighborhood area network (NAN), which is comprised of energy providers, users, and an electricity retailer. Li et al. [8] presents a novel day-ahead power market for distribution systems. Based on the linearized AC power flow model, the distribution locational marginal price for coupled active and reactive power can be calculated. The energy hub at different nodes can trade with each other and optimize their profit based on distribution locational marginal prices. Game theory is applied to solve the energy trading payment problem. Cui et al. [9] develops a risk aversion energy sharing model based on a devised local energy market for addressing the issues caused by uncertain renewable sources from the perspective of community prosumers. In Liu et al. [10], a hybrid energy sharing framework of multiple distribution grids is studied with consideration of combined heat and power and demand response. Fleischhacker et al. [11] proposes a building welfare maximization model as well as a game-theoretical energy sharing model to investigate the effect of sharing PV electricity in an apartment building.

For optimal residential PV panels planning, Senemar et al. [12] proposes a probabilistic dynamic allocation model to optimally determine the capacity of on-site PV generation for a residential energy hub, aiming to minimize the PV energy curtailment as well as consumer cost. In Zhang and Grijalva [13], a data-driven-based method is proposed for the detection, verification, and estimation of residential PV system installations. Authors of Ghiassi-Farrokhfal et al. [14] endeavor to find the optimal installation capacity of PV generation by splitting the budget between PV generation and energy storage for maximizing the revenue from participating in the electricity market. Alhaider and Fan [15] determines the energy storage size and PV panels number for a commercial building considering heating ventilation and air conditioning systems. Li et al. [16] proposes an optimal two-stage placement method for the heterogeneous distributed generators, including PV generators, in a grid-tied multi-energy microgrid with the consideration of the uncertainties from the renewable energy resources. In Cáceres et al. [17], based on a performance indicator and different parameters of the PV system, an economic study is addressed for residential PV panels installation in Santiago, Chile. Yoza et al. [18] presents an expansion planning model of PV and battery systems for the smart house, which considers investment cost, selling price and purchasing price. However, these works only focus on energy sharing or residential PV panels planning. Up to now, the research on integrating energy sharing mechanisms with residential PV panels planning is still at a very early stage.

This chapter develops a novel two-stage game-theoretic framework for residential PV panels planning. In the first stage, Stackelberg game theory is used to model the stochastic bi-level energy sharing problem, which is solved by a proposed descend search algorithm. In the second stage, we develop a stochastic programming-based optimal power flow (OPF) model to optimally allocate residential PV panels for all PV prosumers with minimum expected active power loss. The main contributions of this chapter are threefold,

1) Different from most works related to residential PV panels planning, we innovatively integrate a residential PV panels planning model with the energy sharing mechanism. To our best acknowledge, this has not been studied before. In this way, we can improve economic benefits to PV prosumers and meanwhile facilitate the installation of residential PV panels, which has practical significance.
2) Instead of directly solving the proposed bi-level energy sharing problem by using commercial solvers with Mathematical Program with Equilibrium Constraints (MPEC), we develop an efficient descend search algorithm that can significantly enhance computation efficiency. Moreover, in our proposed solution method, the feasibility of voltage constraint and current constraint is checked to ensure the reliability and security of the distribution grid operation during the energy sharing process. Thus, the proposed solution method can address both economic and operating security concerns.
3) To address the conflict of interests between PV prosumers and normal consumers, the leader–followers Stackelberg game theory is used to model a bi-level PV energy sharing problem. Besides, the uncertainties of PV energy output, load demand, as well as electricity price are simultaneously considered in this energy sharing model.

The rest of this chapter is organized as follows. In Section 7.2, we introduce the system model, including the branch flow model, consumer agents and the coalition of PV prosumer agents. Section 7.3 describes a bi-level energy sharing problem and an efficient search algorithm-based solution method. Section 7.4 presents the optimal PV panels allocation model with the goal of active power loss minimization. Numerical results are given in Section 7.5. Finally, we conclude this chapter in Section 7.6.

7.2 System Modeling

7.2.1 Network Branch Flow Model

In this chapter, the branch flow model (BFM) [19] is used to describe the complex power flow equations of the radial electrical network. Consider a radial distribution grid $\mathcal{M} := (\mathcal{N}, \mathcal{L})$, where $\mathcal{N} := \{0, 1, \ldots, \mathcal{N}\}$ denotes the node set and $\mathcal{L} := \{0, 1, \ldots, \mathcal{L}\}$ denotes the directed line set. Except for the substation node, each node j has a unique ancestor node m and a set of child nodes $\mathcal{C}_j$. We assume the direction of the line connect node j and its ancestor node m is from m to j. Note that the node n is in the set of the child nodes $\mathcal{C}_j$ of the node j. Therefore, the BFM is given as follows:

$$P_{mj} - r_{mj} l_{mj} + p_j = \sum_{n \in \mathcal{C}_j} P_{jn} \quad j \in \mathcal{N} \tag{7.1a}$$

$$Q_{mj} - x_{mj} l_{mj} + q_j = \sum_{n \in \mathcal{C}_j} Q_{jn} \quad j \in \mathcal{N} \tag{7.1b}$$

$$v_j - v_m = 2(r_{mj} P_{mj} + x_{mj} Q_{mj}) - \left(r_{mj}^2 + x_{mj}^2\right) l_{mj} \quad j \in \mathcal{N} \tag{7.1c}$$

$$l_{mj} = \frac{P_{mj}^2 + Q_{mj}^2}{v_j} \quad j \in \mathcal{N} \tag{7.1d}$$

where the active and reactive power balances at each node are described by (7.1a) and (7.1b), respectively; P_{mj} and Q_{mj} represent the active and reactive power flows of line mj (from the ancestor node m to the node j), respectively; p_j and q_j denote the active and reactive power injection (+) or extraction (−) at the node j, respectively; r_{mj} and x_{mj} represent the resistance and reactance of line mj, respectively; l_{mj} is the squared line current magnitude on distribution line mj. The voltage drop/rise on each line can be described by (7.1c). Equation (7.1d) denotes the relationship between the power flow, squared bus voltage magnitude and squared line current magnitude, which is expressed by nonconvex. To convexify (7.1d), it is relaxed into an inequality as (7.2a) and then reformulated into a second-order cone constraint as (7.2b) by the second-order cone programming (SOCP) relaxation method [20].

$$l_{mj} \geq \frac{P_{mj}^2 + Q_{mj}^2}{v_j} \quad j \in \mathcal{N} \tag{7.2a}$$

$$\left\| \left(2P_{mj},\ 2Q_{mj},\ v_j - l_{mj}\right) \right\|_2 \leq v_j + l_{mj} \quad j \in \mathcal{N} \tag{7.2b}$$

It has been proved in Farivar and Low [19] that the relaxation is exact as long as the network is radial and the objective function of the OPF problem is strictly increasing in l_{mj}. In our model, the equality in (7.2b) also holds since we take the network loss into account and it is a strictly increasing function of l_{mj}.

7.2.2 Energy Sharing Agent Model

This chapter proposes a two-stage residential PV panels planning model, where the first stage is to determine the optimal sizing of PV panels by developing a Stackelberg game-based stochastic bi-level energy sharing model. The proposed energy sharing model is performed during the decision-making process of finding the total PV panel installation capacity. Besides, in the practical operation of the distribution system, our proposed bi-level energy sharing model can be used for trading energy among the coalition, consumers and the utility grid.

For better energy sharing coordination, we introduce two kinds of agents, i.e. consumer agent and prosumer agent. Figure 7.1 shows the proposed energy sharing framework between the coalition and consumer agents. Specifically, the consumer agent denotes the aggregation of customers on the same node and the prosumer agent represents the aggregation of PV prosumers (owners of

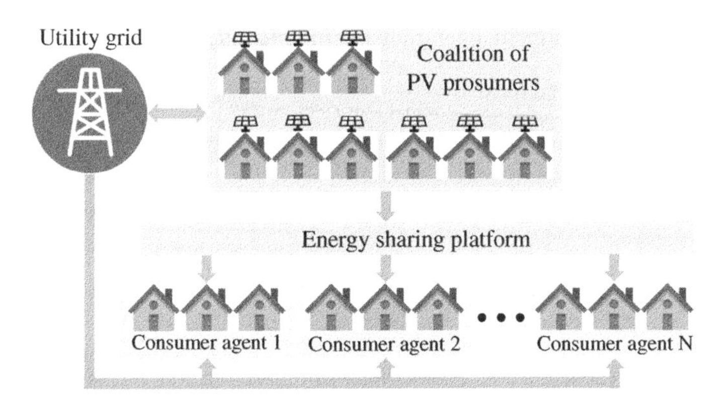

Figure 7.1 The proposed energy sharing framework between the coalition and consumer agents. *Source:* From Xu et al. [A].

PV panels) on the same node. On behalf of its local consumers/prosumers, each agent is allowed to participate in the energy sharing process.

1) Revenue for the coalition of prosumer agents: In our proposed model, all prosumer agents form a coalition to maximize their common benefits by operating their residential PV panels. This assumption is based on the fact that most individual PV prosumers can only provide a small amount of PV energy, posing some challenges to energy management in power girds. Besides, due to the barriers to renewable energy integration [21–24], especially market entry [25], it is efficient and easy to manage the integrated PV energy. Note that the demands of the coalition are satisfied by the self-generated PV energy firstly, then the surplus PV energy is used for energy sharing with consumers. Therefore, before the coalition participates in the energy sharing process, PV energy trading among the prosumers in the coalition has been finished.

The coalition can be a seller or a buyer according to its loads and PV generations. When the coalition becomes a seller, there is a competitive relationship between the coalition and the DSO because they both try to sell electricity to the local consumers. However, when the netload of the coalition is positive, there is a dependency relationship between the coalition and the DSO since the coalition needs to purchase electricity from the utility grid via the DSO. The coalition sells its surplus PV electricity differently to the consumer agents and the utility grid to maximize its revenue. During the energy sharing with local consumer agents, the coalition needs to pay the bill to the energy sharing platform which acts as an intermediary agent to facilitate energy sharing. Therefore, the revenue of the coalition is defined as follows:

$$
\begin{aligned}
Rev^C = & \sum_{i \in \mathcal{N}^C} \sum_{t \in \mathcal{T}} (1 - r^C) c_t^U P_{it}^{PVc} + \sum_{k \in \mathcal{N}^{PV}} \sum_{t \in \mathcal{T}} c_t^{MCP} P_{kt}^{PVg} \\
& - \sum_{k \in \mathcal{N}^{PV}} \sum_{t \in \mathcal{T}} c_t^G \left[\gamma_t^L P_k^L - \eta^{PV} \gamma_t^{PV} \phi^{PV} E_{size}^{PV} \right]^+ \\
& - \left(\alpha c_{inv}^{PV} + c_{o\&m}^{PV} \right) \phi^{PV}
\end{aligned}
\tag{7.3}
$$

where the first term represents the income of selling PV electricity P_{it}^{PVc} to the consumer agents with the uniform price c_t^U. Note that the energy sharing platform charges the coalition intermediary fee in terms of the proportion r^C of this income. The second term is the gain of selling PV electricity P_{it}^{PVg} to the utility grid with the wholesale market clearing price c_t^{MCP}. The third term denotes that the coalition needs to buy electricity from the utility grid when its demand cannot be satisfied by its PV generation. Note that $[\cdot]^+$ represents the projection operator onto the nonnegative orthant, i.e. $[x]^+ = \max(x, 0)$. The last term denotes the investment, operation and maintenance cost of residential PV panels. Note the daily capital recovery factor $\alpha = \dfrac{r^{PV}(1 + r^{PV})^y}{365((1 + r^{PV})^y - 1)}$ is to transform the investment cost c_{inv}^{PV} from the planning horizon into the daily horizon, where r^{PV} is the interest rate and y is the planning horizon.

2) Consumer agent utility function: The utility function of each consumer agent i is composed of three terms as follows:

$$
U_i^C = -\sum_{t \in \mathcal{T}} c_t^U P_{it}^{PVc} - \sum_{t \in \mathcal{T}} c_t^G P_{it}^G - \sum_{t \in \mathcal{T}} w_i^E c^E P_{it}^G \tag{7.4}
$$

where the first term and second term represent the cost of electricity purchased from the coalition and the utility grid, respectively. Generally, the conventional consumer utility function only includes the cost of purchasing electricity. However, it is not rational enough since some consumers have environmental awareness so they are interested in reducing greenhouse gas emissions. Therefore, the third term in (7.4) is for emissions reduction with the individual weight factor w_i^E. The load demand P_{it}^L of each consumer agent i can be satisfied by buying electricity from the coalition (P_{it}^{PVc}) and/or the utility grid (P_{it}^G), i.e. $P_{it}^L = P_{it}^{PVc} + P_{it}^G$. Thus, by replacing P_{it}^G in (7.4) by $P_{it}^L - P_{it}^{PVc}$, the willingness-to-pay (WTP) c_{it}^{WTP} for purchasing local PV electricity of consumer agent i can be characterized as follows:

$$c_{it}^{WTP} = c_t^G + w_i^E c^E \quad i \in \mathcal{N}^C, t \in \mathcal{T} \tag{7.5}$$

Note that the individual weight factor w_i^E is introduced to express the preference for emissions reduction of consumer agent i: 1) $w_i^E > 0$ means that the consumer agent i has strong environmental awareness so his WTP c_{it}^{WTP} is higher than the real-time electricity price c_t^G, 2) $w_i^E = 0$ means that the WTP of the consumer agent i equals to the real-time electricity price, and 3) $w_i^E < 0$ means that the environmental awareness of consumer agent i is weak so his WTP is lower than the real-time electricity price.

7.3 Bi-level Energy Sharing Model for Determining Optimal PV Panels Installation Capacity

7.3.1 Uncertainty Characterization

In this chapter, we consider three sources of uncertainties in sizing the PV panels installation capacity, i.e. PV energy output, load demand, and electricity price. The stochastic programming-based approach is employed to deal with these three uncertainty sources, which are represented by a set of representative scenarios. We assume that local historical public data of the electricity prices, loads and PV generations can be collected by the collation and a well-established backward-reduction algorithm [26] can be used to select the representatives with probabilities, which can distinguish the importance of each scenario. Each representative scenario $w \in \mathcal{W}$ consists of three vectors, given as,

$$w = \{\gamma_{tw}^{PV}, \gamma_{tw}^{L}, c_{tw}^{G}\} \quad t \in \mathcal{T}, w \in \mathcal{W}$$

Besides, p_w is defined as the occurrence probability of scenario w, and the sum of probabilities of all considered representative scenarios equals to one, i.e. $\sum_{w \in \mathcal{W}} p_w = 1$.

7.3.2 Stackelberg Game Model

In a leader-follower structure, the Stackelberg game is suitable for studying the decision-making processes of followers in response to the decision taken by the leaders [27]. In this chapter, we model the coalition as the leader which determines the uniform price of PV electricity and participating consumer agents act as followers to purchase electricity from different sources, e.g. utility grid and coalition, according to the time-varying price. Correspondingly, the Stackelberg game is defined as follows:

$$G = \left\{ \left(Coalition \cup \mathcal{N}^C \right), \{\boldsymbol{c}_{tw}^U\}, \{\boldsymbol{P}_{itw}^{PVc}\}, \{\boldsymbol{P}_{itw}^G\}, \{\boldsymbol{Rev}^C\}, \{\boldsymbol{U}_{itw}^C\} \right\}$$

where $\left(Coalition \cup \mathcal{N}^C \right)$ denote the player sets, the coalition acts as the leader and the consumer agents in the set $\mathcal{N}^C$ take the roles of followers in response to the strategy of the coalition; $\{\boldsymbol{C}_{tw}^U\}$ is the strategy set of the coalition; $\{P_{itw}^{PV_c}\}$ and $\{P_{itw}^{G}\}$ are the strategy sets of consumer agents; $\{Rev^C\}$ and $\{U_{itw}^{C}\}$ are the revenue (7.3) of the coalition and the utility function (7.4) of the consumer agent, respectively.

7.3.3 Bi-level Energy Sharing Model

Stackelberg game theory is adopted to model the energy sharing negotiation [28]. Here, the coalition is the leader who sets the uniform prices for PV electricity and consumer agents are the followers who react to decisions taken by the leader. Thus, the stochastic programming-based bi-level energy sharing model is formulated as follows:

$$max\ Rev^C := \sum_{w \in \mathcal{W}} p_w \left[\sum_{i \in \mathcal{N}^C} \sum_{t \in \mathcal{T}} (1 - r^C) c_{tw}^U P_{itw}^{PVc} + \sum_{k \in \mathcal{N}^{PV}} \sum_{t \in \mathcal{T}} c_{tw}^{MCP} P_{ktw}^{PVg} - \sum_{k \in \mathcal{N}^{PV}} \sum_{t \in \mathcal{T}} c_{tw}^G \left[\gamma_{tw}^L P_k^L - \eta^{PV} \gamma_{tw}^{PV} \phi^{PV} E_{size}^{PV} \right]^+ \right] - \left(\alpha c_{inv}^{PV} + c_{o\&m}^{PV} \right) \phi^{PV} \tag{7.6a}$$

over $\{c_{tw}^U, \phi^{PV}, P_{itw}^{PVc}, P_{itw}^G\}_{i \in \mathcal{N}^C, t \in \mathcal{T}, w \in \mathcal{W}}$

s.t.

$$\sum_{i \in \mathcal{N}^C} P_{itw}^{PVc} + \sum_{k \in \mathcal{N}^{PV}} P_{ktw}^{PVg} + \sum_{k \in \mathcal{N}^{PV}} P_{ktw}^{Curtail} = \left[\eta^{PV} \gamma_{tw}^{PV} \phi^{PV} E_{size}^{PV} - \sum_{k \in \mathcal{N}^{PV}} \gamma_{tw}^L P_k^L \right]^+ \quad t \in \mathcal{T}, w \in \mathcal{W} \tag{7.6b}$$

$$\phi^{PV} E_{size}^{PV} \le E^{PVmax}, \phi^{PV} \in Z^+ \tag{7.6c}$$

$$P_{itw}^{PVc}, P_{ktw}^{PVg}, P_{ktw}^{Curtail} \ge 0 \quad i \in \mathcal{N}^C, k \in \mathcal{N}^{PV}, t \in \mathcal{T}, w \in \mathcal{W} \tag{7.6d}$$

$$max\ U_{itw}^C := -c_{tw}^U P_{itw}^{PVc} - c_{tw}^{WTP} P_{itw}^G \tag{7.6e}$$

$$over\ \{P_{itw}^{PVc}, P_{itw}^G\}_{i \in \mathcal{N}^C, t \in \mathcal{T}, w \in \mathcal{W}}$$

s.t.

$$P_{itw}^{PVc} + P_{itw}^G = \gamma_{tw}^L P_i^L : \lambda_{itw}^L \quad i \in \mathcal{N}^C, t \in \mathcal{T}, w \in \mathcal{W} \tag{7.6f}$$

$$P_{itw}^{PVc} \ge 0 : \mu_{itw}^{PVc} \quad i \in \mathcal{N}^C, t \in \mathcal{T}, w \in \mathcal{W} \tag{7.6g}$$

$$P_{itw}^G \ge 0 : \mu_{itw}^G \quad i \in \mathcal{N}^C, t \in \mathcal{T}, w \in \mathcal{W} \tag{7.6h}$$

where the upper-level problem (7.6a)–(7.6h) is to maximize revenue over considered time periods in all scenarios. Equation (7.6b) describes the dispatch of surplus PV electricity, which can be either fed into consumer agents ($P_{itw}^{PV_c}$), sold to the utility grid ($P_{ktw}^{PV_g}$), or curtailed ($P_{ktw}^{Curtail}$). Note that the right side of (7.6b) can be defined as the surplus PV electricity that can be supplied to the energy sharing process. It should be noted that PV energy curtailment becomes an optimal option in some time periods where all consumer agents are fully provided with local PV generation and the wholesale market clearing prices are negative [29]. For security reasons, (7.6c) is added to ensure that total PV panels installation capacity cannot exceed the PV hosting capacity of the grid. Equation (7.6d) guarantees that the upper-level variables are nonnegative.

The lower-level problem (7.6e)–(7.6h) aims to maximize the utility function of each consumer agent in each time period and scenario. Equation (7.6f) balances the supply and demand for each consumer agent node. In (7.6g) and (7.6h), the decision variables of this lower-level problem are limited to positive values. Since the lower-level problem is linear and continuous, it can be substituted by Karush–Kuhn–Tucker (KKT) conditions [30]. To facilitate the understanding of readers, the proposed bi-level energy sharing model (7.6) can be written as follows:

$$\min_{\{x,\,y,\,\lambda,\,\mu\}} f_1(x, y, \lambda, \mu) \tag{7.7a}$$

$$\text{s.t. } h_1(x, y, \lambda, \mu) = 0 \tag{7.7b}$$

$$g_2(x, y, \lambda, \mu) \geq 0 \tag{7.7c}$$

$$\min_{\{y,\,\lambda,\,\mu\}} f_2(x, y) \tag{7.7d}$$

$$\text{s.t. } h_2(x, y) = 0 : \lambda \tag{7.7e}$$

$$g_2(x, y) \geq 0 : \mu \tag{7.7f}$$

The KKT conditions of the lower-level problem (7.7d)–(7.7f) can be implemented in the upper-level problem (7.7a)–(7.7c), given as follows:

$$\min_{\{x,\,y,\,\lambda,\,\mu\}} f_1(x, y, \lambda, \mu) \tag{7.8a}$$

$$\text{s.t. } h_1(x, y, \lambda, \mu) = 0 \tag{7.8b}$$

$$g_2(x, y, \lambda, \mu) \geq 0 \tag{7.8c}$$

$$\nabla_y f_2(x, y) + \lambda \nabla_y h_2(x, y) + \mu \nabla_y g_2(x, y) = 0 \tag{7.8d}$$

$$h_2(x, y) = 0 \tag{7.8e}$$

$$g_2(x, y) \geq 0 \perp \mu \geq 0 \tag{7.8f}$$

The Lagrangian is introduced as follows:

$$\begin{aligned} L = & -c^{U}_{tw} P^{PVc}_{itw} - c^{WTP}_{tw} P^{G}_{itw} - \lambda^{L}_{itw}\left(P^{PVc}_{itw} + P^{G}_{itw} - P^{L}_{itw}\right) \\ & - \mu^{PVc}_{itw} P^{PVc}_{itw} - \mu^{G}_{itw} P^{G}_{itw} \end{aligned} \tag{7.9}$$

Therefore, the lower-level problem can be replaced by KKT conditions, described as follows:

$$\frac{\partial L}{\partial \lambda^{L}_{itw}} = P^{PVc}_{itw} + P^{G}_{itw} - P^{L}_{itw} = 0 \tag{7.10a}$$

$$\frac{\partial L}{\partial P^{PVc}_{itw}} = c^{WTP}_{itw} - \lambda_{itw} - \mu^{PVc}_{itw} = 0 \tag{7.10b}$$

$$\frac{\partial L}{\partial P^{G}_{itw}} = c^{WTP}_{itw} - \lambda_{itw} - \mu^{G}_{itw} = 0 \tag{7.10c}$$

$$P_{itw}^{PVc} \geq 0 \perp \mu_{itw}^{PVc} \geq 0 \tag{7.10d}$$

$$P_{itw}^{G} \geq 0 \perp \mu_{itw}^{G} \geq 0 \tag{7.10e}$$

Then the stochastic bi-level energy sharing model (7.6) can be formulated as mathematical programming with equilibrium constraints (MPEC) model [31], as follows:

$$\begin{aligned} max\ Rev^C := \sum_{w \in \mathcal{W}} p_w \Bigg[& \sum_{i \in \mathcal{N}^C} \sum_{t \in \mathcal{T}} (1 - r^C) c_{tw}^U P_{itw}^{PVc} + \sum_{k \in \mathcal{N}^{PV}} \sum_{t \in \mathcal{T}} c_{tw}^{MCP} P_{ktw}^{PVg} \\ & - \sum_{k \in \mathcal{N}^{PV}} \sum_{t \in \mathcal{T}} c_{tw}^G \left[\gamma_{tw}^L P_k^L - \eta^{PV} \gamma_{tw}^{PV} \phi^{PV} E_{size}^{PV} \right]^+ \Bigg] - \left(\alpha c_{inv}^{PV} + c_{o\&m}^{PV} \right) \phi^{PV} \end{aligned} \tag{7.11a}$$

$$over\ \left\{ c_{tw}^U, P_{itw}^{PVc}, P_{itw}^G, \phi^{PV}, \lambda_{itw}^L, \mu_{itw}^{PVc}, \mu_{itw}^G \right\}_{i \in \mathcal{N}^C, t \in \mathcal{T}, w \in \mathcal{W}} \tag{7.11b}$$

s.t. (7.6b)-(7.6d), (7.10a)-(7.10e)

7.3.4 Linearization of Bi-level Energy Sharing Model

The MPEC model (7.11) contains two nonlinearities, (i) the nonlinear term $c_{tw}^U P_{itw}^{PVc}$ in the objective function (7.11a); (ii) the complementarity constraints (7.10d) and (7.10e). This may make the problem unsolvable. Therefore, the original model (7.11) needs to be linearized. Firstly, the strong duality condition can be used to obtain a linear expression for $c_{tw}^U P_{itw}^{PVc}$. As stated in the strong duality theorem [32], if a problem is convex, the objective functions of the primal and dual problems have the same value at the optimum. Therefore, the primary objective function (7.6e) of the lower-level problem is equal to its dual objective function, as follows:

$$c_{tw}^U P_{itw}^{PVc} + c_{tw}^{WTP} P_{itw}^G = \lambda_{itw}^L \gamma_{tw}^L P_i^L \tag{7.12}$$

Then we can obtain the following expression for the nonlinear term $c_{tw}^U P_{itw}^{PVc}$,

$$c_{tw}^U P_{itw}^{PVc} = \lambda_{itw}^L \gamma_{tw}^L P_i^L - c_{tw}^{WTP} P_{itw}^G \tag{7.13}$$

Secondly, the complementarity constraints (7.10d) and (7.10e) can be linearized using linear expressions proposed in Fortuny-Amat and McCarl [33], so the following constraints should be added,

$$P_{itw}^{PVc}, P_{itw}^G \geq 0 \quad i \in \mathcal{N}^C, t \in \mathcal{T}, w \in \mathcal{W} \tag{7.14a}$$

$$\mu_{itw}^{PVc}, \mu_{itw}^G \geq 0 \quad i \in \mathcal{N}^C, t \in \mathcal{T}, w \in \mathcal{W} \tag{7.14b}$$

$$P_{itw}^{PVc} \leq (1 - u_{itw}^{PVc}) M \quad i \in \mathcal{N}^C, t \in \mathcal{T}, w \in \mathcal{W} \tag{7.14c}$$

$$P_{itw}^{G} \leq (1 - u_{itw}^{G}) M \quad i \in \mathcal{N}^C, t \in \mathcal{T}, w \in \mathcal{W} \tag{7.14d}$$

$$\mu_{itw}^{PVc} \leq u_{itw}^{PVc} M \quad i \in \mathcal{N}^C, t \in \mathcal{T}, w \in \mathcal{W} \tag{7.14e}$$

$$\mu_{itw}^{G} \leq u_{itw}^{G} M \quad i \in \mathcal{N}^C, t \in \mathcal{T}, w \in \mathcal{W} \tag{7.14f}$$

$$u_{itw}^{PVc}, u_{itw}^G \in \{0, 1\} \tag{7.14g}$$

Finally, the linearized MPEC model can be written as follows:

$$\max \sum_{w\in\mathcal{W}} p_w \Bigg[\sum_{i\in\mathcal{N}^C} \sum_{t\in\mathcal{T}} (1-r^C)\left(\lambda_{itw}^L \gamma_{tw}^L P_i^L - c_{itw}^{WTP} P_{itw}^G\right) + \sum_{k\in\mathcal{N}^{PV}} \sum_{t\in\mathcal{T}} c_{tw}^{MCP} P_{ktw}^{PVg} - \sum_{k\in\mathcal{N}^{PV}} \sum_{t\in\mathcal{T}} c_{tw}^G \left[\gamma_{tw}^L P_k^L - \eta^{PV} \gamma_{tw}^{PV} \phi^{PV} E_{size}^{PV}\right]^+ \Bigg] - \left(\alpha c_{inv}^{PV} + c_{o\&m}^{PV}\right)\phi^{PV} \tag{7.15a}$$

$$\text{over } \left\{c_{tw}^U, P_{itw}^{PVc}, P_{itw}^G, \phi^{PV}, \lambda_{itw}, \mu_{itw}^{PVc}, \mu_{itw}^G\right\}_{i\in\mathcal{N}^C, t\in\mathcal{T}, w\in\mathcal{W}} \tag{7.15b}$$

s.t. (7.6b)–(7.6h), (7.10a)–(7.10c), and (7.14a)–(7.14g)

7.3.5 Descend Search-Based Solution Algorithm

Although the linearized MPEC model (7.15) can be directly solved by cutting-edge solvers, such as CLPEX [34] and Gurobi [35], its computation burden may be quite heavy due to a very large number of scenarios and mixed-integer variables. In this regard, this chapter develops an efficient solution algorithm for solving the linearized MPEC model.

As the leader in the Stackelberg game, the coalition sets the uniform price for its PV electricity sold to the consumer agent. According to the definition stated in Liu et al. [28], the bi-level energy sharing model (7.6) reaches the Stackelberg equilibrium (SE) when all players acquire the optimal solutions. Therefore, our proposed game-theoretic framework reaches a SE as long as the coalition finds the optimal uniform price and meanwhile all consumer agents choose their electricity consumption. In this chapter, it is assumed the coalition knows the load curves of all the consumer agents in the chapter. This assumption is reasonable in practice since the coalition can acquire the load information of each consumer agent by the long-term observation via the nonintrusive load monitoring [36]. Besides, the consumer agent is on behalf of all consumers connected to the same node, so the exact load demand of each consumer cannot be obtained by the coalition. In this regard, the privacy of individual consumers can be protected. Therefore, the optimal uniform price set by the coalition is equal to the WTP offered by the consumer agents. Thus, the coalition prefers to sell its surplus PV electricity to the consumer agent with the highest WTP, followed by the consumer agent with the second-highest WTP, etc. Thus, the surplus PV electricity will be dispatched in the descending order by the WTP of consumer agents. Besides, because the wholesale market clearing price is always lower than the WTP, the coalition prefers to sell its PV electricity first to the consumer agents and then to the utility grid. Note that for the extreme situation where the uniform price equals the electricity price, we assume that the local consumers prefer to purchase the electricity from the collation first due to their environmental awareness. This can facilitate the utilization of local PV energy and reduce the negative impacts on the utility grid caused by PV energy integration.

In this regard, we develop a descend search algorithm to calculate the uniform price, as shown in Algorithm 7.2. The idea

of this search algorithm is to iteratively update the revenue from selling surplus PV electricity to consumer agents, whereby the consumer agents are ranked in descending order by their WTP. Finally, the uniform price that brings about the highest revenue to the coalition will be returned. Therefore, the optimal uniform price can be found by using Algorithm 7.2

Algorithm 7.1 Solution Algorithm for Solving the Proposed Bi-level Energy Sharing Problem

1. Initialize the revenue Rev^C and residential PV panels installation number ϕ^{PV}. Initialize the iteration index $\sigma = 0$ and set the step size ϕ^{PV}_{Step}.
2. **Repeat:**
3. $\sigma = \sigma + 1$;
4. $\phi^{PV(\sigma)} = \phi^{PV(\sigma-1)} + \phi^{PV}_{Step}$;
5. **For** the coalition **do**
6. Execute **Algorithm 7.2**;
6. **End**
7. Receive $Rev^{C(\sigma)}$ from executing **Algorithm 7.2**;
8. **If** $Rev^{C(\sigma)} > Rev^{C(\sigma-1)}$ **then**
9. $Rev^C_{Opt} = Rev^{C(\sigma)}$;
10. $\phi^{PV}_{Opt} = \Phi^{PV(\sigma)}$;
11. **Else** $Rev^C_{Opt} = Rev^{C(\sigma-1)}$;
12. $\phi^{PV}_{Opt} = \Phi^{PV(\sigma-1)}$;
13. **End**
14. **Until** $\phi^{PV} E^{PV}_{per} \geq E^{PVmax}$;
15. **Return** Rev^C_{Opt}, ϕ^{PV}_{Opt};
16. **End procedure**

Algorithm 7.2 Descend Search Algorithm for Seeking the Optimal Internal Uniform Price

1. Receive $\Phi^{PV(\sigma)}$ from **Algorithm 7.1**;
2. **For all** $w \in \mathcal{W}$ **do**
3. **For all** $t \in \mathcal{T}$ **do**
4. **For all** $i \in \mathcal{N}^C$ **do** ▷ Sorted descending by WTP c^{wtp}_{itw}
5. $\mathcal{N}^{*(\sigma)} \leftarrow \mathcal{N}^{*(\sigma)} \cup \{i\}$;
6. **If** $c^{wtp}_{itw} > c^{MCP}_{tw}$ **then continue**
7. **Else** go back to step 3
8. **End**
9. $c^{U(\sigma)}_{tw} \leftarrow c^{wtp}_{itw}$;
10. $P^{PVg(\sigma)}_{tw} \leftarrow \left(\eta^{PV} \gamma^{PV}_{tw} \Phi^{PV(\sigma)} E^{PV}_{per} - \sum_{k \in \mathcal{N}^{PV}} \gamma^L_{tw} P^L_k\right) - \sum_{i \in \mathcal{N}^{*(\sigma)}} \gamma^L_{tw} P^L_i$;
11. **If** $P^{PVg(\sigma)}_{tw} \geq 0$ **then**
12. $Rev^{C(\sigma)}_{tw} \leftarrow \sum_{i \in \mathcal{N}^{*(\sigma)}} (1 - r^C) c^{U(\sigma)}_{tw} \gamma^L_{tw} P^L_i + \left[c^{MCP}_{tw}\right]^+ P^{PVg(\sigma)}_{tw} - \left(\alpha C^{PV}_{inv} + c^{PV}_{o\&m}\right) \Phi^{PV(\sigma)}$;
13. **Else**

14. If $\left(\eta^{PV}\gamma_{tw}^{PV}\Phi^{PV(\sigma)}E_{per}^{PV} - \sum_{k\in\mathcal{N}^{PV}}\gamma_{tw}^{L}P_k^L\right) \geq 0$

15. $Rev_{tw}^{C(\sigma)} \leftarrow (1-r^C)c_{tw}^{U(\sigma)}\left(\eta^{PV}\gamma_{tw}^{PV}\Phi^{PV(\sigma)}E_{per}^{PV} - \sum_{k\in\mathcal{N}^{PV}}\gamma_{tw}^{L}P_k^L\right) - \left(\alpha C_{inv}^{PV} + c_{o\&m}^{PV}\right)\Phi^{PV(\sigma)};$

16. Else $Rev_{tw}^{C(\sigma)} \leftarrow c_{tw}^{G}\left(\sum_{k\in\mathcal{N}^{PV}}\gamma_{tw}^{L}P_k^L - \eta^{PV}\gamma_{tw}^{PV}\Phi^{PV(\sigma)}E_{per}^{PV}\right) - \left(\alpha C_{inv}^{PV} + c_{o\&m}^{PV}\right)\Phi^{PV(\sigma)};$

17. **End**
18. **End**
19. **End**
20. Check power flow by using BFM (1)
21. **If** $l_{itw} \leq l_i^{max}$ && $v^{min} \leq v_{itw} \leq v^{max}$ **then continue**
22. **Else** go back to step 4
23. **End**
24. **End**
25. **End**
26. $b \leftarrow \text{argmax}\left(Rev_{tw}^{C(\sigma)}\right)$; ▷ Find the optimal result
27. $c_{tw}^{U(\sigma)} \leftarrow c_{btw}^{wtp};$
28. $Rev^{C(\sigma)} \leftarrow \left(\sum_{w\in\mathcal{W}} p_w \sum_{t\in\mathcal{T}} Rev_{tw}^{C(\sigma)}\right);$
29. **Return** Rev_{tw}^{C}, $c_{tw}^{U(\sigma)}$;
30. **End** procedure

instead of directly solving the problem (7.8). Moreover, to ensure the reliability and security of the distribution grid during the energy sharing process, the feasibility voltage constraint and current constraint are checked by using the BFM. Our proposed descend search algorithm-based solution method is given as Algorithm 7.1. Besides, Figure 7.2. is plotted to depict the flowchart of our proposed two-stage game-theoretic residential PV panels planning framework.

7.4 Stochastic Optimal PV Panels Allocation in the Coalition of Prosumer Agents

After the first stage, the optimal installation capacity of residential PV panels for the coalition can be obtained. Then the second stage of our proposed two-stage planning framework concerns the optimal residential PV panels allocation among the prosumer agents of the collation. At this stage, the power loss becomes the most important variable since it changes in terms of the allocation result of PV panels. The primary goal of the second-stage model is to minimize active power loss so that economic loss can be reduced. The stochastic optimal residential PV panels allocation model is as follows:

$$min \sum_{w\in\mathcal{W}} p_w \sum_{t\in\mathcal{T}}\sum_{j\in\mathcal{N}} r_{mj} l_{mjtw} \tag{7.16a}$$

$$over \left\{P_{mktw}, Q_{mktw}, P_{mitw}, Q_{mitw}, v_{jtw}, l_{mjtw}, u_k^{PV}\right\}_{k\in\mathcal{N}^{PV}, i\in\mathcal{N}^C, t\in\mathcal{T}, w\in\mathcal{W}, j\in\mathcal{N}}$$

$$s.t. P_{mktw} - r_{mk} l_{mktw} + \eta^{PV}\gamma_{tw}^{PV} u_k^{PV} E_{size}^{PV} - \gamma_{tw}^{L} P_k^L = \sum_{n\in C_k} P_{kntw}$$

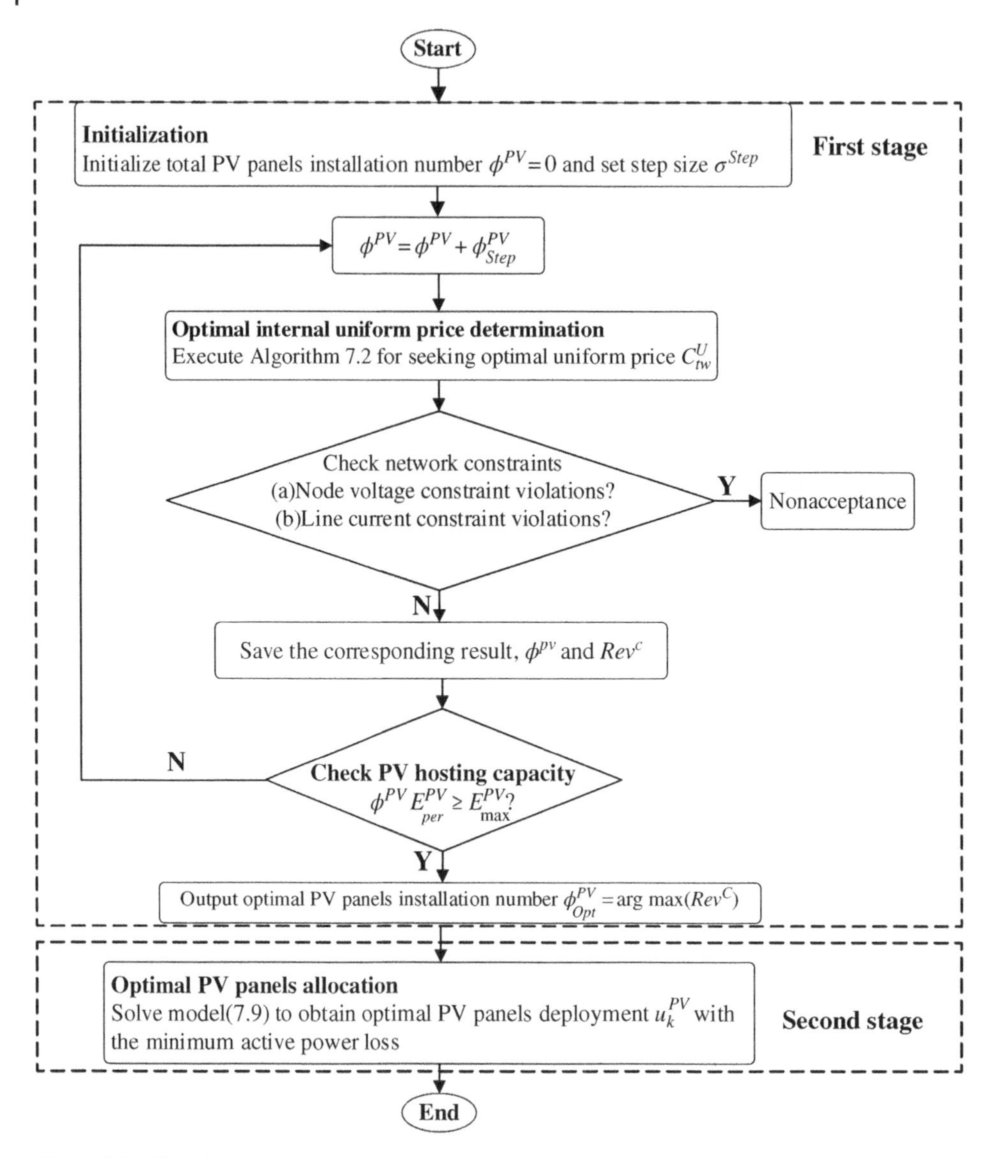

Figure 7.2 Flowchart of our proposed two-stage game-theoretic residential PV panels planning framework. *Source:* From Xu et al. [A].

$$k \in \mathcal{N}^{PV}, t \in \mathcal{T}, w \in \mathcal{W} \tag{7.16b}$$

$$P_{mitw} - r_{mi}l_{mitw} - \gamma^{PV}_{tw}P^L_i = \sum_{n\in\mathcal{C}_i} P_{intw} \quad i \in \mathcal{N}^C, t \in \mathcal{T}, w \in \mathcal{W} \tag{7.16c}$$

$$Q_{mktw} - x_{mk}l_{mktw} - \gamma^L_{tw}Q^L_k = \sum_{n\in\mathcal{C}_k} Q_{kntw} \quad k \in \mathcal{N}^{PV}, t \in \mathcal{T}, w \in \mathcal{W} \tag{7.16d}$$

$$Q_{mitw} - x_{mi}l_{mitw} - \gamma^L_{tw}Q^L_i = \sum_{n\in\mathcal{C}_i} Q_{intw} \quad i \in \mathcal{N}^C, t \in \mathcal{T}, w \in \mathcal{W} \tag{7.16e}$$

$$v_{jtw} - v_{mtw} = 2\left(r_{mj}P_{mjtw} + x_{mj}Q_{mjtw}\right) - \left(r^2_{mj} + x^2_{mj}\right)l_{mjtw}$$
$$j \in \mathcal{N}, t \in \mathcal{T}, w \in \mathcal{W} \tag{7.16f}$$

$$\left\|\left(2P_{mjtw},\ 2Q_{mjtw},\ v_{jtw}-l_{mjtw}\right)\right\|_2 \le v_{jtw}+l_{mjtw}$$
$$j\in\mathcal{N},t\in\mathcal{T},w\in\mathcal{W} \tag{7.16g}$$

$$v^{min} \le v_{jtw} \le v^{max} \quad j\in\mathcal{N},t\in\mathcal{T},w\in\mathcal{W} \tag{7.16h}$$

$$0 \le l_{mjtw} \le l_{mj}^{max} \quad j\in\mathcal{N},t\in\mathcal{T},w\in\mathcal{W} \tag{7.16i}$$

$$u_k^{PV} \le u_k^{PV\max} \quad k\in\mathcal{N}^{PV} \tag{7.16j}$$

$$\sum_{k\in\mathcal{N}^{PV}} u_k^{PV} = \phi^{PV}, u_k^{PV}\in Z^+ \tag{7.16k}$$

where the objective function (7.16a) is to minimize the expected active power loss of all considered uncertainty scenarios. Equations (7.16b) and (7.16c) describe the active power balance on the node with the prosumer agent and consumer agent, respectively. Equations (7.16d) and (7.16e) describe the reactive power balance on each node. Equations (7.16f) and (7.16f) are derived from (7.1c) and (7.2), respectively. Equation (7.16h) is the voltage constraint and (7.16i) is the current constraint. Equation (7.16j) guarantees that the residential PV panels installation number at each prosumer agent node cannot exceed its maximum allowable installation number, which is determined by the installation conditions. Equation (7.16k) ensures that the total installation number of residential PV panels equals the optimal value obtained from the first stage.

7.5 Numerical Results

7.5.1 Implementation on IEEE 33-Node Distribution System

In this section, we employ the modified IEEE 33-node distribution grid as shown in Figure 7.3 to test our proposed residential PV panels planning model. Detailed information on this test system can be referred to [37]. Figure 7.4 depicts 125 ($5*5*5$) uncertainty scenarios, consisting of five load scenarios, five PV output scenarios and five electricity price scenarios. For the sake of simplicity, the electricity price c_{tw}^{G} is assumed to be equal for all agents in the test system and it consists of the market clearing price c_{tw}^{MCP}, generation markup (0.013 \$/kWh) and delivery charge (0.036 \$/kWh) [38], i.e. $c_{tw}^{G}=c_{tw}^{MCP}+0.013+0.036$ /kWh. Due to the characteristics of residential PV panels output, the length of the entire time period $\mathcal{T}$ is thirteen hours (6:00–18:00) and the duration of each time period is one hour. We randomly assume the following weight factors to describe different consumer agent preferences on WTP: $w_i^E=[6.3, 8.2, -7.5, 8.3, 2.6, -8, -4.4, 0.9, 9.2, 9.3,$

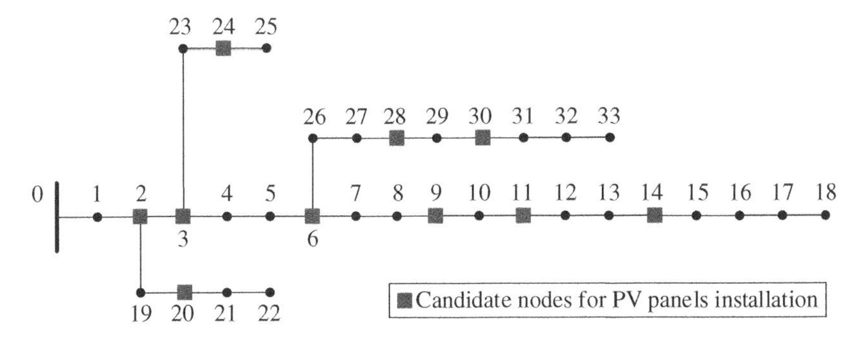

Figure 7.3 IEEE 33-node distribution grid with candidate nodes for PV panels installation. *Source:* From Xu et al. [A].

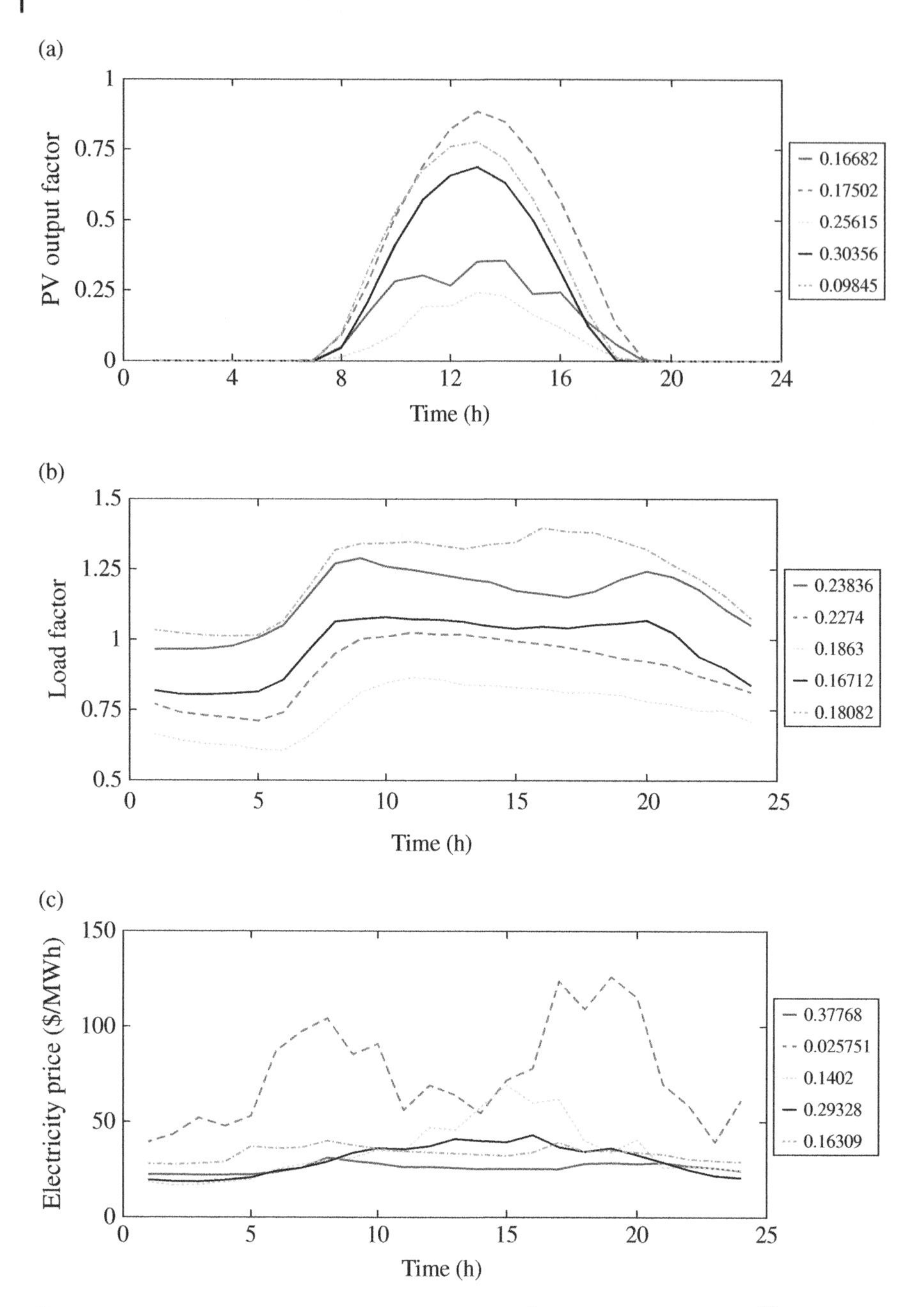

Figure 7.4 Uncertainty scenarios of (a) load factor γ_{tw}^{L}, (b) PV output factor γ_{tw}^{PV}, and (c) electricity price c_{tw}^{G}. *Source:* From Xu et al. [A].

−6.8, 9.4, 9.1, −0.3, 6, −7.2, −1.6, 8.3, 5.8, 9.2, 3.1, −9.3, 7] for consumer agents 1, 4, 5, 7, 8, 10, 12, 13, 15, 16, 17, 18, 19, 21, 22, 23, 25, 26, 27, 29, 31, 32, and 33, respectively. Note that these weight factors have significant effects on optimal residential PV panels planning results and in practice, they can be adjustable depending on the preferences of consumer agents on a case by case basis. Other parameters including greenhouse gas emission price, per residential PV panel size, and residential PV panel investment, maintenance and operation cost are summarized in Table 7.1. Table 7.2 lists the optimal residential PV panels allocation result of our proposed two-stage game-theoretic model.

Table 7.1 Related parameters in the 33-node case.

Parameter name	Parameter value
Greenhouse gas emission price	0.0125 $/kWh [39]
Per residential PV panel size	200 kVA [40]
Residential PV panel planning cost	3.05 $/W [40]
Residential PV panel output efficiency	95% [40]
Planning horizon	20 years
Depreciation rate of PV panels	5%

Source: From Xu et al. [A].

Table 7.2 Optimal residential PV panels planning result in 33-node system.

Candidate locations (nodes)	PV panels capacity (kVA)
2	148.4
3	129
6	259
9	269
11	63.2
14	165.6
20	151.8
24	218.6
28	240
30	255.4

Source: From Xu et al. [A].

Table 7.3 Residential PV panels planning size and cost and daily revenue with and without energy sharing in 33-node test system.

Energy sharing mechanism	No	Yes
Residential PV panels planning size	$6.4 * 10^2$ kVA	$1.9 * 10^3$ kVA
Residential PV panels planning cost	$1.45 * 10^6$ $	$4.3 * 10^6$ $
Daily revenue	32.45 $	110.27 $

Source: From Xu et al. [A].

All case studies are implemented by using MATLAB on a computer with an Intel Core i7 of 2.4 GHz and 12 GB memory.

1) From the Perspective of the Coalition

In this subsection, we demonstrate the results from the perspective of the coalition. Table 7.3 lists comparisons on the PV panels planning size and cost as well as the daily revenue with and without energy sharing. We can see from this table that the PV panels installation capacity with energy sharing is more than that without consideration of energy sharing, so its corresponding PV panels planning cost is higher. This is because local consumer agents with good environmental awareness would like to purchase PV electricity from the coalition, so more PV panels will be invested. Besides, it also can be observed that daily revenue improves significantly after energy sharing between the coalition and consumer agents, which can bring approximately 77.8 $ revenue improvement each day.

Figure 7.5 illustrates the hourly revenue of the coalition in representative scenario 6 and scenario 12. As shown in this figure, hourly revenue is negative in the morning and becomes positive from noon. The main reason is that the residential PV panels only can provide a small amount of PV electricity in the morning, but the PV production is sufficient at midday. Figure 7.6 depicts the dispatch of PV electricity provided by the coalition in representative scenario 4. It can be seen that the coalition sells more PV electricity to the consumer agents in each time period. The coalition allocates the PV electricity in a way to maximize its revenue, so it sells PV electricity differently to the utility grid and the consumer agents in each time period. According to the WTP Eq. (7.5), we can know that the wholesale market clearing price c_{tw}^{MCP} is always lower than the WTP c_{itw}^{WTP}, so the

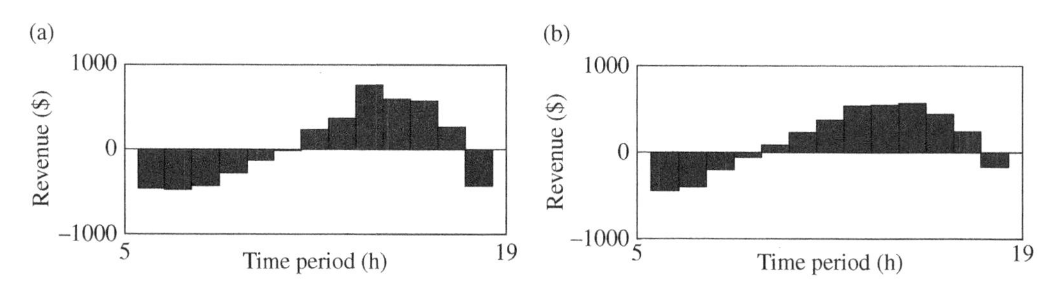

Figure 7.5 Hourly revenue of the coalition in representative (a) scenario 6 and (b) Scenario 12. *Source:* From Xu et al. [A].

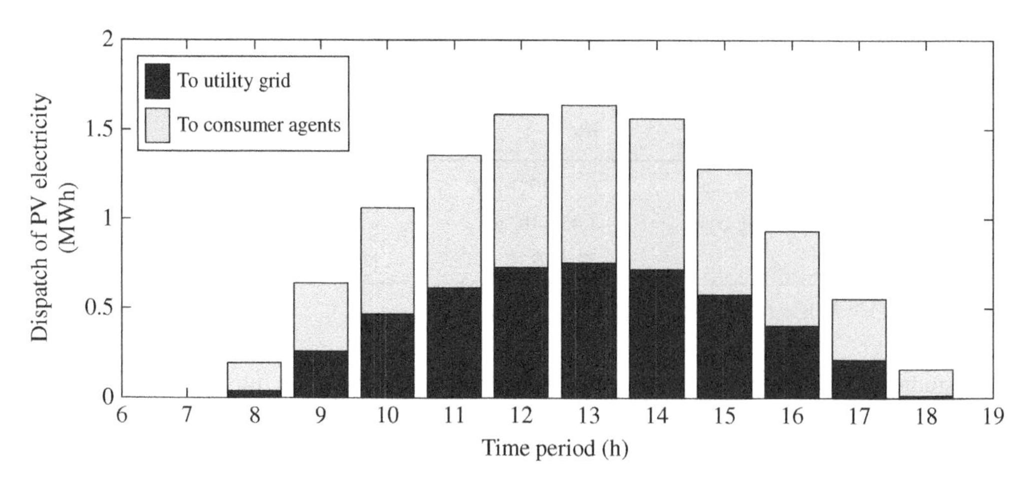

Figure 7.6 PV generation sold to the consumer agents and sold to the utility grid in representative Scenario 4. *Source:* From Xu et al. [A].

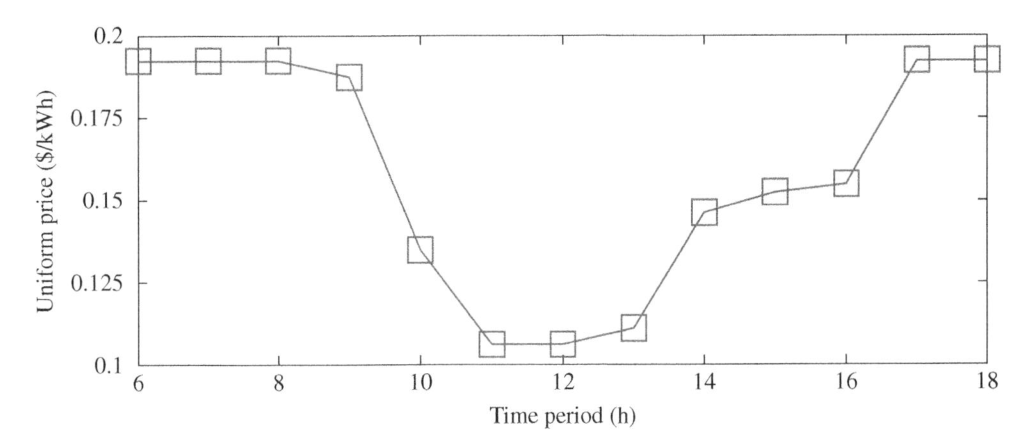

Figure 7.7 Optimal uniform price of each time period in representative Scenario 15. *Source:* From Xu et al. [A].

coalition prefers to sell its PV electricity firstly to the consumer agents and then to the utility grid. Figure 7.7 shows the hourly optimal uniform price in representative scenario 15. It can be observed from this figure that the uniform price is high in time periods of nonpeak PV generation (6:00–10:00 and 16:00–18:00) and relatively low in time periods of peak PV generation (10:00–16:00). This is because the PV electricity can be consumed by local consumers during the daytime, the coalition sets low uniform prices so its PV electricity can be dispatched to more consumer agents. While in the morning or early night, the coalition would like to choose a high uniform price to maximize its revenue with low PV production.

To investigate the computation performance on our proposed solution method (see Algorithm 7.1 and Algorithm 7.2) for solving energy sharing problem (7.6), the conventional mixed-integer linear programming (MILP) method for solving our formulated MPEC problem (7.11) is used as the benchmark. Table 7.4 lists the computation time of solving the proposed bi-level energy sharing problem by these two methods considering the different number of operation scenarios. As shown in this table, both two solution methods can be used to solve the proposed problem with low considered scenario numbers (e.g. 1, 8, and 27) while our proposed method shows a clear advantage over the first method. With high scenarios numbers (e.g. 64, 125), a huge computation burden

Table 7.4 Computation performance on solving proposed bi-level energy sharing problem by different methods considering different number of operation scenarios.

	Computation time (s)	
Number of operation scenarios	**MILP method**	**Proposed method**
$1\ (1*1*1)$	2.91	4.38
$8\ (2*2*2)$	120.57	20.05
$27\ (3*3*3)$	3961.12	67.66
$64\ (4*4*4)$	—	160.38
$125\ (5*5*5)$	—	313.25

Source: From Xu et al. [A].

may be involved so the commercial solver cannot handle the original problem. By contrast, our solution algorithm can solve the same problem with acceptable computation time. Note that the solution obtained by these two methods satisfies all conditions in each considered scenario, so the solution becomes more robust with the consideration of more operation scenarios [41]. In this regard, the robustness and reliability of the solution obtained by our proposed solution method can be more guaranteed since this method can take a large number of uncertainty scenarios into account.

2) The Tradeoff Between Total Residential PV Panels Size and the Daily Coalition Revenue

In this subsection, a sensitivity analysis of the total residential PV panels size with the daily coalition revenue is performed. Figure 7.8 depicts the tradeoff curve between the total residential PV panel size and the daily coalition revenue with and without energy sharing, respectively. With energy sharing, fast-growing daily revenue can be observed with increasing residential PV panels size until the investment capacity reaches about 1900 kVA, then the daily revenue declines after more residential PV panels installation. While without energy sharing, the daily coalition revenue shows a rapid increase at first as increasing residential PV panels installation capacity, then it

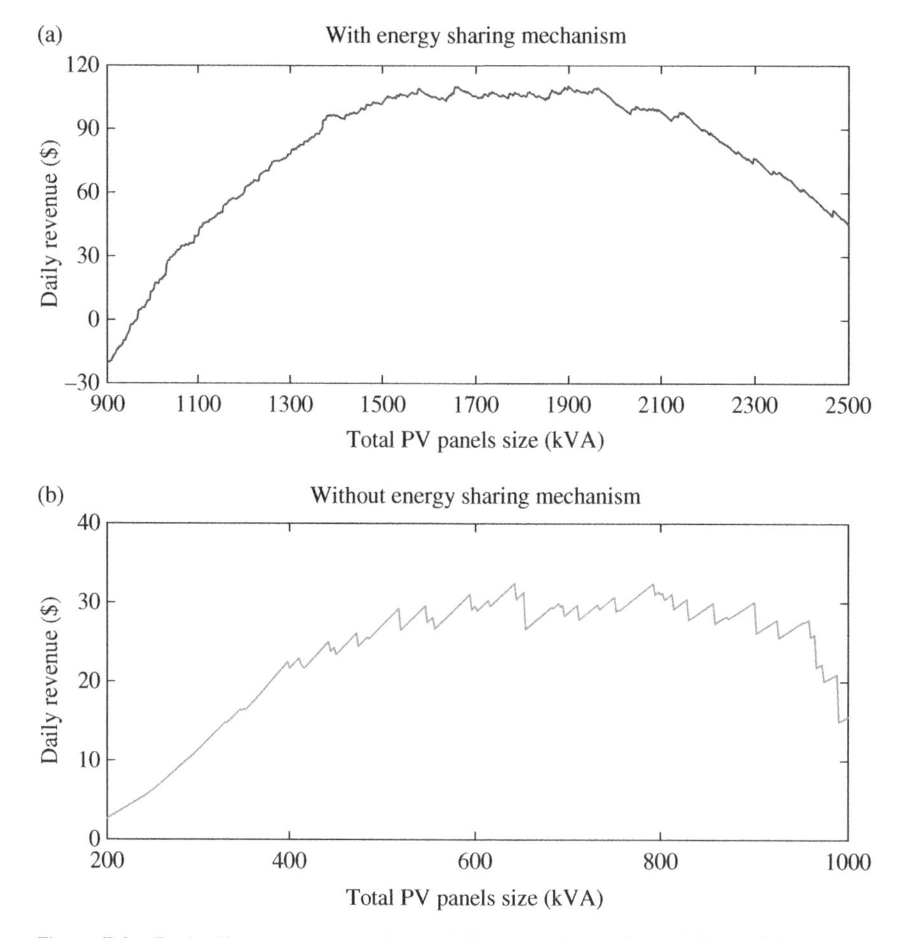

Figure 7.8 Tradeoff curve between the total PV panel size and the daily coalition revenue (a) with and (b) without energy sharing mechanism in the 33-node system. (From [A]).

becomes fluctuant and finally performs a gradual decrease. This is because larger residential PV panels investment leads to high planning costs, which cannot be covered by the economic benefit of PV energy sharing.

7.5.2 Implementation on IEEE 123-Node Distribution System

The IEEE 123-node distribution grid [42] is adopted as the large-scale test system to verify the proposed planning model and solution algorithm. In this case, uncertainty scenarios including load factor, PV energy output factor, and electricity price are the same as those in the 33-node case. Other parameters including greenhouse gas emission price, per residential PV panel size, and residential PV panel investment, maintenance and operation cost are also the same as those in the 33-node case.

Figure 7.9 depicts this modified test system by selecting twenty candidate nodes for PV panels installation, i.e. 5, 14, 19, 26, 35, 38, 43, 47, 51, 55, 63, 75, 77, 84, 86, 93, 103, 112, 119, and 120. Besides, random weight factors in the range of [−10,10] are assumed to define the consumer agent preferences on WTP. Residential PV panels planning result is listed in Table 7.5. Figure 7.10 is plotted to demonstrate the tradeoff between the total residential PV panel size and the daily coalition revenue with the energy sharing mechanism. As shown in this figure, the daily revenue rises rapidly with the increase of the PV panel size. After the PV panel size reaches approximately 6 MVA, the daily revenue decreases with the increasing PV panel size.

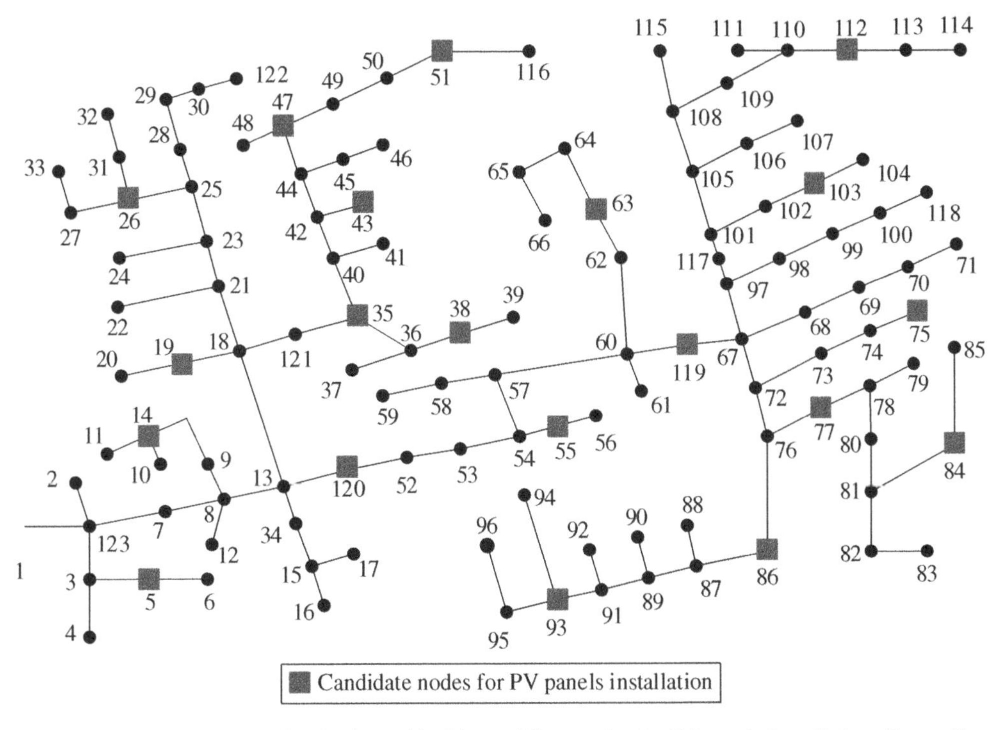

Figure 7.9 IEEE 123-node distribution grid with candidate nodes for PV panels installation. *Source:* From Xu et al. [A].

Table 7.5 Optimal residential PV panels planning result in 123-node system.

Locations (nodes)	PV panels capacity (kVA)	Locations (nodes)	PV panels capacity (kVA)
5	298.67	63	279.10
14	97.37	75	64.01
19	663.21	77	386.75
26	394.31	84	241.06
35	87.65	86	634.42
38	222.52	93	444.33
43	67.30	103	305.76
47	96.56	112	70.35
51	139.54	119	15.68
55	1300.40	120	191.07

Source: From Xu et al. [A].

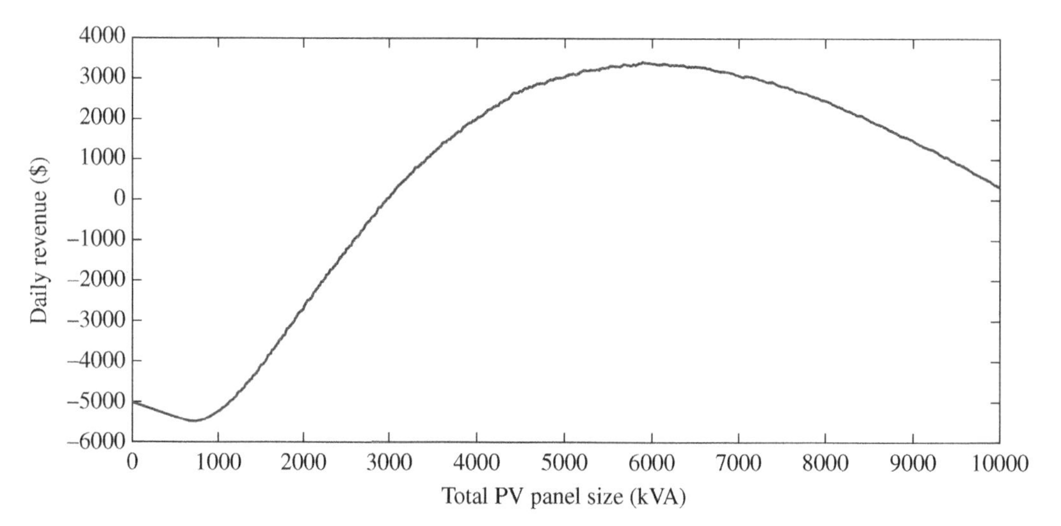

Figure 7.10 Tradeoff curve between the total PV panel size and the daily coalition revenue with energy sharing in the 123-node system. *Source:* From Xu et al. [A].

7.6 Conclusion

In this chapter, we develop a novel two-stage game-theoretic framework of residential PV panels planning. In the first stage, Stackelberg game theory is used to model the stochastic bi-level energy sharing problem, which is solved by our proposed descend search algorithm-based solution method, so the optimal installation capacity of residential PV panels can be obtained. In the second stage, we develop a stochastic programming-based optimal OPF model to optimally allocate residential PV panels for all PV prosumers with minimum expected active power loss. Finally, the IEEE 33-node and 123-node test systems are used to demonstrate the effectiveness of our proposed framework. As shown in numerical results, with the consideration of energy sharing mechanism during the PV panels planning stage, not only economic benefits to PV prosumers can be improved but also the residential PV panels installation is facilitated.

Acknowledgements

The permission given in using the materials from the following paper is very much appreciated.

[A] Xu, X., Li, J., Xu, Y. et al. (2020). A two-stage game-theoretic method for residential PV panels planning considering energy sharing mechanism. *IEEE Transactions on Power Systems* 35 (5): 3562–3573. https://doi.org/10.1109/TPWRS.2020.2985765.

References

1 McKenna, E. and Thomson, M. (2013). Photovoltaic metering configurations, feed-in tariffs and the variable effective electricity prices that result. *IET Renewable Power Generation* 7 (3): 235–245.
2 Li, J., Zhang, C., Xu, Z. et al. (2018). Distributed transactive energy trading framework in distribution networks. *IEEE Transactions on Power Systems* 33 (6): 7215–7227.
3 Wang, Y., Huang, Z., Li, Z. et al. (2020). Transactive energy trading in reconfigurable multi-carrier energy systems. *Journal of Modern Power Systems and Clean Energy* 8 (1): 67–76.
4 Wang, Y.F., Huang, Z.H., Shahidehpour, M. et al. (March 2020). Reconfigurable distribution network for managing transactive energy in a multi-microgrid system. *IEEE Transactions on Smart Grid* 11 (2): 1286–1295. https://doi.org/10.1109/TSG.2019.2935565.
5 Cui, S., Wang, Y.-W., and Xiao, J.-W.J. (2019). Peer-to-peer energy sharing among smart energy buildings by distributed transaction. *IEEE Transactions on Smart Grid* 10 (6): 6491–6501.
6 Zhao, Z., Guo, J., Luo, X. et al. (2020). Energy transaction for multi-microgrids and internal microgrid based on blockchain. *IEEE Access* 8: 144362–144372.
7 Mediwaththe, C.P., Shaw, M., Halgamuge, S.K. et al. (2019). An incentive-compatible energy trading framework for neighborhood area networks with shared energy storage. *IEEE Transactions on Sustainable Energy* 11 (1): 467–476.
8 Li, Z., Lai, C.S., Xu, X. et al. (2021). Electricity trading based on distribution locational marginal price. *International Journal of Electrical Power and Energy Systems* 124: 106322.
9 Cui, S., Wang, Y.-W., Li, C., and Xiao, J.-W.J. (2019). Prosumer community: a risk aversion energy sharing model. *IEEE Transactions on Sustainable Energy* 11 (2): 828–838.
10 Liu, N., Wang, J., and Wang, L.J. (2018). Hybrid energy sharing for multiple microgrids in an integrated heat-electricity energy system. *IEEE Transactions on Sustainable Energy* 10 (3): 1139–1151.
11 Fleischhacker, A., Auer, H., Lettner, G., and Botterud, A.J. (2018). Sharing solar PV and energy storage in apartment buildings: resource allocation and pricing. *IEEE Transactions on Smart Grid* 10 (4): 3963–3973.
12 Senemar, S., Seifi, A.R., Rastegar, M., and Parvania, M.J. (2019). Probabilistic optimal dynamic planning of onsite solar generation for residential energy hubs. *IEEE Systems Journal* 14 (1): 832–841.
13 Zhang, X. and Grijalva, S.J. (2016). A data-driven approach for detection and estimation of residential PV installations. *IEEE Transactions on Smart Grid* 7 (5): 2477–2485.
14 Ghiassi-Farrokhfal, Y., Kazhamiaka, F., Rosenberg, C., and Keshav, S.J. (2015). Optimal design of solar PV farms with storage. *IEEE Transactions on Sustainable energy* 6 (4): 1586–1593.

15 Alhaider, M. and Fan, L.J. (2018). Planning energy storage and photovoltaic panels for demand response with heating ventilation and air conditioning systems. *IEEE Transactions on Industrial Informatics* 14 (11): 5029–5037.

16 Li, Z., Xu, Y., Fang, S., and Mazzoni, S. (2019). Optimal placement of heterogeneous distributed generators in a grid-connected multi-energy microgrid under uncertainties. *IET Renewable Power Generation* 13 (14): 2623–2633.

17 Cáceres, G., Nasirov, S., Zhang, H., and Araya-Letelier, G.J. (2015). Residential solar PV planning in Santiago, Chile: Incorporating the PM10 parameter. *Sustainability* 7 (1): 422–440.

18 Yoza, A., Yona, A., Senjyu, T., and Funabashi, T.J. (2014). Optimal capacity and expansion planning methodology of PV and battery in smart house. *Renewable Energy* 69: 25–33.

19 Farivar, M. and Low, S.H.J. (2013). Branch flow model: relaxations and convexification – Part I. *IEEE Transactions on Power Systems* 28 (3): 2554–2564.

20 Kim, S., Kojima, M.J., and software (2001). Second order cone programming relaxation of nonconvex quadratic optimization problems. *Optimization Methods* 15 (3–4): 201–224.

21 Painuly, J.P. (2001). Barriers to renewable energy penetration; a framework for analysis. *Renewable Energy* 24 (1): 73–89.

22 Foxon, T.J., Gross, R., Chase, A. et al. (2005). UK innovation systems for new and renewable energy technologies: drivers, barriers and systems failures. *Energy Policy* 33 (16): 2123–2137.

23 Byrnes, L., Brown, C., Foster, J., and Wagner, L.D. (2013). Australian renewable energy policy: barriers and challenges. *Renewable Energy* 60: 711–721.

24 Phuangpornpitak, N. and Tia, S. (2013). Opportunities and challenges of integrating renewable energy in smart grid system. *Energy Procedia* 34: 282–290.

25 Union of Concerned Scientists (2014). Barriers to Renewable Energy Technologies. https://www.ucsusa.org/resources/barriers-renewable-energy-technologies (accessed 31 July 2022).

26 Heitsch, H. and Römisch, W. (2003). Scenario reduction algorithms in stochastic programming. *Computational Optimization and Applications* 24 (2–3): 187–206.

27 Myerson, R.B. (2013). *Game Theory*. Harvard University Press.

28 Liu, N., Yu, X., Wang, C., and Wang, J.J. (2017). Energy sharing management for microgrids with PV prosumers: a Stackelberg game approach. *IEEE Transactions on Industrial Informatics* 13 (3): 1088–1098.

29 Li, J., Xu, Z., Zhao, J., and Zhang, C. (2019). Distributed online voltage control in active distribution networks considering PV curtailment. *IEEE Transactions on Industrial Informatics* 15 (10): 5519–5530.

30 Kuhn, H.W. and Tucker, A.W. (1951). Nonlinear programming. In: *Proceedings of the Second Berkeley Symposium on Mathematical Statistics and Probability* (ed. J. Neyman). Berkeley: University of California Press.

31 Luo, Z.-Q., Pang, J.-S., and Ralph, D. (1996). *Mathematical Programs with Equilibrium Constraints*. Cambridge University Press.

32 Dantzig, G. (2016). *Linear Programming and Extensions*. Princeton University Press.

33 Fortuny-Amat, J. and McCarl, B.J. (1981). A representation and economic interpretation of a two-level programming problem. *Journal of the Operational Research Society* 32 (9): 783–792.

34 (2009). V12. 1: User's Manual for CPLEX. *International Business Machines Corporation*, 952 pages. ftp://public.dhe.ibm.com/software/websphere/ilog/docs/optimization/cplex/ps_usrmancplex.pdf (accessed 10 November 2020).

35 Gurobi Optimizer (2020). *Gurobi Optimizer Reference Manual. Version 9.0.* Gurobi Optimization, LLC. 969 pages. https://www.gurobi.com/wp-content/plugins/hd_documentations/documentation/9.0/refman.pdf (accessed 31 July 2022).

36 Shaw, S.R., Leeb, S.B., Norford, L.K., and Cox, R.W. (2008). Nonintrusive load monitoring and diagnostics in power systems. *IEEE Transactions on Instrumentation and Measurement* 57 (7): 1445–1454.

37 Baran, M.E. and Wu, F.F.J. (1989). Network reconfiguration in distribution systems for loss reduction and load balancing. *IEEE Transactions on Power Delivery* 4 (2): 1401–1407.

38 Attar, M., Homaee, O., Falaghi, H., and Siano, P. (2018). A novel strategy for optimal placement of locally controlled voltage regulators in traditional distribution systems. *International Journal of Electrical Power & Energy Systems* 96: 11–22.

39 Kaufmann, R.K. and Vaid, D.J. (2016). Lower electricity prices and greenhouse gas emissions due to rooftop solar: empirical results for Massachusetts. *Energy Policy* 93: 345–352.

40 Sharma, S., Xu, Y., Verma, A., and Panigrahi, B.K. (2019). Time-coordinated multi-energy management of smart buildings under uncertainties. *IEEE Transactions on Industrial Informatics* 15 (8): 4788–4798.

41 Santos, S.F., Fitiwi, D.Z., Shafie-Khah, M. et al. (2016). New multistage and stochastic mathematical model for maximizing RES hosting capacity – Part I: problem formulation. *IEEE Transactions on Sustainable Energy* 8 (1): 304–319.

42 Liu, Z., Wen, F., and Ledwich, G. (2012). Optimal planning of electric-vehicle charging stations in distribution systems. *IEEE Transactions on Power Delivery* 28 (1): 102–110.

8

Networked Microgrids Energy Management Under High Renewable Penetration

Nomenclature

Abbreviations

BESS	Battery energy storage system
CDG	Controllable distributed generator
DSO	Distribution system operator
EMS	Energy management system
MG	Microgrid
MGC	Microgrid community
RESs	Renewable energy sources
SOC	State of charge
VaR	Value-at-risk

Indices

t	Index of time (t=1, 2, ..., T)
i	Index of microgrid (i=1, 2, ..., I)
C	Index of microgrid community
k	Index of scenario (k=1, 2, ..., Ω_K)
$(\hat{\cdot})$	Index of variables in real-time market

Parameters

a^{CG}/b^{CG}	Cost coefficients of CDG
a^{CL}/b^{CL}	Cost coefficients of controllable load
C_t^{CG}	Operation cost of CDG
C_t^{ES}	Operation cost of BESS
C_t^{CL}	The cost of controllable load
C_{it}^{M}	Exchanged power cost of the ith microgrid
$C_t^{C,M}$	Cost of exchanged power in MGC
E_R^{ES}	Rated capacity of BESS
E_t^{ES}	Stored energy in BESS at time t

Smart Energy for Transportation and Health in a Smart City, First Edition. Chun Sing Lai, Loi Lei Lai and Qi Hong Lai.

IC^{ES}	Investment cost of BESS
LCN	BESS total life cycle number
$\eta^{ES,\ Dis}/\eta^{ES,\ Chr}$	BESS discharging/charging efficiencies
$\underline{P^{CG}}/\overline{P^{CG}}$	Lower/upper limits of CDG power output
P_t^L	Electricity load
P_{it}^{RES}	Forecasted renewable power
$\overline{P}^{ES,Dis}/\overline{P}^{ES,Chr}$	Upper limits of BESS discharging/charging power
$\underline{P_i^M}/\overline{P_i^M}$	Lower/upper limits of exchanged power
$\underline{P}^{Exch}/\overline{P}^{Exch}$	Lower/upper limits of interconnection exchange between a MGC and distribution network
ρ_{it}	Price of exchanged power at time t
ρ_t^C	Price of exchanged power between MGC and the distribution network
$Ramp_{CG}^{Up}/Ramp_{CG}^{Up}$	Ramping up/down limits of CDG
$\underline{SOC}/\overline{SOC}$	Lower/upper limits of state of charge
$SUC_{it}^{CG}/SDC_{it}^{CG}$	Start-up/shut-down costs of CDG
γ^{ES}	Battery lifetime depression coefficient
$\underline{\varsigma}/\overline{\varsigma}$	Minimum/maximum ratio of controllable load

Variables

P_t^{CG}	CDG power output
$P_t^{ES,Dis}/P_t^{ES,Chr}$	BESS discharging/charging power
P_t^{CL}	The amount of controllable load
P_{it}^M	Exchanged power of the ith microgrid
$P_t^{C,M}$	Exchanged power amount between MGC and the distribution network
χ_t^{CG}	Commitment status indicator of a CDG
$\chi_t^{ES,Dis}/\chi_t^{ES,Chr}$	BESS discharging/charging indicator
χ_t^{SU}/χ_t^{SD}	Start-up/shut-down indicator of a CDG

8.1 Introduction

Heightened concerns about energy resource limits, climate change, as well as increasing energy prices, have led countries to increased integration of renewable energy sources (RESs) into modern power systems, primarily in the form of solar photovoltaic panels and wind turbines [1]. A transition from fossil-based and nonrenewable fuels to renewable and sustainable energy is occurring around the world [2]. By the end of 2017, the global installed renewable capacity has reached 2180 GW, with solar capacity being around 390 GW and wind power capacity over 500 GW [3]. In such a situation, microgrids (MGs), a cluster of various distributed generators, energy storage systems, loads, and other onsite electric components, are emerging as an effective way to integrate the RESs in distribution networks and satisfy the end-user demands [4]. MGs have a critical role to play in transforming the existing power grid to a future smart grid, which usually operates in grid-connected modes to maximize benefits, and can also operate in islanded modes for enhancing system reliability in grid outage periods [5]. Multiple MGs can be connected to form a networked system. Compared with the traditional individual MG, networked MGs possess the capability of decreasing the network operation cost in grid-connected modes and reducing the amount of load shedding in islanded modes [6].

Energy management system (EMS) is used for optimally scheduling power resources and energy storage systems in MGs to maintain supply-demand balance [4]. Numerous studies have examined the intelligent energy management of networked MGs, which can be categorized into centralized EMS, decentralized EMS, and hybrid EMS based on the architecture. For instance, Olivares et al. present a centralized EMS for isolated MGs, which use model predictive control technique to allow a proper dispatch of the energy storage units [7]. Wang et al. propose a decentralized EMS for the coordinated operation of networked MGs in a distribution system, which aim to minimize the operation cost in the grid-connected mode and maintain a reliable power supply in the island mode [8]. Wang and Mao investigate a hierarchical power scheduling approach to optimally manage power trading, storage, and distribution in a smart grid composed of a macrogrid and cooperative MGs [9]. The merits and demerits of the three prevailing EMSs have been compared and summarized in [10].

Alternately, considering the increasing penetration of RESs, new challenges have been imposed on the scheduling of MGs. RESs (i.e. solar and wind power) are intermittent and stochastic, which highly depend on environmental factors like solar irradiance and wind speed. Due to the uncertainty of renewable energy resources, uncertainty management in scheduling of MGs has become an active research area in recent years. Commonly adopted methods in the literature for MGs uncertainty management are robust optimization [11–14] and stochastic optimization techniques [15, 16]. Kuznetsova et al. present a robust optimization-based optimal energy management strategy to improve system operation performance [11]. In [12], Gupta develops a robust optimization approach to accommodate wind power uncertainty and achieve cost minimization in MGs. In [13], a robust optimization approach is proposed to optimally operate MGs. By collaboratively scheduling energy storage and direct load control, the uncertain outputs of RESs are addressed. By reviewing the literature, it can be found that most works on MGs scheduling by robust optimization method focus on single MG operation. However, the form of networked MGs is emerging given its unprecedented benefits, which requires the optimal operation of MGs with uncertainty management taken into account. Under this circumstance, Hussain et al. design a robust optimization based scheduling method for multi-MGs considering uncertainties in RESs and forecasted electric loads [14].

Stochastic optimization has also been widely used in the planning, operation, and control of MGs. Liang and Zhuang [15] present a detailed survey about stochastic modeling and optimization in a MG. In this survey, the key features of MGs are investigated and a comprehensive review on stochastic modeling and optimization tools for MGs is provided. In [16], a multi time-scale and multi energy-type coordinated MG scheduling solution is proposed. In the day-ahead scheduling model, the uncertainties of RESs are represented by multiple scenarios and the EMS objective is to minimize the MG operation cost. In a real-time dispatch model, the fluctuations of RESs are smoothed out by cooling loads and electrical energy. The prominent defects of applying stochastic optimization on MGs uncertainty management are the high computational requirements when the number of scenarios increases, as well as only providing probabilistic guarantees for constraint satisfaction

[5]. In contrast, robust optimization is immune against all possible realizations of uncertain data within the uncertainty sets. However, shortcomings also exist in this method. Through optimizing the worst-case scenario, robust optimization approaches could result in over-conservative results in MGs operation [14].

Review of the literature has identified that some issues remain open in the scheduling and dispatching of MGs. In [11–13, 16], the uncertainty management is conducted in an individual MG without realizing the emerging trends of networked MGs; and in others, although the uncertainty of RESs are considered in networked MGs, the authors only include it in day-ahead scheduling [14], but ignore RESs dynamic fluctuations in real-time operation. The availability of the high-quality data could potentially facilitate risk hedging and decision making in system operation and planning. The emergence of intelligent, data-driven innovative informatics approaches has the potential to achieve better prognosis and representation of the uncertainties and intermittencies involved in renewable energies [17]. As for the uncertainties in MGs, most of the researchers only consider RESs output uncertainty and electric load uncertainty, and neglect the uncertainty in forecasted day-ahead electricity prices. Moreover, although stochastic optimization has been widely used in MGs energy management, to the best of the authors' knowledge, no researchers have carried out in-depth analysis of risks as well as their impacts on the scheduling and dispatch of networked MGs. Farzan et al. develop stochastic programming optimization models to optimally schedule MGs under uncertainty with risk neutral and risk averse options [18]. Nevertheless, they only focus on the individual MG, without conducting risk analysis on networked MGs. For MGs optimal operation, uncertainties include load and renewable generation uncertainties. Conditional value-at-risk (CVaR) was proposed to represent the potential loss above the mean value [19].

Given that no previous research has considered the above concerns in a single work, a comprehensive research on the noted issues is necessary for the economic and stable operation of networked MGs To close this research gap, this work is focused on developing a two-stage energy management strategy for networked MGs under high penetration of RESs. Compared with previous MG scheduling and dispatching approaches, the distinguished features of the proposed method in this chapter are summarized as follows:

1) A hybrid energy management strategy is adopted for multi MGs. The individual **MG** is networked and regulated by a microgrid community (MGC) for enhancing the capability to accommodate the RESs fluctuations and improving layered privacy.
2) Based on the mean–variance Markowitz theory [20], a risk component is introduced into the optimal energy management of a MGC to estimate profit. The risk-based decision making could greatly influence the scheduling and dispatching results in MGs.
3) A two-stage energy management strategy is proposed for networked **MG**s considering the uncertainties of RESs outputs, electricity load and day-ahead prices. The operation costs are minimized in day-ahead scheduling control based on a hierarchical optimization method. The dynamic fluctuations of RESs and volatility of electricity load are smoothed out in the real-time dispatching stage.

The remainder of this chapter is organized as follows. Section 8.2 presents the problem description. Section 8.3 introduces the components modeling. Section 8.4 provides the mathematical formulation of optimal networked MGs operation. Section 8.5 describes the numerical simulations to demonstrate the effectiveness of the proposed approach. Section 8.6 concludes the chapter and suggests future research challenges.

8.2 Problem Description

In this section, the problem is briefly described, which includes the components and configuration of networked MGs, and the proposed operational strategy in this chapter.

8.2.1 Components and Configuration of Networked MGs

The basic components of a MG consist of RESs (i.e. photovoltaics system and wind turbine), controllable distributed generators (CDGs), battery energy storage systems (BESSs), and electric loads (i.e. both controllable and noncontrollable loads). RESs are able to generate clean and sustainable energy; CDGs, such as micro turbines, can provide stable energy to meet MGs energy demand. BESSs can shift energy to alleviate energy management difficulties through adjusting charging/discharging status; the controllable part in electric loads can help maintain power balance through demand response programs. From the perspective of a MG, the general objective in grid connected mode is to minimize the operation cost or maximize the total benefit under certain operational constraints.

In practical scenarios, some geographically adjacent MGs need to be coordinated controlled as a whole for certain goals, such as economy and reliability [21]. The rapid development of MGs also leads to the emergence of MG community. In view of the merits and promising applications of MGC in recent years [22], the MGs in this chapter are networked under the regulation of MGC. As shown in Figure 8.1, the individual MGs are connected in the MGC with close interaction among each other. MGs in MGC belong to different owners and may have different operation goals. A MGC consists of several MGs and MG community level devices, including community distribution generation units and community BESS. Compared with an individual MG, a MGC has to consider the

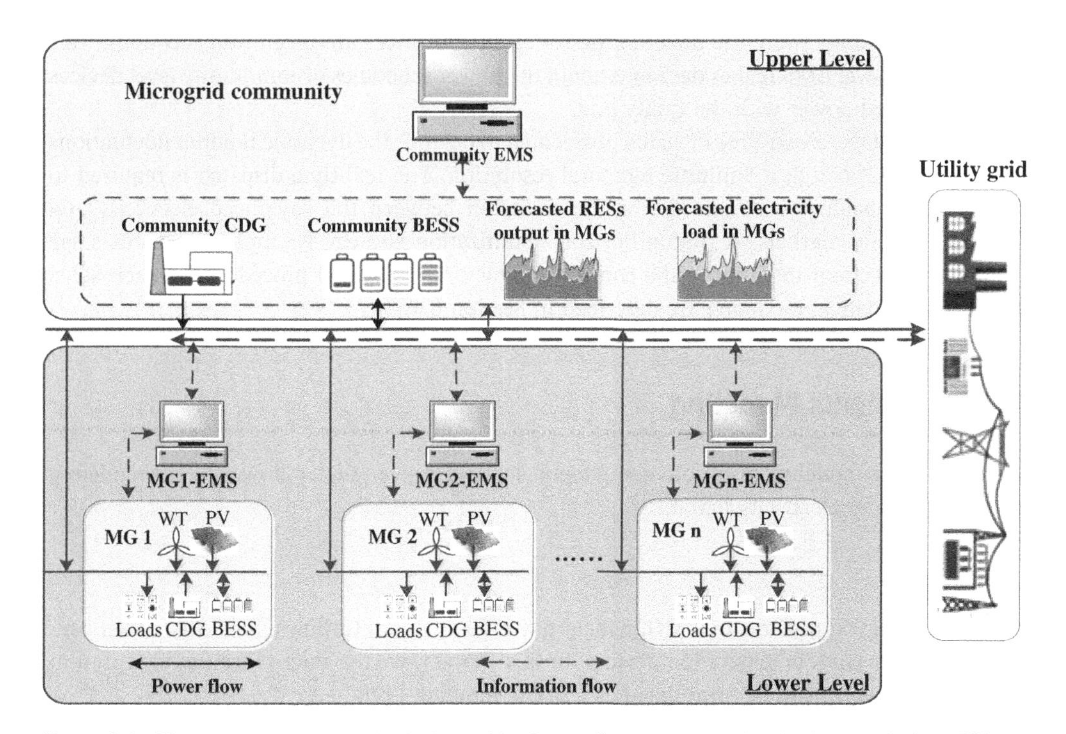

Figure 8.1 The structure of networked microgrids. *Source:* From Wang et al. [A]/with permission of Elsevier.

topology for interconnecting MG and has more levels of control to efficiently manage MGs and community level devices. The distinct features and benefits of a MGC have been detailed in [22]. Note that the energy management in the MGC is achieved in a hybrid way, with better performances than completely decentralized control and centralized control. According to [10], the hybrid EMS is emerging as a trade-off between centralized EMSs and decentralized EMSs. It has better flexibility compared with centralized EMSs and lesser operation cost compared with decentralized EMSs.

8.2.2 Proposed Strategy

In this chapter, a two-stage energy management strategy is proposed for networked MGs with high renewable penetration. The overall objective aims to minimize the operation cost of networked MGs in grid connected modes and predefine the revenue risk into a certain level.

At the first stage, a day-ahead hourly scheduling is formulated for networked MGs. In this stage, a hierarchical optimization strategy is adopted for the energy management of MGC. In the lower level, the optimization is focused on the individual MG and the objective is to minimize its operation cost. In this level, the problem is formulated as a deterministic issue without considering uncertainties in the MG. Through using the forecasted RESs output power, electrical load and electricity price, the lower level EMS determines the commitment status of CDGs, charging/discharging status of BESSs, and the shift or curtailment of controllable loads. Moreover, the exchanged power between the MG and MGC will be determined.

The upper level is to minimize the operation costs of the entire MGC with revenue risk considered. The uncertainties of each MG are collectively considered in the energy scheduling of MG community. Uncertainties including renewable resources and loads in individual MG are broadcast to MG community via information flow channels, and further incorporated into the risk control of networked MGs. In addition, the uncertainties of electricity prices are taken into account at this level. The upper level EMS makes decisions about the power schedules of community level devices and the exchanged power with the utility grid.

At the second stage, a real-time dispatch is executed to balance the dynamic random fluctuations of renewables and load at a 5-minute temporal resolution. The real-time dispatch is required to minimize the imbalance cost considering the deviation between the day-ahead electricity and real-time electricity markets. A rolling horizon optimization strategy is employed in this stage for online optimization to enhance the control accuracy. The detailed procedures in each stage and their mathematical modeling are described in Section 8.4.

8.3 Components Modeling

In this section, the modeling of various components in MGs is given first, followed by the modeling of uncertainties and electricity market.

8.3.1 CDGs

CDGs are flexible components in a MG, which may refer to micro turbines, fuel cells, diesel generators, and other types of generation devices. In this chapter, we use micro turbines to represent CDGs, whose fuel cost can be formulated as a linear function [12]:

$$C_t^{CG} = \chi_t^{CG} a^{CG} + b^{CG} P_t^{CG} \tag{8.1}$$

For other types of CDGs, such as diesel generators and gas-fired power generators, the fuel cost can be formulated as a quadratic function, which can be further approximated using piecewise linear functions.

The operation of a CDG should satisfy the following ramp rate limits and power constraints at each time period at each MG:

$$\underline{P^{CG}}\chi_t^{CG} \leq P_t^{CG} \leq \overline{P^{CG}}\chi_t^{CG};\ \chi_t^{CG} \in \{0, 1\} \tag{8.2}$$

$$\begin{cases} P_t^{CG} - P_{t-1}^{CG} \leq Ramp_{CG}^{Up} \\ P_{t-1}^{CG} - P_t^{CG} \leq Ramp_{CG}^{Down} \end{cases} \tag{8.3}$$

$$\chi_t^{SU} + \chi_t^{SD} \leq 1;\ \chi_t^{SU}, \chi_t^{SD} \in \{0, 1\} \tag{8.4}$$

$$\chi_t^{CG} - \chi_{t-1}^{CG} \leq \chi_t^{SU} - \chi_t^{SD} \tag{8.5}$$

where χ_t^{SU}, χ_t^{SD} are the start-up, shut-down indicator of a CDG (1 means it is in operation and 0 means it is not). Equation (8.2) is the power constraints, Eq. (8.3) is the ramp rate limits, Eq. (8.4) shows that a CDG cannot be started up and shut down simultaneously at any time, Eq. (8.5) shows the relationship between the start-up indicator and shut-down indicator.

8.3.2 BESSs

According to [23], the operation cost of BESSs usually refers to the maintenance cost, which can be formulated as a linear function as:

$$C_t^{ES} = \gamma^{ES} P_t^{ES,Dis}\Delta t + \gamma^{ES} E_t^{ES}\Delta t + \gamma^{ES} P_t^{ES,Chr}\Delta t \tag{8.6}$$

where, Δt is the time duration for converting power to energy. γ^{ES} is calculated as [23]:

$$\gamma^{ES} = \frac{IC^{ES}}{E_R^{ES} \cdot (LCN)} \tag{8.7}$$

BESSs should meet the following constraints during their operation:

$$E_{t+1}^{ES} = E_t^{ES} - P_t^{ES,Dis}\Delta t/\eta^{ES,Dis} + P_t^{ES,Chr}\Delta t\eta^{ES,Chr} \tag{8.8}$$

$$SOC_t = E_t^{ES}/E_R^{ES} \tag{8.9}$$

$$\underline{SOC} \leq SOC_t \leq \overline{SOC} \tag{8.10}$$

$$\begin{cases} 0 \leq P_t^{ES,Dis} \leq \chi_t^{ES,Dis} P^{ES,Dis} \\ 0 \leq P_t^{ES,Chr} \leq \chi_t^{ES,Chr} \overline{P}^{ES,Chr} \end{cases} \tag{8.11}$$

$$\chi_t^{ES,Dis} + \chi_t^{ES,Chr} \leq 1;\ \chi_t^{ES,Dis}, \chi_t^{ES,Chr} \in \{0, 1\} \tag{8.12}$$

$$E_t^{ES} = E_{INIT}^{ES} \text{ if } t = 1 \tag{8.13}$$

where SOC_t is BESS state of charge (SOC) at time t; E_{INIT}^{ES} is the initial stored energy in a BESS. Equation (8.8) shows BESS capacity change, which includes net energy injection and energy losses during charging/discharging process. Equations (8.9) and (8.10) define the BESS SOC constraints. Equation (8.11) limits BESS charging/discharging power capacity. Equation (8.12) means that a

BESS cannot operate in charging mode and discharging mode simultaneously. Equation (8.13) shows the initial energy stored in a BESS.

8.3.3 Controllable Load

Demand response programs are considered in the EMS strategy for adjusting the peak load demand. The controllable load cost is assumed to be a function of controllable load amount and can be represented by a linear function given in [24]:

$$C_t^{CL} = a^{CL} + b^{CL} P_t^{CL} \tag{8.14}$$

The maximum ratio of controllable load is constrained by:

$$\underline{\varsigma} P_t^L \le P_t^{CL} \le \overline{\varsigma} P_t^L \tag{8.15}$$

8.3.4 Uncertain Sets of RESs, Load, and Electricity Prices

In this chapter, RESs output, electricity load, and electricity price are regarded as uncertainties. The historical data in day-ahead market are used as correlated scenarios, hence allowing the correlated probability distributions to be estimated based on the statistical correlations among these uncertainties. Time-series-based methods, such as autoregressive integrated moving average model, are adopted here to generate correlated scenarios [25]. Wind power and solar power forecast errors can be modeled by the Beta distribution [26]:

$$f\left(\Delta P^{RES}; \lambda_1, \lambda_2\right) = \Delta P^{RES\lambda_1 - 1}\left(1 - \Delta P^{RES}\right)^{\lambda_2 - 1} N \tag{8.16}$$

where, ΔP^{RES} is RESs output forecast error, λ_1, λ_2 are the Beta distribution shape parameters, and N is the normalization error. Electricity price and load forecast errors can be modeled by the Gaussian distribution [23]:

$$f\left(\Delta x; \mu_x, \sigma_x^2\right) = \frac{1}{\sqrt{2\pi\sigma_x^2}} \exp\left[-\frac{\left(\Delta x - \mu_x\right)^2}{2\sigma_x^2}\right] \tag{8.17}$$

where, Δx is electricity price or load forecast error, μ_x,σ_x^2 are the mean and standard deviation.

8.3.5 Market Model

In a MGC system, power transactions are conducted between MGs and MGC, MGC and the distribution system operator. Two different electricity price mechanisms are set among MGs, MGC, and distribution system operator. In the view of distribution system operator, MGC is regarded as a price-taker, meaning that the electricity price between MGC and distribution system operator will not be influenced by the scheduling strategy and be determined by the electricity market. The market price is uncertain and denoted by the abovementioned electricity price uncertain set. As for the electricity strike price between MGC and MGs, the bilateral contract is built to reflect market participation. The market bidding strategies include many complex economic problems and are settled based on electricity generation, electricity load, and time of use electricity prices. The detailed steps for making the bilateral contract can be found in [20], which are not detailed here.

8.4 Proposed Two-Stage Operation Model

This section describes the mathematical formulation of the proposed two-stage operation model. The first stage is an hourly day-ahead optimal scheduling model, and the second stage is 5-minute real-time dispatch model.

8.4.1 Hourly Day-Ahead Optimal Scheduling Model

In this stage, considering the hierarchical control structure of an MGC, the hourly day-ahead scheduling is conducted by a two-level hierarchical control for the EMSs. The control structure is composed of the lower level MG energy management and the upper level MGC energy management, which is described below.

8.4.1.1 Lower Level EMS

Objective Function: The objective of the lower level EMS is to minimize the operation cost of individual MGs in the MGC while satisfying some equality and inequality constraints, as shown below:

$$min\ f_1 = \sum_{t=1}^{T} \begin{bmatrix} \left(C_{it}{}^{CG} + SUC_{it}^{CG}\chi_{it}^{SU} + SDC_{it}^{CG}\chi_{it}^{SD}\right) + \\ C_{it}^{ES} + C_{it}{}^{CL} + C_{it}^{M} \end{bmatrix} \tag{8.18}$$

The objective function in the individual MG is in a similar form with previous research [8, 22], which is composed of four terms: the fuel consumption cost, the BESS operation cost, the controllable load cost, and the exchanged power cost. Notably, the fuel consumption cost includes the generation, startup, and shutdown costs of CDG.

The exchanged power cost of the ith MG is calculated as:

$$C_{it}^{M} = \rho_{it} P_{it}^{M} \tag{8.19}$$

where, ρ_{it} is the price of exchanged power at time t, which can be derived from the bilateral contract. It is worth noting that when $P_{it}^{M} > 0$, P_{it}^{M} refers to the surplus power and ρ_{it} refers to the selling price; when $P_{it}^{M} < 0$, P_{it}^{M} refers to the power shortfall and ρ_{it} refers to the buying price.

The calculation of P_{it}^{M} is defined as:

$$P_{it}^{M} = P_{it}^{CG} + P_{it}^{ES,Chr} - P_{it}^{ES,Dis} + P_{it}^{RES} + P_{it}^{CL} - P_{it}^{L} \tag{8.20}$$

Constraints: To guarantee the stable operation of the MG, some equality and inequality constraints should be met.

1) *Power balance constraints:*

   For each MG, the total power generation from local sources and BESS should be equal to the local demand and exchanged power with other MGs.

$$P_{it}^{CG} + P_{it}^{ES,Chr} - P_{it}^{ES,Dis} + P_{it}^{RES} = P_{it}^{M} + P_{it}^{L} - P_{it}^{CL} \tag{8.21}$$

2) *CDG constraints:*

   The operation of a CDG is limited in (2)–(5).

3) *BESS constraints:*

   The operational constraints of a BESS are specified in (8)–(13).

4) *Controllable load constraints:*
   The controllable load amount is constrained by the minimum and maximum ratios defined in (15).
5) *Exchanged power constraints:*
   The exchanged power should be constrained by:

$$\underline{P_i^M} \leq P_{it}^M \leq \overline{P_i^M} \tag{8.22}$$

After the optimization, the lower level EMS can decide the unit start-up/shut-down schedule of CDGs, charging/discharging status of BESS, shift or curtailment amount of controllable load, and the exchanged power at each time interval.

8.4.1.2 Upper Level EMS

Objective Function: Similarly, the objective of the upper level EMS is to minimize the operation cost of a MGC by running a global optimization. In the meantime, given the uncertainties in the whole system, a risk control measure is introduced.

$$min\ f_2 = \sum_{t=1}^{T} \left(\begin{array}{l} (C_t^{C,CG} + SUC_t^{C,CG} \chi_t^{C,SU} + SDC_t^{C,CG} \chi_t^{C,SD}) \\ + C_t^{C,ES} + C_t^{C,M} \end{array} \right) \tag{8.23}$$

where C represents MG community. In the upper level objective function, there are three terms: the MG community fuel consumption cost, community BESS operation cost, and the cost of exchanged power with distribution system operator. Especially, the definitions of $C_t^{C,M}$ and $P_t^{C,M}$ are given by:

$$C_t^{C,M} = \rho_t^C P_t^{C,M} \tag{8.24}$$

$$P_t^{C,M} = P_t^{C,CG} + P_t^{C,ES} + \sum_{i=1}^{I} P_{it}^{RES} + \sum_{i=1}^{I} P_{it}^{M} - \sum_{i=1}^{I} (P_{it}^{L} - P_{it}^{CL}) \tag{8.25}$$

The uncertainties of RESs forecasting and electricity load forecasting in MGs are combined managed in MGC, as observed in (25).

Given the uncertainties in the whole system, we modify Eq. (8.23) into a probabilistic version to mitigate risky decision-making. In addition, to improve the computational efficiency, the initial large set of scenarios is trimmed to a small number of representative scenarios. In this chapter, an efficient scenarios reduction technique, i.e. backward method [26], is adopted to approximate the original scenario set. The risk associated with the cost variability is explicitly captured in the model through the mean–variance Markowitz theory [20]. Equation (8.23) can be rewritten in a probabilistic version as:

$$min\ \ E[O_1] + \varpi \cdot \sigma_{O_1} \tag{8.26}$$

where $E[O_1]$ is the expected operation cost, σ_{O_1} is the standard deviation, and $\varpi \in [0, +\infty)$ is the weighting factor for the inclusion of risk in the objective function. It should be noted that the higher the value of ϖ, the more risk averse. When ϖ=0, the strategy is risk neutral. The calculations of $E[O_1]$ and σ_{O_1} are given by:

$$E[O_1] = \sum_{k \in \Omega_K} \Pr_k \cdot f_{2k} \tag{8.27}$$

$$\sigma_{O_1}^2 = E\left[O_1^2\right] - E^2[O_1] = \sum_{k \in \Omega_K} \Pr_k \cdot f_{2k}^2 - \left(\sum_{k \in \Omega_K} \Pr_k \cdot f_{2k}\right)^2 \tag{8.28}$$

where $\Pr_k$ is the probability of scenario k, f_{2k} refers to the cost function f_2 under scenario k, and Ω_K is the set of reduced representative scenarios. The linearization of quadratic function in (28) has been widely investigated in the literature [27–29], therefore it is not detailed here.

Constraints: In each scenario k, some equality and inequality constraints should be met for the stable operation of an MGC. The CDG constraints and BESS constraints are the same with the lower level EMS, i.e. (2)-(5), (8)-(13). The different parts with the lower level EMS are given below:

1) *Power balance constraints:*

$$P_{tk}^{C,CG} + P_{tk}^{C,ES} + \sum_{i=1}^{I} P_{itk}^{RES} = P_{tk}^{C,M} - \sum_{i=1}^{I} P_{itk}^{M} + \sum_{i=1}^{I} \left(P_{itk}^{L} - P_{itk}^{CL}\right) \tag{8.29}$$

2) *Exchanged power constraints:*

$$\underline{P}^{Exch} \le P_{tk}^{C,M} \le \overline{P}^{Exch} \tag{8.30}$$

After performing the global optimization, the upper level EMS can decide the unit commitment of community CDGs, charging/discharging status of community BESSs, and the power exchange amount with the distribution network.

8.4.2 5-Minute Real-Time Dispatch Model

In real-time dispatch, the dynamic fluctuations of RESs and the volatility of electricity demand are accommodated in the operation of MGs. Note that the real-time dispatch interval could be any short uniform time interval (e.g. 15-minute interval and 5-minute interval). In this chapter, the proposed dispatching interval is assumed to be 5 minutes. For real-time electricity dispatch, a rolling horizon optimization strategy is employed to derive a more accurate value [16]. In rolling horizon optimization scheme, the model inputs are updated at each time step. And at each time step, the model is optimized, with the schedule results derived for all the remaining intervals. Under this circumstance, the schedules in the first interval are mandatory, while the other intervals are only used as references. At the next time step, the control window is moved forward, updating the model inputs and repeating the above procedures until the end of time horizon [30]. As the time scale is 5 minutes, the time window of the dispatch covers 288 intervals (i.e. 24 hours).

In this stage, the objective is to minimize the imbalance cost owing to the deviation between the first stage day-ahead electricity market and the second stage real-time electricity market, defined in a similar form as [31]:

$$min\ \mathrm{f}_3 = \sum_{t=1}^{NT} \left(C_t^{C,M} - \hat{C}_t^{C,M}\right) \tag{8.31}$$

where NT is the total number of dispatch intervals in the real-time market, $\hat{P}_t^{C,M}$ is the real-time dispatched power, and $\hat{C}_t^{C,M}$ is the cost of exchanged power in real-time market. Note that $(\hat{\bullet})$ is used to denote the variables in real-time market. The calculation of $\hat{C}_t^{C,M}$ is given in (32), (33):

$$\hat{C}_t^{C,M} = \hat{\rho}_t^C \cdot \hat{P}_t^{C,M} \tag{8.32}$$

$$\hat{P}_t^{C,M} = \hat{P}_t^{C,CG} + \hat{P}_t^{C,ES} + \sum_{i=1}^{I} \hat{P}_{it}^{RES} + \sum_{i=1}^{I} \hat{P}_{it}^{M} - \sum_{i=1}^{I} \left(\hat{P}_{it}^{L} - \hat{P}_{it}^{CL} \right) \tag{8.33}$$

The complete constraints of (31) include (2), (5), (8)–(13), and (15). In real-time dispatch, slight imbalances in each interval can be handled by automatic generation control or emergency demand response (i.e., instantaneous control).

The step-by-step procedure for carrying out the whole optimization is summarized in Figure 8.2. The formulated problems are based on mixed integer linear programming, which can be easily implemented through commercial software like CPLEX and academic software such as MATLAB.

8.5 Case Studies

8.5.1 Set Up

The proposed approach is tested on an artificial situation, which is based on a modified MGC system composed of three MGs in [32]. The system configuration is denoted in Figure 8.1 as well, specifying the power flow and information flow direction. By verifying the proposed method in an academic MG community, it could verify the feasibility of applying the approach into more complex real MG community applications in future. For instance, a university can be regarded as a MG community connected to a low voltage grid. Hospitals, student apartments, dining rooms, etc., equipped with renewable resources, BESSs, and CDGs can be taken as geographically adjacent MGs in the MG community.

The load in MGs mainly includes domestic appliances, lighting load, air-conditioning load in the buildings, and system devices, such as SCADA device and server. It is assumed that controllable load, such as thermostatically controlled load, exists in every MG and the minimum/maximum ratios are set to be 0 and 20%. The cost coefficients of controllable load a^{CL}, b^{CL} are set as \$0.33/kWh and \$0.05 based on the information provided in [24]. The lower and upper limits for exchanged power capacity in the MG are ±500 kW, and the values are ±1500 kW in the MGC. The parameters related to CDGs of each MG and MGC are tabulated in Table 8.1. BESS parameters in MGs and MGC are shown in Table 8.2. The simulation is coded on a 64-bit PC with 2.40 GHz processor and 8 GB RAM using MOSEK toolbox [33] in MATLAB platform.

Figure 8.3 denotes the day-ahead forecasted electricity load, RESs generation, and electricity prices over 24h on a typical day, which is based on data from the Australian Energy Market Operator [36]. Notably, in Figure 8.3a, the electricity load and RESs generation refer to the cumulative value in three MGs. As clearly observed from Figure 8.3a, there is energy surplus during the daytime and energy shortage during night-time periods. For Figure 8.3b, the electricity price between MGs and the MGC is determined by the bilateral contract, which is a fixed value even in the real-time market. As for the electricity price between the MGC and the distribution network operator (DSO), the value is forecasted and uncertain in the real-time market.

8.5.2 Results and Discussion

To verify the effectiveness of the proposed approach, two cases are compared: i) **Case 1**: An uncoordinated operation strategy. The individual MG operates strategically as the price-taker in the electricity market, aiming to minimize their operation costs. The system total operation cost is the

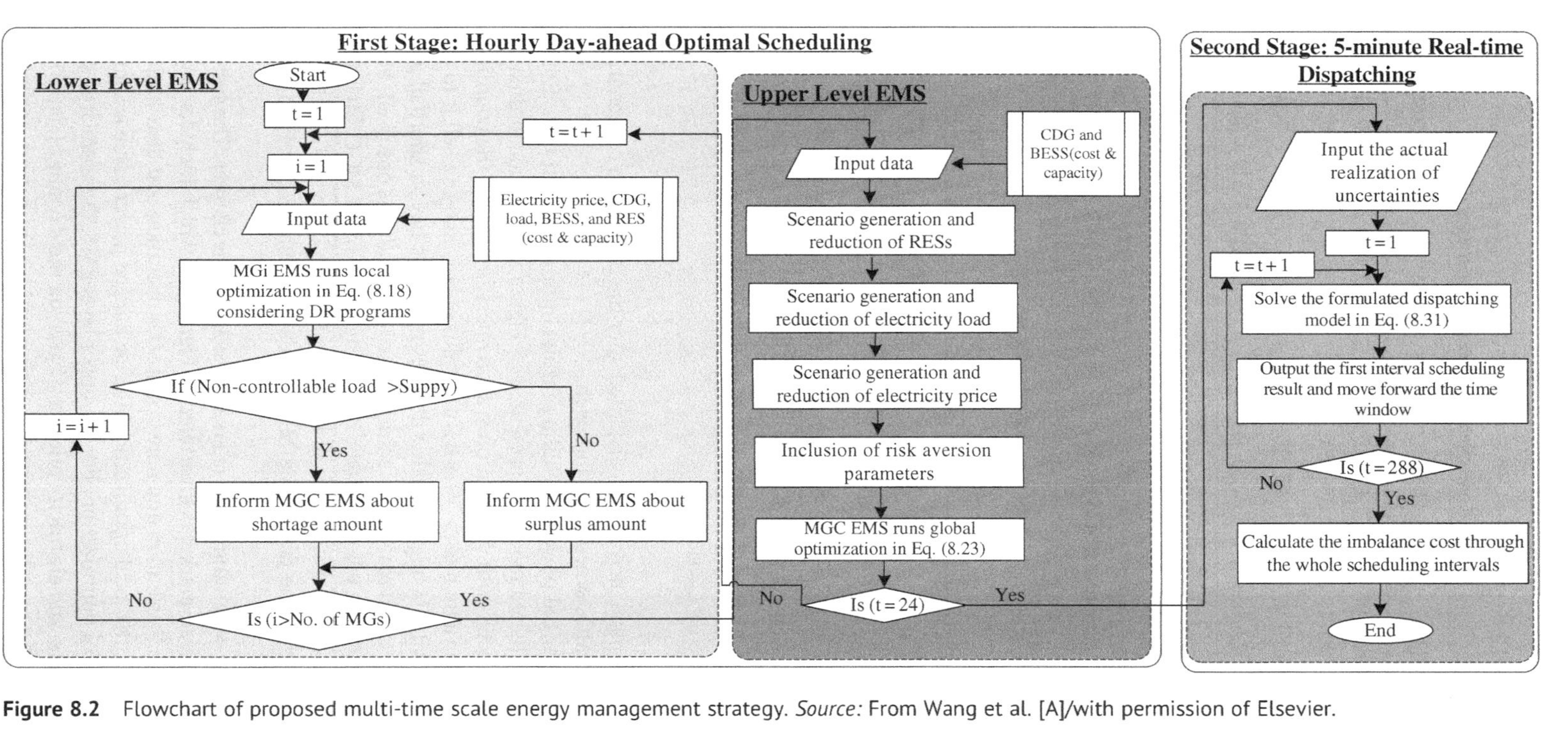

Figure 8.2 Flowchart of proposed multi-time scale energy management strategy. *Source:* From Wang et al. [A]/with permission of Elsevier.

Table 8.1 Parameters of CDGs in each microgrid and MGC [34].

	Controllable distributed generators			
Parameters	**MG1**	**MG2**	**MG3**	**MGC**
a^{CG}, b^{CG}(\$/kWh)	0.30, 0.05	0.22, 0.03	0.43, 0.04	0.31, 0.06
$\underline{P^{CG}}$,$\overline{P^{CG}}$(kW)	0, 200	0, 180	0, 160	0, 500
$Ramp_{CG}^{Up}$, $Ramp_{CG}^{Down}$(kW/h)	80, 85	75, 75	70, 80	220, 230
SUC^{CG},SDC^{CG}(\$)	0.32, 0.15	0.34, 0.18	0.35, 0.15	0.30, 0.20

Source: From Wang et al. [A]/with permission of Elsevier.

Table 8.2 Parameters of BESSs in each microgrid and MGC [35].

	Battery energy storage systems			
Parameters	**MG1**	**MG2**	**MG3**	**MGC**
IC^{ES}(\$), LCN (times)	$80E_R^{ES}$, 2000	$80E_R^{ES}$, 2000	$80E_R^{ES}$, 2000	$80E_R^{ES}$, 2000
E_R^{ES},E_{INIT}^{ES} (kWh)	200, 50	180, 40	220, 60	420, 150
$\overline{P}^{ES,Dis}$, $\overline{P}^{ES,Chr}$(kW)	150	125	160	200
$\underline{SOC}$, $\overline{SOC}$	20%, 80%	20%, 80%	20%, 80%	20%, 80%
$\eta^{ES,\ Dis}$, $\eta^{ES,\ Chr}$	0.95, 0.97	0.98, 0.96	0.95, 0.95	0.98, 0.95

Source: From Wang et al. [A]/with permission of Elsevier.

summation of individual's cost. The imbalance cost is calculated after the realization of scenarios. ii) **Case 2**: The proposed networked MGs scheduling approach.

Figure 8.4a demonstrates the exchanged power scheduling results in three MGs in Case 1. The exchanged power schedules in coordinated mode are presented in Figure 8.4b. In uncoordinated mode, MGs will exchange power directly with distribution system operator without the coordination of MGC. In the figure, a positive value means MGs have surplus energy and can sell it to the distribution system operator/MGC, while a negative value denotes that MGs have a shortfall and need to buy the corresponding amount with real-time electricity price externally. As observed, MGs in Case 2 have more surplus energy over the whole periods. By trading power in the MGC, MGs can support each other with a lower operation cost. As shown in Figure 8.4b, MG1 and MG2 sells the surplus energy most of the time during the day, while MG3 needs to purchase energy at most periods.

By trial and error methods [23], the weighting factor ϖ is set to be 3.6 in this chapter. The system operation cost distributions in day-ahead market is illustrated in Figure 8.5, which includes uncoordinated MGs overall operation cost distribution and networked MGs operation cost distribution. In the proposed control scheme, after the local optimization, the exchanged power value is sent to the upper level as constraints for global optimization. For the upper level EMS, the correlated scenarios of electricity load, RESs outputs, and electricity prices are based on the day-ahead data. The mean operation cost in Case 2 is −\$347.92 and the standard deviation is 93.35. In contrast, the mean

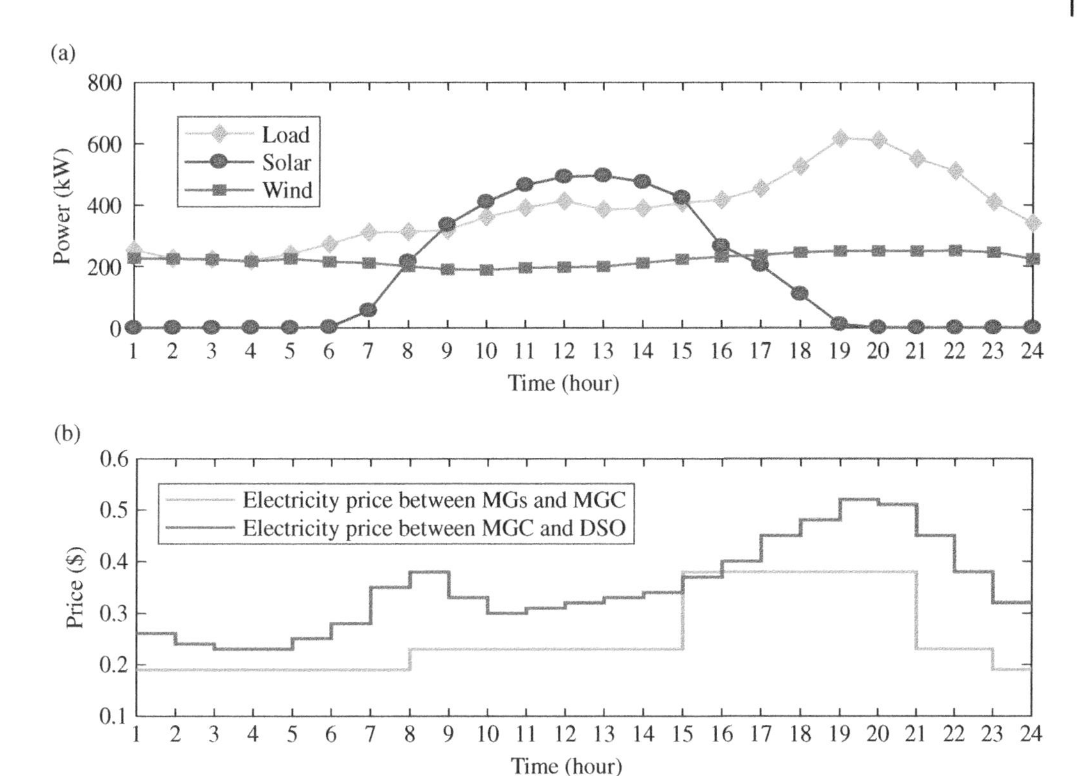

Figure 8.3 Day-ahead forecasted profiles (a) Forecasted electricity load and RESs generation. (b) Forecasted electricity prices. *Source:* From Wang et al. [A]/with permission of Elsevier.

operation cost in Case 1 is only −226.70 and the standard deviation is 97.14, which is higher than the standard deviation in Case 1. In addition, the value-at-risk (VaR) at the 95% confidence level is adopted for evaluating the risk of different operation strategies. The VaR-95% means the expected value of the 5% scenarios with the lowest operation cost, which is -$184.33 in Case 2 and -$52.63 in Case 1.

The power scheduling results of the BESS, CDG, and controllable load in both cases in day-ahead market are shown in Figure 8.6. In both cases, the BESS works in charging mode when electricity prices are relatively lower (i.e. early morning and late afternoon). Instead, BESSs are discharging at morning and evening peak periods, when electricity prices are relatively higher. Therefore, the profits made by BESSs are derived from the differences between peak and off-peak periods. It can be observed that more controllable load and CDG are used in Case 2. This is because for networked MGs operation, CDGs and controllable load are more frequently used to balance the power balance. Through the coordinated operation of various components of the MGs, the total power production follows the load curve.

Figure 8.7 illustrates the exchanged power between the system and the distribution system operator in deterministic day-ahead scheduling, risk-controlled day-ahead scheduling, and real-time dispatch under two cases. Due to the inclusion of risk hedging parameters in risk-controlled day-ahead scheduling, the deterministic scheduling has more surplus energy for selling and less shortage amount for purchasing in both cases. The deviations between real-time exchanged power and risk-controlled day-ahead exchanged power are caused by the dynamic fluctuations of RESs in

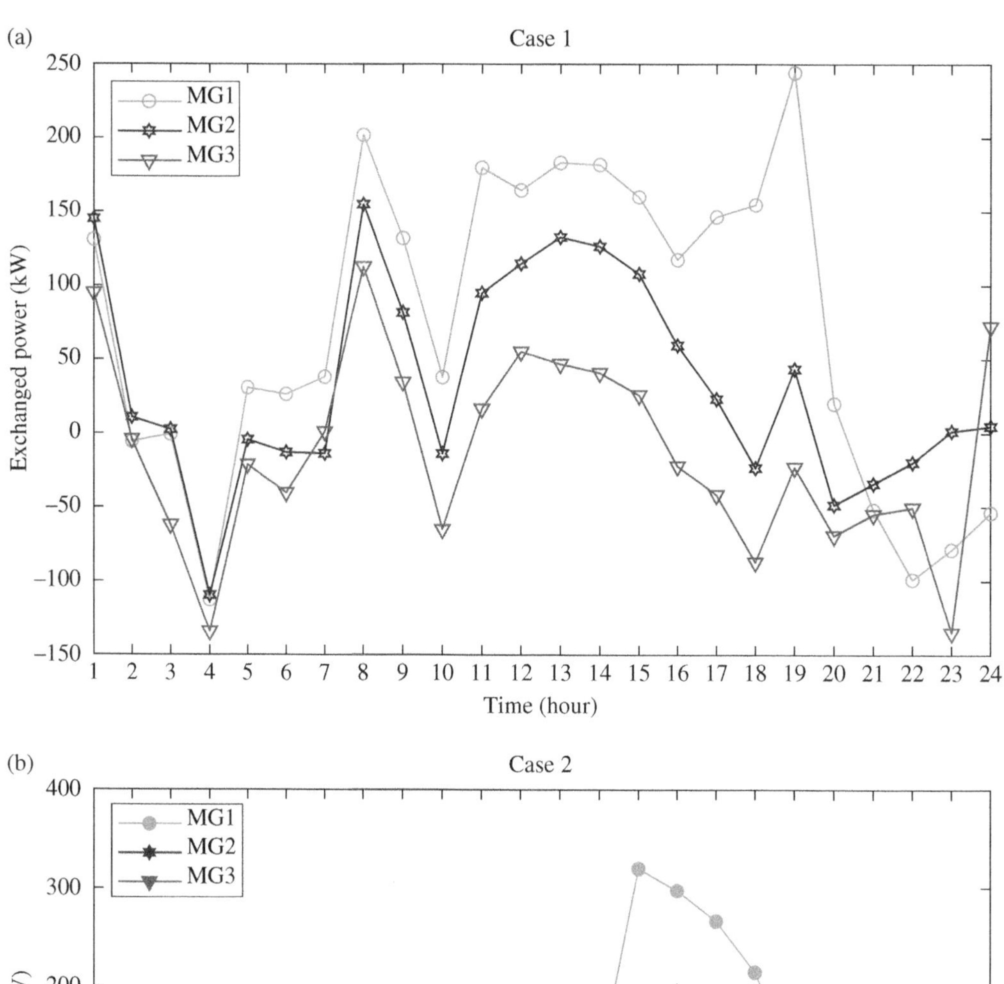

Figure 8.4 Exchanged power results in microgrids in Case 1 and Case 2. (a) With uncoordinated operation strategy. (b) With coordinated operation strategy. *Source:* From Wang et al. [A]/with permission of Elsevier.

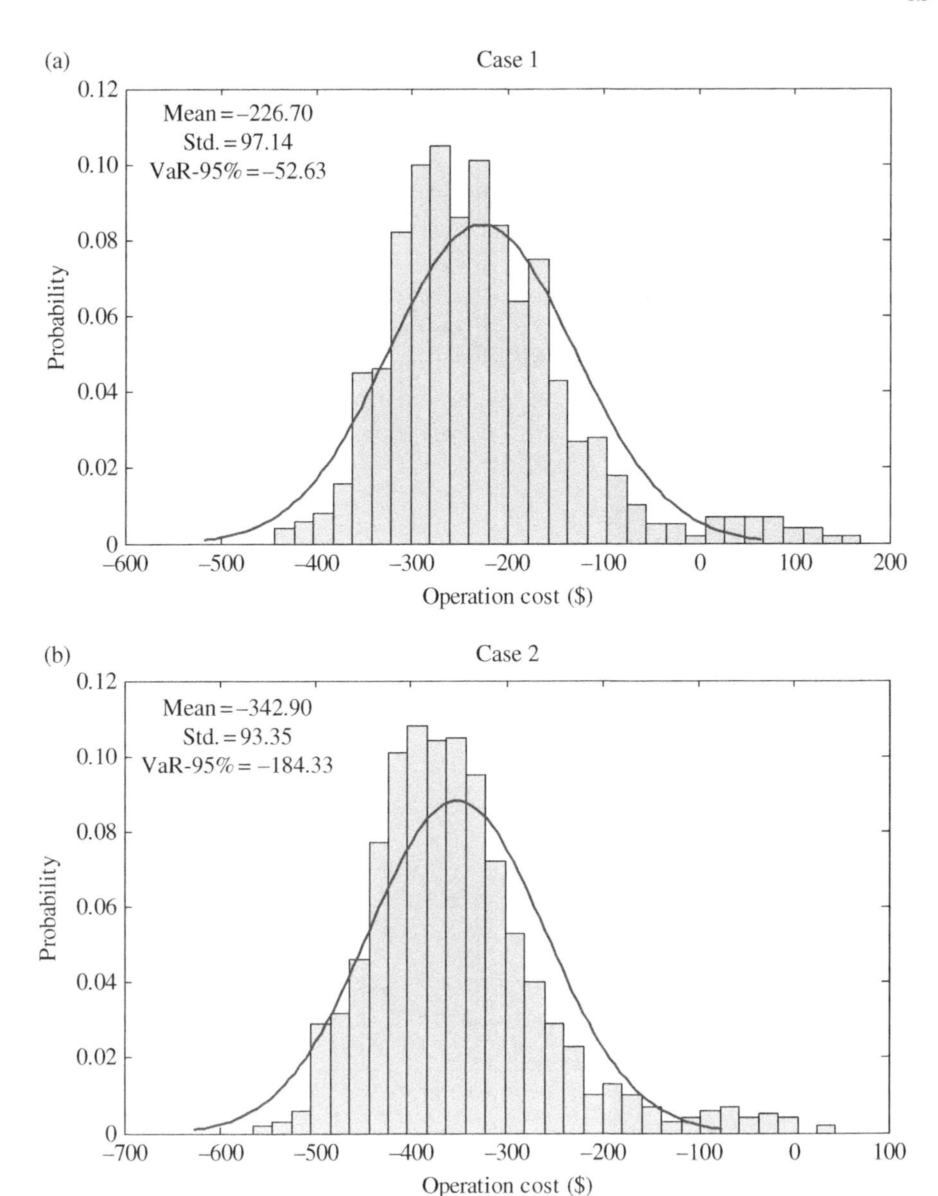

Figure 8.5 Distributions of system operation cost in day-ahead market in Case 1 and Case 2. (a) Uncoordinated microgrids overall operation cost distribution. (b) Networked microgrids operation cost distribution. *Source:* From Wang et al. [A]/with permission of Elsevier.

real time, and the forecast errors in electricity load and electricity price. Compared to Case 1, the results in Case 2 have more surplus power to trade in the electricity market. Hence, more profits are expected via the proposed approach. The real-time community BESS SOC status change is denoted in Figure 8.7b. Similar to day-ahead results, the BESS charging/discharging status corresponds to the real-time electricity price change.

The expected operation costs in the day-ahead and real-time markets for both cases are given in Tables 8.3 and 8.4. It can be found that the net operation cost in Case 1 is -$211.87 and the net

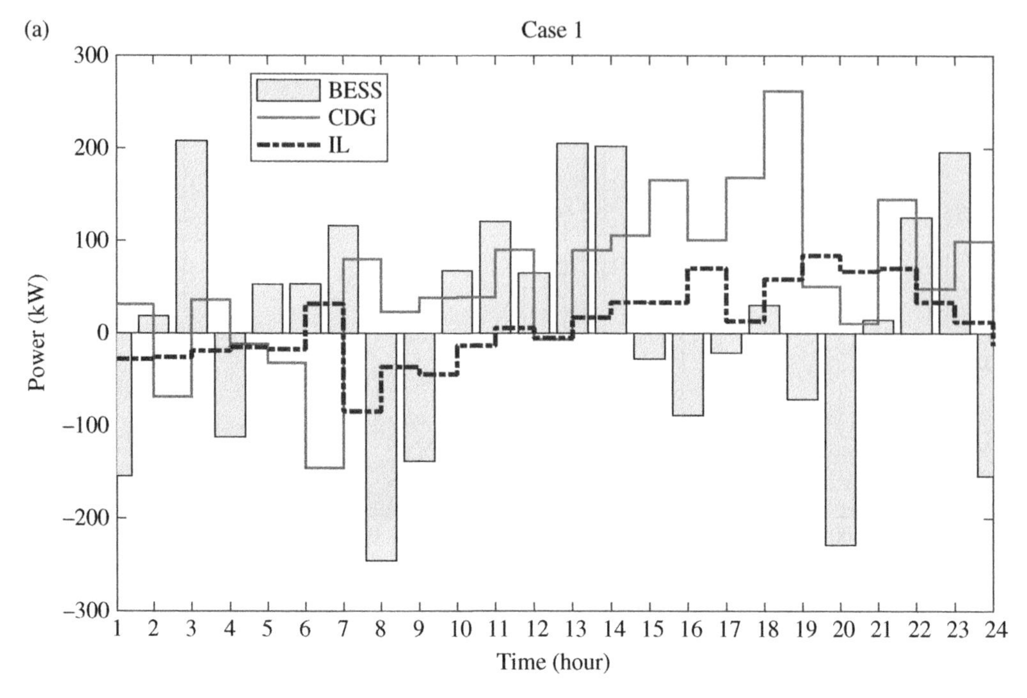

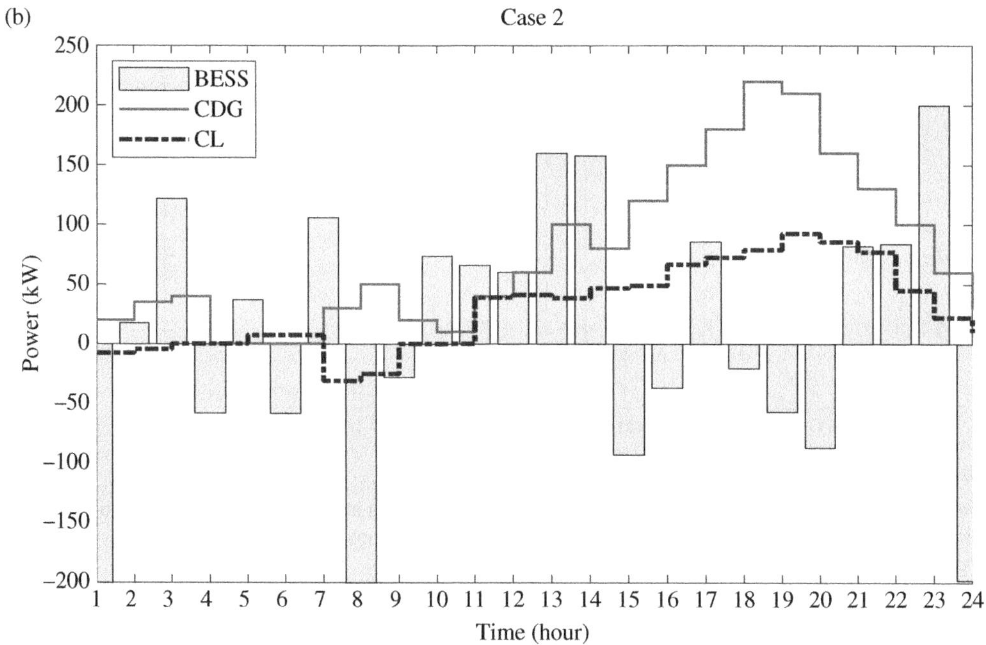

Figure 8.6 (a) Summation of power scheduling of BESS, CDG, and controllable load in three microgrids. (b) Power scheduling of BESS, CDG, and controllable load in MGC. *Source:* From Wang et al. [A]/with permission of Elsevier.

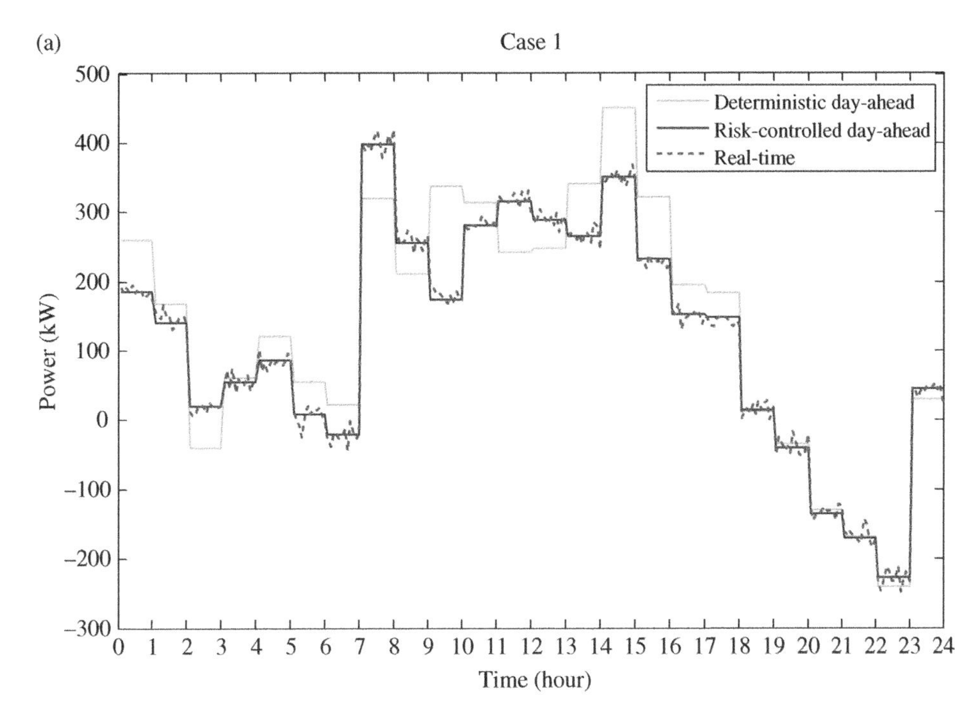

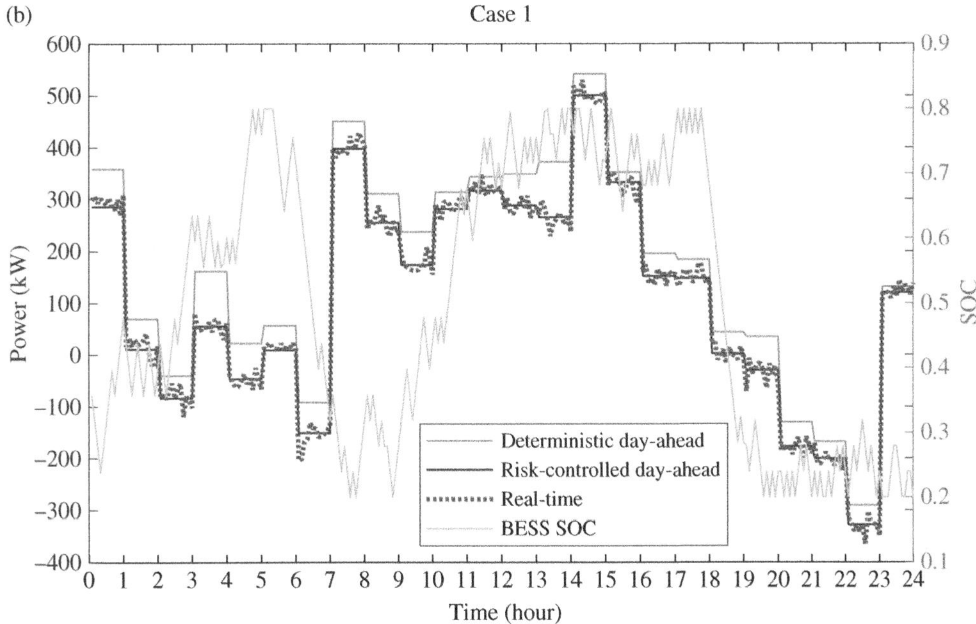

Figure 8.7 (a) Summation of exchanged power in individual microgrids in day-ahead market and real-time market. (b) MGC exchanged power amount in day-ahead market and real-time market and community BESS state of charge change in real-time market. *Source:* From Wang et al. [A]/with permission of Elsevier.

Table 8.3 Operation cost for microgrids components in Case 1.

	CDG	BESS	Controllable load	Total
Expected operation cost ($)	−96.3683	−90.1748	−47.0369	−233.58
Imbalance cost ($)	2.5271	4.4124	14.7697	21.71
Net operation cost ($)	−93.8412	−85.7624	−32.2672	−211.87

Source: From Wang et al. [A]/with permission of Elsevier.

Table 8.4 Operation cost for microgrids components in Case 2.

	CDG	BESS	Controllable load	Total
Expected operation cost ($)	−164.907	−98.8394	−89.6634	−353.41
Imbalance cost ($)	4.2996	1.1404	16.4120	21.85217
Net operation cost ($)	−160.608	−97.699	−73.2513	−331.558

Source: From Wang et al. [A]/with permission of Elsevier.

operation cost in Case 2 is -$331.558. Therefore, more profits can be made by the proposed approach and the economic superiority is verified.

Considering current testing system is in an academic situation, future research can be done toward more complex practical systems. In addition, with more MGs participating in the energy trading system, a blockchain-based transactive energy platform can be designed to efficiently share resources in a peer-to-peer manner [37, 38].

8.6 Conclusions

This chapter proposes a two-stage, i.e. an hourly day-ahead scheduling and 5-minute real-time dispatch, energy management strategy for networked MGs in the presence of high renewable penetration. In the day-ahead scheduling stage, a hybrid EMS control method is adopted considering the hierarchical structure of networked MGs The control objective is to minimize the operation cost on a daily basis and the operation cost variations are captured by incorporating mean–variance Markowitz theory into the objective function. Uncertainties on renewable energy resources output, electricity load, and electricity price are addressed in the first stage. In real-time dispatch stage, the objective is to minimize the imbalance cost given the deviations in the day-ahead and real-time markets. According to the simulation results, the proposed method identifies a techno-economic plan for network MGs under the regulation of MG community as well as provides a risk-hedging strategy. Compared with previous research, it is advantageous in, (i) adopting a hybrid EMS for networked MGs with the presence of MG community, (ii) in depth analyzing the risks of low profit scenarios by incorporating mean–variance Markowitz theory based risk factors, and (iii) comprehensively evaluating uncertainties in the operation system and mitigating dynamic fluctuations of renewable resources in real-time dispatch stage.

Acknowledgements

The permission given to use the materials in the following paper is very much appreciated.

[A] Wang, D., Qiu, J., Reedman, L. et al. (2018). Two stage energy management for networked microgrids with high renewable penetrations. *Applied Energy* 226: 39–48.

References

1 Wang, C., Yan, J., Marnay, G. et al. (2018). Distributed energy and microgrid. *Applied Energy* 210: 685–689.

2 Juraz, J. and Ciapala, B. (2017). Integrating photovoltaics into energy systems by using a run-off-river power plant with pondage to smooth energy exchange with the power grid. *Applied Energy* 198: 21–35.

3 Renewable Capacity Statistics (2017). International Renewable Energy Agency. http://www.irena.org/publications/2017/Mar/Renewable-Capacity-Statistics-2017 (accessed 10 November 2020).

4 Jin, X., Wu, J., Mu, Y. et al. (2017). Hierarchical microgrid energy management in an office building. *Applied Energy* 208: 480–494.

5 Zhang, B., Li, Q., Wang, L., and Feng, W. (2018). Robust optimization for energy transactions in multi-microgrids under uncertainty. *Applied Energy* 217: 346–360.

6 Silvente, J. and Papageorgiou, L.G. (2017). An MILP formulation for the optimal management of microgrids with task interruptions. *Applied Energy* 206: 1131–1146.

7 Olivares, D.E., Canizares, C.A., and Kazerani, M. (July 2014). A centralized energy management system for isolated microgrids. *IEEE Trans Smart Grid* 5 (4): 1864–1875.

8 Wang, Z., Chen, B., and Kim, J. (Mar. 2016). Decentralized energy management system for networked microgrids in grid-connected and islanded modes. *IEEE Trans. Smart Grid* 7 (2): 1097–1105.

9 Wang, Y., Mao, S., and Nelms, R.M. (Dec. 2015). On hierarchical power scheduling for the macrogrid and cooperative microgrids. *IEEE Trans. Ind. Informat.* 11 (6): 1574–1584.

10 Hussain, A., Bui, V.H., and Kim, H.M. (May 2015). A resilient and privacy reserving energy management strategy for networked microgrids. *IEEE Trans. Smart Grid* 9 (3): 2127–2139.

11 Kuznetsova, E., Li, Y., Ruiz, G., and Zio, E. (2014). An integrated framework of agent-based modelling and robust optimization for microgrid energy management. *Applied Energy* 129: 70–88.

12 Gupta, R.A. and Gupta, N.K. (2015). A robust optimization based approach for microgrid operation in deregulated environment. *Energy Conversion and Management* 93: 121–131.

13 Zhang, C., Xu, Y., Dong, Z.Y., and Ma, J. (July 2017). Robust operation of microgrid via two-stage coordinated energy storage and direct load control. *IEEE Trans. Power Syst.* 32 (4): 2858–2868.

14 Hussain, A., Bui, V.H., and Kim, H.M. (April 2016). Robust optimization-based scheduling of multi-microgrids considering uncertainties. *Energies* 9 (4): 278–298.

15 Liang, H. and Zhuang, W. (April 2014). Stochastic modelling and optimization in a microgrid: a survey. *Energies* 7 (4): 2027–2050.

16 Bao, Z., Zhou, Q., Yang, Z. et al. (Sept. 2015). A multi time-scale and multi energy-type coordinated microgrid scheduling solution-part I: model and methodology. *IEEE Trans. Power Syst.* 30 (5): 2257–2266.

17 Xu, Z., Lai, L.L., Wong, K.P. et al. (October 2017). Guest editorial special section on emerging informatics for risk hedging and decision making in smart grids. *IEEE Transactions on Industrial Informatics* 13 (5): 2507–2510.

18 Farzan, F., Jafari, M.A., Masiello, R., and Lu, Y. (March 2015). Toward optimal day-ahead scheduling and operation control of microgrids under uncertainty. *IEEE Trans Smart Grid* 6 (2): 499–507.

19 Wang, D., Wu, R., Lai, C. S. et al., Interactive energy management for networked microgrids with risk aversion. *2019 IEEE International Conference on Systems, Man and Cybernetics (SMC2019)*, Bari, Italy, 6–9 October 2019.

20 Markowitz, H.M. (1959). *Portfolio Selection, Efficient Diversification of Investments*. New Haven, CT: Yale University Press.

21 Ren, L. et al. (2018). Enabling resilient distributed power sharing in networked microgrids through software defined networking. *Applied Energy* 210: 1251–1265.

22 Tian, P., Xiao, X., Wang, K., and Ding, R. (2016). A hierarchical energy management system based on hierarchical optimization for microgrid community economic operation. *IEEE Trans. Smart Grid* 7 (5): 2230–2241.

23 Qiu, J., Meng, K., Zheng, Y., and Dong, Z.Y. (2017). Optimal scheduling of distributed energy resources as a virtual power plant in a transactive energy framework. *IET Gen. Trans. & Dist.* 11 (13): 3417–3427.

24 Luo, F.J., Dong, Z.Y., Meng, K. et al. (2016). Short-term operational planning framework for virtual power plants with high renewable penetrations. *IET Renew. Power Gen.* 10 (5): 623–633.

25 Conejo, A.J., Carrion, M., and Morales, J.M. (2010). *Decision Making Under Uncertainty In Electricity Markets*. New York, USA: Springer.

26 Xu, Y., Dong, Z.Y., Zhang, R., and Hill, D.J. (2017). Multi-time scale coordinated voltage/var control of high renewable-penetrated distribution system. *IEEE Trans. Power Syst.* 32 (6): 4398–4408.

27 Zhang, H., Vittal, V., Heydl, G.T., and Quintero, J. (2012). A mixed integer linear programming approach for multistage security constrained transmission planning. *IEEE Trans. Power Systems* 27 (2): 1125–1133.

28 Alguacil, N., Motto, A.L., and Conejo, A.J. (2003). Transmission expansion planning: a mixed-integer LP approach. *IEEE Trans. Power Systems* 18 (3): 1070–1077.

29 Ahbari, T. and Bina, M.T. (2016). Approximated MILP model for AC transmission expansion planning: global solutions versus local solutions. *IET Generation, Transmission & Distribution* 10 (7): 1563–1569.

30 Wang, D., Meng, K., Gao, X. et al. (2017). Optimal air-conditioning load control in distribution network with intermittent renewables. *Journal of Mod. Power Syst. and Clean Energy* 5 (1): 55–65.

31 Qiu, J., Zhao, J., Wang, D., and Zheng, Y. (2017). Two-stage coordinated operational strategy for distributed energy resources considering wind power curtailment penalty cost. *Energies* 10 (7): 965–983.

32 Bui, V.H., Hussain, A., and Kim, H.M. (2018). A multiagent-based hierarchical energy management strategy for multi-microgrids considering adjustable power and demand response. *IEEE Trans. Smart Grid* 9 (2): 1323–1333.

33 MOSEK ApS (2015). The MOSEK optimization toolbox for MATLAB manual. Version 7.0 [Online]. http://docs.mosek.com/7.0/toolbox/ (accessed 3 August 2017).

34 Li, Z. and Xu, Y. (2018). Optimal coordinated energy dispatch of a multi-energy microgrid in connected and islanded modes. *Applied Energy* 210: 974–986.

35 Zheng, Y., Dong, Z.Y., Luo, F.J. et al. (2014). Optimal allocation of energy storage system for risk mitigation of DISCOs with high renewable penetrations. *IEEE Trans. Power Syst.* 29 (1): 212–220.

36 Australia Energy Market Operator, [Online]: https://www.aemo.com.au/ (Accessed 25/04/2018).

37 Zhang, C., Wu, J., Zhou, Y. et al. (2018). Peer-to-peer energy trading in a microgrid. *Applied Energy* 220: 1–12.

38 Zhao, Z., Guo, J., Luo, X. et al. (2020). Energy transaction for multi-microgrids and internal microgrid based on blockchain. *IEEE Access* 8: 144362–144372.

9

A Multi-agent Reinforcement Learning for Home Energy Management

Nomenclature

i/Ω^{NS}	Index/set of non-shiftable appliances
j/Ω^{PS}	Index/set of power-shiftable appliances
m/Ω^{TS}	Index/set of time-shiftable appliances
n/Ω^{EV}	Index/set of EVs
t/T	Index of time slot
λ_t^G	Electricity price in time slot t
$P_{it}^{d,NS}/P_{jt}^{d,PS}/P_{mt}^{d,TS}$	Energy consumption of non-shiftable appliance i /power-shiftable appliance j/time-shiftable appliance m in time slot t
$P_{nt}^{d,EV}$	Energy consumption of EV n in time slot t
E_t^{PV}/E_t^{PVs}	Solar panel output/surplus solar energy in time slot t
$P_{j,max}^{d,PS}/P_{n,max}^{d,EV}$	Upper bound of Energy consumption for power-shiftable appliance j/EV n

9.1 Introduction

According to the report from the U.S. Energy Information Administration (EIA) [1], in 2019, the U.S. residential energy consumption accounted for 21.2% of the total energy consumption. Due to the concerns on environmental protection and energy crisis, developing renewable energy sources is an effective way for helping us to meet the growing energy demand and reduce the usage of conventional energy sources. It is important to deploy energy renewable energy sources in the residential power system. Among all renewable energy sources, solar photovoltaic (PV) is an easily acceptable technology due to its lower cost and convenient deployment in rooftop. Along with PV power generation, energy storage technologies can be used in conjunction with PV panels to store its excess generation power and then released during periods when solar generation is not available [2, 3]. The combination of PV and energy storage could improve the utilization and robustness of the system. Similar to PV and battery energy storage, electric vehicles (EVs) have also received more attention these days for reducing vehicle CO_2 emissions in comparison with the traditional vehicles. EV sales in China have experienced rapid growth in recent years [4]. EV is different from other household appliances, and it has large capacity and requires higher charging

Smart Energy for Transportation and Health in a Smart City, First Edition. Chun Sing Lai, Loi Lei Lai and Qi Hong Lai.

power. Thus, uncoordinated plug-in elecric vehicle (PEV) charging at home will significantly increase the peak load, which will likely cause the overload of the distribution transformer [5] and will require the expansion of the distribution power capacity. Besides, taking the various household appliances into account, the home energy system is transferred into a complicated system in near future.

This discussion and trends motivate the development of energy management control methods for the home energy management system. Energy storage deployment in residential and commercial applications is an attractive proposition for ensuring proper utilization of solar PV power generation. Energy storage can be controlled and coordinated with PV generation to satisfy electricity demand and minimize electricity purchases from the grid. For optimal energy management, PV generation and load demand uncertainties need to be considered when designing a control method for the PV-based storage system. Another resource available at the residential level is the PEV which also has bi-directional power flow capability. The charging and discharging routines of the PEV can be controlled to help reduce the energy drawn from the power grid during peak hours. There has been extensive existing literature related to this for various objectives such as peak shaving, system cost minimization. For instance, Zheng et al. studied a dispatch strategy of energy storage to assist the power grid when the demand load exceeds a defined threshold [6]. However, the state of charge (SOC) of battery might suffer a low state when the load demand was heavy, e.g. in summer. Sun at al. aimed at minimizing the electricity purchase cost of the home equipped with PV and battery energy storage, and it used the dynamic programming (DP) to find the optimal dispatch between battery and grid. The DP method may suffer the curse of dimensionality and the EV was not considered in the system [7]. Wu at al. considered uncertainty of EV mobility for a stochastic control method to utilize generated solar energy but did not incorporate any PV-based storage [8]. Similarly, DP-based energy management strategy was proposed by Hafiz et al. [9] to minimize the home system cost and smooth the battery SOC, but the detailed results of power dispatch was not presented in the simulation. Di Giorgio and Pimpinella formulated the energy management as an event-driven binary linear programming problem [10]. However, it did not include the energy storage into energy system, and hence the benefits of adopting energy storage were not investigated. Cai et al. introduced the room temperature as a comfort index in the objective function in the home energy management, but the EV was not considered in its studied system [11].

With recent advances in communication technologies and smart metering infrastructures, users can schedule their real-time energy consumption via the home energy management (HEM) system [12]. Such actions of energy consumption scheduling are also referred to as demand response (DR), which balances supply and demand by adjusting elastic loads [13, 14]. Xu et al. had proposed both bottom-up and top-down models for demand response with agent-base approach and neural networks [15]. Many research efforts have been paid on studying HEM system from the demand side perspective. Luo et al. proposed a hierarchical energy management system for home microgrids with consideration of PV energy integration in day-ahead and real-time stages [16]. Luo et al. [17] reported a novel HEM system in finding optimal operation schedules of home energy resources, aiming to minimize daily electricity cost and monthly peak energy consumption penalty. Wu et al. proposed a stochastic programming-based dynamic energy management framework for the smart home with plug-in electric vehicle storage [18]. Pilloni et al. proposed a new smart HEM system in terms of the quality of experience, which depends on the information of consumer's discontent for changing operations of home appliances [19]. In Yu et al. [20] for the smart home equipped with heating, ventilation, and air conditioning, Yu et al. investigated the issue of minimizing electricity bill and thermal discomfort cost simultaneously from the perspective of a long-term time horizon. Sharma et al. proposed a multi-time and multi-energy building energy management system, which is modeled as a nonlinear quadratic programming problem [21]. Keerthisinghe et al. adopted an approximate DP method to develop a computationally efficient HEM system where temporal difference learning is adopted for scheduling distributed energy resources [22]. Huang et al. introduced a new approach for HEM system to solve a DR problem which is formulated using the chance-constrained programming optimization, combining the particle swarm optimization method and the two-point estimation method [23]. Till now, most studies related to HEM system adopt centralized optimization approaches. Generally, due to the assumption of the accurate uncertainty prediction, the optimization method is able to show the perfect performance. However, this assumption is not very reasonable in reality since the optimization model knows all environment information meanwhile removing all prediction errors. Besides, due to a large number of binary or integer variables involved, some of these methods may suffer from expensive computational cost.

As an emerging type of machine learning, reinforcement learning (RL) [24] shows excellent decision-making capability in the absence of initial environment information. The deployment of RL in decision-makings has considerable merits. First, RL seeks the optimal actions by interacting with the environment so it has no requirement for initial knowledge, which may be difficult to acquire in practice. Second, RL can be flexibly employed to different application objects by off-line training and on-line implementation, considering relative uncertainties autonomously. Third, RL is easier to implement in real-life scenarios as compared with conventional optimization methods. The reason is that RL can obtain the optimal results in a look-up table, so its computational efficiency is fairly high. In recent literature, the RL has received growing interests for solving energy management problems. Vázquez-Canteli et al. comprehensively summarizes the algorithms and modeling techniques for reinforcement learning for demand response. Interested readers can further refer to Vázquez-Canteli and Nagy [25]. Ruelens et al. proposed a batch RL-based approach for residential DR of thermostatically controlled loads with predicted exogenous data [26]. Wan et al. used deep RL algorithms to determine the optimal solutions of the EV charging/discharging scheduling problem [27]. RL was adopted by Lu et al. to develop a dynamic pricing DR method based on hierarchical decision-making framework in the electricity market, which considers both profits of the service provider and costs of customers [28]. Lu et al. proposed an hour-ahead DR algorithm to make optimal decisions for different home appliances [29]. Zhou et al. proposed a residential energy management method considering peer-to-peer trading mechanism, where the model-free decision-making process is enhanced by the fuzzy Q-learning algorithm [30]. A model-free DR approach for industrial facilities was presented by Huang et al. based on the actor-critic-based deep reinforcement learning algorithm [31]. Based on RL, Furuzan et al. proposed a multi-agent-based distributed energy management method for distributed energy resources in a microgrid energy market [32]. Mocanu et al. focused on the deep RL-based online optimization of schedules for the building energy management system [33]. Du and Li used the deep neural network and the model-free reinforcement learning to manage the energy in a multi-microgrid system [34]. Remani et al. presented a novel RL-based model for residential load scheduling and load commitment with uncertain renewable sources [35]. In consideration of the state-of-the-art HEM methods in this field, there are still two significant limitations. First, most HEM studies focus only on one category of loads, such as home appliance loads or EV loads, ignoring the coordinated decision-makings for diverse loads. This weakly reflects the operational reality. Second, the integration of renewables, especially solar PV generation, is rarely considered during the decision-making process. With the rapid growth of rooftop installation of residential PV panels [36], allocation of self-generated solar energy should be considered when scheduling residential energy consumption.

To address the above issues, this chapter proposes a novel multi-agent reinforcement learning-based data-driven HEM method. The hour-ahead home energy consumption scheduling problem is formulated as a finite Markov decision process (FMDP) with discrete time steps. The bi-objective of the formulated problem is to minimize the electricity bill as well as DR induced dissatisfaction cost. The main contributions of this chapter are threefold.

1) Under the data-driven framework, we propose a novel model-free and adaptable HEM method based on extreme learning machine (ELM) and *Q*-learning algorithm. To our best knowledge, such method is rarely investigated earlier. The test results show that the proposed HEM method can not only achieve promising performance in terms of reducing electricity cost for householders but also improve the computational efficiency.
2) The conventional HEM methods are based on optimization algorithms with the assumption of perfect uncertainty prediction. However, this assumption is infeasible and unreasonable since the prediction errors are unavoidable. By contrast, our proposed model-free data-driven-based HEM method can overcome the future uncertainties by the ELM-based NN and discover the optimal DR decisions by the learning capability of the *Q*-learning algorithm.
3) In confronting with different types of loads in a residential house (e.g. non-shiftable loads, power-shiftable loads, time-shiftable loads, and EV charging loads), a multi-agent *Q*-learning algorithm-based RL method is developed to tackle the HEM problem involved with multiple loads. In this way, optimal energy consumption scheduling decisions for various home appliances and EV charging can be obtained in a fully decentralized manner.

The remainder of this chapter is organized as follows. Section 9.2 models the home energy consumption scheduling problem as a FMDP. Then our proposed solution approach is presented in Section 9.3. In Section 9.4, test results are given to demonstrate the effectiveness of our proposed methodology. Finally, Section 9.5 concludes the chapter.

9.2 Problem Modeling

As illustrated in Figure 9.1, this chapter considers four agents in a HEM system, which correspond to non-shiftable appliance load, power-shiftable appliance load, time-shiftable appliance load, and EV load, respectively.

In this chapter, the proposed HEM system includes multiple agents, which are virtual to control different kinds of smart home appliances in a decentralized manner. It should note that smart meters are assumed to be installed on smart home appliances to monitor the devices and receive the control command given by the agents. In each time slot, we determine the hour-ahead energy consumption actions for home appliances and EVs. Specifically, in time slot t, the agent observes the state s_t and chooses the action a_t. After taking this action, the agent observes the new state s_{t+1} and chose a new action a_{t+1} for the next time slot $t+1$. This hour-ahead energy consumption scheduling problem can be formulated as a FMDP, where the outcomes are partly controlled by the decision-maker and partly random. The FMDP of our problem contains five tuples, i.e. ($\boldsymbol{S}$, $\boldsymbol{A}$, $\boldsymbol{R}(\cdot, \cdot)$, γ, θ), where $\boldsymbol{S}$ denotes the state set, $\boldsymbol{A}$ denotes the finite action set, $\boldsymbol{R}(\cdot,\cdot)$ denotes the reward set, γ denotes the discount rate, and θ denotes the learning rate. The details about the FMDP formulation are described as follows. Table 9.1 shows the state set, action set, and reward function of each agent.

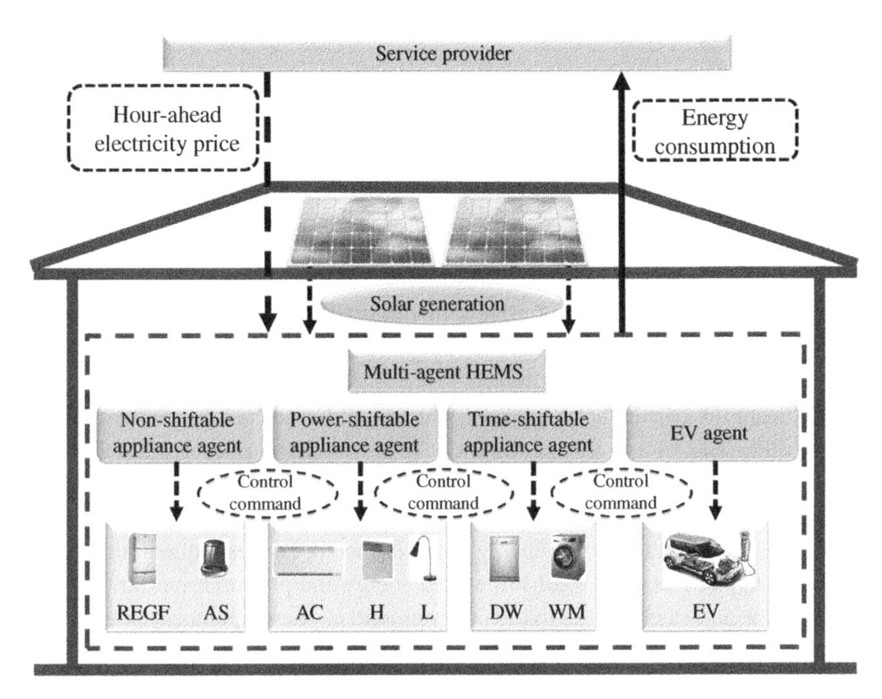

Figure 9.1 Structure of our proposed HEM system (REFG, refrigerator; AS, alarm system; AC, air conditioner; H, heating; L, lighting; WM, washing machine; DW, dishwashing; EV, electric vehicle). *Source:* From Xu et al. [A].

Table 9.1 State set, action set, and reward function of each agent.

Item ID	State set	Action set	Reward function
REFG	$\{(\lambda_t^G, \lambda_{t+1}^G, \dots, \lambda_T^G), (E_t^{PV}, E_{t+1}^{PV}, \dots, E_T^{PV})\}$	1 – "on"	Eq. (9.1)
AS	$\{(\lambda_t^G, \lambda_{t+1}^G, \dots, \lambda_T^G), (E_t^{PV}, E_{t+1}^{PV}, \dots, E_T^{PV})\}$	1 – "on"	Eq. (9.1)
AC1	$\{(\lambda_t^G, \lambda_{t+1}^G, \dots, \lambda_T^G), (E_t^{PV}, E_{t+1}^{PV}, \dots, E_T^{PV})\}$	{0.7, 0.8, ..., 1.4} – "power ratings"	Eq. (9.2)
AC2	$\{(\lambda_t^G, \lambda_{t+1}^G, \dots, \lambda_T^G), (E_t^{PV}, E_{t+1}^{PV}, \dots, E_T^{PV})\}$	{0.7, 0.8, ..., 1.4} – "power ratings"	Eq. (9.2)
H	$\{(\lambda_t^G, \lambda_{t+1}^G, \dots, \lambda_T^G), (E_t^{PV}, E_{t+1}^{PV}, \dots, E_T^{PV})\}$	{0.5, 0.6, ..., 1.5} – "power ratings"	Eq. (9.2)
L1	$\{(\lambda_t^G, \lambda_{t+1}^G, \dots, \lambda_T^G), (E_t^{PV}, E_{t+1}^{PV}, \dots, E_T^{PV})\}$	{0.2, 0.3, ..., 0.6} – "power ratings"	Eq. (9.2)
L2	$\{(\lambda_t^G, \lambda_{t+1}^G, \dots, \lambda_T^G), (E_t^{PV}, E_{t+1}^{PV}, \dots, E_T^{PV})\}$	{0.2, 0.3, ..., 0.6} – "power ratings"	Eq. (9.2)
WM	$\{(\lambda_t^G, \lambda_{t+1}^G, \dots, \lambda_T^G), (E_t^{PV}, E_{t+1}^{PV}, \dots, E_T^{PV})\}$	0 – "off" 1 – "on"	Eq. (9.3)
DW	$\{(\lambda_t^G, \lambda_{t+1}^G, \dots, \lambda_T^G), (E_t^{PV}, E_{t+1}^{PV}, \dots, E_T^{PV})\}$	0 – "off" 1 – "on"	Eq. (9.3)
EV	$\{(\lambda_t^G, \lambda_{t+1}^G, \dots, \lambda_T^G), (E_t^{PV}, E_{t+1}^{PV}, \dots, E_T^{PV})\}$	{0, 3, 6} – "charging rates"	Eq. (9.4)

REFG, refrigerator; AS, alarm system; AC, air conditioner; H, heating; L, light; WM, wash machine; DW, dishwashing; EV, electric vehicle.
Source: From Xu et al. [A].

9.2.1 State

The state s_t can describe the current situation in the FMDP. In this chapter, the state s_t in time slot t can be defined as a vector, defined as,

$$s_t = \{(\lambda_t^G, \lambda_{t+1}^G, \ldots, \lambda_T^G), (E_t^{PV}, E_{t+1}^{PV}, \ldots, E_T^{PV})\}$$

where s_t consists of two types of information,

1) $(\lambda_t^G, \lambda_{t+1}^G, \ldots, \lambda_T^G)$ indicate the current electricity price λ_t in and predicted future electricity prices from the next time slot $t+1$ to the end time slot T. In each time slot, the hour-ahead electricity price can be informed by a service provider.
2) $(E_t^{PV}, E_{t+1}^{PV}, \ldots, E_T^{PV})$ indicate the current solar panel output E_t^{PV} and predicted future solar panel outputs from the next time slot $t+1$ to the end time slot T. In this chapter, we assume that the householder owns a residential solar system, including the solar panels and the inverter systems [37], operating at the maximum power point (MPP) [38]. In each intraday time slot, the householder needs to allocate the generated solar energy to the home appliances sequentially. Consider that non-shiftable appliances always consume fixed energy and play an important role in ensuring the convenience and safety of the living environment, so these appliances are served first. Also, the comfort level of the living environment should be taken into account, so the surplus self-generated solar energy E_t^{PVs} is delivered according to the descending order of the dissatisfaction coefficients of the remaining home appliances. Finally, the surplus solar energy will be curtailed or sold to the utility grid with wholesale market clearing price.

9.2.2 Action

In this study, the action denotes the energy consumption scheduling of each home appliance as well as EV battery charging, described as follows:

1) *Action set for non-shiftable appliance agent*: Non-shiftable appliances, e.g. refrigerator and alarm system, require high reliability to ensure daily-life convenience and safety, so their demands must be satisfied and cannot be scheduled. Therefore, only one action, i.e. "on," can be taken by the non-shiftable appliance agent.
2) *Action set for power-shiftable appliance agent:* Power-shiftable appliances, such as air conditioner, heating, and light, can operate flexibly by consuming energy in a predefined range. Hence, power-shiftable agents can choose discrete actions, i.e. 1,2,3, ..., which indicate the power ratings at different levels.
3) *Action set for time-shiftable appliance agent:* The time-shiftable loads can be scheduled from peak periods to off-peak periods to reduce the electricity cost and avoid peak energy usage. Time-shiftable appliances, including wash machine and dishwasher, have two operating points, "on" and "off."
4) *Action set for EV agent:* As the EV user, the householder would like to reduce electricity cost by scheduling EV battery charging. It should be noted that in this chapter, EV battery discharging is not considered since it can significantly shorten the useful lifetime of EV battery [39]. Le Floch et al. suggested that by EV charger can provide discrete charging rates [40].

9.2.3 Reward

The reward represents the inverse utility cost of each agent, described as follows:

1) The reward of non-shiftable appliance agent

$$r_{it}^{NS} = -\lambda_t^G \left[P_{it}^{d,NS} - E_{it}^{PVs}\right]^+ \quad i \in \Omega^{NS}, t = \{1, 2, \ldots, T\} \tag{9.1}$$

The reward of non-shiftable appliance agent only concerns on electricity cost since the non-shiftable loads are immutable. Note that $[\cdot]^+$. represents the projection operator onto the nonnegative orthant, i.e. $[x]^+ = max(x, 0)$

2) The reward of power-shiftable appliance agent

$$\begin{aligned} r_{jt}^{PS} &= -\lambda_t^G \left[P_{jt}^{d,PS} - E_{jt}^{PVs}\right]^+ - \alpha_j^{PS}\left(P_{j,max}^{d,PS} - P_{jt}^{d,PS}\right)^2 \\ j &\in \Omega^{PS}, t = \{1, 2, \ldots, T\} \end{aligned} \tag{9.2}$$

where the first term denotes the electricity cost and the second term is the dissatisfaction cost caused by reducing power ratings of power-shiftable appliances. This dissatisfaction cost is defined by a quadratic function [41] with an appliance-dependent coefficient α_j^{PS}, which can be adjusted to achieve a trade-off between the electricity cost and the satisfaction level.

3) The reward of time-shiftable appliance agent

$$\begin{aligned} r_{mt}^{TS} &= -\lambda_t^G \left[u_{mt} P_{mt}^{d,TS} - E_{mt}^{PVs}\right]^+ - \alpha_m^{TS}\left(t_m^s - t_m^{ini}\right)^2 \\ m &\in \Omega^{TS}, t = \left[t_m^{ini}, t_m^{end}\right] \end{aligned} \tag{9.3}$$

where u_{mt}.. is the binary variable representing the operating point of the time-shiftable appliance m in time slot t, i.e. $u_{mt} = 1$ (on) or $u_{mt} = 0$ (off). When the time-shiftable loads are scheduled, dissatisfaction cost of householder would be raised due to the waiting time for them to start. Therefore, electricity bill (first term) and dissatisfaction cost (second term) should be taken into account simultaneously for operating time-shiftable appliancesα_m^{TS} is the dissatisfaction coefficient describing the tolerance of waiting time for the appliance m and it is determined by personal dependence on devices. Thus, a higher α_m^{TS} means that the wait for appliance m start is more likely to cause dissatisfaction. Note that time-shiftable appliance m should start to operate during its normal working period $\left[t_m^{ini}, t_m^{end}\right]$

4) The reward of EV agent

$$\begin{aligned} r_{nt}^{EV} &= -\lambda_t^G P_{nt}^{d,EV} - \alpha_n^{EV}\left(P_{n,max}^{d,EV} - P_{nt}^{d,EV}\right)^2 \\ n &\in \Omega^{EV}, t \in \left[t_n^{arr}, t_n^{dep}\right] \end{aligned} \tag{9.4}$$

where the first two terms of (9.4) describe that the EV owner needs to pay the electricity cost ($\lambda_t^G P_{nt}^{d,EV}$) during the period $[t_n^{arr}, t_n^{dep}]$. Besides, the second term of (9.4) represents the cost of "charging anxiety" with the dissatisfaction coefficient α_n^{EV}, which describes the fear that the EV has insufficient energy to get its destination without underfilled EV battery.

9.2.4 Total Reward of HEM System

After giving the rewards (see Eqs. (9.1)–(9.4)) of all agents in the proposed HEM system, the total reward R can be acquired, described as follows,

$$R = -\sum_{t \in T} \left\{ \begin{array}{l} \lambda_t^G \left(\begin{array}{l} \left[P_{it}^{d,NS} - E_{it}^{PVs}\right]^+ - \left[P_{jt}^{d,PS} - E_{jt}^{PVs}\right]^+ \\ - \left[u_{mt} P_{mt}^{d,TS} - E_{mt}^{PVs}\right]^+ - P_{nt}^{d,EV} \end{array} \right) \\ - \left(\begin{array}{l} \alpha_j^{PS} \left(P_{j,max}^{d,PS} - P_{jt}^{d,PS}\right)^2 - \alpha_m^{TS} \left(t_m^s - t_m^{ini}\right)^2 \\ - \alpha_n^{EV} \left(P_{n,max}^{d,EV} - P_{nt}^{d,EV}\right)^2 \end{array} \right) \end{array} \right\} \tag{9.5}$$

9.2.5 Action-value Function

The quality of the action a_t under the state s_t, i.e. energy consumption scheduling in time slot t, can be evaluated by the expected sum of future rewards for the horizon of K time steps, given as follows,

$$Q_\pi(s, a) = \mathrm{E}_\pi \left[\sum_{k=0}^{K} \gamma^k \cdot r_{t+1} \mid s_t = s, a_t = a \right] \tag{9.6}$$

where $Q_\pi(s, a)$ represents the action-value function and π is the policy mapping from a system state to an energy consumption schedule. $\gamma \in [0, 1]$ is the discount rate denoting the relative importance of future rewards for the current reward. When $\gamma = 0$, the agent seems to be shortsighted since it only cares about the current reward, while $\gamma = 1$ indicates that the agent is foresighted and it considers future rewards. To balance the trade-off between current reward and future reward, setting a fraction in the range [0, 1] for γ is suggested.

The objective of the energy consumption scheduling problem is to find the optimal policy π^*, i.e. a sequence of optimal operating actions for each home appliance and EV battery, to maximize the action-value function.

9.3 Proposed Data-Driven-Based Solution Method

In this chapter, the proposed reinforcement learning-based data-driven method is comprised of two parts (see Figure 9.2), (i) an ELM-based feedforward NN is trained for predicting the future trends of electricity price and PV generation, (ii) a multi-agent Q-learning algorithm-based RL method is developed for making hour-ahead energy consumption decisions. Details of this data-driven-based solution method are given in the following subsections.

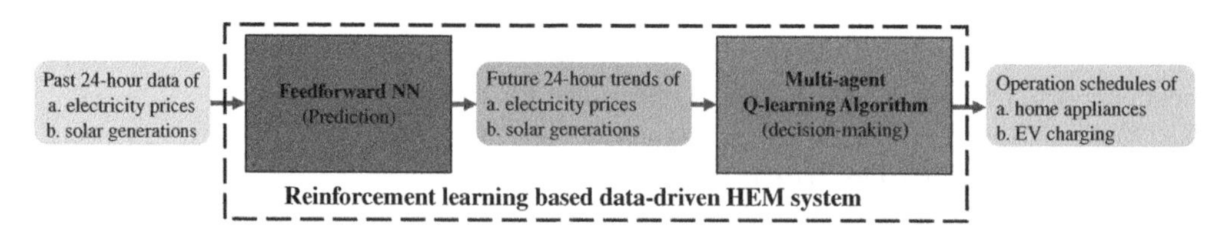

Figure 9.2 Schematic of the reinforcement learning-based data-driven HEM system.

9.3.1 ELM-Based Feedforward NN for Uncertainty Prediction

As a well-studied training algorithm, ELM algorithm has become a popular topic in the fields of load forecasting [42], electricity price forecasting [43], renewable generation forecasting [44], and probabilistic PV power prediction [45]. Since the input weights and biases of the hidden layer are randomly assigned and free to be tuned further when using ELM algorithm, some exceptional features can be obtained, e.g. fast learning speed and good generalization. To deal with the uncertainties of electricity prices and solar generations, we propose an ELM-based feedforward NN to dynamically predict future trends of these two uncertainties. Specifically, at each hour, the inputs of the trained feedforward NN are past 24-hour electricity price data and solar generation data, and its outputs are the forecasted future 24-hour trends of electricity prices and solar generations. This predicted information will be fed into the decision-making process of energy consumption scheduling, as described in the following subsection.

9.3.2 Multi-Agent *Q*-Learning Algorithm for Decision-Making

After acquiring the predicted future electricity prices and solar panel outputs, we employ the *Q*-learning algorithm to use this information to find the optimal policy π^*. As an emerging machine learning algorithm, *Q*-learning algorithm is widely used for the decision-making process to gain the maximum cumulative rewards [46]. The basic mechanism of this algorithm is to construct a *Q*-table where *Q*-value $Q(s_t, a_t)$ of each state-action pair is updated in each iteration until the convergence condition is satisfied. In this way, the optimal action with optimal *Q*-value in each state can be selected. The optimal *Q*-value $Q_\pi^*(s_t, a_t)$ can be obtained by using Bellman equation [47], given as below:

$$Q_\pi^*(s_t, a_t) = r(s_t, a_t) + \gamma * max\, Q(s_{t+1}, a_{t+1}) \tag{9.7}$$

The Q-value can be updated i terms of reward, learning rate, and discount factor, described as follows,

$$Q(s_t, a_t) \leftarrow Q(s_t, a_t) + \theta \begin{bmatrix} r(s_t, a_t) \\ + \gamma * max\,(Q(s_{t+1}, a_{t+1})) \\ - Q(s_t, a_t) \end{bmatrix} \tag{9.8}$$

where $\theta \in [0, 1]$ denotes the learning rate indicating to what extent the new *Q*-value can overturn the old one. When $\theta = 0$, the agent exploits the prior information exclusively, whereas $\theta = 1$ indicates that the agent considers only the current estimate and overlooks the prior information. A value of a decimal between 0 and 1 should be applied to θ, trading off the new *Q*-value and old *Q*-value.

9.3.3 Implementation Process of Proposed Solution Method

Algorithm 9.1 demonstrates the implementation process of our proposed solution approach for solving the FMDP problem as described in Section 9.2. Specifically, in the initial time slot, i.e. $t = 1$, the HEM system initializes power rating, dissatisfaction coefficient, discount rate, and learning rate. In each time slot, the trained DFM is used to forecast future 24-hour electricity prices as well as solar panel outputs, as shown in Algorithm 9.2. Upon obtaining the predicted information, the multi-agent *Q*-learning algorithm is adopted to make ideal energy scheduling decisions for different residential appliances and EV battery charging iteratively, as shown in Algorithm 9.3. Specifically, in each episode σ, the agent observes the state s_t and then chooses an action a_t using the exploration

Algorithm 9.1 Proposed Data-driven-based Solution Method

1. Initialize power rating, using time, dissatisfaction coefficient α, discount factor γ and learning rate θ
2. **For** time slot $t = 1 : T$ **do**
3. **For** HEM system **do**
4. Execute **Algorithm 9.2**
5. **End for**
6. **Receive** extracted information about future electricity prices and solar generations
7. **For** each agent **do** ▷ Sort descending by α
8. Execute **Algorithm 9.3**
9. **End for**
10. **End for**

Algorithm 9.2 Feedforward NN (Features Extraction)

1. Update the input electricity price data $\{\lambda_{t-23}^{G}, \ldots, \lambda_{t}^{G}\}$ and solar generation data $\{E_{t-23}^{PV}, \ldots, E_{t}^{PV}\}$
2. Extract the future trends of electricity prices and solar generations

$$\{\lambda_{t+1}^{G}, \lambda_{t+2}^{G}, \ldots, \lambda_{T}^{G}\} \leftarrow \mathrm{NN}\ (\{\lambda_{t-23}^{G}, \ldots, \lambda_{t}^{G}\})$$

$$\{E_{t+1}^{PV}, E_{t+2}^{PV} \ldots, E_{T}^{PV}\} \leftarrow \mathrm{NN}\ (\{E_{t-23}^{PV}, \ldots, E_{t}^{PV}\})$$

3. Output the extracted information

Algorithm 9.3 Q-learning Algorithm (Decision-making)

1. Initialize Q-value Q arbitrarily
2. **Repeat** for each episode σ
3. Initialize the state s_t
4. **Repeat**
5. Choose the action a_t for the current state s_t by using ε-greedy policy
6. Observe the current reward $r_t(s_t, a_t)$ and the next state s_{t+1}
7. Update the Q-value $Q(s_t, a_t)$ via Eq. (9.8)
8. **Until** s_{t+1} is terminal
9. **Until** termination criterion, i.e., $|Q^{(\sigma)} - Q^{(\sigma-1)}| \leq \tau$, is satisfied
10. Output the optimal policy π^*, i.e., $\{a_t^*, a_{t+1}^*, \ldots, a_T^*\}$
11. Execute the optimal action a_t^* for current time slot t

and exploitation mechanism. To realize the exploration and exploitation, the ε-greedy policy ($\varepsilon \in [0, 1]$) [48] is adopted so the agent can either execute a random action form the set of available actions with probability ε or select an action whose current Q-value is maximum, with probability $1 - \varepsilon$. After taking an action, the agent acquires an immediate reward $r(s_t, a_t)$, observes the next state s_{t+1} and updates the Q-value $Q(s_t, a_t)$ via Eq. (9.8). This process is repeated until the state s_{t+1} is terminal. After one episode, the agent checks the episode termination criterion, i.e. $|Q^{\sigma} - Q^{\sigma-1}| \leq \tau$, where τ is a system-dependent parameter to control the accuracy of the convergence. If this termination criterion is not satisfied, the agent will move to the next episode and repeat the above process. Finally, each agent will gain optimal actions for each coming hour, i.e. $h = 1, 2, \ldots, 24$. Note that only the optimal action for the current hour is taken. The above procedure will be repeated until the end hour, namely, $h = 24$. Besides, the flowchart in Figure 9.3 clearly depicts the process.

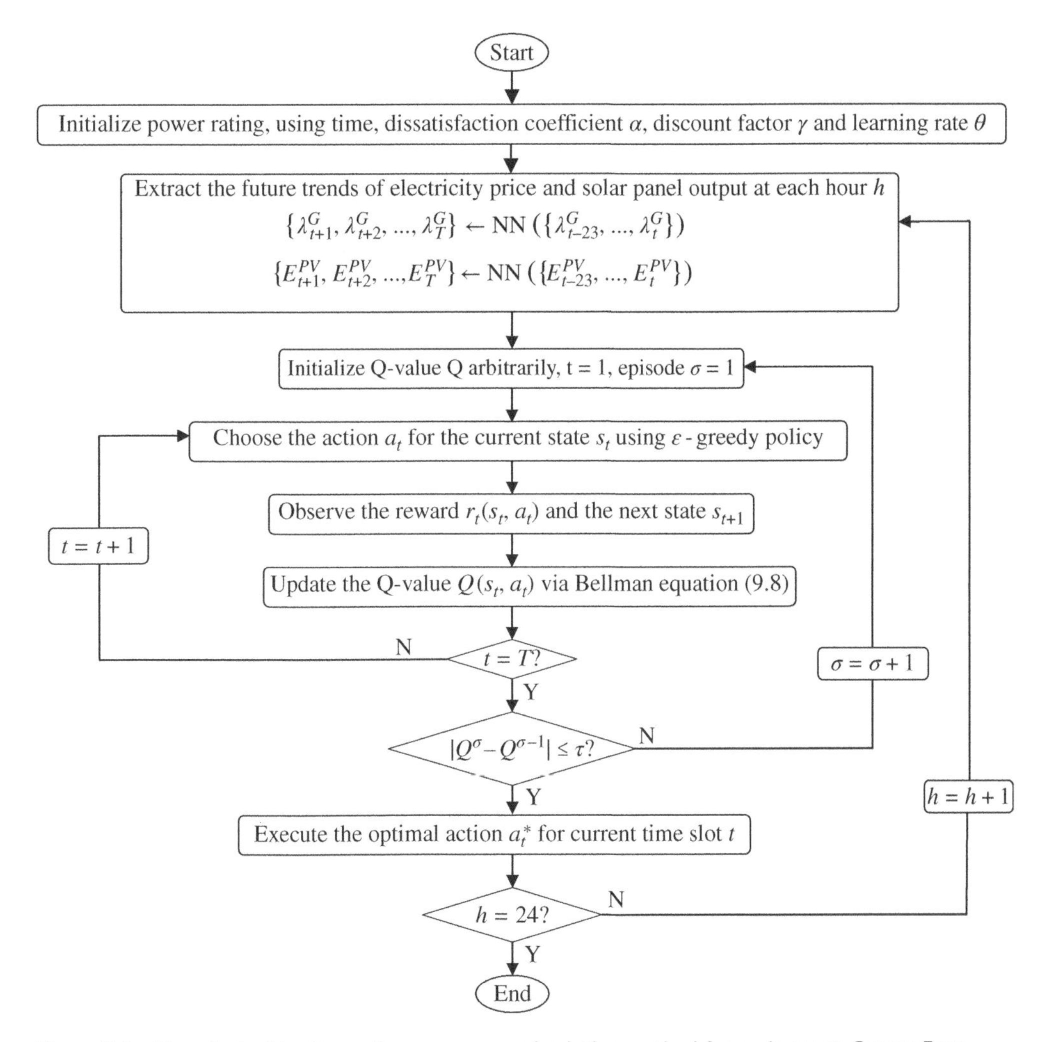

Figure 9.3 Flowchart of implementing our proposed solution method for each agent. *Source:* From Xu et al. [A].

9.4 Test Results

9.4.1 Case Study Setup

In this study, real-world data is utilized for training our proposed feedforward NN. The hourly data electricity prices and solar generations from 1 January 2017 to 31 December 2018 lasting 730 days are collected from PJM [49]. After a number of accuracy tests, the trained feedforward NN for electricity price data consists of three layers, i.e. one input layer with 24 neurons, one hidden layer with 40 neurons and one output layer with 24 neurons, and the trained feedforward NN for solar generation data also includes three layers, i.e. one input layer with 24 neurons, one hidden layer with 20 neurons and one output layer with 24 neurons. The number of training episode is 50 000. As for the parameters related to the Q-learning algorithm. The discount rate γ is set to 0.9, so the obtained strategy is foresighted. To ensure that the agent can call all state-action pairs and learn new knowledge from the system, the learning rate θ as well as turning parameter ε are both set to 0.1.

In this chapter, simulations are conducted on a detached residential house with two same solar panels, two non-shiftable appliances (REFG and AS), five power-shiftable appliances (AC1, AC2, H, L1, and L2), two time-shiftable appliances (WM and DW) and one EV. Detailed parameters of these home appliances and the EV battery are listed in Table 9.2. Besides, our proposed HEM method can be applied to residential houses with more home appliances and renewable resources. All simulations are implemented by using MATLAB with an Intel Core i7 of 2.4 GHz and 12 GB memory.

9.4.2 Performance of the Proposed Feedforward NN

Figures 9.4 and 9.5 show performance of the proposed feedforward NN for extracting features of electricity prices as well as solar generations, respectively. For the hour-ahead horizon, the mean absolute percentage error (MAPE) of the forecasted PV output is 8.82%, and the MAPE of the forecasted electricity price is 9.34%. In these two figures, the blue line represents the extracted future

Table 9.2 Parameters of each house appliance and EV battery.

Item ID	Dissatisfaction coefficient	Power rating (kWh)	Using time
REFG	—	0.5	24 h
AS	—	0.1	24 h
AC1	0.05	0–1.4	24 h
AC2	0.08	0–1.4	24 h
H	0.12	0–1.5	24 h
L1	0.02	0–0.6	6 p.m.–11 p.m.
L2	0.03	0–0.6	6 p.m. – 11 p.m.
WM	0.1	0.7	7 p.m. – 10 p.m.
DW	0.06	0.3	8 p.m. – 10 p.m.
EV	0.04	0–6	11 p.m. – 7 a.m.

REFG, refrigerator; AS, alarm system; AC, air conditioner; H, heating; L, light; WM, wash machine; DW, dishwashing; EV, electric vehicle.
Source: From Xu et al. [A].

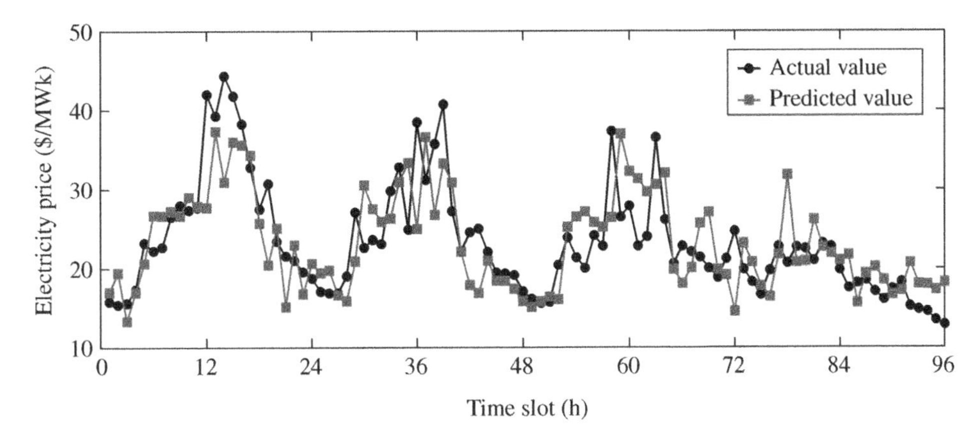

Figure 9.4 Comparison of the actual and predicted electricity prices on 1–4 January 2019. *Source:* From Xu et al. [A].

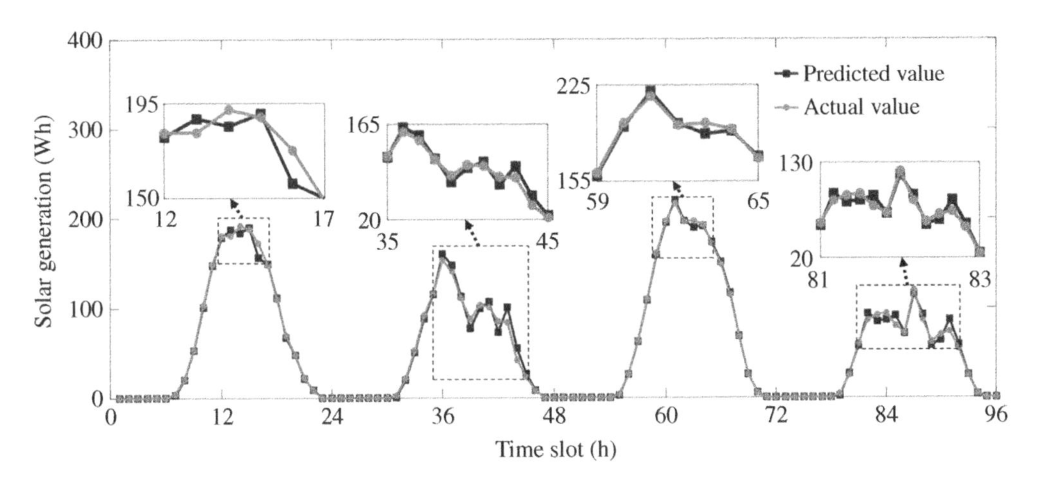

Figure 9.5 Comparison of the actual and predicted solar generations on 1–4 January 2019. *Source:* From Xu et al. [A].

values, and the red line indicates the actual values. It can be observed that both extracted trends of electricity prices and solar generations are generally similar to actual ones, though small errors can be observed from some time slots. Therefore, the proposed feedforward NN can generate accurate and reasonable forecasting values, which can benefit the following decision-making process for energy consumption scheduling.

To investigate the impact of the prediction accuracy on the solution results, we compare the performance on operation cost between cases 1–3 and optimal solution. Note that the optimal solution can be obtained by conventional optimization method based on the perfect prediction. Each case includes two kinds of predicted information, e.g. PV generation and electricity price. Table 9.3 lists the MAPE of the prediction result in Cases 1–3. Figure 9.6 is plotted to demonstrate the comparison result. As shown in this figure, with the increase of the prediction accuracy, the operation cost obtained by the *Q*-learning algorithm-based RL method is lower, which becomes closer to the optimal solution. Therefore, the prediction accuracy has a direct effect on the optimal result. In this

Table 9.3 MAPE of prediction result in cases 1–3.

	Case 1 (%)	Case 2 (%)	Case 3 (%)
MAPE of PV generation prediction	4.41	8.82	17.64
MAPE of electricity price prediction	4.67	9.34	18.68

Source: From Xu et al. [A].

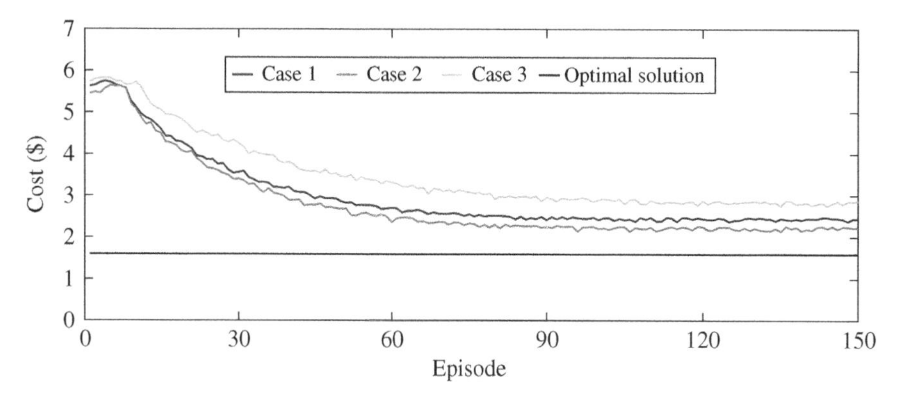

Figure 9.6 Comparison of operation cost with different prediction accuracy. *Source:* From Xu et al. [A].

chapter, we introduce the ELM-based NN to dynamically produce the future uncertainties to provide state statutes for the agent in Q-learning algorithm. However, developing more accurate prediction model to reduce the prediction error is out of the scope of this chapter, since more explanatory variable should be included, e.g. load demand, historical data, power trading direction, energy policy, etc.

9.4.3 Performance of Multi-Agent Q-Learning Algorithm

Table 9.4 lists the performance on computation time, considering a different number of state-action pair. As shown in this table, with the increase of the state-action pair, the Q-learning algorithm takes more time to fill up the Q-table and find the optimal Q-value. It should be noted that the state space is fixed (24 state statuses), so the state-action pair increases with larger considered action space (power ratings of home appliance). For example, 15 action statues correspond to 15 power ratings, i.e. 0, 0.1, 0.2, ..., 1.4 kWh, and 150 action statues correspond to 150 power ratings, i.e. 0, 0.01, 0.02, ..., 1.4 kWh. Therefore, more accurate energy consumption scheduling poses a minor effect on optimal results. Besides, it is difficult to achieve precise control of the power rating for most home appliances. In this regard, it is reasonable to consider a small number of state-action pairs for each home appliance in each time slot, resulting in short search time.

Figure 9.7 depicts the convergence of the Q-value for each power-shiftable agent on 1 January 2019. It can be seen from this figure that each power-shiftable appliance agent converges to the maximum Q-value. In the beginning, the Q-value is low since the agent takes poor actions, then it becomes high as the agent discovers the actions by learning them through trials and errors, finally reaching the maximum Q-value.

Table 9.4 Computational efficiency performance with different number of state-action pair.

No. of state-action pair (No. of state status * No. of action status)	Computation time (s)
$3.6 * 10^2$ (24 * 15)	1.315
$3.6 * 10^3$ (24 * 150)	2.067
$3.6 * 10^4$ (24 * 1500)	7.385
$3.6 * 10^5$ (24 * 15 000)	317.992

Source: From Xu et al. [A].

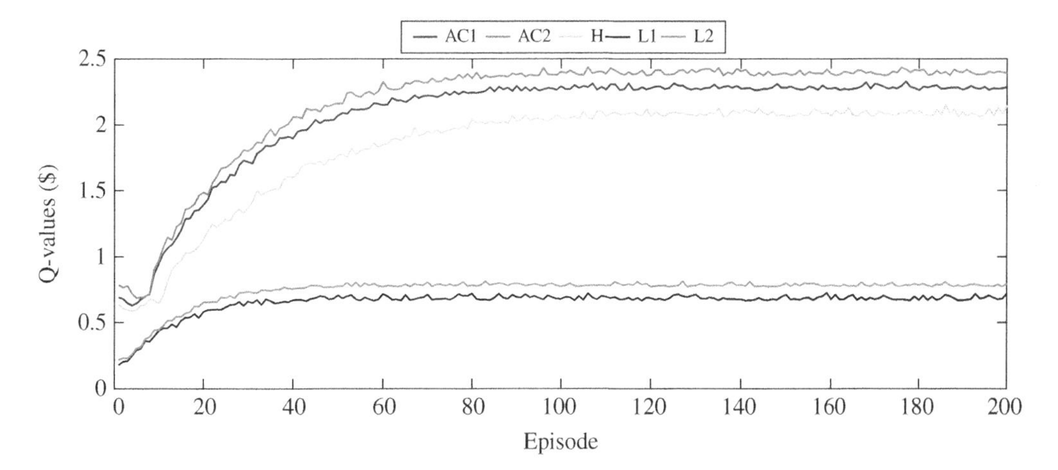

Figure 9.7 The convergence of Q-value for power-shiftable agents on 1 January 2019. *Source:* From Xu et al. [A].

To illustrate the effectiveness of our proposed HEM algorithm, Figure 9.8 is plotted to show the energy consumption of all power-shiftable appliances in each time slot. As shown in this figure, the energy consumption is high during the first five time slots. Then these five appliances reduce their energy consumption since the electricity price increases from the time slot 6. As the electricity price reaches its maximum in around time slot 13, the energy consumption of each appliance decreases to its minimum value. Finally, with the price goes down, the energy consumption starts to increase.

Figure 9.9 demonstrates the results of daily energy consumption for AC1 with different five dissatisfaction coefficients. We can see that as the dissatisfaction coefficient increases, the daily energy consumption goes up since the dissatisfaction coefficient can be regarded as a penalty factor. This creates a trade-off between saving electricity bill and decreasing dissatisfaction caused by reducing power rating of AC1. Besides, this figure also shows that the agent learns to increase energy consumption for low dissatisfaction during the off-peak time slots and decrease energy consumption for low electricity cost during the on-peak time slots. These observations verify that the proposed method can be applied to consumers for helping them manage their individual energy consumption.

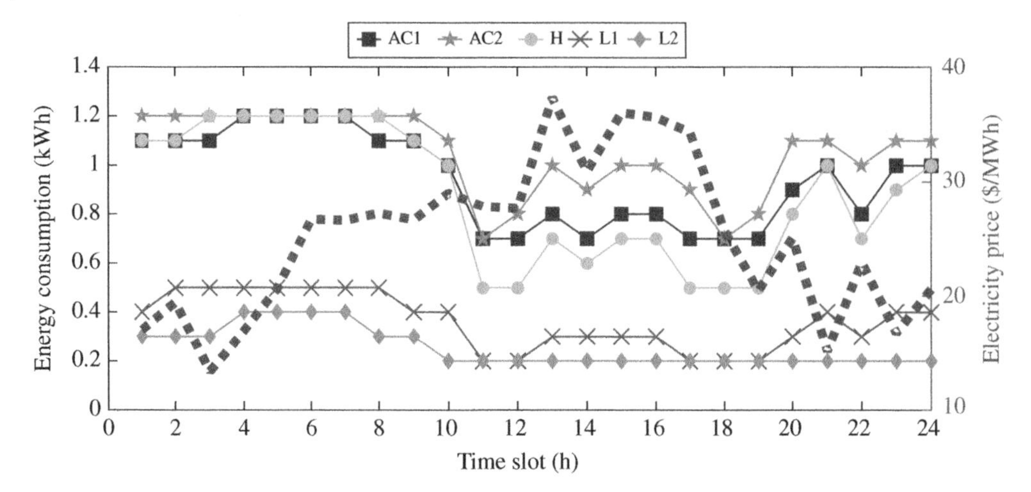

Figure 9.8 Energy consumption of five power-shiftable appliances in each time slot. *Source:* From Xu et al. [A].

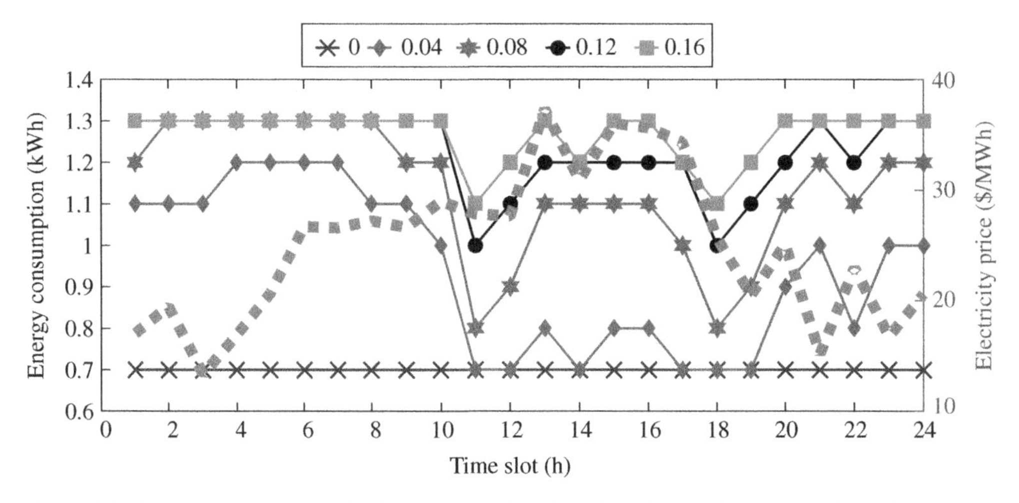

Figure 9.9 Energy consumption of AC1 with changing dissatisfaction coefficient α during 24-time slots without consideration of solar generation. *Source:* From Xu et al. [A].

Figure 9.10. gives the daily energy consumption of each home appliance and EV in two different cases with and without DR, along with the electricity prices and solar panel outputs. With DR mechanism, more energy is consumed when the price is low, and the load demand is reduced when the price is high, as shown in Figure 9.10a. Thus, the power-shiftable or time-shiftable loads can be reduced or scheduled to off-peak periods, maintaining the overall energy consumption at a low level during the on-peak periods. By contrast, for the case without DR, as shown in Figure 9.10b, no reduction or shift on energy consumption can be observed. The comparison of electricity costs in these two cases is listed in Table 9.5, which shows that the electricity cost can be significantly reduced with DR.

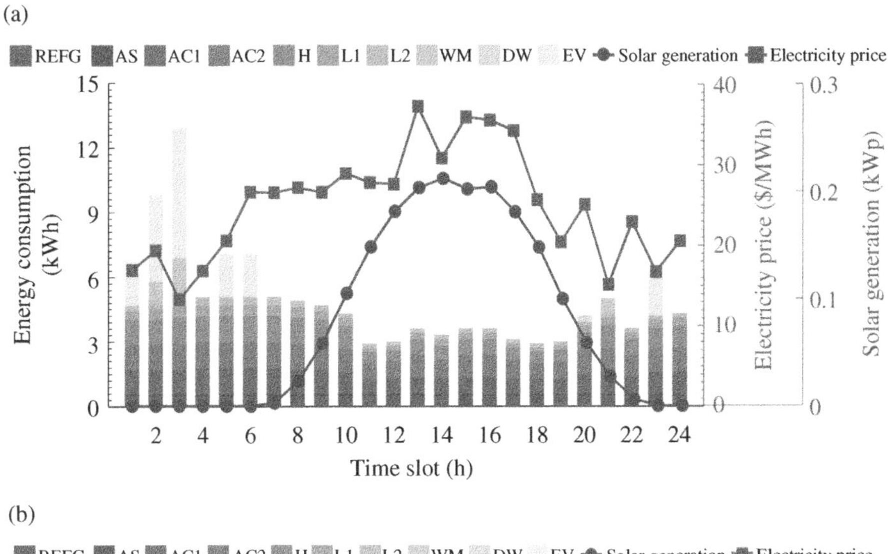

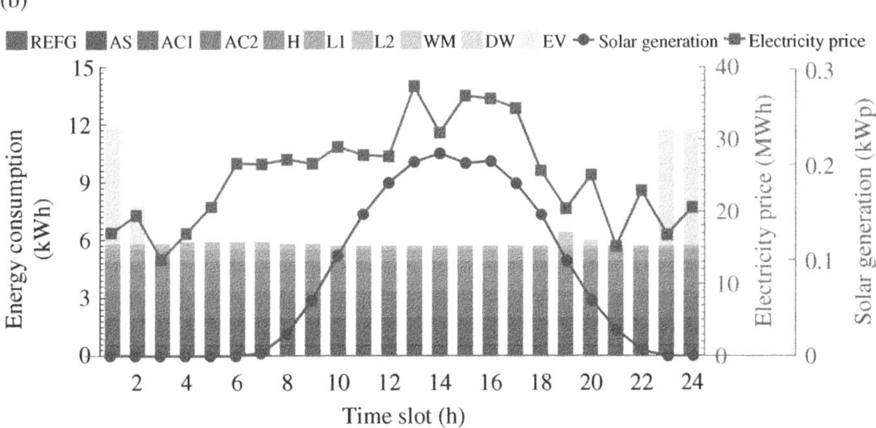

Figure 9.10 Energy consumption of all appliances on 1 January 2019, (a) with and (b) without DR. *Source:* From Xu et al. [A].

9.4.4 Numerical Comparison with Genetic Algorithm

To evaluate the performance of our proposed *Q*-learning algorithm-based solution method, the genetic algorithm (GA)-based [50] solution method is compared as a benchmark. The benchmark problem is a mixed-integer nonlinear programming problem, and its objective function is to minimize the electricity cost and dissatisfaction cost, as given by Eqs. (9.1)–(9.4). We can see from Figure 9.11 that our proposed solution method (red line) shows a poor performance at the initial training stage since it is undergoing trials and errors. However, after experiencing more iterations, our method adapts to the learning environment and adjusts its policy via exploration and exploitation mechanism. For a longer time, it outperforms the GA-based solution method (blue line). The reason is that the RL agent not only considers the current reward but also the future rewards so it can learn from the environment while the GA algorithm has low learning capability. Note that the black dashed line in Figure 9.11 is plotted as the benchmark to show the optimal result obtained by the conventional optimization method, which knows all environment information and removes prediction errors.

Table 9.5 Comparison of electricity cost with and without.

	Electricity cost ($)	
Item ID	**With DR**	**Without DR**
REFG	0.492	0.492
AS	0.098	0.098
AC1	0.836	1.378
AC2	0.942	1.378
H	0.731	1.476
L1	0.301	0.591
L2	0.223	0.591
WM	0.023	0.051
DW	0.012	0.012
EV	0.399	1.262
Total	4.057	7.329

REFG, refrigerator; AS, alarm system; AC, air conditioner, H: heating; L, light; WM, wash machine; DW, dishwashing; EV, electric vehicle.
Source: From Xu et al. [A].

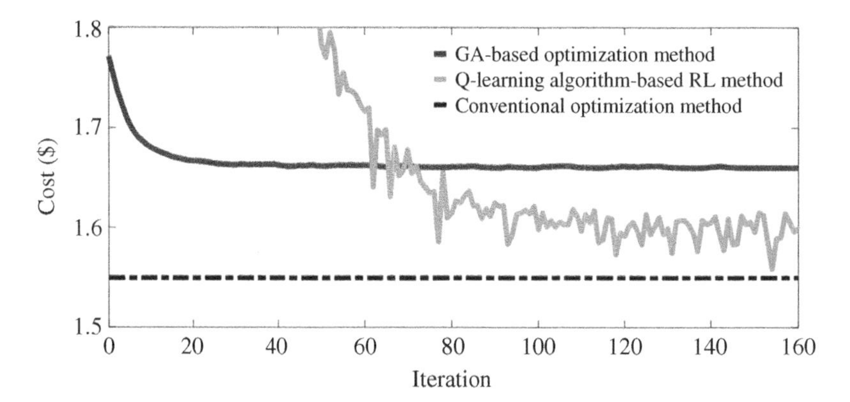

Figure 9.11 Optimization performance comparison of the three methods for scheduling AC1 loads. *Source:* From Xu et al. [A].

Besides, Table 9.6 is added to compare the computational efficiency between the proposed solution method and benchmark. It can be observed that our proposed solution method is able to significantly reduce the computation time. The reasons can be summarized into two aspects: (i) GA algorithm is based on Darwin's theory of evolution, which is a slow gradual process that works by making changes to the making slight and slow changes. Moreover, GA usually makes slight changes to its solutions slowly until getting the best solution. (ii) In the *Q*-learning algorithm, the agent chooses an action using the exploration and exploitation mechanism, so it is fast by employing the ε-greedy policy to explore and exploit the optimum from the look-up table. Note that only a small number of state-action pairs need to be searched by the *Q*-learning algorithm, resulting

Table 9.6 Comparison on computation efficiency by GA optimization method and Q-learning algorithm-based RL method.

	Average computation time of running 1000 times (s)
GA-based optimization method	46.296
Q-learning algorithm-based RL method	1.107

Source: From Xu et al. [A].

in a high computation efficiency. In this regard, considering the adaptivity of model-free RL to the external environment, it is suggested to accept our proposed well-performing solution method for HEM system.

9.5 Conclusion

Based on a feedforward NN and Q-learning algorithm, this chapter proposes a new multi-agent RL-based data-driven method for HEM system. Specifically, ELM is employed to train the feedforward NN to predict future trends of electricity price and solar generation according to real-world data. Then, the predicted information is fed into the multi-agent Q-learning algorithm-based decision-making process for scheduling the energy consumption of different home appliances and EV charging. To implement the proposed HEM method, the FMDP is utilized to model the hour-ahead energy consumption scheduling problem with the objective of minimizing the electricity bill as well as DR included dissatisfaction. Simulations are performed on a residential house with multiple home appliances, an EV and several PV panels. The test results show that the proposed HEM method can not only achieve promising performance in terms of reducing electricity cost for householders but also improve the computational efficiency. In the future, energy storage for rooftop solar PV systems will also be considered in the HEM system. Besides, more effective uncertainty prediction model will be developed to facilitate the decision-making process of DR.

Acknowledgements

The permission given to use the materials in the following paper is very much appreciated.

[A] Xu, X., Jia, Y., Xu, Y. et al. (2020). A multi-agent reinforcement learning based data-driven method for home energy management. *IEEE Transactions on Smart Grid*. https://doi.org/10.1109/TSG.2020.2971427.

References

1 U.S. Energy Information Administration (n.d.). Energy consumption estimates by sector. https://www.eia.gov/consumption/ (accessed 20 August 2020).

2 Lai, C.S. and McCulloch, M.D. (2017). Levelized cost of electricity for solar photovoltaic and electrical energy storage. *Applied Energy* 190: 191–203.

3 Liu, N., Chen, Q., Lu, X. et al. (2015). A charging strategy forpv-based battery switch stations considering service availability and self-consumption of pv energy. *IEEE Transactions on Industrial Electronics* 62 (8): 4878–4889.
4 Zheng, J., Sun, X., Jia, L., and Zhou, Y. (2020). Electric passenger vehicles sales and carbon dioxide emission reduction potential in Chinas leading markets. *Journal of Cleaner Production* 243: 118607.
5 Moses, P.S., Masoum, M.A.S., and Hajforoosh, S. (2012). Overloading of distribution transformers in smart grid due to uncoordinated charging of plug-in electric vehicles. *2012 IEEE PES Innovative Smart Grid Technologies (ISGT)*, pp. 1–6.
6 Zheng, M., Meinrenken, C.J., and Lackner, K.S. (2015). Smart households: dispatch strategies and economic analysis of distributed energy storage for residential peak shaving. *Applied Energy* 147: 246–257.
7 Sun, C., Sun, F., and Moura, S.J. (2016). Nonlinear predictive energy management of residential buildings with photovoltaics & batteries. *Journal of Power Sources* 325: 723–731.
8 Wu, X., Hu, X., Moura, S. et al. (2016). Stochastic control of smart home energy management with plug-in electric vehicle battery energy storage and photovoltaic array. *Journal of Power Sources* 333: 203–212.
9 Hafiz, F., de Queiroz, A.R., and Husain, I. (2019). Coordinated control of pev and pv-based storages in residential systems under generation and load uncertainties. *IEEE Transactions on Industry Applications* 55 (6): 5524–5532.
10 Di Giorgio, A. and Pimpinella, L. (2012). An event driven smart home controller enabling consumer economic saving and automated demand side management. *Applied Energy* 96: 92–103.
11 Cai, J., Zhang, H., and Jin, X. (2019). Aging-aware predictive control of pv-battery assets in buildings. *Applied Energy* 236: 478–488.
12 Yan, D., Li, T., Ma, C. et al. (2020). Cost effective energy management of home energy system with photovoltaic-battery and electric vehicle. *The 46th Annual Conference of the IEEE Industrial Electronics Society (IECON 2020)*, Singapore (18–21 October 2020).
13 Shareef, H., Ahmed, M.S., Mohamed, A., and Al Hassan, E. (2018). Review on home energy management system considering demand responses, smart technologies, and intelligent controllers. *IEEE Access* 6: 24498–24509.
14 Chen, Y., Xu, Y., Li, Z., and Feng, X. (2019). Optimally coordinated dispatch of combined-heat-and-electrical network with demand response. *IET Generation, Transmission & Distribution* 13 (11): 2216–2225.
15 Xu, F.Y., Wang, X., Lai, L.L., and Lai, C.S. (2014). Agent-based modeling and neural network for residential customer demand response. *Proceedings of the 2013 IEEE International Conference on Systems, Man and Cybernetics*, Manchester, UK (13–16 October 2013).
16 Luo, F., Ranzi, G., Wang, S., and Dong, Z.Y. (2018). Hierarchical energy management system for home microgrids. *IEEE Transactions on Smart Grid.*
17 Luo, F., Kong, W., Ranzi, G., and Dong, Z.Y. (2019). Optimal Home Energy Management System with Demand Charge Tariff and Appliance Operational Dependencies. *IEEE Transactions on Smart Grid* 10 (5): 5536–5546.
18 Wu, X., Hu, X., Yin, X., and Moura, S. (2016). Stochastic optimal energy management of smart home with PEV energy storage. *IEEE Transactions on Smart Grid* 9 (3): 2065–2075.
19 Pilloni, V., Floris, A., Meloni, A., and Atzori, L. (2016). Smart home energy management including renewable sources: a qoe-driven approach. *IEEE Transactions on Smart Grid* 9 (3): 2006–2018.
20 Yu, L., Jiang, T., and Zou, Y. (2017). Online energy management for a sustainable smart home with an HVAC load and random occupancy. *IEEE Transactions on Smart Grid* 10 (2): 1646–1659.

21 Sharma, S., Xu, Y., Verma, A., and Panigrahi, B.K. (2019). Time-coordinated multi-energy management of smart buildings under Uncertainties. *IEEE Transactions on Industrial Informatics* 15 (8): 4788–4798.

22 Keerthisinghe, C., Verbič, G., and Chapman, A. (2016). A fast technique for smart home management: ADP with temporal difference learning. *IEEE Transactions on Smart Grid* 9 (4): 3291–3303.

23 Huang, Y., Wang, L., Guo, W. et al. (2016). Chance constrained optimization in a home energy management system. *IEEE Transactions on Smart Grid* 9 (1): 252–260.

24 Sutton, R.S. and Barto, A.G. (1998). *Introduction to reinforcement learning*, vol. 4. Cambridge: MIT Press.

25 Vázquez-Canteli, J.R. and Nagy, Z. (2019). Reinforcement learning for demand response: a review of algorithms and modeling techniques. *Applied Energy* 235: 1072–1089.

26 Ruelens, F., Claessens, B.J., Vandael, S. et al. (2016). Residential demand response of thermostatically controlled loads using batch reinforcement learning. *IEEE Transactions on Smart Grid* 8 (5): 2149–2159.

27 Wan, Z., Li, H., He, H., and Prokhorov, D. (2019). Model-free real-time EV charging scheduling based on deep reinforcement learning. *IEEE Transactions on Smart Grid* 10 (5): 5246–5257.

28 Lu, R., Hong, S.H., and Zhang, X. (2018). A dynamic pricing demand response algorithm for smart grid: reinforcement learning approach. *Applied Energy* 220: 220–230.

29 Lu, R., Hong, S.H., and Yu, M. (2019). Demand response for home energy management using reinforcement learning and artificial neural network. *IEEE Transactions on Smart Grid* 10 (6): 6629–6639.

30 Zhou, S., Hu, Z., Gu, W. et al. (2019). Artificial intelligence based smart energy community management: a reinforcement learning approach. *CSEE Journal of Power and Energy Systems* 5 (1): 1–10.

31 Huang, X., Hong, S.H., Yu, M. et al. (2019). Demand response management for industrial facilities: a deep reinforcement learning approach. *IEEE Access* 7: 82194–82205.

32 Foruzan, E., Soh, L.-K., and Asgarpoor, S. (2018). Reinforcement learning approach for optimal distributed energy management in a microgrid. *IEEE Transactions on Power Systems* 33 (5): 5749–5758.

33 Mocanu, E., Mocanu, D.C., Nguyen, P.H. et al. (2019). On-line building energy optimization using deep reinforcement learning. *IEEE Transactions on Smart Grid* 10 (4): 3698–3708.

34 Du, Y. and Li, F. (2020). Intelligent multi-microgrid energy management based on deep neural network and model-free reinforcement learning. *IEEE Transactions on Smart Grid* 11 (2): 1066–1076.

35 Remani, T., Jasmin, E., and Ahamed, T.I. (2019). Residential load scheduling with renewable generation in the smart grid: a reinforcement learning approach. *IEEE Systems Journal* 13 (3): 3283–3294.

36 McCabe, A., Pojani, D., and van Groenou, A.B. (2018). Social housing and renewable energy: community energy in a supporting role. *Energy Research Social Science* 38: 110–113.

37 Green, M.A. (1982). *Solar Cells: Operating Principles, Technology, and System Applications*, 288. Prentice-Hall, Inc.

38 Walker, G. (2001). Evaluating MPPT converter topologies using a MATLAB PV model. *Journal of Electrical Electronics Engineering, Australia* 21 (1): 49.

39 Rezvanizaniani, S.M., Liu, Z., Chen, Y., and Lee, J. (2014). Review and recent advances in battery health monitoring and prognostics technologies for electric vehicle (EV) safety and mobility. *Journal of Power Sources* 256: 110–124.

40 Le Floch, C., Kara, E.C., and Moura, S. (2016). PDE modeling and control of electric vehicle fleets for ancillary services: a discrete charging case. *IEEE Transactions on Smart Grid* 9 (2): 573–581.

41 Yu, M. and Hong, S.H. (2017). Incentive-based demand response considering hierarchical electricity market: a Stackelberg game approach. *Applied Energy* 203: 267–279.

42 Rafiei, M., Niknam, T., Aghaei, J. et al. (2018). Probabilistic load forecasting using an improved wavelet neural network trained by generalized extreme learning machine. *IEEE Transactions on Smart Grid* 9 (6): 6961–6971.

43 Chai, S., Xu, Z., and Jia, Y. (2019). Conditional density forecast of electricity price based on ensemble ELM and logistic EMOS. *IEEE Transactions on Smart Grid* 10 (3): 3031–3043.

44 Fu, W., Wang, K., Li, C., and Tan, J. (2019). Multi-step short-term wind speed forecasting approach based on multi-scale dominant ingredient chaotic analysis, improved hybrid GWO-SCA optimization and ELM. *Energy Conversion and Management* 187: 356–377.

45 Wu, X., Lai, C.S., Bai, C. et al. (2020). Optimal kernel ELM and variational mode decomposition for probabilistic PV power prediction. *Energies* 13: 3592.

46 Watkins, C.J. and Dayan, P. (1992). Q-learning. *Machine Learning* 8 (3–4): 279–292.

47 Kappen, H.J. (2011). Optimal control theory and the linear bellman equation. In: *Inference Learning Dynamic Models* (ed. D. Barber, A.T. Cemgil and S. Chiappa), 36387. Cambridge University Press.

48 Tokic, M. and Palm, G. (2011). Value-difference based exploration: adaptive control between epsilon-greedy and softmax. *Annual Conference on Artificial Intelligence*, Berlin, Germany (4–7 October 2011), 335–346. Springer.

49 Attar, M., Homaee, O., Falaghi, H., and Siano, P. (2018). A novel strategy for optimal placement of locally controlled voltage regulators in traditional distribution systems. *International Journal of Electrical Power & Energy Systems* 96: 11–22.

50 Koza, J.R. (1994). *Genetic Programming*, vol. 17. Cambridge, MA, USA: MIT Press.

10

Virtual Energy Storage Systems Smart Coordination

10.1 Introduction

As an effective solution to future energy crisis, renewable energy resources are playing a vital role in current power systems. Based on the electricity forecast of International Energy Agency (IEA), the share of renewable energy in meeting global power demand would reach almost 30% in 2023, up from 24% in 2017 [1]. During this period, more than 70% of global electricity generation growth is met by renewables, led by solar PV. Solar PV brings unprecedented environmental and technical benefits, such as low-carbon emissions and congestion management. Nevertheless, the large-scale penetration of PV energy in distribution network causes many power quality issues as well, such as harmonic pollution [2] and voltage rise [3]. According to [3], voltage rise is the most significant one among the power quality issues in distribution network. Overvoltage problem usually occurs at the time of high PV penetration periods and light load periods [4]. In contrast, undervoltage happens at the time of low PV penetration periods and heavy load periods.

10.1.1 Related Work

Previous researchers have proposed a variety of techniques on regulating voltage in distribution networks, such as installing voltage regulators [5], changing line impedances and conductor size [6], modulating the tap set points of secondary transformer [7], applying reactive power compensation in PV inverters [8], reducing solar power generation amount [9, 10], and employing battery energy storage systems (BESSs) [4, 11]. Owing to the randomness of PV energy and customer load, the former three methods require frequent changes of set point and are not flexible for the utility side. By contrast, the latter three methods are based on end-user side and more promising to regulate voltage. However, unavoidable shortcomings still exist here. For example: (i) the distribution network usually has high R/X ratio, therefore, reactive power compensation is not effective enough [3], (ii) in terms of PV generation reduction, the energy efficiency is reduced via this method, and (iii) although the BESS cost has dropped, customers still bear financial burden on installing large energy storage systems [12–14].

Except for voltage regulation, loading management is another important issue in distribution network due to the increasing energy-hungry appliances [15]. Especially in recent years, air conditioners are rapidly making the way to households because of summer heat and falling upfront cost. In line with an investigation led by Ausgrid [16], air conditioners contribute more than half

Smart Energy for Transportation and Health in a Smart City, First Edition. Chun Sing Lai, Loi Lei Lai and Qi Hong Lai.

of the load in some of their substations in summer days. If such demand comes up to certain ratio of feeder load, challenges would be imposed to system operation. Network infrastructure capacity needs to be upgraded by system operator to maintain reliable electricity supply. Consequently, billions of dollars would be spent for network upgrading to deal with the short but sharp peak load period. Nevertheless, the infrastructure upgrading is only used for short periods of the year to meet peak demand, which is not cost-effective for system operators.

How to address the noted issues (i.e. voltage regulation and loading management) in a technically and economically efficient manner needs to be considered by the researchers. With the popularity of demand response (DR) technologies, an alternative way to address peak load is through DR programs by shaping the load curve for optimal use of energy and improving asset investment overall efficiency [17, 18]. In fact, the pressure on integrating large-scale PV resources into distribution network can be alleviated as well with the help of DR programs. Owing to rapid progress in control and communication techniques, thermostatically controlled loads (TCLs) in end-user side can be equipped with control modules to be better involved in DR programs. Among various types of TCLs, air conditioners receive researchers' increasing attentions because they have relatively fast response time with least end-user disruptions and mainly contribute the summer peak load. When air conditioners are turned on/off, the room temperature can maintain within certain range by storing large amount of heat/cold air. This phenomenon is referred as thermal inertia, which is defined as a thermal mass being capable of resisting the change on its temperature faced with the fluctuation of ambient temperature [19]. Consequently, a household can shift its energy consumption over the planning horizon to help consume the peak PV generation amount or reduce the peak load periods. The thermal buffering capacity in an air-conditioned household can imitate the energy buffering characteristics of physical energy storage systems, such as batteries, and hence can be viewed as virtual energy storage systems (VESSs). In fact, the air-conditioned household becomes a battery.

The concept of VESSs is not new. Researchers in [20, 21] have taken the flexible loads in power systems as VESSs. In [20], Meng et al. coordinated DR from domestic refrigerator to form the VESS, aiming to provide frequency service for the system. In [21], the author employed virtual energy storage through distributed control of flexible loads, which was believed to be innovative solutions for integrating renewables. VESSs can be integrated with other energy resources to provide desired services for the system operator. Therefore, distribution utilities are encountering new opportunities in the situation of growing number of air conditioners and increasing penetration of PV resources at customer side. By coordinating VESSs, network voltage regulation and overloading issues can be solved in an efficient and economical way.

This work is inspired by the fact that air conditioners at end-user side are contributing summer peak loads in distribution network and the voltage regulation is necessary with growing PV penetration in customer side. If a large number of air-conditioned households is coordinated by DR programs, the households can be viewed as VESSs and offer significant system support. Hence, the objective of this work is to put forward an innovative method to regulate distribution network with the help of VESSs. Previous research mainly focused on the adoption of battery storage for supporting stable operation of power systems, which is inferior in terms of economics and flexibility [22, 23]. By utilizing virtual storage systems instead of practical battery storage systems, distribution system operator would reduce the infrastructure investment to a large extent, as well as gain much more flexibility given the fast growth of customer demand and renewable resources penetration.

10.1.2 Main Contributions

In this chapter, a two-level dispatch strategy is proposed to share the required active power adjustment among VESSs for voltage regulation and overloading management. Specifically, in the lower level, the more precise VESS model is built to reflect the dynamic thermal process of air-conditioned households. Given a single VESS has limited capacity to participate in DR program, a group of VESSs in a residential district is aggregated to an aggregator for effective involvement in the control scheme. In the upper level, a consensus-driven distributed control strategy is adopted to fairly regulate system voltage and loading. Compared with centralized control scheme, distributed control shares information through a limited communication network, thus it is more robust against communication failure and is more efficient in coordinating the available units in the system. Among the various types of distributed control schemes, consensus control is a typical approach to achieve the common objective by operating in the same manner via information exchange [24]. Consensus control has been widely applied in microgrids and distribution systems, such as load shedding [25], frequency regulation [26], and power sharing [27]. In this chapter, if the voltage and network loading violate certain limits, the consensus-driven distributed control will coordinate the active power to support system voltage and loading. Compared with existing works in the literature, the contributions of this chapter are detailed in two aspects:

1) The first contribution is the application of VESS, i.e. air-conditioned households, on voltage regulation and overloading management in distribution network. In contrast to conventional energy storage systems, VESSs effectively support system operation at a far less cost.
2) The other contribution is applying a distributed consensus control on distribution network management with high robustness and scalability. The influences of dynamic communication network topology changes on system performance are investigated. Through exchanging information with neighboring aggregators via sparse communication networks, active power support is fairly shared among participating aggregators.

The remaining parts of the chapter are as follows. VESS modeling, aggregation, and coordination strategy are introduced in Section 10.2. Section 10.3 presents the proposed control approach, including overloading management and voltage regulation strategies. In Section 10.4, case studies are carried out and simulation results are analyzed to demonstrate the performance of the proposed method scheme. Conclusions and future works are given in Section 10.5.

10.2 VESS Modeling, Aggregation, and Coordination Strategy

In this section, the thermal modeling of VESS is firstly introduced, followed by the aggregation method. In the last subsection, the VESS coordination strategies are given.

10.2.1 VESS Modeling

In order to capture the thermal behavior of VESS, the fundamental part is to build a comprehensive thermal model for air-conditioned household. Previous researchers have developed different complexities of thermal models to represent the thermal process of TCLs. In [28, 29], the authors proposed the first order differential equation to build the individual TCL model. However, this method cannot reflect the actual thermal process of heating, ventilating, and air-conditioning (HVAC)

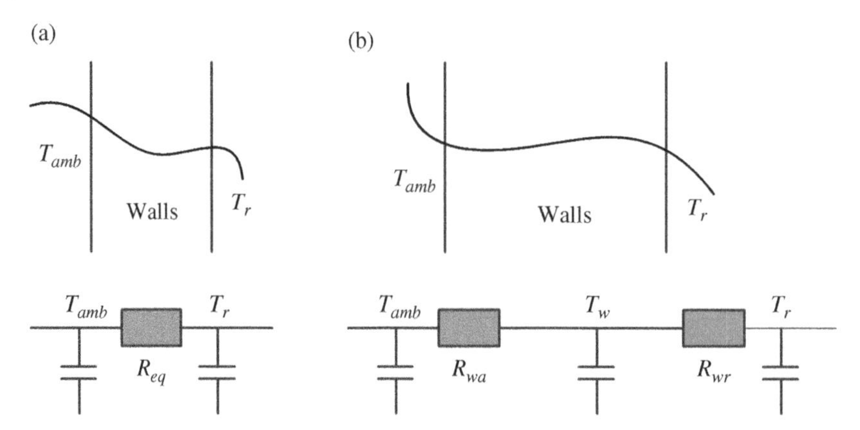

Figure 10.1 Schematic of VESS models: (a). one-parameter model (b). two-parameter model. *Source:* From Wang et al. [A].

systems owing to the inherent large thermal mass temperature dynamics. In this simplified model, as shown in Figure 10.1a, only ambient temperature, indoor air temperature, and thermal resistance are taken into account, while neglecting explicit model of building materials (especially walls). In fact, indoor air temperature is not only subject to the difference between indoor and outdoor air, but also depends on the thermal energy exchange with internal walls and participations. The thermal mass inside the building has significant temperature variation influenced by external environment, hence could greatly impact the indoor air temperature.

In this chapter, a two-parameter thermal model is built to more precisely capture the thermal dynamics of an air-conditioned building, as shown in Figure 10.1b. In this model, the building is divided into two inter-connected parts. One is in-house part, and the other one is the additional thermal mass part with significant thermal capacity, such as walls. The efficiency of a two-parameter thermal model has been proved by [30], where the authors compared the impacts of different complexities of thermal model on indoor air calculation. Thermal dynamic process of VESS can be depicted by:

$$\frac{dT_r(t)}{dt} = \frac{1}{M_a \times Cp_a}\left[\frac{dQ_{gain\_a}(t)}{dt} - \frac{dQ_{ex\_w\_r}(t)}{dt} - \frac{dQ_{ac}(t)}{dt}\right] \tag{10.1}$$

$$\frac{dT_w(t)}{dt} = \frac{1}{M_w \times Cp_w}\left[\frac{dQ_{gain\_w}(t)}{dt} + \frac{dQ_{ex\_w\_r}(t)}{dt}\right] \tag{10.2}$$

$$\frac{dQ_{gain\_a}(t)}{dt} = \frac{T_{amb} - T_r}{R_{eq}} \tag{10.3}$$

$$\frac{dQ_{ex\_w\_r}(t)}{dt} = \frac{T_w - T_r}{R_{wr}} \tag{10.4}$$

$$\frac{dQ_{gain\_w}(t)}{dt} = \frac{T_{amb} - T_w}{R_{wa}} \tag{10.5}$$

$$\frac{dQ_{ac}(t)}{dt} = \frac{P_{ac}}{\eta} \tag{10.6}$$

where, t refers to time; T_r, T_w, T_{amb} are indoor air temperature, wall temperature, and ambient temperature, respectively; M_a, M_w are mass of indoor air and wall; Cp_a, Cp_w are the thermal capacity of indoor air and wall; Q_{gain_a}, Q_{gain_w} are the heat absorbed by indoor air from the ambient and heat absorbed by the wall from the ambient; $Q_{ex\_w\_r}$ is heat exchange between indoor air and the wall; Q_{ac} is the cooling energy from air conditioners; R_{eq}, R_{wa}, R_{wr} are the equivalent thermal resistance of house envelope, the equivalent thermal resistance between ambient and wall outer surface, and the equivalent thermal resistance between indoor air and wall inner surface, respectively; η is air conditioner performance coefficient; P_{ac} is air conditioner rated power.

Equations (10.1) and (10.2) represent indoor air temperature and wall temperature change rate. Equations (10.3)–(10.6) represent heat absorption rate between ambient and indoor air, heat exchange rate between wall and indoor air, heat absorption rate between wall and indoor air, and air conditioner cooling rate respectively. To conveniently calculate indoor air temperature and build a flexible operational planning scheme, the proposed dynamic thermal model can be linearized as below [19]:

$$T_r(t) = \left(1 - \frac{1}{M_a Cp_a R_{eq}}\right) T_r(t-1) + \frac{1}{M_a Cp_a R_{eq}} T_{amb}(t-1) + \frac{T_w(t-1) - T_r(t-1)}{M_a Cp_a R_{wr}} - S_{ac}(t)\frac{Q_{ac}(t-1)}{M_a Cp_a}, \quad \forall t \in [1, N] \tag{10.7}$$

$$T_w(t) = T_w(t-1) + \frac{T_{amb}(t-1) - T_w(t-1)}{M_w Cp_w R_{wa}} + \frac{T_r(t-1) - T_w(t-1)}{M_w Cp_w R_{wr}}, \forall t \in [1, N] \tag{10.8}$$

where N is the total time steps; S_{ac} is air conditioner operation status. 1 means air conditioner is ON, 0 means air conditioner is OFF. The operation status of air conditioners is determined by a thermostatic switching law with the predetermined temperature deadband:

$$S_{ac}(t) = \begin{cases} 0 & \text{if } S_{ac}(t-1) = 1 \;\&\; T_r < T_r^{\min} \\ 1 & \text{if } S_{ac}(t-1) = 0 \;\&\; T_r > T_r^{\max} \\ S_{ac}(t-1) & \textit{otherwise} \end{cases} \tag{10.9}$$

$$T_r^{\min} = T_{set} - \frac{T_{db}}{2}; \quad T_r^{\max} = T_{set} + \frac{T_{db}}{2} \tag{10.10}$$

where, $T_r^{\min}$, $T_r^{\max}$ refer to the lower and upper limits of indoor air temperature; T_{set} is customer preferred temperature set point; T_{db} represents the temperature deadband scope. To guarantee the thermal comfort of end-users, the indoor air temperature should be restricted as:

$$T_r^{min} \le T_r(t) \le T_r^{max} \tag{10.11}$$

In addition, wall temperature should meet the lower and upper limits:

$$T_w^{min} \le T_w(t) \le T_w^{max} \tag{10.12}$$

10.2.2 VESS Aggregation

Due to the limited TCL amount an individual VESS can provide, its contribution volume usually cannot meet the minimum load requirement for participating demand response programs. Therefore, it is a good practice to aggregate the individual VESS through an aggregator, so that the benefits of DR programs can be effectively reaped. In this work, each residential community comprised of hundreds of air-conditioned households is regulated by an aggregator. The aggregated energy consumption of VESSs is denoted as:

$$P_{VESSs}(t) = \sum_{j=1}^{N_{air}} \frac{1}{\eta_j} P_{ac,j} S_{ac,j}(t) \tag{10.13}$$

where P_{VESSs} means the overall power consumption of aggregated VESSs, N_{air} is the total air conditioner number in the aggregator, and j means the jth air conditioner.

To characterize a group of VESSs and further aggregate them, heterogeneous operating scenarios (i.e. different T_{set}, T_{db}, $P_{rate,j}$, etc.) are generated by Monte-Carlo simulation [31]. By generating multiple building model cases within the intermittency range, the practical air-conditioned household scenarios are modeled. For example, the parameters of an air-conditioned household model can be: house length is 13 m, house width is 12 m, house height is 5 m, wall width is 0.25 m, set point of thermostat is 24.5 °C, number of windows is 4, and air conditioner rated power is 4 kW.

After building the aggregation model, the maximum controllable capacity of each VESSs aggregator should be estimated for the upper level control.

Objective:

$$Maximize\ \ P_i^{max}(t) = \sum_{j=1}^{N_{air}} \frac{1}{\eta_j} \left(1 - S_{ac,j}(t)\right) P_{rate,j} \tag{10.14}$$

where $P_i^{\max}$ refers to the maximum controllable capacity in aggregator i. It is assumed that reactive power is fully compensated in air conditioner side, hence $P_i^{\max}$ specifically means the maximum controllable active power in the aggregator. It is worth noting that the decision variable in (10.14) is $S_{ac,j}(t)$, and hundreds of VESSs are controlled by the same aggregator here.

The calculation of (10.14) should meet following conditions:

Subject to:

$$S_{ac}(t) = \begin{cases} 0 & \text{if } S_{ac}(t-1) = 1\ \&\ T_r < T_r^{min} \\ 1 & \text{if } S_{ac}(t-1) = 0\ \&\ T_r > T_r^{max} \\ S_{ac}(t-1) & \quad otherwise \end{cases} \tag{10.15a}$$

$$T_r^{min} \leq T_r(t) \leq T_r^{max} \tag{10.15b}$$

$$T_w^{min} \leq T_w(t) \leq T_w^{max} \tag{10.15c}$$

The maximum capacity estimation model in (10.14) is a nonlinear mix-integer programming problem, which is quite difficult to be solved by conventional mathematical methods. Therefore, heuristic-based programming methods are employed in this chapter to solve the capacity estimation model. After getting the maximum controllable active power in the aggregator, it can be further utilized in the upper level control for voltage regulation and loading management.

10.2.3 VESS Coordination Strategies

In this part, the VESS coordination strategies are presented to demonstrate the VESS monitoring, communication and dispatching procedures. The detailed procedures are given below.

1) **Initialization.** End users' thermal comfort can be obtained via collecting historical data based on different weather scenarios or via the collection of household surveys regarding end users' preferences on indoor temperature, relative humidity, etc.
2) **Monitoring**. Home energy management system (HEMS) in the air-conditioned household can monitor the room temperature and air conditioner status in real-time, and interact with the smart meter wirelessly.
3) **Estimation**. After receiving the relevant information, smart meters then send it to the aggregator for estimating the maximum controllable active power at the current status through the heuristic approach. The estimated maximum controllable power information can be further transferred to the upper level for participating in the distribution network management.
4) **Communication and Dispatch**. In the upper level control scheme, the required active power support information will be shared among interconnected aggregators through the communication network in an iterative manner, the detailed control strategy will be illustrated in the next section. Once an equilibrium point is achieved, the aggregator would meet its power adjustment commitment to the utility by informing its HEMSs to selectively turn ON/OFF air conditioners.

The proposed hierarchical coordination strategy of aggregated VESSs for distribution network is illustrated in Figure 10.2. In order to encourage the participation of end customers on DR programs, participants will receive financial reimbursement from the utility, such as cash reward and reduced electricity price. Note that the frequent motor starting has been recognized as a major cause of flicker, especially for large-scale motors connected to distribution network. To keep the system dynamic stability, especially mitigate voltage fluctuations, reducing the starting current of a motor is one effective measure. A series of starting techniques can be employed, such as using power electronic soft starter and full inverter control of motor.

10.3 Proposed Approach for Network Loading and Voltage Management by VESSs

In this section, through the coordination of aggregated VESSs, a new method to regulate distribution network loading and voltage is proposed. By reducing the active power consumption in peak demand time, the overloading is curtailed, as well as preventing the occurrence of undervoltage. For overvoltage regulation, VESS helps increase the active power consumption in peak PV generation periods, thus over generated solar power can be consumed to eliminate voltage rise.

To guarantee the active participation in the control scheme, except for the financial incentives for VESSs, the active power curtailment and consumption should be implemented fairly among participating VESSs. Therefore, a consensus decision-making method is needed to reach an agreement among all aggregators without a bidding strategy [32, 33]. The detailed control schemes will be introduced in the following subsections.

10.3.1 Network Loading Management Strategy

For a multi-agent system, the information exchange among agents can be denoted by a weighted graph $G = (V, E, A)$. Introduction for graph theory is omitted here, which can be found in many previous research [15, 34].

Considering the disturbance in communication channels, link failures are likely to happen, thus the static communication network is not appropriate to reflect its dynamics. Here, a time varying

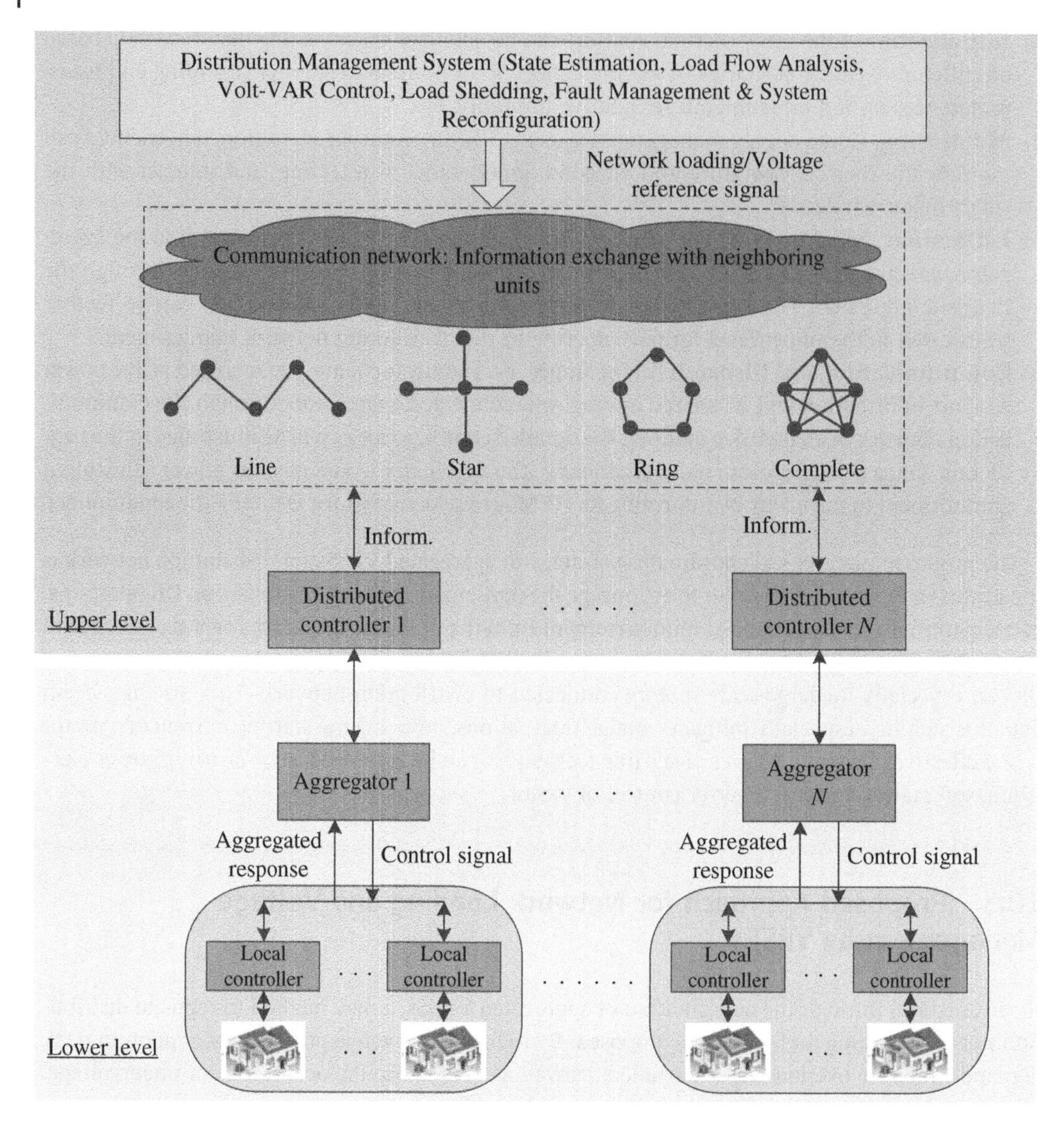

Figure 10.2 Proposed hierarchical coordination strategy of aggregated VESSs for distribution network. *Source:* From Wang et al. [A].

coefficient matrix is given to model the complete network topology between aggregators and disclose the dynamic change characteristic. The communication matrix is defined as:

$$\Psi(t) = \begin{bmatrix} \psi_{11}(t) & \psi_{12}(t) & \cdots & \psi_{1m}(t) \\ \psi_{21}(t) & \psi_{22}(t) & \cdots & \psi_{2m}(t) \\ \vdots & \vdots & \vdots & \vdots \\ \psi_{m1}(t) & \psi_{m2}(t) & \cdots & \psi_{mm}(t) \end{bmatrix} \tag{10.16}$$

where $\psi_{ij}(t)$ reflects the communication link between ith and jth aggregator at time t. $\psi_{ij} = 1$ if there is directed communication link between aggregator i and aggregator j, otherwise $\psi_{ij} = 0$. Moreover,

$\psi_{ii} = 1$ for all aggregators. Hence, for a given connected network, the dynamic network topology can be reflected in the adjacency matrix by its values change with time. According to [33], the consensus state will converge to the same value if a spanning tree exists in the communication graph, and the Laplace matrix of the communication graph has a simple zero eigenvalue, and all the other eigenvalues with positive real parts.

In the proposed control scheme, a virtual leader is assigned to define the primal information state via the measurement of critical point [32]. The power exchange between the substation and main grid is continually monitored by the virtual leader, which is further used as the input control signal of distributed controller for network loading management. The distributed controller will be initiated once the apparent power constraint $S(t) \leq S^{max}$ is violated. The information state of the virtual leader is defined as:

$$\eta_0(t) = K_p m(t) + K_i \sum_{k=0}^{t} m(t) + K_d[m(t) - m(t-1)] \tag{10.17}$$

where $m(t) = S^{max} - S(t)$; $\eta_0(t)$ is the virtual leader initial information state; K_p,K_i,K_d are the proportional, integral, and derivative gains, respectively, whose values are given as $K_p = 0.66$, $K_i = 0.001$, $K_d = 31$ by trial and error methods [4]. The adoption of virtual leader can greatly relieve the heavy computational burden in the control center given that only virtual leader needs to communicate with the distribution network management system. The information state from the virtual leader will be sent to the connected aggregators at discrete time step.

By using the communication matrix in (10.16), a transition weight between aggregator i and aggregator j can be defined as:

$$\Gamma_{ij}(t) = \frac{\psi_{ij}(t)}{\sum\limits_{j \in I_i} \psi_{ij}(t)} \tag{10.18}$$

Γ_{ij} mathematically represents the communication relevance between i and j, and directly determines the convergence speed.

In terms of followers, the consensus-based control scheme will be achieved iteratively. The information state of follower aggregators will be decided based on the states of virtual leader and neighboring aggregators. According to consensus algorithm [35], the consensus state of follower aggregators is updated as:

$$\eta_i(t) = \eta_i(t-1) + \sum_{j \in I_i} \Gamma_{ij}(t-1)\left[\eta_j(t-1) - \eta_i(t-1)\right] \tag{10.19}$$

where I_i is the neighbors of aggregator i. By (10.19), each aggregator could update its consensus state to adjust the active power commitment and finally converge to a unique equilibrium point, denoted as:

$$\frac{P_1}{P_1^{max}} = \frac{P_2}{P_2^{max}} = \cdots = \frac{P_i}{P_i^{max}} = \cdots = \frac{P_n}{P_n^{max}} = \eta_i \tag{10.20}$$

where P_i is the required active power commitment for aggregator i, P_i^{max} is the maximum controllable active power in aggregator i, which is derived by the aggregation model in (10.14).

Therefore, the required active power commitment for network loading management for aggregator i is calculated as:

$$P_i(t) = \eta_i(t) P_i^{max}(t) \tag{10.21}$$

10.3.2 Voltage Regulation Strategy

During normal hours, the bus voltage is kept within normal voltage range considering the constant balance between electricity load and electricity generation. However, bus voltages are easily drifting out of the standard voltage deviation ranges (i.e. normally 5–10% deviations based on different national standards) at high PV generation and light load periods. Therefore, the objective for voltage regulation is to control the local bus voltage within the acceptable range:

$$V^{min} \leq V_i(t) \leq V^{max} \tag{10.22}$$

where $V_i(t)$ is bus i voltage at time t; V^{min} and V^{max} are the minimum and maximum acceptable voltage limits.

Once the voltage limitation occurs, the distributed control scheme will be initiated:

$$\begin{cases} \varepsilon_i(t) = g_i\left[V_i(t) - V_i^{max}\right] \\ \varepsilon_i(t) = g_i\left[V_i(t) - V_i^{min}\right] \end{cases} \tag{10.23}$$

where ε_i is the information state for aggregator i and g_i is the weight for bus i. Note that the voltage limit violation is a localized problem, not a network wide one. Therefore, the distributed control strategy is well suited [36].

The relationship between bus voltage change and power change can be represented by Jacobian matrix:

$$\begin{bmatrix} \Delta P \\ \Delta Q \end{bmatrix} = \begin{bmatrix} J_1 & J_2 \\ J_3 & J_4 \end{bmatrix} \cdot \begin{bmatrix} \Delta \theta \\ \Delta |V| \end{bmatrix} \tag{10.24}$$

where

$$\begin{bmatrix} \Delta \theta \\ \Delta |V| \end{bmatrix} = \begin{bmatrix} J_1 & J_2 \\ J_3 & J_4 \end{bmatrix}^{-1} \begin{bmatrix} \Delta P \\ \Delta Q \end{bmatrix} = \begin{bmatrix} A & B \\ C & D \end{bmatrix} \begin{bmatrix} \Delta P \\ \Delta Q \end{bmatrix} \tag{10.25}$$

Given that reactive power is fully compensated in this chapter and active power considerably affects the voltage magnitude in distribution networks, the approximate sensitivity of bus voltage to power can be given by:

$$\frac{\partial V}{\partial P} = C \tag{10.26}$$

With the proposed communication structure, the aggregator is only aware of C_{ji} corresponding to their neighbors. Hence, the sensitivity matrix is defined as:

$$\overline{C_{ji}} = \begin{cases} C_{ji} & i \in \{I_j \cup j\} \\ 0 & i \notin \{I_j \cup j\} \end{cases} \tag{10.27}$$

where, I_j is the neighbors of aggregator j, with which j can communicate with. The transition weight in voltage regulation is given by:

$$\Gamma_{ij}(t) = \frac{\overline{C_{ji}} \psi_{ij}(t)}{\sum_{j=1}^{n} \overline{C_{ji}} \psi_{ji}(t)} \tag{10.28}$$

The control actions for the aggregator are determined subject to local voltage and its neighbors' information state. The information state of each aggregator is updated at discrete time step as:

$$\varepsilon_i(t) = \Gamma_{ii}(t)\varepsilon_i(t) + \sum_{j \in I_i} \Gamma_{ij}(t)\varepsilon_j(t-1) \tag{10.29}$$

Finally, the required active power commitment for voltage regulation for aggregator i is defined as:

$$P_i(t) = \varepsilon_i(t)P_i^{max}(t) \tag{10.30}$$

Comparing the active power commitment in Eq. (10.30) with the active power in Eq. (10.21), the larger value should be chosen in the control scheme in case that undervoltage violation and network overloading occur simultaneously. By choosing a larger value, both violations can be refrained and system normal operation can be guaranteed. To better understand the operation conditions for network loading management and voltage regulation, Figure 10.3 has been given.

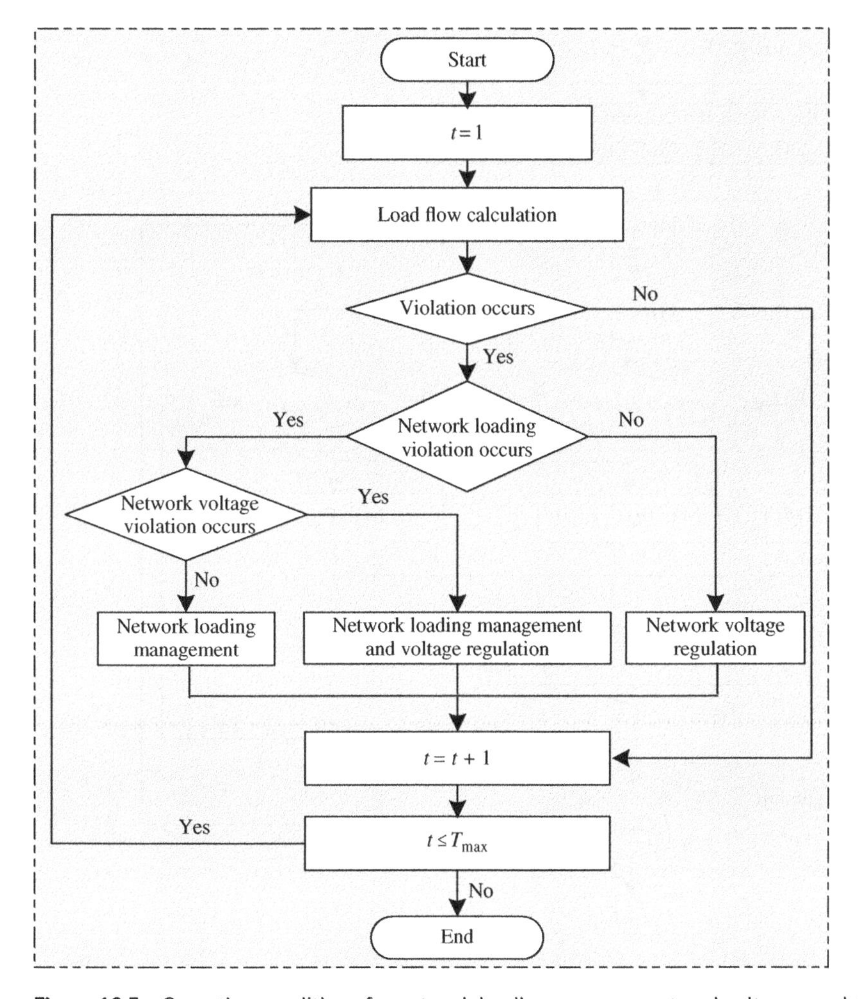

Figure 10.3 Operating conditions for network loading management and voltage regulation. *Source:* From Wang et al. [A].

The control flowchart for network loading management and voltage regulation is further shown in Figure 10.4. The solid blue line means the aggregated maximum controllable VESSs power is transferred from lower level to upper level. In contrast, the dotted blue line means the control signal from upper level is transferred to lower level for selectively charging/discharging VESSs. It should be noted that the consensus algorithm convergence time depends on system size and

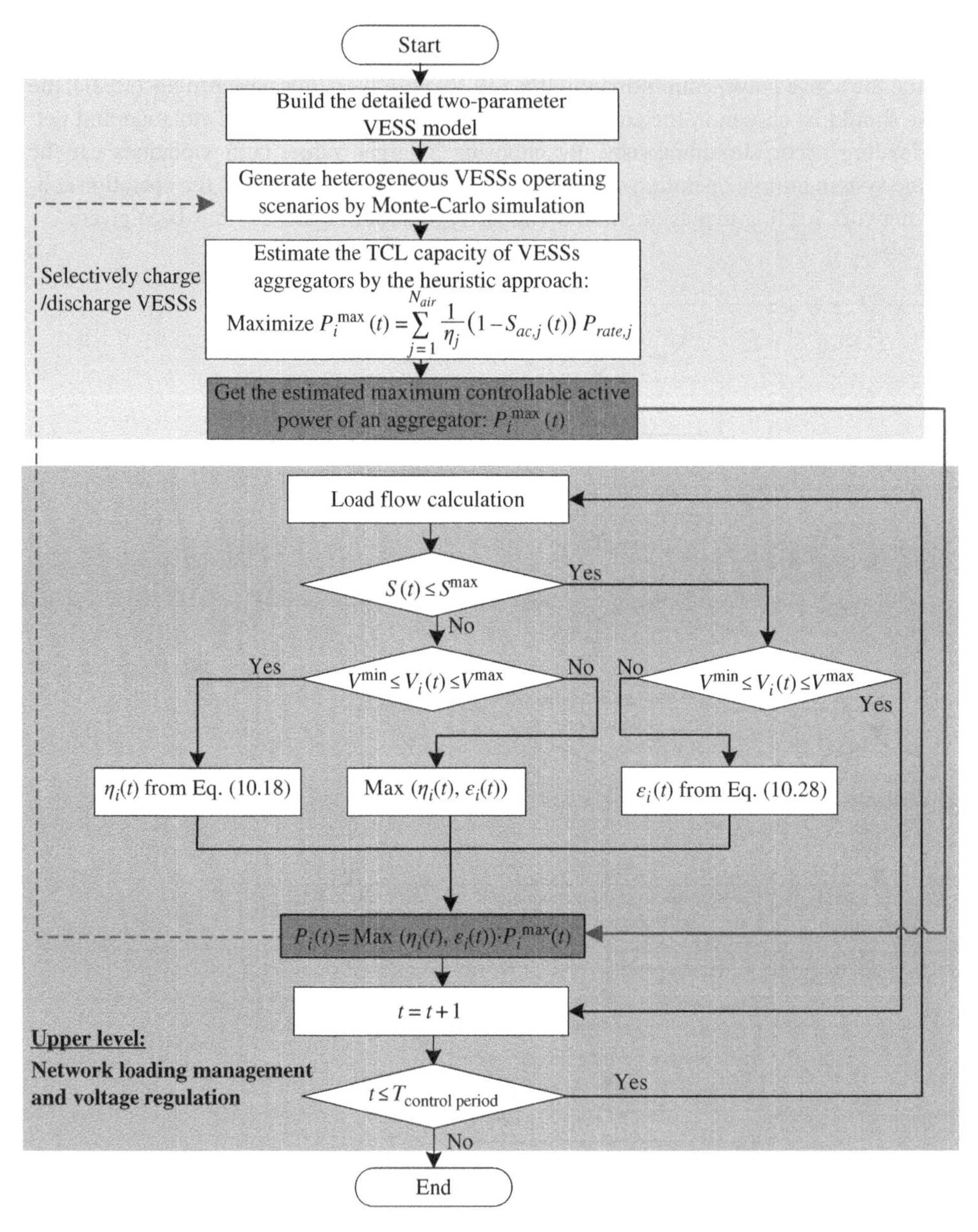

Figure 10.4 Flowchart of proposed control scheme for network loading management and voltage regulation. *Source:* From Wang et al. [A].

communication networks. The updating interval in this work is much more than enough for the proposed algorithm to achieve equilibrium.

10.4 Case Studies

One 11 kV nine-node feeder, which approximately supplies 1000 residential and commercial customers in New South Wales, Australia [37], is used in this case study to demonstrate the performance of proposed control scheme. It is estimated that more than 75% of households along the feeder have at least one air-conditioner installed. The one-line diagram of nine-node test feeder is shown in Figure 10.5, with connected PV systems and aggregators in the network.

There is one virtual leader and five aggregators, with 150 air conditioners regulated by each aggregator, distributed across the system. The heterogeneous VESS models are generated by Monte-Carlo simulation method, with parameters sampling ranges provided in Table 10.1. In this work, the thermal resistance of VESS includes glass windows and walls, which is calculated as a lump-sum resistance. The thermal capacitance of walls is taken into account as well.

Differential evolution (DE) algorithm is employed in this chapter to solve the maximum active power estimation model in Eq. (10.14). The maximum iteration time and population size are both fixed as 100 in DE algorithm. The detailed DE algorithm process is omitted here, which can be found in a similar work in [38]. Different temperature deadband scopes are given for the aggregators to simulate different households thermal comfort preferences. In this work, the temperature deadband scopes for five aggregators are 1, 2, 3, 4, and 5 °C, respectively. By solving the model in Eq. (10.14), the maximum controllable capacity of each VESSs aggregator in one day is denoted in Figure 10.6. In actual situations, a variety of factors influence the capacity of VESSs, such as building parameters (e.g. building material characteristics and room volume), meteorological parameters (e.g. indoor temperature and ambient temperature), and human parameters (e.g. temperature deadband scopes and ideal temperature set points). As observed from Figure 10.6, the larger temperature deadband scopes are, the more maximum controllable capacity each VESSs aggregator has.

The substation contains one transformer with summer rating being 3.3 MVA and the secondary transformer voltage at local nodes is 230 V, which is regarded as 1.0 p.u. (per unit). The maximum

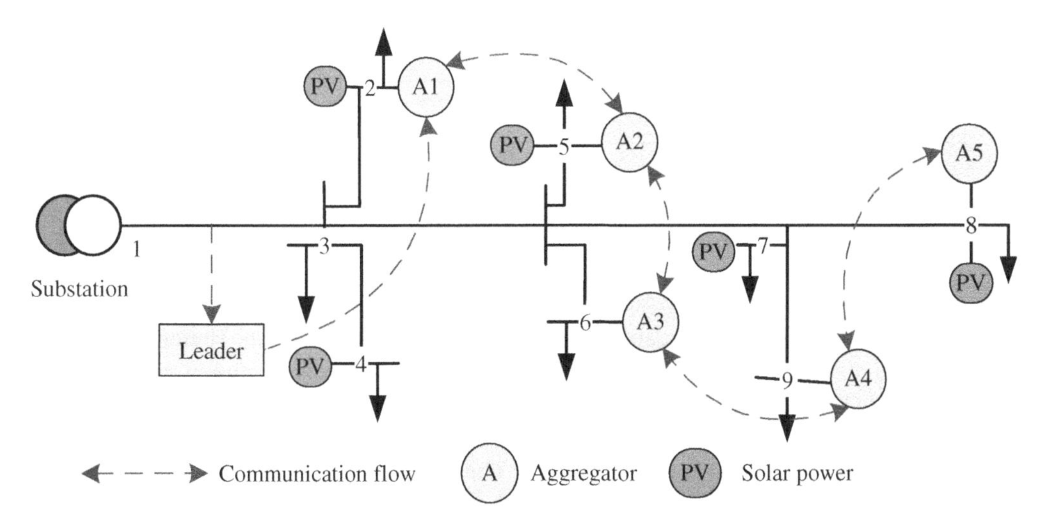

Figure 10.5 One-line diagram of nine-node test feeder. *Source:* From Wang et al. [A].

Table 10.1 VESS parameters sampling ranges for Monte-Carlo simulation.

House length/m	House width/m	House Height/m
8–22	10–20	3–9
Wall width/m	**Set point of thermostat/ °C**	**No. of Windows**
0.2–0.4	23–26	3–8
	Air conditioners rated power/kW	
	1–7	

Source: From Wang et al. [A].

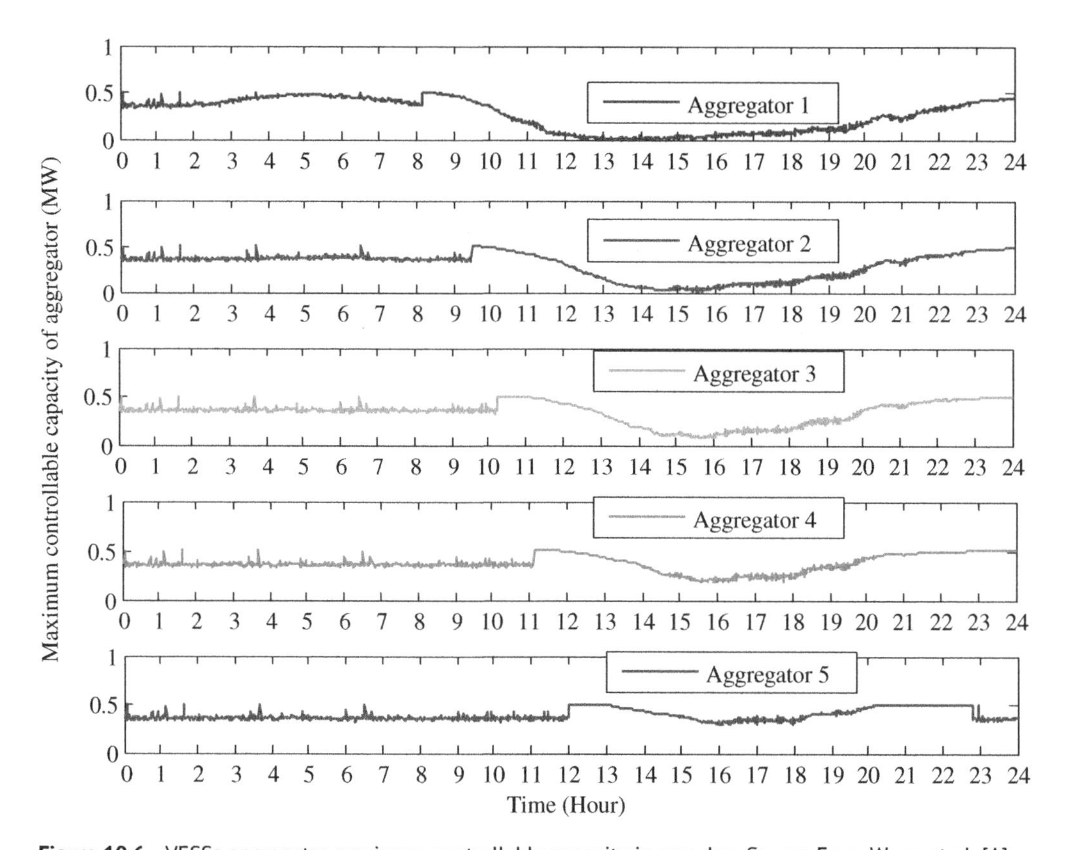

Figure 10.6 VESSs aggregator maximum controllable capacity in one day. *Source:* From Wang et al. [A].

voltage limit is set as 1.05 p.u. and the minimum voltage limit is set as 0.96 p.u. Therefore, to guarantee the system normal operation, the system voltage should be controlled between the minimum and the maximum limits, and the network loading should be no larger than 3.3 MVA. The rated PV capacity in each node is denoted in Table 10.2, where one-minute solar radiance resolution data is derived based on a typical summer data in UQ Solar Photovoltaic Centre to get the daily PV power profile [39]. The system load is predicted based on the classical residential load demand in [40].

Table 10.2 Rated capacity in each node of the system.

Node	2	4	5	7	8
PV Capacity (MW)	0.36	0.45	0.52	0.54	0.48

Source: From Wang et al. [A].

10.4.1 Case 1

In this case study, the performance of proposed control scheme is verified. The simulation program is conducted in MATLAB on a 4 core, 64-bit DELL Desktop with Intel Core i5-3570S CPU and RAM 8 Gb.

Figure 10.7a demonstrates the system node voltage change in one day under high PV power penetration and peak residential electricity demand. It can be found that the voltage in local nodes keeps rising because of PV penetration into the system and crosses the maximal voltage limit during 8:30–14:30. Compared with overvoltage, undervoltage occurs during 19:00–21:00 because of the high residential electricity demand. Once the voltage violation happens, the proposed control scheme is initiated to regulate system voltage by the smart coordination of aggregated VESSs.

Observing the grey line in Figure 10.8, it can be clearly seen that system apparent power crosses summer rating value from 18:30 to 21:30. This is caused by the peak electricity demand in hot summer days when residents are back to home from work, especially caused by turning on the large amount of air conditioners. Similar to voltage regulation, once overloading occurs, the proposed control strategy is implemented to manage system loading.

Figure 10.7b demonstrates the voltage control performance by the proposed strategy. With the proposed control scheme, the nodes voltage is maintained within the desirable range during the day. Specifically, from 8:30 to 14:30, voltage rise is mitigated and is controlled to be no more than 1.05 p.u. This control effect is achieved by the extra consumption of PV generation by turning on VESSs, hence the reverse power from PV inverters into distribution network are reduced. From 19:00 to 21:00, voltage drop is eliminated as well and kept larger than the minimal voltage limit, i.e. 0.96 p.u. During these periods, VESSs are turned off to help reduce electricity demand.

Figure 10.8 shows the network loading control performance. It can be observed that the system overloading is well controlled below the summer rating value from 18:30 to 21:30, which is achieved by the reduced electricity consumption via selectively turning off VESSs. Given that VESSs are required to turn off during both undervoltage periods and overloading periods, a larger active power control value should be executed during these two periods, which can guarantee the overall system control performance.

The contribution amount of each VESSs aggregator is denoted in Figure 10.9. The VESSs work in charging mode, i.e. VESSs are turned on, during high PV penetration periods to help lower nodal voltage. During 18:30–21:30, VESSs work in discharging mode, i.e. VESSs are turned off, to compensate voltage drop and network overloading. It should be noted that by considering indoor air temperature restrictions as shown in (10.11), the thermal comfort in residential household is not compromised during the whole control period.

10.4.2 Case 2

Given the communication channel disturbance in Eq. (10.16), different communication topology networks can be formed among aggregators, which could cause impacts on the algorithm

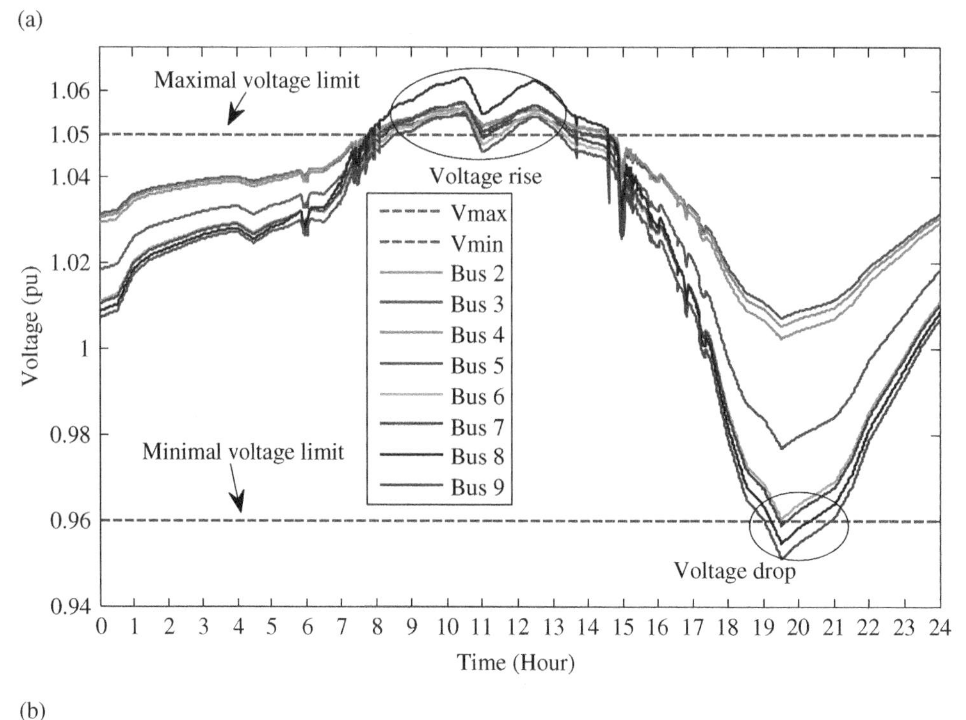

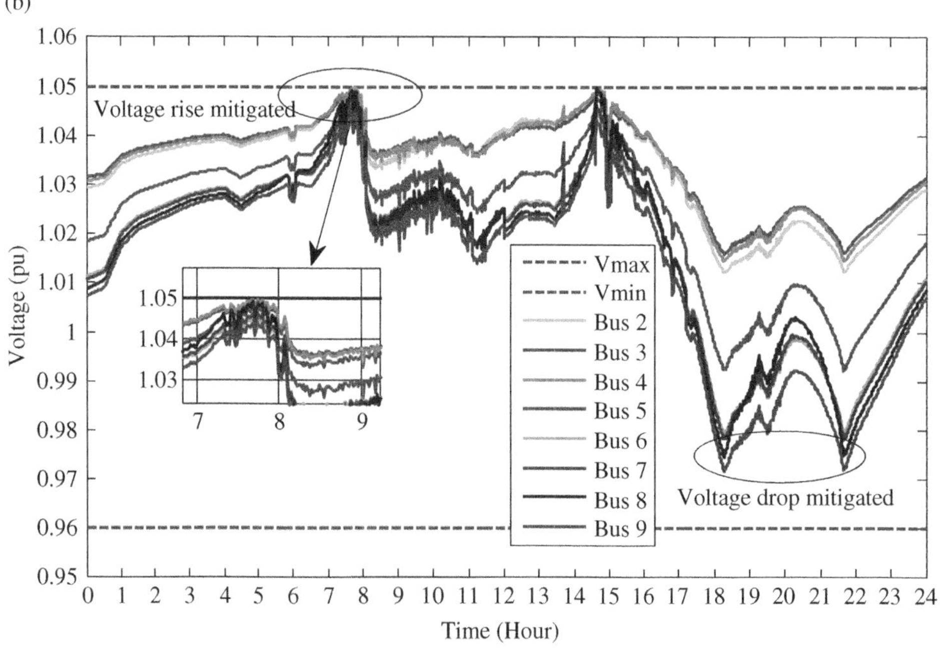

Figure 10.7 Voltage profile in one day (a) Voltage profile before control. (b) Voltage profile after adopting proposed control scheme. *Source:* From Wang et al. [A].

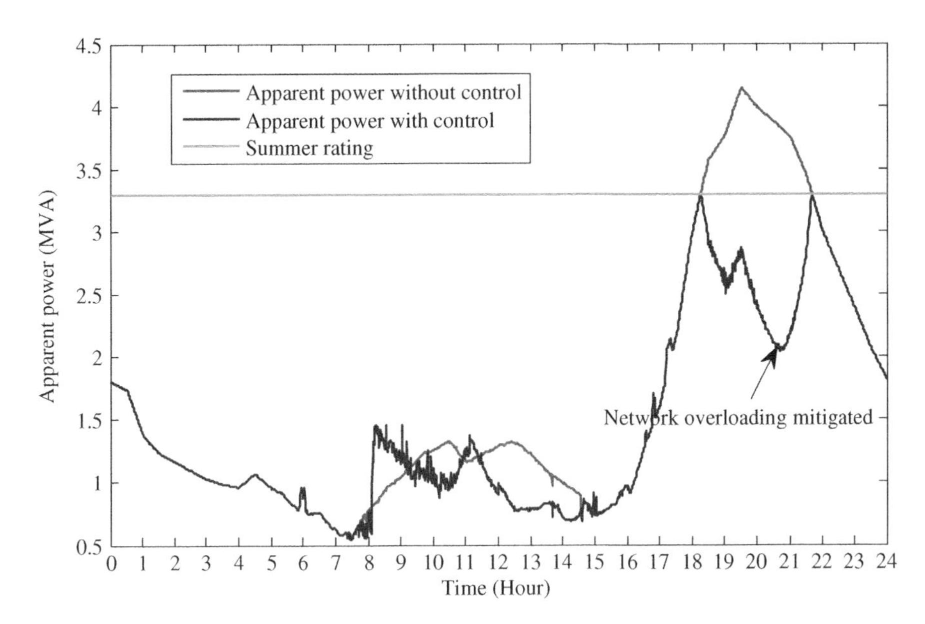

Figure 10.8 Network loading change in one day without and with control. *Source:* From Wang et al. [A].

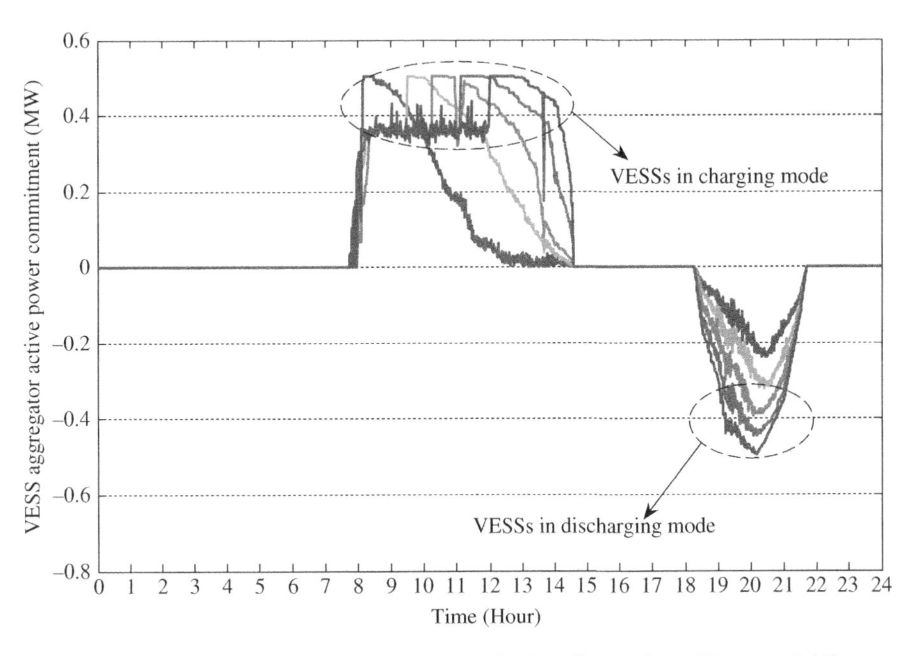

Figure 10.9 VESSs aggregator active power contribution. *Source:* From Wang et al. [A].

convergence. Therefore, the influence of dynamic communication topology on system performance is investigated in this part to demonstrate the robustness of proposed control scheme.

The common adopted communication topologies for the aggregators are line shape, star shape, ring shape, and complete shape. Figures 10.10–10.12 separately present the star-shape topology, complete-shape topology, and ring-shape topology, together with the voltage control

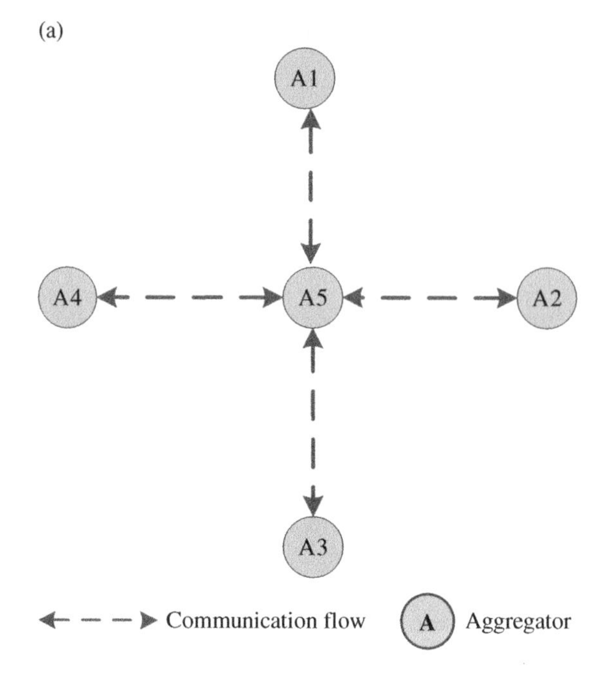

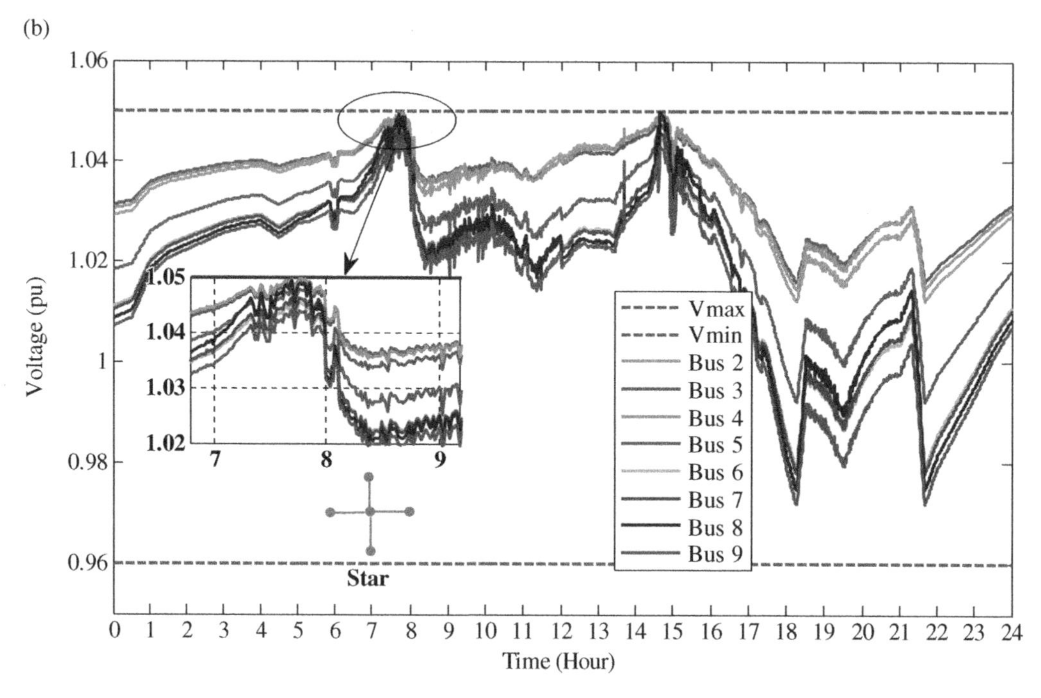

Figure 10.10 Voltage performance under Star-shape topology (a) Star-shape topology. (b) Voltage regulation performance under Star-shape topology. *Source:* From Wang et al. [A].

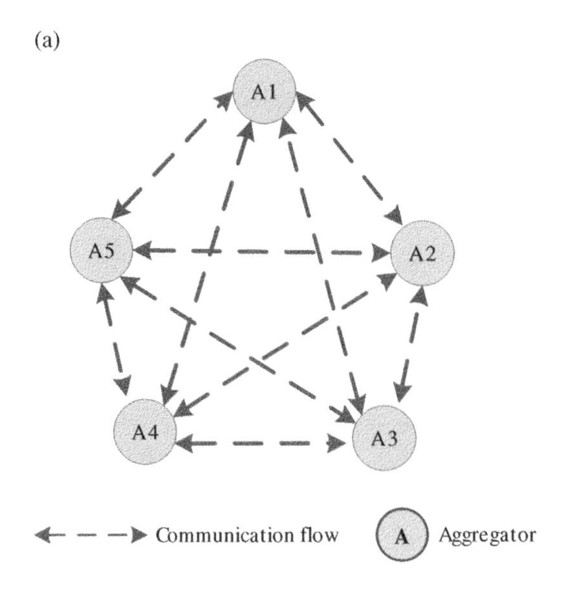

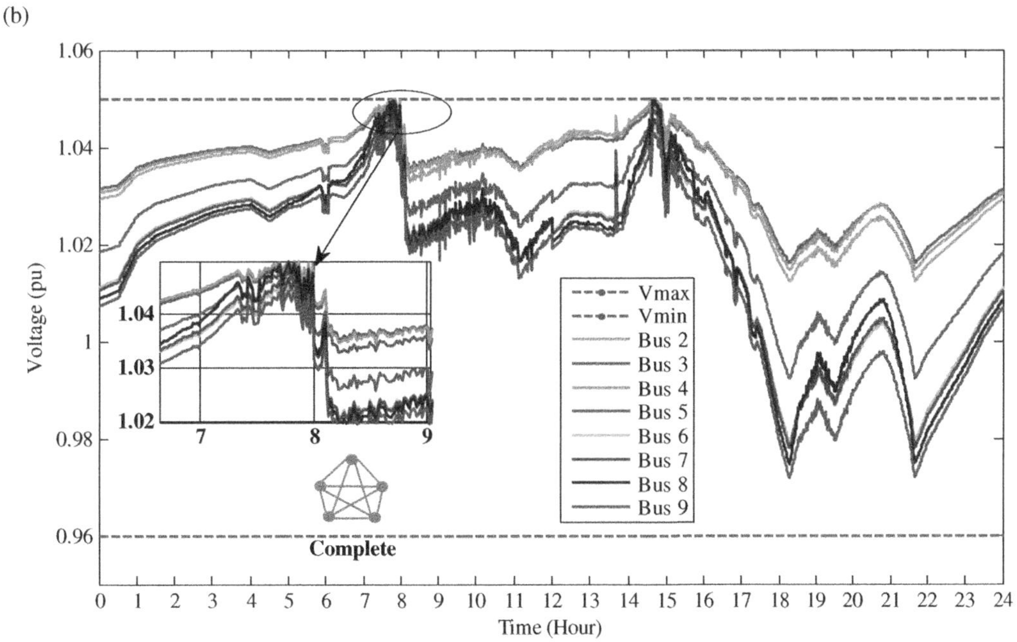

Figure 10.11 Voltage regulation performance under Complete-shape topology (a) Complete-shape topology, (b) Voltage regulation performance under Complete-shape topology. *Source:* From Wang et al. [A].

performances. For network loading effects under different topologies, these three topologies have similar control results to line-shape topology shown in Figure 10.8. Therefore, only the network loading management performance under ring-shape topology is given here for reference. The control results under line-shape topology have been given in Case 1. Four different communication network topologies all demonstrate satisfying system control performances, which verify the control scheme robustness in terms of dynamic network topologies. The control scheme

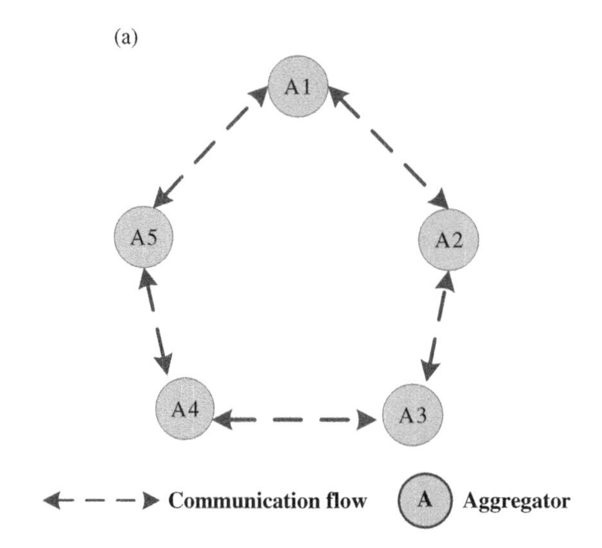

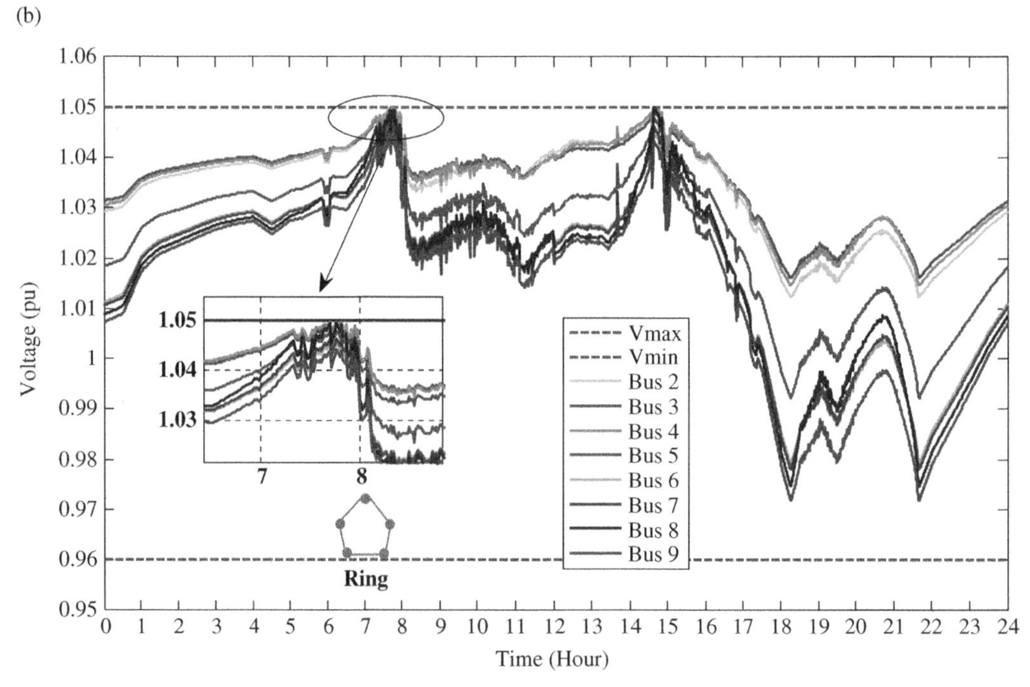

Figure 10.12 Voltage regulation performance and network loading management performance under Ring-shape topology (a) Ring-shape topology, (b) Voltage regulation performance under Ring-shape topology, (c) Network loading management performance under Ring-shape topology. *Source:* From Wang et al. [A].

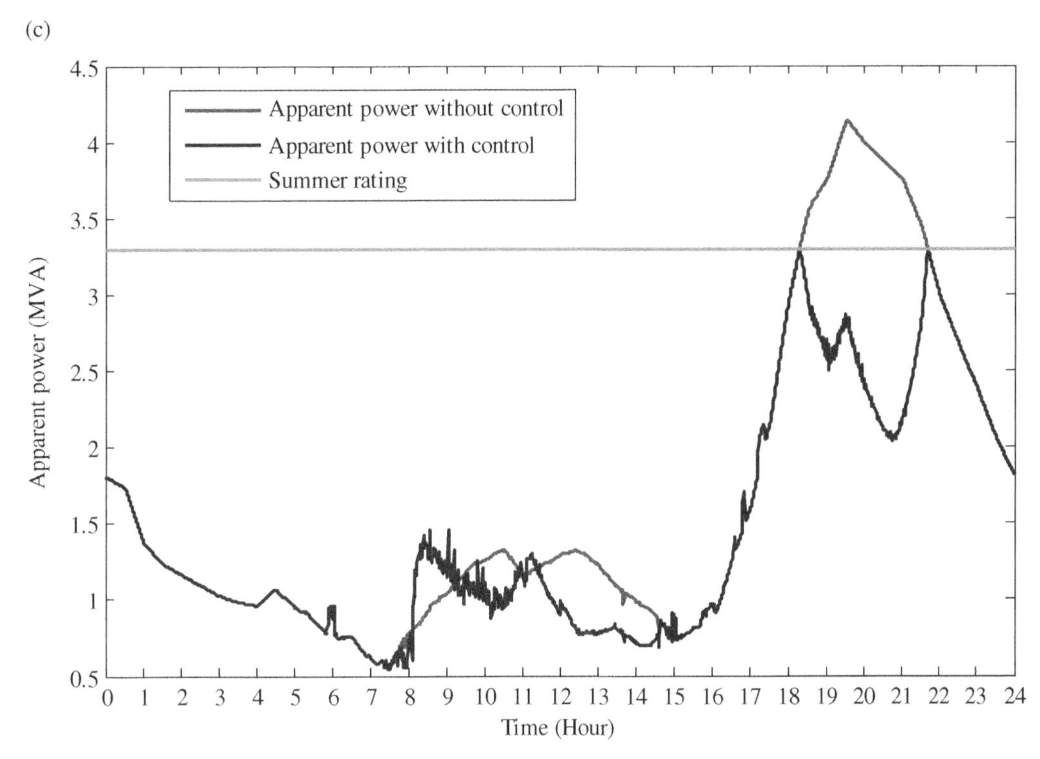

Figure 10.12 (Continued)

Table 10.3 Calculation speed and system performance for different network topologies.

Network topology	Calculation speed (s)	Voltage regulation effectiveness	Network loading management effectiveness	Robust to communication failure
Line	29.02	Yes	Yes	No
Star	23.24	Yes	Yes	No
Ring	17.95	Yes	Yes	Yes
Complete	15.74	Yes	Yes	Yes

Source: From Wang et al. [A].

calculation speed and system performance of these four communication network topologies are summarized in Table 10.3.

It is worth noting that complete network topology shows the fastest calculation speed, owing to faster consensus algorithm convergence speed compared with the other three topologies. On the other hand, complete topology has the most sophisticated communication channels, which also increase the communication cost dramatically. It can also be observed that ring-shape topology and complete topology are robust to communication failures. When one communication link is removed among aggregators, the system control performance will not be influenced. Combining calculation speed and communication costs, ring-shape topology is more ideal from system operator's perspective, with relatively fast computation speed and less communication line investment.

10.5 Conclusions and Future Work

A smart coordination strategy is proposed in this chapter to coordinate aggregated VESSs for distribution network, voltage regulation, and loading management. Through the lower level control, the VESS model is built and the aggregated amount of maximum controllable active power is calculated. Through the upper level control strategy, a consensus-driven distributed control strategy is employed to fairly share the active power commitment among aggregators for distribution network management. The simulation results on a nine-node test feeder in NSW prove the effectiveness of proposed control scheme, as well as demonstrate the robustness in case of dynamic communication network topologies and communication failure. The proposed method is superior over other methods in terms of, (i) utilizing the thermal buffering capacity of air-conditioned households to form VESSs, with far less cost than hard energy storage techniques, (ii) reaping demand response programs benefits via aggregators, with end-users' thermal comfort guaranteed, and (iii) adapting to dynamic communication network topologies, in the meanwhile ensuring the equilibrium point is reached among participating aggregators. With the increasing penetration of renewable resources and deployment of smart household energy management systems, the proposed method is believed to be a promising method on managing practical distribution networks in a techno-economic manner.

Future work will focus on improving the thermal modeling of virtual energy storage system via taking into account other important factors, including solar irradiance, window area, etc. In addition, the VESSs control in large-scale systems with complex communication network will be investigated.

Acknowledgements

The permission given in using materials form the following paper is very much appreciated.

[A] Wang, D., Lai, C.S., Li, X. et al. Smart coordination of virtual energy storage systems for distribution network management. *International Journal of Electrical Power and Energy Systems* 129: 106816.

References

1 International Energy Agency (n.d.). Renewables 2018 [Online]. https://www.iea.org/renewables2018/ (accessed 21 November 2020).

2 Akagi, H. and Isozaki, K. (Apr. 2012). A hybrid active filter for a three-phase 12-pulse diode rectifier used as the front end of a medium-voltage motor drive. *IEEE Trans. Power Electron.* 27 (4): 1758–1772.

3 Zeraati, M., Golshan, M.E.H., and Guerrero, J.M. (2018). Distributed control of battery energy storage systems for voltage regulation in distribution networks with high PV penetration. *IEEE Trans. Smart Grid* 9 (4): 3582–3593.

4 Rafi, F.H.M., Hossain, M.J., and Lu, J. (2016). Hierarchical controls selection based on PV penetrations for voltage rise mitigation in a LV distribution network. *International Journal of Electrical Power & Energy Systems* 81: 123–139.

5 Salem, M.R., Talat, L.A., and Soliman, H.M. (1997). Voltage control by tap changing transformers for a radial distribution network. *Proc. Inst. Elect. Eng., Generation Transmission and Distribution* 144: 517–520.

6 Masters, C.L. (2002). Voltage rise: the big issue when connecting embedded generation to long 11 kV overhead lines. *Power Engineering Journal* 16: 5–12.

7 Tonkoski, M. (2014). Transformer voltage regulation-compact expression dependent on tap position and primary /secondary voltage. *IEEE Trans. Power Delivery* 29 (3): 1516–1517.

8 Chen, C.S., Lin, C., Hsieh, W. et al. (2013). Enhancement of PV penetration with DSTATCOM in Taipower distribution system. *IEEE Trans. Power Syst.* 28 (2): 1560–1567.

9 Tonkoski, R. and Lopes, L.A.C. (2011). Impact of active power curtailment on overvoltage prevention and energy production of PV inverters connected to low voltage residential feeders. *Renewable Energy* 36 (12): 3566–3574.

10 Ghosh, S., Rahman, S., and Pipattanasomporn, M. (2017). Distribution voltage regulation through active power curtailment with PV inverters and solar generation forecasts. *IEEE Trans Sustainable Energy* 8 (1): 13–22.

11 Wang, Y., Tan, K., Peng, X., and So, P. (2016). Coordinated control of distributed energy-storage systems for voltage regulation in distribution networks. *IEEE Trans. Power Delivery* 31 (3): 1132–1141.

12 Yan, X., Gu, C., Li, F., and Xiang, Y. (Feb. 2018). Network pricing for customer-operated energy storage in distribution networks. *Applied Energy* 212: 283–292.

13 Sidhu, A.S., Pollitt, M.G., and Anaya, K.L. (2018). A social cost benefit analysis of grid-scale electrical energy storage projects: a case study. *Applied Energy* 212: 861–894.

14 Lai, C.S. and McCulloch, M.D. (2017). Levelized cost of electricity for solar photovoltaic and electrical energy storage. *Applied Energy* 190: 191–203.

15 Wang, D., Meng, K., Luo, F. et al. (2016). Coordinated dispatch of networked energy storage systems for loading management in active distribution networks. *IET Renewable Power Generation* 10 (9): 1374–1381.

16 Smith, R., Meng, K., Dong, Z.Y., and Simpson, R. (2013). Demand response: a strategy to address residential air-conditioning peak load in Australia. *J. Mod. Power Syst. Clean Energy* 1 (3): 223–230.

17 Rahiman, F.A., Zeineldin, H.H., Khadkikar, V. et al. (2014). Demand response mismatch (DRM): concept, impact analysis, and solution. *IEEE Trans. Smart Grid* 5 (4): 1734–1743.

18 Ihsan, A., Jeppesen, M., and Brear, M.J. (2019). Impact of demand response on the optimal, techno-economic performance of a hybrid, renewable energy power plant. *Applied Energy* 238: 972–984.

19 Wang, D., Meng, K., Gao, X. et al. (2017). Optimal air-conditioning load control in distribution network with intermittent renewables. *J. Mod. Power Syst. Clean Energy* 5 (1): 55–65.

20 Cheng, M., Sami, S.S., and Wu, J. (2017). Benefits of using virtual energy storage system for power system frequency response. *Applied Energy* 194: 376–385.

21 Busic, A. (2015). *Virtual Energy Storage Through Distributed Control of Flexible Loads*. San Francisco, CA: *California France Forum on Energy Efficiency Technologies*.

22 Akagi, S., Yoshizawa, S., Ito, M., and Fujimoto, Y. (2020). Multipurpose control and planning method for battery energy storage systems in distribution network with photovoltaic plant. *International Journal of Electrical Power & Energy Systems* 116: https://doi.org/10.1016/j.ijepes.2019.105485.

23 Ranaweera, I., Midtgard, O.M., and Korpas, M. (Dec. 2017). Distributed control scheme for residential battery energy storage units coupled with PV systems. *Renewable Energy* 113: 1099–1110.

24 Wang, L. and Chen, B. (July 2019). Distributed control for large-scale plug-in electric vehicle charging with a consensus algorithm. *International Journal of Electrical Power & Energy Systems* 109: 369–383.

25 Liu, W., Gu, W., Xu, Y. et al. (2015). Improved average consensus algorithm based distributed cost optimization for loading shedding of autonomous microgrids. *International Journal of Electrical Power & Energy Systems* 73: 89–96.

26 Wu, X., Shen, C., and Iravani, R. (2018). A distributed cooperative frequency and voltage control for microgrids. *IEEE Trans. Smart Grid* 9 (4): 2764–2776.

27 Xin, H., Liu, Y., Qu, Z., and Gan, D. (2014). Distributed control and generation estimation method for integrating high-density photovoltaic systems. *IEEE Trans. Energy Convers.* 29 (4): 988–996.

28 Bashash, S. and Fathy, H.K. (2013). Modeling and control of aggregate air conditioning loads for robust renewable power management. *IEEE Trans. Cont. Syst. Tech.* 21 (4): 1318–1327.

29 Lu, N. and Zhang, Y. (2013). Design considerations of a centralized load controller using thermostatically controlled appliances for continuous regulation reserves. *IEEE Trans. Smart Grid* 4 (2): 914–921.

30 Wang, H., Meng, K., Dong, Z. et al. (2013). Demand response through smart home energy management using thermal inertia. *Australian Universities Power Engineering Conference,* Hobart, Australia (29 September 2013 to 3 October 2013).

31 Talwariya, A., Singh, P., and Kolhe, M. (2019). A stepwise power tariff model with game theory based on Monte-Carlo simulation and its applications for household, agricultural, commercial and industrial consumers. *International Journal of Electrical Power & Energy Systems* 111: 14–24.

32 Karavas, C.-S., Kyriakarakos, G., Arvanitis, K.G., and Papadakis, G. (2015). A multi-agent decentralized energy management system based on distributed intelligence for the design and control of autonomous polygeneration microgrids. *Energy Conversion and Management* 103: 166–179.

33 Saber, R.O. and Murray, R.M. (2004). Consensus problems in networks of agents with switching topology and time-delays. *IEEE Transactions on Automatic Control* 49 (9): 1520–1533.

34 He, W., Chen, G., Han, Q., and Qian, F. (2017). Network-based leader-following consensus of nonlinear multi-agent systems via distributed impulsive control. *Information Sciences* 380: 145–158.

35 Olfati-Saber, R., Fax, J., and Murray, R.M. (2007). Consensus and cooperation in networked multi-agent systems. *Proc. IEEE* 95 (1): 215–233.

36 Mehmood, K.K., Khan, S.U., Lee, S.J. et al. (2018). A real-time optimal coordination scheme for the voltage regulation of a distribution network including an OLTC, capacitor banks, and multiple distributed energy resources. *International Journal of Electrical Power & Energy Systems* 94: 1–14.

37 Meng, K., Dong, Z.Y., Xu, Z. et al. (2019). Coordinated dispatch of virtual energy storage systems in smart distribution networks for loading management. *IEEE Trans. Systems, Man, and Cybernetics: Systems* 49 (4): 776–786.

38 Luo, F., Dong, Z.Y., Meng, K. et al. (2017). An operational planning framework for large scale thermostatically controlled load dispatch. *IEEE Trans. Ind. Informat.* 13 (1): 217–227.

39 The University of Queensland (2015). UQ SOLAR Photovoltaic Data. The University of Queensland, Australia [Online]. http://solar.uq.edu.au/user/reportPower.php?dtra=day&dts=2015-11-26 (accessed 21 November 2020).

40 Jardini, A.J., Tahan, C.M.V., Gouvea, M.R. et al. (2000). Daily load profiles for residential, commercial, and industrial low voltage consumers. *IEEE Trans. Power Delivery* 15 (1): 375–380.

11

Reliability Modeling and Assessment of Cyber-Physical Power Systems

Nomenclature

e_{xy}	Edge from node x to node y in the directed graph for the power information system
f_{xy}	Flow in e_{xy}
C_{xy}	Capacity of e_{xy}
C_x	Capacity of node x residing in the current state in the information layer
$C_{x.\ max}$	Maximum capacity of node x
Ψ	Total inflow injected into sink node
V_t	Set of sink nodes
V_s	Set of source nodes
V_{inter}	Set of intermediate nodes
ls_x	Loss of information after coming across node x
ub_{xy}	Upper bound of flow in e_{xy}
lb_{xy}	Lower bound of flow in e_{xy}

11.1 Introduction

With the technological advances in Information and Communication Technology (ICT), urbanization becomes a global phenomenon. By 2030, 60% of the population will live in urban. Modern cities today compete with each other and need to stay smart and be smarter with high-quality living standard. Cyber-physical systems (CPSs) support our day-to-day activities and is becoming more important to smart cities deployment.

CPSs are defined as transformative technologies for managing interconnected systems between its physical assets and computational capabilities [1]. The ability to interact with, and expand the capabilities of, the physical world through computation, communication, and control is a key enabler for future technology developments. CPSs research aims to integrate knowledge and engineering principles across the computational and engineering disciplines, namely, networking, control, software, human interaction, learning theory, as well as electrical, mechanical, biomedical, and material science to develop new supporting technology with a small impact from cyber risk.

Embedded computers and networks monitor and control the physical processes, usually with feedback loops where physical processes affect computations and vice versa. There are considerable

Smart Energy for Transportation and Health in a Smart City, First Edition. Chun Sing Lai, Loi Lei Lai and Qi Hong Lai.

challenges, particularly because the physical components of such systems introduce safety and reliability requirements qualitatively different from those in general purpose computing [2]. With recent developments that have resulted in higher availability and affordability of sensors, data acquisition systems and computer networks. Consequently, the ever-growing use of sensors and networked intelligent electronic devices have resulted in the continuous generation of lots of data. In such an environment, CPS can be further developed for managing the data and leveraging the interconnectivity of devices to reach the goal of intelligence, resilience, and self-adaptability [3].

The CPSs must be networked. The most widely used networking techniques today introduce a great deal of timing variability and stochastic behavior. Successful applications include communication systems and home appliances. Such networking poses considerable technical challenges. For example, embedded software relies on bench testing for concurrency and timing properties. In a networked environment, it becomes impossible to test the software under all possible conditions. Moreover, general purpose networking techniques make program behavior much more unpredictable. To be specific, recent advances in time synchronization across networks promise networked platforms that share a common notion of time to a known precision [4].

The use of parallel and distributed computing is pervasive for a wide range of cyber-physical applications. The simulation of a CPS includes physical plants, end-nodes, i.e. the computers that execute distributed programs, and the intermediate networks. The proposed model comes with the notion of external events to allow the co-simulation of interactions between end-nodes and plants/network [5]. In the transition to CPS, the expectation of reliability will only increase. In fact, without improved predictability and reliability, CPS will not be deployed into such applications as power system operation. The physical world, however, is not totally predictable. CPSs will not be operating in a controlled environment and must be robust to unexpected conditions and adaptable to subsystem failures.

The future of smart grid is facing considerable challenges due to significant increases in its scale and complexity which lays additional stress on power system operation and control. In this arena, it is important to realize online utilizations and real-time exchanges of energy with the support of information technologies. Smart grid is generally characterized by more sensors, additional communication channels, higher computational efforts, and added control hierarchies. As a result, smart grid is evolved into a cyber-physical power system in which embedded computers and communication networks closely monitor and control the physical processes for distributed energy delivery [6–8].

The strong coupling of cyber and physical systems in power systems has tied the performance of cyber systems to the operating characteristics of physical power systems [9]. However, the conventional power system reliability literature has focused on the failure of physical system components, in which cyber element functionalities are considered perfectly reliable. With the expansion in the deployment of cyber infrastructure, the reliability analyses of cyber-physical power systems have attracted more attention among researchers in recent years [10–12].

Falahati et al. defined cyber-power interdependency as direct and indirect types which are based on impacts of cyber failures on power systems. The state mapping or state updating models were proposed to quantitatively evaluate the cyber-power reliability [13, 14]. Lei et al. presented cyber-physical interface matrix to implement the protection system reliability at substation level without the necessity of considering the details of cyber elements [15, 16]. Xin et al. proposed an information-energy flow model for cyber-physical power systems [17]. It introduced a path-branch incidence matrix to describe the cyber network in cyber-physical sensitivity analyses for evaluating the coupling among cyber/physical quantities. Marashi et al. populated a Markovian reliability model for smart grids that can be substantiated with failure data collected from field measurements or simulations [18]. Guo et al. investigated the reliability assessment of cyber-physical microgrid systems in island mode based on networks connectivity [19]. However, these valuable cyber-physical reliability studies do not include cyber system analyses in any significant details.

Reliability is a key metric to measure how consistently the power information system (i.e. cyber space in power system) can perform data collection, transfer, and processing according to specific power system requirements. The research on communication network reliability, especially that of network connectivity, has achieved promising results. There are many measures to evaluate the communication network reliability, such as reliability block diagram, fault tree analysis, and state space method. Wang et al. analyzed the communication network reliability in wide-area measurement system (WAMS) using Markov modeling and state enumeration techniques [20]. Dai et al. discussed the reliability of communication systems in wide-area protection using fault tree analysis and Monte-Carlo simulation [21]. Liu et al. developed the validity of cyber link to measure the connectivity of network topology [22].

However, these methods may not be directly applicable to examine the reliability of a power information system in which cyber components are linked via information flows. That is because these studies only consider binary-state operation-failure representations of physical systems, in which the information flow can pass through only when components are in operation state. By only considering the binary-state of power information equipment one can over-estimate the reliability of power information system. This is because the mean-time-to-failure (MTTF) of power information equipment can change over time in normal operating conditions and corresponding state transition intervals can be several orders higher than those of information flow. Accordingly, the stochastic-flow network (SFN) model is suitable for the reliability analysis of information flow which uses multiple states to represent the degree of deterioration of information performance. The SFN reliability can be computed in terms of level d which means the maximum network flow is not less than d units [23].

Generally, analytical methods and Monte-Carlo simulation are adopted for the reliability evaluation of SFN model [24]. However, the analytical method, which uses the graph theory and the probability theory to compute the network reliability accurately, is often applicable to small-scale networks [25, 26]. With the expansion of network size and the increase in the number of states, the computational complexity of analytical methods will rise exponentially, which will lead to NP-hard solution difficulties [27]. The Monte-Carlo simulation has demonstrated several advantages in system state extractions, which lends itself to evaluate multi-state large-scale networks. Ramirez-Marquez and Coit introduced the nonsequential Monte-Carlo simulation approach combined with d-minimal path set to calculate the multi-state reliability [28]. However, the process of forming the d-minimal path set requires the enumeration of all possible system states, which could result in a lower efficiency for evaluating large complex systems [29].

The information and energy flows interact only when they form a close-loop. Therefore, certain power system applications can be regarded as coupling processes for information and energy, such as wide area protection and control system, which cover wide area information collection, communication network for transmitting the information, and control center for information processing, decision-making, and control output implementation. Therefore, wide area protection and control system in smart grid links the power information system with the power system to realize the closed-loop.

Based on the previous research results, in this chapter an innovative idea for the reliability analyses of cyber-physical power systems is proposed. The reliability of cyber space at the d-level us studied. The scope of the previous work on quantitative reliability modeling of communication networks in intelligent substations is also broadened [23] and the work is extended to the information system of smart grid. A reliability evaluation methodology is presented for the information space in cyber-physical power system, which is expressed by the composite Markov process using the Monte-Carlo simulation and the maximum-flow technique.

First, a composite Markov model is formed, which contains not only the binary-state representation of physical components, such as equipment failure and disabled communication links, but also a multi-state representation of information flow. In this way, causes and consequences of cyber-physical power system failures are depicted clearly, which can stem out of either the physical layer or the information layer in power information system. For example, the failure can occur when all devices are working normally, but the information system lacks the stated communication quality. Then, according to the functional duties, the components in the power information system are classified into several types of logical nodes and the corresponding flow constraint description is stated. In order to improve the solution efficiency, the linear programming model is used to obtain maximum flows, instead of enumerating all system states to obtain information flow distributions.

Finally, the sequential Monte-Carlo simulation with a two-level sampling is applied to the proposed model and reliability indices are defined for evaluating the power information system. The proposed sequential Monte-Carlo approach offers significant advantages for calculating reliability indices related to frequency and duration of failures in complex systems, because it simulates stochastic characteristics according to time lapses. In each more refined sampling scenario, maximum flow is computed which determines whether the flows in the current system state meet functional requirements. The proposed cyber space reliability solution can also be extended to evaluate the cyber-physical power system reliability and the robustness in smart grid.

The remainder of this chapter is organized as follows. Section 2 establishes the Markov models for both information flow performance and physical system operation, and combines them to form a composite Markov model for the power information system. Section 3 gives the node classifications and their flow constraints in the power information system. It also presents a linear programming model to obtain maximum flows. Section 4 provides the definition of reliability indices and presents the corresponding calculation procedure using the sequential Monte-Carlo simulation with a two-level sampling. Section 5 implements the proposed reliability evaluation method on the case study based on the IEEE 14-bus system and analyzes the results in different scenarios. Section 6 concludes the chapter.

11.2 Composite Markov Model

In this chapter, the topology of power information system is described by a directed graph, in which all components such as sources, sinks, and switches are abstracted as nodes, while communication channels are presented as edges in the physical layer. The information layer represents the information flow through the cyber space in power information system.

11.2.1 Multistate Markov Chain of Information Layer

In the existing literature, the multistate division of components adopts a static partitioning and Markov chain is introduced in [23] to form the multistate of random flows through the information layer.

The information flow through a component can be arbitrary at any time which is stated as a discrete-state stochastic process $\{X_c(t),\ t \in [0, \infty)\}$ with state space Ω_c^0. Since the number of states is excessively large in practice, typical values are usually used to reflect fluctuations in information transmission rates. Thus, the original state space Ω_c^0 is discretized here.

Let the historical information flow of a component be represented as $v(t_k) \in [v_{min}, v_{max}]$ with sampling time series $\{t_k\}_{k=0}^{m}$, where v_{max} and v_{min} represent the maximum and the minimum of recorded flow during the total observation period, respectively. To discretize the information flow of a component into N levels, the total interval is divided into N intervals $[v_i, v_{i+1})$, $i \in \{1, 2, \ldots, N\}$ where the length of each interval is $\Delta v = (v_{max} - v_{min})/N$. The information flow of the component in state i is calculated as:

$$f_i = \sum_{v(t_k) \in [v_i,\ v_{i+1})} v(t_k)/k_i \tag{11.1}$$

where k_i is the number of sampling data in interval $[v_i, v_{i+1})$. The discretized state space is $\Omega_c = \{1, 2, \ldots, N\}$.

The information flow of a component is assumed to be time-homogeneous; then, the stochastic process $\{X_c(t),\ t \in [0, \infty)\}$ within state space Ω_c is a time-homogeneous Markov chain with a transition probability which satisfies

$$p_{ij}(t) = \Pr\{X(t_s + t) = j \mid X(t_s) = i\}; i,j \in \Omega_c \tag{11.2}$$

where p_{ij} represents the probability of transferring from state i to state j in the information layer, and t_s stands for the initial time. The probability distribution of each state within the Markov chain can be figured out by analyzing the historical data of state transitions. To solve the Markov chain, the following designations should be noted.

Let T_i^k be the kth sojourn time in state i of the target component in the information layer of power information system which obeys the exponential distribution, K_i be the accumulated number of occurrences in state i, and K_{ij} be the counting number of state transitions from state i to state j. Then, the transition rate matrix A_c in the information layer can be obtained. Among the elements in the matrix, the unbiased estimation of state transition rates can be obtained as follows:

$$\hat{a}_{ij} = \frac{K_{ij}}{\sum_{k=1}^{K_i} T_i^k}, i \neq j;\ \hat{a}_{ii} = -\sum_{j=1}^{N} a_{ij}, j \neq i \tag{11.3}$$

Then, the mean time of duration in state i is calculated by

$$T_{c.mean}^{i} = \sum_{k=1}^{K_i} T_i^k / K_i = 1/\sum_{\substack{j=1 \\ i \neq j}}^{N} \hat{a}_{ij} \tag{11.4}$$

The multistate transition in the information layer is shown in Figure 11.1. Let $p_c^i(t)$, $i \in \Omega_c$ be the probability of the component being in state i and P_c be the stochastic matrix in the information layer. Both Kolmogorov forward and Kolmogorov backward equations can describe the characteristics of the probability time-evolution in Markov chain. When the initial value is given, the Kolmogorov backward equation $\dot{P}_c = A_c P_c$ can be worked out in which $\dot{P}_c$ is the derivative of P_c, and the probability in each state $p_c^i(t)$, $i \in \Omega_c$ can be figured out. The steady-state probability distribution $\Pi_c := \left[\pi_c^1, \pi_c^2, \ldots, \pi_c^N\right]^T$ can be obtained when t is increased to infinity, where

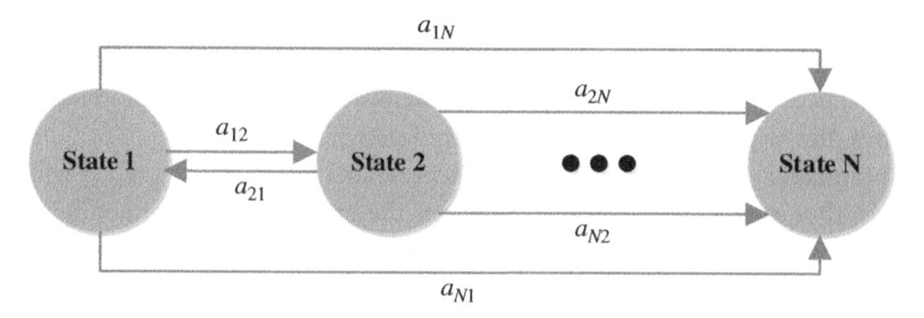

Figure 11.1 Multistate transition in the information layer. *Source:* From He et al. [A].

$$\pi_c^i = \lim_{t \to \infty} p_c^i(t); i \in \{1, 2, \ldots, N\} \tag{11.5}$$

$$s.t. \quad \sum_{i=1}^{N} \pi_c^i = 1 \tag{11.6}$$

11.2.2 Two-state Markov Chain of Physical Layer

A component may encounter failure due to electromagnetic interference, insufficient power supply, partial damage, etc., and would not be restored to working state without repair, as shown in Figure 11.2. The two-state Markov chain of a component at the physical layer is described as follows:

Let $\{X_p(t), t \in [0, \infty)\}$ be the two-state Markov chain of the physical layer where $\Omega_p = \{0, 1\}$ is the state space, 0 for operation state and 1 for failure state. Let $p_p^i(t)$, $i \in \Omega_p$ be the probability of the component residing in state i, A_p be the transition rate matrix and P_P be the stochastic matrix in the physical layer. Then, the Kolmogorov backward equation $\dot{P}_p = A_p P_p$ can be solved with the steady-state distribution stated as

$$\Pi_p := \left[\pi_p^0, \pi_p^1\right]^{\mathrm{T}} \tag{11.7}$$

where

$$\pi_p^i = \lim_{t \to \infty} p_p^i(t); i \in \{0, 1\} \tag{11.8}$$

$$s.t. \quad \pi_p^0 + \pi_p^1 = 1 \tag{11.9}$$

Designating λ as the failure rate and μ as repair rate of a component, the mean duration in operation state is $T_{p.mean}^0 = 1/\lambda$, and the mean duration in failure state is identified as $T_{p.mean}^1 = 1/\mu$. By solving the Kolmogorov backward equation of the binary-state Markov chain with the initial conditions $p_p^0(0) = 1$, the steady-state probability is obtained as

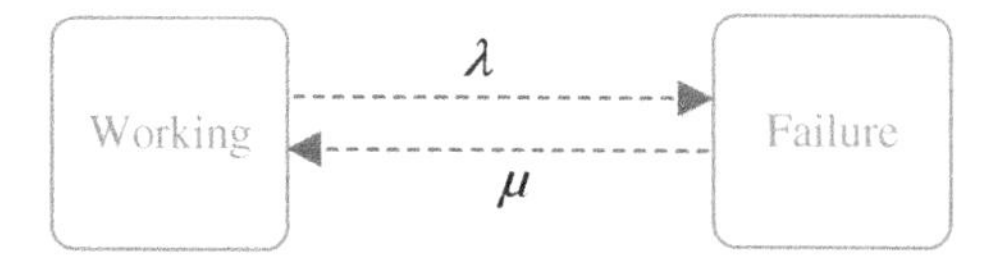

Figure 11.2 Two-state transition of the physical layer. *Source:* From He et al. [A].

$$\begin{bmatrix} p_p^0(t) \\ p_p^1(t) \end{bmatrix} = \begin{bmatrix} \dfrac{\mu}{\lambda + \mu} + \dfrac{\lambda}{\lambda + \mu} e^{-(\lambda + \mu)t} \\ \dfrac{\lambda}{\lambda + \mu} - \dfrac{\lambda}{\lambda + \mu} e^{-(\lambda + \mu)t} \end{bmatrix} \tag{11.10}$$

and

$$\begin{bmatrix} \pi_p^0 \\ \pi_p^1 \end{bmatrix} = \begin{bmatrix} p_p^0(+\infty) \\ p_p^1(+\infty) \end{bmatrix} = \begin{bmatrix} \dfrac{\mu}{\lambda + \mu} \\ \dfrac{\lambda}{\lambda + \mu} \end{bmatrix} \tag{11.11}$$

11.2.3 Coupling Model of Physical and Information Layers

Theoretically, the reliability model of power information system can be regarded as that of the coupling of physical and information flow layers. The component reliability model in power information system represented the performance of both information and physical layers in predefined conditions. In this chapter, it is proposed to build a composite Markov model (CMM) that correlates the two component layers.

The operation state of physical layer is the basis of the state transition in the information layer. For instance, when discussing a component's information, it is essentially implied that the component is at working state. No state transition will occur in information layer when a component is physically in failure with a 0 Mbps information, which is defined as no information state. After repair, the component is restored to working state in the physical layer with the corresponding normal state in the information layer, as shown in Figure 11.3.

According to previous discussions, the multi-state Markov chain $\{X_c(t), t \in [0, \infty)\}$ in the information layer is lower in rank than that of the two-state Markov chain $\{X_p(t), t \in [0, \infty)\}$ in the physical layer. The composite Markov chain in the form of $\{X_{c.p}(t), t \in [0, \infty)\}$ is expressed as

$$X_{c.p}(t) = \begin{cases} X_c(t), & X_p(t) = 0 \\ 0, & X_p(t) = 1 \end{cases} \tag{11.12}$$

The state space $\Omega_{c.p} = \{0, 1, 2, \ldots, N\}$ is the intersection of the state space in the information layer and the no information state caused by the physical layer failure. The probability vector of being in any states of composite Markov chain is designated as $P_{c.p} = \left[p_{c.p}^0(t), p_{c.p}^1(t), \ldots, p_{c.p}^N(t)\right]^T$.

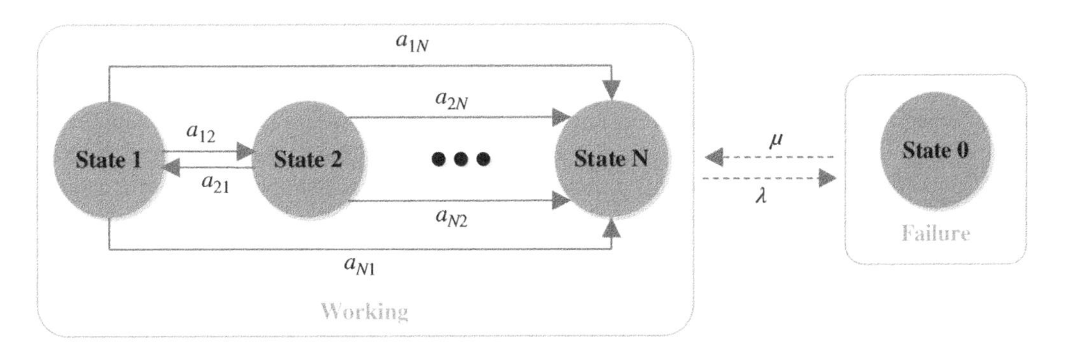

Figure 11.3 Coupling model in both information and physical layers. *Source:* From He et al. [A].

The probability vector $P^{*}_{c.p} = \left[p^{1}_{c.p}(t), \dots, p^{N}_{c.p}(t)\right]^{T}$ excludes the no information state. According to the inherent logic between the Markov chain in the information layer and its counterpart in the physical layer, there are

$$p^{i}_{c.p}(t) = \begin{cases} \left(\dfrac{\mu}{\lambda+\mu} + \dfrac{\lambda}{\lambda+\mu}e^{-(\lambda+\mu)t}\right)p^{i}_{c}(t), & i \in \Omega_c \\ \dfrac{\lambda}{\lambda+\mu} - \dfrac{\lambda}{\lambda+\mu}e^{-(\lambda+\mu)t}, & i = 0 \end{cases} \tag{11.13}$$

Thus, the steady-state probability $\Pi_{c.p} := \left[\pi^{0}_{c.p}, \pi^{1}_{c.p}, \pi^{2}_{c.p}, \dots, \pi^{N}_{c.p}\right]^{T}$ is obtained, where,

$$\sum_{i=1}^{N} \pi^{i}_{c.p} = \pi^{0}_{p};\ \sum_{i=0}^{N} -\pi^{i}_{c.p} = 1 \tag{11.14}$$

The mean duration time in each state is determined by

$$T^{i}_{cp.mean} = \begin{cases} T^{0}_{p} \times T^{i}_{c}/\sum_{j\in\Omega_c} T^{j}_{c} = T^{i}_{c}/\left(\lambda\sum_{j\in\Omega_c} T^{j}_{c}\right), & i \in \Omega_c \\ T^{1}_{p} = 1/\mu, & i = 0 \end{cases} \tag{11.15}$$

The composite Markov model of a component is then obtained.

11.3 Linear Programming Model for Maximum Flow

The reliability of the power information system in CMM can be evaluated by the probability in terms of level d. Therefore, the maximum flow through the network needs to be computed, in order to detect whether such a transmitted flow can meet the minimum demand d, and further examine whether the current system state is valid. In [23], the improved depth-first state-tree is used to search the target space for intelligent substations. In this chapter, in order to reduce the workload, the linear programming model is presented to acquire the maximum flow for the cyber space in smart grid.

11.3.1 Node Classification and Flow Constraint Model

The power information system can be depicted by a directed graph consisting of nodes and edges. The nodes in the directed graph represent various communication devices in the power information system while the edges represent the links among various communication devices. Usually, edges are considered to be perfectly reliable which means edges do not fail and there is no information loss on edges. If a link is considered to be unreliable during modeling, it should be treated as a node with a single inlet/outlet. The nodes are classified into six types as follows, which are based on functional duties assumed by communication devices.

1) Acquisition node. The acquisition nodes represent sources of information flow within the power information system. Usually, an acquisition node uploads data at a constant rate. However, a variable rate may occur when the power system fails.

2) Central node. Generally, there is only one central node in a power information system. Central node represents the control center in cyber space, which is the destination of the data collection for absorbing information flows.
3) Action node. These nodes represent action objects of relay protection devices such as circuit breakers and intelligent operation units. An action node is the terminal of control commands, which can also be regarded as absorbing flows.
4) Convergence node. The total outflow of a convergence node may or may not be equal to its total inflow, which depends on its role in the system. Devices with the preceding property such as routers and switches that serve as information aggregators are considered convergence nodes.
5) Switching node. A switching node performs optional data exchanges between its inlets and outlets. That is, outflows are associated with certain inflows in a given configuration, such as partition constraints in the virtual local area network. Switches can be abstracted as switching nodes.
6) Routing node. In a routing node, an outflow can be initiated from several inlets, but an inflow is only transmitted to a certain outlet. Therefore, the total inflow of a routing node is equal to its total outflow. A routing node usually represents a router.

The six types of nodes are summarized in Figure 11.4, which can further be summarized as source, intermediate, and sink nodes. The first three types of nodes, listed above, are source and sink points of information flow, and the direction of flows is either receiving or sending, so there is no flow conservation between input and output ports of nodes. The second three types of nodes belong to intermediate nodes. The task of intermediate node is to forward data, which neither generates nor absorbs information flows. For convergence and switching nodes, inflows may be

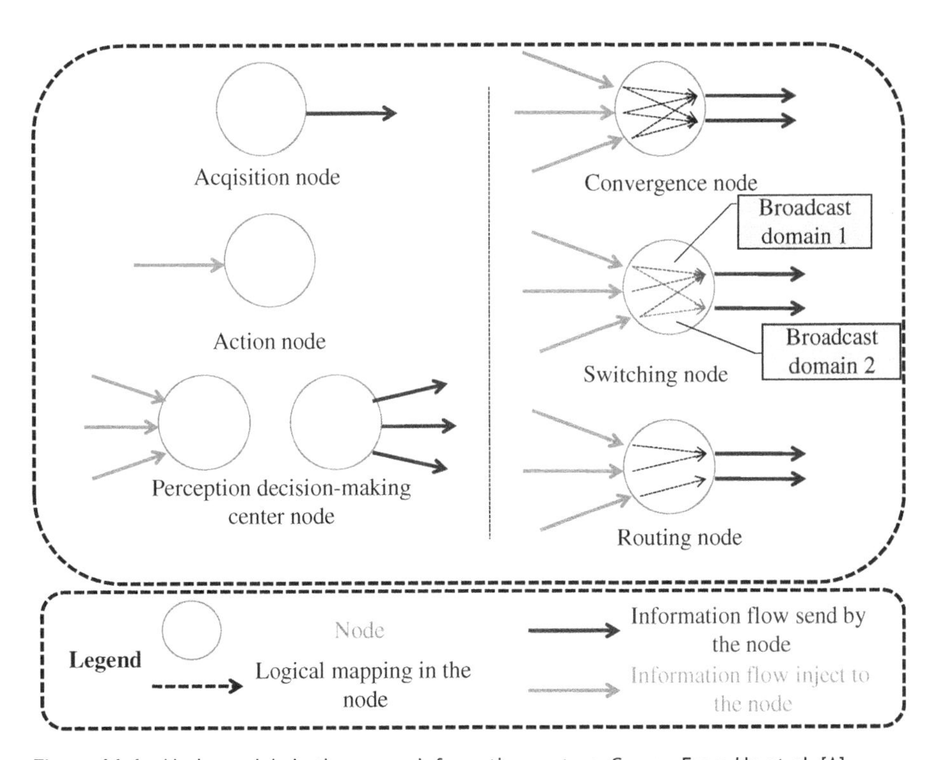

Figure 11.4 Node models in the power information system. *Source:* From He et al. [A].

replicated, resulting in additional traffic. However, no additional traffic is created in routing nodes because an inflow is only exchanged with a certain outlet.

Assume there is a constraint matrix Z^x for intermediate node x, which reflects the relation between inflows and outflows. Let Φ_x^+ be the set of nodes connected to the inlets of node x; Φ_x^- be the set of nodes connected to the outlets of node x; $F_{\Phi_x^+}$ be the column vector recording all inflows of node x; $F_{\Phi_x^-}$ be the column vector recording all outflows of node x. The flow constraint in an intermediate node x is stated as

$$F_{\Phi_x^-} = Z^x F_{\Phi_x^+} \tag{11.16}$$

The general node descriptions for the power information system are provided here. In fact, when considering different levels of communication networks or different power system implementations, the included nodes may be different. For example, the switching nodes will be included when considering the communication network at the substation level. On the contrary switches inside the substation are generally excluded when studying WAMS. And action nodes are needed when focusing on wide area protection; whereas they are not required in WAMS.

11.3.2 Programming Model for Network Flow

The maximum flow problem in power information system can be solved by a linear programming model.

The linear programming model is stated as follows:

$$MAX\ \Psi = \sum_{x \in \Phi_y^+} \sum_{y \in V_t} f_{xy} \tag{11.17}$$

$$s.t.\ \sum F_{\Phi_x^-} = \sum Z^x F_{\Phi_x^+} - ls_x, \forall x \in V_{inter} \tag{11.18}$$

$$\sum F_{\Phi_x^+} \leq C_x, \forall x \in V_{inter} \cup V_t \tag{11.19}$$

$$\sum F_{\Phi_x^-} \leq C_x, \forall x \in V_{inter} \cup V_s \tag{11.20}$$

$$lb_{xy} \leq f_{xy}, \leq ub_{xy}, \forall x,y \in V_s \cup V_{inter} \cup V_t \tag{11.21}$$

$$lb_{xy} = 0 \tag{11.22}$$

$$ub_{xy} = min\left(C_{x.max}, C_{y.max}, C_{xy}\right) \tag{11.23}$$

Equation (11.17) maximizes the flow received by sink node. Equation (11.18) is the flow constraints of intermediate nodes by taking information loss into consideration. Equations (11.19) and (11.20) indicate that the total inflow and outflow of a node cannot exceed the current capacity of the node. Equations (11.21)–(11.23) limit the flow of edges.

The above optimization model can be solved by the simplex or the interior point method. Simplex method transfers the linear program to standard form with the help of slack variables and builds a simplex tableau. Then pivot operations will be performed on the simplex tableau until the optimum value for the original linear program is obtained. Simplex method is available in the optimization toolbox of MATLAB. The maximum flow Ψ_m is obtained after performing the simplex method on the linear programming model.

11.4 Reliability Analysis Method

11.4.1 Definition and Measures of System Reliability

Power information system is repairable. The repair in CMM means not only the physical repair of devices, but also making a state transition from information blockage to normal level by means of process regulation and routing strategies. Similarly, the failure in CMM includes the physical failure of devices and the blockage of information flow.

A reliable power information system includes a reliable physical layer in which a flow from the source node to the sink node can satisfy the minimum demand d at sink. That is, there is at least one minimum path between source and sink devices to ensure the topological connectivity in physical layer. In addition, an adequate flow margin is necessary for the terminal to acquire at least d units of information flow from the source.

For specific smart grid applications, the setting of the demand flow threshold d could vary. Considering the state estimation as an example, suppose there are i measurements and n nodes in the power network. Thereby, the threshold is determined according to the minimum required measurement quantities that satisfy the system observability. That is, the rank of the augmented Jacobian matrix in state estimation must be larger than n. Hence, the setting of the information flow threshold is closely related to the algorithm or the function adopted by the follow-up application in a control center.

To better assess the reliability of power information system, several indices are considered in this chapter, including system availability expressed as A_{sys}, $MTTF$, and mean-time-to-repair ($MTTR$). $MTTF$ and $MTTR$ are calculated as

$$MTTF = \frac{T_W}{K_W};\ MTTR = \frac{T_F}{K_F} \tag{11.24}$$

where T_W is the total sojourn time and K_W is the accumulated number of intervals for being in the working state of the information system. T_F is the total sojourn time and K_F is the accumulated number of intervals in failure state of the power information system.

A_{sys} is stated as

$$A_{sys} = \frac{MTTF}{MTTF + MTTR} \tag{11.25}$$

11.4.2 Sequential Monte-Carlo Simulation

11.4.2.1 System State Sampling

The traditional Monte-Carlo simulation is based on the binary-state sampling of components assuming that the system state remains unchanged during a sampling period, which will lead to large errors when applied directly to the sampling of multi-state components in CMM. Therefore, the two-level sampling for the sequential Monte-Carlo simulation is utilized in the work to calculate the reliability indices of the power information system.

Suppose there are n nodes in cyber space. Let $i_{sys} = (i_1, i_2, ..., i_n)$ be the power information system state vector, in which the elements are the current state of components.

For a multistate component, the first to be sampled is the current state, followed by the sojourn time in the state. Assume that there are $N+1$ independent states for component x marked as 0, 1, ..., N, and *rand* is a random number with uniform distribution in [0,1]. The state of the component x at time t can be obtained as follows:

$$i_x = \begin{cases} 0 & 0 \leq rand \leq \pi^0_{c.p}(t) \\ 1 & \pi^0_{c.p}(t) < rand \leq \pi^0_{c.p}(t) + \pi^1_{c.p}(t) \\ \ldots & \\ N & \sum_{i=1}^{N-1} \pi^i_{c.p}(t) < rand \leq 1 \end{cases} \tag{11.26}$$

Let τ_x be the sojourn time of component x residing in state i. According to the property of the continuous-time Markov chain, τ_x obeys an exponential distribution which is stated as

$$D(\tau_x) = 1 - e^{-\frac{1}{T^i_{x.mean}}\tau_x} \tag{11.27}$$

where $T^i_{x.mean}$ is the mean sojourn time of component x residing in state i. Let $D(\tau_x) = rand$. Then,

$$\tau_x = -T^i_{x.mean} \ln(1 - rand) \tag{11.28}$$

Note that $(1 - rand)$ is equivalent to *rand*. Therefore, (11.28) can also be expressed as

$$\tau_x = -T^i_{x.rand} \ln(rand) \tag{11.29}$$

The power information system state vector $i_{sys} = (i_1, i_2, ..., i_n)$ and the sojourn time vector $\tau_{sys} = [\tau_1, \tau_2, ..., \tau_n]$ are obtained by sampling all components.

11.4.2.2 Reliability Computing Procedure

The flow chart for the reliability calculation is shown in Figure 11.5 and the steps are discussed as follows.

1) Input basic data of the power information system, including the historical flow data of each component, the total simulation time T_{sim}, and the demand flow threshold $d\%$.
2) Establish the composite Markov model of each component.
3) Initialize the power information system state i_{sys} and the sojourn time τ_{sys} according to (11.26) and (11.29).
4) Calculate the maximum flow corresponding to the current system state by the linear programming model. If the flow received by the terminal is less than $d\%$, the current network is unreliable. Update the reliability indices.
5) Update the system simulation time $t_{sim} = t_{sim} + min(\tau_{sys})$. If the total simulation time is not reached, update the sojourn time of each component as $\tau_x = \tau_x - min(\tau_{sys})$. Perform the two-level component sampling with the shortest duration and update the system state and the corresponding sojourn time.
6) Calculate the reliability indices after the preset total simulation time is reached ($t_{sim} = T_{sim}$).

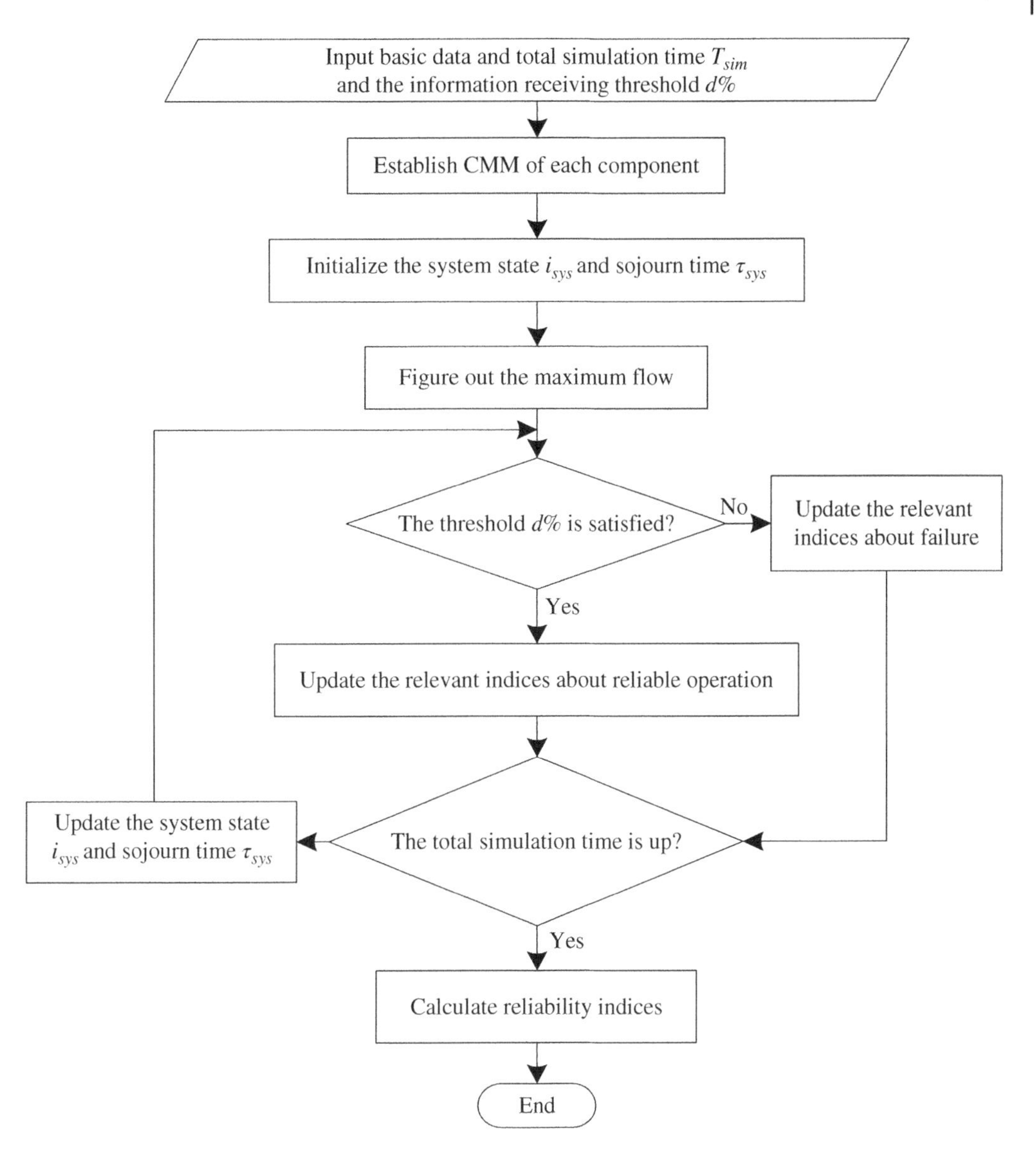

Figure 11.5 Reliability calculating flow chart. *Source:* From He et al. [A].

11.5 Case Analysis

11.5.1 Case Description

The reliability of the power information system constructed in the IEEE 14-bus system, shown in Figure 11.6, will be discussed. The power information system includes a 10-node and 15-link model among which Nodes 5 and 6 belong to the same substation, and Nodes 4, 7, 8, and 9 belong to another same substation. Node 5 is the master station where the control center is located and other nodes are slave stations [30]. Synchronous digital hierarchy (SDH) optical fiber communication is adopted in the overall architecture. The Synchronous Transport Module level-1 (STM-1) mode is

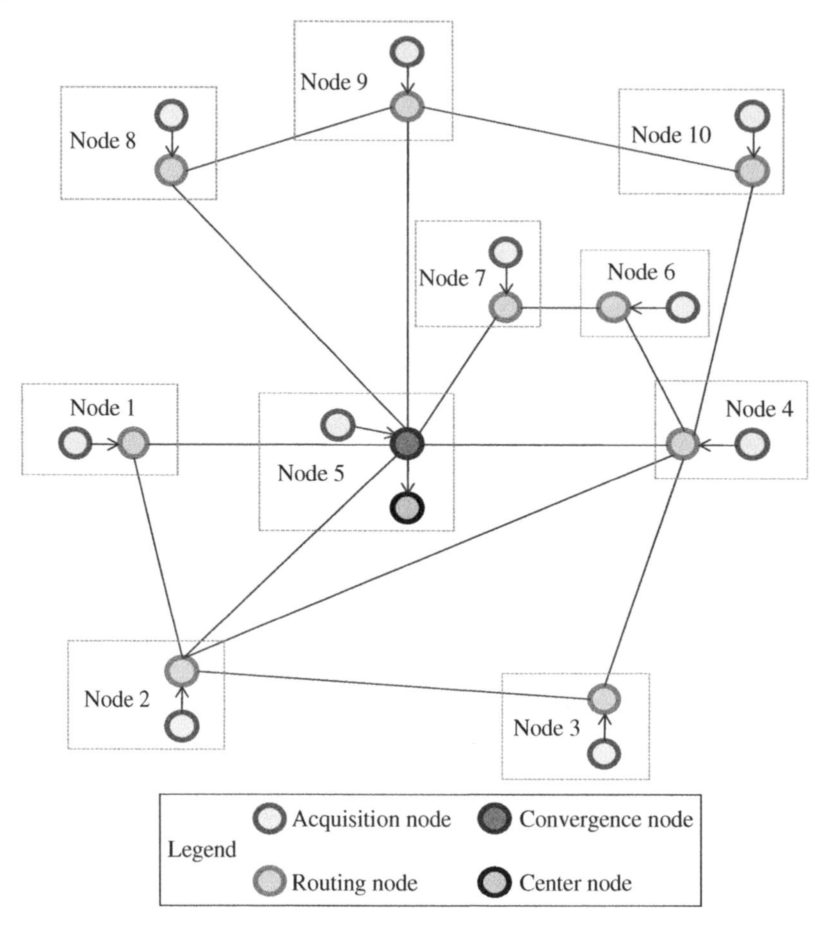

Figure 11.6 Network topography of the power information system based on IEEE 14-bus system. *Source:* From He et al. [A].

used to connect the nodes and all routers support STM-1 rate. The capacity of intermediate nodes is 155 Mb/s while the capacity of central node is unlimited. Links are reliable and full-duplex with sufficient capacity is used to meet the information transmission needs of the corresponding nodes. The various business flows uploaded by the server in each substation are represented by a single acquisition node.

The parameters of various nodes in the physical layer are given as follows. The failure rates of acquisition nodes, routing nodes, and center node are $\lambda_{ac} = 1.801 \times 10^{-5}$ time/h, $\lambda_{ru} = 1.202 \times 10^{-5}$ time/h and $\lambda_{ct} = 7.982 \times 10^{-6}$ times/h, respectively [31]. The repair time of all nodes is 24 hours. The convergence node is treated as a routing node.

In the information layer, the flow level is divided into four states of each acquisition node, namely, 10, 20, 30, and 40 Mb/s. The transition rate matrix is

$$A_c = \begin{bmatrix} -0.4002 & 0.2002 & 0.1499 & 0.0500 \\ 0.6003 & -0.8010 & 0.1503 & 0.0504 \\ 0.6002 & 0.1997 & -0.8500 & 0.0501 \\ 0.5995 & 0.2000 & 0.1501 & -0.9495 \end{bmatrix}$$

Table 11.1 Parameters of composite markov model of acquisition nodes.

State	0	1	2	3	4
Stationary distribution	0.0003	0.5998	0.1998	0.1500	0.0501
Mean sojourn time	1.0000	2.4979	1.2484	1.1764	1.0531

Source: From He et al. [A].

Table 11.2 Parameters of binary-state markov model of routing nodes.

State	0	1
Stationary distribution	0.9997	0.0003
Mean sojourn time	83 195	24

Source: From He et al. [A].

Table 11.3 Parameters of binary-state markov model of central nodes.

State	0	1
Stationary distribution	0.9998	0.0002
Mean sojourn time	111 099	24

Source: From He et al. [A].

For the given parameters, the composite Markov model of acquisition nodes and the binary-state Markov model of routing nodes and the central node are established. The stationary distributions and the mean sojourn time for each state are shown in Tables 11.1–11.3. Using the sequential Monte-Carlo simulation, the capacity of each node in the current system state is obtained after each sampling and updating. The flowchart using a linear programming model is shown in Figure 11.7. The maximum flow in the current system state is calculated using the optimization toolbox in MATLAB and the reliability indices are computed.

11.5.2 Calculation Results and Analysis

11.5.2.1 Effect of Demand Flow on Reliability

Set the total simulation time to 100 years. In order to reflect some congestion due to insufficient bandwidth, the acquisition node has three high-level with low distribution states, namely 20 Mb/s-20%, 30 Mb/s-15%, and 40 Mb/s-5%, which may generate flow gap between the sink at the control center and the sources at substations. Because of the random sampling of the Monte-Carlo simulation, the flow gaps are different in different system states, which can reflect the congestion degree of communication network stochastically. In this case, the system reliability is related to the demand *d*. The Monte-Carlo simulation is carried out with different demand flow thresholds and the reliability indices are shown in Table 11.4. By considering the dual

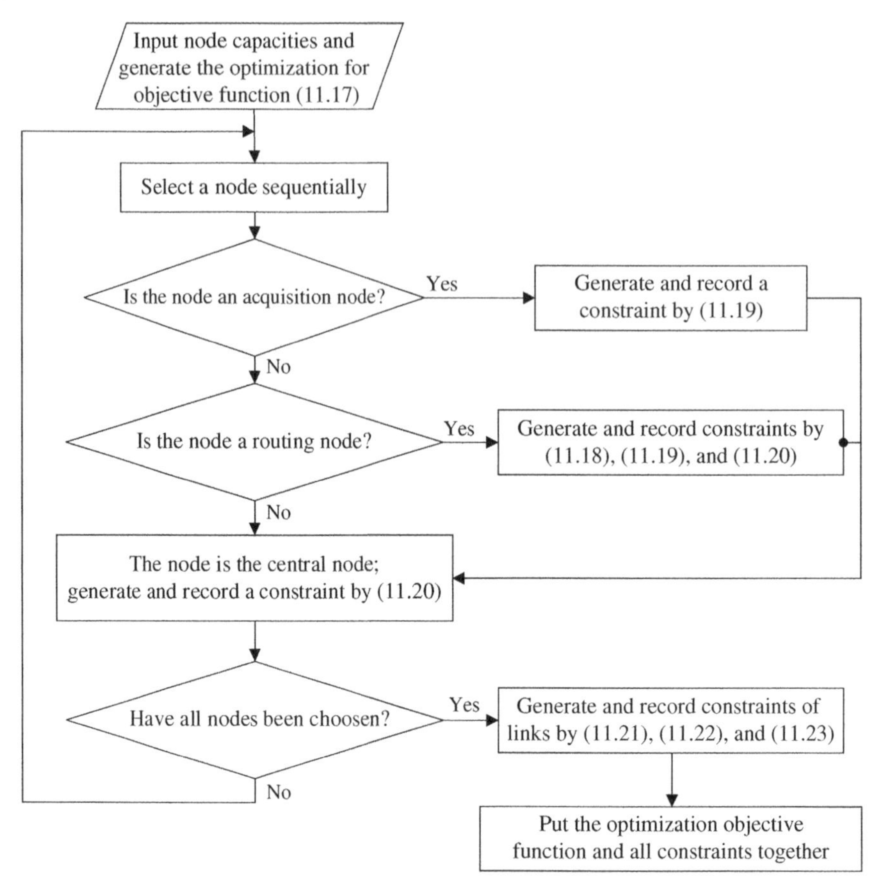

Figure 11.7 The proposed flowchart using a linear programming model. *Source:* From He et al. [A].

Table 11.4 Reliability indices by sequential Monte-Carlo simulations with different information receiving threshold.

Demand flow threshold (%)	Availability (%)	MTTF(h)	MTTR(h)
100	76.70	2.5	0.76
95	85.96	3.8	0.62
90	92.21	6.1	0.51
85	95.91	10.4	0.44
80	98.00	19.1	0.39
75	99.08	37.3	0.35
70	99.82	179.7	0.32
65	99.90	463.3	0.44
60	99.97	2762.5	0.91
55	99.97	39 808	10.59
50	99.97	58 358	15.82

Source: From He et al. [A].

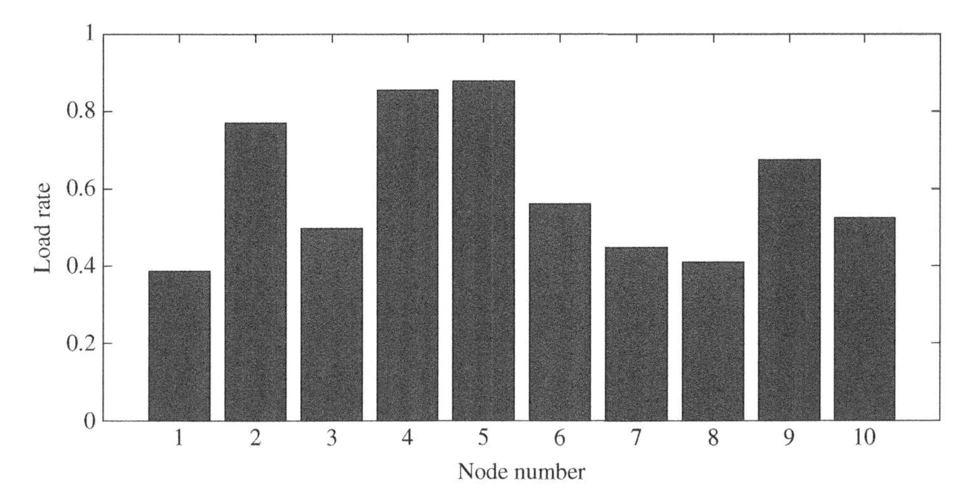

Figure 11.8 Information load rate of intermediate nodes. *Source:* From He et al. [A].

configuration, only one half of the information is sent in the lowest case, which is enough to ensure the normal implementation of the control function. It is obvious that the network availability decreases as the demand flow threshold rises. Specifically, the availability is nearly 100% when the demand is lower than 70%. Once the threshold exceeds 80%, the availability plummets. When the demand is 100%, MTTF lasts only 2.5 hours and the availability drops to 76.70%, indicating that the system availability is insufficient.

The information load rate of the intermediate nodes is shown in Figure 11.8, in which the serial number is consistent with that of the associated acquisition node and the load rate of the entire network exceeds 40%. Nodes 2, 4, and 5 are hub nodes which exchange data with at least four links whose load rates are 77.0, 85.6, and 87.8%, respectively.

Figure 11.9 shows the network flow distribution in a certain scenario. At present, the total traffic flow reaches 170 Mb/s while the maximum flow that can be absorbed by the sink node is 155 Mb/s. So, the convergence node has reached the performance upper bound. Therefore, there exists 15 Mb/s flow gap and the information receiving rate of the central node is 91.18% in this scenario. If the demand flow threshold is set to 100%, the communication network in this scenario will be unreliable.

11.5.2.2 Effect of Node Capacity on Reliability

The capacity of each hub node is upgraded to a higher level, i.e. 310 Mb/s, while d is still set to 100%, and the system availability is shown in Table 11.5. The four scenarios in the table refer to Scenario 1 as the reference basis, and Scenarios 2–4 are the availability level when the bandwidths of routing nodes 2, 4, and 5 are upgraded to 310 Mb/s. It can be seen that the system availability jumps to 99.67 from 76.70% after improving the capacity of Node 5. However, the availability in Scenarios 2 and 3 has almost no change compared with that of Scenario 1.

The information load rate of routing nodes corresponding to four scenarios is shown in Figure 11.10. Taking Scenario 2 as an example, when the capacity of routing Node 2 is raised, the information load rate of routing Node 2 decreases considerably. However, the information load rate of the connecting nodes increases to varying degrees, as the overall network availability

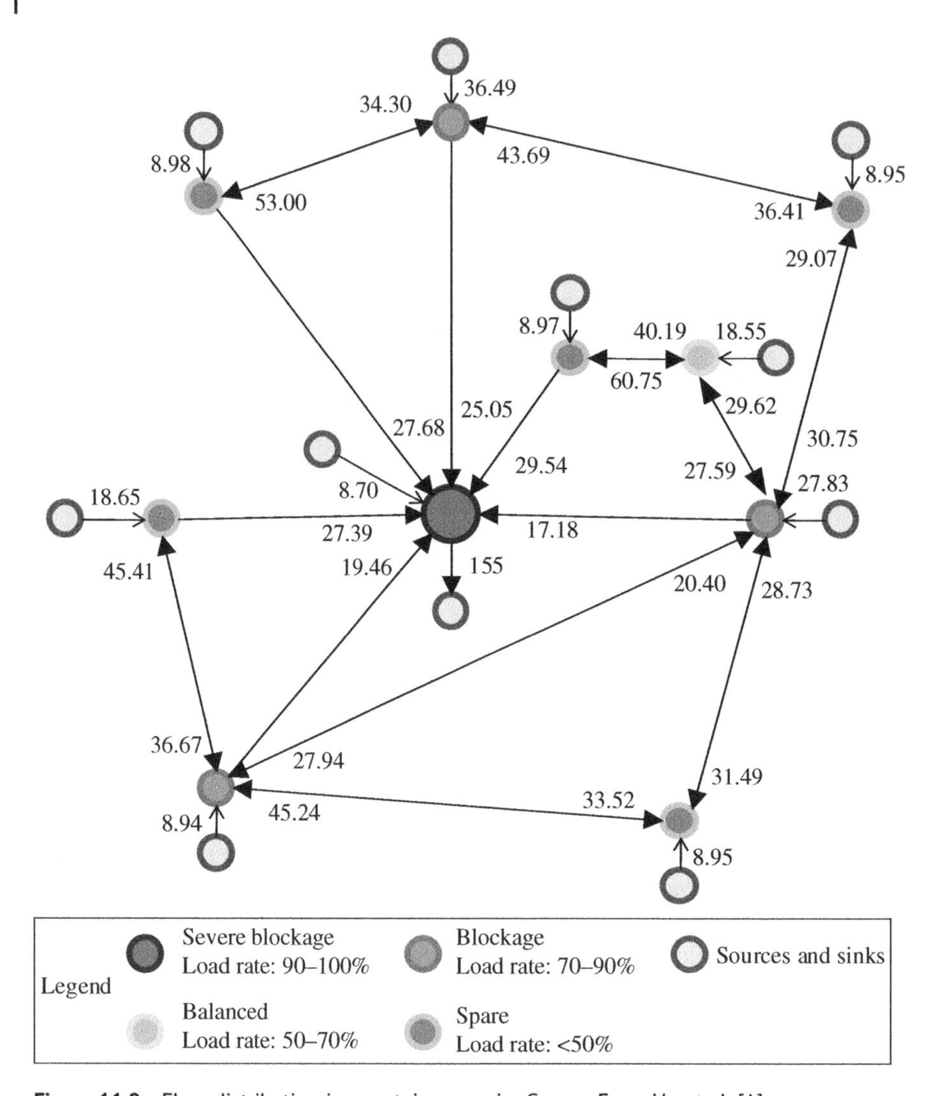

Figure 11.9 Flow distribution in a certain scenario. *Source:* From He et al. [A].

Table 11.5 Availability in different scenarios.

Scenario	1	2	3	4
Availability (%)	76.70	76.57	76.57	99.67

Source: From He et al. [A].

is not improved substantially. In Scenario 4, all network data streams should go through the convergence node before entering the control center. That is, the maximum flow that can be absorbed by the sink node depends largely on the capacity of the convergence node. Therefore, upgrading the convergence node bandwidth can significantly improve the availability of the overall information system.

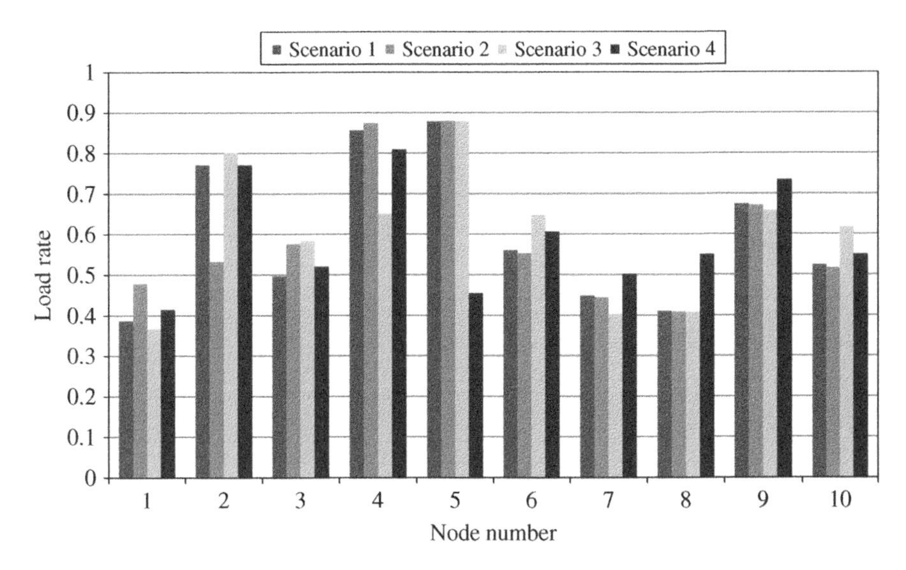

Figure 11.10 Information load rate under different scenarios. *Source:* From He et al. [A].

11.5.2.3 Effect of the Information Flow Level on Reliability

The reliability difference between using the proposed composite Markov model which provides the coupling of physical and information layers and the traditional operation-failure model in the physical layer is discussed. If the flow in the information layer is not considered, only the physical network connectivity will be considered; that is, the system is considered to be reliable if there is a minimum path between source and sink nodes. The system reliability indices for using a two-state expression are shown in Table 11.6. Compared with Table 11.6, it is found that the MTTF for using the physical connectivity is several orders of magnitude lower than that of integrating data transmission conditions when the demand flow threshold is set at 60–100%; however, there is a minute difference between them when the threshold is less than 50%.

In addition, *MTTR* is quite low when the demand is in the range of 60–100%. Under these cases, the system failure is mainly caused by the jamming of information flow. Since the state transition rate in the information layer is much faster than that in the physical layer, the information repair process is very short, which leads to a low *MTTR*. Once d is set below a certain threshold, the network is sufficient regardless of injected flow fluctuations. Therefore, the information flow level will no longer affect the system reliability and *MTTR* is considered to have mainly originated from the physical repair time. Similarly, *MTTR* may exceed 20 hours when $d\%$ is less than a certain level because the repair time is much longer than that of flow state transition.

In order to ensure the reliable operation of the power system, it is necessary to ensure that the control center can acquire enough information by lowering the demand flow threshold even if there

Table 11.6 Reliability indices of the physical network.

Availability (%)	MTTF	MTTR
99.69	7703.2	24.05

Source: From He et al. [A].

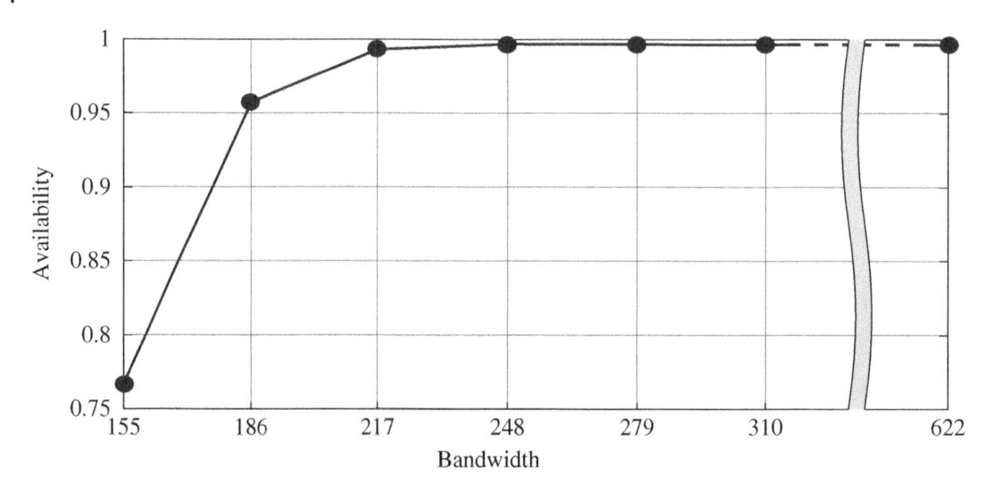

Figure 11.11 Impact of the overall bandwidth on system reliability. *Source:* From He et al. [A].

is a large sum of redundant data in the power information network. The impact of improving the system bandwidth on availability is illustrated in Figure 11.11. If the power information network bandwidth is upgraded from STM-1 to STM-4, the influence of the information flow level on reliability can be neglected, while the availability is entirely inclined to reflect the physical characteristics.

Therefore, the reliability of power information systems can be improved by increasing the network bandwidth, which of course has a cost burden. If the demand for data collection is reduced, there is no guarantee that the key information can be fully transmitted to the control center because of the data redundancy; so, there may exist hidden risks.

In summary, when the power information system is insufficient, the system failure is dominated by information flow fluctuations, and the physical failure will have almost no effect on the reliability of the power information system. With an increase in capacity or decrease in demand, failure events will be reduced in the information layer, and the system reliability will be determined by not only the blockage of information flow but also the physical system failures. When the bandwidth is raised to an ideal level or the receiving rate is set at a low level, there will be no system congestion as the information flow fluctuates; hence, the power information system reliability depends only on the state transition of the physical layer. For a similar communication network arrangement, based on the IEEE 14-bus system, the reliability analyses using the Markov model with two states in physical will present over-optimistic results which cannot bring out the deeper system characteristics [20].

11.6 Conclusion

The existing reliability evaluation methods of the power information system focus on the operation-failure characteristics of the physical system, in which there are hardly any details of cyber space models in the reliability analysis of power systems. A composite cyber-physical power system model is proposed for coupling the information flow with the physical performance to evaluate the power information system reliability. The multistate transition of information flow and the two-state transition of the physical system are considered. Moreover, A linear programming model is established for calculating the maximum flow using the power information system, which is much more efficient than using the state enumerating technique. Also, the sequential Monte-Carlo simulation with a two-level sampling and the corresponding indices are applied to evaluate the power information system. In cyber-physical power system, the reliable operation of a power system depends highly on the situational awareness in the information system. The present study will be extended to a wide-area communication network which will allow the realization of the situational awareness and reliability analyses in a much larger regional power system.

Acknowledgements

The permission given in using materials form the following paper is very much appreciated.

[A] He, R., Xie, H., Deng, J. et al. (2020). Reliability modeling and assessment of cyber space in cyber-physical power systems. *IEEE Transactions on Smart Grid* 11 (5): 3763–3773. https://doi.org/10.1109/TSG.2020.2982566.

References

1 Samad, T. and Annaswamy, A.M. (ed.) (2014). Resilient cyber-physical systems. In: *The Impact of Control Technology*, Report. IEEE Control Systems Society.
2 Edward, A.L. (2008). Cyber physical systems: design challenges. Technical Report No. UCB/EECS-2008-8, University of California at Berkeley. http://citeseerx.ist.psu.edu/viewdoc/download?doi=10.1.1.156.9348&rep=rep1&type=pdf (accessed 11 November 2020).
3 National Institute of Standards and Technology (2013). Foundations for innovation in cyber-physical systems. Workshop Report, Prepared by Energetics Incorporated Columbia, Maryland, For the National Institute of Standards and Technology. https://www.nist.gov/system/files/documents/el/CPS-WorkshopReport-1-30-13-Final.pdf (accessed 11 November 2020).
4 Johannessen, S. (2004). Time synchronization in a local area network. *IEEE Control Systems Magazine* 24 (2): 61–69.
5 Shum, C., Lau, W.H., Mao, T. et al. (2018). DecompositionJ: parallel and deterministic simulation of concurrent Java executions in cyber-physical systems. *IEEE Access* 6: 21991–22010.
6 Bose, A. (2010). Smart transmission grid applications and their supporting infrastructure. *IEEE Transactions on Smart Grid* 1 (1): 11–19.
7 Chen, S.Z., Lu, J., Zhang, G., and Zhang, Y. (2019). Immunizing variable frequency transformer from dual-side asymmetrical grid faults via a single-converter-based novel control strategy. *IEEE Transactions on Power Delivery* https://doi.org/10.1109/TPWRD.2019.2940771.
8 Sha, L., Gopalakrishnan, S., Liu, X., and Wang, Q. (2008). Cyber-physical systems: a new frontier. *2008 IEEE International Conference on Sensor Networks, Ubiquitous, and Trustworthy Computing (SUTC 2008)*, Taichung (11–13 June 2008), 1–9.
9 Shi, L., Dai, Q., and Ni, Y. (Oct. 2018). Cyber-physical interactions in power systems: A review of models, methods, and applications. *Electric Power Systems Research* 163: 396–412.
10 Singh, C. and Sprintson, A. (2010). Reliability assurance of cyber-physical power systems. *IEEE Power Energy Society General Meeting*, Minneapolis, MN (25–29 July 2010), 1–6.

11 Zhang, G., Yuan, J., Li, Z. et al. (2020). Forming a reliable hybrid microgrid using electric spring coupled with non-sensitive loads and ESS. *IEEE Transactions on Smart Grid*, https://doi.org/10.1109/TSG.2020.2970486.

12 Lei, H., Chen, B., Butler-Purry, K.L., and Singh, C. (2018). Security and reliability perspectives in cyber-physical smart grids. *Proceedings of IEEE Innovative Smart Grid Technologies – Asia (ISGT Asia)*, Singapore (22–25 May 2018), 42–47.

13 Falahati, B., Fu, Y., and Wu, L. (2012). Reliability assessment of smart grid considering direct cyber-power interdependencies. *IEEE Transactions on Smart Grid* 3 (3): 1515–1524.

14 Falahati, B. and Fu, Y. (2014). Reliability assessment of smart grids considering indirect cyber-power interdependencies. *IEEE Transactions on Smart Grid* 5 (4): 1677–1685.

15 Lei, H., Singh, C., and Sprintson, A. (2014). Reliability modeling and analysis of IEC 61850 based substation protection systems. *IEEE Transactions on Smart Grid* 5 (5): 2194–2202.

16 Lei, H. and Singh, C. (2015). Power system reliability evaluation considering cyber-malfunctions in substations. *Electric Power Systems Research* 129: 160–169.

17 Xin, S., Guo, Q., Sun, H. et al. (2017). Information-energy flow computation and cyber-physical sensitivity analysis for power systems. *IEEE Journal on Emerging and Selected Topics in Circuits and Systems* 7 (2): 329–341.

18 Marashi, K., Sarvestani, S.S., and Hurson, A.R. (2018). Consideration of cyber-physical interdependencies in reliability modeling of smart grids. *IEEE Transactions on Sustainable Computing* 3 (2): 73–83.

19 Guo, J., Liu, W., Syed, F.R., and Zhang, J. (2019). Reliability assessment of a cyber physical microgrid system in island mode. *CSEE Journal of Power and Energy Systems* 5 (1): 46–55.

20 Wang, Y., Li, W., and Lu, J. (2010). Reliability analysis of wide-area measurement system. *IEEE Transactions on Power Delivery* 25 (3): 1483–1491.

21 Dai, Z., Wang, Z., and Jiao, Y. (2011). Reliability evaluation of the communication network in wide-area protection. *IEEE Transactions on Power Delivery* 26 (4): 2523–2530.

22 Liu, W., Gong, Q., Han, H. et al. (2018). Reliability modeling and evaluation of active cyber physical distribution system. *IEEE Transactions on Power Systems* 33 (6): 7096–7108.

23 Ruiwen, H., Jianhua, D., and Lai, L.L. (2018). Reliability evaluation of communication-constrained protection systems using stochastic-flow network models. *IEEE Transactions on Smart Grid* 9 (3): 2371–2381.

24 Yingkui, G. and Jing, L. (2012). Multi-state system reliability: a new and systematic review. *Procedia Engineering* 29: 531–536.

25 Yeh, W. (2015). An improved sum-of-disjoint-products technique for symbolic multi-state flow network reliability. *IEEE Transactions on Reliability* 64 (4): 1185–1193.

26 Lisnianski, A., Elmakias, D., Laredo, D., and Haim, H.B. (2012). A multi-state Markov model for a short-term reliability analysis of a power generating unit. *Reliability Engineering & System Safety* 98 (1): 1–6.

27 Lin, Y.K. (2001). A simple algorithm for reliability evaluation of a stochastic-flow network with node failure. *Computers & Operations Research* 28 (13): 1277–1285.

28 Ramirez-Marquez, J.E. and Coit, D.W. (2005). A Monte-Carlo simulation approach for approximating multi-state two-terminal reliability. *Reliability Engineering & System Safety* 87 (2): 253–264.

29 George-William, H. and Patelli, E. (2016). A hybrid load flow and event driven simulation approach to multi-state system reliability evaluation. *Reliability Engineering & System Safety* 152: 351–367.

30 Giovanini, R., Hopkinson, K., Coury, D.V., and Thorp, J.S. (2006). A primary and backup cooperative protection system based on wide area agents. *IEEE Transactions on Power Delivery* 21 (3): 1222–1230.

31 Zhihui, D. (2012). Research on Reliability and Risk Assessment of Protection Systems. Ph.D. dissertation, North China Electric Power University, Beijing, China.

12

A Vehicle-To-Grid Voltage Support Co-simulation Platform

12.1 Introduction

The rapid decentralization of power system is driving revolutionary changes in its underlying ICT infrastructures. Various smart grid applications require distributed computation and pervasive data communication for coordinating millions of automated devices. The study of these smart grid applications often requires performance evaluation, either via experimentations on real-world testbed or via simulation. Unfortunately, real-world platforms are rarely available for the purpose of experiments due to the costs and the critical nature of power stability. Even if a testbed is available, experiments are usually bounded to a small-scale with a limited set of scenarios. Furthermore, experimental results are often difficult to repeat due to varying operational conditions. In contrast, simulation enables fully configurable experiments with arbitrary hypothetical scenarios at a low cost and the results are easily repeatable. It is therefore of no surprise that most published results are obtained via software-based simulations.

For a comprehensive simulation of smart grid, all three aspects of electrical, communication, and distributed computation, as well as their interdependencies should be modeled. Given the complexity and heterogeneity of the smart grid, creating a comprehensive simulator from scratch would be costly and time consuming. Therefore, the co-simulation designs are often implemented by combining existing well-developed and validated domain-specific simulators, e.g. power system and communication network simulators, with each of them responsible for modeling and simulating some aspects of the smart grid [1, 2].

This smart grid co-simulation paradigm was pioneered by the work of Electrical Power and Communication Synchronizing Simulator [3] (EPOCHS), and later works [4–16] contributed to this field mainly in three directions:

1) Exploring various combinations of power system and communication simulators for integrating and expanding the tool sets to support different smart grid applications.
2) Improving time synchronization mechanisms. This is to ensure that simulation events and messages across all simulators are processed in correct timestamp order, such that simulation outcomes are causally correct and repeatable.
3) Improving the interoperability of co-simulation plat- forms by conforming to standardized co-simulation frame-works such as Distributed Interactive Simulation (DIS, IEEE 1278 Std.) and High Level Architecture (HLA, IEEE 1516 Std.).

Smart Energy for Transportation and Health in a Smart City, First Edition. Chun Sing Lai, Loi Lei Lai and Qi Hong Lai.

Despite previous advancements, discussions are lacked on the modeling and simulation of distributed computation. Most previous works simply delegated such responsibility to either electrical or communication simulators, leading to several problems:

- Electrical/communication simulators allow creation of custom models via "user-codes." However, user code development is not as flexible as general programming, typically subject to many limitations such as reduction of usable language features, libraries, and interfaces. In particular, multi-threading and blocking calls are generally forbidden to prevent nondeterministic execution and deadlocks.
- Due to user code limitations, existing software cannot be simulated in its deployable form. Researchers need to rewrite the software with respect to user-code constraints. Depending on the complexity of target software, development process can be time-consuming and labor-intensive.
- Due to user-code constraints and development costs, researchers may need to discard nonessential parts of the software which may result in loss of functional fidelity.
- Electrical/communication simulators do not provide mechanisms to model computation delays which cause loss of timing fidelity.
- Separation of deployment and simulation codes complicates version control and code maintenance.

To mitigate these problems, we delegate the simulation of distributed software to *direct-execution simulators*, in conjunction with the typical integration of electrical and communication simulators. In a direct-execution simulation, the original code of the target program will be executed to emulate its own functional behavior. Additional simulation control codes are inserted to execute alongside the target codes, which are responsible for: (i) redirect program's inter-actions with real systems, i.e. I/O, system clock, timers, to their simulated counterparts, (ii) determine computation delays and track the logical timestamp of program's actions, and (iii) control the order of execution over actions performed by different threads within the simulator, and the order of execution over events across co-simulators.

In this chapter, we discuss the challenges and operating principles of a direct-execution simulation, then provide an overview on the design of DecompositionJ simulation framework (DEterministic, COncurrent Multi-PrOcesssing SImulaTION for Java programs). This framework performs compiler-based source analysis on a target Java program, then automatically transforms the original program into a direct-execution simulator by instrumenting simulation control codes. The proposed framework eliminates the need for manual code modification, hence significantly reduces the cost for developing and maintaining a model for the target software, yet at the same time fully retains its functional fidelity. Furthermore, this framework does not require new hardware support and programmer annotation, therefore is highly compatible with existing Java execution environments and development tools. This allows researchers to use mainstream IDEs and debuggers to investigate and debug the target software during simulation. Further details on the framework, such as definitions of the simulation model, exploitation of parallelism, scalability, and performance evaluations, can be found in [17].

To demonstrate the usefulness of our framework, we apply the direct-execution simulation techniques on a popular multi-agent platform JADE (Java Agent Development Frame-work) [18]. The DecompositionJ framework is used to analyze and transform JADE source code to produce direct-execution simulators, which are then integrated with power system simulator PSCAD and network simulator OPNET via a runtime infrastructure (RTI). The entire co-simulation platform adheres to the HLA standard. Using this co-simulation platform, a case study on agent-based fault location, isolation, and service restoration (FLISR) has been conducted. Through the analysis of simulation results, it is shown that the proposed direct-execution simulation framework is able to facilitate the understanding, evaluation, and debugging of distributed smart grid software. The rest of this chapter is organized as follows: Section 12.2 discusses and compares related works in smart grid co-simulation with our proposed framework. Section 12.3 presents the operating principle, design, and implementations of DecompositionJ framework. Section 12.4 discusses an HLA-based co-simulation platform that integrates electrical simulator – PSCAD, communication simulator – OPNET, and DecompositionJ simulators. Section 12.5 presents the FLISR case study, and results are shown in Section 12.6. Section 12.7 gives a case study on V2G for voltage support application. Conclusion is given in Section 12.8.

12.2 Related Works

The software-based co-simulation paradigm in smart grid research was pioneered by the work of Electrical Power and Communication Synchronizing Simulator [3] (EPOCHS), which combined power system simulators PSCAD (for electromagnetic transients) and PSLF (for electro-mechanical dynamics) with communication simulator NS2 (for packet-level simulation of wired/wireless communications). Simulators are interfaced to a runtime-infrastructure (RTI) which facilitates the exchange of data and the synchronization of event execution across simulators such that simulation events are executed in correct order according to their timestamps. The capability of EPOCHS was demonstrated with an example case study in agent-based wide-area monitoring, protection, and control (WAMPC).

Many other co-simulation frameworks and platforms have since been proposed. Table 12.1 summarizes their characteristics with respect to several aspects: (i) target use cases, (ii) selection of simulation tools for power systems, communication networks, and distributed software, (iii) time synchronization schemes, and (iv) co-simulation interfaces. Each of these aspects is discussed in the following.

12.2.1 Simulation of Power Systems

The selection of power system simulator depends primarily on the target use case. In general, power system simulation models can be classified into two types:

1) Steady state models of which the stable state of the power network is solved using power flow analysis. The model is typically used in power network planning, demand side management (DSM), energy markets, and optimization studies. Example simulation tools include Adevs, OpenDSS, GridLab-D, PowerWorld, PSSE, DigSilent, and MATLAB.
2) Transient dynamic models of which the power network is characterized at circuit level by differential equations. The trapezoidal rule is then applied to discretize the equations such that sampling and switching events can be modeled. Simulator then solves the system equations repeatedly for each time step to obtain numerical time-domain solutions. These models are typically used to study power system control and protection, where transitions between stable states occur due to the changing of operation point. Example tools include PSCAD, PSLF, Adevs-THYMS, DIgSILENT, and MATLAB.

Since the proposed co-simulation framework targets on delay-sensitive applications that operate during state transitions (e.g. power system restoration), the PSCAD simulator is selected to model the fast transients in the electrical system.

Table 12.1 Characteristics of related works in smart grid co-simulation.

Related works	Year	Example use case	Power	Commun.	Computation	Synchronization	Interface
EPOCHS [3]	2006	Agent-based WAMPAC	PSCAD, PSLF (Transient)	NS-2	External Process	Time-stepped	Adhoe
Nutaro et al. [4]	2007	WAMPAC	Adevs (steady state)	NS-2	NS-2	Master-slave	Adhoe
Nutaro [5]	2011	WAMPAC	Adevs-THYMS (Transient)	OMNET ++	Adevs, OMNET ++	Master-slave	Adhoe
PowerNet [6]	2011	Networked Control	Modelica (Transient)	NS-2	NS-2	Time-stepped	Adhoe
Mets et al. [7]	2011	DSM, DR	VTB (Transient)	OMNET ++	OMNET++	Master-slave	Adhoe
MAPNET [8]	2011	WAMPAC	PSLF (Transient)	OPNET	MATLAB	Master-slave	Adhoe
VPNET [9]	2011	Distributed Control	VTB (Transient)	OPNET	VTB	Time-stepped	Adhoe
GECO [10]	2012	WAMPAC	DIgSILENT (Transient)	NS-2	NS-2	Event driven	Adhoe
Levesque et al. [11]	2012	DSM, DR	OpenDSS (steady state)	OMNET ++	OMNET++	Master-slave	Adhoe
INSPIRE [12]	2014	WAMPAC, IEC 61850	DIgSILENT (Transient)	OPNET +	External Process	Event-driven (Parallel)	Adhoe
Celli et al. [13]	2014	DSM, DR	OpenDSS (steady state)	NS-2	MATLAB	Event-driven	Adhoe
Perkonigg et al. [14]	2015	Agent-based protection and control	-	OPNET	Extended JADE	Event-driven (Parallel)	HLA
Li et al. [15]	2017	DSM, DR, AMI	GridLabD (Steady state)	CORE	GridLabD	Real-time	Adhoe
Garau et al. [16]	2017	Distribution protection and control	OpenDSS (Steady state)	NS-3	MATLAB	Event-driven	Adhoe
This Chapter	2018	Agent-based protection and control	PSCAD (Transient)	OPNET	DecompositionJ	Event-driven (Parallel)	HLA

Source: From Shum et al. [A]/IEEE.

12.2.2 Simulation of Communication Network

Packet-level network simulators such as OPNET, NS-2, NS-3 and OMNET++ are commonly used for simulating smart grid communications. The principles for these tools are similar, that is, the processing and transmission of messages are modeled as a sequence of discrete events along the logical time-line. Selection of suitable simulator depends on model availability, development flexibility, and licensing types (e.g. open source or proprietary).

For this chapter, OPNET is selected for its rich set of wired and wireless communication models, and its built-in HLA co-simulation interface.

12.2.3 Simulation of Distributed Software

Distributed software programs are usually modeled within the network simulator [4–7, 10, 11], or within power system simulator [5, 8, 9, 15], and their drawbacks have been discussed in the Introduction section. Other approaches [3, 12, 13, 16] allow the target software to be modeled as external processes and interfaced to the simulators as slaves. We do not consider these slave processes to be co-simulators because they are not time-regulating; rather, they simply perform computation and return results upon receiving triggering signals from their master simulators. Modeling using slave processes suffers from the following limitations: (i) blocking operations are forbidden within slave codes since master simulators need to wait for slave results, otherwise deadlocks will occur; and (ii) multi-threading within slave process should be avoided to prevent nondeterministic results. These limitations significantly hindered the modeling of distributed software as multi-threading and blocking operations are very common.

Here, the work of Perkonigg et al. [14] is discussed in particular, which extended the JADE platform to support the simulation of unmodified agent codes. This is achieved by using a wrapper agent class to override the standard JADE agent class. The wrapper agent class provides the same APIs as the standard agent class, and in addition, implements simulation control codes to (i) redirect agent communications events to OPNET simulator, (ii) track the computation delay and logical timestamp of the agent, and (iii) control agent code executions to achieve synchronization. This design philosophy is similar to that of a direct-execution simulation, but the implementation is limited to JADE. In this chapter, a generalized design at the language level to support simulation of Java programs is proposed.

12.2.4 Time Synchronization

Note that during a co-simulation session, simulators execute local events and advance local logical clocks in parallel. Synchronization mechanisms must be installed to ensure events and messages across all simulators are executed in correct timestamp order. The methods used by previous works can be classified into three types: master-slave, time-stepped, and event-driven.

1) In a master-slave synchronization scheme, one of the simulators is chosen as the master that triggers slave simulators to execute events and exchange information. The exact logical time for interaction is only known by the master. This synchronization method can often be found in co-simulations with a steady state power system simulator (slave), which calculates new system states when being triggered by the communication simulator (master). The master-slave approach should only be used if the slave simulators do not actively generate events that affect the master.
2) In a time-stepped synchronization scheme, all co-simulators execute independently until a specific logical time is reached. When all co-simulators have reached the time step, messages accumulated during this period can be exchanged between simulators. The simulation is then resumed until the next step. Since messages accumulated within a time step are delayed until the next synchronization point, system error may occur and accumulate throughout the

simulation. This error can be reduced by decreasing step-size statically or allowing changes of step-size adaptively, but cannot be completely eliminated.

3) In an event-driven synchronization scheme, a simulator may send and receive messages at arbitrary time. During simulation, a time bound is assigned to each simulator. By limiting the execution to events within this bound, it can be guaranteed that local events and external messages are processed in a correct order. This eliminates system error that may occur in a time-stepped approach. Depending on the implementation, the synchronization protocol can be sequential or parallel.

Since the proposed co-simulation platform adheres to the HLA standard, its time synchronization mechanism is inherently parallel and event-driven which (i) fully eliminates system error due to synchronization and (ii) allows the exploitation of parallelism in the host computer.

12.2.5 Co-Simulation Interface

Since most proprietary simulation tools do not provide direct interfaces with standardized co-simulation frame-works, the implementation of adhoc interfaces to integrate communication and power system simulators is prominent in practice, especially in earlier works. Recent developments [12, 14] and this chapter, however, devote efforts to the conformity of standardized IEEE 1516 High-Level Architecture (HLA) [19] interface to achieve better interoperability, reusability and scalability in co-simulator design. The HLA standard specifies (i) object model templates to define co-simulation message structures understandable by all federates and (ii) the runtime infrastructure (RTI) to coordinate the message exchange, time synchronization, and management of a co-simulation session. Details on HLA-based co-simulation architecture is presented in Section 12.4.

12.3 Direct-Execution Simulation

The use of distributed software is pervasive for a wide range of smart grid applications. Their designs typically utilize multi-threading, with dedicated threads for handling network events. In addition, synchronization operations are often performed by threads as a means to achieve inter-thread coordination. However, these kinds of design patterns are not supported by smart grid co-simulators reported in the literature because the unsupervised execution of multi-threading and blocking codes within a electrical/communication simulator may result in problems such as nondeterministic outcomes and deadlocks. In this section, we explain how direct-execution simulators are designed to eliminate this kind of problems. The following discussions are focused on the Java language, due to its popularity in distributed software development.

The major challenges in simulating a multi-thread Java execution is to ensure deterministic results. To achieve this, we first explore the reasons for nondeterministic behaviors, and then describe the corresponding counter-measures.

According to Java memory model (JMM) [20], a single-thread Java program performs a sequence of actions. Given a particular control flow path, the order for action execution is uniquely defined according to the thread-local semantics. Therefore, the execution of a single-thread program is deterministic and repeatable as long as it does not contain unspecified actions such as reading object references and external inputs.

However, for a multi-threaded program, actions issued by different threads are not necessarily in order, hence their execution order may vary in different runs and produce different timing and functional behaviors. These differences accumulate as the simulation proceeds and lead to diverged results. The causes of nondeterministic/un-repeatable behaviors are summarized as below:

1) Actions' execution order is not deterministically defined and enforced as explained above. Thread scheduling is not defined nor controlled by the program, which contributes to the non-determinism of action's execution order.
2) The outcomes of Java synchronization operations (e.g. lock/unlock, wait/notify) may be non-repeatable even if their order of execution are exactly repeated. For example, when multiple threads contend for a lock, the order of lock acquisition is not defined in JMM even if the order of contention is the same.
3) Interactions between a program and external systems may not be exactly repeatable.

Corresponding countermeasures are employed by the DecompositionJ framework to eliminate these four sources of nondeterminism.

1) In the simulation, actions are executed in an order determined by their timestamps. To enforce this timestamp ordering of action execution, a *logical time barrier* is inserted ahead of each thread action. This barrier postpones an action execution until all actions with lesser timestamps are completed.
2) Thread scheduling is modeled by introducing logical processors, logical threads, logical scheduler, and scheduling actions in the simulation. Logical threads participate in the scheduling by executing *processor-contend, acquire,* and *release* actions. The outcomes of these actions are determined by the logical scheduler based on the first-come-first-served principle. A logical thread must acquire a logical processor to continue action execution. This is enforced by inserting a *logical processor barrier* after a *processor release*, which postpones action execution until the thread has re-acquired a logical processor.
3) Java synchronization operations in the program are replaced with their deterministic versions, which interacts with the logical scheduler.
4) External systems are replaced with their simulated counterparts, i.e. co-simulators. External interactions with other co-simulators are performed by exchanging times-tamped messages. External messages and thread actions are processed according to their timestamp order. This is enforced by inserting *logical time barrier* before the processing of an external message.

12.3.1 Operation of a Direct-Execution Simulation

The design of a direct-execution simulator is in itself a very complicated matter that deserves dedicated paper [17] for discussion. Reference [17] provides detailed discussions on (i) concurrent execution model, (ii) definitions of actions, interaction relationships, and constraints, (iii) exploitation of parallelism, and (iv) performance and scalability evaluation. In the following subsections, we provide an overview on the runtime operation of direct-execution simulation by first describing the data structures used in the simulation, then followed by the discussion on the tracking of timestamps and the enforcing of timestamp order.

12.3.1.1 Simulation Metadata

The simulation *metadata* is a global data structure that tracks the states of a simulated execution. Metadata provides the necessary information for evaluating action timestamps and calculating the outcomes of scheduling and synchronization operations, and it consists of:

- Processor States: For each processor, metadata maintains its ID, logical clock, the latest dispatched thread, operating frequency, context switching delay, and an idle flag.

- Logical Threads: For each logical thread, metadata maintains its ID, the latest acquired processor and time-slice, the lock it contends, and the wait set it resides in.
- Logical Scheduler: This includes a set of active threads and a processor contention queue.
- External Event Queue: This includes messages received from external processes, which are sorted according to their timestamps.
- Locks: For each lock, metadata keeps track of its locked/unlocked state, latest acquiring thread, and a contention queue.
- Wait sets: For each wait set, metadata contains its wait queue and the associated lock.

Since the metadata is a shared data structure, a metadata lock is used to prevent concurrent access that leads to inconsistent states. A thread must hold the metadata lock when accessing metadata.

12.3.1.2 Enforcing Simulated Thread Scheduling

To simulate the effect of thread scheduling, actions can only be executed when the logical thread has acquired a logical processor. Therefore, processor barrier codes are inserted at the beginning of threads and at the end of every processor releasing operation. When a thread reaches a processor barrier, it will wait until a logical processor is assigned to it by the logical scheduler. Note that the metadata lock must be (i) acquired before entering processor barrier and (ii) released while waiting inside or leaving the barrier.

12.3.1.3 Tracking Action Timestamps

The timestamps of thread actions are tracked by using the processors' logical clock in the metadata. When a thread is being released from processor barriers, the clock of the newly acquired processor is updated. After leaving the processor barrier, the processor's clock will be incremented by the control codes each time a new action is reached. Essentially, processor clock tracks the timestamp of the next action to be performed by the running thread.

12.3.1.4 Enforcing Timestamp Order

The timestamp order of event execution is enforced by inserting logical time barrier ahead of synchronization actions or external messages. Note that in a co-simulation environment, it is also necessary to ensure that no external messages with lesser timestamps will arrive after leaving a logical time barrier. Therefore, the barrier contains a loop, in which a thread invokes RTI's *NextMessageRequest* service to obtain the next external message with a timestamp less than that of the next action, and waits until a *TimeAdvanceGrant* is issued by the RTI. This request-grant cycle repeats until no further messages with lesser timestamp is received.

12.3.1.5 Handling External Events

Messages received from co-simulators will be stored in the external event queue in the metadata. Events in the queue are sorted according to their timestamps and await to be processed by an external event thread (EET). The main body of EET is a loop. In each iteration, EET first waits at a logical time barrier until the next external message has the least timestamp. It then executes the handler associated with that message. If the event queue is empty, EET will wait at an *external event barrier* until new events arrive.

The operation principle of an external event thread is similar to that of a logical thread but with the following differences: (i) since external actions are not performed in the context of a logical thread, handlers must not contain blocking operations such that EET will not be blocked as a result of executing external operations and (ii) external actions do not require logical processors for execution and therefore EET will not be blocked by processor barriers.

12.3.2 DecompositionJ Framework

The DecompositionJ framework is, in essence, a runtime library and a compiler-based code analyzer and transformer. Data structure and various mechanisms used in a direct-execution simulation are implemented in a purposely designed runtime library, such as:

1) The simulation metadata.
2) Execution order enforcing mechanisms. These include logical scheduler, external event thread, barriers, and time tracking mechanisms that are inserted into the target source code.
3) Deterministic versions of Java synchronization operations and time related operations. These include: *Object.wait/notify/notifyAll, Thread.start/sleep/yield/ join/interrupt/interrupted/ isInterrupted/ isAlive/getState*, monitor *enter/exit, System.currentTimeMillis/ nanoTime.*
4) Simulated versions of external interactions. For example, a simulated version of *java.net* package has been implemented to facilitate the simulation of TCP communications between Java programs. The simulation package translates various network events into co-simulation messages which are exchanged with the OPNET network simulator.

With the aid of the runtime library, the compiler-based analyzer and transformer can then convert a target Java program into a direct-execution simulator of itself. The analysis and conversion procedures are as follows:

The parser generates abstract syntax trees (AST) for all the compilation units, i.e. *.java* source files of the target program.

1) AST nodes representing Java synchronization operations are instrumented with barriers or replaced by their deterministic counterparts in the runtime library. Specifically, (i) for a volatile variable access, metadata lock and logical time barrier are inserted before the access, and metadata unlock is inserted after the access, (ii) for Synchronization-Blocks and Synchronization-Methods, they are replaced by a try-finally block that begins with a monitor enter and ends with a monitor exit operation, and (iii) for commonly used synchronization operations, they are replaced by their counter-parts in the library.
2) AST nodes representing external interactions are replaced by their simulated versions, e.g. the *java.net* package.
3) Time tracking codes are then inserted between actions. Note that it is not necessary to inject a tracking code between every action since multiple intra-thread actions can be lumped into one if there is no branching in their control flow, i.e. they belong to the same basic block. Therefore, basic blocks are identified by using precise exceptional intra-procedural control flow analysis [21] and tracking codes are only inserted before each basic block.
4) The transformed ASTs are then rewritten into Java source which can then be compiled using JDKs to produce simulator executables.

The JastaddJ/ExtendJ compiler framework is used to automatically perform the above analysis and source-to-source transformation. It is important to note that manual modification of the original program source code is not required. The simulator produced by DecompositionJ is compatible with any JDK, JVM, and their associated development tools such as debuggers, profilers, and IDEs. To begin the simulation, users are only required to write a startup code to specify parameters for the simulation environment, e.g. number of logical processors and their operating frequencies, and then redirect the control flow to the simulator produced by DecompositionJ. Therefore, the entire process is convenient and almost fully automatic.

12.4 Co-Simulation Platform for Agent-Based Smart Grid Applications

As noted in recent literatures, the multi-agent software platform JADE [18] has become a popular choice for implementing and studying distributed smart grid applications. JADE features full compliance with the FIPA [22] (Foundation for Intelligent Physical Agents) specification designed by IEEE Computer Society for promoting agent interoperability. The platform provides Agent Management System, Directory Facilitator, Agent Interaction Protocols (AIPs), and Message Transport Services (MTS). The entire software is composed of more than 400 K lines of codes and continue to evolve under active developments.

The multi-threading nature and complexity of JADE make it a good candidate for demonstrating the capability and usefulness of the proposed framework. As shown in Figure 12.1, the source code of JADE and agents are automatically analyzed and transformed using the DecompositionJ framework to produce Virtual JADE (V-JADE) simulators. V-JADE simulators then participate in a co-simulation with electrical system simulator PSCAD and network simulator OPNET.

As illustrated in Figure 12.2, the co-simulation is achieved by interfacing each simulator with a common HLA-compliant Runtime Infrastructure. For V-JADE, the RTI interface has been developed as part of the external event handling mechanism inside the DecompositionJ runtime library. For PSCAD, the RTI interface has been developed as an external library linked to user-defined modules. For OPNET, the RTI interface is already built into the software. Note that in the JADE architecture, the multi-agent platform is formed by multiple containers distributed over networked

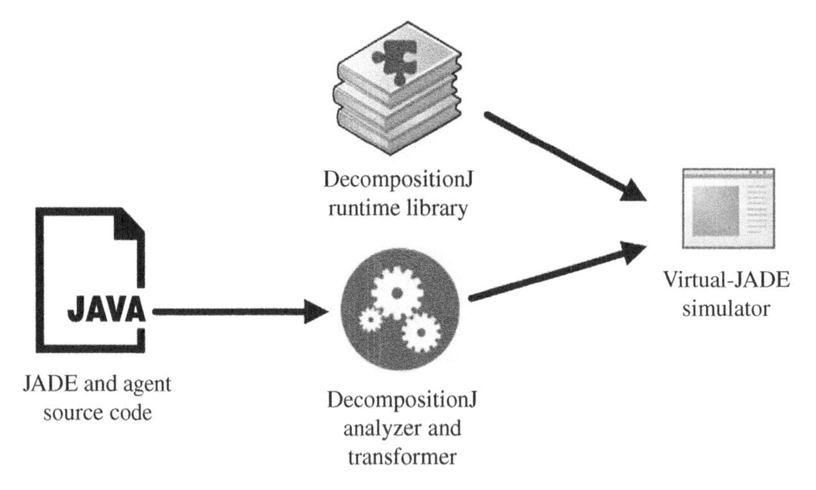

Figure 12.1 Converting JADE and agent code into direct-execution simulators using DecompositionJ. *Source:* From Shum et al. [A]/IEEE.

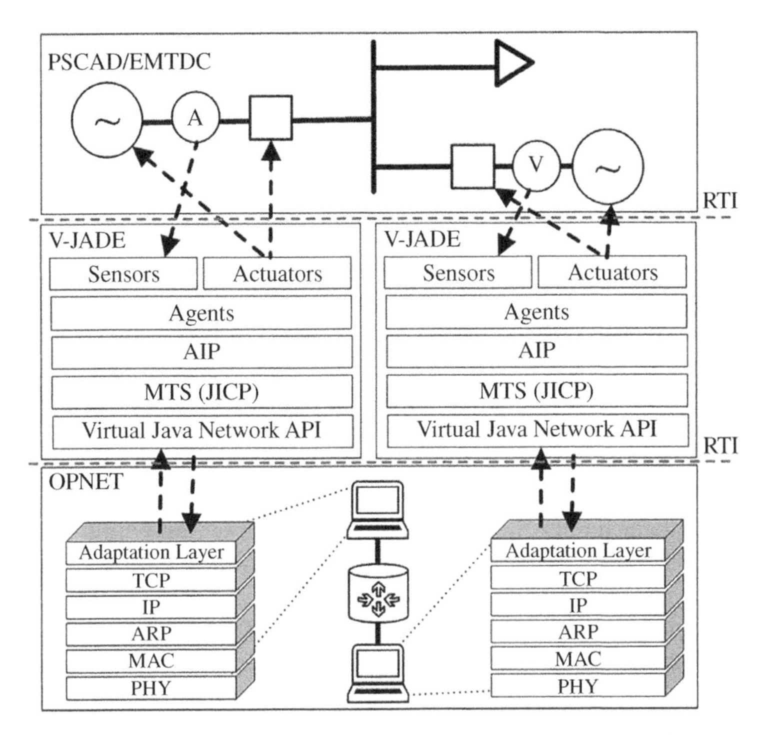

Figure 12.2 Co-simulation architecture. *Source:* From Shum et al. [A]/IEEE.

computers. Each container may encapsulate more than one agent. Therefore, multiple V-JADE simulators are used in the co-simulation.

12.4.1 Co-Simulation Message Exchange

To develop an HLA co-simulation, it is crucial that the format of exchanged messages is made known to all simulators. Therefore, a Federate Object Model (FOM) must be designed to specify the structures and attributes of all objects and interactions being exchanged. The FOM specification file is then shared between simulators.

In our co-simulation design, V-JADEs and PSCAD exchange information via *sensor* and *actuator* objects. At the initialization phase of a co-simulation, *sensor* and *actuator* objects are created and registered on the RTI by PSCAD. During simulation, sensor readings from the electrical system are periodically published to the RTI by PSCAD. The RTI then delivers the messages to V-JADEs according to time- stamp order. For *actuator* objects, their values are published by V-JADE and subscribed by PSCAD.

The message exchange between V-JADEs and OPNET is considerably more complicated. Firstly, each V-JADE simulator is associated with a *workstation* object which is in turn associated with multiple *network interface* objects. Each *network interface* contains simulated network attributes exposed to V-JADE, which include *workstation ID/network interface ID / MAC address/hostname/IP version/IP address/subnet mask* and *MTU*. During the co-simulation initialization phase, OPNET creates, registers, and publishes *workstation* and *network interface* objects to V-JADE simulators.

Secondly, when TCP socket operations are performed by V-JADE during simulation, RTI Interaction messages are sent to OPNET with reference to particular *workstation* and *interface* objects. Different types of interactions are used for different types of the operation, which include socket *create/open/listen/send/close* and *abort*. Conversely, when TCP socket events occur in OPNET, interactions will be sent to V-JADE with reference to particular *workstation* and *inter- face*. The interaction types include *open confirmation / close indication / close confirmation / data reception / error Indication / Abort indication* and *FIN reception*. The above types cover most functionalities of the *java.net* package, which is sufficient for simulating the communications between JADE containers.

12.4.2 Co-Simulation Time Synchronization

Proper ordering of events across co-simulators is necessary to ensure causality and repeatability of the results. Time Management Service provided by HLA compliant RTIs handles this ordering using conservative time synchronization algorithms.

During the co-simulation, a simulator refrains from processing the next local event until it is guaranteed by the RTI that no messages with timestamps less than the next event will be received. This stop-and-wait procedure is achieved by first issuing a *NextMessageRequest* with t_{next_event} as parameter to the RTI, and then wait until a *TimeAdvanceGrant* is received from RTI. By issuing the *NextMessageRequest*, the simulator also guarantees that it will not generate a message with timestamp less than $\min(t_{next\_msg}, t_{next\_event}) + LA$, where t_{next_msg} refers to the timestamp of next message that will be delivered to the simulator, and *LA* refers to simulator's look ahead. The lookahead is a nonnegative value that allows other simulators to execute beyond this simulator's local time. Using large lookahead values improve parallelism and hence enhance the performance of the co-simulation. Deriving a suitable lookahead value depends on the physical limitations of the simulated system, i.e. how quickly a simulated system can react to events received from another simulated system. For OPNET and Virtual JADE simulators, the overhead is assumed to be 1 μs with the consideration of the processing delays between JVM and underlaying protocol stacks. For PSCAD, the lookahead is the same as the time-step size since PSCAD will not send out messages between simulation steps.

12.5 Agent-Based FLISR Case Study

Agent-based Fault Location, Isolation, and Service Restoration is a delay sensitive application that relies on fast decision response to reduce service interruption time. With this simulation case study, we demonstrate that the timeliness of service restoration can be influenced by the configurations of the JADE software and background traffics. In addition, we investigate a scenario in which both electrical and communication system failures occur simultaneously.

12.5.1 The Restoration Problem

When fault occurs in a radially configured power distribution network, the protection system isolates the faulted section and consequently blocks the power flow toward downstream sections. Switches are reconfigured to restore service for the affected sections with a network topology to minimize de-energized loads. Optimal network reconfiguration problem is traditionally solved off-line and the solutions are statically programmed to react upon events. This approach does not adapt well to the dynamic operation of smart distribution grids. Agent-based solutions were proposed in [23–28] to calculate optimal configuration based on dynamic information in a distributed manner.

In a reconfiguration problem, the power distribution network is considered as a graph $\mathbb{G} = \{V, E\}$, where each edge corresponds to a switch, and each node corresponds to a set of feeder buses and lines bounded by a common set of switches. For a node v, its power generation, generation capacity, and load consumption are given by $G(v)$, $G^{max}(v)$ and $L(v)$ respectively; and for an edge $\{v_i, v_j\}$, its power flow (from node v_i to v_j), and line capacity are given by $P(v_i, v_j)$ and $P^{max}(v_i, v_j)$. Note that power flow is directional, i.e. $P(v_i, v_j) = -P(v_j, v_i)$. A modified IEEE 34-bus system is shown in Figure 12.3a and its graph model is shown in Figure 12.3b which illustrates the graph representation of distribution network. This system is also used in this FLISR case study.

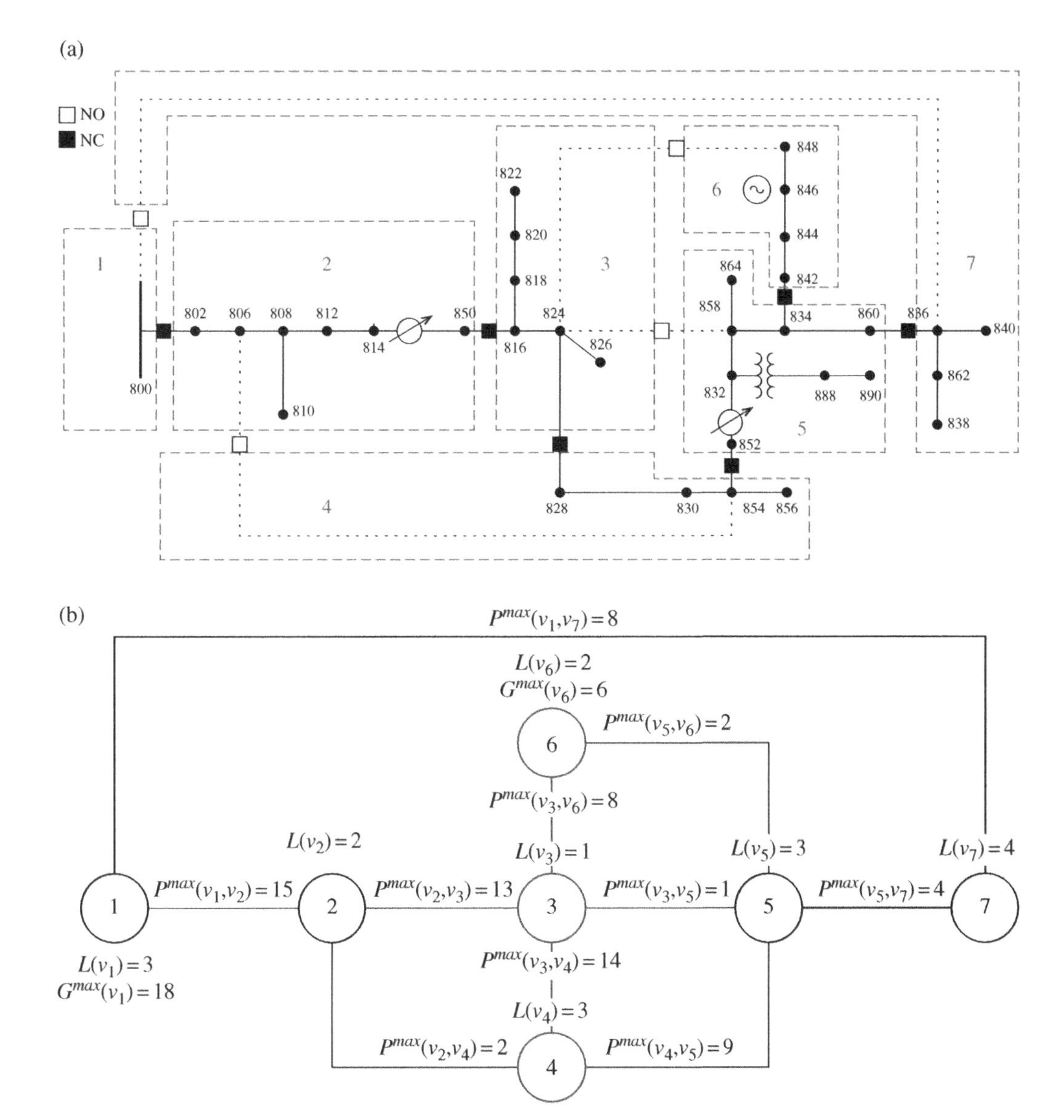

Figure 12.3 (a) A modified IEEE 34-bus distribution network, and feeder buses bounded by a common set of switches are grouped in dashed boxes. (b) The graph model for the distribution network. *Source:* From Shum et al. [A]/IEEE.

When a fault occurs, the nodes are classified into three disjoint subsets: supplier nodes V^s(nodes with positive net power output), consumer nodes V^c, and faulted nodes V^f. Consider a graph $\mathbb{G}$with N supplier nodes, the reconfiguration algorithm aims to find a set of N trees $H' = \{T1, T2, TN\}$ for restoration, such that the total restored load is maximized:

$$H' = \underset{H=\{T1,\ T2,\ TN\}}{\text{argmax}} \sum_{T\in H}\sum_{v\in T} L(v) \tag{12.1}$$

The optimization subjects to the following constraints:

1) Each tree must contain a supplier node as the root and exclude all the faulted nodes.

$$\forall T_n = \{V_n, E_n\} \in H. \quad |V_n \cap V^s| = 1 \tag{12.2}$$

$$\forall T_n = \{V_n, E_n\} \in H. \quad V_n \cap V^f = \emptyset \tag{12.3}$$

2) The trees are disjoint.

$$\forall T_n, T_m \in H. \quad T_n \neq T_m \Rightarrow T_n \cap T_m = \emptyset \tag{12.4}$$

3) Power balance must be satisfied, i.e. the input power of a node is equal to the total output power for its successor nodes plus local load. For consumer node, the net input power equals load consumption.

$$\forall v \in V^c \cap H. \quad \sum_{\{v',\ v\}\in E} P(v', v) = L(v) \tag{12.5}$$

For supplier node, the sum of generation and net input power equals consumption.

$$\forall v \in V^s. \quad G(v) + \sum_{\{v',\ v\}\in E} P(v', v) = L(v) \tag{12.6}$$

1) Power generation of each DER is bounded by its maximum capacity.
2) Power flow of each line is bounded by maximum line capacity.

$$\forall v \in V^s. \quad 0 \le G(v) \le G^{max}(v) \tag{12.7}$$

$$\forall \{v_i, v_j\} \in H. \quad |P(v_i,\ v_j)| \le P^{max}(v_i, v_j) \tag{12.8}$$

12.5.2 Reconfiguration Algorithm

Due to the distributed nature of multi-agent systems and the dynamic nature of smart grid, each node on the network only possesses local information. Therefore, after the fault, the agents at each node explore the graph and obtain information from agents in other nodes in order to determine the restoration trees. Since each restoration tree must contain a supplier node, the exploration and tree growth start from the supplier node. The procedures of the reconfiguration algorithm are as follows:

1) Initially, a restoration tree T_n is created for each supplier node $\{v_i, v_j\}$, which contains v_n^s as its root.
2) For each restoration tree, frontier edges F_n is defined as the edges between nodes inside T_n and nodes outside T_n.F_n will be updated when a new node is recruited to the tree.
3) For each tree node v other than the root supplier node, $in(v)$ represents the parent tree node that supplies power to v, and $out(v)$ represents the set of child tree nodes which draws power from v.

4) For each tree node v, $R(v)$ represents the surplus power that can be drawn from the node if new nodes are added to the tree through v. By constraining $R(v) \geq 0$ for all nodes, Eqs. (12.5)–(12.8) can be enforced.

$$\forall v \in V^c. \quad R(v) = min\,(R(\mathrm{pr}(v)), P^{max}(\mathrm{pr}(v), v)) - L(v) - \sum_{v' \in out\,(v)} P(v, v') \tag{12.9}$$

$$\forall v \in V^s. \quad R(v) = G^{max}(v) - L(v) - \sum_{v' \in out\,(v)} P(v, v') \tag{12.10}$$

5) A weight w, representing the load of the neighboring node reachable through the edge, is assigned to each frontier edge.

$$\forall \{v_i, v_j\} \in F_n, \quad w(v_i, v_j) = w(v_j, v_i) = L(v_j) \tag{12.11}$$

6) For as long as F_n is not empty, the algorithm greedily explores the frontier edge with the largest weight and attempts to recruit the neighboring node through that edge.
7) To ensure that a neighboring node is eligible for recruitment, the supplier node's agent enquires its state through communications. The neighboring node will be recruited unless it (i) is faulted or (ii) is already included in another restoration tree, or (iii) has a load exceeding the R value of its parent node.
8) New frontier edges F_{new} are added to F_n when a new node is recruited. Supplier node communicates with the new node in order to acquire F_{new} and their associated weights. Additionally, R value of all tree nodes will be updated to account for the new load.
9) The algorithm ends when all the frontier edges are explored. The distribution network will be restored according to the topology of the restoration trees.

12.5.3 Restoration Agents

The FLISR mechanisms including the reconfiguration algorithm are implemented using a two-tier multi-agent system (MAS) with four agent types. As shown in Figure 12.4, the upper tier consists of Node Agents (NA) and Switch Agents (SA), while the lower tier consists of Load Agents (LA) and

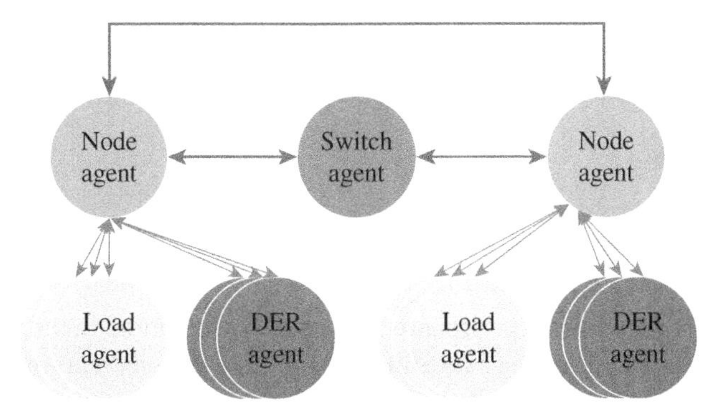

Figure 12.4 Two-tiered MAS hierarchy. *Source:* From Shum et al. [A]/IEEE.

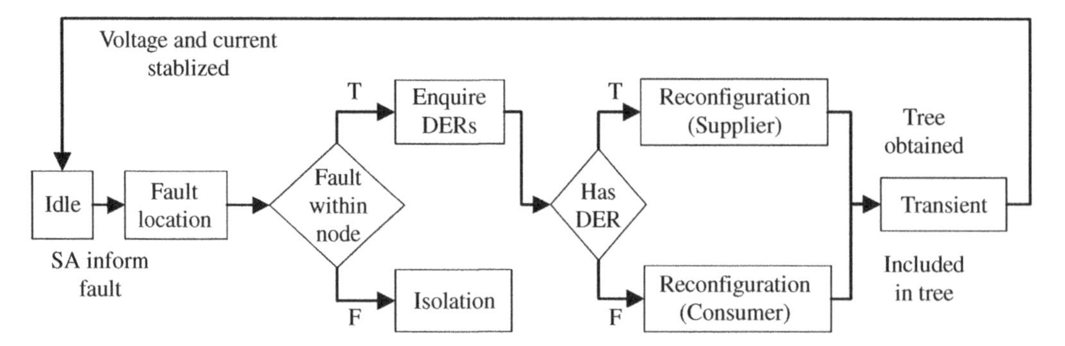

Figure 12.5 Operation flow of a node agent. *Source:* From Shum et al. [A]/IEEE.

DER Agents (DA). Lower tier agents can only communicate with their supervising NAs, while upper tier agents can communicate with one another. The operation of each type of agent will be presented in the following.

SAs, LAs, and DAs are connected to their corresponding physical switches, loads, and DERs through sensors and actuators. These agents control and monitor their corresponding physical devices and respond to their supervising NAs' requests, such as adjusting operating points and reporting device status. In addition, SAs are also responsible for over current protection and fault detection. NAs have no direct interaction with the electrical system, but they are responsible for coordinating the FLISR mechanism. Figure 12.5 depicts its operating states. Initially, NA stays in idle state until an anomaly is reported by a neighboring SA. Upon which the NA will enquire current and voltage readings from all neighboring SAs. If a fault is detected within its segment, NA will isolate the fault by requesting all neighboring SA to open their switches and it will then remain isolated and not be restored. On the contrary, NA will proceed to the inquiry phase to identify its local generations and consumptions through subordinate DAs and LAs. After that, NAs with local generations (i.e. supplier node) will initiate the reconfiguration algorithm, while consumer NAs will wait and respond to inquiries. Supplier NAs will enter transient state and request both DERs and switches to adjust to new network configuration upon the completion of the reconfiguration algorithm. Finally, NAs return to idle state when stabilized bus voltages is observed.

12.5.4 Communication Network Configurations

For communication infrastructure, we consider gateway routers being installed at all buses, which are connected by optical carriers running alongside with distribution lines. The distances between routers are stipulated in the IEEE 34-bus system dataset. Each agent runs in a separate JADE container which is mapped to a workstation that connects to a router. Further details are listed in Table 12.2.

12.6 Simulation Results

The agent-based FLISR case has been simulated using the proposed co-simulation platform. In this section, we first explain the MAS operation by presenting a trace of significant simulation events in the FILSR process. Then the performance of MAS is evaluated with different network and software configurations: (i) background traffics, (ii) communication link failure, (iii) communication link

Table 12.2 Characteristics of simulated network.

Characteristics	Configuration
Transport layer	Background traffic: UDP. JADE traffic: TCP
TCP Configurations	TCP Reno, Karn's Algo., initial retransmission timeout: 0.5 s
Network layer	IPv4
Routing	EIGRP
Link layer	Ethernet. MTU: 1500 Bytes. Speed grade: OC-1 (51.8 Mbps)
Router processing	Switching rate: 200 000 pkt/s. Forwarding: 20 000 pkt/s
Background traffic	From all buses to bus 800 (substation)

Source: From Shum et al. [A]/IEEE.

failure time, and (iv) location of JADE main container responsible for maintaining all agents' network address.

12.6.1 Agent Actions and Events

The event trace in Table 12.3 is produced without simulating background traffic or link failure. This table only shows relevant information from supplier Node agents, i.e. NA1 and NA6.

12.6.1.1 Phase 1 – Fault Detection

At $t = 0$, a three-phase ground fault occurs between node 806 and 808 (inside v_2), which causes fault current to flow through Bus 800 toward node 808 and voltage sag in the distribution network. The fault current is first detected by SA12 and reported to NA1 at $t = 0.00931$ (#1), using the *Switch Anomaly Inform: Inform* message. This triggers NA1 to change its state from Idle to Fault Location (#2). Similarly, NA6 enters Fault Location state after receiving the message from SA56 (#15, 16). Note that only the first Switch Anomaly Inform messages trigger the state change, later messages are neglected and not shown in the trace.

12.6.1.2 Phase 2 – Fault Location

During Fault Location phase, NA1 sends *Switch Status Query: Query-ref* messages to all its neighboring SAs (#3), and awaits their responses. Responses (*Switch Status Query: Inform*) containing recent current readings are received at $t = 0.01303$ (#4). The readings indicate that the fault is outside v_1 (#5). Consequently, NA1 proceeds to Enquire DERs and Loads state (#6). Similarly operations are also performed by NA6 (#17, 24–26).

12.6.1.3 Phase 3 – Enquire DERs

To identify the total power generation capacity and load consumption of the node. NA1 sends a query message (*DER/Load Status Query: Query-ref*) to DA1 and LA1 (#7). The corresponding response messages are received at #8, indicating that v_1 has a generation capacity of 18.0 and load of 3.0, and hence 15.0 units are available for restoring other nodes. NA1 then enters Restoration Supplier state (#9). Similarly, NA6 follows the same routine (#27, 29–30) and has a surplus power of four units. For nodes without DER, their NAs will enter Restoration Consumer state after receiving the response messages from their subordinate LAs.

Table 12.3 Simulation event trace for NA1 and NA6.

#	Time(s)	Agt	Type	Event	Remarks
1	0.00931	NA1	Recv	Switch Anomaly: Inform	From SA12
2	0.00931	NA1	State	Idle → Fault Location	
3	0.00935	NA1	Send	Switch Status Query: Query-ref	To SA12, SA17
4	0.01303	NA1	Recv	Switch Status Query: Inform	SA12 (over cur.), SA17 (under vlt.)
5	0.01303	NA1	Algo	Team Status Confirmed	Not Faulted
6	0.01304	NA1	State	Fault Location → Enquire	
7	0.01307	NA1	Send	DA/LA Status Query: Query-ref	To DA1, LA1
8	0.01571	NA1	Recv	DA/LA Status Query: Inform	From DA1, LA1
9	0.01572	NA1	State	Enquire → Reconfig. Supplier	
10	0.01573	NA1	Algo	Initialize frontier edge set	$\{v_1, v_2\}\{v_1, v_7\}$
11	0.01573	NA1	Algo	Initialize $R(v_1)$	N/A → 15.0
12	0.01573	NA1	Algo	Attempt to restore frontier edge	$\{v_1, v_7\}$
13	0.01576	NA1	Send	Restoration Query: Query-If	To NA7
14	0.02143	NA1	Recv	Restoration Query: Agree	From NA7
15	0.02872	NA6	Recv	Switch Anomaly: Inform	From SA56
16	0.02874	NA6	State	Idle → Fault Location	
17	0.02878	NA6	Send	Switch Status Query: Query-ref	To SA56. SA36
18	0.03239	NA1	Recv	Restoration Query: Inform	From NA7
19	0.03240	NA1	Algo	New frontier edge discovered	$\{v_1, v_7\}$
20	0.03240	NA1	Algo	Update $R(v_7)$	N/A → 4.0
21	0.03240	NA1	Algo	Update $R(v_1)$	15.0 → 11.0
22	0.03241	NA1	Algo	Attempt to restore frontier edge	$\{v_7, v_5\}$
23	0.03243	NA1	Send	Restoration Query: Query-If	To NA5
24	0.04143	NA6	Recv	Switch Status Query: Inform	SA56 (under vlt.), SA36 (under vlt.)
25	0.04143	NA6	Algo	Team Status Confirmed	Not Faulted
26	0.04144	NA6	State	Fault Location → Enquire	
27	0.04147	NA6	Send	DA/I.A Status Query: Query-ref	To DA6, LA6
28	0.04599	NA1	Recv	Restoration Query: Agree	From NA5
29	0.04916	NA6	Recv	DA/LA Status Query: Inform	From DA6, LA6
30	0.04917	NA6	State	Enquire → Reconfig. Supplier	
31	0.04918	NA6	Algo	Initialize frontier edge set	$\{v_6, v_3\}\{v_6, v_5\}$
32	0.04918	NA6	Algo	Initialize $R(v_6)$	N/A → 4.0
33	0.04918	NA6	Algo	Attempt to restore frontier edge	$\{v_6, v_5\}$
34	0.04918	NA6	Algo	Violate constraint	$\omega(v_6, v_5) \leq P^{max}_{(v_6, v_5)}$
35	0.04918	NA6	Algo	Attempt to restore frontier edge	$\{v_6, v_3\}$
36	0.04921	NA6	Send	Restoration query: query-if	To NA3
37	0.05302	NA1	Recv	Restoration query: inform	From NA5
38	0.05303	NA1	Algo	New frontier edge discovered	$\{v_5, v_3\}\{v_5, v_4\}\ \{v_5, v_6\}$

Table 12.3 (Continued)

#	Time(s)	Agt	Type	Event	Remarks
39	0.05303	NA1	Algo	Update $R(v_5)$	N/A → 1.0
40	0.05303	NA1	Algo	Update $R(v_7)$	4.0 → 1.0
41	0.05303	NA1	Algo	Update $R(v_1)$	11.0 → 8.0
42	0.05304	NA1	Algo	Attempt to restore frontier edge	$\{v_5, v_4\}$
43	0.05305	NA1	Algo	Violate constraint	$\omega\{v_5, v_4\} \leq R\{v_5\}$
44	0.05305	NA1	Algo	Attempt to restore frontier edge	$\{v_5, v_6\}$
45	0.05305	NA1	Algo	Violate constraint	$\omega\{v_5, v_6\} \leq R\{v_5\}$
46	0.05306	NA1	Algo	Attempt to restore frontier edge	$\{v_1, v_2\}$
47	0.05307	NA1	Send	Restoration query: query-if	To NA2
48	0.05679	NA6	Recv	Restoration query: agree	From NA3
49	0.05689	NA1	Recv	Restoration query: refuse	From NA2, faulted
50	0.05695	NA1	Algo	Attempt to restore frontier edge	$\{v_5, v_{32}\}$
51	0.05697	NA1	Send	Restoration query: query-if	To NA3
52	0.06244	NA6	Recv	Restoration query: inform	From NA3
53	0.06245	NA6	Algo	New frontier edge discovered	$\{v_3, v_2\}\{v_3, v_4\}$ $\{v_3, v_5\}$
54	0.06245	NA6	Algo	Update $R(v_6)$	4.0 → 3.0
55	0.06246	NA6	Algo	Attempt to restore frontier edge	$\{v_3, v_5\}$
56	0.06246	NA6	Algo	Violate constraint	$\omega(v_3, v_5) \leq P^{max}_{(v_3,\ v_5)}$
57	0.06246	NA6	Algo	Attempt to restore frontier edge	$\{v_3, v_4\}$
58	0.06248	NA6	Send	Restoration query: query-if	To NA4
59	0.06557	NA1	Recv	Restoration query: refuse	From NA3, restored
60	0.06733	NA1	Algo	No restorable frontier edge left	
61	0.06733	NA1	State	Reconfig. supplier → Transient	
62	0.06735	NA1	Send	DA operation request: request	To DA1 ($P = 10$)
63	0.06788	NA1	Recv	DA operation request: agree	From DA1
64	0.07050	NA6	Recv	Restoration query: agree	From NA4
65	0.07650	NA6	Recv	Restoration query: inform	From NA4
66	0.07651	NA6	Algo	New frontier edge discovered	$\{v_4, v_5\}$,$\{v_4, v_2\}$
67	0.07651	NA6	Algo	Update $R(v_6)$	3.0 → 0.0
68	0.07651	NA6	Algo	Update $R(v_3)$	3.0 → 0.0
69	0.07652	NA6	Algo	No restorable frontier edge left	$R(v_6) = 0$
70	0.07652	NA6	State	Reconfig. supplier → Transient	
71	0.07654	NA6	Send	DA operation request: request	To DA6 ($P = 6$)
72	0.07708	NA6	Recv	DA operation request: agree	From DA6

Source: From Shum et al. [A]/IEEE.

12.6.1.4 Phase 4 – Reconfiguration

After entering the reconfiguration phase, each supplier node becomes the root of a restoration tree and its node agent executes reconfiguration algorithm to expand the tree. Consumer nodes also participate in the process by reacting to the messages from supplier nodes. Initially, T_1 only consists of the root node v_1 with two frontier edges v_1, v_2 and v_1, v_7 (#10). The edge with the largest weight (i.e. v_1, v_7) is selected for restoration. NA1 sends a *Restoration Query: Query-If* message to NA7 (#13) to verify that the constraints mentioned in step 7) of the reconfiguration algorithm are satisfied. Note that v_7 is restorable because it is not faulted, nor has it been included in any restoration tree. Therefore, the recruitment of v_7 is confirmed with a *Restoration Query: Agree* message replied to NA1 (#14). Upon recruitment, NA7 issues *Switch Operation Request: Request* to SA17 to establish connection to *T*1 by closing the switch. Following that, NA7 proceeds to enquire its neighboring SAs and NAs for a list of new frontier edges. These edges are then identified and reported back to NA1 through a *Restoration Query: Inform* message (#18,19). Finally, NA1 updates the surplus power of v_1 and v_7 (#20,21). The above process represents one iteration of the reconfiguration algorithm. Each restoration supplier will repeat this process to explore its neighborhood and attempt to expand the tree until all eligible frontier edges are exhausted. For NA1 and NA6, their reconfiguration phases end at $t = 0.06733$(#61) and $t = 0.07652$ (#70), respectively. Figure 12.6 shows the final network configuration.

12.6.1.5 Phase 5 – Transient

Once the algorithm ends, supplier node enters transient state and issues an *Operation Request: Request* to its DA (#62 for NA1, #71 for NA6). The DA then replies with an *Operation Request: Agree* message and starts power generation. The communication and computational delays of MAS operations are observable from the electrical system, which is shown in Figure 12.7. The fault current is detected at #1 by SA12 which controls the switch to open at the next zero crossing point. Voltages of node 836 and node 824 are restored shortly after the DERs agree to power generation (#63, 72).

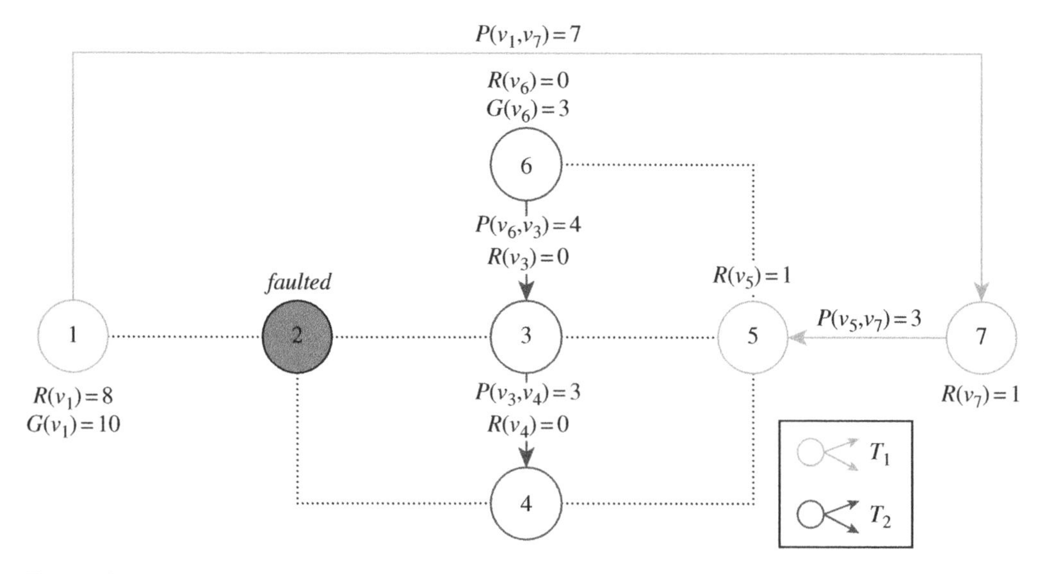

Figure 12.6 New network configuration, with two restoration trees rooted at node-1 and node-6. *Source:* From Shum et al. [A]/IEEE.

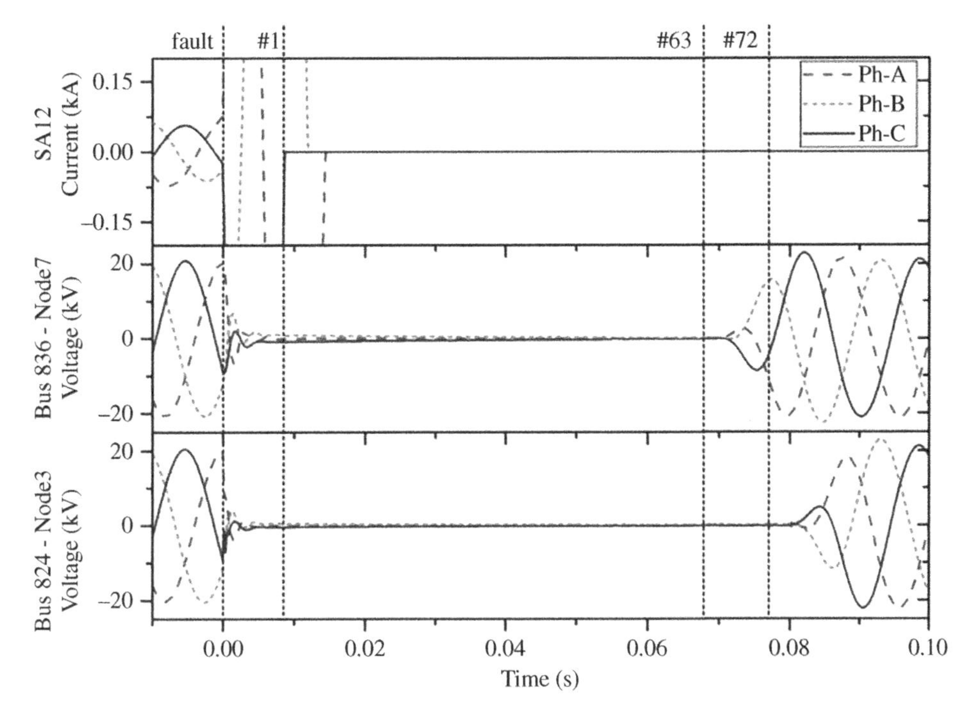

Figure 12.7 Simulation of electrical transients during fault and restoration process. *Source:* From Shum et al. [A]/IEEE.

12.6.2 Effects of Background Traffics and Link Failure

Background traffic is sent from all feeder buses to a data aggregator at bus 800 using IP-based protocols. Note that two OC-1 links are connected to bus 800, i.e. from node 836 to 802, allowing a maximum throughput of 100.224 Mbps (payload) which is equivalent to 3.037 Mbps per bus. Figure 12.8 shows the solution time, i.e. time for supplier NA to enter transient state, of NA1

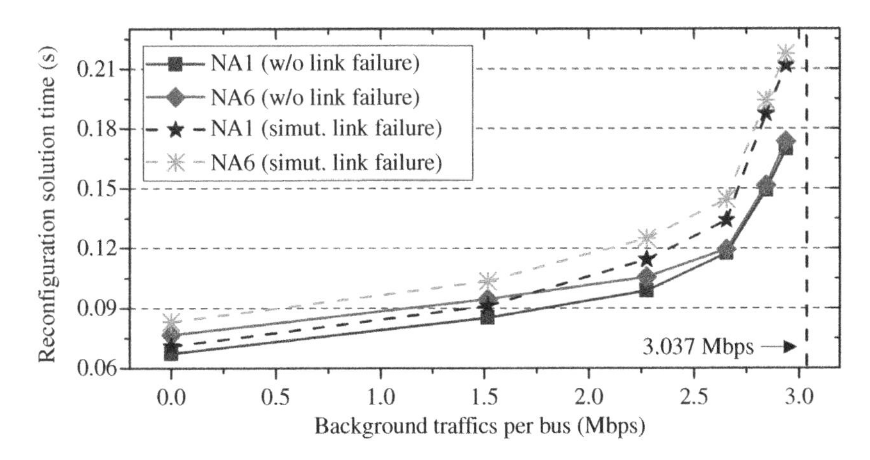

Figure 12.8 Reconfiguration solution time under different levels of background traffic. *Source:* From Shum et al. [A]/IEEE.

Table 12.4 Percentage change of reconfiguration time.

Data rate per bus / 3.037 Mbps	0%	50%	75%	87.5%	93.75%
NA1 solution lime % change	+5.33%	6.98%	15.78%	14.21%	25.70%
NA6solution lime % change	+8.61%	9.43%	18.51%	21.00%	28.13%

Source: From Shum et al. [A]/IEEE.

and NA6 with different levels of traffics. As the configuration of background traffic rate is increased toward the maximum throughput, the average packet queuing delay approaches infinity, which explains the exponentially increasing solution time.

Since communication links run along with power distribution lines, it is likely for an electrical fault to correlate with communication link failures. To study the impact, a link failure is scheduled to occur simultaneously with the electrical fault between nodes 806 and 808. As shown in Figure 12.8, the reconfiguration time increases due to more background traffics being distributed over the remaining links. Table 12.4 shows the percentage increment of reconfiguration time caused by simultaneous link failure and the results indicate a more significant impact under higher traffics.

12.6.3 Effects of Link Failure Time

Since JADE message transport service benefits from the retransmission mechanisms provided by the underlaying TCP protocol, application programmers may often assume reliable and timely agent communications without concerning packet losses in the network so long as the reachability between JADE containers is maintained. However, the co-simulation reveals that when both electrical fault and communication link failures occur, their temporal proximity greatly affects the reconfiguration solution time.

In Figure 12.9, an electrical fault occurs at $t = 0$. The X-axis represents the link failure time and Y-axis represents the reconfiguration solution time of node agents.

If link failure occurs in zone I, i.e. it occurs less than 7.5 ms after the electrical fault, the effect of link failure on solution time is insignificant. This is because agent communications only begin at 9.31 ms (#1 in Table 12.3) after the fault, hence routers adjacent to the failed link have sufficient time to detect it and update their forwarding tables before the agent communications. As a result, no MAS packet is lost.

However, a dramatic increment of solution time is observed if link failure occurs in zone II, i.e. between 7.5 and 27.5 ms after the fault. This is because agents have begun establishing TCP connections with each other via three-way handshakes before alternative routes are updated in router's forwarding tables. The loss of handshake packets is not timely re-transmitted by TCP because the session's round-trip time (RTT) has yet to be measured and a rather long initial re-transmission timeout of 0.5 s is used, leading to a significant impact on the solution time.

If link failure occurs in zone III, i.e. 27.5 ms after the fault, it will not have a significant impact on the solution time. Because TCP handshake between agents has been completed before the link failure and RTT measurements can be used to perform timely retransmission with a negligible timeout delay.

A possible solution to achieve short reconfiguration time is to modify JADE or agent codes which include (i) TCP connections are preestablished between every pair of agents, and (ii) heartbeat

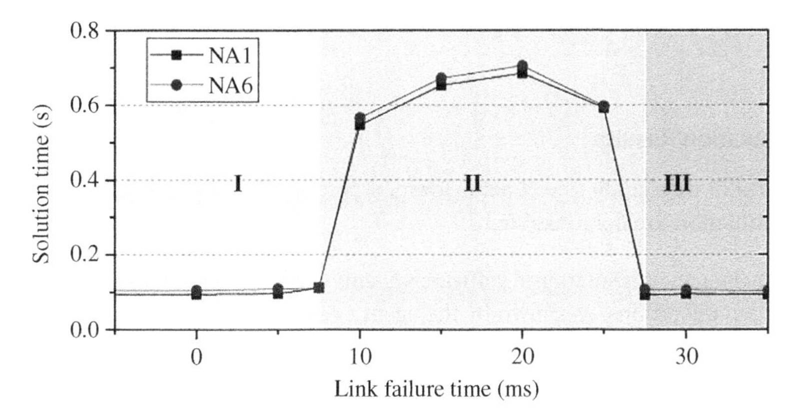

Figure 12.9 Relationship between reconfiguration solution time and link failure time. *Source:* From Shum et al. [A]/IEEE.

signals are sent between agents to periodically update the RTT. However, this is outside the scope of this chapter.

The above analysis of simulation results demonstrates the capability of the proposed framework to capture the long solution time when communication failure and electrical fault occurred in quick succession.

12.6.4 Effects of Main-Container Location Configuration

Under the JADE architecture, a *Global Agent Descriptor Table* storing the network addresses of all agents is maintained by the main container (MC). The network location of MC has an impact on the solution time since agents must first enquire MC to resolve each other's network address before they attempt to communicate for the first time.

As shown in Figure 12.10, the solution time for different node agents varies depending on the location of the MC. When NA1's container is selected as MC, its solution time is benefited because all agent interactions originated from NA1 can be resolved locally. Similar benefit for NA6 can be

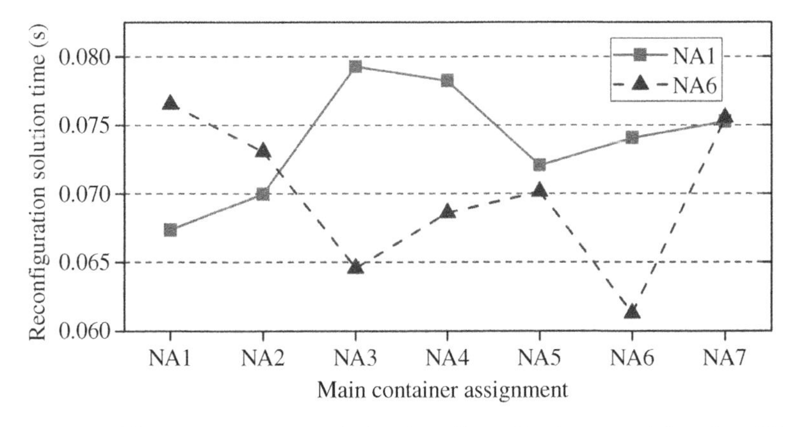

Figure 12.10 Reconfiguration solution time affected by the network location of Main Container. *Source:* From Shum et al. [A]/IEEE.

obtained if its container is assigned as MC. The timing information is useful for finding the optimal MC location.

12.6.5 Summary on Simulation Results

Through this agent-based FLISR case study, the benefits for incorporating direct-execution simulators in a smart grid co-simulation are demonstrated:

- The detailed and authentic behavior of smart grid software systems can be generated to facilitate the understanding of system operations, as shown by the agent event trace.
- Interdependencies between communication systems and software systems can be simulated, as shown by the effects of background traffics on reconfiguration solution time.
- Inherent design concerns in software and communication systems can be revealed by co-simulation, as demonstrated in the case where electrical fault and link failure occurred in quick succession.
- Software designers may fine-tune the JADE main container location to improve solution time.

12.7 Case Study on V2G for Voltage Support

A scenario of V2G voltage support during fault disruption was developed, and the power quality performance under different network infrastructures (Wi-Fi and WiMAX) was investigated. Wi-Fi and WiMAX are considered to be potential technologies for the development of vehicular communication facilities supporting vehicle-to-vehicle (V2V) and vehicle-to-infrastructure (V2I) applications [29]. The modeling of electrical and communication network is described in this section.

12.7.1 Modeling of Electrical Grid and EVs

A modified IEEE 14-bus system (Figure 12.11) was used to model the transmission and distribution grid of a "virtual" city. Taken into account the total of 85 MW loads on the distribution network (bus 6, 9–14) and the typical U.S. electricity consumption of 1400 W per capita, we assumed the city has a population of 60 000. In consideration of a 5% EV penetration of America/Europe automobile market (0.8 vehicles per capita), a fleet of 2400 EV was estimated. Further assume that, 80% of the fleet (1920 EVs) are plugged and ready to provide V2G service during the fault. The distribution of EVs at bus 9–14 is proportional to the resistive load on each bus, given by standard IEEE 14-bus system data, Table 12.5 shows the numbers.

A distribution management system (DMS) is responsible for fault management, voltage & frequency monitoring and V2G services coordination. On the one hand, it collects phasor data from each bus at a rate of 120 Hz for system monitoring. On the other hand, it commands V2G aggregators at bus 9–14 to manage individual EVs on that bus. In this case study, EV operates in either normal or voltage support mode. In normal mode, when an EV is plugged, it sends a charging request message to the aggregator and consumes active power according to a charging schedule provided in the received response message. In voltage support mode, EV injects specified amount of active and reactive power into the bus according to the discharging schedule given by the aggregator. EV stops discharging after battery being drained or EV being unplugged. In PSCAD, we modeled EVs with a set of parameters such as battery capacity, (dis)charging power, initial/targeted

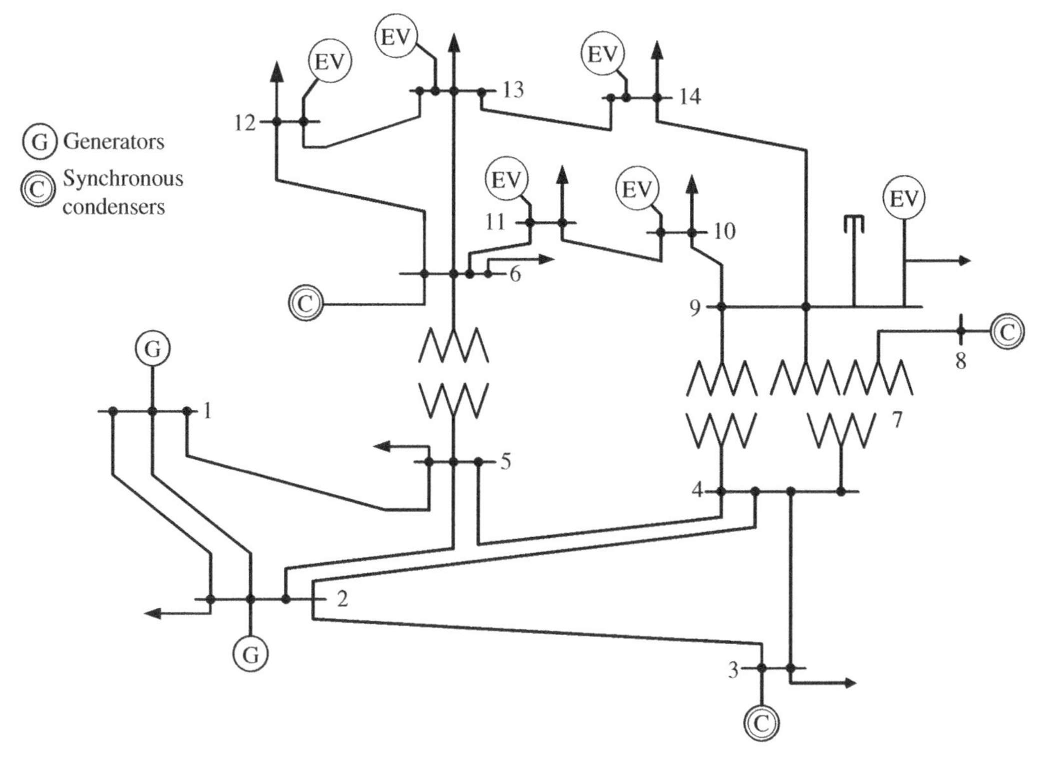

Figure 12.11 IEEE 14-bus system topology and EV locations.

Table 12.5 Amount of EV and total (dis)charging power at each bus.

BUS	9	10	11	12	13	14
Amount of EVs	730	220	90	160	350	370
Max (dis)charging power (MW)	11	3.3	1.35	2.4	5.25	5.55

Source: From Shum et al. [B]. © 2013 IEEE.

state-of-charge (SoC) and charging deadline. Table 12.6 shows the distributions of these parameters, in conformity with realistic EV data (e.g. Tesla Model S [30]).

During voltage support mode, the amount of power injected by the EVs into each bus is optimized for minimal overall generation cost. The MATPOWER toolbox developed by Zimmerman, R.D et al. [31] was used for solving optimal AC power flow problems in an offline fashion. The optimization problem is formulated as:

$$min_{\theta,V_m,P_g,Q_g} \sum_{i=1}^{14} \left[a_0^i + a_1^i \left(p_g^i \right) + a_2^i \left(p_g^i \right)^2 \right] \tag{12.12}$$

Subject to constraints

$$P_{bus}(\theta, V_m) + P_d - C_g P_g = 0 \tag{12.13}$$

Table 12.6 Electrical characteristics of EV.

Electrical characteristics	Distribution	Range
Max (Dis)charging power	Uniform	10/20 kW
Battery capacity	Uniform	60–85 kWh
Current state-of-charge	Gaussian $\mu = 90\%, \sigma^2 = 0.2$	0–100%
Targeted state-of-charge (SoC)	Delta	100%
Power max ramp rate	Delta	100% per 30 ms

Source: From Shum et al. [B]. © 2013 IEEE.

$$Q_{bus}(\theta, V_m) + Q_d - C_g Q_g = 0 \tag{12.14}$$

$$V_m^{min} \leq V_m \leq V_m^{max} \tag{12.15}$$

$$P_g^{min} \leq P_g \leq P_g^{max} \tag{12.16}$$

$$Q_g^{min} \leq Q_g \leq Q_g^{max} \tag{12.17}$$

The objective function (12.12) is to minimize the sum of the 14 2nd-order polynomial cost functions for real power generation; with respect to four parameters θ, V_m, P_g, Q_g. Those are vectors representing the voltage angles, magnitudes, and active and reactive power generations respectively. In this study, the generation-cost-polynomial coefficients for EVs are twice of those for generators, to reflect the expensive financial rewards or compensations pay to the EV owners.

In Eqs. (12.13) and (12.14), P_{bus}, Q_{bus}, P_d, Q_d, C_gP_g, C_gQ_g are 14×1 vectors representing the real and reactive power injection, load injection and generation injection, respectively. Equations (12.13) and (12.14) together constitute the traditional formulation of power balancing equations to ensure the consumed power of each bus equivalents to injected power plus generated power [31].

Equations (12.15)–(12.17) set the upper and lower bounds for voltage magnitude, real and reactive generation of each bus. For bus voltages, we tolerated steady state voltage in range of 0.94–1.06 pu. For generators and synchronous compensators, we allowed maximum power output 20% higher than that in standard IEEE 14-bus system's solution, in conformity with the observation that many real-world power systems have 20% reserved operational margin. For the EVs, we allow maximum discharging power to be the same as max charging power.

12.7.2 Modeling of Communication Network

The communication network for V2G application is illustrated in Figure 12.12. The PMUs are connected to DMS through dedicated FDDI links to provide real-time phasor information. The aggregators are connected to DMS through public network, and EVs communicate with aggregators through public-network-connected wireless base stations. In this case study, three network infrastructures were considered. (i) An Ideal network (control case). (ii) Wi-Fi/IEEE 802.11 g being used by the base stations (or APs). (iii) WiMAX/IEEE 802.16 e being used by the base stations. In order to provide realistic simulation results, round-trip-time delays close to the measurements obtained from real-world WiMAX and Wi-Fi networks are adopted, as reported by Kim et al. [32–34]. Table 12.7 shows the RTT of each link. In all cases, IP-based connectionless protocol is used.

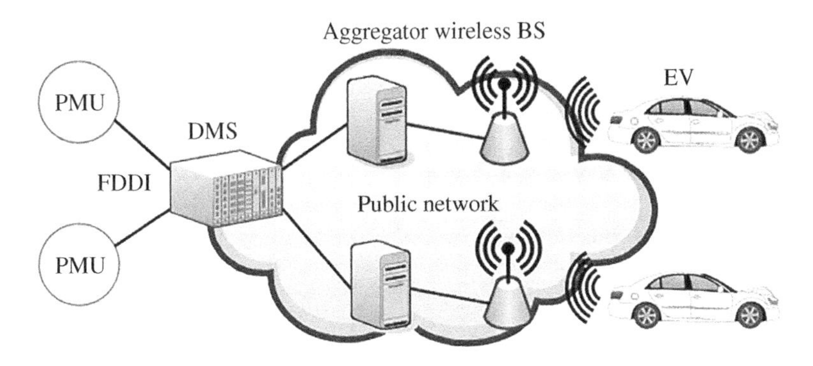

Figure 12.12 Connection of PMUs, DMS, aggregators, and EVs. *Source:* From Shum et al. [B]. © 2020 IEEE.

Table 12.7 RTT delay of each link.

Link	Min	Max	Mean	Std. dev
PMU – DMS	500 μs	500 μs	500 μs	0
DMS – Aggregator	8	12	10	0.7
Aggregator – Base Station	8	12	10	0.7
Base Station – EV (Wi-Fi)	5	25	10	7
Base Station - EV (WiMAX)	45	465	140	15

Unless otherwise specified, all values are in milliseconds.
Source: From Shum et al. [B]. © 2013 IEEE.

12.7.3 Simulation Events

Initially, EVs are in charging mode. Then, a scheduled fault occurs at bus 5 at $t = 10$ s, triggering the circuit breakers to isolate bus 5 from the rest of the grid. Next, as the power flow is changed, voltages on bus 9–14 begin to drop. Once the measured bus voltages drop below a threshold of 0.96 pu, DMS will command the aggregators to start voltage support service. After that, EVs switch to discharging mode to maintain an acceptable voltage level of 0.94–1.06 pu. Later, an automatic reclosing is scheduled at $t = 20$ s, causing the voltage to raise above the threshold of 1.06 pu. Finally, DMS commands the aggregators to switch EVs back to normal operation.

12.7.4 Co-simulation Results

The effects of V2G voltage support can be verified by comparing the Vrms waveforms in Figures 12.13 and 12.14. Figure 12.13 shows the per unit Vrms fluctuations on bus 9–14 with V2G support over an ideal communication network. The voltages experienced a sudden dip to about 0.85 pu immediately after the fault and quickly recovered to over 0.95 pu as the result of fault isolation and V2G support. The EV fleet continued to support the grid until automatic reclosing was executed at $t = 20$ s, by then the system returned to normal operation. In contrast, Figure 12.14 shows the waveforms when V2G is not available. All bus voltages dropped below 0.9 pu until

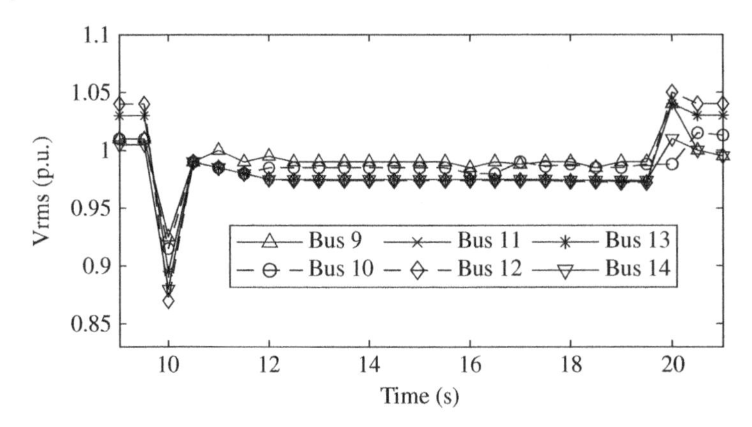

Figure 12.13 Voltage waveform of bus 9–14 (Ideal). *Source:* From Shum et al. [B]. © 2020 IEEE.

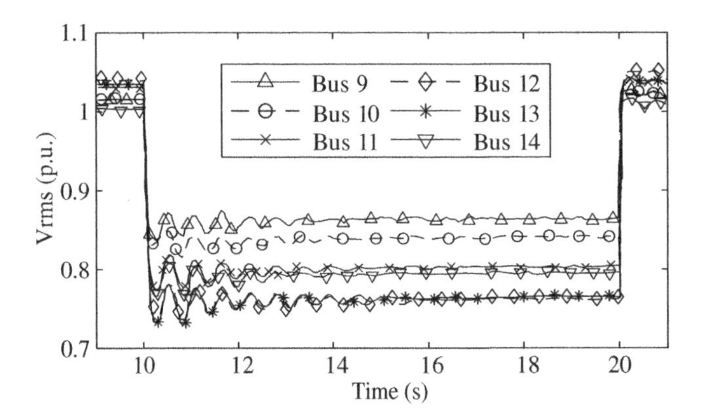

Figure 12.14 Voltage waveform of bus 9–14 (No V2G). *Source:* From Shum et al. [B]. © 2020 IEEE.

the grid was healed. Since bus 9 was the only bridge for power to transfer from generators to the distribution network, buses located further away from bus 9 experienced larger sags.

To see the impacts on power quality with respect to Wi-Fi and WiMAX network, we picked bus 12 for the comparisons of the voltage and frequency stability. Figure 12.15 shows the voltage waveform shortly after the fault. With an ideal network, the bus voltage recovered back to an acceptable level (0.94 pu) within 70 ms (4.2 cycles) and the lowest transient voltage was around 0.85 pu. With WiMAX network, recovering to 0.95 pu required 165 ms (9.9 cycles), and a deeper sag to 0.8 pu was recorded. Using the Wi-Fi network gave voltage waveforms similar to the ideal case. Moreover, Figure 12.16 shows the voltage after automatic reclosing; Figures 12.17 and 12.18 show the frequency fluctuation. In addition, Table 12.8 provides power quality measurements with respect to the duration of disruptions perceived by bus 9–14.

It can be drawn that lower latency links (e.g. Wi-Fi) can result in better power qualities. Nevertheless, there were other design considerations such as signal coverage and in-band interference not being considered. Therefore, in practice, it is better to make judgments about the tradeoffs on a case-by-case basis, and this is the time when co-simulation technology comes in handy for the evaluation of specific scenarios.

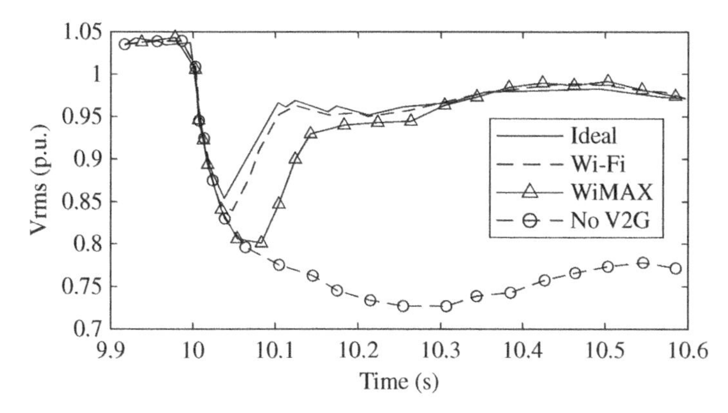

Figure 12.15 Voltage waveform of bus 12 (after fault). *Source:* From Shum et al. [B]. © 2020 IEEE.

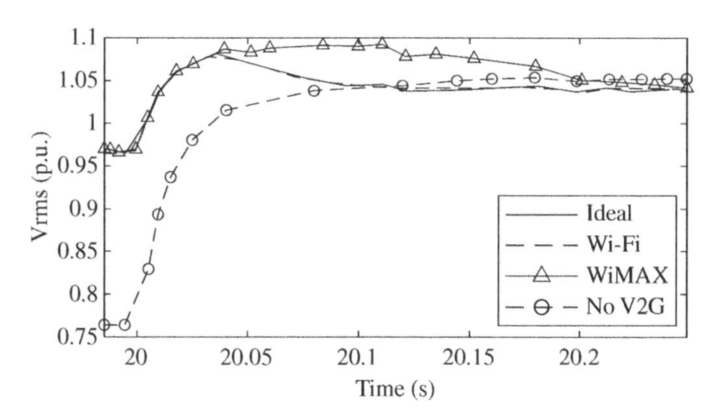

Figure 12.16 Voltage waveform of bus 12 (after reclosing). *Source:* From Shum et al. [B]. © 2020 IEEE.

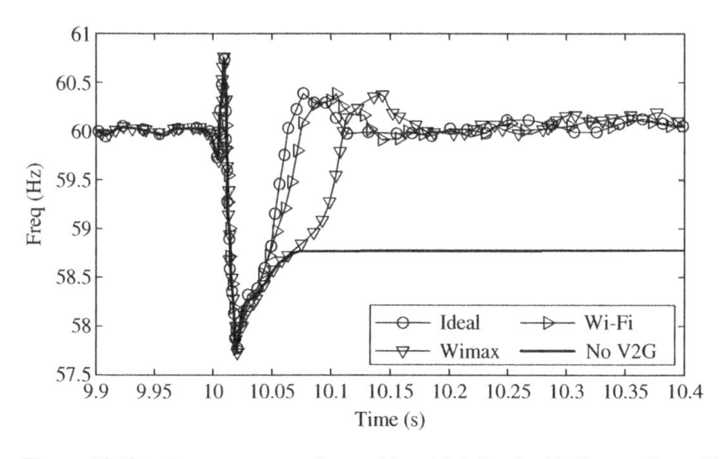

Figure 12.17 Frequency waveform of bus 12 (after fault). *Source:* From Shum et al. [B]. © 2020 IEEE.

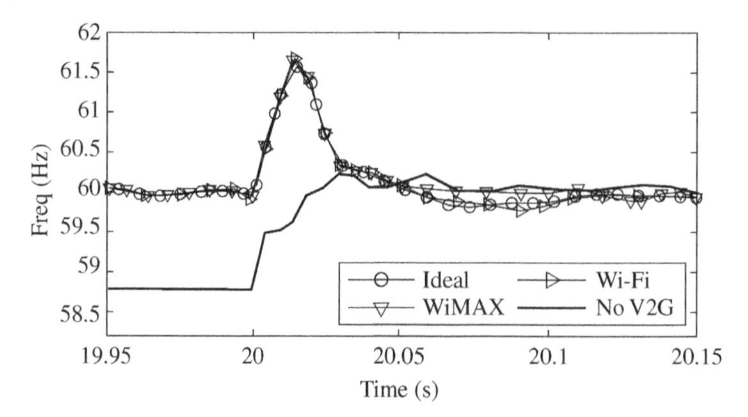

Figure 12.18 Frequency waveform of bus 12 (after reclosing). *Source:* From Shum et al. [B]. © 2020 IEEE.

Table 12.8 Duration of disruption.

		Duration of disruption on bus 9–14 (cycles)					
Disruptions	**Network**	**9**	**10**	**11**	**12**	**13**	**14**
Vrms < 0.94 pu	Ideal	2.82	3.48	3.84	4.2	4.14	4.02
	Wi-Fi	3.48	4.14	4.62	4.98	4.98	4.86
	WiMAX	6.54	7.32	7.92	9.9	8.58	8.1
Vrms > 1.06 pu	Ideal	0	0	0.99	2.73	2.34	0
	Wi-Fi	0	0	1.86	3.6	3.15	0
	WiMAX	0.06	0.36	6.87	9.93	8.76	2.16
Freq < 0.97 pu	Ideal	0	0.06	1.32	0.6	0.6	1.38
	Wi-Fi	0	0.06	1.98	0.66	0.6	1.92
	WiMAX	0	0.06	2.52	0.84	0.6	2.04
Freq > 1.03 pu	Ideal	0	0	0	0.06	0	0
	Wi-Fi	0	0	0	0.06	0	0
	WiMAX	0	0	0	0.06	0	0

Source: From Shum et al. [B]. © 2013 IEEE.

12.8 Conclusions

This chapter presents a novel integration of direct-execution simulators to a Smart Grid co-simulation platform that adheres to the HLA standard. The DecompositionJ framework automatically performs static analysis and source-to-source transformation to convert a target program into its own simulator. This removes the burden of manual modeling and code development, and simplifies version control and maintenance. The transformed simulator is compatible with existing Java environments and tools, which facilitates researchers to debug and study a target program during simulation. A complex multi-agent platform JADE is simulated and used in an FLISR case study. Results illustrate that low-level details in software and network configuration can affect the time required for restoration, which would otherwise be overlooked. This convenient and effective direct-execution simulation framework with high-fidelity modeling capability is believed to be important for enabling simulation studies on a wide range of smart grid applications. A case study on V2G voltage support was reported, demonstrating the capability of our co-simulator in evaluating the performance and adequacy of communication infrastructures with respect to a smart grid application. The co-simulation results show that by using low latency Wi-Fi/IEEE 802.11g in V2I communication instead of WiMAX/IEEE 802.16e, better power quality performance can be obtained.

Acknowledgements

The permission given in using materials from the following papers is very much appreciated.

[A] Shum, C., Lau, W.H., Mao, T. et al. (2018). Co-simulation of distributed smart grid software using direct-execution simulation. *IEEE Access* 6: 20531–20544.

[B] Shum, C., Lau, W.H., Lam, K.A. et al. (2013). The development of a smart grid co-simulation platform and case study on Vehicle-to-Grid voltage support application. *IEEE SmartGird Comm 2013 Symposium – Smart Grid Standards, Co-simulation, Test-beds and Field Trails*, USA (21–24 October 2013).

References

1 Mets, K., Ojea, J.A., and Develder, C. (2014). Combining power and communication network simulation for cost-effective smart grid analysis. *IEEE Commun. Surveys Tuts.* 16 (3): 1771–1796, 3rd Quarter.

2 Li, W., Ferdowsi, M., Stevic, M. et al. (2014). Cosimulation for smart grid communications. *IEEE Transactions on Industrial Informatics* 10 (4): 2374–2384.

3 Hopkinson, K., Wang, X., Giovanini, R. et al. (2006). EPOCHS: a platform for agent-based electric power and communication simulation built from commercial off-the-shelf components. *IEEE Transactions on Power Apparatus and Systems* 21 (2): 548–558.

4 Nutaro, J., Kuruganti, P.T., Miller, L. et al. (2007). Integrated hybrid-simulation of electric power and communications systems. *Proceedings of the IEEE Power Engineering Society General Meeting*, Tampa, FL, USA (24–28 June 2007), 1–8.

5 Nutaro, J. (2011). Designing power system simulators for the smart grid: Combining controls, communications, and electro-mechanical dynamics. *Proceedings of the IEEE Power Engineering Society General Meeting*, San Diego, CA, USA (24–28 July 2011), 1–5.

6 Liberatore, V. and Al-Hammouri, A. (n.d.). Smart grid communication and co- simulation. *Proceedings of the IEEE Energytech*, Cleveland, OH, USA (25–26 May 2011), 1–5.

7 Mets, K., Verschueren, T., Develder, C. et al. (2011). Integrated simulation of power and communication networks for smart grid applications. *Proceedings of the IEEE 16th International Workshop Computer-Aided Modeling Design Communication Links and Networks (CAMAD)*, Cleveland, Kyoto (10–11 June 2011), 61–65.

8 Li, W., Li, H., and Monti, A. (2011). Using co-simulation method to analyze the communication delay impact in agent-based wide area power system stabilizing control. *Proceedings of the Grand Challenges Modeling Simulation Conference*, Vista, CA, USA (June 2011), 356–361.

9 Li, W., Monti, A., Luo, M., and Dougal, R.A. (2011). VPNET: a co-simulation framework for analyzing communication channel effects on power systems. *Proceedings of the. IEEE Electrics Ship Technologies Symp.osium*, Alexandria, VA, USA (10–13 April 2011), 143–149.

10 Lin, H., Veda, S.S., Shukla, S.S. et al. (2012). GECO: global event-driven co-simulation framework for interconnected power system and communication network. *IEEE Transactions on Smart Grid* 3 (3): 1444–1456.

11 Lévesque, M., Xu, D.Q., Joós, G., and Maier, M. (2012). Communications and power distribution network co-simulation for multidisciplinary smart grid experimentations. *Proceedings of the 45th Annual Simulation Symposium*, San Diego, CA, USA (March 2012), 2:1–2:7.

12 Georg, H., Müller, S.C., Rehtanz, C., and Wietfeld, C. (2014). Analyzing cyber-physical energy systems: the INSPIRE cosimulation of power and ICT systems using HLA. *IEEE Transactions on Industrial Informatics* 10 (4): 2364–2373.

13 Celli, G., Pegoraro, P.A., Pilo, F. et al. (2014). DMS cyber- physical simulation for assessing the impact of state estimation and communication media in smart grid operation. *IEEE Transactions on Power Apparatus and Systems* 29 (5): 2436–2446.

14 Perkonigg, F., Brujic, D., and Ristic, M. (2015). Platform for development and validation agent-based smart grid applications incorporating accurate communications modelling. *IEEE Transactions on Industrial Informatics* 11 (3): 728–736.

15 Li, X., Huang, Q., and Wu, D. (2017). Distributed large-scale co-simulation for IoT-aided smart grid control. *IEEE Access* 5: 19951–19960.

16 Garau, M., Celli, G., Ghiani, E. et al. (2017). Evaluation of smart grid communication technologies with a co-simulation platform. *IEEE Wireless Communications* 24 (2): 42–49.

17 Shum, C., Lau, W.H., Mao, T. et al. (2018). DecompositionJ: parallel and deterministic simulation of concurrent Java executions in cyber-physical systems. *IEEE Access* 6: 21991–22010.

18 Bellifemine, F., Bergenti, F., Caire, G., and Poggi, A. (2005). JADE – A Java agent development framework. In: *Multi-Agent Programming* (ed. R.H. Bordini, M. Dastani, J. Dix and A. El Fallah Seghrouchni), 125–147. New York, USA: Springer.

19 (2010). *IEEE Standard for Modeling and Simulation (M&S) High Level Architecture (HLA)—Framework and Rules*. IEEE Standard 1516–2010 (Revision of IEEE Standard 1516–2000). (18 August 2010), 1–38.

20 Manson, J., Pugh, W., and Adve, S.V. (2005). The Java memory model. *Proceedings of the 32nd ACM SIGPLAN-SIGACT Symp. Principles Programming Language*, Long Beach, CA, USA (January 2005), 378–391.

21 Söderberg, E., Ekman, T., Hedin, G., and Magnusson, E. (2013). Extensible intraprocedual flow analysis at the abstract syntax tree level. *Science of Computer Programming* 78 (10): 1809–1827.

22 FIPA (2014). *Foundation for Intelligent Physical Agents. A Standards Organization of the IEEE Computer Society*. [Online]. http://www.fipa.org/ (accessed 7 February 2016).

23 Nguyen, C.P. and Flueck, A.J. (2012). Agent based restoration with distributed energy storage support in smart grids. *IEEE Transactions on Smart Grid* 3 (2): 1029–1038.

24 Eriksson, M., Armendariz, M., Vasilenko, O.O. et al. (2015). Multiagent-based distribution automation solution for self- healing grids. *IEEE Transactions on Industrial Electronics* 62 (4): 2620–2628.

25 Solanki, J.M., Khushalani, S., and Schulz, N.N. (2007). A multi-agent solution to distribution systems restoration. *IEEE Transactions on Power Apparatus and Systems* 22 (3): 1026–1034.

26 Ren, F., Zhang, M., Soetanto, D., and Su, X. (2012). Conceptual design of a multi- agent system for interconnected power systems restoration. *IEEE Transactions on Power Apparatus and Systems* 27 (2): 732–740.

27 Ye, D., Zhang, M., and Sutanto, D. (2011). A hybrid multiagent framework with *Q*-learning for power grid systems restoration. *IEEE Transactions on Power Apparatus and Systems* 26 (4): 2434–2441.

28 Sharma, A., Srinivasan, D., and Trivedi, A. (2015). A decentralized multiagent system approach for service restoration using DG islanding. *IEEE Transactions on Smart Grid* 6 (6): 2784–2793.

29 Zhong, F., Kulkarni, P., Gormus, S. et al. (2013). Smart grid communications: overview of research challenges, solutions, and standardization activities. *Communications Surveys & Tutorials, IEEE* 15 (1): 21–38.

30 TESLA MOTORS (n.d.). TESLA MODELS FACTSl TESLA MOTORS. (TESLA MOTORS WEBSITE), [online]. http://www.teslamotors.-com/models/facts (accessed 1 May 2013).

31 Immerman, R.D., Murillo-Sanchez, C.E., and Thomas, R.J. (2011). MATPOWER: steady-state operations, planning, and analysis tools for power systems research and education. *IEEE Transactions on Power Systems* 26 (1): 12–19.

32 Kim, D., Cai, H., Na, M., and Choi, S. (2008). Performance measurement over Mobile WiMAX/IEEE 802.16e network. *2008 International Symposium on a World of Wireless, Mobile and Multimedia Networks* (23–26 June 2008).

33 Cosma, R., Cabellos-Aparicio, A., Domenech-Benlloch, J. et al. (2008). Measurement-based analysis of the performance of several wireless technologies. *2008 16th IEEE Workshop on Local and Metropolitan Area Networks*, (3–6 September 2008).

34 Chou, C.-M., Li, C.-Y., Chien, W.-M., and Lan, K.-C. (2009). A feasibility study on vehicle-to-infrastructure communication: WiFi vs. WiMAX. *2009 Tenth International Conference on Mobile Data Management: Systems, Services and Middleware* (18–20 May 2009).

13

Advanced Metering Infrastructure for Electric Vehicle Charging

13.1 Introduction

Rising environmental concern for pollution and greenhouse gas emission, favorable government policies for using electric vehicles (EVs), and significant investment by EV manufacturers are some of the major factors driving the global EV market. Some of the manufacturers are also promoting workplace and residential charging stations to minimize the charging constraints. For instance, in December 2017, Electrify America LLC announced to install more than 2800 residential and workplace charging stations by June 2019 in 17 different metropolitan cities of the United States [1]. However, lack of global standard for the charging infrastructure is one of the major reasons that slow down the market growth. But technological advancement in EV charging stations powered by renewable energy provides new opportunities in the market growth.

The COVID-19 pandemic will affect global EV markets. Based on car sales data during January to April 2020, it was estimated that the passenger car market will contract by 15% over the year relative to 2019. Overall, electric car sales would account for about 3% of global car sales in 2020 due to supporting policies, particularly in China and Europe. China has extended its subsidy scheme until 2022. China and Europe have also strengthened and extended their new energy vehicle mandate and CO_2 emissions standards, respectively [2].

Also according to reference [2], only about 17 000 electric cars were on the world's roads in 2010. By 2019, that number had increased to 7.2 million, 47% of which were in The People's Republic of China. Nine countries had more than 100 000 electric cars on the road. At least 20 countries reached market shares above 1%. The global EV market volume was 2265.5 thousand units in 2019 and is expected to register a compound annual growth rate (CAGR) of 40.7% from 2020 to 2027 [3].

EV is a key technology to reduce air pollution in densely populated areas and a promising option to contribute to energy diversification and carbon emission reduction. To date, 17 countries have announced 100% zero-emission vehicle targets or the phase-out of internal combustion engine vehicles through 2050 [2]. France, in December 2019, was the first country to put this intention into law, with a 2040 timeframe.

In 2019, there were about 7.3 million chargers worldwide, of which about 6.5 million were private, light-duty vehicle, slow chargers at home, multi-dwelling buildings, and workplaces. Convenience, cost-effectiveness and a variety of support policies at all governance levels, such as preferential rates, equipment purchase incentives, and rebates, are the main drivers for the growth of private charging [2].

Smart Energy for Transportation and Health in a Smart City, First Edition. Chun Sing Lai, Loi Lei Lai and Qi Hong Lai.

Policy actions for EV s depend on the status of the EV market or technology. Setting vehicle and charger standards are prerequisites for wide EV adoption. In the early stages of deployment, public procurement schemes, for buses and municipal vehicles, have the double benefit of demonstrating the technology to the public and providing the opportunity for public authorities. Importantly, they also allow the industry to produce and deliver bulk orders to foster economies of scale.

The vast majority of car markets offer some form of subsidy or tax reduction for the purchase of an individual or company electric car as well as support schemes for deploying charging infrastructure to maximize power system security, quality and reliability. Provisions in building codes to encourage charging facilities are becoming more common. So are mandates to build charging infrastructure along road corridors.

It is expected that electricity demand on slow chargers represent the majority of EV electricity demand, mainly due to a continuing dominance of private charging. Fast-charging infrastructure is gradually deployed to respond to the growth in relative shares of EV s with higher battery capacity and power requirements, e.g. buses and trucks. With the projected size of the global EV market, expansion of battery manufacturing capacity will largely be driven by electrification in the car market. The electrification of cars is a crucial driver in cutting unit costs of automotive battery packs.

A proper integration of EV s with power systems is essential to benefit both. Balancing electricity demand and supply will become an increasing challenge to ensure the smooth integration of variable renewables-based energy generation and the electrification of multiple end-use sectors. Over the coming decade, managing EV charging patterns will be key to encourage charging at periods of low electricity demand or high renewables-based electricity generation. With 250 million EV s on the road by 2030 in the Sustainable Development Scenario [2], the share of EV charging in the average evening peak demand could rise to as high as 4–10% in the main EV markets in China, European Union, and the United States, assuming unmanaged charging. A range of ready options with various degrees of complexity can be introduced to reduce EV charging at peak system demand, thereby reducing the need for upgrades to generation, transmission, and distribution assets to maximize the use of resources. While off-peak charging at night with simple demand response and/or night time tariffs would more than halve the contribution of EVs to peak demand, controlled charging in response to real-time price signals from utilities could further expand variable renewable electricity generation and enhance the range of services that EVs could offer to the grid.

According to [2], not only are there means to relieve the negative impact of EV charging on power systems, but the 16 000 GWh of energy that can be stored in EV batteries globally in 2030 could actively supply energy to the grid at suitable times via vehicle-to-grid (V2G) solutions. The V2G possibility depends on availability of vehicles to participate in such services at suitable times, consumer acceptance, and the ability for participants to generate revenues, as well as other technical constraints related to battery discharge rates or impacts on battery lifetime. It is estimated that 5% of the total EV battery capacity could be made available for V2G applications during peak times. This could provide about 600 GW of flexible capacity globally by 2030 across China, the United States, Europe, and India, contributing to offset reduced renewable electricity generation during peaks as well as the increase of capacity needs to meet peak demand.

According to the objectives of the 'Guidance on Accelerating the Construction of Electric Vehicle Charging Infrastructure' by the State Council of China, this guideline seeks to set out a framework through which to systematically improve the charging infrastructure of EVs and to promote the healthy and rapid development of the EV industry [4].

The guideline sets targets in 2020 to add more than 12 000 new centralized charging and battery replacement stations, 4.8 million decentralized charging stations and develop public transport and rental services. The main areas are promoting the construction of charging infrastructure;

strengthening the capacity of the power grid to support increased demand from EV charging facilities; accelerating the improvement of standardized specifications and technological innovation; exploring sustainable business models and developing relevant pilot projects.

Simple solutions should be implemented, such as the promotion of workplace charging or the use of off-peak tariffs. However, releasing the full flexibility potential of EVs through dynamic controlled charging and V2G services would require the adaptation of regulatory and market frameworks. Currently, flexible EV integration is not on track for power systems to accommodate the distributed loads that EV batteries represent in a co-ordinated way and on a large scale. Specific stakeholders such as aggregators, along with business models that make use of new regulatory frameworks to reward EV owners for providing flexibility services are also needed for EV batteries to contribute to the power system stability on a significant scale [5].

The carbon release due to EV charging is arousing critical environmental concerns. Apart from environmental and economical reasons, the most significant driving force of EV market is strong policy, including legislations, benefits, and rebates from various governments as mentioned before. Regardless of the initiative, the rapid growth of EV definitely increases the energy demand. It was reported that the extra energy demand of EV may further increase the peak demand and cause the distribution circuit congestion [6–10]. More importantly, it was pointed out that the EV hype may reduce CO_2 emission if appropriate demand side management (DSM) strategy is adopted. In [6], it was highlighted that the U.S. electric infrastructure is designed to meet the highest expected demand for power, which only occurs a few 100 hours a year (at most 5% of the time). For 95% of the time, the electric grid is seriously underutilized and therefore huge amount of greenhouse gas is generated unnecessarily. Unfortunately, the U.S. electric grid is not a special case. Most of the developed areas have adopted a similar approach to stabilize the electricity supplies. To fully utilize the electric infrastructure, researchers have proposed numerous DSM strategies [11, 12] for EV charging. It is also well documented that by adopting centralized control of EV charging, peak demand will be more accurately estimated and hence will significantly benefit the utility [13–15]. The centralized control EV charging may be achieved by enabling dynamic demand response, load profile flattening and improved generation resource utilization. On the other hand, a prepaid smart metering scheme for smart-grid application based on centralized authentication and charging using the WiMAX prepaid accounting model was developed [16]. However, this smart metering scheme was designed for payment only and other important services such as demand monitoring, customer service, and meter management service were ignored.

The significant development of energy networks toward a more intelligent direction leads smart meters as one of the most fundamental devices in the intelligent energy networks. Smart energy meters can exchange the information on energy consumption and the status of energy networks between utilities and users. Smart energy meters can also be used to monitor and control home appliances and other devices according to the individual request [17].

Intelligent metering systems are being rolled-out on a large-scale worldwide; however, the lack of established international standards represents a serious issue to be overcome for a complete development of an efficient, affordable, and profitable market. The identification of suitable communication protocols and cost-effective network architectures represent a challenging aspect. Barbiroli et al. [18] reported different network design solutions for wireless smart metering systems for practical applications.

Although, the role and proposed features of EV Advanced Metering Infrastructure (AMI) have been discussed in [19], the practical design of the entire system is missing. It is thus notified that a full feature AMI for EV charging service is desperately required. Similar to typical AMI system, the EVAMI measures, collects, and analyses energy consumption from EV charging station regularly via communication links. "Infrastructure of AMI" refers to a network that connects the measurement devices to utility servers allowing data collection and distribution of information to customers, suppliers, utility companies, and service providers. Interactive communication carries distinctive meaning of AMI because the influence of real-time energy consumption on users' charging behavior was documented. With the timely information, utility easily monitors energy demand on time and location basis. In this chapter, an AMI platform for EV charging (EVAMI), enabling data exchange between automobile owners, utility information systems (e.g. MDMS, billing), and the EV charging stations, are introduced. Owing to the requirement for high availability, the proposed solution adopts Power Line Communication (PLC) for network of onsite charging system (OCS) which transmits data via the power line. Therefore, no extra effort is necessary for rewiring and yet high scalability is achieved. As a result, EV charging may take place in a wide range of locations such as multistory car parks. This chapter is organized as follows: the demand response EVAMI system is discussed in Section 13.2. The system architecture, protocol design, and implementation are discussed in Section 13.3. The performance evaluation is discussed in Section 13.4. Finally, a conclusion is drawn in Section 13.5.

13.2 EVAMI Overview

In this section, the system overview of EVAMI is presented. EVAMI is a total solution which serves utility employee and EV owners. It should be noted that EV owners possess varying characters ranging from normal energy consumers to huge energy consumption users and yet it is anticipated that EV owners may charge their vehicles at different locations. Thus PLC seems one of the potential candidates. The versatility of the PLC system is that the OCS can be connected to the utility management system by collaborating with ordinary internet access technologies such as WiMAX, LTE, Optical fiber, etc. to provide TCP/IP communications.

13.2.1 Advantage of Adopting EVAMI

In general, EV charging consumers will receive electricity bills from multiple utilities. In such a circumstance, it is difficult to obtain the personal consumption profile of individuals from a single utility. One of the salient features of the developed EVAMI is that a third party customer information platform (a neutral platform) is implemented to serve various utilities. The key contribution in this chapter is that the proposed EVAMI facilitates an owner to obtain a single charging bill that includes the overall charging profile. The EVAMI (Figure 13.1) is divided into three parts: OCS, Third Party Customer Service platform (TPCSP), and Utilities Information Management System.

13.2.2 Choice of Signal Transmission Platform

Since signals are typically transmitted over a long distance, an inexpensive, and yet reliable transmission medium must be employed. Wireless transmission may be a good choice but the skills required may be complex. On the other hand, wireline transmission requires preinstallation which may be tedious and sometimes impossible because of the need for renovation. Alternatively, PLC may be used since it facilitates ease of deployment and flexibility of modification. The cost and weight of the power line are also comparatively lower [20]. For short to medium distance where signals do not have to change phase or to go through transformers, PLC is a potential technology

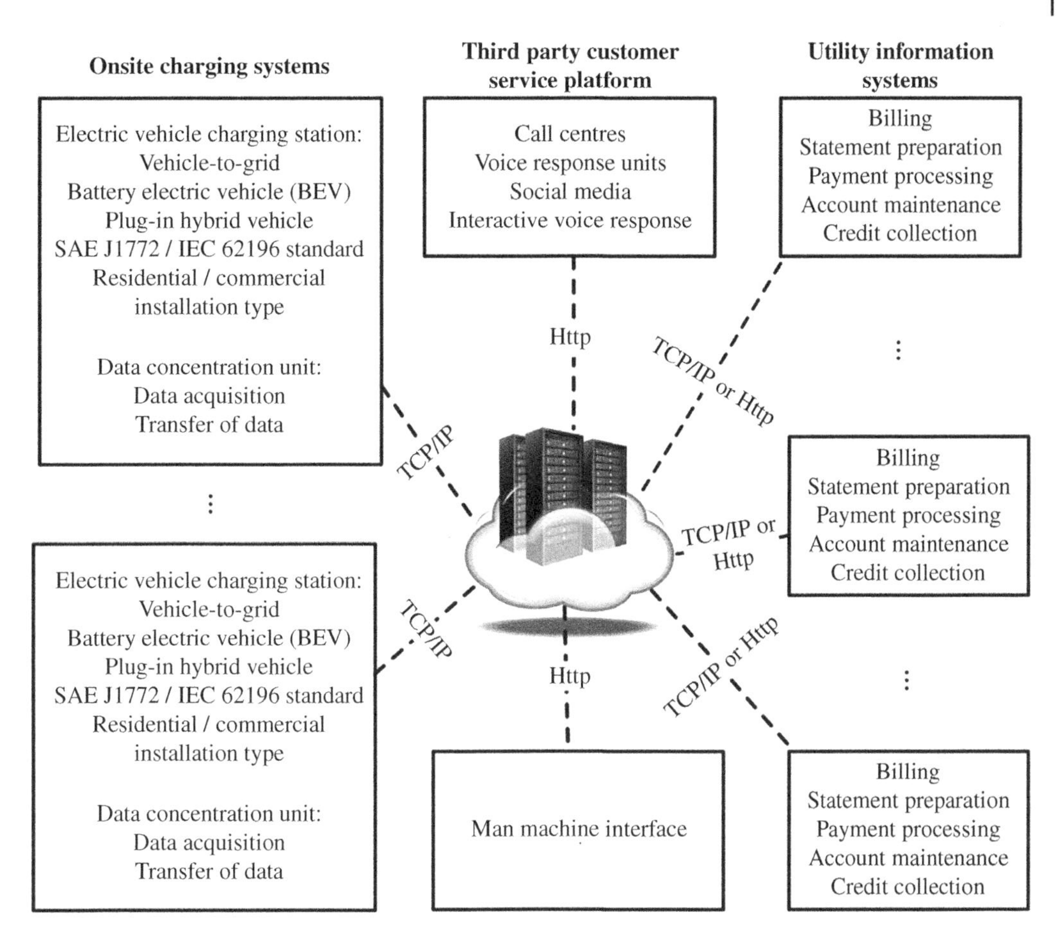

Figure 13.1 EVAMI overview. *Source:* From Lam et al. [A]/Scientific Research Publishing/CC BY 4.0.

candidate for local area network (LAN). Typically, a PLC LAN should be no more than 10 km. PLC, specified by the HomePlug 1.0/1.1 standard [21], provides a 14/85 Mb/s raw data rate. This is adequate for most electronic device communications. The PLC systems operate by imposing a modulated carrier signal into the power line. Depending on the signal transmission characteristics of the power line used, PLC may operate in different frequency bands. In this work, PLC is adopted in order to furnish the captioned features into the EVAMI. The specification of the PLC module is listed in Table 13.1.

Table 13.1 Specification of PLC module.

Used band	2–32 MHz
Modulation	OFDM
Data rate	Up to 200 Mbps
Encryption	Triple DES

Source: From Lam et al. [A]/Scientific Research Publishing/CC BY 4.0.

13.2.3 Onsite Charging System

OCS consists of an EV charging station (EVCS) and a Data Concentration Unit (DCU). The function of CS is to serve EV owners in car parks. Generally, it handles the initialization and termination of the charging session. During the charging period, the OCS monitors the charging status and updates the meter reading. It also monitors the health status of the CS. The DCU is a PLC internet gateway which supports TCP/IP communication and connects the OCS to the Utility Information Management System (UIMS). In this work, PLC is adopted in order to furnish the captioned features into both the EVCS and DCU.

13.2.4 EV Charging Station

EVCS (refer to Figure 13.1) is composed of an electric power plug and a smart meter. Basically, the EVCS informs the DCU whether an EV has been plugged in or not. The charging process is designed not to start immediately after plug-in. To avoid unauthorized use of service, the EVCS starts the charging session only after receiving the activation command from the DCU. It then authorizes the charging and monitors the entire charging session and provides detailed report of the charging session. Furthermore, the EVCS equips self-healing ability and reports to the DCU when fault or error is detected. The operation flow of the EVCS is summarized in Figure 13.2.

13.2.5 Utility Information Management System

Utility Information Management System (UIMS) bridges the OCS, the TPCSP, and other smart grids applications in order to provide billing services and perform demand response operations. Based on TCP/IP protocol, the UIMS collects the real time charging service data from numerous car parks. As a result, the UIMS not only provides the real time personal consumption profile to customers but also timely location based energy demand to utilities. By incorporating such

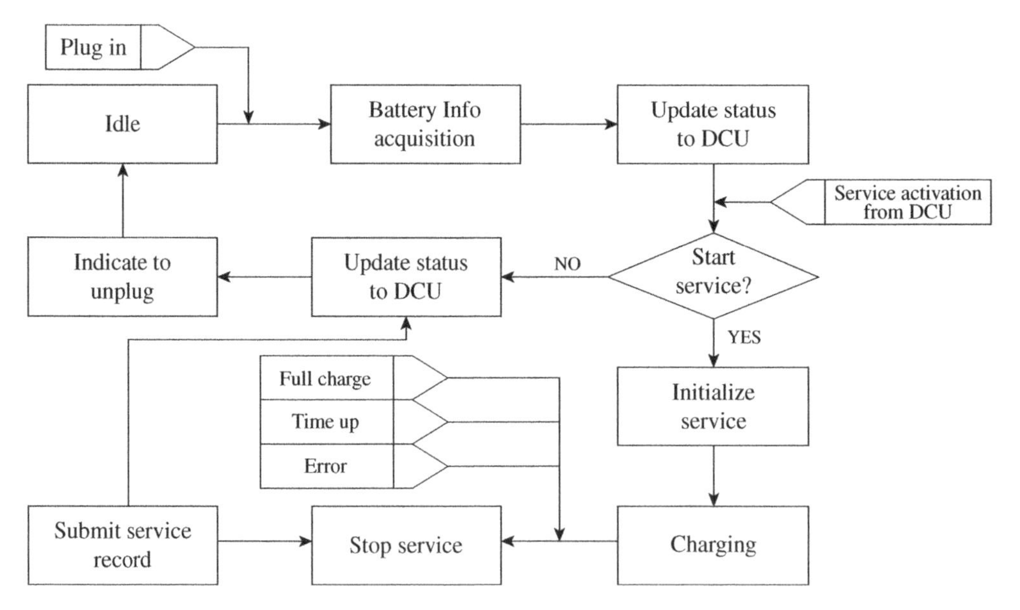

Figure 13.2 Operation flow of the EVCS. *Source:* From Lam et al. [A]/Scientific Research Publishing/CC BY 4.0.

information, the DR policy can be defined with slight effort. The UIMS forwards the defined policy to the OCS and the TPCSP. In addition to demand response (DR), UIMS also provides information to MDMS of utilities for billing. By updating the service record of TPCSP, a single bill can be issued to customers. To facilitate data processing, an administrative web portal is designed for the UIMS. The web portal provides EV energy demand reports to utility employees and manages customer data.

13.2.6 Third Party Customer Service Platform

TPCSP is a neutral platform which interacts with multiple UIMS. Differentiated from UIMS, TPCSP serves the EV owner through a web portal for charging service. TPCSP collects service records from numerous utilities and combines them to form a single bill disclosing the overall personal consumption profile. Therefore, an EV owner may record an overall consumption habit due to EV charging. More importantly, TPCSP enables the communication between EV owners and utilities so that utilities can deliver their message to EV at any time. Therefore, the TPCSP does not only benefit EV owners but also utilities by fostering an efficient DR platform.

13.3 System Architecture, Protocol Design, and Implementation

EVAMI adopts a three-tier architecture including the web tier, the application tier and the enterprise information tier. The web tier handles the presentation and communication function while the application tier supports the business intelligence and the enterprise information tier so that data may be organized systematically. The system architecture of EVAMI is illustrated in Figure 13.3.

By considering the system functions, EVAMI can be divided into five subsystems: Car Park Management Subsystem (CPMS), Meter Management Subsystem (MMS), User Management Subsystem (UMS), Demand Response Service Subsystem, and Billing Service Subsystem (BSS). Each of the subsystems serves clients for different purposes.

Car Park Management Subsystem

The CPMS records, handles, and processes the data of car park. The CPMS is comprised of two major components: Car Park Management Service Module (CPMSM) and Car Park Information Database (CPIDB). CPMSM helps service managers of utilities to manage the information of car park, thus an EV owner may obtain the updated information via the portal.

Meter Management Subsystem

The MMS records, handles, and processes the data of a meter. The MMS is comprised of two major components: Meter Management Service Module (MMSM) and Meter Information Database (MIDB). The MMS collects meter reading and status from OCS to enable service mangers and EV owners to obtain the energy consumption data through web portals. Furthermore, the MMS monitors the status of the meter and generates meter error report automatically should operation failure be detected. The report will be delivered to the service manager via emails and warning messages will also be displayed on the web portal.

User Management Subsystem

The UMS records, handles, and processes the data of users. The UMS is comprised of two major components: the User Management Service Module (UMSM) and the User Information Database (UIDB). The UMSM identifies the role of users and helps service mangers to manage the information of EV owners. The UIDB stores the personal details of EV owner.

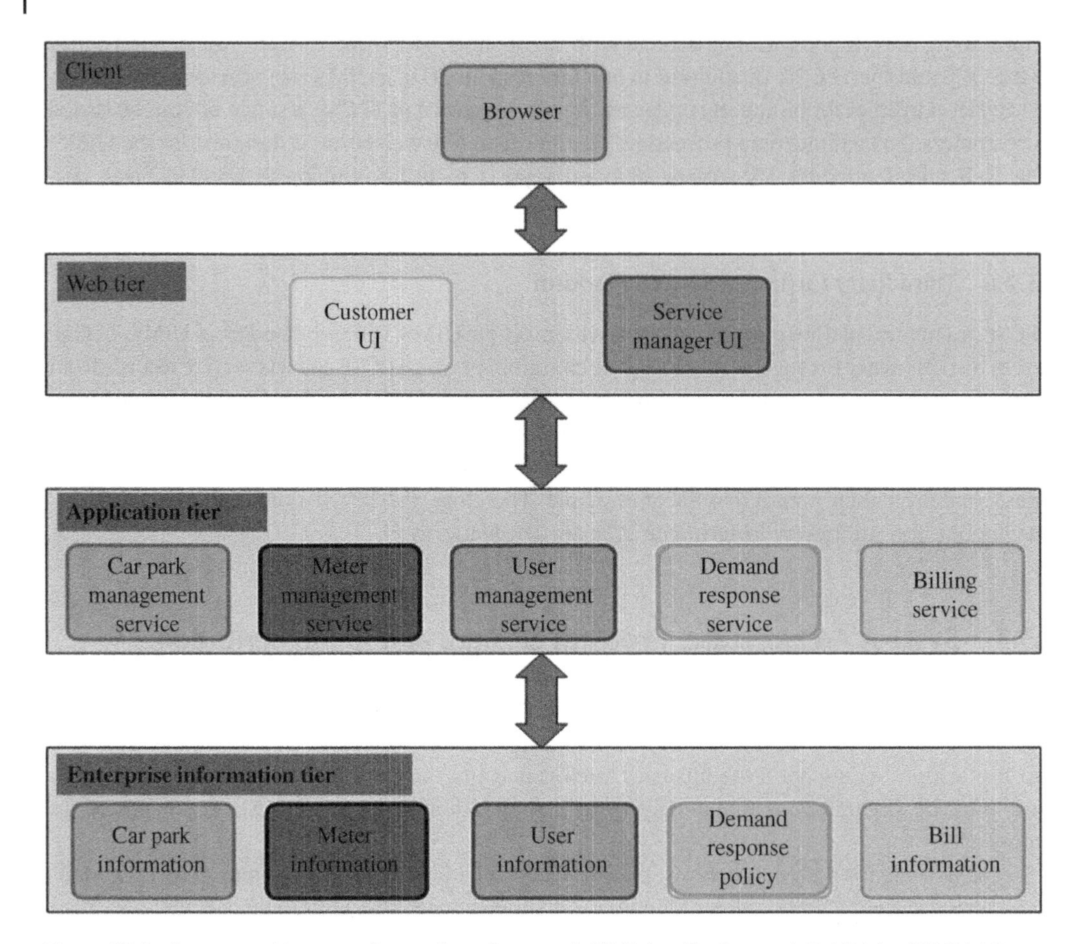

Figure 13.3 System architecture. *Source:* From Lam et al. [A]/Scientific Research Publishing/CC BY 4.0.

Demand Response Subsystem

The Demand Response Subsystem (DRS) records, handles, and processes the data of users. The DRS is comprised of two major components: Demand Response Service Module (DRSM) and Demand Response policy database (DRPDB). Under normal circumstance, the UIMS updates the DR policy regularly in order to fulfill the energy demand. The most up-to-date policy is stored in the DRPDB. The DRSM generates DR signals and transmits them to the DCU and the TPCSP. The TPCSP then informs EV owners by displaying messages in "welcome page" of the web portal or during the charging service subscription.

Billing Service Subsystem

The BSS records, handles, and processes the billing data. The BSS is comprised of two major components, namely, the Billing Service Module (BSM) and the Billing Information Database (BIDB). The BSS cooperates with MMS and UMS to generate the most updated payment record.

13.3.1 Communication Protocol

To support the operation of above subsystems, a series of communication protocol has been defined to cater for the Charging Service Session Management (CSSM), Demand Response, and Device Management.

13.3.1.1 Charging Service Session Management

The CSSM consists of two parts: Charging Service Session Initialization (CSSI) and Charging Service Session Termination.

13.3.1.1.1 Charging Service Session Initialization

The CSSI process involves mobile users, TPCSP, UIMS, and OCS. The detailed procedure is summarized in Figures 13.4 and 13.5. Figure 13.4 describes the interaction between the EVCS and the DCU during charging service initialization. Once the smart EVCS is plugged in, it updates its current status to the DCU with updated commands. The DCU will start the timer and wait for service activation from the utility server. After the timeout period or when the service activation is received, the DCU sends the "decision" service command to the EVCS. The EVCS will start and stop service corresponding to the decision. The EVCS then sends the service command and the updated commands to the DCU. The Service Command consists of the charger information, the consumption, and the time stamp to facilitate the DCU to upload the service record to the utility server. The EVCS then updates its current status to the DCU with updated commands. It should be pointed out that the DCU is only the intermediate node of the communication. The communication between the DCU, TPCS platform, UIMS, and user is illustrated in Figure 13.5.

When an EV starts a charging session, the DCU receives "Plug in" status from the EVCS. The user is required to complete the authentication process with the third party server before "requests for service" is allowed. The third-party server then receives the service request including user information, target charger information, and service content. The third-party server will then set up connection with the utility server. Authentication between two servers is needed and the third party

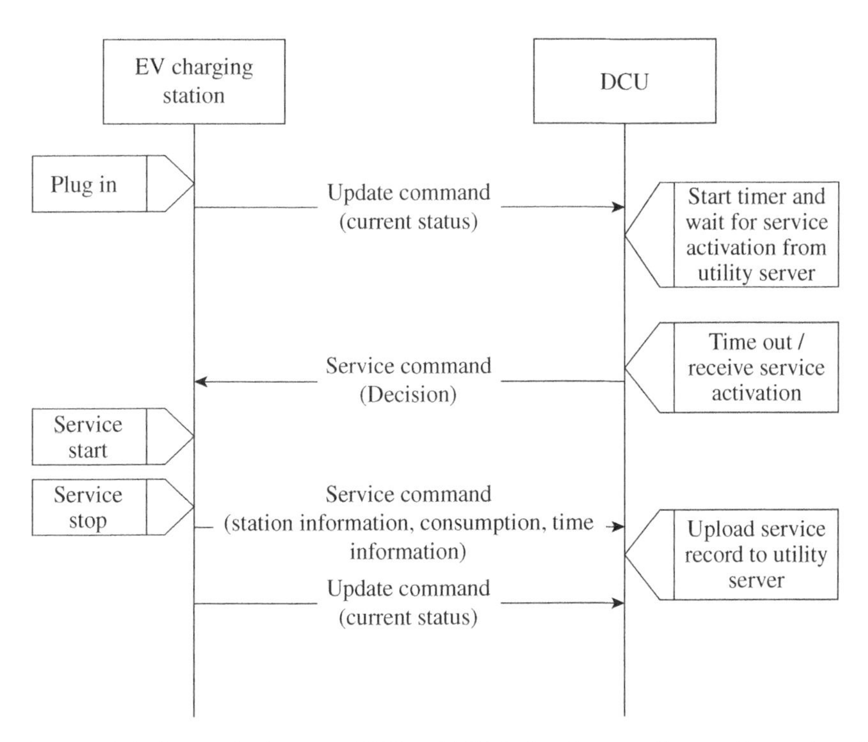

Figure 13.4 Interaction between EVCS and DCU during Charging Process. *Source:* From Lam et al. [A]/Scientific Research Publishing/CC BY 4.0.

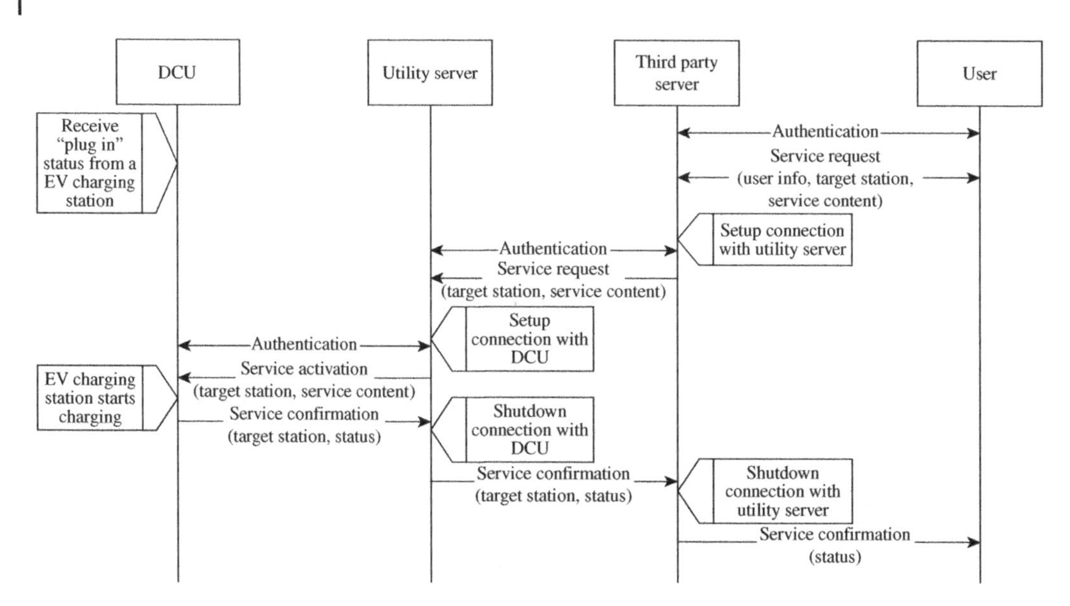

Figure 13.5 Data flow during service initialization process. *Source:* From Lam et al. [A]/Scientific Research Publishing/CC BY 4.0.

sends the service request containing the target charger information and the service content to the utility server. The utility server sets up connection with the DCU and carries out the authentication process. After authentication is completed, service activation including target charger information and service content will be received by DCU. The EVCS then starts charging and the DCU will send a service confirmation with target charger information and status to the utility server. The connection between the utility server and the DCU will be shut down. The utility server then forwards the service confirmation to the third-party server and the third party server will shut down the connection with the utility server. Finally, the user then receives the status from the third party server.

13.3.1.1.2 Charging Service Session Termination

The Charging Service Session Termination (CSST) process involves mobile users, TPCSP, UIMS, and OCS. The detailed procedure is summarized in Figure 13.6.

When the charging process is finished, the DCU receives the service termination from the EVCS. The DCU then sets up connection with the utility server. When the authentication process is completed, the DCU forwards the service termination commands including the target charger information and the service summary to the utility server. The connection between the DCU and the utility server is then shut down and the utility server will set up connection with the third-party server. The two servers will complete the authentication process. The utility server then sends the service termination to the third-party server and the utility server shuts down the connection with the third-party server. After the third-party server has received the service termination, a service status alert with service summary and status will be sent to users.

13.3.1.2 Device Management

13.3.1.2.1 EVCS Failure Report

The EVCS failure report procedure is described in Figure 13.7.

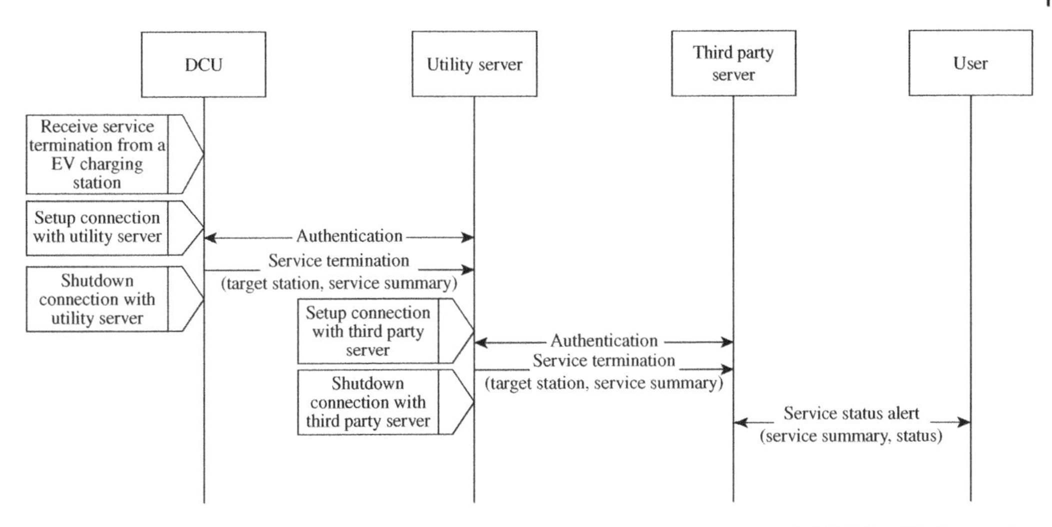

Figure 13.6 Data flow during service termination process. *Source:* From Lam et al. [A]/Scientific Research Publishing/CC BY 4.0.

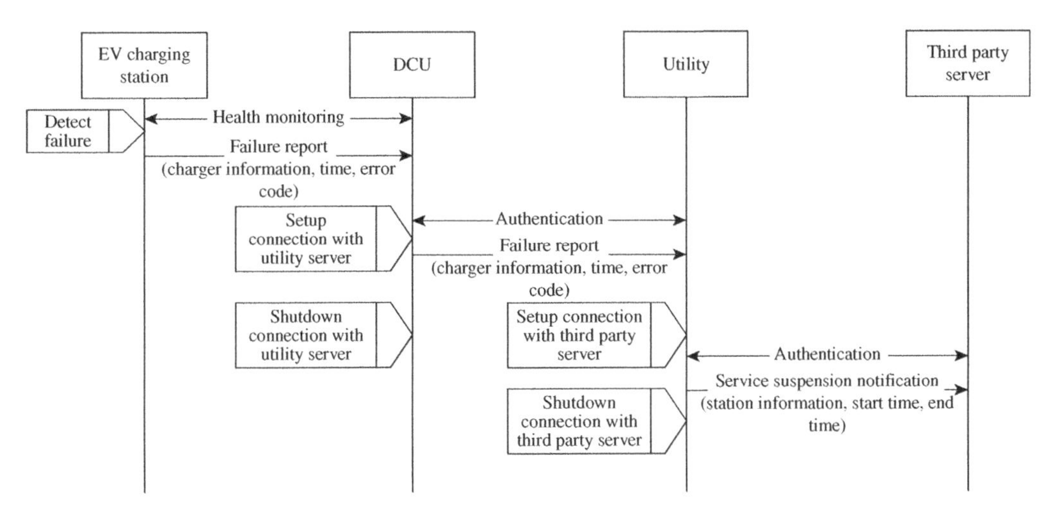

Figure 13.7 EVCS failure report procedure. *Source:* From Lam et al. [A]/Scientific Research Publishing/CC BY 4.0.

The EVCS and the DCU carry out health monitoring periodically. When the EVCS detects a failure, it sends a failure report to the DCU to report the charger information, time, and error code. Then the DCU will set up a connection with the utility server. After the authentication process is completed, the DCU forwards the failure report to the utility server, and the connection between the DCU and the utility server will be shut down. The utility server then sets up the connection with the third-party server and performs authentication. Afterward, the utility sends the "service suspension notification" which includes charger information, start time, and end time to the third party server. The connection between the two servers will then be shutdown.

13.3.1.3 Demand Response Management

The EVAMI performs the demand response in two approaches: dynamic pricing and charging session scheduler. The operations of two demand response approaches are presented in Figures 13.8 and 13.9.

13.3.1.3.1 Energy Price Modification

Dynamic pricing has well been recognized as an effective approach to reshape the load profile [11–15]. By offering attractive energy price, utility increases the energy demand during the off-peak hours. Meanwhile, utility reduces the energy demand by raising the energy price in the rush hours. As a result, the load profile is flattened. To update the time of use (TOU) price, the utility server will set up a connection with the DCU and complete the authentication process. The utility sends the updated setting of time period and price to the DCU, and the connection is shut down afterward. The DCU then forwards the most up-to-date setting to the EVCS to facilitate the DCU to set up a connection with utility server. After the authentication process is completed, the

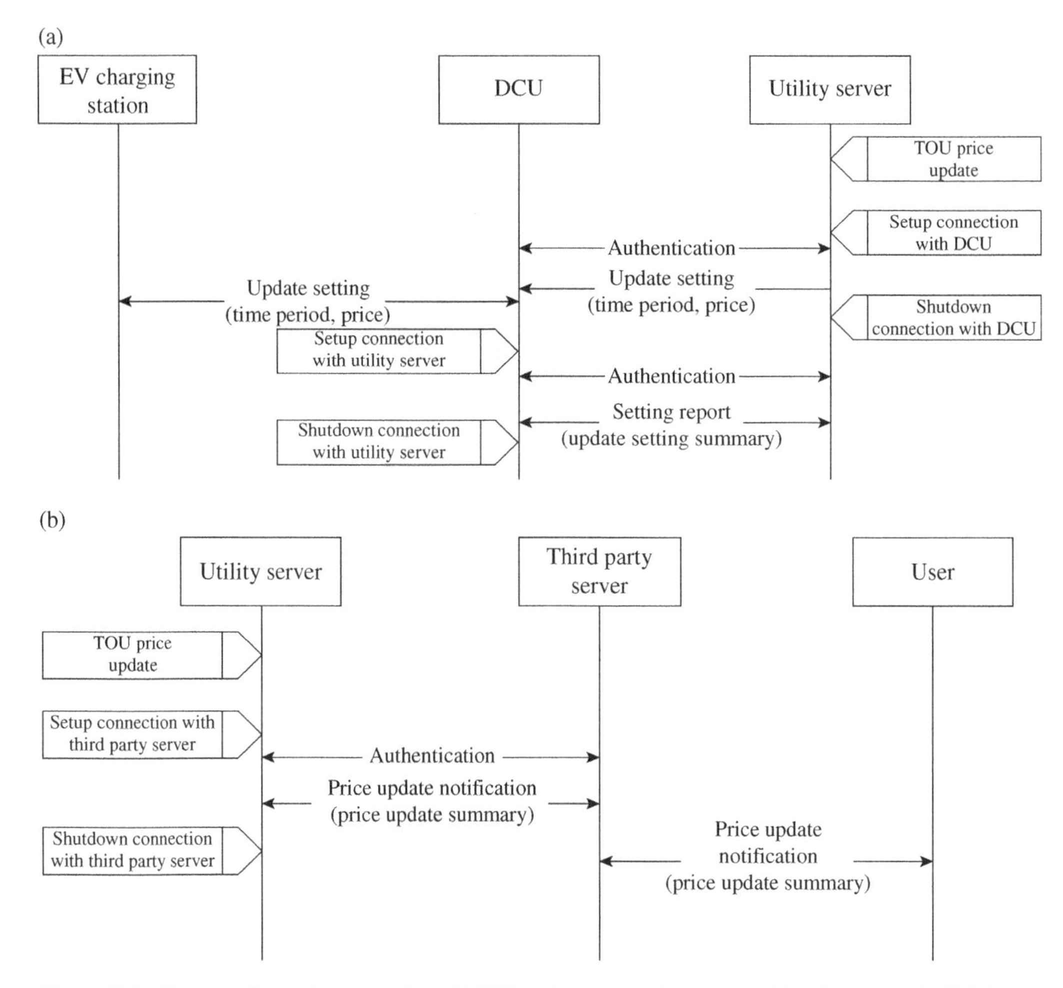

Figure 13.8 Energy price update procedure. (a) DCU exchanges setting report with utilityserver. (b) Third party server exchanges price update notification with user. *Source:* From Lam et al. [A]/Scientific Research Publishing/CC BY 4.0.

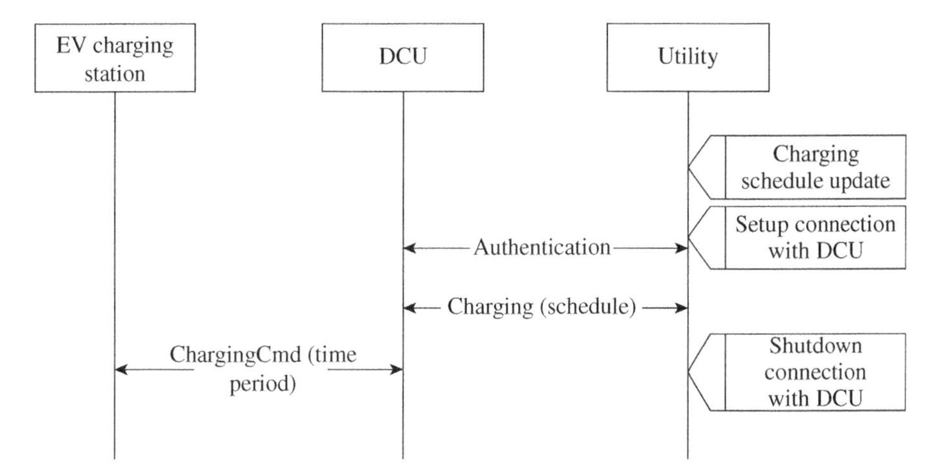

Figure 13.9 Charging session schedule update. *Source:* From Lam et al. [A]/Scientific Research Publishing/CC BY 4.0.

DCU exchanges the setting report with update setting summary to the utility server. The DCU shuts down the connection after then.

When the utility server has received the TOU price update, it also sets up a connection with the third party server. The authentication process is carried out and the two servers exchange the price update notification with price update summary. The connection between the two servers will be shut down. Finally, the third-party server exchanges the price update notification with user.

13.3.1.3.2 Charging Session Schedule update

The charging session scheduler is invoked when a new charging session is initialized. In general, the scheduler will schedule the charging service at off-peak hours to minimize loadings due to EV charging. The procedure of updating the charging scheduler is illustrated in Figure 13.9. To update charging schedule, the utility server sets up connection with the DCU. Once the authentication is completed, the utility server sends the charging schedule to the DCU. The connection will shut down afterward. The DCU then forwards the charging schedule comprising the time period data log to the EVCS.

13.3.2 Web Portal

Two web portals have been developed to serve the service manager and the EV owner. The user interfaces are shown in Figures 13.10 and 13.11, respectively. By incorporating the web portal, the service manager may capture the full EV charging service via the portal. Some information includes the overall energy profile as shown in Figure 13.10, customer information and status of each charger. Such timely information shifts the paradigm of dynamic pricing from single dimension (time series) to two dimensions (time and geography). In contrast, EV owners may obtain the energy profile of their EV charging, electronic bill (Figure 13.11a) as well as monitoring the charging status of their vehicles (Figure 13.11b). More importantly, EV owners may obtain the real-time EV parking information which ensures EV owners to enjoy charging service.

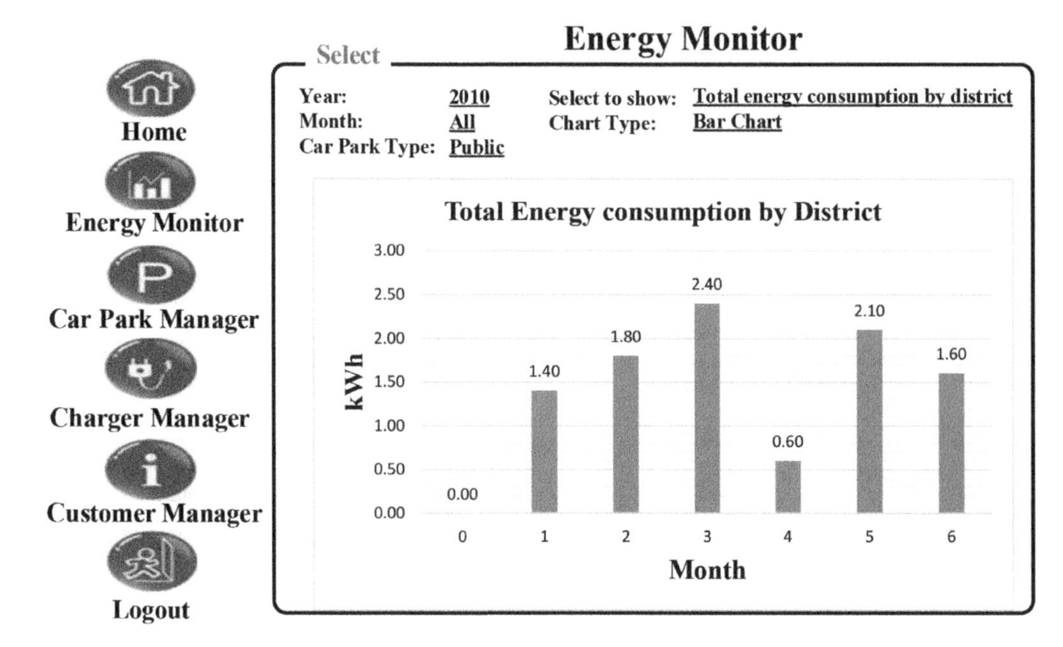

Figure 13.10 User interface for service manager. *Source:* Scientific Research Publishing Inc.

13.4 Performance Evaluation

In this section, the performance of EVAMI is reported. There are two parts: The first part investigates the network performance of OCS and the second part evaluates the effectiveness of EVAMI for DR applications.

13.4.1 Network Performance of OCS

An experiment was conducted to investigate the network performance of OCS. Without losing generality [20], OCS based on six EVCSs was devised during experiment. The network throughput and end to end delay were measured under three different system occupancies: (1) 0%, (2) 50%, and (3) 100% implying (1) all EVCSs are idle, (2) three EV are charging, and (6) all EV are charging respectively. The results are shown in Figure 13.12.

In Figure 13.12, The end-to-end delay of OCS is bounded by 11.2 ms while the system throughput ranges from 4.6 to 5.1 Mbps. In general, End to END delay increases as the system occupancy increases because the traffic loadings of the system increase when more charging service sessions are established. Apart from the traffic loadings, interference is another important factor yielding performance degradation. When an EV is plugged into the system and the charging process is started, interference is also injected into the PLC network. Therefore, the packet error may increase during transmission. As a result, the system throughput will drop.

13.4.2 Effectiveness of EVAMI on Demand Response

To investigate the effectiveness of EVAMI for DR applications, a case study has been conducted. An efficient AMI system normally carries high traffics; hence heavy loadings are studied. As a result, the load profiles of residential buildings are collected by recording the meter readings every

(a)

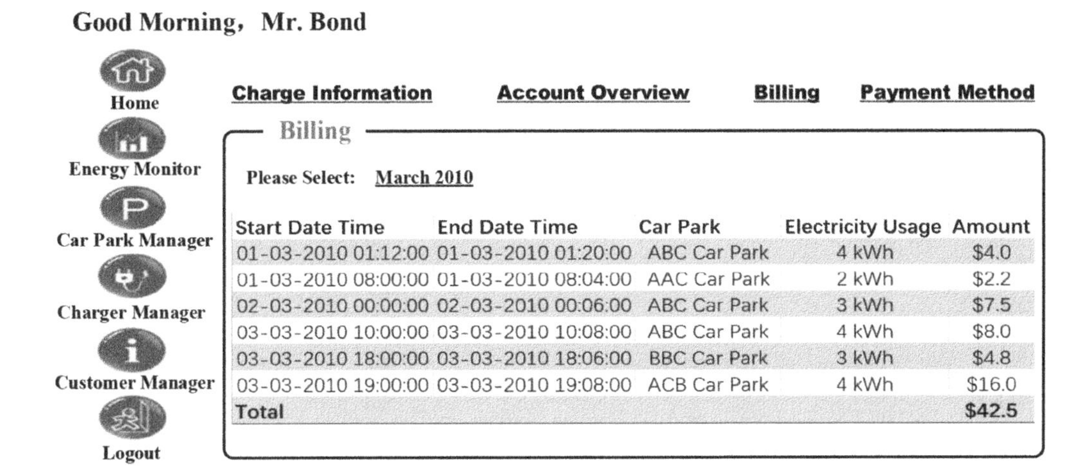

Start Date Time	End Date Time	Car Park	Electricity Usage	Amount
01-03-2010 01:12:00	01-03-2010 01:20:00	ABC Car Park	4 kWh	$4.0
01-03-2010 08:00:00	01-03-2010 08:04:00	AAC Car Park	2 kWh	$2.2
02-03-2010 00:00:00	02-03-2010 00:06:00	ABC Car Park	3 kWh	$7.5
03-03-2010 10:00:00	03-03-2010 10:08:00	ABC Car Park	4 kWh	$8.0
03-03-2010 18:00:00	03-03-2010 18:06:00	BBC Car Park	3 kWh	$4.8
03-03-2010 19:00:00	03-03-2010 19:08:00	ACB Car Park	4 kWh	$16.0
Total				$42.5

(b)

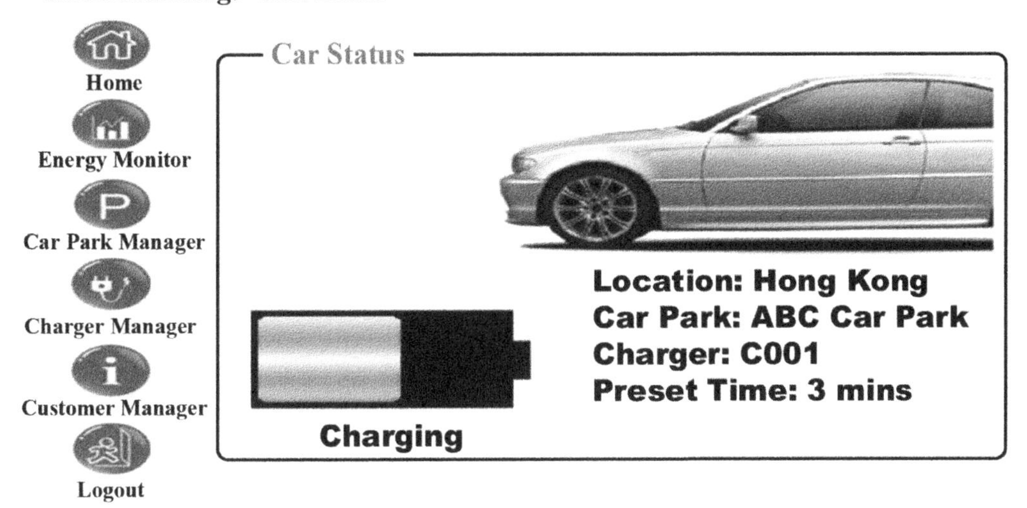

Figure 13.11 User interface for EV owner. (a) Charging profile and billing information. (b) Vehicle charging status. *Source:* From Lam et al. [A]/Scientific Research Publishing/CC BY 4.0.

30 minutes. The monitored residential building was 60 floors tall and there were eight apartments on each floor. The energy consumption for 420 apartments was measured and the study lasted for two months. The average 30 minutes energy consumption of the entire residential building is indicated as normal in Figures 13.13 and 13.14. It is assumed that the study has an EV scenario that is based on the average driving requirement of 25 km/day which is estimated from the Hong Kong government data [22, 23] at the time.

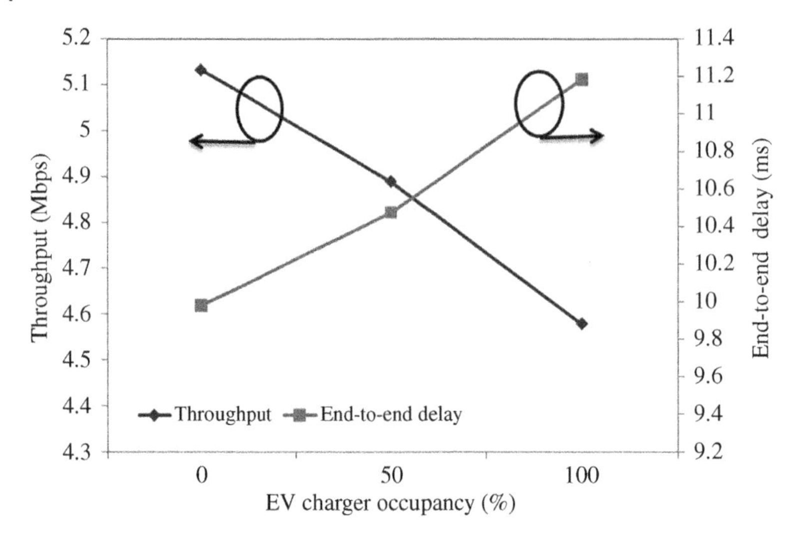

Figure 13.12 Network performance of OCS. *Source:* From Lam et al. [A]/Scientific Research Publishing/CC BY 4.0.

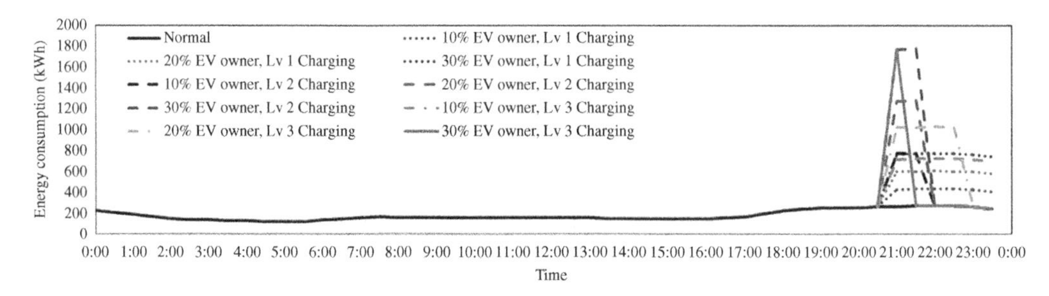

Figure 13.13 Impact of EV charging level on load profile of a residential building. *Source:* From Lam et al. [A]/Scientific Research Publishing/CC BY 4.0.

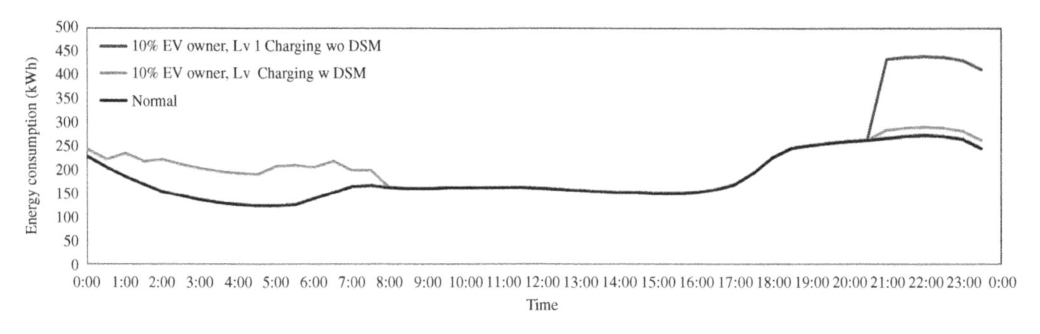

Figure 13.14 Impact of EV AMI on load profile of a residential building. *Source:* From Lam et al. [A]/Scientific Research Publishing/CC BY 4.0.

Table 13.2 Definition of charging level [24].

Level 1 "Basic" / "Trickle" Charging	110 V (volts)-120 V alternating current (AC), single phase, 15 A (amps) continuous.
Level 2 "Quick" / "Standard" Charging	208–240 V alternating current (AC), single phase, typically 220 V/30 A continuous but up to 80 A specified.
Level 3 "Fast" / "DC" / "Commercial" Charging	408—480 V 120 A Converts three phase alternating current (AC) to direct current (DC) for ~80% to full charging in 15–30 min.

Source: From Lam et al. [A]/Scientific Research Publishing/CC BY 4.0.

The definition of the three charging level is recalled and is listed in Table 13.2.

In Figure 13.14, the impact due to different EV charging level on the load profile is investigated. In additional to the Charging level, Figure 13.14 also considers the effect of EV penetration, from 10 to 30%, on the load profile, Nevertheless, the increase of EV penetration yields an increase of energy demand. For normal practice, customers charge their EV at around 20:00. Without any demand response policy, peak consumption raises 65% if 10% of residences charge their EV with level 1 charging mode. It should be noted that the peak energy demand dramatically jumps to 1778 kWh (560% of normal case) if 30% of residential charge their EV with level 3 charging mode. It is thus seen that the penetration of EV and charging level significantly affect the peak energy demand. Therefore, level 1 charging mode is concluded as the most appropriate charging method for charging scheduler of EVAMI. For this study, the global penetration of hybrid EV was estimated to be over 10% from 2012 to 2015 [25, 26]. To reflect a realistic projection, the EVAMI performance is evaluated based on the assumption that 10% of residents charge their EV with Level 1 charging. The findings are shown in Figure 13.14. Without the EVAMI, the peak energy demand reaches 440 kWh when 10% of residents charge their EV with level 1 charging. After applying scheduler of EVAMI, the peak energy consumption is reduced by 36% (which falls to 280 kWh). More importantly, EVAMI successfully increases the energy demand at "off-peak" by 54%. This investigation reveals that the EVAMI does not only reduce the peak consumption but also relocate the energy demand effectively. From [27], saving 1 Wh peak demand reduces 50 Wh non-coincidental peak load. If EVAMI is widely adopted in Hong Kong, 3.3 TWh electricity consumptions are saved annually when penetration of EV has reached 10%. At the same time, 2.2G tons of CO_2 emission is reduced which is worth HK$3.3G. Apart from benefiting the utilities, EVAMI also provides a full function web portal to the EV owners. TPCPS interacts with different utilities' information systems to collect all EV charging service transactions and therefore a single bill can be issued to EV owners. As a result, EV owners may obtain the overall profile of their EV charging usage.

13.5 Conclusion

In this chapter, an AMI platform for EV charging (EVAMI) is proposed to furnish a wide range of features to both EV owners and utility service managers. By enabling the interactive and efficient communication platform, the proposed EVAMI facilitates utilities to perform demand response effectively. Apart from the traditional demand response approach, a charging scheduler is introduced to advice the charging mode and the schedule to EV charging station in order to reshape the load profile. As such, the proposed solution not only provides the time based load profile but also the location-based load profile. Such timely information shifts the paradigm of dynamic pricing from a single dimension (time series) to two dimensions (time and geographic). It must be stressed that the incorporation of TPCSP facilitates a single bill to be issued to EV owners. Hence, EV owners understand their energy usage and thus may perform energy saving activity.

In this investigation, the Power Line Communication network has been proposed. Evaluation of the performance shows that a data rate of 4.6 to 5.1 Mbps with an end to end delay of 9.8–11 ms is achieved. It must be stressed that the invention of TPCPS in the EVAMI allows a single bill to be issued to EV owners.

Furthermore, an insight into the impact of charging level on the load profile is also provided. In general, level 1 charging has less effect on the load profile than level 2 and level 3 charging. In particular, the level 3 charging shapes the load profile in a "sharp peak". In this work, the charging scheduler only carries out demand response by adopting level 1 charging and an appropriate schedule. As a result, the peak energy consumption is reduced by 36%, i.e. from 440 to 280 kWh. More importantly, the adoption of EVAMI successfully increases the energy demand at "off-peak" by 54%. It is important to point out that the EVAMI does not only reduce the peak consumption but also relocates the energy demand effectively.

Acknowledgements

The permission given in using materials from the following paper is very much appreciated.

[A] Lam, L.K., Ko, K.T., Tung, H.Y. et al. (2011). Advanced metering infrastructure for electric vehicle charging. *Smart Grid and Renewable Energy, Scientific Research* 2(4): 312–323.

References

1 Bebon, J. (2017). VW's Electrify America Plans to Install 2800 Charging Stations. https://ngtnews.com/vws-electrify-america-plans-install-2800-charging-stations (accessed 28 February 2021)
2 Global EV Outlook (2020). Technology Report, IEA. https://www.iea.org/reports/global-ev-outlook-2020 (accessed 31 July 2022).
3 (2021). Electric Vehicle Market (By Product: Battery Electric Vehicles (BEV) and Plug-in Hybrid Electric Vehicles (PHEV)) – Global Market Size, Trends Analysis, Segment Forecasts, Regional Outlook 2020 – 2027. Precedence Research. https://www.precedenceresearch.com/electric-vehicle-market (accessed 25 February 2021).
4 (n.d.). Guidelines for the development of electric vehicles charging infrastructure. Last updated: 16 April 2021. https://www.iea.org/policies/2695-guidelines-for-the-development-of-electric-vehicles-charging-infrastructure (accessed 31 July 2022).
5 Xie, C., Wang, D., Lai, C.S. et al. (2021). Optimal sizing of battery energy storage system in smart microgrid considering virtual energy storage system and high photovoltaic penetration. *Journal of Cleaner Production, Journal of Cleaner Production* 281: 125308.
6 Ipakchi, A. and Albuyeh, F. (2009). Grid of the future. *IEEE Power and Energy Magazine* 7 (2): 52–62.
7 Kintner-Meyer, M.C., Schneider, K., Pratt, R. et al. (2007). Impacts Assessment of Plug-in Hybrid Vehicles on Electric Utilities and Regional US Power Grids: Part 1: Technical Analysis. Pacific Northwest National Laboratory (PNNL) (PNNL-SA-61669).

8 (2007). Environmental Assessment of Plug-in Hybrid Electric Vehicles. Volume 1: Nationwide Greenhouse Gas Emissions, EPRI, Palo Alto, CA 1015325.
9 Duvall, M. (2008). How many plug-in hybrids can a smart grid handle? *Plug-in 2008 Conference and Exposition*, San Jose, CA.
10 Hadley, S.W. and Tsvetkova, A.A. (2007). Potential Impacts of Plug-in Hybrid Electric Vehicles on Regional Power Generation. Oak Ridge National Laboratory (ORNL) (ORNL/TM-2007/150).
11 Putrus, G.A., Suwanapingkarl, P., Johnston, D. et al. (2009). Impact of electric vehicle on power distribution network. *The Proceeding of IEEE Vehicle Power and Propulsion Conference 2009*, Dearborn (7–10 September 2009), 827–831.
12 Galus, M.D. and Andersson, G. (2008). Demand management of grid connected plug-in hybrid electric vehicles (PHEV). *IEEE Energy 2030 Conference 2008*, Atlanta (17–18 November 2008).
13 Short, W. and Denholm, P. (2006). Preliminary assessment of plug-in hybrid electric vehicles on wind energy markets. National Renewable Energy Lab (NREL), (NREL/TP-620-39729).
14 Denholm, P. and Short, W. (2006). Evaluation of utility system impacts and benefits of optimally dispatched plug-in hybrid electric vehicles. National Renewable Energy Laboratory (NREL), (NREL/TP-620-40293).
15 K. Parks, P. Denholm, and T. Markel (2007). Costs and emissions associated with plug-in hybrid electric vehicle charging in the Xcel Energy Colorado Service Territory. National Renewable Energy Laboratory (NREL), (NREL/TP-640-41410).
16 Khan, R.H., Aditi, T.F., Sreeram, V., and Iu, H.H.C. (2010). A prepaid smart metering scheme based on WiMAX prepaid accounting model. *Smart Grids and Renewable Energy* 1 (2): 63–69.
17 Sun, Q., Li, H., Ma, Z. et al. (2016). A comprehensive review of smart energy meters in intelligent energy networks. *IEEE Internet Things Journal.* 3 (4): 464–479.
18 Barbiroli, M., Fuschini, F., Tartarini, G., and Corazza, G.E. (2017). Smart metering wireless networks at 169 MHz. *IEEE Access* 5: 8357–8368.
19 Rua, D., Issicaba, D., Soares, F.J. et al. (2010). Advance metering infrastructure functionalities for electric mobility. *The Proceeding of 2010 IEEE PES Innovative Smart Grid Technologies Conference Europe (ISGT Europe)*, Gothenburg (11–13 October 2010).
20 Lin, Y.-J., Latchman, H.A., Lee, M., and Katar, S. (Dec. 2002). A power line communication network infrastructure for the smart home. *IEEE Wireless Communications* 9 (6): 104–111.
21 Intellon (n.d.), http://www.intellon.com (accessed 20 November 2001).
22 Anonymous (2010). Hong Kong Energy End-use Data 2010. Electrical & Mechanical Services Department of Hong Kong Government. http://www.emsd.gov.hk/emsd/e_download/pee/HKEEUD2010.pdf (accessed 21 June 2011).
23 Anonymous (2011). Vehicles Registration & Licensing. Transportation Department of Hong Kong government. http://www.td.gov.hk/filemanager/en/content_281/table41a.pdf (accessed 21 October 2012).
24 Earley, M.W., Sargent, J.S., Sheehan, J.V., and Caloggero, J.M. (1999). *National Electrical Code Handbook*. Long Beach, CA: Nat. Fire Protection Assoc.
25 Niedermeyer, E. (2011). Plotting the electrified future: BCG downgrades EV penetration, pacific crest offers bear and bull cases. The Truths about cars. http://www.thetruthaboutcars.com/2011/06/plotting-the-electrified-future-bcg-downgrades-ev-penetration-pacific-crest-offers-bear-and-bull-cases/ (accessed 3 August 2013).
26 Anonymous (2010). Electric cars and hybrids set to impact negatively on Powder Metallurgy (PM) part production. ipmd.net. http://www.ipmd.net/articles/articles/001057.html (accessed 22 April 2012).
27 Raynolds, N. (2000). The contribution of energy efficiency to the reliability of the U.S. electric system. Alliance to Save Energy.

14

Power System Dispatching with Plug-In Hybrid Electric Vehicles

Nomenclature

Index/Set

N_g	Set of units
T	Set of dispatch periods
K	Set of buses

Deterministic Data

F_C	Production cost function
F_e	CO_2 emission function
C_e	Cost coefficient of CO_2 emission
$\rho_c(t)$	Charging price in period t
$\rho_d(t)$	Discharging price in period t
P_{dc}	Average discharging power
$L(t)$	Active power loss in period t
$D(t)$	Load demand in period t
P_c	Average charging power
$R(t)$	Reserve requirement in period t
R_{di}	Ramp down rate of unit i
R_{ui}	Ramp-up rate of unit i
S_{di}	Shutdown rate of unit i
S_{ui}	Start-up rate of unit i
$X_i^{on}(t)$	Duration of continuously on state of unit i in period t
$X_i^{off}(t)$	Duration of continuously off state of unit i in period t
T_i^{on}	Minimum up time of unit i
T_i^{off}	Minimum down time of unit i
P_i^{min}	Minimum active power of unit i
P_i^{max}	Maximum active power of unit i
N_e^{max}	Maximum number of charging and discharging
N_c^{sch}	Scheduled number of charging PHEVs for the whole day

Smart Energy for Transportation and Health in a Smart City, First Edition. Chun Sing Lai, Loi Lei Lai and Qi Hong Lai.

N_d^{sch}	Scheduled number of discharging PHEVs for the whole day
Y	Bus-unit incidence matrix
P_d	Active load vector
Q_d	Reactive load vector
P_L^{max}	Maximum capacity vector of transmission line
V_{min}	Lower limit vector of bus voltage
V_{max}	Upper limit vector of bus voltage
Q_{min}	Diagonal matrix of minimum reactive power output
Q_{max}	Diagonal matrix of maximum reactive power output
T_{min}	Lower limit vector of transformer tap
T_{max}	Upper limit vector of transformer tap
γ_{min}	Lower limit vector of phase shifter angle
γ_{max}	Upper limit vector of phase shifter angle

Variables

$P_i(t)$	Active power output of unit i in period t
$u_i(t)$	On/off state of unit i in period t
P_g	Active power output vector
Q_g	Reactive power output vector
$N_c(t)$	Charging number in period t
$N_d(t)$	Discharging number in period t
S_{p1}, S_{q1}	Slack variables
S_{p2}, S_{q2}	Slack variables
$S_D(t)$	Discharging number at each bus in period t
$S_C(t)$	Charging number at each bus in period t
P_{in}	Vector of injection active power, $P_{in,j}$ for the injection active power of bus j
Q_{in}	Vector of injection reactive power, $Q_{in,j}$ for the injection reactive power of bus j
$G_{jk}+jB_{jk}$	Mutual admittance between bus j and bus k
δ_{jk}	Voltage angle between bus j and bus k
V	Bus voltage vector, V_j for voltage of bus j
H,N,E,F	Jacobi matrices
M,J,R,S	Jacobi matrices
A,B,C,D	Jacobi matrices
ΔP_L	Increment vector of active power flow
$\Delta\delta$	Increment vector of voltage angle
ΔV	Increment vector of bus voltage
ΔT	Increment vector of transformer tap
$\Delta\gamma$	Increment vector of phase shifter angle
ΔQ	Increment vector of reactive power output
P_L	Power flow vector
T	Transformer tap vector
γ	Phase shifter angle vector

14.1 Introduction

It is getting clear that the deployment of electric vehicles has real benefits to the society, for example reference [1] discussed real CO_2 emissions benefits and end user's operating costs of a plug-in hybrid electric vehicle (PHEV). Various means could be used to maximize the benefits. However, EV charging needs to be connected to the power grid to obtain electricity, large-scale charging may cause overload and safety to the power grid. Therefore, it is importance to study the impact of EV charging on the power grid. Although researchers have done some related work on the impact of EV charging on the power grid [2, 3], there is no generally applicable method. References [4, 5] reported the impact of different charging strategies on the distribution networks in Germany and New Zealand, respectively. In [6], an unsupervised algorithm is proposed to extract the local EV charging load. This algorithm can effectively reduce the interference of other similar charging behaviors and make power grid decisions much smarter. In [7], the authors proposed a charging strategy based on cost-benefits analysis, which aims to minimize daily production costs and peak-to-average ratio (PAR). The method can be applied to small and medium distribution networks. In [8], some current controlled charging methods and strategies were reviewed, and their advantages and disadvantages were analyzed. Reference [9] proposed a model to quantify the benefits of controlled charging to the power grid. The model is verified through the load data of three million electric vehicles in California, which proves that the model is effective. Reference [10] compared the cost and benefits of decentralized and centralized algorithms in large-scale power systems, so as to improve the strategies of plug-in electric vehicle (PEV) smart charging. In [11], the authors proposed a new PEV smart charging method, which can determine different charging modes based on factors such as vehicle type and driver habits, so that both the power grid and vehicle owners can benefit from it.

Investigation has shown that PHEVs have the potential to contribute 1.68% out of the 10% renewable energy in transportation sector targeted by the Republic of Ireland [12]. Suppose that the oil price is 2.57 $/gallon, and the electrical price is 8.6 cent/kWh, the vehicle-to-grid (V2G) technology could save $450 for the PHEV users per year in America [13]. Besides, the capability of optimizing the fluctuation of load curve and improving the penetration of renewable energy has also been proved [14–16].

Though the popularization of PHEVs provides opportunities for the improvement of traffic system, large penetration of PHEVs is a tough problem not to be ignored for the power system [17, 18]. Therefore, interest of the dispatch issue with PHEVs taken into account is growing gradually among researchers as well as the system operators [19–22]. Of these researches, References [19, 20] tackle the same topic as this chapter. A unit commitment (UC) model with "grid-connected" vehicles is proposed with the vehicles treated as small portable power plants, and finally solved by means of the Particle Swarm Optimization [19]. On that basis, reliability consideration is described in [20], in which the desired level of reliability is demonstrated by two indexes named the loss of load probability and the expected unserved energy.

Although many efforts have been spent, there are still some aspects to be tackled especially the lack of security considerations in system modeling. Hence, the basic aim of this work is the intelligent scheduling of existing generating units and large number of PHEVs by means of SCUC. A two-stage model is developed for the day-ahead schedule of power system considering both the power system benefit and security constraints. The main contributions of this chapter are three-fold.

14.1.1 Model Decoupling

By virtue of the controllable charging and discharging, the mathematical model is decoupled into two submodels based on the assumption that the V2G facility is well served.

14.1.2 Security Reinforcement

When committing the charging and discharging capacity into each access bus, security constraints should be satisfied to ensure the feasibility of the final solution. In this way, the Newton–Raphson algorithm is employed for iterative optimization, where AC power flow equations are calculated in pursuit of the precise solution.

14.1.3 Potential for Practical Application

A case study based on real-life data will be used to demonstrate the benefit gained in the integration of EV with the grid to improve power system security.

Because the energy storage of other EVs may not be available for 24 hours dispatching, only the schedulable PHEVs are discussed in this chapter. Schedulable here means the PHEV is registered to be dispatched, and whose owner should follow the EV aggregator's order to charge or discharge at the right time and the right place.

This chapter is organized as follows: the framework of PHEV dispatching is discussed in Section 14.2. The framework for the two-stage model is given in Section 14.3. The charging and discharging modes are explained in Section 14.4. The optimal dispatching model with PHEVs is developed in Section 14.5. Numerical examples are reported in Section 14.6. Practical applications are included in Section 14.7. Finally, a conclusion is drawn in Section 14.8.

14.2 Framework of PHEVs Dispatching

The conventional centralized dispatching approach seems improper for the dispersive PHEVs, as a result, a two-layer dispatching model based on the EV aggregator is proposed. Serving as a middleman between the drivers and the electrical utilities or electricity market, the concept of EV aggregator is now receiving growing interests [23, 24]. Specifically, the bottom-layer dispatch is managed by the system dispatcher, in which PHEVs within the same fleet are regarded as a big virtual electric vehicle (also known as the virtual power plant), though the formulation of virtual PHEV will be taken care by the aggregator. The top-layer dispatch deals with decomposing and allocating the specified charging and discharging schedules for individual PHEV entities, which is taken care by the aggregator. This chapter focuses on the top layer dispatching model within the EV aggregator framework. Other issues such as the designs of the charging facilities and the communication platform, etc. will be investigated in the future scope. The proposed dispatching model should perform following tasks in providing charging services.

1) The charging and discharging prices, as well as the permissible charging periods for the second day are published through the communication platform.
2) According to their realistic travel patterns, the PHEV users submit the information involving the available capacity and periods for dispatching/discharging to the corresponding aggregator at the prescriptive time.
3) The aggregator of PHEV fleet manages all the information received, and then forecasts the total available capacity and period for charging or discharging from the standpoint of a virtual vehicle with main objective of maximizing the energy trading profits. Note that all the fast charging station, the normal charging station and the battery exchange stations should be taken into account to determine the bid capacity and bid price, as shown in Figure 14.1.

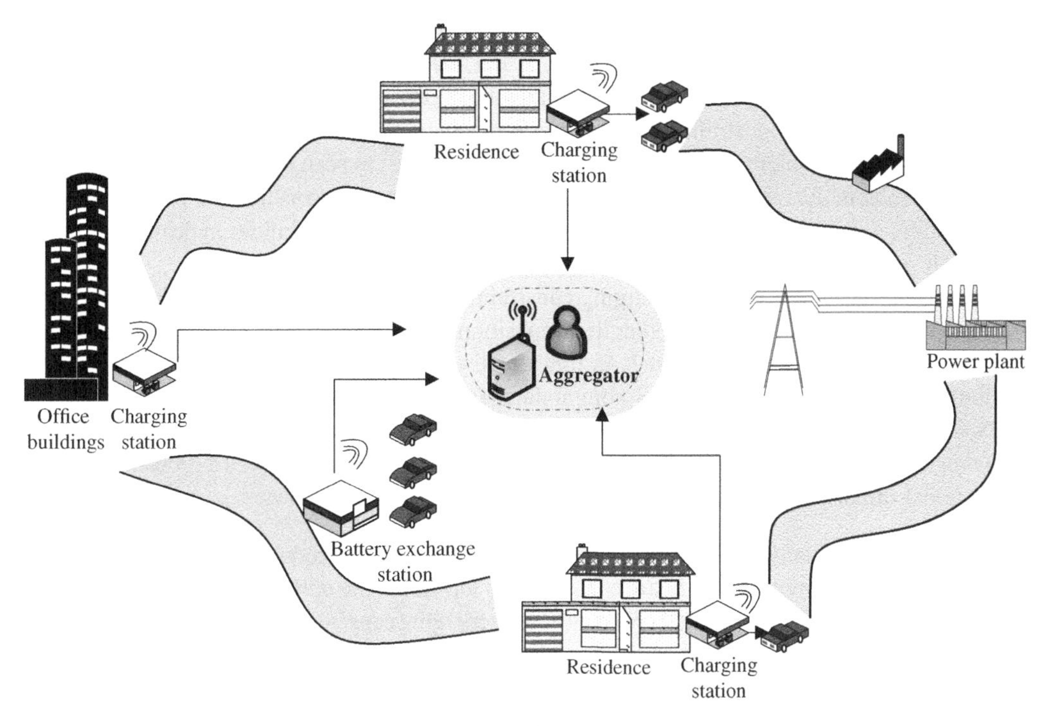

Figure 14.1 EV aggregator illustration. *Source:* From Cai et al. [A] © 2015 IEEE.

4) According to the collected bid information, the day-ahead market is cleared with the dispatching model developed in this chapter. Thus, the final charging and discharging scheme and states/power output of units can correctly fulfill the network secure constraints while achieving the economic benefits.
5) After the charging and discharging schemes are sent back to power plants and aggregators, the aggregator decomposes the determined schedule into specific schemes for individual PHEVs, during which the clients could be classified into different groups. Certainly, the aggregator is responsible for balancing themselves in operation according to the determined schedules from the bottom layer.

With the help of aggregator, power system dispatcher will deal only with the aggregated PHEVs load. The proposed two-layer framework therefore provides an effective grid integration solution for many PHEVs based on both centralized and distributed control.

14.3 Framework for the Two-Stage Model

Power system dispatching with PHEVs should consider not only economic benefit but also system security constraints. Consequently, like the SCUC, it is a discrete optimization problem of high complexities and nonlinearities, and difficult to be solved. Traditionally, the Benders' decomposition approach was applied to solve the SCUC [25]. The basic idea is to decompose the mixed integer problem into a master problem and a sub problem. Firstly, the integer variables such as on/off states of units are solved in the master problem. Secondly, based on the known integer variables the continuous variables such as power output are calculated in the sub problem. Finally, checkup whether the termination condition is satisfied. If not, new constraint named Bender's cut will be added into the master problem, then repeat the above process until the termination condition is satisfied. However, the constraints in the master problem may increase with iterations carried out between the master problem and the sub problem. The extreme case can involve 24 cuts added into the master problem during one iteration, indicating that the security constraint is not fulfilled for any scheduling interval. Thus, even the Benders' decomposition has difficulties in dealing with the basic SCUC of large-scale systems.

With additional charging and discharging constraints, the SCUC will become even more challenging. To tackle this challenge, another decomposition approach based on the adequate capacity of charging and discharging is presented. Assume that the charging/discharging facilities are widely distributed with sufficient capacity to accommodate large number of PHEVs, so that the available PHEV storage can play a role in power flow optimization. To model this problem, we develop a two-stage program to optimally schedule the PHEV charging and discharging as well as power output of conventional units, of which details of the framework are given as follows:

1) Sub-model 1 is actually standard UC model, in which the power production cost, CO_2 emission cost, and PHEV economic benefit are considered to formulate the optimization objective, i.e. the total social cost. Note that the total charging and discharging capacity as well as generators output in each dispatching interval are attained, but the charging locations cannot be determined yet.
2) The sub-model 2 optimally allocates the optimal total charging and discharging capacity to different locations of PHEV charging stations.

14.4 The Charging and Discharging Mode

14.4.1 PHEV Charging Mode

The fundamental work of PHEVs charging research is to build the charging load model. Single battery charging load curve is different from each other for different voltage and capacity. Reference [13] has made great efforts on this subject, where the study area is the Xcel Energy Colorado service territory. During the simulation, the average charging power ranges from 1.8 to 2.0 kW with a full charging time of six hours. In addition, four vehicle-charging scenarios are developed for evaluation, namely the uncontrolled charging, the delayed charging, the off-peak charging, and the continuous charging, respectively. In each of the four scenarios, aggregated hourly charging profile for a fleet of vehicles is developed. The hourly load was then added to the base case load to evaluate the incremental system impact. The details of the four charging scenarios and the simulation of load impact are shown in the Appendix 14.A. Figures 14.A.1 and 14.A.2 give the load patterns with PHEV charging for summer and winter respectively for the uncontrolled, delayed, off-peak and continuous charging.

According to the investigation of Reference [13], it can easily come to the conclusion that the off-peak charging mode can not only provide a potential mechanism for reducing the probability of system capacity increase resulted from reserve shortage, but also effectively drive up the load curve during low demand. Therefore, the following research is carried out based on the off-peak charging mode.

14.4.2 PHEV Discharging Mode

Discharging characteristic of the PHEV depends on various factors, e.g. the battery type, battery size, inserting state, power of discharging circuit, discharging cycle and so on. For instance, the full discharging time of Prius PHEV is about 5.5 hours with an average discharging power of 1.2 kW; the SUV PHEV-40 has an electricity consumption around 0.26 kWh/km, whose battery capacity is up to 14.4 kWh; while the economic PHEV-20 consumes about 0.16 kWh/km with a total capacity of 5.0 kWh. According to the existing battery storage technology, the maximum power of discharging circuit is mainly 9.6 kW and the discharging time of which is six hours under normal circumstances [26].

14.4.3 PHEV Charging and Discharging Power

The difficulty of charging and discharging modeling is how to take vehicles' dispersion and drivers' habit into account. Nevertheless, the uncertainty of charging and discharging power for PHEV may arise from the charging and discharging behavior, battery type and some other external factors, among which charging and discharging behavior is considered the most important. The indeterminacy used to be handled upon some assumptions, for example, the PHEV should only charge or discharge once in a day. Whereas, such kind of constraint seems a little bit rigid for those PHEVs with long driving distance per day. Besides, the mathematical statistics theory and stochastic simulation method (such as Monte-Carlo) are frequently employed for the simulations of charging/V2G power or probability distributions investigation [27]. However, it seems appropriate for the load forecast of un-schedulable PHEVs only.

Commonly, there are two control approaches for PHEV charging and discharging, called direct control and indirect control respectively. Both the two aim to reduce the bad influence brought by the indeterminate charging and discharging habits. The direct control is essentially equivalent to the concentrated dispatch of PHEVs, in which when and where the charging or discharging occur are entirely scheduled by the system operator. On the contrary, the electricity prices (such as peak-valley price) are regarded as the guidance for dispatching control to achieve the goal of leading reasonable charging and discharging.

In order to maximize the social benefit while maintaining the secure operation of the power system, both the direct and indirect control are integrated in this chapter. Furthermore, some assumed conditions are summarized as follows. (i) The available charging periods are limited to avoid the accretion of peak demand, as well as to limit the number of variables calculated in the sub-model 1 for time saving. (ii) The discharging price varies with charging periods, but no restriction of the discharging period is built in order to make full use of the V2G technology.

Though utilization of the specific charging/discharging power of each PHEV may improve the modeling accuracy, it becomes complicated or even infeasible for the power system with high penetration of PHEVs. As a result, the average charging and discharging power is applied instead. The average charging and discharging power may change with the progress of battery storage technology, and in such case, it is only needed to update the power data in the two-stage model.

14.5 The Optimal Dispatching Model with PHEVs

14.5.1 Sub-Model 1

Power systems benefits are calculated in terms of production cost reduction, CO_2 emission reduction, and charging/discharging profit. To unify the unit of each objective, the CO_2 emission cost is firstly defined and supposed to be the linear function of emission quantity. These three sub objectives are expected to be taken into account in sub-model 1. Actually, a multi-objective problem could hardly obtain the genuine optimal solution but a set of Pareto solutions instead. In this way, the single objective programming is built with the social cost defined as the production cost pluses the CO_2 emission cost then subtracts the PHEV users' profit due to V2G services.

The sub-model 1 determines the units' state/active power output and the charging/discharging capacity of PHEVs in each period while minimizing the social cost. Meanwhile, the constraints of the system power balance, reserve demand, unit ramp up/down rate, minimum on/off time and active power output limits are calculated. The objective function and constraints are expressed as (14.1)–(14.11).

$$\min \sum_{i=1}^{N_g}\sum_{t=1}^{T}[F_c(P_i(t))u_i(t) + C_e \times F_e(P_i(t))u_i(t) - (\rho_d(t)N_d(t) - \rho_c(t)N_c(t))] \tag{14.1}$$

1) System power balance

$$\sum_{i=1}^{N_g}[P_i(t)u_i(t)] + P_{dc}N_d(t) = L(t) + D(t) + P_cN_c(t) \tag{14.2}$$

2) Reserve demand

$$P_{dc}N_d(t) + \sum_{i=1}^{N_g} P_i^{max}(t)u_i(t) \geq L(t) + D(t) + P_cN_c(t) + R(t) \tag{14.3}$$

3) Ramp up/down rate

$$R_{di} \leq P_i(t+1) - P_i(t) \leq R_{ui} \tag{14.4}$$

4) Start up and shut down rate

$$S_{di} \leq P_i(t+1) - P_i(t) \leq S_{ui} \tag{14.5}$$

5) Mini up/down time

$$\left(X_i^{on}(t) - T_i^{on}\right)(u_i(t) - u_i(t+1)) \geq 0 \tag{14.6}$$

$$\left(X_i^{off}(t) - T_i^{off}\right)(u_i(t+1) - u_i(t)) \geq 0 \tag{14.7}$$

6) Generation limit

$$P_i^{min}u_i(t) \leq P_i(t) \leq P_i^{max}u_i(t) \tag{14.8}$$

7) Number of charging and discharging limit

$$N_c(t) + N_d(t) \leq N_e^{max} \tag{14.9}$$

$$\sum_{t=1}^{T} N_c(t) = N_c^{sch} \tag{14.10}$$

$$\sum_{t=1}^{T} N_d(t) = N_d^{sch} \tag{14.11}$$

The constraints (14.9)–(14.11) state that except for the traditional UC constraints, the schedule should satisfy a planned bound on the storage capacity. For the sake of computation time, the piecewise linear approach introduced in Reference [28] is applied to convert the quadratic production cost function and the quadratic CO_2 emission function into linear ones.

14.5.2 Sub-Model 2

Since the charging and discharging operation can have great effects on the power flow distribution, it is highly important to arrange well, the PHEVs network access locations. Essentially, the key issue of sub-model 2 is the optimization of charging and discharging schedule subject to power flow constraints.

The total charging and discharging capacity of each time interval has been determined in the sub-model 1, then the main work left for the sub-model 2 is to determine the specific capacity at each access node and whether to charge or discharge. In order to guarantee the feasibility and accuracy of the dispatching solution, the AC power flow constraints are considered. It is assumed that each load node has the charging and discharging station, also known as the network access facility. In addition, PHEVs that connected to the same access node are combined as one vehicle-to-grid (V2G) power resource.

We use AC power flow equations for modeling flow in the power grid. Conventionally, the power flow of the network is calculated through multiple iterations until convergence. Nevertheless, it is hard for the power flow calculation to converge with the control variables (such as reactive power output, transformer tap, and phase shifter angle) kept within their limits during the first few iterations. Accordingly, the slack variables are added to the standard power flow modified equations, as described in (14.15). It is assumed that the slack variables are all non-negative, and the physical meanings of the slack variables vector S_{p1} and S_{q1} could be understood as the virtual injection of active power and reactive power at each bus, while S_{p2} and S_{q2} can be understood as virtual active and reactive load demand. Since any load flow violations in transmission line is not allowed, no slack variable is appended for the transmission power flow equation as in (14.16).

The economic and environmental benefits have been considered in sub-model 1, and thus the sub-model 2 is designed with summation of the slack variables as objective function, which aims to minimize the additional virtual power generation or load demand, shown in (14.12).

Constraints (14.13) and (14.14) are the standard ones for power network with additional charging and discharging capacity limits: (14.13) is the network active power balance constraint at each bus while (14.14) is for reactive power balance. Even the advanced solver (such as CPLEX) seems powerless for the basic SCUC problem with AC power flow taken into account. Actually, security check accounts for most of the calculation time. If Newton-Raphson algorithm is employed for iterative optimization, then the control variables could be updated and the parametric matrixes (the admittance matrix as well as the Jacobian matrix) can bccomc known.

Constraints (14.17)–(14.21) specify the ranges of control variables. Constraints (14.22) and (14.23) fulfill the original scheme resulting from the sub-model 1. The objective function and the constraints are listed as follows:

$$\min \sum S_{p1} + S_{p2} + S_{q1} + S_{q2} \tag{14.12}$$

1) Power balance

$$d_P = YP_g + P_{dc}S_D(t) - P_d - P_cS_C(t) - P_{in} \tag{14.13}$$

$$d_Q = YQ_g - Q_d - Q_{in} \tag{14.14}$$

2) Modified power flow function

$$\begin{bmatrix} d_P \\ d_Q \end{bmatrix} + \begin{bmatrix} 0 \\ Y\Delta Qu \end{bmatrix} = -\begin{bmatrix} H & N & E & F \\ M & J & R & S \end{bmatrix} \begin{bmatrix} \Delta\delta \\ \Delta V \\ \Delta T \\ \Delta\gamma \end{bmatrix} + \begin{bmatrix} S_{p1} \\ S_{q1} \end{bmatrix} - \begin{bmatrix} S_{p2} \\ S_{q2} \end{bmatrix} \tag{14.15}$$

$$\Delta P_L = [A \ B \ C \ D][\Delta\delta \quad \Delta V \quad \Delta T \quad \Delta\gamma]^T \tag{14.16}$$

3) Power flow limit

$$-P_L^{max} \le P_L + \Delta P_L \le P_L^{max} \tag{14.17}$$

4) Bus voltage limit

$$V_{min} \le V + \Delta V \le V_{max} \tag{14.18}$$

5) Reactive power bound of units

$$Q_{min}u(t) \le Q_g + \Delta Q \le Q_{max}u(t) \tag{14.19}$$

6) Transformer tap limit

$$T_{min} \le T + \Delta T \le T_{max} \tag{14.20}$$

7) Phase shifter angle limit

$$\gamma_{min} \le \gamma + \Delta\gamma \le \gamma_{max} \tag{14.21}$$

8) PHEV charging and discharging amount limit

$$\sum_{j \in K} S_{Cj}(t) = N_c(t) \tag{14.22}$$

$$\sum_{j \in K} S_{Dj}(t) = N_d(t) \tag{14.23}$$

The computational procedure of the sub-model 2 is shown in Figure 14.2.

14.6 Numerical Examples

According to the simulation results of Reference [13]: (i) the average charging power of PHEV is set at 1.8 kW with the full charging time of six hours, so that the total charging capacity adds up to 10.8 kWh, (ii) the average discharging power of PHEVs is 1.7 kW, with a total discharging capacity of 10.2 kWh. Suppose that N_e^{max} accounts for 90% of the total PHEVs. The reserve requirement is assumed to be 10% of the load demand and the total scheduling period is 24 hours. All calculations have been solved using CPLEX solver.

The 6-bus system is used for case study and Figure 14.3 presents the diagram [25]. The load buses including bus 3, bus 4, and bus 5 are considered as EV network access nodes. Parameters of the units, load demand, bus data, line data, and charging/discharging prices are listed in Tables 14.1–14.5. The 6-bus system is considered for simulation with 6000 "grid-connected" PHEVs while N_c^{sch} and N_d^{sch} are all set at 18 000 hours, and the CO_2 emission cost coefficient is set at 1 \$/kg.

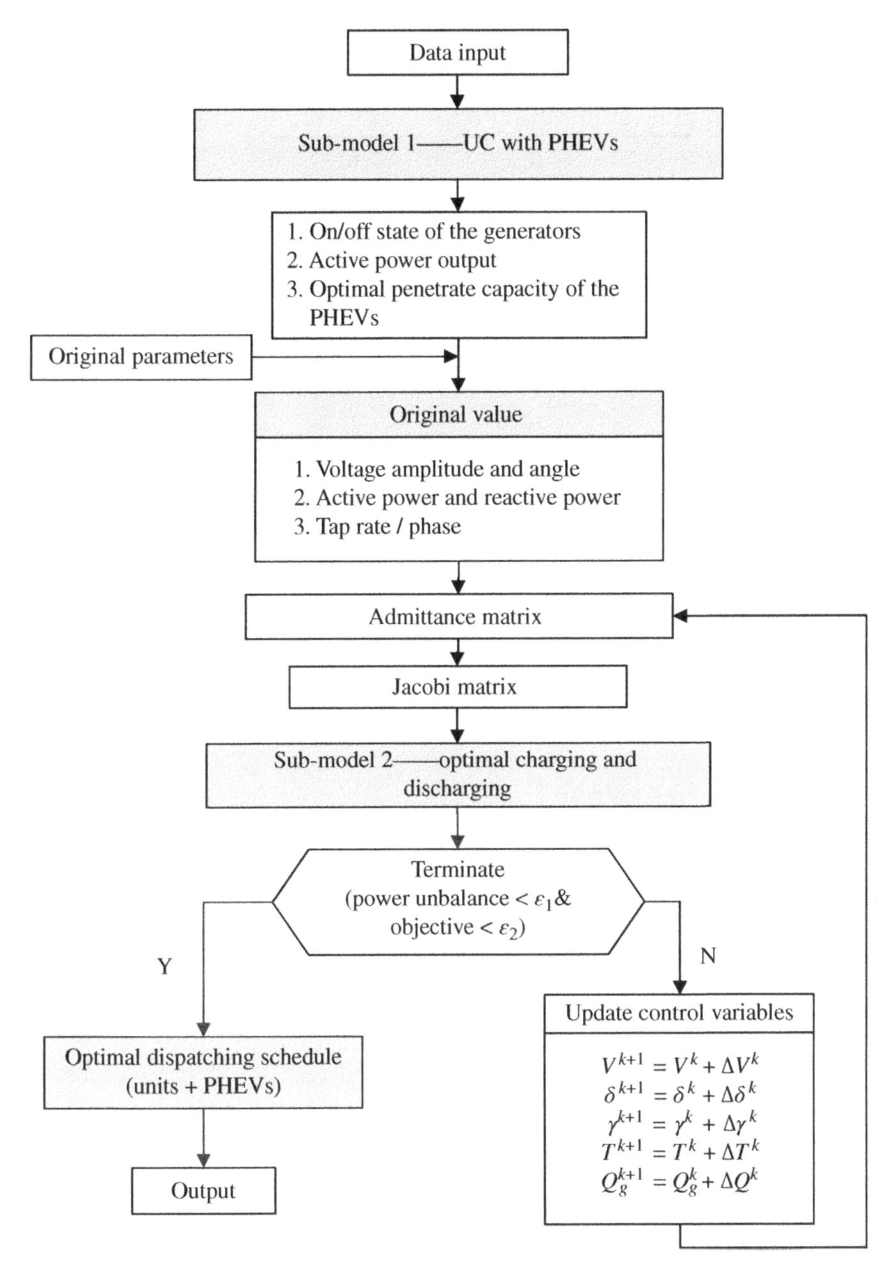

Figure 14.2 The flowchart of the computational procedure. *Source:* From Cai et al. [A] © 2015 IEEE.

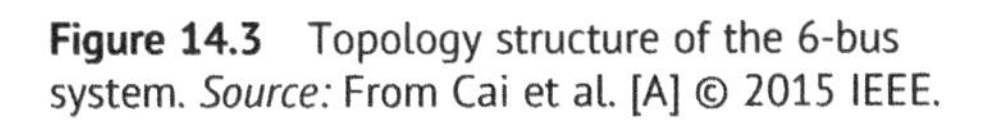

Figure 14.3 Topology structure of the 6-bus system. *Source:* From Cai et al. [A] © 2015 IEEE.

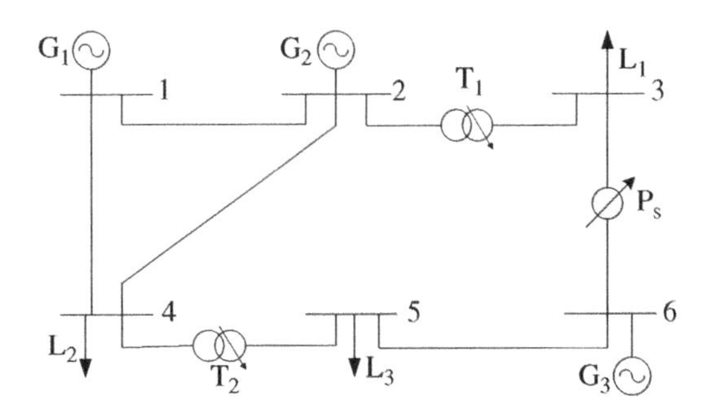

Table 14.1 Parameters of generators of the 6-bus system.

No.	G_1	G_2	G_3
P^{min}(MW)	100	10	10
P^{max}(MW)	220	100	20
Q_{min}(MVAr)	−80	−40	−40
Q_{min}(MVAr)	200	70	50
α(kg/h)	130.00	137.00	160.00
γ(kg/(MW²·h))	0.022	0.044	0.084
a($/h)	1000	700	670
b($/(MW·h))	16.19	16.60	27.79
c($/(MW²·h))	0.00048	0.00200	0.00173
$R_u R_d$(MW/min	55	50	20
S_u(MW/min)	100	80	30
Initial online time(h)	4	1	−1
T^{on}(h)	4	2	1
T^{off} (h)	4	3	1

Source: From Cai et al. [A] © 2015 IEEE.

Table 14.2 Hourly load data of the 6-bus system.

Period	Active load (MW)	Reactive load (MVAr)	Period	Active load (MW)	Reactive load (MVAr)
1	178	50	13	247	69
2	168	47	14	248	70
3	161	45	15	253	71
4	157	44	16	260	73
5	158	44	17	261	73
6	163	46	18	251	70
7	176	49	19	250	70
8	194	51	20	242	68
9	209	53	21	242	68
10	221	59	22	231	66
11	233	65	23	205	56
12	240	67	24	200	56

Source: From Cai et al. [A] © 2015 IEEE.

Both the cases with and without PHEVs consideration are calculated for comparisons, and the results are summarized in Table 14.6.

PHEVs are taken into account to calculate the generation units' schedule. The production cost, CO_2 emission cost, and PHEVs' economic benefit are also obtained as $122 949, $8741 and $1019, respectively and the social cost is $130 671 in total. Test results suggest that the time dependent

Table 14.3 Periods and prices of charging and discharging.

Period	Charging price (cent/kWh)	Period	Discharging price (cent/kWh)
23	8.8	10	10
24	8.6	11	12
1	8.2	12	9.8
2	8.2	14	12
3	8.0	15	12
4	8.0	16	12
5	8.2	17	10
6	8.5	20/21	9.8

Note: Discharging price of any other period is 8.5 cent/kWh.
Source: From Cai et al. [A] © 2015 IEEE.

Table 14.4 Bus data of the 6-bus system.

Bus No.	Upper voltage	Lower voltage	Initial voltage
1	1.05	0.95	1
2	1.15	0.85	1
3	1.15	0.85	0.96
4	1.05	0.91	1
5	1.15	0.85	0.96
6	1.15	0.85	0.96

Source: From Cai et al. [A] © 2015 IEEE.

Table 14.5 Branch data of the 6-bus system.

Line No.	Type	From bus	To bus	R (pu)	X (pu)	Flow limit (MW)	Max tap	Min tap
1	L	1	2	0.0050	0.170	200	-	-
2	L	1	4	0.0030	0.258	100	-	-
3	L	2	4	0.0070	0.197	100	-	-
4	L	5	6	0.0020	0.140	100	-	-
5	P	3	6	0	0.018	100	30^0	-30^0
6	T	2	3	0	0.037	100	1.053	1.02
7	T	4	5	0	0.037	100	1.053	1.02

Note: L, T and P represent for transmission line, transformer and phase shifter respectively.
Source: From Cai et al. [A] © 2015 IEEE.

Table 14.6 The optimal dispatching schemes with and without PHEVs.

	Case 1: Schedule with PHEVs					Case 2: Schedule without PHEVs		
	Units (MW)			PHEVs (MW)		Units (MW)		
T	G_1	G_2	G_3	Charging	Discharging	G_1	G_2	G_3
1	100	80	-	-	-	100	79.07	-
2	120.73	58	-	9.72	-	116	53.01	-
3	161.69	10	-	9.72	-	151.97	10	-
4	167.67	-	-	9.72	-	157.94	-	-
5	168.67	-	-	3.24	-	158.95	-	-
6	168.45	-	-	-	-	163.98	-	-
7	176	-	-	-	1.06	119.06	58	-
8	132	63.16	-	-	-	132	63.16	-
9	146.25	64	-	-	-	146.25	64	-
10	152.33	70	-	-	-	152.33	70	-
11	156.40	76	-	-	2.00	158.40	76	-
12	165.44	76	-	-	-	165.44	76	-
13	172	76.48	-	-	-	172	77.49	-
14	164.31	76	-	-	9.18	172.52	82	-
15	169.34	76	-	-	9.18	172.52	82	-
16	172	80.38	-	-	9.18	176	86.56	-
17	176	86.57	-	-	-	176	86.57	-
18	172	80.51	-	-	-	172	80.51	-
19	172	79.5	-	-	-	172	79.5	-
20	167.45	76	-	-	-	167.45	76	-
21	167.45	76	-	-	-	167.45	76	-
22	156.39	76	-	-	-	156.39	76	-
23	142.23	64	-	-	-	142.23	64	-
24	146.92	64	-	-	-	137.2	64	-

Source: From Cai et al. [A] © 2015 IEEE.

electricity prices succeed in guiding the charging in periods of low demand while discharging during high demand periods. Case 2 is the conventional SCUC with only the generators considered, the production cost is $123 647, while the CO_2 emission cost is $8663, and hence the social cost is up to $132 310. Due to additional PHEVs charging, the CO_2 emission of case 1 is a little bit higher than case 2. But the production cost of case 1 is lower than case 2 owing to the energy storage of PHEVs.

Based on the sub-model 2, the optimal charging and discharging schedule is shown in Table 14.7. Without the PHEVs taken into account, the power flows of the transmission lines are almost the same as the model considering PHEVs, displayed in Table 14.8.

Different cost coefficients of CO_2 emission are considered, and the corresponding schedules are listed in Table 14.9. For the case applying 1$/kg as cost coefficient in Table 14.9, the CO_2 emission is

Table 14.7 The optimal charging and discharging schedules for PHEVs of the 6-bus system.

	Charging PHEVs (10^3)				Discharging PHEVs (10^3)				
	t				t				
Bus	**2**	**3**	**4**	**5**	**7**	**11**	**14**	**15**	**16**
3	2	2	2	1.8	0.622	1.178	2	2	2
4	2	2	1.4	-	-	-	2	2	2
5	1.4	1.4	2	-	-	-	1.4	1.4	1.4
Total	5.4	5.4	5.4	1.8	0.622	1.178	5.4	5.4	5.4

Source: From Cai et al. [A] © 2015 IEEE.

Table 14.8 The line flow of the 6-bus system.

Branch No.	Power flow (MW)	Branch No.	Power flow (MW)
1→2	40.54	2→1	−40.42
1→4	96.66	4→1	−96.31
2→4	90.66	4→2	−89.97
5→6	38.28	6→5	−38.24
3→6	−38.24	6→3	38.24
2→3	13.76	3→2	−13.76
4→5	63.28	5→4	−63.28

Source: From Cai et al. [A] © 2015 IEEE.

Table 14.9 Optimal dispatching schemes under different cost coefficients of CO_2 emission.

CO_2 emission cost coefficient ($/kg)	Production cost ($)	CO_2 emission cost ($)	PHEVs' economic benefit ($)
1	122 949	8741	1019
2	122 972	17 443	1019
3	122 972	26 165	1019
4	122 986	34 855	1006

Source: From Cai et al. [A] © 2015 IEEE.

19.5, 19.7, and 27.3 kg higher than those with the cost coefficients set at 2, 3, and 4 $/kg, respectively. Test results have shown that the higher the cost coefficient is, the lower the total CO_2 emission would be. However, the production cost increases with the decrease of CO_2 emission. According to the units' parameters shown in Table 14.1, once the power output increases one unit, the emission increment of the three units diminish in the sequence of G_1, G_2, and G_3, which means that the power output of G_2 and G_3 will increase while G_1 decreases, causing increment of the total production cost.

14.7 Practical Application – The Impact of Electric Vehicles on Distribution Network

In most existing strategies which consider the impact of large-scale uncontrolled charging of electric vehicles on the power grid, only comparatively analyzed the impact of controlled and uncontrolled charging on the entire load curve of the power grid. This chapter provides a controlled charging management of electric vehicles plugged in the power grid. We firstly use the Monte-Carlo method to simulate the charging load of EVs. And then, the peak and valley time periods are divided by the membership function. Next, the consumer psychology model is used to optimize the time-of-use electricity price, with the minimum peak-valley load difference as the objective. Finally, the impact of electric vehicles on the entire load curve under controlled and uncontrolled charging is analyzed.

14.7.1 Modeling of Electric Vehicles

Tables 14.10 and 14.11 show the statistical results of plug-in time and charging demand of electric private vehicles respectively where the collecting number of chargers is 141 and the number of vehicles is 7135. Charging data of private vehicles come from public charging station and charging spots in residential areas. Some of these charging stations are located near residential areas, while most of

Table 14.10 Electric private vehicle plug-in time.

Duration	Charging time (s)	Percent (%)
0:00–8:00	7723	19.08
8:00–14:00	10 746	26.55
14:00–17:00	7438	18.38
17:00–19:00	4654	11.50
19:00–22:00	6205	15.33
22:00–24:00	3701	9.15

Source: Based on Feng et al. [B].

Table 14.11 Electric private vehicle charging quantity.

Charging quantity	Charging time (s)	Percent (%)
0–10 kWh	5277	12.92
10–20 kWh	115 554	28.55
20–30 kWh	8941	22.09
30–40 kWh	8181	20.22
40–60 kWh	5737	14.18
60–80 kWh	827	2.04

Source: Based on Feng et al. [B].

them are located in power stations near enterprises and public institutions. It can be seen that in Yangjiang residential areas, many users choose to charge their cars at work during the day.

Since charging spots have not been installed in many residential areas, users mainly choose to charge their cars after work or at night. At present, the battery capacity of private vehicles is mainly between 50 and 70 kWh, while a small number of private vehicles have a battery capacity of 80 kWh, so the charging quantity of private vehicles is mainly between 10 and 60 kWh.

The probability density function (PDF) of charging demand for electric vehicles is not described accurately by a single distribution function. In this chapter, multiple distribution functions are considered to jointly describe the PDF characteristics of charging demand.

The charging demand analysis of EV based on PDF method can refer to the gray image processing technology. The image gray histogram reflects the frequency of gray value at a certain level in the image, which can also be considered as the estimation of the gray probability density of the image. The histogram of charging demand variation reflects the probability density estimation of plug-in time and charging quantity. The Gaussian Mixture Model (GMM) is widely used in the modeling of grayscale image processing. In this chapter, the GMM is used to describe the probability density distribution characteristics of charging demand.

The mathematical model of GMM is

$$f(x) = \sum_{i=1}^{n} a_i \cdot e^{\left(\frac{x-b_i}{c_i}\right)^2} \tag{14.24}$$

where a_i, b_i, and c_i are the distribution parameters of the model, which can be obtained by maximum likelihood estimation.

According to the summary of plug-in time and charging quantity discussed above, Figure 14.4a presents the probability distribution of the probability distributions of plug-in time of electric private vehicles and Figure 14.4b presents the probability distribution of charging quantity of electric private vehicles. Figure 14.4a can be modelled by GMM, and Figure 14.4b can be modeled by Gamma distribution as follows:

$$f(x, \beta, \alpha) = \frac{\beta^{\alpha}}{\Gamma(\alpha)} x^{\alpha-1} e^{\beta x} \tag{14.25}$$

where α and β are the distribution parameters of the model. The corresponding PDFs parameters are summarized in Table 14.12.

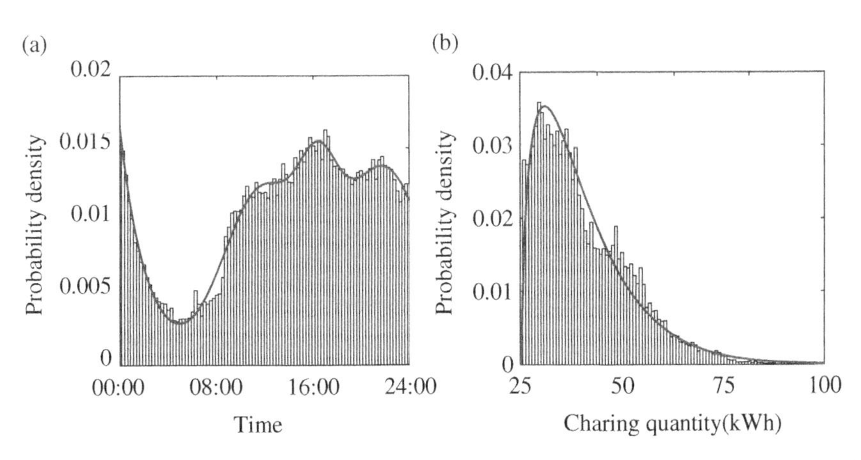

Figure 14.4 The plug-in time and charging quantity of electric private vehicles. (a) plug-in time, (b) charging quantity. *Source:* Based on Feng et al. [B].

Table 14.12 Plug-in-time and charging quantity of private vehicles and electric buses.

	Charging period	Plug-in time	Charging quantity
Electric private vehicles	00:00–24:00	$a_1 = 0.00728, b_1 = 65.63, c_1 = 9.539$	Gamma distribution $\alpha = 1.6163$ $\beta = 12.6565$
		$a_2 = 3.181 * 10^{11}, b_2 = -545.3, c_2 = 98.58$	
		$a_3 = 0.01365, b_3 = 87.38, c_3 = 20.21$	
		$a_4 = 0.01202, b_4 = 46.81, c_4 = 17.75$	
Electric buses	00:00–06:00	$a_1 = 1.267, b_1 = -184.1, c_1 = 111.7$	$a_1 = 0.01396, b_1 = 39.29, c_1 = 93796$ $a_2 = 0.03079, b_2 = 23.85, c_2 = 3.621$ $a_3 = 0.23590, b_3 = 12.70, c_3 = 11.08$ $a_4 = 0.20810, b_4 = 12.09, c_4 = 11.04$
		$a_2 = 0.02907, b_2 = 120.6, c_2 = 68.26$	
		$a_3 = 0.02299, b_3 = 236.2, c_3 = 83.03$	
	06:00–12:00	$a_1 = 0.01576, b_1 = 698.8, c_1 = 62.70$	
		$a_2 = 0.01086, b_2 = 523.9, c_2 = 86.85$	
	12:00–16:00	$a_1 = 0.01480, b_1 = 762.1, c_1 = 78.89$	
		$a_2 = 0.01060, b_2 = 862.0, c_2 = 116.5$	
	16:00–24:00	$a_1 = 0.12290, b_1 = 1328, c_1 = 10.64$	
		$a_2 = 0.01547, b_2 = 1439, c_2 = 51.90$	
		$a_3 = 0.00787, b_3 = 1272, c_3 = 12.58$	
		$a_4 = 0.00457, b_4 = 1068, c_4 = 128.9$	

Source: Based on Feng et al. [B].

Table 14.13 Electric buses plug-in time.

Duration	Charging time (s)	Percent (%)
0:00–8:00	39 595	48.35
8:00–14:00	19 125	23.35
14:00–17:00	6207	7.58
17:00–19:00	2070	2.53
19:00–22:00	1541	1.88
22:00–24:00	13 359	16.31

Source: Based on Feng et al. [B].

Tables 14.13 and 14.14 show the statistical results of plug-in time and charging quantity of electric buses, respectively.

Since most buses in Yangjiang operate until 22:00, the number of charging buses increases sharply after 22:00. Some buses that are not charged at night choose to be charged during the day to maintain normal daily operation. As the battery capacity of buses is larger than that of private vehicles, it can be seen from Tables 14.11 to 14.13 that the charging quantity of buses is larger, mainly within the range of 30–120 kWh. Unlike private vehicles, buses have specific operating time and time headways, so the probability distribution of their plug-in time is relatively complicated,

Table 14.14 Electric buses charging quantity.

Charging quantity	Charging time (s)	Percent (%)
10–30 kWh	5652	6.90
30–60 kWh	24 999	30.52
60–90 kWh	31 340	38.27
90–120 kWh	11 044	13.48
120–150 kWh	8088	9.88
150–180 kWh	774	0.95

Source: Based on Feng et al. [B].

which requires piecewise fitting. The result of piecewise fitting for the probability density related to the time of the whole day is shown in Figure 14.5. Figures 14.5 (a), (b), (c) and (d) show the distribution for plug-in time 00.00–06:00, 06.00–12:00, 12.00–16:00 and 16.00–24:00 respectively.

The probability density functions of charging quantity for electric buses is shown in Figure 14.6.

According to the summary of plug-in time and charging quantity discussed earlier, Figures 14.5 and 14.6 present the probability distributions of electric buses. The corresponding PDFs parameters are summarized in Table 14.12.

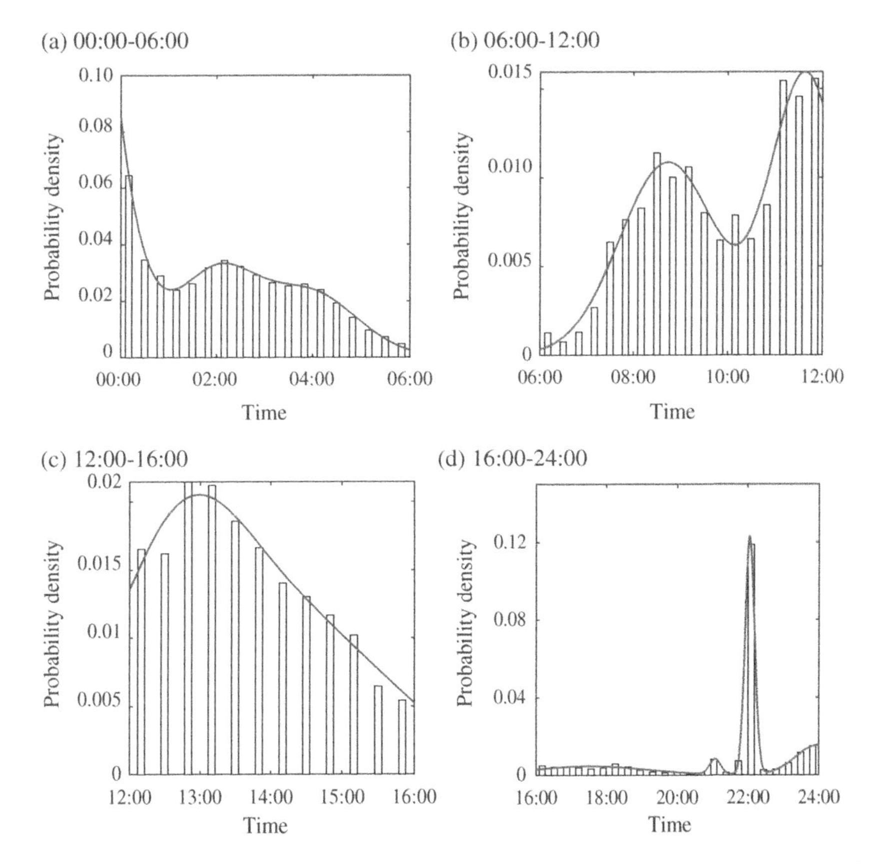

Figure 14.5 The plug-in time of buses. (a) Plug-in time 00.00–06:00. (b) Plug-in time 06.00–12:00. (c) Plug-in time 12.00–16:00. (d) Plug-in time 16.00–24:00. *Source:* Based on Feng et al. [B].

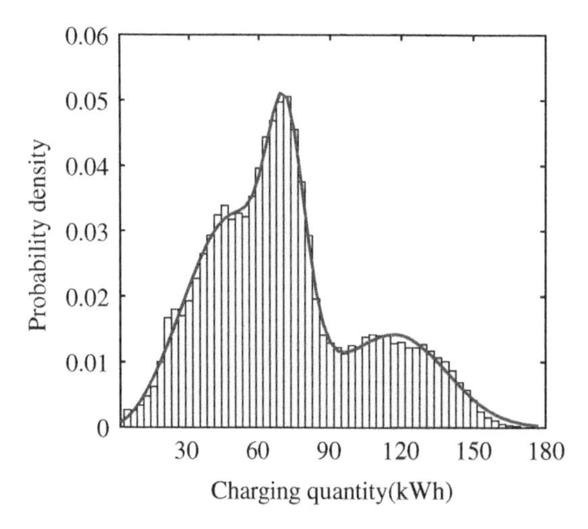

Figure 14.6 The charging quantity of electric buses. *Source:* Based on Feng et al. [B].

14.7.2 Uncontrolled Charging

1.Calculation of EV Charging Load Based on Monte–Carlo simulation (MCS)

MCS is a random simulation method to study probabilistic phenomena. The approximate results are calculated on the basis of random sampling. With the increase of sampling times, the probability of getting correct results increases gradually. The stochastic process of Monte-Carlo method can be divided into three steps:

1) *Model:* Construct a probabilistic model.
2) *Sample:* Take a random sample from a given probability distribution.
3) *Result: Obtain an approximate result.*

According to the Statistical Yearbook of Guangdong Province 2019, the vehicle PARC of Yangjiang in 2018 was 371 917, including 352 874 private cars and 449 public buses [29]. In its 13th Five-Year Plan, Yangjiang Development and Reform Bureau predicted that the number of vehicles in 2018 would be 394 120, where 389 177 private cars and 460 public buses. By 2020, the population will reach 618 035, including 612 667 private cars and 610 public buses [30]. As shown in Table 14.15, the vehicle PARC of Yangjiang during the 13th Five-Year Plan period has been growing rapidly in recent years, of which the main increment is from private passenger cars, which will increase by nearly twice in 2020 compared to that in 2015.

Table 14.15 Prediction of vehicle PARC in Yangjiang.

Vehicle types	2015	2016	2017	2018	2019	2020
Bus	302	348	400	460	530	610
Taxi	677	719	719	719	719	719
Goods truck/ Sanitation vehicle	922	1010	1107	1213	1329	1456
Utility vehicle	2500	2517	2533	2550	2567	2584
Private vehicle	197 030	247 213	310 177	389 177	488 299	612 677
Total	201 431	251 806	314 936	394 120	493 444	618 035

Source: Based on Feng et al. [B].

Table 14.16 Prediction of EV PARC in Yangjiang (From [B]).

Vehicle types	2015	2016	2017	2018	2019	2020
Bus	99	140	198	261	334	417
Taxi	0	5	5	5	405	405
Goods truck/ Sanitation vehicle	0	13	31	54	86	131
Utility vehicle	0	3	7	12	18	25
Private vehicle	1	177	680	1747	3680	6913
Total	100	339	922	2080	4524	7892

Source: Based on Feng et al. [B].

Electric vehicles will be fully promoted as a national target in the future in China. As shown in Table 14.16, the electric vehicles PARC predicted by Yangjiang Development and Reform Commission during the 13th Five-Year Plan period shows that Yangjiang will firstly promote and develop public transport and official vehicles, and then shift its focus to private vehicles [30].

It is predicted that the penetration rate of electric vehicles on public buses in Yangjiang will reach 68.4% in 2020, much higher than that of private cars, which is 1.1%. Since the base number of private cars is much higher than that of other types of vehicles, the larger penetration rate of electric private cars in the future will impact power grid greater.

Assuming that the total number of private vehicles and public buses will be 610 000 and 1000, respectively in 2020.The charging power of the fast charging spot for private vehicles is 42, and 7 kW for slow charging spot. Also, the charging power of the public buses is 70 kW. The fast charging ratio of the private vehicles' charging spots is 13.5% and the slow charging is 76.5%. Using MCS, the charging load curve of the private vehicles is shown in Figure 14.7. If all buses use fast charging, the obtained charging load curve of the bus is shown in Figure 14.8.

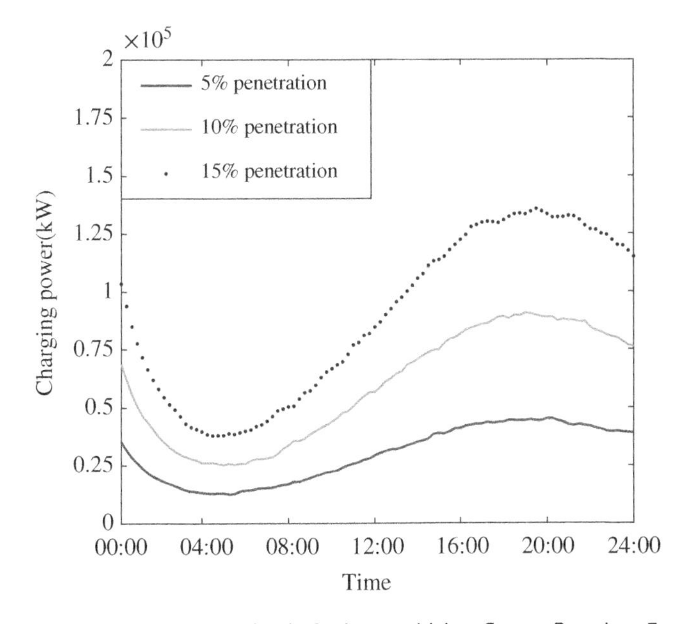

Figure 14.7 Charging load of private vehicles. *Source:* Based on Feng et al. [B].

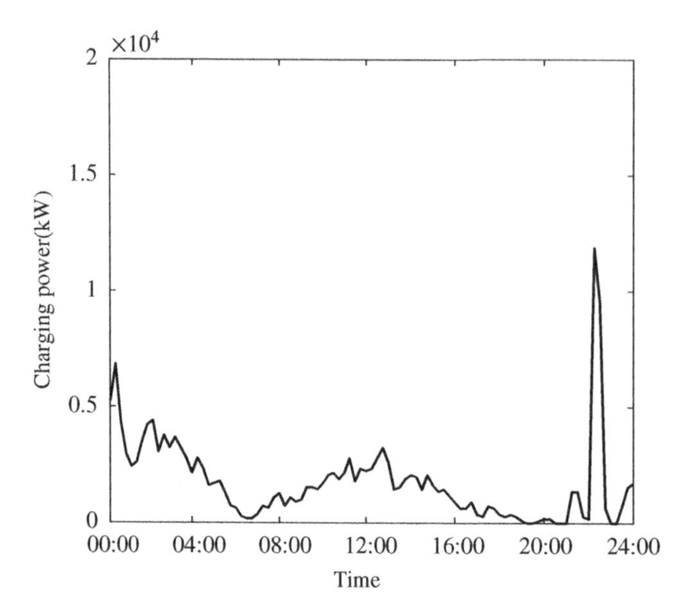

Figure 14.8 Charging load of buses. *Source:* Based on Feng et al. [B].

14.7.3 Results

As seen from Figure 14.8, the charging peak demand of the private vehicle is between 19:00 and 21:00. During this period, most owners of electric private cars go home from work then the electric vehicles may be charged. The charging demand of the private vehicles is low between 7:00 and 9:00, when most of the electric vehicles have been out of charging for going to work. This shows that the charging demand of private vehicles is affected by the human's activity. The charging demand of the buses is relatively large during the daytime, and the charging peak demand appears between 22:00 and 23:00, which can be seen from Figure 14.8. Adding the uncontrolled charging load of electric private vehicles and electric buses to the typical daily load profile in Yangjiang, the results are shown in Figure 14.9. Three cases are conducted, in which electric private vehicle penetration rates are 5, 10, and 15%.

As seen from Figure 14.9, the uncontrolled charging of electric vehicles will bring an extra peak load to the regional power grid, which increases the weakness of the power supply. With the penetration rate of electric private vehicles increases, this phenomenon will become more obvious.

14.8 Conclusions

With the rapid development of electric vehicles, the large-scale charging behavior of electric vehicles will inevitably affect the distribution network. The optimal plug-in capacities for PHEVs and the schemes, including when and where charging and discharging occurred has been studied with Security Constrained unit commitment scheme. In this chapter, the calculation method of electric vehicle charging load has been investigated. The uncontrolled charging of electric vehicles will

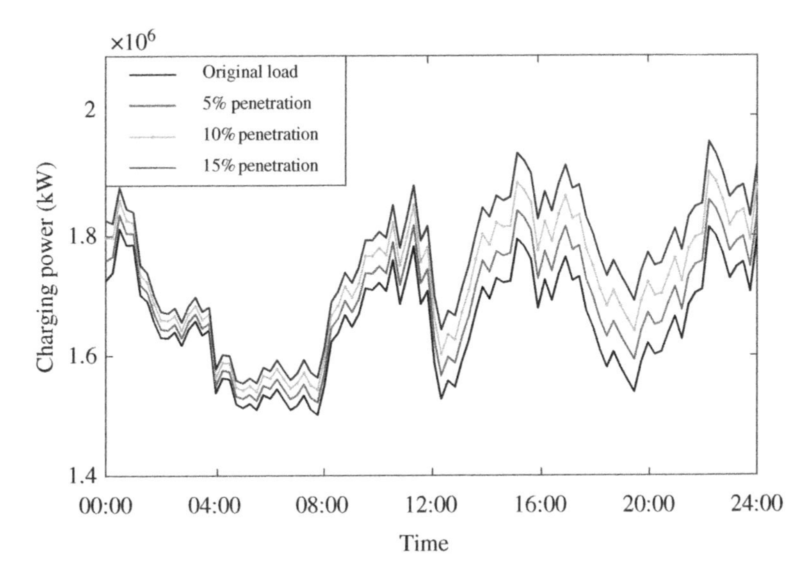

Figure 14.9 Load curve of the region considering the uncontrolled charging load during the day.
Source: Based on Feng et al. [B].

bring an extra peak load to the regional power grid, which increases the power system instability and can affect smart city operation.

Acknowledgements

The permission given in using materials from the following papers is very much appreciated.

[A] Cai, Q., Xu, Z., Wen, F. et al. (2015). Security constrained unit commitment-based power system dispatching with plug-in hybrid electric vehicles. *2015 IEEE 13th International Conference on Industrial Informatics (INDIN)*, Cambridge, UK (22–24 July 2015).

[B] Feng, K., Zhong, Y., Hong, B. et al. (2020). The impact of plug-in electric vehicles on distribution network. *Proceedings of the 2020 IEEE International Smart Cities Conference,* Cambridge, UK (28 September to 1 October 2020).

References

1 Millo, F., Rolando, L., Fuso, R., and Mallamo, F. (2014). Real CO_2 emissions benefits and end user's operating costs of a plug-in Hybrid Electric Vehicle. *Applied Energy* 114: 563–571.
2 Iana, V. and Campillo, J. (2017). Adoption barriers for electric vehicles: experiences from early adopters in Sweden. *Energy* 120: 632–641.
3 Farahani, H.F. (2017). Improving voltage unbalance of low-voltage distribution networks using plug-in electric vehicles. *Journal of Cleaner Production* 148: 336–346.
4 Hanemann, P., Behnert, M., and Bruckner, T. (2017). Effects of electric vehicle charging strategies on the German power system. *Applied Energy* 203: 608–622.

5 Su, S.J., Lie, T.T., and Zamora, R. (2019). Modelling of large-scale electric vehicles charging demand: a New Zealand case study. *Electric Power Systems Research* 167: 171–182.
6 Munshi, A.A. and Mohamed, Y. (2019). Unsupervised nonintrusive extraction of electrical vehicle charging load patterns. *IEEE Transactions on Industrial Informatics* 15: 266–279.
7 Mehta, R., Srinivasan, D., Tivedi, A., and Yang, J. (2019). Hybrid planning method based on cost-benefit analysis for smart charging of plug-in electric vehicles in distribution systems. *IEEE Transactions on Smart Grid* 10: 523–534.
8 Solanke, T.U. et al. (2020). A review of strategic charging–discharging control of grid-connected electric vehicles. *Journal of Energy Storage* 28: 101193.
9 Zhang, J., Jorgenson, J., Markel, T. et al. (2019). Value to the grid from managed charging based on California's high renewables study. *IEEE Transactions on Power Apparatus and Systems* 34: 831–840.
10 Cheng, A., Tarroja, B., Shaffer, B., and Samuelsen, S. (2018). Comparing the emissions benefits of centralized vs. decentralized electric vehicle smart charging approaches: a case study of the year 2030 California electric grid. *Journal of Power Sources* 401: 175–185.
11 Humayd, A. and Bhattacharya, K. (Mar. 2019). Design of optimal incentives for smart charging considering utility-customer interactions and distribution systems impact. *IEEE Transactions on Smart Grid* 10: 1521–1531.
12 Foley, A.M., Leahy, P.G., McKeogh, E.J., and Gallachóir, B.P.O. (n.d.). Electric vehicles and displaced gaseous emissions. Presented at the IEEE Vehicle Power and Propulsion Conference, Lille, France (1–3 September 2010), 1–6.
13 Parks, K., Denholm, P., and Markel, T. (2007). Costs and emissions associated with plug-in hybrid electric vehicle charging in the Xcel energy Colorado service territory. National Renewable Energy Laboratory Technical Report, [Online]. http://www.nrel.gov/analysis/pdfs/41410.pdf.
14 Vlachogiannis, J.G. (2009). Probabilistic constrained load flow considering integration of wind power generation and electric vehicles. *IEEE Transactions on Power Systems* 24 (4): 1808–1817. https://doi.org/10.1109/TPWRS.2009.2030420.
15 Hernandez, J.E., Kreikebaum, F., and Divan, D. (n.d.). Flexible electric vehicle (EV) charging to meet renewable portfolio standard (RPS) mandates and minimize green house Gas emissions. Presented at the 2010 IEEE Energy Conversion Congress and Exposition (ECCE), Atlanta, USA (11–16 September 2010), 4270–4277.
16 Qian, K.J., Zhou, C.K., Allan, M., and Yuan, Y. (2011). Modeling of load demand due to EV battery charging in distribution systems. *IEEE Transactions on Power Systems* 26 (2): 802–810.
17 Sortomme, E., Hindi, M.M., Macpherson, S.D.J., and Venkata, S.S. (2011). Coordinated charging of plug-in hybrid electric vehicles to minimize distribution system losses. *IEEE Transactions on Smart Grid* 1 (1): 198–205.
18 Clement, N.K., Haesen, E., and Driesen, J. (2010). The impact of charging plug-in hybrid electric vehicles on a residential distribution grid. *IEEE Transactions on Power Systems* 25 (3): 371–380.
19 Saber, A.Y. and Veneyagamoor, G.K. (2010). Intelligent unit commitment with vehicle-to-grid–a cost emission optimization. *Journal of Power Sources* 195 (3): 898–911.
20 Ghanbarzadeh, T., Goleijani, S., and Moghaddam, M.P. (n.d.). Reliability constrained unit commitment with electric vehicle to grid using hybrid Particle Swarm Optimization and Ant Colony Optimization. Presented at the 2011 IEEE Power and Energy Society General Meeting, Detroit, USA (24–29 July 2011), 1–7.
21 Saber, A.Y. and Venayagamoorthy, G.K. (n.d.). Optimization of vehicle-to-grid scheduling in constrained parking lots. Presented at the IEEE Power & Energy Society General Meeting, Calgary, Canada (26–30 July 2009), 1–8.
22 Pillai, J.R. and Bak-Jensen, B. (2011). Integration of vehicle-to-grid in the Western Danish power system. *IEEE Transactions on Sustainable Energy* 2 (1): 12–19.

23 Han, S., Han, S., and Sezaki, K. (2010). Development of an optimal vehicle-to-grid aggregator for frequency regulation. *IEEE Transactions on Smart Grid* 1 (1): 65–72.

24 Bessa, R.J. and Matos, M.A. (n.d.). The role of an aggregator agent for EV in the electricity market. 7th Mediterranean Conference and Exhibition on Power Generation, Transmission, Distribution and Energy Conversion, Napa, Cyprus (7–10 November 2010), 1–9.

25 Fu, Y., Shahidehpour, M., and Li, Z. (2005). Security-constrained unit commitment with AC constraints. *IEEE Transactions on Power Systems* 20 (3): 1538–1550.

26 Denholm, P. and Short, W. (n.d.). An evaluation of utility system impacts and benefits of optimally dispatched plug-in hybrid electric vehicles. [Online]. National Renewable Energy Laboratory Technical Report. [2006-10]. http://www.nrel.gov/docs/fy07osti/40293.pdf.

27 Zhao, L., Prousch, S., Hubner, M., and Moser, A. (n.d.). Simulation methods for assessing electric vehicle impact on distribution grid. Presented at the 2010 IEEE PES transmission and distribution conference and exposition, New Orleans, US (19–22 April 2010), 1–7.

28 Carrion, M. and Arroyo, J.M. (2006). A computationally efficient mixed-integer linear formulation for the thermal unit commitment problem. *IEEE Transactions on Power Systems* 21 (3): 1371–1377.

29 (2019). *Guangdong Statistical Yearbook-2019. Statistics Bureau of Guangdong Province.* Beijing: China Statistics Press.

30 (2016). *Yangjiang Electric Vehicle charging Infrastructure Special planning (2016–2020).* Yangjiang: Yangjiang Development and Reform Bureau.

Appendix 14.A

Figures 14.A.1 and 14.A.2 show the summer and winter time load patterns with PHEV charging.

1) Uncontrolled Charging

The uncontrolled charging considers a simple PHEV scenario where vehicle owners charge their vehicles exclusively at home in an uncontrolled manner. The PHEV begins charging as

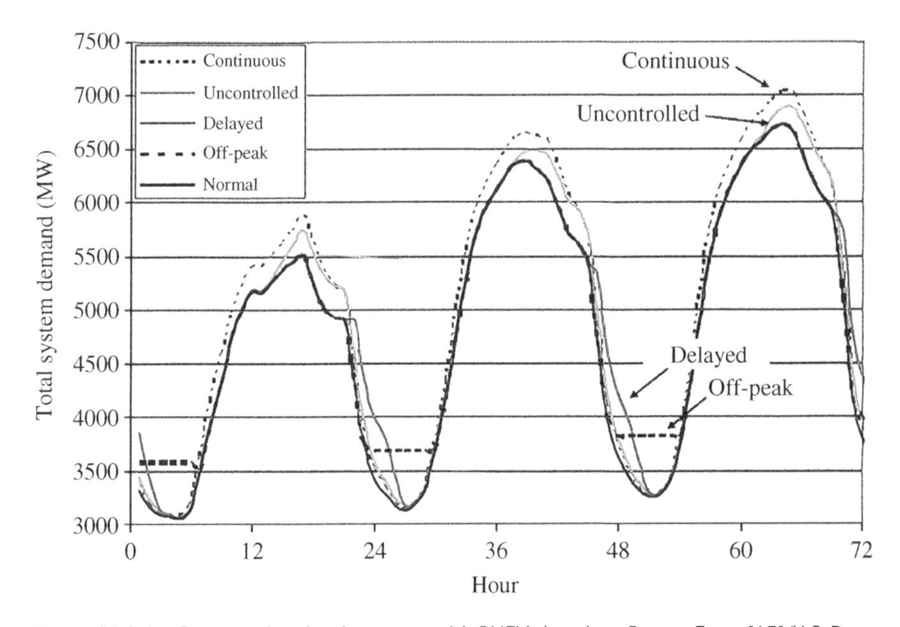

Figure 14.A.1 Summer time load patterns with PHEV charging. *Source:* From [13]/U.S. Department of Energy/ Public Domain.

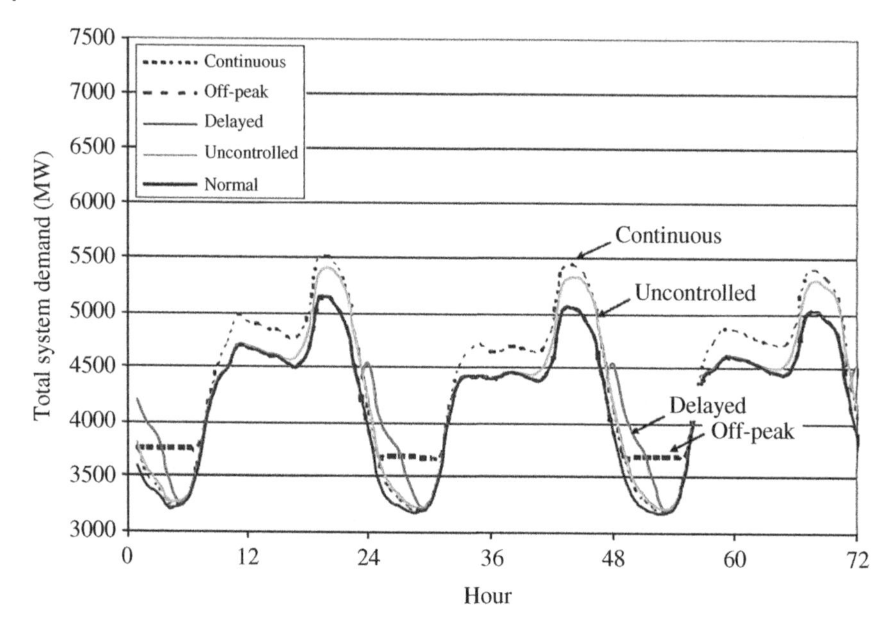

Figure 14.A.2 Winter time load patterns with PHEV charging. *Source:* From [13]/U.S. Department of Energy/ Public Domain.

soon as it is plugged in, and stops when the battery is fully charged. This can be considered a reference or "do nothing" case, because it assumes a business-as-usual infrastructure requirement (no charging stations at work or other public locations).

2) Delayed Charging

The delayed case is similar to uncontrolled charging, in that all charging occurs at home. However, it attempts to better optimize the utilization of low-cost off-peak energy by delaying initiation of household charging until 10 p.m.

3) Off-Peak Charging

The off-peak charging scenario also assumes that all charging occurs at home in the overnight hours. However, it attempts to provide the most optimal, low-cost charging electricity by assuming that vehicle charging can be controlled directly or indirectly by the local utility. This allows the utility to precisely match the vehicle charging to periods of minimum demand, allowing the use of lowest-cost electricity, and improving overall utility system performance.

4) Continuous Charging

The continuous charging scenario is similar to uncontrolled charging, in that it assumes that charging occurs in an uncontrolled fashion whenever the vehicle is plugged in. However, it also assumes that public charging stations are available wherever the vehicle is parked. As a result, the vehicle is continuously charged whenever it is not in motion, (limited by the battery capacity).

15

Machine Learning for Electric Bus Fast-Charging Stations Deployment

Nomenclature

Abbreviation

e-bus	Electric Bus
AP	Affinity Propagation algorithm
BPSO	Binary Particle Swarm Optimization algorithm
GA	Genetic algorithm
SOC	State of Charge

Parameters

D	Dimension of the particles
c_1, c_2	Learning factors
λ	The damping factor
ω	Inertia weight
P	Charging power of the equipped charging facilities, kWh
$C_{battery}$	Battery capacity of e-bus, kWh
u	Energy consumption per kilometer, kWh/km
α	Fluctuation coefficient of charging demand
k_s	Operation simultaneous rate of charging facilities
k_{eff}	Charging efficiency of the charging facility, %
TR_j	The operation time of route j, minute
TS_j	The spacing interval of route j, minute
SOC_{max}	Maximum *SOC* of e-bus battery
$SOC_{warning}$	Minimum *SOC* of e-bus battery
$L_{i,j}$	Length of route j operated by the ith terminus, km
η_b	Assuming scaling factor, %
g_t	Electricity price of charging, ¥/kWh
N	Number of distribution network nodes
μ	Charging capacity redundancy of the fast-charging station
λ_d	Influencing factor on the operation routes
r_0	Discount rate, %

Smart Energy for Transportation and Health in a Smart City, First Edition. Chun Sing Lai, Loi Lei Lai and Qi Hong Lai.

γ	Operating life of the charging station, year
p_1	Unit price of the charging facility, ¥
ω_c	Construction cost of charging station c, ¥
α_{cn}	Equipment and installation cost of power distribution line per kilometer, ¥/km
p_2	Unit price of the transformer, ¥
g_p	Average tariff including tax, ¥/kWh
L	Service radius of fast-charging station, km
U_n^{min}, U_n^{max}	Upper/lower margins of voltage magnitude, kV
NS_{min}	Minimum number of charging facilities
NS_{max}	Maximum number of charging facilities
P_n^{max}	Maximum power allowed for node n of distribution network, kW
P_{loss1}	Active power loss before fast-charging station access, kW
P_n	Load power of node n of distribution network, kW

Variables

F	Objective function
p_{id}	Best position of particle in BPSO
p_{gd}	Best position of global in BPSO
v_{id}	Particle velocity in BPSO
$d_{i,c}$	Distance from charging demand point i to fast-charging station c, km
$var1_{i,j,b,t}$	The operating state of e-bus b in route j of terminus i at the time of its t-th operation
$var2_{i,j,b,t}$	The charging state of e-bus b in route j of terminus i at the time of its t-th charging
$SOC1_{i,j,b,t}$	SOC of e-bus b in route j of terminus i at the time of its t-th departure
$SOC2_{i,j,b,t}$	SOC of e-bus b in route j of terminus i at the time of its t-th arrival
$E_{i,j}$	Energy consumption of route j operated by the i-th terminus, kWh
$E_{i,c}$	Energy consumption from bus terminus i to charging station c, kWh
$TD_{i,j,b,t}$	Dwelling time of e-bus b in route j operated by terminus i at the time of its tth dwell, minute
$TC_{i,j,b,t}$	Charging time of e-bus b in route j operated by terminus i at the time of its tth dwell, minute
$T_{i,c}$	Driving time from bus terminus i to charging station c, minute
$e_{i,j,b,t}$	Electric energy obtained of e-bus b in route j of terminus i at the time of its t-th charging, kWh
NSD_c	Number of charging spots required for daytime charging of fast-charging station c
TCD_c	Effective charging time of fast-charging station c during the daytime, minute
$\Delta tc_{i,j,b}$	Time required for e-bus b in route j operated by terminus i to be fully charged at night, minute
TCN_c	Effective charging time of fast-charging station c at night, minute
NSN_c	Number of charging spots required for night charging of fast-charging station c
NS_c	Number of charging spots to be built for the fast-charging station c
NT_c	Number of transformers in the charging station c
$n_{c,i}$	Number of charging times of terminus i of charging station c
$Y_{c,n}$	State variable representing the connection between charging station c and node n in distribution network.

$L_{c,n}$	Length of the power distribution line connecting charging station c to node n of distribution network, km
x_{id}	State variable representing whether the bus terminus is the charging station or not
NB_j	Number of e-buses of route j
NI_c	Number of charging demand points belonging to the same cluster of fast-charging station c
NJ_i	Number of bus routes operated by charging demand point i
NT_b	Number of arrivals of e-bus b
NC	Number of charging stations
C_{1c}	Equipment and installation cost of charging station c, ¥
C_{2c}	Construction cost of power distribution line, ¥
C_{3c}	Operation and maintenance cost of charging station c, ¥
C_{4c}	Travel cost of e-bus to charging station c, ¥
C_5	Power loss cost, ¥
P_{loss2}	Active power loss after fast-charging station access, kW
U_n	Voltage magnitude of bus n, kV
$P_{c,n}$	Charging power of the charging station c access node n of distribution network, kW

Sets and Indices

$\vec{x}_i$	The set of bus terminus
$S(i, j)$	The set of "similarity"
$R(i, k)$	The set of "responsibility"
$A(i, k)$	The set of "availability"
b	The e-buses index, where $b = 1, 2, ..., NB_j$
c	The fast-charging station index, where $c = 1, 2, ..., NC$

15.1 Introduction

With the depletion of fossil energy resources and increased environmental pollution, low-carbon energy usage and the reduction of carbon emissions attract people's awareness. As a major energy producer and consumer, China has also been committed to energy transition and reducing carbon emissions [1]. Electric vehicles, as a relatively new mode of transportation, are considered as one of the solutions for China to reduce its carbon emissions and dependence from oil. If electric vehicles are developed rapidly, based on 140 million of Chinese automobiles in 2020, it can save 32.29 million tons of oil and replace 31.1 million tons of oil, that is, a total of 63.39 million tons of oil, which is equivalent to cutting the demand for automobile oil by 22.7%. The battery electric buses (BEBs) play an important role in urban transportation in many cities. By analyzing the life cycle costs and carbon emissions of different type of electric buses, some studies point out that electric vehicles produce less carbon emissions than the traditional ones, which is important to improving air quality and environmental pollution [2, 3]. With the development of new energy technology and charging management technology, the environmental friendliness of electric vehicles is expected to be further improved [4, 5]. In addition, the use of electric vehicles can also contribute to the consumption of new energy and the stable operation of distribution network [6]. For example, electric vehicles can be used to mitigate photovoltaic output fluctuations in low-voltage grids [7]. The demand response of electric vehicles can also be used to smooth the wind power and limit the ramp rate, so as to solve the problem of high penetration of wind farms [8]. Some studies have shown that electric vehicles can be used as portable generators, which supply power to critical loads in emergency condition by using vehicle-to-grid (V2G) technology [9].

However, with the growth of electric vehicles, problems such as lack of charging infrastructure and unreasonable distribution are gradually exposed [10]. To better understand the interaction between the promotion of electric vehicles and construction of charging infrastructure, China has issued a series of policies to promote the development of a national charging network [11].

Therefore, reasonable charging station deployment may help in avoiding noneconomic investment and further promoting the penetration of electric vehicles in the market. Most of the current studies focus on private electric vehicles or taxis, and fewer consider the optimization of electric bus charging stations. For example, Yang et al. described the decision-making process of EV users and the driving characteristics of EVs, and further analyzed EV charging demand variation curve [12]. Hosseini, Seyedmohsen, and MD Sarder proposed a Bayesian Network (BN) model that considered uncertainty, quantitative, factors and qualitative factors, further assessed the site selection of charging stations from a sustainability and technical point of view [13]. From the perspective of distribution network, Cheng et al. described a distributed test network model, which combined active and reactive power optimization methods to determine the optimal placement of charging stations to reduce power losses [14]. From the driver's point of view, according to the trip success ratio of EVs, Alhazmi et al. selected the charging stations to optimize the trip success ratio from the existing candidate charging stations [15]. Morro-mello et al. also proposed a method to optimize the allocation of fast charging stations for urban electric taxis which met the planning requirements of all urban planners [16].

Compared with the randomness of private cars and taxis, electric bus has a fixed operation route and a systemic operation schedule every day. This is not only beneficial to the countries to reduce carbon emissions as well as the transition to low-carbon energy, but also to the charge of grid for V2G scheduling. From the technological development history of electric buses in China, electric buses occupy a large share of the Chinese new energy vehicle market and will play a key role in urban transportation electrification [17]. From the recent announcements of the National Development and Reform Commission and Ministry of Industry and Information Technology, it is suggested that China will vigorously promote electrification of public sector in the future, including public transportation, sanitation vehicles, and taxis [18–20]. Mahmoud et al. reviewed the development history of electric bus technology and pointed out that electrification of public buses in cities is feasible [21]. Gao et al. indicated that high-power charging technology can make the service reliability of electric buses in operation consistent with that of traditional diesel buses [22].

Thus, the electrification of urban bus network is very significant for a country. In the process of popularization of electric buses, the construction of charging infrastructure is very important, but there are few researches on the optimization of charging infrastructure of electric buses. To date, there are few literatures documenting the optimization of charging infrastructure for e-buses. Some of these studies are briefly summarized in Table 15.1. You et al. studied the optimal charging scheme for the battery swapping station of electric buses to meet the battery swapping demand and minimize the total cost of the battery swapping station [23]. Bi et al. proposed a framework of multi-objective optimization model based on life cycle assessment for siting the location of wireless charger in multi-route electric bus system [24]. However, the research of Chen et al. pointed out

Table 15.1 Research on e-bus charging stations deployment.

Literature		Bi et al. [24]	Ke et al. [26]	He et al. [27]	Wang et al. [28]	Xylia et al. [29]	Rogge et al. [32]	This work
Location of study		University of Michigan	Penghu, Taiwan	Salt Lake City, Utah	Qingdao, China	Stockholm, Sweden	European Cities: Aachen, German; Roskilde, Danish	Yangjiang, China
Optimization method		Genetic algorithm	Genetic algorithm	Mixed-integer linear programming	Linear programming relaxation algorithm; Multiple backtracking and greedy algorithm	Mixed-integer linear programming	Grouping genetic algorithm; Mixed-integer non-linear programming	Affinity propagation algorithm; Binary particle swarm optimization
Charging station deployment		Deploy large-scale wireless charging infrastructure at bus stops	Build charging station in parking lots	Install fast-charging stations at an on-street bus stop or a bus terminal	Install electric vehicle charging stations at selected bus stops	Deploy charging stations at major transport hubs	Plan depot charging station	Deploy fast-charging station in bus terminus
Bus dispatching schedule		Y	Y	Y	Y	N	N	Y
Charging station sharing for different bus routes		N	N	N	Y	Y	N	Y
Objective function	Construction cost of power distribution line	N	N	N	N	N	N	Y
	Operation and maintenance cost of charging station	N	N	N	N	Y	N	Y

(*Continued*)

Table 15.1 (Continued)

Literature		Bi et al. [24]	Ke et al. [26]	He et al. [27]	Wang et al. [28]	Xylia et al. [29]	Rogge et al. [32]	This work
	Travel cost of e-bus to charging station	Y	Y	N	N	Y	Y	Y
	Power loss cost	N	N	N	N	N	N	Y
	Greenhouse gas emissions	Y	N	N	N	N	N	N
	Installation cost of energy storage systems	N	N	Y	N	N	N	N
Constraints	Bus voltages	N	N	N	N	N	N	Y
	Line flows	N	N	N	N	N	N	Y

Source: From [A]/with permission of Elsevier.

that although wireless charging and swapping stations can relieve drivers' range anxiety, their construction costs are too high [25]. The bus transportation system in Penghu was studied by Ke et al. The impact of day and night charging on the construction cost of the electric bus transportation system was examined, so as to improve the practicability of electric buses [26]. He et al. proposed a mathematical model for the optimal planning of fast charging stations to alleviate the problem of high charges caused by high-speed charging, and applied the model to the public transport system in Salt Lake City. The optimization results showed that the fast charging stations could be built at on-street bus stops that are shared by many bus lines [27]. Wang et al. minimized the total installation cost of EV charging stations to determine which bus stops would be selected to build charging stations [28]. Xylia et al. minimized the total costs or the total energy consumption of the electrification e-bus system, and the charging stations were deployed at major transport hubs [29]. Lajunen's research also pointed out that, compared with charging at night and charging at bus stops along the way, charging stations at bus terminals are cheaper and more suitable for bus electrification during the whole life cycle [30]. Gallet et al. pointed out that the power consumption from terminal to terminal of most lines was less than 40 kW/h, making it possible to conduct fast charging at the bus terminal [31]. However, the study of Rogge et al. assumed that all bus terminals should be equipped with charging stations, which would only lead to high investment costs and redundant equipment including transformers and power converters [32]. Therefore, based on the study of the above literature, this chapter proposes an optimization model of electric bus fast charging station considering the bus network and the distribution network. The contributions of this chapter are as follows:

1) With the affinity propagation (AP) clustering algorithm, we proposed clustering the adjacent terminuses into the same class according to the geographical location of each bus terminuses in order to share resources. AP is a clustering algorithm based on the information transfer mechanism between data points, which can avoid determining the number of clusters and setting the initial values.
2) A real-life bus dispatching schedule is adopted to simulate daily charging load for a city's bus transit system.
3) The optimal cost model of charging station is proposed, where the bus transit system and power distribution network are considered as well.
4) With binary particle swarm optimization, we optimize the deployment of fast-charging stations due to discrete site selection, also the charging capacities such that the total bus transit system cost is minimized.
5) A methodology is developed to solve the fast-charging stations deployment problem. The convergence behavior and the total cost are investigated.

The remainder of the chapter is organized as the following. Section 15.2 provides operating characteristics of electric buses and the clustering of bus stations. Section 15.3 formulates the optimal model, followed by a numerical study to demonstrate the effectiveness of the proposed model in Section 15.4. Section 15.5 concludes the chapter.

15.2 Problem Description and Assumptions

In this section, the description and assumptions regarding bus lines and fast-charging stations of an EB system are presented. In addition, an analytical model for clustering bus stations is proposed.

15.2.1 Operating Characteristics of Electric Buses

Assuming a bus network is with multiple bus lines. Then each bus line has a fixed loop line with the same start point and end point. Due to the operational requirements of electric buses, they require larger onboard batteries than that of other types of electric vehicles, so even using fast charging mode, it will take dozens of minutes to several hours for charging up to a reasonable amount of stored energy. Therefore, electric buses can only be recharged near the start stations in the period of waiting for the next departure only after they have run the operating lines. However, because of the large number of bus terminals, and the charging demand of each station is related to the number of vehicles, shift changing time, and driving time of each line, the investment cost will be too high to build charging stations at each bus station. And the equipment in the station is often redundant. Therefore, by clustering the adjacent bus terminal stops and building relatively centralized fast-charging station, the investment of the fast-charging station can be reduced and the usage effectiveness can be improved.

15.2.2 Affinity Propagation Algorithm

Affinity propagation (AP) is a clustering algorithm based on the information transfer mechanism between data points [11]. This algorithm can avoid determining the number of clusters and being sensitive to the initial values compared to traditional clustering algorithms. In this chapter, each bus terminal stations are regarded as potential clustering centers $\vec{X} = \{\vec{x_1}, \vec{x_2}, ..., \vec{x_n}\}$. According to the geographical location information of the terminal stations, the similarity matrix S between the stations is constructed, where the similarity $S(i, j)$ indicates how well the terminal stations with index j is suited to be the exemplar for data point i.

$$S(i,j) = -\|\vec{x_i} - \vec{x_j}\|^2 \tag{15.1}$$

The terminal stations with larger values of $S(k, k)$ are more likely to be chosen as an exemplar. These values are referred to as "preferences." The iterative process is to perform an exemplar competition according to the "availability" and "responsibility" between the terminal stations. The "responsibility" $R(i, k)$, sent from data point i to candidate exemplar point k, reflects the accumulated evidence for how well-suited point k is to serve as the exemplar for point i, taking into account other potential exemplars for point i. The "availability" $A(i, k)$, sent from candidate exemplar point k to point i, reflects the accumulated evidence for how appropriate it would be for point i to choose point k as its exemplar, taking into account the support from other points that point k should be an exemplar. The responsibilities are computed using the following rules [33]:

$$R_{t+1}(i,k) = \begin{cases} S(i,k) - \max\limits_{j \neq k} \{A_t(i,j) + R_t(i,j)\}, i \neq k \\ S(i,k) - \max\limits_{j \neq k} \{S(i,j)\}, i = k \end{cases} \tag{15.2}$$

$$A_{t+1}(i,k) = \begin{cases} \min\left\{0, R_{t+1}(k,k) + \sum\limits_{j \neq \{i,k\}} \max\{0, R_{t+1}(j,k)\}\right\}, i \neq k \\ \sum\limits_{j \neq k} \max\{0, R_{t+1}(j,k)\}, i = k \end{cases} \tag{15.3}$$

The damping factor λ is introduced to avoid numerical oscillation and adjust the convergence rate of AP clustering algorithm in the iterative updating process. Then the above equations are updated as follows:

$$R_{t+1}(i,k) = (1-\lambda)\cdot R_{t+1}(i,k) + \lambda\cdot R_t(i,k) \tag{15.4}$$

$$A_{t+1}(i,k) = (1-\lambda)\cdot A_{t+1}(i,k) + \lambda\cdot A_t(i,k) \tag{15.5}$$

The clustering division of the terminal stations can be obtained through the AP clustering algorithm. Since the fast-charging site selection is discrete, Binary Particle Swarm Optimization algorithm is used to solve the optimization problem. Based on the results of clustering, the binary code is used to generate initial particle populations. Each population represents a combination mode, and the fast-charging stations in each combination mode are selected from the terminal stations. The advantage is that it can ensure that each charging station and its charging demands can be classified into the same category, and moreover, the combination of fast-charging station sites updated in each iteration is guaranteed to be a feasible solution, which reduces the search space.

In this chapter, $\vec{x_i}$ in BPSO represents the ith particle position, and each particle represents a solution of the planning problem.

$$\vec{x_i} = [x_{i1}, x_{i2}, \ldots, x_{id}, \ldots, x_{iD}] \tag{15.6}$$

where D is the dimension of the particles, which corresponds to the number of terminal stations of each cluster. The value of x_{id} indicates whether the ith particle selects the dth station as the fast-charging station, and its values are {0,1}. Its particle velocity updates as follows [34]:

$$v_{id}^{t+1} = \omega\cdot v_{id}^t + c_1\cdot rand\left(p_{id}^t - x_{id}^t\right) + c_2\cdot rand\left(p_{gd}^t - x_{id}^t\right) \tag{15.7}$$

where ω is inertia weight, c_1 and c_2 are learning factors, *rand*() is a random positive number between 0 and 1, p_{id} is the best individual position, and p_{gd} is the best position, globally.

After updating, each particle velocity v_{id} will be mapped to the value probability of x_{id} by the *sigmoid*() function, and its position will be updated by Eq. (15.9) [35]:

$$Sig(v_{id}) = \frac{1}{1+\exp(-v_{id})} \tag{15.8}$$

$$x_{id} = \begin{cases} 1 & rand() \le Sig(v_{id}) \\ 0 & otherwise \end{cases} \tag{15.9}$$

where the function $Sig(v_{id})$ is a sigmoid limiting transformation, which represents the probability that the position x_{id} takes 1.

15.3 Model Formulation

In the following section, a mathematical program is developed to optimize deployment of fast-charging stations, as well as the capacity in order to minimize the total social cost. The detailed model solution and optimal flow chart is shown in Figure 15.1.

15.3.1 Capacity Model of Electric Bus Fast-Charging Station

Compared with other types of electric vehicles, electric buses have a fixed operation mode. Therefore, this chapter obtains the spatiotemporal distribution characteristics of buses according to its

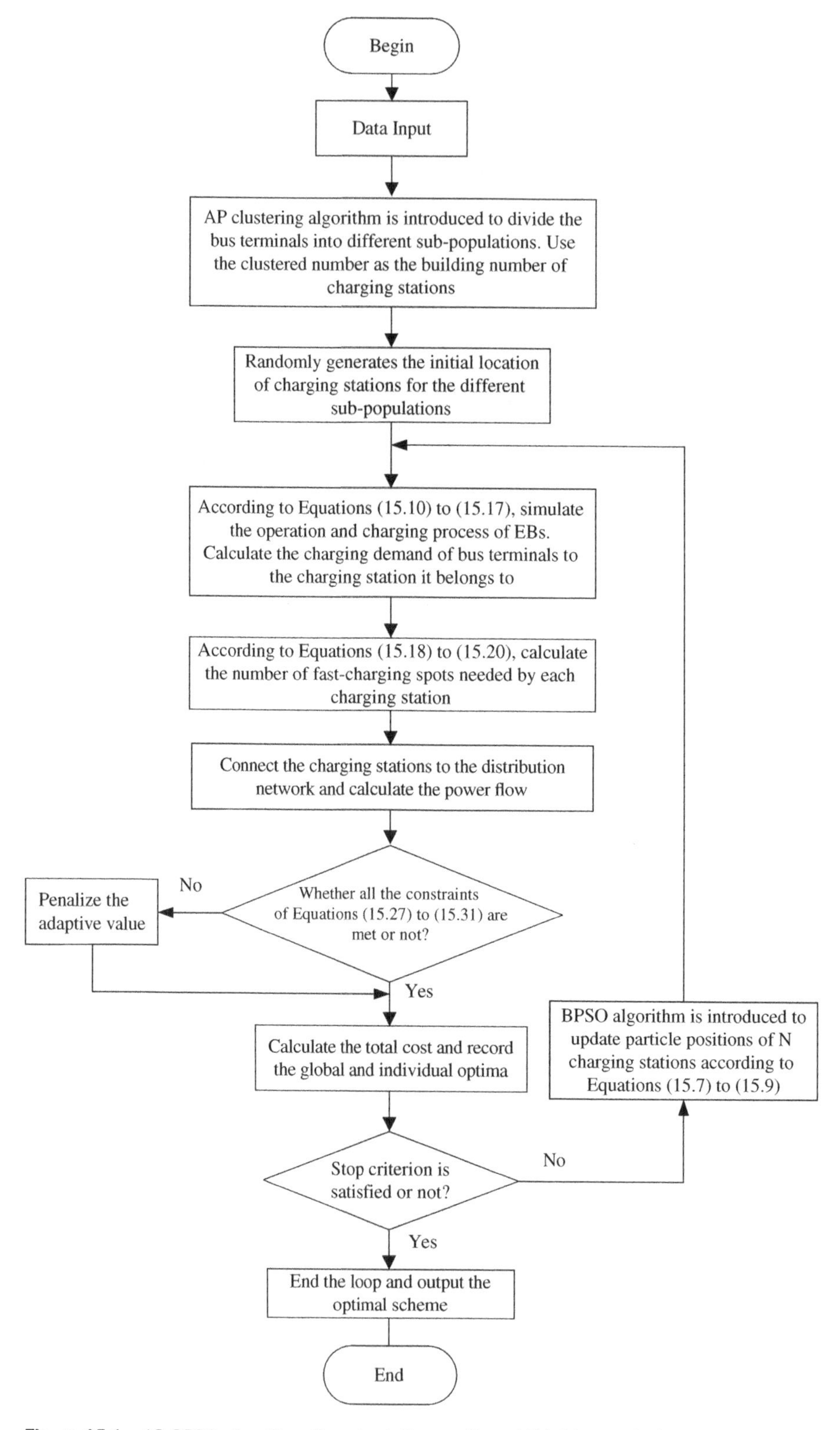

Figure 15.1 AP-BPSO algorithm flowchart. *Source:* From [A]/with permission of Elsevier.

operation scheduling plan, and simulates the operational task according to the spatio-temporal distribution characteristics. According to the characteristics of the bus line, to ensure the demand of the line, the number of electric buses needs to be reasonably configured. The number of electric buses required for each line is determined by Eq. (15.10):

$$N_b = ceil\left(\frac{T_{run} + T_{fc}}{T_{shift}}\right) \tag{15.10}$$

where N_b is the number of electric buses that need to be configured, T_{run} is the operation time of a bus line, T_{fc} is the recharging time, and T_{shift} is the spacing interval of a bus line, and ceil() is the function round up to an integer.

The simulation model needs to assume some state variables to track the EB state throughout the whole process. Assume that $\text{var}_{1,i,j,b,t}$, $\text{var}_{2,i,j,b,t}$, $SOC_{1,i,j,b,t}$, and $SOC_{2,i,j,b,t}$ are the operating state, the charging state, the state of charge when the vehicle is departing, and the state of charge when the vehicle is arriving at bus b in line j of terminal I at the time of its tth departure or arrival, respectively. $SOC_{\max}$ is the redundancy of the state of charge to prevent overcharging, $SOC_{warning}$ is the warning value of the state of charge, $C_{battery}$ is the battery capacity of EB, and P is the rated value of charging spot.

1) The first bus departure is scheduled according to the operation schedule, at this moment $SOC1_{i,j,b,t} = 1$ and $\text{var1}_{i,j,b,t} = 1$.
2) After the T_{run} time, the vehicle arrives at the station, at this moment $\text{var}_{1,i,j,b,t} = 0$, the arrival time should be recorded and $SOC_{2,i,j,b,t}$ are recorded as follows:

$$SOC_{2,i,j,b,t} = SOC_{1,i,j,b,t} - \frac{E_{i,j}}{C_{battery}} \tag{15.11}$$

where the power consumption of line j operated by the ith station:

$$E_{i,j} = uL_{i,j}\lambda_d \tag{15.12}$$

where u is the power consumption per kilometer, $L_{i,j}$ is the length of line j operated by the ith station, λ_d is the influencing factor on the operation lines, for example, slope, rugged degree, etc., which together named as a comprehensive factor, and generally taken between 1.1 and 1.3 [5].
3) According to Eq. (15.13), to determine whether the vehicle needs to be charged, by setting $\text{var}_{2,i,j,b,t}$ at 1; otherwise, setting as 0 and wait for the next departure time:

$$SOC_{2,i,j,b,t} - \frac{E_{i,j} + E_{i,c}}{C_{battery}} \leq SOC_{warning} \tag{15.13}$$

where $E_{i,c}$ is the power consumption from bus terminal i to charging station c.
4) If the vehicle needs to be charged, the charging time for the vehicle is calculated according to Eq. (15.14). Meanwhile, in order to ensure that the SOC of the vehicle can meet the next operation task and not exceed the spacing interval, the charging time T_{fc} needs to meet the constraint of Eq. (15.15). The electric energy obtained from charging is given by Eq. (15.16).

$$T_{fc,i,j,b,t} = \begin{cases} \dfrac{60\left(SOC_{max} - SOC_{2,i,j,t}\right)C_{battery} + E_{i,c}}{P} & if\,\text{var}_{2,i,j,t} = 1 \\ 0 & \text{else} \end{cases} \tag{15.14}$$

$$\begin{cases} T_{fc,i,j,b,t} \geq \dfrac{60\left(SOC_{warning} - SOC_{2,i,j,t}\right)C_{battery} + E_{i,j} + E_{i,c}}{P} \\ T_{fc,i,j,b,t} \leq T_{dwell,i,j,t} - 2T_{i,c} \end{cases} \tag{15.15}$$

$$e_{i,j,b,t} = \frac{T_{fc,i,j,t} \cdot P}{C_{battery} \cdot 60} \tag{15.16}$$

where $T_{dwell,i,j,t}$ is the dwelling time of an electric bus in bus terminal, $T_{i,c}$ is the driving time from bus terminal i to charging station c.

5) After a period of $T_{dwell,i,j,t}$, it is time for departure, at this moment $SOC_{1,i,j,b,t}$ is:

$$SOC_{1,i,j,b,t} = \begin{cases} SOC_{2,i,j,b,t} + e_{i,j,b,t} - 2E_{i,c} & if\,\mathrm{var}_{2,i,j,t} = 1 \\ \mathrm{SOC}_{2,i,j,b,t} & \text{else} \end{cases} \tag{15.17}$$

6) Repeat steps 2–5 until all vehicles, lines, and terminals have been traversed to obtain the total charging demand and effective charging time of each bus terminal during the daytime operation. The number of charging spots required for charging during the day is as follows:

$$m_{daytime} = ceil\left(\frac{\sum_{i=1}^{N_i}\sum_{j=1}^{N_j}\sum_{b=1}^{N_b}\sum_{t=1}^{N_t} e_{i,j,b,t} \cdot (1+\mu)}{T_{daytime} \cdot P \cdot k_s \cdot k_{eff}} \cdot \alpha\right) \tag{15.18}$$

where N_i is the number of charging demand points belonging to the same cluster of fast-charging station c, N_j is the number of bus lines operated by charging point i, N_t is the number of arrivals of vehicle b; μ is the charging capacity redundancy of the fast-charging station, P is the charging power of the equipped charging facilities, $T_{daytime}$ is the effective charging time of the fast-charging station, k_s is the operation simultaneous rate of charging facilities, k_{eff} is the charging efficiency of the charging facility, and α is the fluctuation coefficient of charging demand.

The period from the end of operation on one day to the beginning of operation on the next day is the charging time at night. During this period, the number of charging spots should meet the operation requirements of all the vehicles on the next day. According to the SOC when the vehicles are fully charged, the number of charging spots required for night charging is as follows:

$$m_{night} = \mathrm{ceil}\left(\frac{\sum_{i=1}^{N_i}\sum_{j=1}^{N_j}\sum_{b=1}^{N_b} \Delta t_{fc,i,j,b}}{T_{night}}\right) \tag{15.19}$$

where $\Delta t_{fc,i,j,b}$ is the time required for vehicle b to be fully charged at night, T_{night} is the effective charging time at night.

The number of charging facilities to be built for the fast-charging station c is given as follows:

$$m_c = \max\left(m_{daytime}, m_{night}\right) \tag{15.20}$$

15.3.2 Deployment Model of Electric Bus Fast-Charging Station

The objective function for EB fast-charging station planning is to minimize the cost of charging station and the cost of distribution network. The costs of charging station include the investment and construction cost of charging station C_{1c}, the operation and maintenance cost of charging station C_{3c}, and the travel cost of EB to charging station C_{4c}. The cost of distribution network includes construction cost of distribution line C_{2c} and power loss C_5.

$$\min F = \sum_{c=1}^{N_c} (C_{1c} + C_{2c} + C_{3c} + C_{4c}) + C_5 \tag{15.21}$$

where N_c is the number of charging stations.

1) Investment cost of charging station

$$C_{1c} = (m_c p_1 + e_c p_2 + \omega_c) \frac{r_0 \cdot (1 + r_0)^z}{(1 + r_0)^z - 1} \tag{15.22}$$

where r_0 is the discount rate, z is the operating life of the charging station, m_c is the number of charging facilities at the charging station c, p_1 is the unit price of the charging facility, e_c is the number of transformers in the charging station c, p_2 is the unit price of the transformer, and ω_c is the construction cost of charging station c.

2) Construction cost of distribution line

$$C_{2c} = \alpha_{cn} Y_{cn} L_{cn} \frac{r_0 \cdot (1 + r_0)^z}{(1 + r_0)^z - 1} \tag{15.23}$$

where Y_{cn} is the state variable representing the connection between charging station c and node n in distribution network. If the charging station c is connected to node n of the distribution network, then $Y_{cn} = 1$. Otherwise, it is 0. α_{cn} is the investment cost per kilometer; L_{cn} is the length of the distribution line connecting charging station c to node n of the distribution network.

3) Annual operation and maintenance cost of charging station

$$C_{3c} = (C_{1c} + C_{2c}) \cdot \eta_b \tag{15.24}$$

where η_b is a ratio which depends on utilities practice.

4) Travel cost to charging station

$$C_{4c} = 365 \sum_{c=1}^{N_c} \sum_{i=1}^{N_i} 2 g_t E_{i,c} n_{ci} \tag{15.25}$$

where g_t is the electricity price of charging (per kWh), and n_{ci} is the number of charging times.

5) Power loss cost

After the charging station is connected to the distribution network, the active power loss of the distribution network will increase. The increased annual power loss cost is shown as follows:

$$C_5 = g_p \int_0^{8760} [P_{\text{loss2}}(t) - P_{\text{loss1}}(t)]\, dt \tag{15.26}$$

where g_p is the average tariff including tax, P_{loss1} is the active power loss of distribution network before fast-charging station access and P_{loss2} is the active power loss of distribution network after fast-charging station access.

15.3.3 Constraints

1) When the charging facilities are configured, the limitation of the disposal area should be taken into account. The limitation of the disposal area is converted into the maximum number of charging facilities that can be installed in the station as shown below:

$$m_c \in [S_{min}, S_{max}] \tag{15.27}$$

2) In order to satisfy the charging reliability, the distance from the charging demand point to the fast-charging station needs to be met as given below:

$$d_{ij} \leq 0.5L \tag{15.28}$$

3) Each charging station can only access to one distribution network node:

$$\sum_{n=1}^{N} Y_{cn} = 1 \tag{15.29}$$

4) The constraints of the distribution network include load constraints and voltage constraints:

$$U_n^{min} < U_n < U_n^{max}, \ \mathrm{n} = 1,2,\ldots,N \tag{15.30}$$

$$P_{cn} + P_n \leq P_n^{max} \tag{15.31}$$

where U_n is the voltage magnitude of node n of distribution network, $U_n{}^{min}$ and $U_n{}^{max}$ are the upper and lower margins of voltage magnitude of node n of distribution network, P_{cn} is the charging power of the charging station c access node n of distribution network, P_n is the load power of node n of distribution network, and $P_n{}^{max}$ is the maximum power allowed for node n of distribution network.

15.4 Results and Discussion

A numerical study to demonstrate the effectiveness of the proposed model is given. The numerical study is based on a real bus transit system in urban area Yangjiang City, which is a coastal city in South China.

15.4.1 Spatio-temporal Distribution of Buses

A bus transit system with 26 bus routes is utilized in this numerical study. The routes of the bus transit system cover 510.8 km of road segments in urban and suburb areas of Yangjiang City. The 26 bus routes serve 388 bus stops, where 34 terminuses are included. In this chapter, only the urban area of the city is considered, where the geographical distribution of the terminuses is shown in Figure 15.2. The simulation parameters utilized for bus dispatching schedule is listed in Table 15.2.

For simplicity, it is assumed that the e-buses used for the 26 bus routes are with the same model. In the future, the whole subnetwork in Figure 15.2 will be served by the Yutong E6 e-bus with an on-board battery capacity of 85.85 kWh, and the charging power is 120 kW. The proposed optimization model can help Yangjiang city to determine the locations and number of fast-charging stations, the numbers of charging spots within fast-charging stations, and the cost of the electrified bus network.

15.4.2 Optimized Deployment of EB Fast-Charging Stations

The AP clustering algorithm is used to obtain the charging demand location of each terminus and the number of fast-charging stations needed to be built for the bus network. Based on the model proposed in Section 15.3, the optimal solution is obtained for e-bus fast-charging stations by using

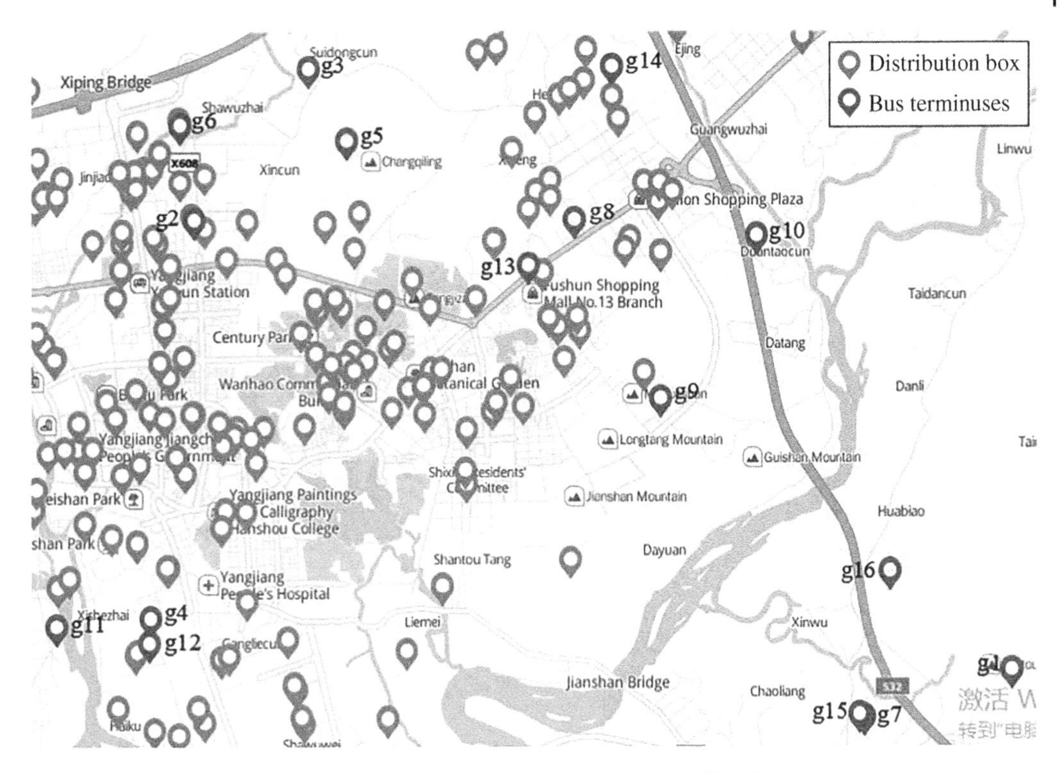

Figure 15.2 Schematic diagram of the planning area. *Source:* From [A]/Elsevier.

the BPSO algorithm. As a result, the planning scheme obtained by the proposed optimization model is 4 fast-charging stations to serve 16 bus terminuses and 26 bus routes. The detailed charging station planning scheme is shown in Table 15.3. The optimal cost of the model is reported in Table 15.4. The location and service of the fast-charging stations is shown in Figure 15.3, where the selected bus terminus is marked blue. The variation of the *SOC* of e-buses in the daytime operation is shown in Figure 15.4. Due to the large number of e-buses involved in the study, only the bus route 8 is taken as an example, and the rest of the e-buses are similar.

15.4.3 Comparison of Different Planning Methods

To present the economic benefits of the proposed optimization model, a comparison of placing fast-charging stations at selected terminuses is conducted as shown in Figure 15.5a. Conventional planning scheme is to build charging stations in each bus terminus. Although the travel cost on the way to charging station should not be considered in conventional planning scheme, other costs might be much higher and the equipment will be redundant because fast-charging stations are built in each bus terminus. For AP clustering only, the obtained location of fast-charging stations is the closest to other bus terminuses. But the difference in the number of routes and e-buses operated by different terminuses is ignored, resulting in a larger total travel cost. It can be seen that the deployment of charging stations cannot be determined only by the distance between bus terminuses. The calculated costs of GA algorithm and BPSO algorithm are slightly higher than that of proposed

Table 15.2 Bus dispatching schedule.

Station	Line	Number of buses	First bus	Last bus	Time headway (min)	Route length (km)
g1	2	7	7:00	20:00	20	20
g2	3	15	6:10	21:15	15	27
	5	3	6:35	18:35	60	23
	8	22	6:05	22:00	9	30
	11	2	7:15	19:00	180	29
	12	9	6:05	21:00	13	25
	14	3	7:30	22:30	40	12
g3[a]	7	2	8:00	16:30	180	22
	4	7	7:00	20:00	30	23.5
g4	6	2	6:55	17:35	140	21
g5	9	1	8:30	17:30	390	30
g6	10	1	7:30	17:00	120	8.3
g7[a]	25	1	7:30	17:30	120	7
	26	1	7:30	17:30	120	3
g8	1	5	7:00	20:00	40	21
g9	4	7	7:00	20:00	30	23.5
g10	6	2	6:55	17:35	140	21
g11	7	2	8:00	16:30	180	22
g12	10	1	7:30	17:00	120	8.3
g13[a]	15	2	9:40	19:30	80	15
	20	1	7:00	17:00	480	26
g14	16	1	9:00	15:30	180	21
g15	18	1	10:20	13:50	120	10
g16	26	1	7:30	17:30	120	3

[a] Note: One bus line has two terminal stops. Some terminals are far away from the planned area, which will not be included in the table.
Source: From [A]/with permission of Elsevier.

Table 15.3 Planning scheme of fast-charging stations.

Terminuses used for charging stations	Number of fast-charging spots	Served e-buses	Number of charging times	Charging demand (kWh)
g1	2	11	18	767
g2	9	72	171	7554
g9	3	18	29	1226
g12	1	5	5	183

Source: From [A]/with permission of Elsevier.

Table 15.4 Cost of fast-charging stations.

Result	Value (¥[a] × 10^6)
Equipment and installation cost	0.442
Distribution line construction cost	0.096
Operation and maintenance cost	0.061
Total travel cost	0.093
Total power loss cost	0.114

[a] Chinese RMB ¥1 ≈USD $0.1501.
Source: From [A]/with permission of Elsevier.

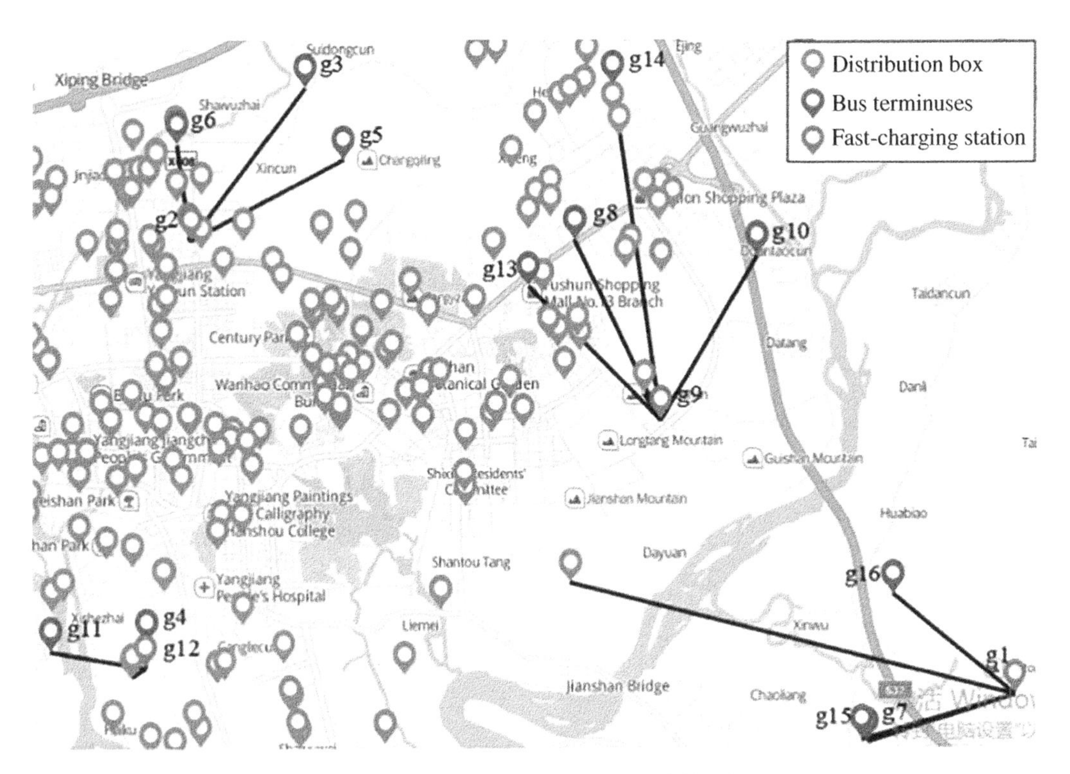

Figure 15.3 Deployment of fast-charging stations. *Source:* From [A]/Elsevier.

method in this paper. The convergence of these three algorithms is compared, as shown in Figure 15.5b. The proposed method has a faster rate of convergence and a reduced total cost than BPSO algorithm. The reason is that the terminuses have been classified before optimization. After the classification, the number of terminuses within each class decreases. When the BPSO algorithm is used for each class, its optimization range becomes smaller. BPSO needs to be calculated several times for several classes, but the calculation speed is shorter than that of non-clustering. Therefore, for the electrification of public transport network in Yangjiang city, the fast-charging stations deployment based on the proposed optimal model is more economical.

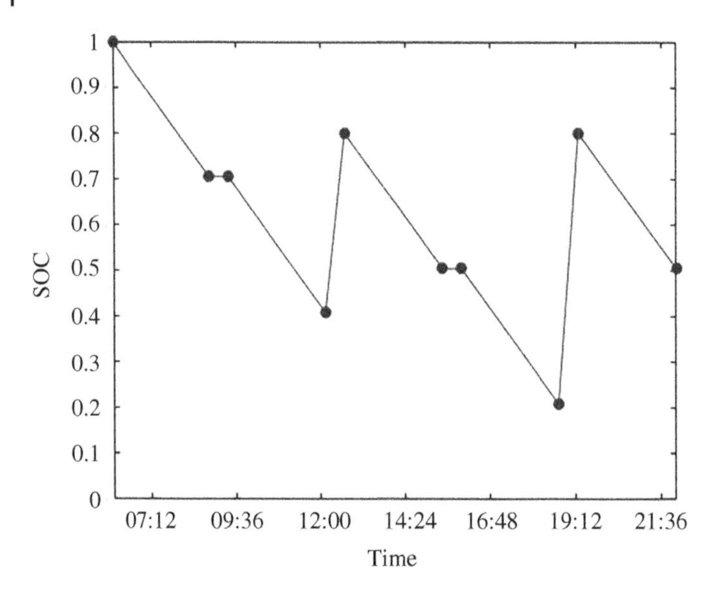

Figure 15.4 Changes in SOC of bus line 8. *Source:* From [A]/with permission of Elsevier.

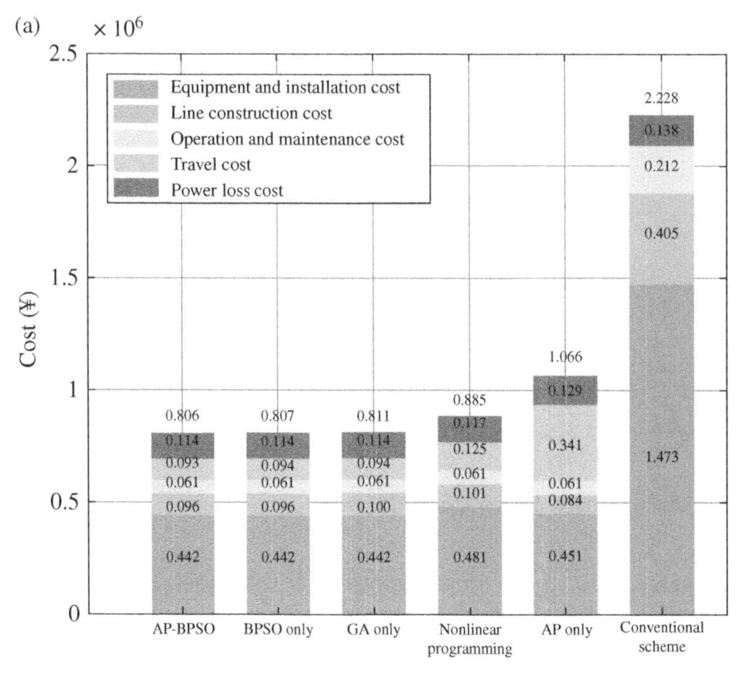

Figure 15.5 Cost and convergence comparison of different optimization planning methods. (a) Cost comparison of different optimization planning methods, (b) Convergence comparison of different optimization planning methods. *Source:* From [A]/with permission of Elsevier.

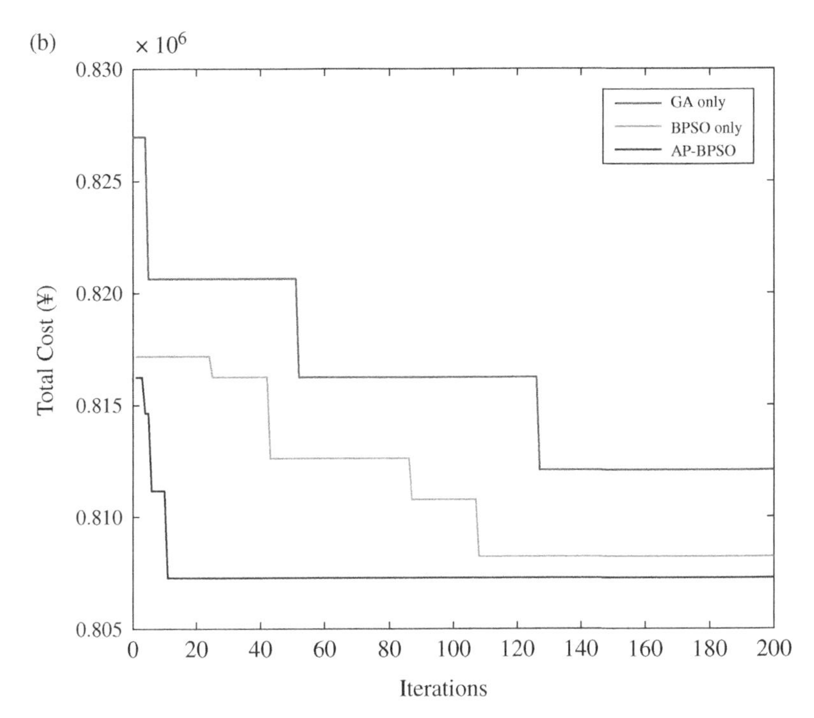

Figure 15.5 (Continued)

15.4.4 Comparison Under Different Time Headways

Compared to transit operation in big cities, the time headway for most the bus routes in Yangjiang City are relatively long. The further impact of different time headway on the result of the model is examined and analyzed. According to the bus operation schedule of Yangjiang City, the time headway is modified for bus routes with other model parameters fixed, and the results on deployment and cost are shown in Figures 15.6 and 15.7, respectively. The minimum fleet size of each bus route is given by Eq. (15.10). Three groups of time headways, namely, 20 minutes (Scenario 1), 15 minutes (Scenario 2), and 10 minutes (Scenario 3) are considered. It can be seen that with the decrease of departure interval, the need for fast-charging spots, charging times of vehicles, the number of vehicles in service and charging load are all increased in the public transport network. The result is predictable, as the reduction in departure intervals leads to an increase in the number of missions for the vehicle to perform, leading to an increase in the need for charging.

15.4.5 Comparison Under Different Battery Size and Charging Power

In addition, we also studied the impact of different types of e-buses on the proposed model. For simplicity, only the difference in battery capacity is considered. The e-bus models used are Yutong bus E6 (Scenario 1), E8 (Scenario 2), and E10 (Scenario 3), with battery parameter values illustrated in Table 15.5 [36]. Figure 15.8 shows the comparison of charging details for different capacities of buses. It is assumed that when e-buses have enough power, they can operate longer and perform more tasks, and they can even meet the needs of a day's circular operation for a bus route with no

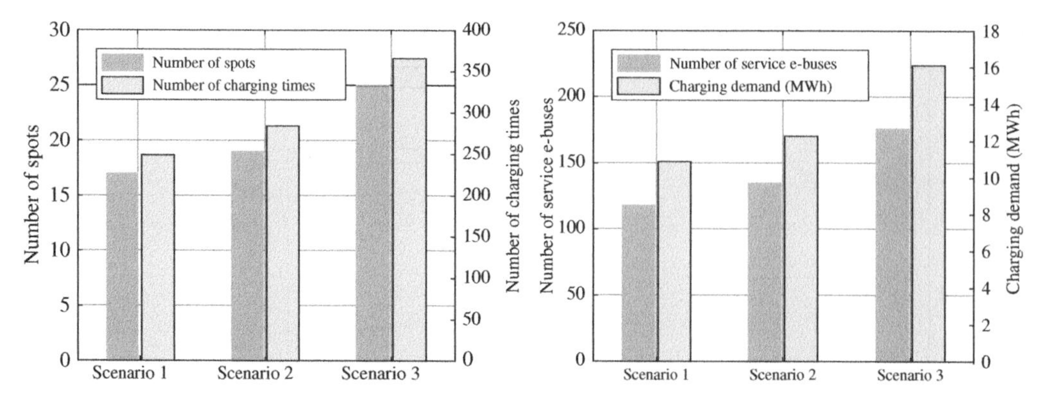

Figure 15.6 Deployment comparison under different time headways. *Source:* From [A]/with permission of Elsevier.

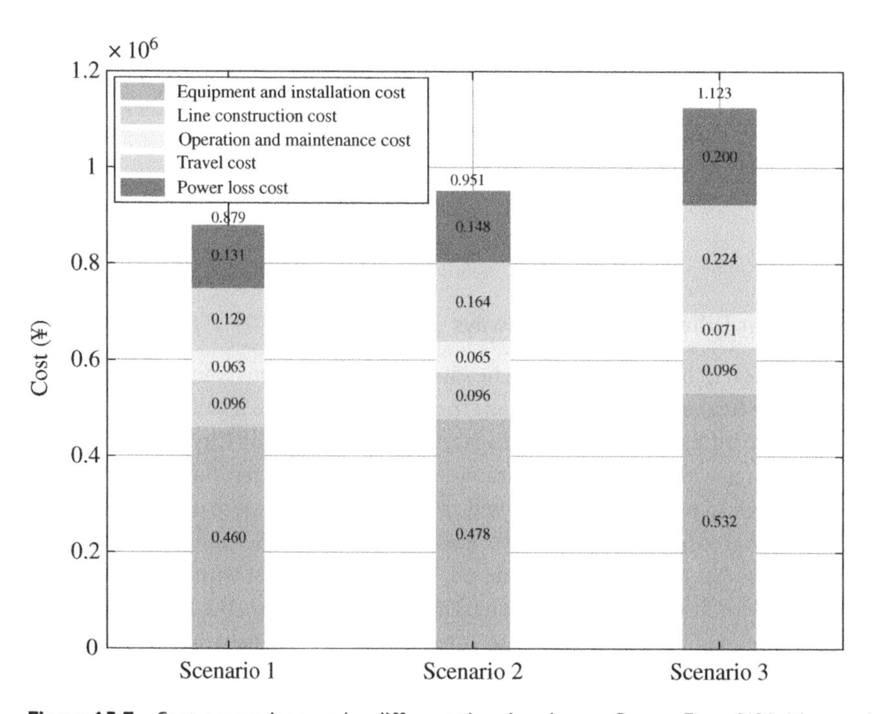

Figure 15.7 Cost comparison under different time headways. *Source:* From [A]/with permission of Elsevier.

need to be recharged. However, as seen from Figure 15.8, the duration of charging time decrease when the battery capacity of the buses is increased, while the number of fast-charging spots, service e-buses, and charging demands increase. This is due to the fact that increase in the battery capacity would lead to increase in the required charging time. When a battery of e-bus is not fully charged, new e-buses are coming, resulting in an increase in the number of e-buses to be charged per unit of time. As a result, the number of fast-charging facilities, operating e-buses and charging demand are also increased. Furthermore, in order to investigate the influence of charging power, three groups of

Table 15.5 Parameter values of the battery sizes.

E-bus type	Battery capacity (kWh)
E6	85.85
E8	122.93
E10	202.93

Source: From [A]/with permission of Elsevier.

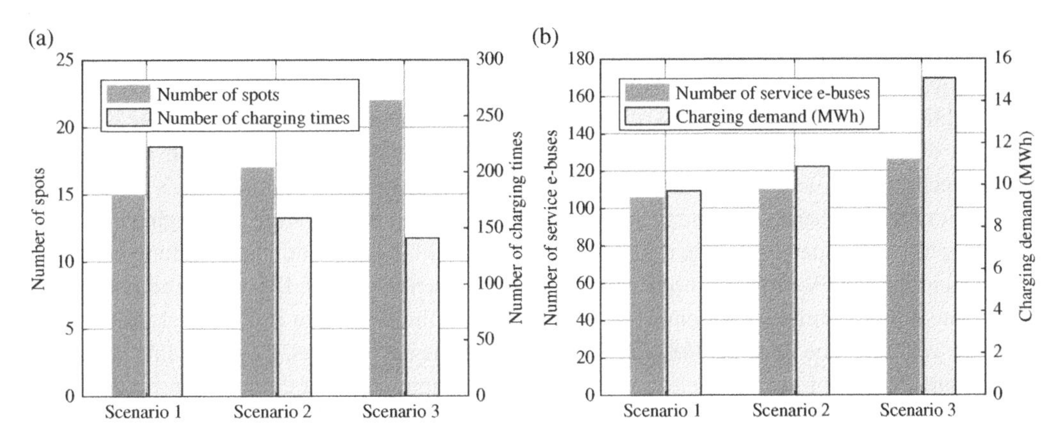

Figure 15.8 Deployment comparison under different battery size. (a) Number of spots and charging times, (b) Number of service e-buses and charging demand (MWh). *Source:* From [A]/with permission of Elsevier.

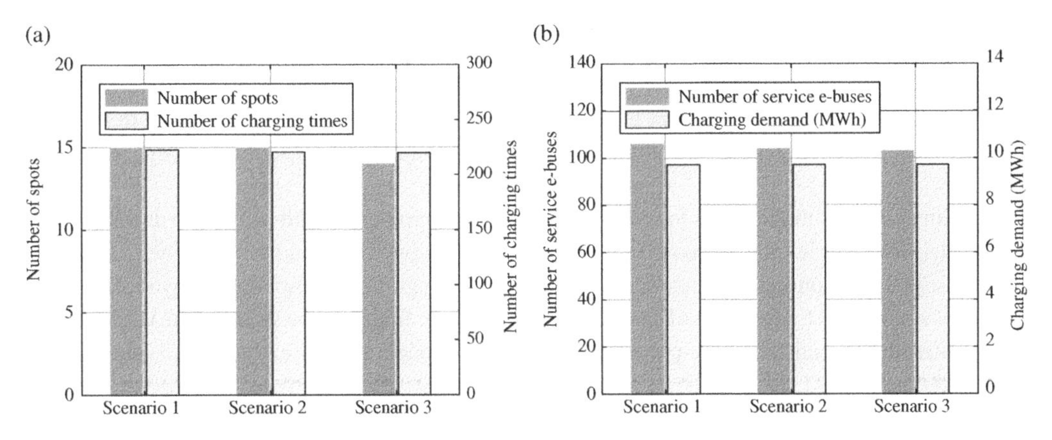

Figure 15.9 Deployment comparison under different charging power. (a) Number of spots and charging times, (b) Number of service e-buses and charging demand (MWh). *Source:* From [A]/with permission of Elsevier.

fast-charging facilities were set for comparison, namely 120 kW (Scenario 1), 150 kW (Scenario 2), and 180 kW (Scenario 3) of Star Charge brand [37], as shown in Figure 15.9. The minimum fleet size for each bus route should satisfy Eq. (15.10). Figure 15.9 shows that with the increase of charging power, there is a reduction in the number of fast-charging spots, charging times, and operating e-buses. This is because as the charging power increases, e-buses need less charging time.

15.4.6 Policy and Business Model Implications

The main stakeholders in electric bus public transportation include the central and local governments, electric bus manufacturers, users/bus companies, and providers of charging infrastructure. Unlike conventional buses, electric buses need charging facilities. The availability of charging station facilities is important to the development of the electric bus public transportation infrastructures. This immediately raises questions on the exact location of those facilities and the role of energy providers, for example, State Grid and Southern Grid in China. Also, manufacturers need workable strategies when they enter the market. The government subsidies will need a discussion or debate too. Naturally, a number of issues in the electric bus industry has to be considered, such as policy implementation, technology innovation, business model, and the whole supply chain.

Recently, new policies focused more on the construction, operation, and business models of charging facilities. For example, in China, there is the incentive policies on EV charging facility construction during the 13th five-year-plan [38]. In addition, new energy generation and energy storage technologies were considered to be an important part of the EV industry's strategy in 2016. A number of policies continue to support the construction and operation of charging facilities, especially those intended for public transportation systems [39]. In addition to this, there are some government funded EV projects in other countries that contribute to the EV policy and regulatory development. For example, Reference [40] reports that in the UK, there is plug-in grant for low-emission vehicles. The Electric Vehicle Homecharge Scheme provides grant funding of up to 75% toward the cost of installing electric vehicle smart charge points at domestic properties. Reference [41] reports that Singapore aims to phase out Internal Combustion Engine vehicles and have all vehicles run on cleaner energy by 2040. National Electric Vehicle Centre was formed to promote wider EV adoption. This includes accelerating the deployment of a nationwide EV charging infrastructure, focusing on vehicle taxes and incentives, building EV regulations and standards, and cultivating a robust EV ecosystem, with a target of 60 000 charging points by 2030. Reference [42] reports that the New Zealand Government are considering additional measures to increase the uptake of electric and low emissions vehicles. The existing exemption from road user charges has been extended until 31 March 2024 to encourage the uptake of electric vehicles. Also, in January 2021, the Government agreed to introduce the Clean Car Standard for imported new and used light vehicles.

Regarding business models, it is foreseen that a holistic approach should be used to develop business models for electric mobility based on electric buses on one hand and on the other hand give decision support to concerning organizations. With the integration of mobile energy storage into the power system and the buildup of charging infrastructure, there will be shifts in the value chain, the revenue model, and the value proposition. Business models seek to explain how value is created [43].

Rapid urbanization with increasing population density has caused multiple problems in urban areas. One of the most critical problems is transportation regarding new infrastructure. Traffic congestion threatens accessibility to destination points, especially those that are in the city center. Public transport is an essential element of urban life since it reduces car traffic and provides city residents with mobility. Buses play a significant role in the provision of public transportation, being cheap, flexible, and, in many cases, superior in terms of capacity and speed. Therefore, the bus remains the most suitable solution from an economic, environmental, and social point of view regarding the balanced and sustainable urban development [44, 45].

A cleaner power grid and an increase in system charging efficiency would enhance the future benefits of electric buses [46]. Urban public transportation is a multicriteria decision-making (MCDM) problem because multiple actors are involved in the design and operation. The decision-making process regarding public transportation is a very complicated task involving multiple economic, environmental, and sociopolitical issues [47].

Sun et al. looked at fast-charging station choices and behaviors [48]. Wang et al., and An studied electric vehicle charging infrastructure planning and design [49, 50]. Gao et al. studied battery capacity and recharging [22]. Panchal et al. studied degradation testing and battery modeling [51], design and simulation of batteries [52] and financial model of lithium-ion storage [53].

A business model aims to identify appropriate specific evaluation criteria for electric buses transportation under clean technology and to develop and propose methods to select the transportation structures and elements such as charging station and batteries to maximize the contribution to sustainability and profit. Many methods could be considered such as computationally intelligent methods; for example, deep neural network [54], genetic algorithm, and particle swarm optimization are used in the present paper. Electric buses could also help to address air and noise pollution issues [55]

In summary, the present proposed method and studied system could be used as examples to further carry out sensitivity study and identify more parameters to develop business model and energy policy. This could be considered as a future work.

15.5 Conclusions

In this chapter, the problem of deploying fast-charging stations for an electric bus system is studied to ensure that buses on each bus line will have enough energy to complete the operation. The purpose is to build fast-charging stations at selected bus terminals to minimize the total cost of the system for deploying fast-charging stations. This chapter proposes a planning model for locating and sizing of electric bus fast-charging station based on the consideration of both the bus operation network and distribution network. The Affinity Propagation clustering algorithm is used to cluster the bus terminals, and then the Binary Particle Swarm Optimization algorithm is used to find the optimal solution of the deployed fast charging station. The case study based on a real-world bus network is provided to demonstrate the effectiveness of the model and to avoid the situation of equipment redundancy in the station and excessive costs caused by too many charging stations. The decrease of the time headway will lead to the increase of fleet size and charging demand, thus increasing the total cost of fast-charging stations.

In the future, the optimization model will be extended to different bus operation networks by customizing model data to assist the deployment of fast-charging stations in the electrification of bus networks in practice.

Acknowledgements

The permission given is using materials from the following paper is very much appreciated.

[A] Wu, X., Feng, Q., Bai, C. et al. (2021). A novel fast-charging stations locational planning model for electric bus transit system, *Energy*. https://doi.org/10.1016/j.energy.2021.120106.

References

1 Wen, Y., Cai, B., Yang, X. et al. (2020). Quantitative analysis of China's low-carbon energy transition. *International Journal of Electrical Power & Energy Systems* 119: 105854.

2 Lajunen, A. and Lipman, T. (2016). Lifecycle cost assessment and carbon dioxide emissions of diesel, natural gas, hybrid electric, fuel cell hybrid and electric transit buses. *Energy* 106: 329–342.

3 Correa, G., Muñoz, P.M., and Rodriguez, C.R. (2019). A comparative energy and environmental analysis of a diesel, hybrid, hydrogen and electric urban bus. *Energy* 187: 115906.

4 Rupp, M., Handschuh, N., Rieke, C. et al. (2019). Contribution of country-specific electricity mix and charging time to environmental impact of battery electric vehicles: a case study of electric buses in Germany. *Applied Energy* 237: 618–634.

5 Zhou, B., Ye, W., Zhou, B. et al. (2016). Real-world performance of battery electric buses and their life-cycle benefits with respect to energy consumption and carbon dioxide emissions. *Energy* 96: 603–613.

6 Nikoobakhta, A., Aghaei, J., Niknam, T. et al. (2019). Electric vehicle mobility and optimal grid reconfiguration as flexibility tools in wind integrated power systems. *International Journal of Electrical Power & Energy Systems* 110: 83–94.

7 Brinkel, N.B.G., Gerritsma, M.K., AlSkaif, T.A. et al. (2020). Impact of rapid PV fluctuations on power quality in the low-voltage grid and mitigation strategies using electric vehicles. *International Journal of Electrical Power & Energy Systems* 118: 105741.

8 Raoofat, M., Saad, M., Lefebvre, S. et al. (2018). Wind power smoothing using demand response of electric vehicles. *International Journal of Electrical Power & Energy Systems* 99: 164–174.

9 Suhail Hussain, S.M., Aftab, M.A., Ali, I. et al. (2020). IEC 61850 based energy management system using plug-in electric vehicles and distributed generators during emergencies. *International Journal of Electrical Power & Energy Systems* 119: 105873.

10 Capuder, T., Sprčić, D.M., Zoričić, D. et al. (2020). Review of challenges and assessment of electric vehicles integration policy goals: integrated risk analysis approach. *International Journal of Electrical Power & Energy Systems* 119: 105894.

11 Ji, Z. and Huang, X. (2018). Plug-in electric vehicle charging infrastructure deployment of China towards 2020: policies, methodologies, and challenges. *Renewable and Sustainable Energy Reviews* 90: 710–727.

12 Yang, J., Fuzhang, W., Yan, J. et al. (2020). Charging demand analysis framework for electric vehicles considering the bounded rationality behavior of users. *International Journal of Electrical Power & Energy Systems* 119: 105952.

13 Hosseini, S. and Sarder, M.D. (2019). Development of a Bayesian network model for optimal site selection of electric vehicle charging station. *International Journal of Electrical Power & Energy Systems* 105: 110–122.

14 Wang, C., Dunn, R., Robinson, F. et al. (2017). Active-reactive power approaches for optimal placement of charge stations in power systems. *International Journal of Electrical Power & Energy Systems* 84: 87–98.

15 Alhazmi, Y.A., Mostafa, H.A., and Salama, M.M.A. (2017). Optimal allocation for electric vehicle charging stations using trip success ratio. *International Journal of Electrical Power & Energy Systems* 91: 101–116.

16 Morro-Mello, I., Padilha-Feltrin, A., Melo, J.D. et al. (2019). Fast charging stations placement methodology for electric taxis in urban zones. *Energy* 188: 116032.

17 Du, J., Li, F., Li, J. et al. (2019). Evaluating the technological evolution of battery electric buses: China as a case. *Energy* 176: 309–319.

18 National Development and Reform Commission (n.d.). Implementation advices on promoting consumption, expanding capacity, improving quality and accelerating the formation of a strong domestic market [EB/OL] [28 February 2020]. https://www.ndrc.gov.cn/xxgk/zcfb/tz/202003/t20200313_1223046.html (accessed 8 March 2020).

19 Ministry of Industry and Information Technology (n.d.). The equipment industry division I organized a symposium on the promotion of electric vehicles in the public sector 2020 [EB/OL]. http://www.miit.gov.cn/n1146290/n1146402/n1146440/c7824104/content.html (accessed 8 March 2020).

20 National Development and Reform Commission (n.d.). Opinions on accelerating the establishment of a system of green production and consumption laws and policies 2020 [EB/OL]. https://www.ndrc.gov.cn/xxgk/zcfb/tz/202003/t20200317_1223470.html (accessed 22 April 2020).

21 Mahmoud, M., Garnett, R., Ferguson, M. et al. (2016). Electric buses: a review of alternative powertrains. *Renewable and Sustainable Energy Reviews* 62: 673–684.

22 Gao, Z., Lin, Z., LaClair, T.J. et al. (2017). Battery capacity and recharging needs for electric buses in city transit service. *Energy* 122: 588–600.

23 You, P., Yang, Z., Zhang, Y. et al. (2016). Optimal charging schedule for a battery switching station serving electric buses. *IEEE Transactions on Power Systems* 31 (5): 3473–3483.

24 Bi, Z., Keoleian, G.A., and Ersal, T. (2018). Wireless charger deployment for an electric bus network: a multi-objective life cycle optimization. *Applied Energy* 225: 1090–1101.

25 Chen, Z., Yin, Y., and Song, Z. (2018). A cost-competitiveness analysis of charging infrastructure for electric bus operations. *Transportation Research Part C Emerging Technologies* 93: 351–366.

26 Ke, B.-R., Chung, C.-Y., and Chen, Y.-C. (2016). Minimizing the costs of constructing an all plug-in electric bus transportation system: a case study in Penghu. *Applied Energy* 177: 649–660.

27 He, Y., Song, Z., and Liu, Z. (2019). Fast-charging station deployment for battery electric bus systems considering electricity demand charges. *Sustainable Cities and Society* 48: 101530.

28 Wang, X., Yuen, C., Hassan, N.U. et al. (2017). Electric vehicle charging station placement for urban public bus systems. *IEEE Transactions on Intelligent Transportation Systems* 18 (1): 128–139.

29 Xylia, M., Leduc, S., Patrizio, P. et al. (2017). Locating charging infrastructure for electric buses in Stockholm. *Transp Res Part C Emerg Technol* 78: 183–200.

30 Lajunen, A. (2018). Lifecycle costs and charging requirements of electric buses with different charging methods. *Journal of Cleaner Production* 172: 56–67.

31 Gallet, M., Massier, T., and Hamacher, T. (2018). Estimation of the energy demand of electric buses based on real-world data for large-scale public transport networks. *Applied Energy* 230: 344–356.

32 Rogge, M., Wollny, S., and Sauer, D.U. (2015). Fast charging battery buses for the electrification of urban public transport – a feasibility study focusing on charging infrastructure and energy storage requirements. *Energies* 8: 4587–4606.

33 Frey, B.J. and Dueck, D. (2007). Clustering by passing messages between data points. *Science* 315: 972–976.

34 Kennedy, J. and Eberhart, R.C. (1997). Discrete binary version of the particle swarm algorithm. *IEEE International Conference on System, Man, and Cybernetics. Computational Cybernetics and Simulation*, volume 5 4104-8; Orlando, USA (12–15 October 1997), vol. 5, 4104–4108.

35 Gómez, M., López, A., and Jurado, F. (2010). Optimal placement and sizing from standpoint of the investor of photovoltaics grid-connected systems using binary particle swarm optimization. *Applied Energy* 87: 1911–1918.

36 Yutong (n.d.). Yutong: e-bus. https://en.yutong.com/products/new-energy-ehicles/ (accessed 13 August 2022).

37 Star Charge (n.d.). Wanbang Digital Energy (Star Charge), September 2020. http://www.starcharge.com/.

38 MOST (n.d.). MOST: Policies on EV charging facility construction. http://www.most.gov.cn/tztg/201601/t20160120_123772.htm (accessed 12 November 2020).
39 State Council (2015). "Internet + Energy" Strategy (In Chinese). http://www.gov.cn/xinwen/2015-07/04/content_2890205.htm (accessed 12 November 2020).
40 Department for Transport (n.d.). Low-emission vehicles eligible for a plug-in grant. https://www.gov.uk/plug-in-car-van-grants (accessed 8 December 2021).
41 Land Transport Authority (n.d.). Electric vehicles. https://www.lta.gov.sg/content/ltagov/en/industry_innovations/technologies/electric_vehicles.html (accessed 8 December 2021).
42 Ministry of Transport (n.d.). Electric vehicles programme. https://www.transport.govt.nz/area-of-interest/environment-and-climate-change/electric-vehicles-programme/ (accessed 8 December 2021).
43 Zot, C., Amit, R.H., and Massa, L. (2011). The business model: recent developments and future research. *Journal of Management* http://doi.org/10.1177/0149206311406265.
44 Lai, C.S., Jia, Y., Dong, Z. et al. (2020). A review of technical standards for smart cities. *Clean Technologies, MDPI* 2 (3): 290–310.
45 Hamurcu, M. and Eren, T. (2020). Electric bus selection with multicriteria decision analysis for green transportation. *Sustainability, MDPI* 12: 2777. https://doi.org/10.3390/su12072777.
46 Zhou, B., Wu, Y., Zhou, B. et al. (2016). Real-world performance of battery electric buses and their life-cycle benefits with respect to energy consumption and carbon dioxide emissions. *Energy, Elsevier* 96: 603–613.
47 Perez, J.C., Carrillo, M.H., and Montoya-Torres, J.R. (2015). Multi-criteria approaches for urban passenger transport systems: a literature review. *Annals of Operations Research* 226: 69–87.
48 Sun, X.-H., Yamamoto, T., and Morikawa, T. (2016). Fast-charging station choice behavior among battery electric vehicle users. *Transportation Research Part D: Transport and Environment* 46: 26–39.
49 Wang, H., Zhao, D., Meng, Q. et al. (2019). A four-step method for electric-vehicle charging facility deployment in a dense city: an empirical study in Singapore. *Transportaion Research Part A: Policy Practice* 119: 224–237.
50 An, K. (2020). Battery electric bus infrastructure planning under demand uncertainty. *Transportation Research Part C: Emerging Technologies* 111: 572–587.
51 Panchal, S., Rashid, M., Long, F. et al. (2018). Degradation testing and modeling of 200 Ah LiFePO4 battery. SAE Technical Paper; SAE International: Detroit, MI, USA.
52 Panchal, S., Dincer, I., Agelin-Chaab, M. et al. (2018). Design and simulation of a lithium-ion battery at large C-rates and varying boundary conditions through heat flux distributions. *Measurement* 116: 382–390.
53 Lai, C.S., Locatelli, G., Pimm, A. et al. (October 2019). A financial model for lithium-ion storage in a photovoltaic and biogas energy system. *Applied Energy* 251 (1): 113179.
54 Lai, C.S., Mo, Z., Wang, T. et al. (2020). Load forecasting based on deep neural network and historical data augmentation. *IET Generation Transmission and Distribution* https://doi.org/10.1049/iet-gtd.2020.0842.
55 Teoh, L.E., Khoo, H.L., Goh, S.Y., and Chong, L.M. (2018). Scenario-based electric bus operation: a case study of Putrajaya. *Malaysia International Journal of Transportation Science Technology* 7: 10–25.

16

Best Practice for Parking Vehicles with Low-power Wide-Area Network

16.1 Introduction

Internet of Things (IoT) applications are one of the emerged technology applications that are considered a key digital transformation enabler for several industries and utilities, including smart cities, factories, electricity grids, health, education, transportation, and many other services. IoT is playing an increasingly important role in smart city construction. The number of IoT devices is facing a hugely increasing trend from 2012 to 2020 as shown in Figure 16.1.

Infrastructure aging, smart mobility, energy efficiency, and personal safety are some of the issues that put serious strains on the economic sustainability of human activities, as well as the quality of life in modern urban areas. As a matter of fact, the living quality in medium and big cities has been deteriorating significantly in the last several decades, and it is expected to get worse in the future. For example, Figure 16.2 shows how a beautiful city could be affected under air pollution. The people within the city will breath with poor air and cannot see things properly.

Projections indicate that there will be more than 41 mega-cities (with more than 10 million people) by 2030, with a relevant increase with respect to the current 28 mega-cities. At the same time, the percentage of population residing in urban areas is expected to reach 66% by 2050, compared to a figure of 54% in 2014 [1]. In recent years, the smart city idea has emerged as a revolutionary approach to tackle the challenges posed by the increasing complexity of urban environments [2]. By utilising Information and Communication Technologies (ICTs) smartly, smart cities promise to bring wide-ranging improvements to urban living, making the use of physical infrastructures such as residential and public buildings, roads, utilities, more efficient, properly adapting to evolving circumstances, collecting valuable information for decision-making processes and effectively engaging citizens in local governance. The expected reward of such a process is an inclusive environment where, with innovative technical solutions, citizens experience healthy, safer, and more efficient conditions. Turning this vision into reality would require the integration of urban infrastructures, citizens and governance in a comprehensive and holistic way, which envisions the different components of the smart cities. ICTs play the role by connecting the different parts for example infrastructures, people, and governance and providing the computational capability for data analysis, features extraction, and decision-making support. As plenty of connected devices will operate in order to sense the physical world as well as to adapt the context to changing circumstances, the IoT was proposed. In most cases, connectivity is ensured by wireless networks, as they enable the exchange of information with a flexible and low-cost deployment. In this regard,

Smart Energy for Transportation and Health in a Smart City, First Edition. Chun Sing Lai, Loi Lei Lai and Qi Hong Lai.

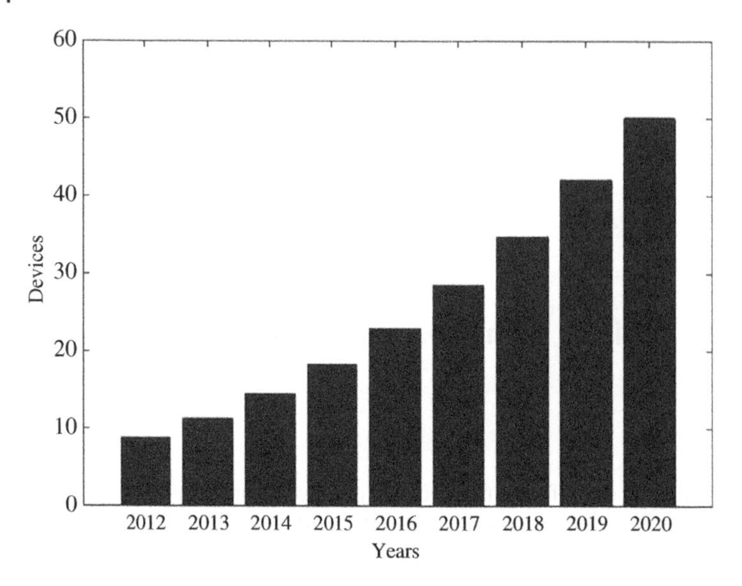

Figure 16.1 Number of IoT devices from 2012 to 2020.

a heterogeneous communication technology and different network architectures should be adopted, depending on the characteristics of the service to be implemented, e.g. low bit rate or high bit rate, and with coverage such as long-range or low-range links.

In the IDTechEx report [3], it showed that the adoption of low power wireless network technology in four different areas, namely, smart cities, assest tracking, agriculatuer, and smart homes in 2018.

For smart cities, governments worldwide are investing heavily by increasing connectivity to the infrastructure and the environments, for example, in street lighting and environmental monitoring to reduce crime rates among other applications.

Regarding asset tracking, low power networks provide a new business model in the form of subscriptions for tracking of things.

Turning to agriculture, there are new ways to monitor crops, water usage, environmental conditions and other aspects designed to ensure production uniformity and good yields on farms and vineyards to promote the so-called smart agriculture. The long range and low power requirements of low-power wide-area networks (LPWAN) are useful to some of the applications.

As for smart homes, wireless networks help buildings connect effectively. Government regulation and environmental management are adopted to make homes and building more energy efficient.

The total number of connections by applications and technologies used are given in Figure 16.3:

It was mentioned that there will be 2.7 billion LPWAN connections in 2029. LPWA networks have developed in multiple ways. On one hand, there are low power, unlicensed communication technologies, some of which are highly proprietary and focused on a particular application; on the other hand, there are large telecommunication companies that have added low power versions as extensions of their cellular network, with the leading cellular protocols, with NB-IoT and LTE, wrapped up into the 5G standard. There are many types available with different governments and territories pushing different types [3].

The increased population density worldwide in cities and villages and the evolution of human and community lifestyles have led to the rapid development of mobile communication technology.

(a)

(b)

Figure 16.2 A beautiful city affected by air pollution for (a) summer and (b) spring.

Thus, the 3rd Generation Partnership Project (3GPP) has standardized the vision for 5th generation mobile communication technology (5G), which comprises the traditional LTE network along side the future 5G network. The 5G mobile networks get to overcome the previous cellular standards limitations and to become an enabling technology for future IoT applications. The 3GPP releases of 5G, starting from release 15 and beyond, introduced enhanced cellular communications services such as enhanced mobile broadband (eMBB) and massive machine type communications (mMTC), and they also introduced new services such as ultrareliable low latency communication (URLLC) and time-sensitive communications (TSC) [4–6]. The URLLC and mMTC in 5G are closely related to IoT [6]. Compared to 4G, it is expected that 5G mobile networks will efficiently support the basic IoT requirements, such as good coverage, high data throughput, low latency, high scalability, high

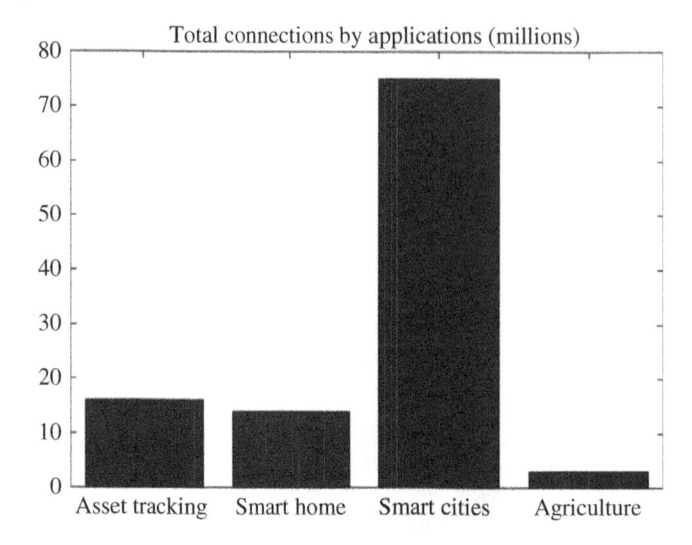

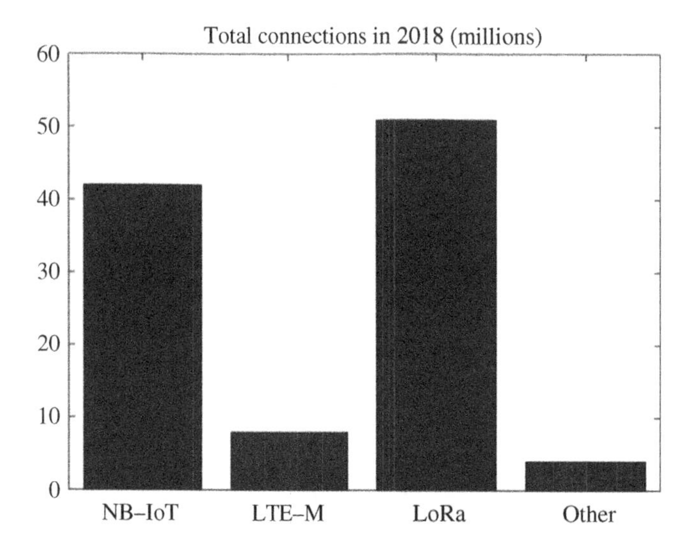

Figure 16.3 Number of connections by applications and technologies used.

energy efficiency, and the connectivity provision for IoT users [4]. 5G aims to maximize the capacity of the network to be at least 1000 times larger than 4G [7].

The IoT 5G includes current technologies such as unlicensed bands, for example, short-range communication technologies including Bluetooth, Zigbee, Wi-Fi, and the unused spectrum band, namely, cognitive radio [8] and LPWANs.The latter are divided into non-3GPP networks such as SigFox, LoRa, [9, 10], and 3GPP networks such as long-term evolution (LTE) – Min Release 12 and narrowband-IoT (NB-IoT) in Release 13 [10–18]. Recently, Fog radio access network (Fog-RAN) and Cloud radio access network (C-RAN) have been proposed for fifth-generation (5G) wireless communications that serve IoT applications [19]. For smart city applications, LPWANs are the most suitable alternatives [9]. Therefore, only 5G cellular IoT is discussed here. NB-IoT achieves better LPWA performance in terms of coverage and the variety of data rates and deployment scenarios compared to other LPWA technologies [20]. This technology is a modified design of LTE that aims to serve IoT traffic. Besides, it can provide enhanced coverage compared to LTE. It is used in three modes, namely standalone, in-band, and guard-band [21]. The subcarrier bandwidth is 180 kHz in the case of coexistence with current LTE networks to provide long periods of connections over mobile operator networks since various objects may have small amounts of data [21]. The aim of NB-IoT is globally covering low-cost or complex ubiquitous IoT devices worldwide [22]. To cater for the large amount of IoT requirements, the 3GPP has standardized that the mMTC services will be served by further evolving NB-IoT and LTE-M as integral parts of the 5G specifications, such that there will be no need for new standards in the near future. Although they are known as 4G technologies, the 3GPP has started a working group in the 5G core network to support LTE-M and NB-IoT radio access networks in order to guarantee the compatibility of these technologies with 5G [23].

Efficient and profitable IoT communication systems should be low cost, size, weight, and power directed. With the increased dependence on NB-IoT technology in human lives, a large number of NB-IoT terminals need to be connected to the network in order to cater for the user needs. According to the predictions by Ericsson, the number of connected IoT devices will reach 1.5 billion by 2022 [24]. The rapid increase in the number of massive IoT devices has added a growing challenge to 5G design owing to its contrasting requirements [7] from two main perspectives. First, the narrow band width may restrict the transmission performance of the IoT in addition to the communication overhead that is required before data transmission [7]. Therefore, it is important to achieve continuous improvements concerning the efficient use of these resources to support the continued growth of IoT applications [25]. Second, access reservation protocols designed for 5G networks face massive connectivity for IoT applications [9], because the device density is expected to be larger than the capability of the techniques to realize access reservation and random access (RA) procedures, and may significantly affect end-to-end delay.This issue would be critical, especially for the scenarios of burst arrival [26]. The access requests and the signalling associated with each periodic IoT transmission process are a burden to both IoT device and network resources. These problems are further transferred with uplink traffic and remain opened issues, as discussed in [21, 27]. Until now, 5G systems have been considered as an evolution of the LTE-advanced (LTE-A) standard, where the 5G new radio (NR) access reservation is also based on LTE-A random access (RA) procedure [26, 28]. Access class barring (ACB) and its modifications is a standardized technique to process burst access in LTE and NRRA [26]. However, it may not be appropriate for IoT applications [29]. Most adopted solutions focused on random-access phase optimization and other optimization techniques should be considered. Currently, studies and developments for the NB-IoT have been widely adopted by global mobile operators and manufacturing companies [25].

Reference [30] adopted the random accessing stage in Media Access Control (MAC) to improve the system throughput by using Markov chain modelling technique. In [31], feasible strategies for resource sharing between the multimedia and sensory data in a hybrid LTE/NB-IoT wireless deployment was introduced. Dynamic resource sharing with reservation strategy was used as a preferred solution for reliable collection of heterogeneous data in large-scale 5G-grade IoT deployments.

The NB-IoT group communication technique was adopted clearly in [32]. Their proposal was based on adapting Multimedia Broadcast Multicast Service (MBMS) protocol for serving downlink traffic of IoTs. The cashing based group access was adopted in [7] for Fog-RAN and in [33] for Industrial IoT covered by small base stations.

The fast development and implementation of IoT-based technologies have allowed for various technological achievements in different aspects of life. The main goal of IoT technologies is to simplify processes in different fields, to ensure a better efficiency and resiliency of systems and finally to improve quality of life. Sustainability and adativitiy have become key issues for population where the dynamic development of IoT technologies is bringing different benefits, but this fast development must be carefully monitored and evaluated from an environmental and societal point of view to limit the presence of harmful impacts and ensure the smart utilization of limited global resources. Significant research efforts are needed in the previous sense to carefully investigate the pros and cons of IoT technologies. Reference [34] gave a review on Opportunities, issues, and challenges on IoT toward a smart and sustainable future. It covers the latest advancements in the fields: (i) IoT technologies in Sustainable Energy and Environment, (ii) IoT enabled Smart City, (iii) E-health – Ambient assisted living systems, and (iv) IoT technologies in Transportation and Low Carbon Products. The review provided an understanding of current technological progress in IoT application areas as well as the environmental implications linked with the increased application of IoT products. For example, according to [35], in 2020, US data centers are expected to consume 73 billion kilowatt-hours of energy, according to one energy-usage report. As if that was not bad enough, each kilowatt-hour of energy requires two gallons of water to cool it. So for 2020 alone, that equates to some 220 000 Olympic-size swimming pools of water. Obvious, this gives a pressure on the use of natural resources.

The IoT technology has been a fundamental one that facilitates the construction of smart applications, aiming at improving the production efficiency and human living quality, such as smart city, smart healthcare, smart industry, smart agriculture, etc. Larger coverage of wireless technology is required to achieve the best performance of IoT applications which are in the trend of increase these days. The LPWAN technologies are suitable alternatives to fulfill the requirements. The LPWAN is proposed to replenish the vacancy of current wireless technologies including Wi-Fi and Bluetooth, which provide service in a short distance with high bandwidth. LPWAN become an applicable and essential part in smart city. Table 16.1 gives a comparison between LPWAN and other types of wireless technologies helow:

The LPWAN earns its popularity around the world due to its low power, long-distance, and low cost in communications. It could provide up to 10–40 km in rural zones and 1–5 km in urban zones with high energy efficiency [36]. The cost of a radio chipset is about 2 USD and operating cost per device is about 1 USD for one year [37]. These characteristics of LPWAN had attracted researchers to perform numerous studies on applications, and indoor and outdoor evaluation of LPWAN. Lots

Table 16.1 Comparison between LPWAN and other types of wireless technologies.

	Local area network Short range communication	**Low power wide area (LPWAN) Internet of Things**	**Cellular network Traditional M2M**
Market	40%	45%	15%
Advantages	Well-established standards	Low power consumption Low cost	High capacity Less transmission power
Disadvantages	Privacy violations High setup cost	Not situable for high data rate applications Not situable for low latency applications	Infrastructure neededHandover needed
Examples	Bluetooth 4.0 Zigbee, Wi-Fi	LoRa	GS 3G+/H+, 4G

of LPWAN applications are developed accordingly and they have proved the reliability and appropriateness of LPWAN technologies. LPWAN technology is one of the most suitable choices for these applications which need to transmit data over long distances. Currently, popular LPWAN technologies include Sigfox [38], LoRaWAN [37, 39], and NB-IoT [40].

The governments all over the world are pacing rapidly on smart city initiative and LPWAN deployment is, nonetheless, one of the focuses of the Government. Since there are various technologies for LPWAN and a global IEEE standard is not in place, it is desperate to develop a new standard to aid sensor interoperability and combat IoT cyber threat. This investigation intends to compile best practices for the technologies to achieve the best performance. To investigate the appropriate technology to be applied to different IoT applications, a guideline on the usage of these technologies should be introduced. Thus, a guideline would be proposed for the best practices of LPWAN to promote IoT development. In this chapter, LPWAN technologies including Sigfox, LoRaWAN, and NB-IoT were evaluated. Technical features among these technologies would be compared. Then the success factors such as coverage, latency, etc would be compared to further define the characteristics of each technology. At last, an IoT index, LP-INDEX would be developed for LPWAN technologies to evaluate their performance in different scenarios. The LPWA-index evaluates the LPWAN in different aspects including Latency, Data Capacity, Power and Cost, Coverage, Scalability, and Security, and it provides a final score for each LPWAN technique based on application requirements. Besides, the LPWAN network was applied to better explore the characteristics through experiments.

The contributions are summarized as follows:

1) An LPWAN index for the evaluation of LPWAN, LP-INDEX, is proposed to provide a guideline for finding the best practices of LPWAN. The developed LP-INDEX guides LPWAN based applications and leads the LPWAN deployment in smart city construction.
2) Testing for three LPWAN comparisons based on a parking detection sensor was implemented as a use case to illustrate how LP-INDEX works. To the best of our knowledge, there are no similar testings in related work.

The remaining of this chapter is organized as follows: Section 16.2 summarizes the state-of-art works in LPWAN. Section 16.3 introduces LP-INDEX for best practices of LPWAN technologies. Section 16.4 gives a case study. Finally, Section 16.5 provides a conclusion and future work.

16.2 Related Work

The unique advantages of LPWAN have attracted a lot of researchers to perform further studies on them. This section discusses the previous studies related to the conducted LPWAN technologies.

Nowadays numerous researchers concentrate on LPWAN-based applications, such as monitoring, localization, sensing, etc. These works contribute to facilitating the implementation of LPWAN technologies in reality and attract more attention from investigators. Meanwhile, since the development of LPWAN is still in its infancy, more researches focus on the evaluation of LPWANs.

Except for the works listed above which evaluate LPWANs technologies individually, some researchers studied them together and compared various LPWANs in different terms. In a related work [41], the authors discussed different LPWAN technologies in Media Access Control layer. They compared Sigfox, LoRa, NB-IoT with their features of the radio frequency and evaluated them in terms of packet error rate. In another study [42], the authors explored the main features of LoRa and Sigfox and conducted a coverage estimation test case. They also performed experiments to estimate the largest coverage of LPWAN. Besides, in [43] the authors reviewed the main LPWANs and list their technical features such as modulation, frequency, bandwidth, etc. The authors further compared them in terms of performance factors such as coverage, latency, etc. It was available to know the strengths and drawbacks of different LPWANs from this paper, but it lacked experiments and original algorithm.

The mentioned works all contribute to the research of LPWAN technology. However, among all current researches, studies related to the best practices of LPWAN are lacking behind. The lack of related guideline on LPWAN choice based on specific scenarios causes a confusion for users. For decision-makers, it is hard for them to select the most applicable one from so many LPWAN technologies. For this reason, IEEE P2668 is proposed based on LPWAN index to evaluate different LPWANs in various applications.

16.2.1 LoRaWAN

LoRa (Long Range) is a low-power wide-area network (LPWAN) protocol developed by Semtech. It was developed by Cycleo of Grenoble, France and acquired by Semtech, the founding member of the LoRa Alliance.

LoRa Technology LoRa [44] adopts a proprietary modulation based on chirp spread spectrum (CSS) technology, which exploits chirps whose frequency increases or decreases linearly over a certain amount of time; information is inserted in chirps by introducing a frequency discontinuity at different time offsets [45]. These chirps occupy a bandwidth of 125, 250, or 500 kHz. One of the most important parameters of the physical layer is the Spreading Factor (SF), which is the ratio between the signal bandwidth to the symbol rate. By maintaining the bandwidth constant, it is possible to improve the receiver sensitivity by increasing the airtime, that is duration of a packet transmission. To describe it in a more accurate way, each increment by one unit of the SF will correspond to a doubling of the airtime and a decrement of the receiver sensitivity by about 3 dB. At MAC layer LoRa uses the LoRaWAN protocol that describes three Classes: (i) Class A: end devices, after the transmission of a packet, open two receive windows to get an acknowledgement (ACK) or received at from the gateway, then they stay in idle mode until the next transmission, (ii) Class B: end devices have more receive windows synchronized with a beacon provided by the gateway, and (iii) Class C: end devices stay continuously in reception mode. Any device has to be compliant with,at least, ClassA, where ALOHA protocol is used in uplink. As for network aspects, LoRaWANs are generally laid out in a star topology and the central node is usually called gateway.

LoRa is a physical layer technology that modulates the signals in the sub-GHz Industrial, Scientific, and Medical (ISM) band [37]. LoRa uses unlicensed ISM bands, 923.2 MHz in Hong Kong. The CSS modulation spreads a narrow-band signal over a wider channel bandwidth in LoRa. This results in a signal with low noise levels, high interference resilience, and it is difficult to be detected or jammed. One unique feature of LoRaWAN is the flexible establishment without Internet Service Providers (ISP), which makes it cost less and has an extra chance to ensure security. It replaces the cost of subscription and management fees by gateway purchase and installation fees. Though the cost of the LoRa gateway is not a small number, the whole cost of LoRaWAN network establishment is decreased if the cheap gateways are selected. Besides, the flexibility brought by the private establishment of LoRaWAN network becomes another advantage. Comparing to other LPWANs whose base stations are fixed, it is more convenient to provide a good cover for applications. Users can deploy the gateway to the most appropriate site, which enables most devices to send and receive high-quality signals. In addition, security can be further ensured if users do not connect to the public network. The ability of flexible network establishment is a unique advantage of LoRa. Another unique feature of LoRa is that it adopts an adaptable data rate (ADR) algorithm through six spreading factors (SF7–SF12) to make tradeoff with coverage. A higher spreading factor indicates larger coverage at the expense of a lower data rate. The LoRa data rate varies between 300 bps and 50 kbps based on spreading factor and channel bandwidth. Also, messages transmitted through various spreading factors can be received simultaneously by LoRa base stations. In 2015, LoRaWAN, a LoRa-based communication protocol is first proposed by LoRa-Alliance. Based on LoRaWAN, signals transmitted by end devices can be received by all the base stations which are covered. The success transmission ratio is thus increased due to this redundant reception. Since multiple base stations are needed nearby, this capability incurs additional network deployment costs. However, there are more advantages than disadvantages. For example, a time difference of arrival (TDOA)-based localization technique was developed, supported by very precise time synchronization between different base stations [39]. Therefore, LoRaWAN is applied to localization by analyzing the strength of the signal received from the same terminal device. This feature also prevents handoffs because the terminal always sends signals to all available base stations, so no handoffs are required.

16.2.2 NB-IoT

NB-IoT is specified in Release 13 of the Third Generation Partnership Project (3GPP) as a Narrow Band IoT technology in 2016 [40]. NB-IoT occupies a frequency bandwidth of 200 kHz, which corresponds to one resource block in Global System for Mobile Communications (GSM) and LTE transmission. NB-IoT is developed based on the LTE protocol. It can be seen as a simple version of LTE without complex protocol functionalities. It employs the quadrature phase-shift keying modulation (QPSK). The NB-IoT is further improved to be more suitable for IoT applications. When the LTE backend transmits signals to available end devices in a cell, it will decrease these necessary steps adopted in LTE protocols such as channel quality monitoring, carrier aggregation, and dual connectivity, which are unnecessary for IoT applications. It decreases the power and resource consumption to a minimum. For each cell, at most 100 000 devices are allowed to connect to the NB-IoT carrier. It could break through the number limit by adding more carriers. There are 12 subcarriers inside the channel and each subcarrier is separated by 15 kHz. NB-IoT uses single carrier frequency division multiple access (SC-FDMA) modulation and orthogonal frequency division multiple (OFDM) access modulation for uplink and downlink transmissions. This makes large connections and reliable two-way communication possible. Since it is deployed in the licensed band, NB-IoT has a relatively large throughput, which enables device firmware to be updated over the air. The NB-IoT uplink effective data rate is 0.5–140 kbps, and the downlink effective data rate is 0.3–125 kbps. In addition, NB-IoT benefits from a licensed band with no duty cycle restrictions. But the disadvantage is the high deployment cost of narrowband IoT. NB-IoT has three network deployment methods, namely in-band, guard-band, and stand-alone [38]. NB-IoT spectrum is deployed inside the LTE spectrum band with 180 kHz bandwidth which is one resource block of an LTE channel in in-band mode. For guard-band deployment, 180 kHz NB-IoT spectrum is placed by ISPs in the existing LTE signal's guard-bands. NB-IoT spectrum can also be entirely separated from the existing LTE spectrum in stand-alone method. Reference [46] gives technical decision-makers an overview of NB-IoT, the communications technology that will underpin industrial-grade IoT deployments.

16.2.3 Sigfox

Sigfox is a French global network operator founded in 2010 that builds wireless networks to connect low-power objects such as electricity meters and smart watches, which need to be continuously on and emitting small amounts of data. Sigfox employs the differential binary phase-shift keying (DBPSK) and the Gaussian frequency shift keying (GFSK) that enables communication using the Industrial, Scientific and Medical ISM radio band which uses 868 MHz in Europe and 902 MHz in the US. It utilizes a wide-reaching signal that passes freely through solid objects, called ultra narrow band and requires little energy, being termed LPWAN. Sigfox has partnered with a number of firms in the LPWAN industry such as Texas Instruments. The ISM radio bands support limited bidirectional communication. The network is based on one-hop star topology and requires a mobile operator to carry the generated traffic. The signal can also be used to easily cover large areas and to reach underground objects. As of October 2018, the Sigfox IoT network has covered a total of 4.2 million square kilometers in a total of 50 countries.

Sigfox connects each customer with end-to-end service. The Sigfox base stations are deployed with software-defined radios. They are connected to the backend server through an IP-based network [38]. The application of the ultra-narrow band makes bandwidth more efficient and noise levels lower. This further leads to lower power consumption and cost. In Sigfox, data from the base stations to the end devices as downlink communication will be sent following an uplink communication. Due to the number of messages in the downlink, it is impossible to acknowledge each uplink message. Then, the uplink communication reliability is ensured by time and frequency diversity, and transmission duplication. Each end-device message will be transmitted three times (by default) over different frequency channels. Sigfox divides the global region into seven regions from RC1 to RC7. Each region specifies different operating rules for the Sigfox device, including frequency range, data rate, multiple access mechanisms, and hardware specifications. Sigfox uses a lightweight protocol for short message transmission to ensure low power consumption. This lightweight protocol typically limits up to 140 upstream transfers per day, with a maximum payload of 12 bytes and a maximum downstream transfer of 28 bytes, and is used only for upstream identification. However, this prevents the Sigfox network from responding to most upstream messages. The frequency-hopping used in RC4 allows each message frame to be broadcast three times on a different frequency. Besides, the second send can be executed after 20 seconds. Therefore, the delivery of packaging can be ensured. The use of 100 Hz bandwidth ultra-narrow band (UNB) modulation has achieved ultra-low noise level, thus the long-distance transmission and high sensitivity at the receiver end can be realized.

16.3 LP-INDEX for Best Practices of LPWAN Technologies

The IEEE Standard Working Group P2668 [47–49] is developing an indicator, namely the IoT Index (IDex), to measure the maturity of IoT objects and produce quantifiable evaluation results. The IoT objects refer to a variety of physical and cyber objects in the IoT ecosystem, such as devices, machines, networks, systems, services, and infrastructure. The IDex indicates the integral performance of IoT objects pertinent to their applications. An IoT object can have different IDex levels when it is implemented to various applications and scenarios. For instance, a wireless protocol can achieve a high IDex level in noncritical use cases, but it may attain a low IDex level in critical use cases. With the salient features offered by the IDex, the IoT stakeholders can compare, adapt, and develop IoT solutions that fit the applications' requirements and comply with the global standards. An example of the IDex use case on evaluating is provided as follows:

As mentioned, LPWAN is an emerging wireless technology that aims to address the IoT's needs for high-scalability, low-power, and large-coverage connections. As a trade-off, the LPWANs are usually latency-tolerant and lack advanced security measures. For IDex applications in LPWAN, a sub-trail named LP-INDEX is proposed.

The LP-INDEX can manifest high-level guidance on the wise selection (and integration) of IoT solutions. It evaluates the LPWAN based applications based on the following six aspects, namely, Latency, Data Capacity, Power and Cost, Coverage, Scalability, and Security. For each application, a set of preferences can be defined according to the application requirement in terms of these aspects. For instance, an intelligent transport system requires 10–50 Mbps of data rate, 10 ms of latency, and a connection density of 4000/km^2. Then, the weightings are distributed for all the aspects. The weightings are considered mainly based on the application requirement. Special requirements of users will be focused in particular. Evaluations based on these preferences, and representative grades for the six aspects are made. Finally, a total score is given according to Equation (16.1) as given below:

$$S_{total} = \sum\nolimits_{1}^{i} w_i S_i \tag{16.1}$$

In Eq. (16.1), $\boldsymbol{w}$ and S represent the weighting and evaluation score for each preference.

The following sections introduce the specific content for the six considered aspects of LPWAN.

16.3.1 Latency

Latency indicates the time it takes for a request to travel from the sender to the receiver and for the receiver to process that request. In other words, latency is a measure of delay. In a network, latency measures the time it takes for data to get to its destination across the network. It is usually measured as a round trip delay. It is desirable to have this time to remain as close to zero as possible, however, in practice, there are constraints preventing the latency times to be zero but we try our best to keep it as low as possbile. Therefore, it is an essential parameter for these applications with restricted limitations in real-time. However, LPWAN generally does not perform well on latency due to the network characteristic. Thus, other wireless networks are recommended to replace LPWAN for those applications with ultra-low latency requirements. Though the low latency is not the advantage of LPWAN, it still deserves to be compared among different LPWANs. Some applications for LPWAN still require as low latency as possible due to the application need. Latency is another element that contributes to network speed. A so-called low latency network connection is one that generally experiences small delay times, while a high latency connection generally suffers from long delays. Latency is also referred to as a ping rate and typically measured in milliseconds (ms). The ping is the request sent to the server, and the ping rate is how long it takes for that request to transmit and come back with the result. Low latency means there is a strong, reliable network connection, which reduces the chance for a connection loss or delay. The delayed messages may transmit incorrect information and confuse at peak hours.

16.3.2 Data Capacity

Data capacity refers to the amount of data that can be stored in a tag. Increased data capacity increases the usefulness of the tag and its cost. In general, large data capacity will need more data storages such as data bank. Data capacity reflects the available transmitted message in one turn, which is essential in such applications with a large packet size requirement. Data rate and payload length are thought to be the main parameters to reflect the performance of data capacity in this part. Higher data rates enable more devices to transmit more information, thus improving data transmission capacity. The larger the payload is, the more information is sent around. Due to the various LPWAN protocols, the three LPWANs have various data capacity characteristics. It is worth mentioning that Sigfox has the largest data transfer limitation daily. This may bring a negative effect for these applications in asking for frequent communication.

16.3.3 Power and Cost

Low power is one of the main characteristics of the three LPWAN technologies. The performance of power consumption is critical for battery-powered IoT end devices because unrealistic costs would be incurred by frequently replacing large network batteries. Sensors in LPWAN applications, such as temperature and humidity sensors, are dedicated to maintaining minimum power consumption to extend the battery life of sensor devices. Thus, the power consumption for different operation modes should be considered. In addition, the working period of the different patterns determines the duty cycle of each technology. For example, longer sleep patterns can reduce power consumption. Among these three, NB-IoT consumes a larger power due to its communication protocol. Unlike, Sigfox and NB-IoT, LoRaWAN provides an extra choice, Class C with more power consumption but supporting bidirectional communication.

Regarding cost, it is the basic issue that is considered in all applications. Usually, the user should choose the most suitable one within the budget. In terms of cost, several parameters including costs for sensor, gateway, installation, subscription, management, and sensor recurring should be considered. As discussed previously, usage fees are needed for Sigfox and NB-IoT service, while for LoRaWAN it is free. For LoRaWAN, since the network could be established freely, the additional cost is generated for LoRaWAN gateways. Besides, the fees for deployment, the sensor, and maintenance of the three LPWAN technologies should be considered in practice.

Low Power and Low Cost makes the products have a longer lifetime and this usually decreases the overall cost requirement.

16.3.4 Coverage

The coverage indicates the operation range of the LPWAN technology. In general, the large coverage is a common characteristic of LPWANs due to their protocol. This characteristic fulfills the gap of the traditional wireless network. The traditional wireless network needs hundreds or thousands of base stations to cover a city while for LPWAN only several base stations are needed to achieve similar performance. For LPWAN, though they all have large coverage, the range is a little different due to the protocol used. Besides, the flexible network establishment is another considered element. As discussed previously, LoRaWAN could be established conveniently by deploying LoRaWAN gateways. Sigfox and NB-IoT end devices need to be deployed based on service base station location from Internet Service Provider. The flexible network establishment could bring stable communication of good quality for some applications while for others this function may not be useful. Large coverage provides stable signals with less base stations. Sufficient coverage ensures the continuous operation of sensors.

16.3.5 Scalability

Scalability measures the capacity to be changed in size or scale. That is, scalability is the measure of a system's ability to increase or decrease in performance and cost in response to changes in application and system processing demands. Larger scalability can support more users.

Supporting thousands of end devices is a common feature of Sigfox, LoRaWAN, and NB-IoT. At the design stage, these LPWAN technologies should provide high scalability. This results from the topology structure and operation mechanism of LPWAN. For instance, strategies like optimal channel selection make the same available channels to accommodate more communications. For applications with limited users, scalability is not so important because enough channels are usually available. However, for large-scale systems, the scalability limitation could influence the performance of LPWAN.

16.3.6 Security

Security mechanism ensures the privacy of data. With high security, the LPWAN could prevent the user's privacy from hacking. LoRaWAN supports the security by a unique 128-bit Advanced Encryption Standard (AES) key, called AppKey and a globally unique identifier, EUI-64-based DevEUI, both of which are used during the device authentication process [50]. For Sigfox, It uses the counter (CTR) encryption key in the AES128 algorithm mode, which is unique for each device. The air security of the uplink implements several mechanisms: a message counter for replay attack protection, AES-128 in Cipher block chaining (CBC) mode for authentication and integrity checking, and CRC-16 for error detection. NB-IoT inherits the security mechanisms for confidentiality and authentication from LTE networks. LTE provides symmetric encryption and signature mechanisms to prevent data leakage and uses SIM cards to authenticate and identify devices in the network.

16.4 Case Study

16.4.1 Experimental Setup

In this part, the experimental setup and the evaluation results of LPWAN are introduced. The setup of parking sensors is shown in Figure 16.4.

16.4.2 Depolyment of Car Park Sensors

In this test, parking detection sensors are used as a use case. The parking detection sensor could detect the occupied/free status of park lots and transmit the data through the LPWAN network. In the test, different parking detection sensors with the three LPWAN techniques were adopted. Based on the experiments, the best applicable technique was selected for this application.

16.4.3 Evaluation Matrices and Results

The evaluation results of the three IoT technologies are given in Table 16.2. Parking sensors detect the approach of vehicles and send notice through the wireless network. In this test, three sensors were used and they are installed with LoRaWAN, Sigfox, and NB-IoT.

Due to the large number of sensors, the power and cost of each parking detection sensor should be as low as possible. The numbers from 1 to 5 represent the performance and weighting level for

Figure 16.4 The setup of parking detection sensors. *Source:* From [A].

Table 16.2 The weightings used for the case study to calculate LP-INDEX.

Item	Weighting (%)	LoRaWAN	Sigfox	NB-IoT
Latency	30	4	2	5
Data Capacity	2	2	1	5
Power and Cost	20	5	4	3
Coverage	30	4	5	1
Scalability	5	4	4	5
Security	13	5	4	5
LP-INDEX	100%	4.29	3.64	3.4

Source: From [A] © 2020 IEEE.

the six parameters. The larger number indicates higher weighting, that is, a better performance. The numbers for LPWAN represent the performance of various LPWAN techniques in this application. The weightings are determined by experts such as project owners according to the application requirement. In the smart parking system, latency, coverage, and power and cost are considered as the most significant ones. The low latency ensures the normal operation of the system. The delayed messages may transmit incorrect information and cause congestion at peak hours. The coverage ensures the communication quality with limited gateway numbers and locations for NB-IoT and Sigfox, because their locations are fixed. The poor quality communications may cause a high packet loss rate. Low power and cost are also important due to a large number of parking sensors. With the values of performance and weighting, the comprehensive score could be calculated for each LPWAN by adding the weighted performance scores. The final score represents a comprehensive evaluation for the LPWAN for such an application. The LPWAN with the highest score is considered as the most applicable one in the application.

For this application, the weightings for the six aspects Latency, Data Capacity, Power and Cost, Coverage, Scalability, and Security are set as 30, 2, 20, 30, 5, and 13%, respectively. Among them, latency, and coverage are distributed with more weighting by considering their importance. The monitoring latency should be set as low as possible. The delayed messages may transmit incorrect information and confuse at peak hours. Sufficient coverage ensures the continuous operation of sensors. If the communication is in poor quality due to a weak signal, it is easy to lose important detection results. The aspect in power and cost, is given a weight of 20%. Due to the large number, the power and cost of each parking detection sensor should be as low as possible so that it will not cause unaffordable situations. Security will depend on the consequences caused and this will be influenced by some social factors. It should be mentioned that the results in Table 16.2 only represent a particular performance of three LPWANs in this application rather than a generalized one. For other applications, new preferences and weightings should be set to satisfy requirements and achieve the best performance. With these constraints in mind, the following technologies have been analyzed and compared following the captioned evaluation criteria.

The number (1–5) represents the ranking from low to high.

Based on LP-INDEX, LoRaWAN is considered to be the best network to be used in the parking detection sensor application.

16.5 Conclusion and Future Work

LP-INDEX is proposed to explore the most suitable LPWAN techniques in different applications. The work leads to the provision of guidelines in helping smart cities depolyment. In the present study, the LP-INDEX considers the best practice of LPWAN in terms of six aspects according to the specific requirements of the applications Each aspect is assigned with a reasonable weighting which in general is obtained from experience. A parking detection sensor testing was implemented as a use case to illustrate how LP-INDEX works. In the future, more aspects should be included for LP-INDEX determination. It is foreseen that this work will have a very high practical significance to the society.

Acknowledgements

The persmission given in using the materials in the following paper is very much appreciated.

[A] Wang, H., Liu, Y., Wei, Y. et al. (2020). LP-INDEX: explore the best practice of LPWAN technologies in smart city. *Proceedings of the 2020 IEEE International Smart Cities Conference (ISC2)*, Piscataway, NJ, USA (28 September 2020 to 1 October 2020).

References

1 Department of Economic and Social Affairs (2012). World Urbanization Prospects—The 2011 Revision. http://www.un.org/en/development/desa/population/publications/pdf/urbanization/WUP2011_Report.pdf (accessed 5 April 2018).

2 Andrisano, O., Bartolini, I., Bellavista, P. et al. (2018). The need of multidisciplinary approaches and engineering tools for the development and implementation of the smart city paradigm. *Proceedings of the IEEE* 4: 738–760.

3 Jiang, L. and Das, R. (n.d.). Low-Power Wide-Area Networks 2019–2029: Global Forecasts, Technologies, Applications, Report, IDTechEx, UK. https://www.idtechex.com/en/research-report/low-power-wide-area-networks-2019-2029-global-forecasts-technologies-applications/614 (accessed 3 March 2021).

4 de Almeida, I.B.F., Mendes, L.L., and Rodrigues, J. (2019). 5G waveforms for IoT applications. *IEEE Communication Surveys and Tutorials* 21 (3): 2554–2567.

5 Ghosh, A., Maeder, A., and Baker, M. (2019). 5G evolution: a view on 5G cellular technology beyond 3GPP release15. *IEEE Access* 7: 127639–127651.

6 Li, S., Ni, Q., Sun, Y. et al. (2018). Energy-efficient resource allocation for industrial cyber-physical IoT systems in 5G era. *IEEE Transactions on Industrial Informatics* 14 (6): 2618–2628.

7 Wang, Q., Chen, D., Zhang, N. et al. (2017). LACS: a light weight label based access control scheme in IoT-based 5G caching context. *IEEE Access* 5: 4018–4027.

8 Alzahrani, B. and Ejaz, W. (2018). Resource management for cognitive IoT systems with RF energy harvesting in smart cities. *IEEE Access* 6: 62717–62727.

9 Akpakwu, G.A., Silva, B.J., and Hanck, G.P. (2017). A survey on 5G networks for the internet of things: communication technologies and challenges. *IEEE Access* 6: 3619–3647.

10 Chettri, L. and Bera, R. (2020). A comprehensive survey on internet of things (IoT) towards 5G wireless systems. *IEEE Internet of Things Journal* 7 (1): 16–32.

11 3GPP (2015). 3rd Generation Partnership Project; Technical Specification Group GSM/EDGE Radio Access Network; Cellular System Support for Ultra Low Complexity and Low Throughput Internet of Things. GPP Technical Report document (TR)45.820/r1(v1.1.0).

12 3GPP (2016). 3rd Generation Partnership Project; Evolved Universal Terrestrial Radio Access (E-UTRA); NB-IoT; Technical Report for BS and UE Radio Transmission and Reception. GPP Technical Specifications document (TS)6.802/r1(v1.0.0).

13 3GPP (2017). 3rd Generation Partnership Project; Technical Specification Group Radio Access Network; Evolved Universal Terrestrial Radio Access (E-UTRA) and Evolved Universal Terrestrial Radio Access Network (E-UTRAN); Overall description; Stage 2. GPP Technical Specifications document (TS)6.00/r14(14.4.0).

14 3GPP (2017). 3rd Generation Partnership Project; Technical Specification Group Radio Access Network; Evolved Universal Terrestrial Radio Access (E-UTRA); User Equipment (UE) procedures in idle mode. GPP Technical Specifications document (TS)6.04/r14(v14.4.0).

15 3GPP (2017). 3rd Generation Partnership Project; Technical Specification Group Radio Access Network; Evolved Universal Terrestrial Radio Access (E-UTRA); User Equipment (UE) radio access capabilities. GPP Technical Specifications document (TS)6.06/r14(v14.4.0).

16 3GPP (2017). 3rd Generation Partnership Project; Technical Specification Group Radio Access Network; Evolved Universal Terrestrial Radio Access (E-UTRA); User Equipment (UE) Radio Transmissionand Reception. GPP Technical Specifications document (TS)6.101/r15(v15.0.0).

17 3GPP (2017). 3rd Generation Partnership Project; Technical Specification Group Radio Access Network; Evolved Universal Terrestrial Radio Access (E-UTRA); Base Station (BS) Radio Transmissionand Reception.GPP Technical Specifications document(TS) 6.104/r15(v15.0.0).

18 3GPP (2017). 3rd Generation Partnership Project; General Packet Radio Service (GPRS) Enhancements for Evolved Universal Terrestrial Radio Access Network (EUTRAN) access. GPP Technical Specifications document (TS) 2.401/r15(v15.2.0).

19 Nassar, A. and Yilmaz, Y. (2019). Reinforcement learning for adaptive resource allocation in Fog RAN for IoT with heterogeneous latency requirements. *IEEE Access* 7: 128014–128025.

20 Cao, J., Yu, P., Xiang, X. et al. (2019). Anti-quantum fast authentication and data transmission scheme for massive devices in 5G NB-IoT system. *IEEE Internet of Things Journal* 6 (6): 9794–9805.

21 Xu, J., Yao, J., Wang, L. et al. (2018). Narrow band internet of things: evolutions, technologies, and openissues. *IEEE Internet of Things Journal* 5 (3): 1449–1462.

22 Zhang, Y., Ren, F., Wu, A. et al. (2019). Certificate less multi-party authenticated encryption for NB-IoT terminals in 5G networks. *IEEE Access* 7: 114721–114730.

23 Kodheli, O., Andrenacci, S., Maturo, N. et al. (2019). An uplink UE group-based scheduling technique for 5G mMTC systems over LEO satellite. *IEEE Access* 7: 67413–67427.

24 Goudos, S.K., Deruyck, M., Plets, D. et al. (2019). A novel design approach for 5G massive MIMO and NB-IoT green networks using a hybrid Jaya-differential evolution algorithm. *IEEE Access* 7: 105687–105700.

25 Al-Turjman, F., Every, E., and Zahmatkesh, H. (2019). Small cells in the forth coming 5G/IoT: traffic modelling and deployment overview. *IEEE Communication Surveys and Tutorials* 21 (1): 28–65.

26 Vilgelm, M., Linares, S.R., and Kellerer, W. (2019). Dynamic binary countdown for massive IoT random access in dense 5G networks. *IEEE Internet of Things Journal* 6 (4): 6896–6908.

27 Xia, N., Chen, H., and Yang, C. (2018). Radio resource management in machine-to-machine communications – a survey. *IEEE Communication Surveys and Tutorials* 20 (1): 791–828.

28 Centenaro, M., Vangelista, L., Saur, S. et al. (2017). Comparison of collision-free and contention-based radio access protocols for the internet of things. *IEEE Transactions on Communications* 65 (9): 3832–3846.

29 Verma, S., Kawamoto, Y., and Kato, N. (2019). Energy-efficient group paging mechanism for QoS constrained mobile IoT devices over LTE-A pro networks under 5G. *IEEE Internet of Things Journal* 6 (5): 9187–9199.

30 Sun, Y., Tong, F., and Zhang, Z. (2018). Throughput modeling and analysis of random access in narrow-band internet of things. *IEEE Internet of Things Journal* 5 (3): 1485–1493.

31 Begishev, V., Petrov, V., Samuylov, A. et al. (2018). Resource allocation and sharing for heterogeneous data collection over conventional 3GPP LTE and emerging NB-IoT technologies. *Computer Communications* 120: 93–101.

32 Tsoukaneri, G., Condoluci, M., Mahmoodi, T. et al. (2018). Group communications in narrowband-IoT: architecture, procedures, and evaluation. *IEEE Internet of Things Journal* 1 (1): 1–10.

33 Duan, P., Jia, Y., Liang, L. et al. (2018). Space-reserved cooperative caching in 5G heterogeneous networks for industrial IoT. *IEEE Transactions on Industrial Informatics* 14 (6): 2715–2724.

34 Nižetić, S., Šolić, P., González-de-Artaza, D.L.-d.-I., and Patrono, L. (2020). Internet of things (IoT): opportunities, issues and challenges towards a smart and sustainable future. *Journal of Cleaner Production* 274: 122877.

35 Office of Scientific and Technical Information (2016). United States Data Center Energy Usage Report, 1 June. https://www.osti.gov/biblio/1372902/ (accessed 17 October 2020).

36 Centenaro, M., Vangelista, L., Zanella, A., and Zorzi, M. (2016). Long-range communications in unlicensed bands: the rising stars in the IoT and smart city scenarios. *IEEE Wireless Communications* 23 (5): 60–67.

37 Raza, U., Kulkarni, P., and Sooriyabandara, M. (2017). Low power wide area networks: an overview. *IEEE Communication Surveys and Tutorials* 19 (2): 855–873.

38 Sigfox (2020). Sigfox connected objects: radio specifications. https://storage.sbg.cloud.ovh.net/v1/AUTH_669d7dfced0b44518cb186841d7cbd75/prod_medias/b2be6c79-4841-4811-b9ee-61060512ecf8.pdf (accessed 1 December 2020).

39 Pospisil, J., Fujdiak, R., and Mikhaylov, K. (2020). Investigation of the performance of TDoA-based localization over LoRaWAN in theory and practice. *Sensors* 20: 5464. https://doi.org/10.3390/s20195464.

40 Vodafone Limited (2017). Narrowband-IoT: pushing the boundaries of IoT. White paper. https://www.vodafone.com/business/news-and-insights/white-paper/narrowband-iot-pushing-the-boundaries-of-iot (accessed 1 December 2020).

41 Tan, C. and Tan, H.-P. (2020). Evaluation of Sigfox LPWAN for sensor-enabled homes to identify at risk community dwelling seniors. *2019 IEEE 44th Conference on Local Computer Networks (LCN)*, IEEE, Osnabrueck, Germany (14–17 October 2019).

42 Mekki, K., Bajjic, E., Chaxel, F., Meyer, F. (2018). Overview of cellular LPWAN technologies for IoT deployment: Sigfox, LoRaWAN, and NB-IoT. *2018 IEEE International Conference on Pervasive Computing and Communications Workshops (PerCom Workshops)*, IEEE, Athens, Greece (19–23 March 2018).

43 Mermer, G.B. and Zeydan, E. (n.d.). A comparison of LP-WAN technologies: an overview from a mobile operators' perspective. *2017 25th Signal Processing and Communications Applications Conference (SIU)*, IEEE, Antalya, Turkey (May 2017).

44 Augustin, A., Clausen, T., and Townsley, W. (2016). A study of LoRa: long range & low power networks for the internet of things. *Sensors* 16: https://doi.org/10.3390/s16091466.

45 Vangelista, L. (2017). Frequency shift chirp modulation: the LoRa modulation. *IEEE Signal Processing Letters* 24: 1818–1821.

46 (2017). Narrowband-IoT (NB-IoT) is a new standard connecting internet of things projects. White papers (February). https://www.vodafone.com/business/news-and-insights/white-paper/narrowband-iot-pushing-the-boundaries-of-iot (accessed 2 August 2019).

47 Wu, C.K., Tsang, K.F., Liu, Y. et al. (2019). Supply chain of things: a connected solution to enhance supply chain productivity. *IEEE Communications Magazine* 57 (8): 78–83.

48 Working Group Chair: Tsang, K.F. (2018). IEEE P2668 – Standard for Maturity Index of Internet-of-things: Evaluation, Grading and Ranking. https://standards.ieee.org/project/2668.html (accessed 1 October 2020). The scope of this standard is to measure the maturity of objects in Internet-of-Things (IoT) environment.

49 Tsang, K.F. and Huang, V. (2019). Conference on sensors and internet of things standard for smart city and inauguration of IEEE P2668 internet of things maturity index [Chapter News]. *IEEE Industrial Electronics Magazine* 13 (4): 130–131.

50 LoRaWAN Security Whitepaper (2017). LoRaWAN™ security: full end-to-end encryption for IoT application providers. February [Online]. https://lora-alliance.org/sites/default/files/2019-05/lorawan_security_whitepaper.pdf (accessed 1 December 2020).

17

Smart Health Based on Internet of Things (IoT) and Smart Devices

17.1 Introduction

According to World Population Prospects 2019, by 2050, 1 in 6 people in the world will be over the age of 65, up from 1 in 11 in 2019 [1]. There has been great progress in the fight against communicable diseases, with better hygiene conditions and access to treatment improving quality of life. However, the prevalence of obesity is increasing and is linked to an increase in chronic conditions, such as diabetes. It is expected that 642 million people world-wide will be diagnosed with diabetes by 2040 and the number of dementia sufferers will double over the next 20 years [2].

More than 20 years ago, the World Health Organization (WHO) defined telemedicine as the delivery of healthcare services, where distance is a critical factor, by all healthcare professionals using information and communication technologies for the exchange of valid information for diagnosis, treatment and prevention of disease and injuries, research and evaluation, and for the continuing education of healthcare providers, all in the interests of advancing the health of individuals and their communities [3].

Although today doctors and patients can communicate via online platforms such as email or WhatsApp, physical consultations remain necessary for most primary care. This is changing with the development of home-use technologies. Patients suffering chronic conditions can be equipped with smart devices that easily measure parameters such as heart rate, blood pressure, blood glucose, and body temperature. In the event of readings that are outside the normal set range, the device will alert the user.

This information can be synced to a platform where doctors will be able to review the information remotely and the same software can be used to give patients medication reminders and advice on maintaining a healthy lifestyle. This will lead to shorter queues in the waiting room, better health outcomes and reduced workload as data collected will be processed by computers. From patient profiling, we can infer the likelihood of the development of certain diseases and prevent them before they occur.

Amongst wearable devices, blood pressure monitors are relatively mature in their development. Hypertension is estimated to result in 10 million deaths per year, and 80% of cases could have improved outcomes with regular monitoring. The majority of blood pressure monitors available consist of an inflatable cuff connected to an electronic device that measures blood flow. This simple, noninvasive design allows regular monitoring of blood pressure. Although the device is easy to use, factors such as mental state and activity can affect measurements, so patients must be well informed, and the data should be reviewed by a medical practitioner. More recently, novel cuffless

Smart Energy for Transportation and Health in a Smart City, First Edition. Chun Sing Lai, Loi Lei Lai and Qi Hong Lai.

pocket-sized and wearable finger ring blood pressure monitors have been developed with some already available on the market [4, 5]. These novel devices provide a more portable and convenient method of monitoring to allow measurement of real-time changes in blood pressure, further improving health outcomes [6].

Global health spending has been increasing due to labor costs, increased life expectancy, emerging market growth and developing new advanced treatments. Technological developments, such as artificial intelligence (AI), 3D printing, synthetic biology, and data analytics, can make healthcare more effective and accessible. Synthetic biology is an interdisciplinary branch that brings together biology and engineering. It involves the redesigning of biological parts and systems found in nature to have new abilities. An example of this is the in vitro synthesis of DNA, combined with cognitive calculation, AI, 3D printing and nanotechnology to print biological tissues [7] or harnessing bacteria to attack cancer cells [8]. On the molecular level, RNA molecules have been engineered with new biological functions to regulate gene expression in vivo. Although largely done in yeast and bacteria, efforts have now been directed to mammalian systems with the hope of curing genetic disorders by reprogramming gene circuits and cell behavior. Direct alterations to remove defective genes or replace missing genes can be achieved through engineering RNA to direct the CRISPR/Cas gene editing system [9].

Computational techniques have aided molecule's design and evolution, speeding up the process from design to product significantly [10], while advances in nanotechnology has aided delivery of these molecules to target cells [11]. An example of this on the global scale is the COVID-19 vaccines produced by Pfizer-BioNTech and Moderna. They consist of messenger RNA (mRNA) coding for the virus spike protein and uses lipid nanoparticles for delivery into cells, and are arguably the most effective vaccines in the fight against COVID-19 to date.

Health systems are increasingly oriented toward a smart approach, from which the term smart healthcare is derived. Smart healthcare can be defined as using mobile and electronic technology for better disease diagnosis, improved treatment, and enhanced quality of life for patients.

In the last decade or so, Mobile health (mHealth) and Electronic health (eHealth) have been harnessed in medicine. mHealth is defined by the World Health Organization's (WHO) Global Observatory as medical and public health practice supported by mobile devices, such as mobile phones, patient monitoring devices, personal digital assistants (PDAs), and other wireless devices. The most common application of mHealth is the use of mobile phones and other wireless technology in medical care to educate consumers about preventive healthcare services. However, mHealth is also used for disease surveillance, treatment support, epidemic outbreak tracking and chronic disease management. Despite the lack of safety and quality validation of these technologies by medical regulatory bodies, individuals have adopted mHealth devices as self-management aids, while medical professionals are at a loss for words on how to relate to them [12, 13].

Electronic health (eHealth) is an emerging field in the intersection of medical informatics, public health and business, referring to health services and information delivered. It is the use of emerging information and communication technology, especially the Internet, to improve or enable healthcare without providers and patients being directly in contact with each other.

Both mHealth and eHealth play a role in supporting healthcare with electronics. They perform similar functions, however, the means by which the information is provided is the primary difference. With mHealth services, patients are able to log, store, and monitor their health records on their personal mobile devices. mHealth and eHealth are thus vital components of the smart healthcare system.

A common framework for all mHealth initiatives around the world will be useful in order to assess whatever mHealth solution is desirable in different areas, adapting it to the specifics of each context, to bridge the gap between health authorities, patients, and mHealth developers. Due to this consumer-based and rapid introduction within the world of patient health aids, mHealth solutions present unique and stakeholder-specific challenges to the medical environment. Patients, healthcare providers, administrators, authorities, and mHealth developers alike are operating without clear direction from potentially improper use of mHealth apps by individuals, to medical systems' inability to react due to lack of technological and organizational support. Therefore, health authorities should push for mHealth evaluation and certification. Furthermore, policies and strategies should be introduced to medical professionals and their practices. Methods should also be developed to ensure patients are properly informed on how to select and use mHealth solutions.

The COVID-19 pandemic has caused extensive concern and economic catastrophe for consumers and businesses across the globe. The increase in bandwidth requirements on networks, as millions of employees and students became homebound, has significantly impacted the telecom sector. The current scenario has confirmed that the telecom infrastructure is critically essential, and telecommunication services during the lockdown have enabled many to recommence work, study, shop, and access healthcare. Policymakers are reconsidering regulations for telecom networks to enhance network bandwidth. Telecom companies are now also considerably focused on ensuring that there is adequate network capacity throughout the entire day, and not only during peak times, as progress in static traffic has increased significantly.

With all the complexities that COVID-19 has brought upon telecom service providers, the urgency to achieve even better and faster network operation is one of the most crucial aspects faced by the sector. The ultimate solution is to automate the process of identifying issues in cellular networks and imminent analytical tasks, in order to significantly diminish the effort required to maintain the network in good shape.

With the ongoing COVID-19 pandemic, there have been new demands placed on healthcare, such as the need for solutions that can aid in providing care remotely. Indeed, a number of foundational shifts are arising from and being exacerbated by the spread of COVID-19. Examples include increasing consumer involvement in health care decision-making; the rapid adoption of virtual health and other digital innovations; the push for interoperable data and data analytics use; and unprecedented public–private collaborations in vaccine and therapeutics development. Amid these dynamics, governments, healthcare providers, funding bodies, and other stakeholders around the globe are being challenged to quickly pivot, adapt, and innovate. An increasing demographic of underserved consumers and communities are leading to health inequities. Today's socioeconomic, mental, and behavioral health crises have made it clear that players across the healthcare landscape need to innovate to better serve the needs of people across the world. In [14], issues driving change in the healthcare sector were looked at and questions and actions to health leaders have been suggested for consideration in the near future. How stakeholders analyze, understand, and respond to these issues will shape their ability to navigate from recovering to thriving in the post-pandemic and advance their journey along the path to the Future of Health.

The concept of health smart homes (HSH) has been discussed for more than two decades or so. Based on the use of tele-medical information systems and communication technologies, HSH systems provide healthcare services for people with special needs who wish to remain independent and living in their own home. The large diversity of needs in a home-based patient population requires complex technology. Meeting these needs technically requires the use of a distributed approach and the combination of many hardware and software techniques [15].

Smart healthcare, that measures users' living conditions and health status using small sensing devices and collecting their data over a network under daily life, is a new trend. It is getting paid more attention along with the increase of demands of preventive care which in daily life has become more important against the increase in medical costs associated with aging and the increase in lifestyle-related diseases. Smart healthcare, that is measuring user's living conditions and health status using small sensing devices and collecting their data over a network under daily life, can be expected as a new trend along with an increase in demands of preventive care. The technologies of low power large-scale integrated circuit (LSI), wireless communications and small vital sensors have been rapidly advanced. It implies that the time to market of a wearable vital monitoring device has been certainly reducing. This device could be an effective tool for realizing a smart healthcare world where people can enjoy a safe and healthy life with unplanned medical ICT support [16].

When compared with other sectors, such as financial services, retail and travel; pharma and healthcare has been relatively slow in its digital transformation journey. This is about technology, data, process, and organizational change. However, in recent years, organizations in the sector have made large steps when it comes to using digital to improve the customer experience. Because of social media and the innovations of young companies in pharma and healthcare sector, this leads to healthcare providers and patients with greater access to knowledge about health conditions and making it possible to better manage treatments with digital tools.

Reference [17] examines how digital is transforming the pharma and healthcare industry and consider how different groups or companies in the pharma and healthcare sector can better serve their customers including patients, doctors, and healthcare providers. Digital transformation will depend on communications, content, data and culture, and this will make sure it is useful for the business as the end consumer becomes increasingly knowledgeable and demanding with advice and examples of best practice.

Healthcare touches our lives and is essential to our communities. Together with governments and healthcare organizations, modern technology is used to improve care today and help shape a healthier future. New standards of care with robots to assist health workers in their day-to-day tasks are developed for a smarter healthcare for a healthier future. For example, [18] reported the use of Robotic Nurse Assistant to provide valuable assistance to nurses by carrying out essential but repetitive tasks, such as measuring patients' vital signs, synchronizing patient data for real-time follow-up and delivering medicine and snacks. Some of the technologies help people from being regularly in hospital and giving back the quality of life. More control is possible for both medical professionals and patients.

Smart healthcare has been a field of incredible change and growth in the new age of smart houses, smart communities, and all things digital. It has become among the most important ingredients of citizens. A total of 50 billion health-related things are projected to be wired by 2021, and 5G would increase the Internet of Health Things (IoHT) standard of operation. The IoHT includes a network linked with many types of medical sensors and devices such as image digitizer, vital sign sensing devices such as ECG, blood pressure, wound [19, 20] and oxygen monitor devices, display, modems, storage devices, and communication networks. 5G promises a drastic extension of the spectrum, expected to be 100 times quicker than 4G. In brief, the convergence of high-speed broadband, reduced latency, and larger capacity would boost modern healthcare.

The healthcare industry is no longer an isolated zone untouched by technological innovations. Top healthcare technologies such as IoT, Big Data, advanced analytics, and many other technological innovations have turned traditional healthcare into smart healthcare. The newest research conducted by our industry experts show how the market for smart healthcare solutions is growing at a tremendous pace. With an expected market value of close to USD 251 billion by 2022 [21]. The key factors driving smart healthcare market growth are growing demand for remote health monitoring. The smart healthcare market is expected to grow at a CAGR of 24.11% during 2019–2024.

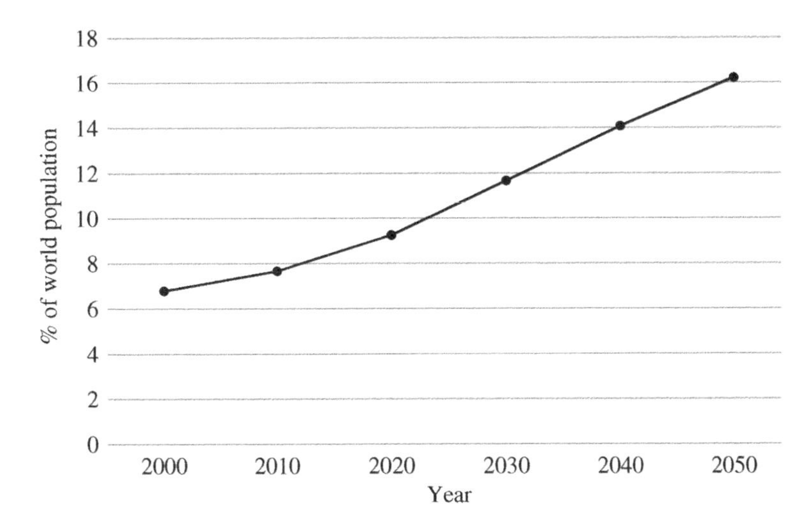

Figure 17.1 Percentage of the world population over 65 from 2020 to 2050. *Source:* UN World Population Prospect 2008.

IoT in healthcare technologies is also popularly known as Internet of Medical Things (IoMT) which is one major technological innovation added to the element of smartness in the healthcare industry [21]. Along with identifying, monitoring, and informing caregivers about the patient's vitals – it also provides the much-needed critical data to the healthcare providers so that issues can be identified at an early stage leading to better delivery of care services.

IoMT is particularly a boom for a burdened healthcare system. By 2050, the elderly population (aged 65 or older) is expected to be close to 1.5 billion [22]. As reported in [23], two billion people will be aged 60 and over by 2050, 22% of the world's population. Figure 17.1 shows the percentage over 65 from 2020 to 2050. High incidence of health issues makes healthcare solutions for the elderly a very costly affair. IoMT devices not only help the elderly to keep a close eye on their vitals, such as heart rate, glucose levels, and sleep patterns – but also reminds them of medication. The portable IoMT devices make it a lot easier for the aging population to conduct their routine blood and urine tests as well.

It has been noticed that the governments across the globe are working toward building an effective healthcare paradigm – and technology, especially Big Data, is the very foundation on which this infrastructure can be built. The global healthcare big data market size is expected to grow by USD 5.78 billion between 2018 and 2022. This report offers an analysis of the market based on end-users, including hospitals and clinics, finance and insurance agencies, and research organizations; solution such as services and software and geography, namely, the Americas, APAC – Asia Pacific; EMEA – Europe, Middle East and Africa. The big data spending market research report analyzes the market's competitive landscape and offers information on several companies including Hewlett Packard Enterprise, Huawei, IBM, and Microsoft [24].

In the United Kingdom, the number of people living with progressive neurodegenerative disorders, such as dementia, is increasing. Supporting their ongoing needs places significant strain on national health and social care resources. Providing 24-hour monitoring for patients is a significant challenge, which is set to increase further due to the aging population. A key philosophy in dementia care is supporting people to live well with dementia by promoting independence and enabling them to continue living at home for as long as possible. Over 80% of patients prefer to stay in their own home in the later years of life, and evidence shows that moving people into residential care can hasten the progression of dementia. However, allowing patients to live at home safely is often quite challenging as there are limitations in current care provision, and the responsibility of caring often falls to informal carers [25].

The impact of carer burnout is also well documented in literature and, given that community care provision relies on feedback from informal carers to raise concerns, those who do not have a social support network are at increased risk of adverse events. This often means that opportunities for early intervention in dementia care are missed.

Currently, there are no technological solutions capable of monitoring the progression of dementia 24-hours a day, seven days a week. Although telehealth solutions are available, they are often expensive, intrusive, and increase the cost of overall standard care plans by 10%, with little or no benefit to the patient. Where such technologies are being used, they are incapable of being personalized to the individual. This significantly limits the effectiveness of most current solutions. Consequently, large-scale usage within NHS trusts, councils, and social services in the United Kingdom is not feasible.

However, with the introduction of the smart metering infrastructure, new and exciting opportunities for a variety of emerging applications such as health and social care are possible. Smart meters can continually monitor household electricity consumption, and the technology utilizes this infrastructure to capture the detailed habits of an individual's interactions with electrical devices, identifying anomalies or changes in a person's routine to facilitate early intervention practice (EIP) by monitoring deviations in behavior that correlates with disease progression.

17.2 Technology Used in Healthcare

The market for smart healthcare is steadily growing around the world. Statistics show that global spending on smart healthcare services between 2016 and 2018 increased at a compound annual growth rate (CAGR) of 60% [26].

Japan has a large aging population, and elderly people aged over 60 account for 20.5% of its total population. Therefore, Japan has a huge demand for smart healthcare products related to senile diseases. Europe also has an aging population and a large amount of unexpected immigration; as a result, much of its medical equipment needs to be urgently upgraded. To meet these demands, many European countries have seen rapid growth in their smart healthcare equipment market, and Germany has become the largest producer and exporter in Europe.

Emerging economies such as India, China, and Brazil are expected to provide a wide range of opportunities for players in the industry. Furthermore, the growing adoption of mobile platforms, AI and 5G, and growing awareness and preference for home healthcare will also boost the growth of this market.

Nearly 100 hospitals are exploring different use cases of 5G+ healthcare, such as remote group consultation, remote surgery, mobile ward rounds, emergency rescue, and intrahospital monitoring. These hospitals are among the many first-movers to apply 5G in their industry [26]. The urgent needs of smart healthcare will mean huge value in 5G+ healthcare.

China has huge prospects for 5G+ healthcare. This is because China's medical resources are not evenly allocated, making telemedicine a pressing need. With a shortage of medical resources, the healthcare industry needs to go digital to boost efficiency. Given the current situation facing China's healthcare industry, digital healthcare is key to addressing these challenges, with 5G smart healthcare at the core of it. With high bandwidth, low latency, and massive connections, 5G will make mobile remote diagnoses and remote surgery possible, and increase the management efficiency within hospitals.

In combination with cloud computing, mobile edge computing, big data, AI, and blockchain, 5G will drive upgrades and transformations in medical information systems and remote medical platforms. If anything has made the healthcare technologies way smarter, then it is cloud computing technology. Cloud computing and edge computing encourage cost savings [27, 28], scalability, and system flexibility; now with the increased use of cloud-assisted medical collaboration, the demand for cloud computed healthcare solutions has grown exponentially.

Edge or cloud computing can detect, monitor, and notify the occurrence of illness in a patient to its caregiver earlier and easier. An edge is a computing location at the edge of a network, along with the hardware and software at those physical locations. Cloud computing is the act of running workloads within clouds, while edge computing is the act of running workloads on edge devices. Cloud computing has made it easier for even small hospitals to gain access to information and store it in a cost-effective manner. These small hospitals can now deliver the best services to their patients.

It seems that the main benefits of edge computing over cloud computing are better data management, reliable, uninterrupted connection, lower connectivity costs, and better security practices. However, cloud computing has made it easier for even small hospitals to gain access to information and store it in a cost-effective manner. These small hospitals can now deliver the best services to their patients [29]. The cloud-based software solutions facilitate more detailed case discussions from any place – enabling confident, timely, and individualized treatment decisions. For patients this may increase chances of being reviewed and starting treatment sooner.

The healthcare sector has jumped on the wearable devices bandwagon, as these devices allow for more comprehensive real-time monitoring and data collection without requiring extra work or effort. Additionally, because wearables are convenient, easy to use, and ergonomic, they are attractive to consumers and ideal for health professionals. The global wearable healthcare devices market is projected to reach USD 46.6 billion by 2025 from USD 18.4 billion in 2020, at a CAGR of 20.5% from 2020 to 2025 [30].

The wearable healthcare devices market in the Asia Pacific is expected to grow at a high pace in the coming years. The factors attributing to the high growth of the Asia Pacific region include the presence of a large patient population, growing penetration of smartphones, increasing disposable income, improving the standard of living and economic conditions, and the presence of several low-cost device manufacturers in the Asia Pacific.

Over time, the thousands of data points patients create throughout their journey not only help identify changes in their health but also enable tracking of similar patient cohorts across institutions and geographies. This will eventually allow for comparison of outcomes as well as predictive measures through AI and machine learning. Digital solutions that can capture the performance of therapies have the potential to support value-based care models.

Technology has now made it easier for citizens to make informed choices. There is a lot of data that has been collected and collated that reveals the truth to us. Using data analytics, we can tell what we eat affects our longevity, how our surrounding affects our health, and what lifestyle changes we can make to boost our immunity from diseases. Smart technology can help us live a healthier life [31].

We can shrink the number of deaths caused by noncommunicable diseases. This can be achieved by leveraging the power of data. Data analytics can help us focus more on preventive measures rather than treating illnesses. With enough data on these diseases available, organizations are able to formulate policies that will help prevent their wide spreading.

To help change the focus from curative measures to preventive measures, individuals need to take more responsibility. More often, individuals are aware of what is beneficial to them; however, they have difficulty in altering their daily routines. Smart technology can help the change of daily routine. For instance, you can get a mobile phone app that helps you to measure your physical activity. You can also get apps that will provide you with health-related advice.

Wearable technology has also contributed significantly in improving our healthcare systems. With wearable devices such as smartwatches, smart bracelets, among others, we can keep track of our health. The devices will alert you about where you are going. Combined with data analysis, this technology is already changing lives. Wearable technologies can assist the initiation of positive behavioral changes. Poor air quality and lighting will make our homes unhealthy. With smart home technology such as wireless communication, you can accurately measure all the essential variables and allow building homes that are healthy and sustainable.

In precision medicine, it refers to medical care designed to optimize efficiency or therapeutic benefit for specific groups of patients. This is achieved using genetic or molecular profiling and taking into account individual variability in genes, environment, and lifestyle. That is, precision medicine uses your genetic code to create individualized medical treatments, and is a prime example of translating fundamental science into medical value. DNA, or deoxyribonucleic acid, is the hereditary material in humans and almost all organisms. Ribonucleic acid (RNA) is a molecule transcribed from DNA. Both DNA and RNA consist of nucleotides, each containing a five-carbon sugar backbone, a phosphate group, and a nitrogen base. DNA provides the code for the cell's activities, while RNA converts that code into proteins to carry out cellular functions. Unlike DNA, RNA is single-stranded. An RNA strand has a backbone made of alternating sugar (ribose) and phosphate groups. Transcription is the process of copying DNA into RNA. For protein coding genes to be expressed, this process requires RNA polymerase II (RNAPII) to be loaded onto the DNA template as part of transcription initiation, followed by productive elongation and then termination. There is tight regulation at each step to ensure that genes are expressed in a timely and faithful manner. The recruitment of accessory proteins is needed for this, such as transcription elongation factors, mRNA processing factors and chromatin regulators. DNA is constantly exposed to both endogenous and exogenous damaging agents. For example, ultraviolet (UV) radiation from sunlight is the leading cause of skin cancer in humans [32, 33]. Exposure to UV light produces bulky adducts – such as cyclobutane pyrimidine dimers (CPDs) and pyrimidine-(6,4)-pyrimidine photoproducts (6–4PPs) – which are transcription-blocking lesions (TBLs), that cause transcriptional stress and impediment of RNAPII elongation [34]. This too is observed with camptothecin (CPT) treatment, a topoisomerase I inhibitor and chemotherapy drug [35], which can also cause transcriptional read through onto neighboring genes [36]. If there is prolonged pausing due to the presence of transcription blocking DNA lesions, DNA damage repair pathways are triggered [37]. To preserve genomic integrity, cells have an arsenal of repair proteins that engage the appropriate repair pathway [38]. If damage is unrepaired and transcription not restarted, there can be catastrophic cellular outcomes including triggered programmed cell death and genomic instability leading to the development of disease [39]. As aberrant gene expression is the fundamental cause of most diseases, understanding the key players involved using high throughput sequencing and other technological advancements will enable us to modulate these processes, with implications for regulating ageing and cancer through targeted therapy. With precision medicine, it is possible to treat diseases that were previously costly and difficult to manage. Examples of such diseases include cancer, neurological disorders, and other rare genetic conditions. Novel treatments are created with an individual's genetics and other variables in mind. This technology will significantly help us live healthier lives. For those individuals that are already living with conditions such as diabetes, cancer, among others, this technology will help them manage the conditions better.

Healthcare is becoming smart due to the devices connected to patients with bracelets, watches, and more to collect data on their health status, even remotely, and prevent critical situations. Healthcare and IoT applied to medicine require the use of objects integrated by sensors that communicate with each other and exchange information. This exchange of information is necessary to prevent illness and diagnose disease in advance. AI can now accurately predict heart attack or stroke. With re-elaboration of big data, machine learning can save lives, and enable a diagnosis to be made even before human intervention.

Reference [40] reported some technologies such as nanotechnology, blockchain, and augmented reality (AR) to make medical care less expensive and more accessible to patients. In particular, the younger generation of patients expect personalized and comfortable services that require the use of multichannel retail communication, the adoption of social media and mobile apps. The smart healthcare workforce requires expert staff with additional skills. In addition to making work of professionals more efficient, staff members are supported by digital or robotic assistants.

Sustaining technology relies on incremental improvements to an already established technology. Disruptive technology lacks refinement, often has performance problems because it is new, appeals to a limited audience and may not yet have a proven practical application. However, it can create a new industry. Some most disruptive technologies are AI, AR, and Internet of Things (IoT).

AI has been around for decades. Today, it is being used to learn more about you in order to make better, more accurate suggestions for you, AI is used to collect data from searching histories, products purchased, and by discovering your preferences. Healthcare is seeing numerous possibilities of integrating AI into existing systems. One application of AI in healthcare involves connecting AI with an amputee's brain so that they can better communicate with and control the attached prosthetic leg [41]. Previous work with powered prosthetics relied on preprogrammed behaviors based on the movements of non-amputee individuals, but this approach limits the use of the prosthetic to pre-mapped areas and obstacles. This would work great if you were only walking to the same places every day, but it is not practical to retrain your prosthetic every time you want to go someplace new. Instead of relying on preprogrammed movements to drive the prosthetic, AI could be adopted to better learn how to move more effectively. A new AI-based approach has been proposed for prosthetic limb movement to mimic the motion of the user's leg to make walking smoother and more intuitive. The integration of both human and AI should create a better world for all. Reference [42] reported that AI presented huge opportunities for the pharma and healthcare industries. Big data strategies could save the US healthcare system up to $100 billion a year thanks to AI-assisted efficiencies in trials, research, and clinical practice.

AR is the use of displays, cameras, and sensors to overlay digital information onto the real world. AR allows us to bring the most useful information from the digital realm into our perception of the environment around us. AR is not a new concept, but over the last few years, advances in camera and sensor technology and AR-focused software research have made it practical – we are still in the early stages of the AR revolution, but this year and into the future, we can expect to see an explosion of AR devices and applications enter the market. Indeed, healthcare and medical fields will be among the first to embrace AR in a big way. In fact, today there are many nurses and doctors interacting with AR applications every day to improve patient education and outcomes.

Healthcare workers have been quick to realize the benefits of AR technologies. Education is an obvious application of AR in the healthcare field. Healthcare workers have to learn a huge amount of information about anatomy and the way the body functions. AR applications give learners the ability to visualize and interact with three-dimensional representations of bodies. AR allows medical professionals to help patients understand surgical procedures and the way medicines work. Today, surgeons use several techniques to visualize the area on which they are to operate, but AR, which can project three dimensional representations of the patient's anatomy into the surgeon's field of view, is likely to improve accuracy and outcomes for patients. A practical application of augmented reality which is in use today is vein visualization. Needless to say, patients are uncomfortable with being injected or having blood taken, and the experience is much worse when it is difficult to find a vein and the needle has to be inserted several times [43].

Augmented reality is used in healthcare facilities across the world today for applications that include vein visualization, surgical visualization and education. As the hardware and software have reduced the cost of AR, healthcare providers are investigating the potential benefits of AR to their customers and their business. The global AR, virtual reality (VR), and mixed reality (MR) market is forecast to reach 30.7 billion U.S. dollars in 2021, rising to close to 300 billion U.S. dollars by 2024 [44]. An overview and look to the future of how AR is being widely adopted in the healthcare industry, creating business opportunities for companies with AR expertise.

17.2.1 Internet of Things

As the sophistication of both hardware and software in the consumer electronics industry is occurring at an astounding rate, there is a large increase in the share of electronic devices produced around the world that are manufactured with internet connectivity. The Internet of Things (IoT) is an expansive network of things or devices that are connected to the internet, which facilitates their intercommunication. IoT is another technology that will help to bridge the gap between the physical and digital spheres. Table 17.1 shows the number of IoT connected devices worldwide from 2020 to 2030 [45]. Forecasts suggest that by 2030, around 50 billion of these IoT connected devices will be in use around the world, creating a massive web of interconnected devices spanning everything from smartphones to kitchen appliances [46].

Table 17.1 Number of IoT connected devices worldwide from 2020 to 2030.

Year	Number in billions
2020	8.74
2021	10.07
2022	11.57
2023	13.15
2024	14.76
2025	16.44
2026	18.15
2027	19.91
2028	21.72
2029	23.57
2030	25.44

The use of smart technology in healthcare has been advancing steadily over the past several years, putting powerful devices such as connected inhalers in the hands of everyday consumers and allowing them to better manage and address their own health needs, as well as to quickly access help if something goes wrong. Wearable devices such as smart watches can also allow healthcare professionals to remotely monitor ongoing conditions and gather data, allowing observation and treatment.

Sustainable health and social care are an important elements of a smart city. Health and social care programs need to be developed and supported in a manner that will satisfy the demands both of present population and future generations. In the digital society, massive quantities of data are created daily from sensors, personnel records, social networks, the IoT, businesses, and the Internet in numerous scales and formats. This certainly poses a major challenge to the efficiency of our built systems and infrastructures; improving the efficiency and sustainability of devices, cloud, big data networks, and software has become increasingly important.

The IoT has opened up a world of possibilities in treatment and diagnosis of disease. When connected to the internet, ordinary medical devices can collect invaluable additional data, give extra insight into symptoms and trends, enable remote care, and generally give patients more control over their lives and treatment.

There were 17 billion connected devices in 2016 and the projection for 2020 spans anywhere from 28 to 100+ billion [47]. The ability to connect devices to the internet is nothing new, but we are now connecting more things to the internet than ever before. Imagine your alarm going off in the morning and prompting your coffee maker to start brewing your morning cup before your self-driving car drives you to a smart office environment in which your personal space is perfectly adapted to your needs.

IoT develops new relationships between things and other things, things and people, and people and other people to make our lives easier, more efficient, and more effective. For example, we can control smart thermostats from our phones so that the temperature is ideally suited to you when you enter your home. Future developments could see our cars connect with our calendars to navigate automatically to our destination along the optimal route, or our fridge ordering groceries when it detects a given food item is close to depletion.

On a more global scale, IoT will significantly transition us into smart cities. With the help of sensors, IoT will make our cities more efficient, cost-effective, and safer places in which to live. Of course, much of this is a long way off. It would require some of our existing infrastructure to be replaced. With a projected global worth of 6.2 trillion USD by 2025 [48], we can expect most industries to be impacted by IoT, most notably the healthcare sector, which will see improvements in diagnoses, treatments, and predictive health monitoring.

With so much of our lives relying on emerging technologies, we must be aware of the vulnerabilities opened by so much of our data being stored in opaque, private enterprise databases on which nontransparent AI algorithms such as neural networks are adopted for training. There are many different opinions on the future of technology and how it will impact our lives. It may drive productivity, help us live longer, and increase efficiencies. However, it may introduce security risk to our daily life too. Clearly it is up to us to shape the final outcome.

The long-term effects of COVID-19 are increasingly hitting home. Everybody seems to know someone who has been affected by it. But there is increasing hope for improved care due to IoT technology. Even before COVID-19, IoT technology was beginning to drive telehealth, including remote medical patient monitoring. But now, during the pandemic, telehealth has taken on a whole new urgency. Today, everything from convention centers to tent villages are being converted into medical centers with demanding healthcare requirements. In order to function, these makeshift healthcare sites need advanced, high performance, high security, machine-to-machine, and machine-to-control center communications over dedicated mobile networks.

Healthcare providers need to be able to upload and download high-definition images, like radiological scans, to traditional medical centers such as hospitals to better treat patients. This is something that cannot be done successfully without Wi-Fi. Wi-Fi does not have the stability and reliability of the communication channel, nor the end-to-end security necessary to handle sensitive patient data.

Fortunately, with the emergence of new 5G technologies, healthcare-related IT services and applications are set to become better connected than ever, resulting in an enormous impact on both healthcare providers and patients alike. But right now, COVID-19, as a global emergency, is bringing the urgency of medical IoT deployments, front and center. For the industry to be successful, the speed of the enabling network technology, both in set up and operations, is as key as the network edge devices themselves.

Enter private Gigabit LTE networks which, through wireless wide area network (WAN) routers, enable temporary healthcare centers to rapidly come online and provide the high-performance, high-security communications required between the remote facility and the medical center. It is important to note that deploying a wired internet connection to these pop-up clinics would require a lot more time as well as costly information technology (IT) infrastructure equipment.

Recent modern IoT devices and network technology support this due to high-speed gigabit-class LTE connections, dual path wireless modems with link redundancy and traffic load management, edge security systems, firewall protection, and internet content filtering. Fast, secure, highly sophisticated, technologically cutting-edge, and enterprise scalable private mobile IoT networks are being built [49]. Gigabit LTE links give remote doctors reliable and secure at-home and remote office connections that can provide the bandwidth required to enable real-time video and extremely large data files. Remote medical care, diagnosis, and even surgery has been advised as one of the potential applications for 5G network technology.

5G technology will help turn antiquated healthcare systems in hospitals into smart hospitals where telehealth services can be successfully provisioned for patients globally. 5G opens up entirely new horizons for telehealth, the technology that allows patients to connect virtually with doctors and other healthcare providers, communicating via real-time video or live chat. Telehealth allows chronically ill patients to obtain critical healthcare where they might otherwise have difficulty leaving their home to travel to their doctor for care. And as 5G promises to bring ultra-fast speeds with low latency, telehealth applications will improve dramatically.

With 5G-enabled devices such as remote wearables, healthcare providers can monitor patients remotely and gather real-time data, patient records, inventory, health alerts, that can be used for preventative care and other individually tailored healthcare.

All IoT projects rely on devices. Any large-scale IoT deployment will likely have a variety of new and legacy devices that use different technologies and serve multiple purposes ultimately reflecting the evolution of the IoT field and the scope of your deployment. There are five pillars to any IoT project, namely, devices, connectivity, device management, data processing, and application [50].

Many businesses do not give connectivity a second thought and instead, settle for Wi-Fi or any other popular wireless technology. Furthermore, the devices might be too resource-constrained to even use Wi-Fi. At this point, it may be in a business's best interest to consider cellular connectivity, turning attention to cellular IoT. Low power consumption and long battery life of cellular connectivity will ensure devices are accessible almost everywhere.

When it comes to device management, it is important to discuss the importance of connectivity and industry standards. The protocol stack chosen for device management can make or break a project. When it comes to protocols, regulated open standards are your safest option. They have been developed with the IoT industry in mind, address its needs and grow as the industry grows. The bigger your deployment, the more robust and flexible your device management solution needs to be so that you can handle anything coming your way and this can only be secured with proper protocols.

Devices are constantly collecting data. However, data is useless unless it is aggregated and properly processed. When information does not make its way through where it needs to, it leads to misinformation and wrong conclusions, so make sure all your data is in place. When you have hundreds of thousands of devices with multiple sensors and actuators that collect heaps of data every second of every day, you are bound to lose control quickly if you do not have a platform to manage it all. The platform you choose needs to be versatile to accommodate different solutions and flexible to adapt to future changes which are the only constants in the IoT industry.

There has been a lot of progress in introducing IoT to the production environment including defined actions with clear workflow and products with a specific use. IoT has been stealthily infiltrating research labs and is beginning to show significant results. Compared to manufacturing, laboratories have smaller numbers of a wider variety of equipment, and this equipment is used in several different ways. The employees tend to be more educated, skilled, and thus have a stronger influence on how operational decisions are made. No matter what field or industry the lab serves, it is in everyone's best interest to speed up product discovery and development without increasing costs. IoT enables a unified solution that monitors everything. It eliminates the large number of custom solutions that have been required to get individual pieces of equipment in specific labs to work efficiently.

Many lab procedures require large numbers of similar procedures with tight tolerances, as well as continual checking on ongoing processes over many days, 24 hours a day. Automation and remote tracking can both reduce load and stress on researchers and improve accuracy. The ability to track processes remotely makes experimentalists' lives easier and more flexible while ensuring that any variation is caught and responded to quickly. Stir plates can be turned off and temperatures modified remotely. High-throughput assays with a lot of pipetting can be done more quickly and with greater accuracy. Monitoring equipment such as freezers and incubators can show great benefits. The samples being kept at a given temperature are usually the product of many hours of skilled labor, and are being prepared for the next step in the procedure. An unacceptable temperature variation, not to mention complete failure, can set things back by weeks, with attendant costs.

Any lab with experimental animals knows that keeping cages clean is a significant task. IoT-enabled sensors can allow for real-time monitoring of cage conditions such as ammonia level, humidity, temperature, and animal movement, via sound and light sensors. Cages can be cleaned when as needed, minimizing disruption to animals, while also keeping an eye on individual animal health. It can also reveal when ventilation is malfunctioning, a water bottle is empty, or an animal is manifesting the early signs of illness.

There are many standards in place or under development to promote and give recommendations, guidelines for IoT applications and some of them are related to smart healthcare.

Data networks are used throughout industry. Often, a user must turn to a single manufacturer for networks, sensors and controls, to insure compatible operation. It would make tremendous economic sense to enable any sensor to communicate using any data network. Smart sensors make this a reality. IEEE 1451 is a set of smart transducer interface standards describing a set of open, common, network-independent communication interfaces for connecting transducers (sensors or actuators) to microprocessors, instrumentation systems, and control/field networks [51].

IEEE P2668 is a standard for maturity index of IoT: Evaluation, grading, and ranking. The scope of this standard is to measure the maturity of objects in IoT environment, namely IoT objects. IoT objects shall represent things or devices or the entire system such as health infrastructure. The standard defines the mechanism and specifications for evaluation, grading and ranking of the performance of IoT objects by using indicator values, referred as IoT Index (IDex). IDex shall classify the objects into multiple levels of performance and give a quantitative representation and indication of the performance of objects. IDex shall manifest guidance on blending of IoT objects to evolve into better performance [52].

Reference [53] reported that the European Standard aims to establish a common baseline across the European and wider global market, raising the security bar for all consumer IoT devices from near-zero to a good level. It also contains outcome-focused provisions to create the necessary flexibility and cover all consumer IoT. By providing the foundation for consumer IoT assurance supported by evolving consumer IoT document set in ETSI TC CYBER, an overview of the initiatives in consumer IoT security can be found in [54]. As more devices in our homes connect to the internet and as people entrust their personal data to an increasing number of services, the cyber security of the IoT has become a growing concern. Poorly secured products threaten consumer's privacy, and some devices are exploited by attackers. For Information Security Management Systems (ISMS) Standards, there is ISO27000-series. The ISO 27000 family of information security management standards is a series of mutually supporting information security standards that can be combined to provide a globally recognized framework for best practice information security management [55].

IEC 80001 series is for Medical Devices Cybernetics consideration. IEC 80001-1:2021 is about safety, effectiveness and security in the implementation and use of connected medical devices or connected health software – Part 1: Application of risk management. It specifies general requirements for organizations in the application of risk management before, during and after the connection of a health IT system within a health IT infrastructure whilst engaging appropriate stakeholders [56]. Also, in the IT Applications in healthcare technology, lots of information on standards are listed in [57].

Turning to an insulin pump which is a medical device used for the administration of insulin in the treatment of diabetes mellitus. There is a standard [58] that established a normative definition of communication between personal telehealth insulin pump devices and personal computers in a manner that enables plug-and-play interoperability. While [59–61] reported standards that define a common core of communication functionality for personal telehealth glucose meters, blood pressure monitors and electrocardiograph (ECG) devices.

17.2.2 Smart Meters

Smart metering (SM) supports distributed technologies and consumer participation and extracts energy data using two-way communication. Load monitoring has great potential in many useful applications, such as human behavior, energy conservation, controllable load quantitative evaluation, energy awareness, and load prediction.

A modern smart city is normally full of civilians with enriched lives that normally demand communication using Wi-Fi or Bluetooth for wireless delivery in the same frequency band. Thus, the application of ZigBee to advanced metering infrastructure (AMI) in high-traffic areas needs to be handled with special consideration to mitigate the potentially hostile interferences [62].

In the design, there are multiple parameters that are indicative for consideration, for instance, high power and high throughput for fast data transmission and low latency. However, these factors may produce different effects, e.g. the high-power transmission that causes the feeling of potential health hazard versus the well accepted low power, the high throughput demanded by users versus the low through put generally achieved in a hostile environment, the low latency commonly requested versus the high latency normally occurring in noisy communications. A practical solution is needed by optimizing these key parameters.

Since the deployment of the metering infrastructures is unlikely to grow over time with the same hardware and software architectures, and the number of grid sensors is expected to increase significantly. Hence, it is essential to properly represent and discover the intrinsic semantic of the measured data in order to have a full understanding of the information context, which allows assessing the degree of confidence of the corresponding content. The idea of converting massive data acquired by pervasive field sensors into high-level information and eventually into actionable intelligence at different application domain could represent a strategic tool in solving further operation problems [63].

Applications of many new metering and measurement devices, capable of closely monitoring and sensing grid operation in real-time, result in overwhelming amount of measurement data of high precision and resolution. By far, how to make the best use of the massive data remains quite a challenging task facing power system researchers and practitioners [64]. Some of the difficulties in smart meter data could be summarised as below:

For data quality, as the size of data set is very large, sometimes in the region of several gigabytes or more and also the data origin is from many sources, real-world databases include inconsistent, incomplete, and noisy data. Therefore, a number of data preprocessing techniques, including data cleaning, integration, transformation and reduction need to be applied to minimize noise, inconsistencies and incompleteness in data. In many cases, the current techniques will be too slow to achieve a workable solution for real-life problems.

Turning to data explosion, when smart metering is fully deployed and operated at 30-minute sampling rate, energy suppliers will need to ingest, store, and process at least around 4500–9000 times more of the current data size, reaching 50 terabytes. To manage data sets in such a large volume, the main problem of using relational database management systems is its low scalability [65] and this require a scalable solution that can grow for practical use.

Another urgent problem is due to the lack of standards. To take advantage of these large new data sets, it is essential to gain access to data; develop the data management and programming capabilities to work with large-scale data sets. New approaches to summarize, describe, and analyze the information contained in big data must be developed [66]. Integration of many different forms of data from systems like meter data management, outage management, customer management, billing platforms, and asset management is required. Standards for data description and communication are essential. These facilitate data reuse by making it easier to import, export, compare, combine, and understand data. Standards also eliminate the need for each data originators to develop unique descriptive practices [67]. The lack of worldwide industry standards around data from smart grids and meters could lead to concerns about sharing data with competitors, it also brings worry around data ownership, and concerns about data accuracy. A further worry is on data privacy and data protection.

One of the potential application areas of big data in smart grid is consumer behavior analysis. Many problems need to be fixed, including communication network congestions, increased complexity of optimization problems, management of large data uncertainties, vulnerability of centralized computing systems. In solving all these issues, the conceptualization of decentralized, self-organizing, proactive, and holistic computing framework for decision support in a data-rich, but information-limited domain, represents one of the most relevant research directions [68].

Nonintrusive load monitoring (NILM) and semi-intrusive load monitoring (SILM) are fast developing techniques for devices operation recognition in system monitoring. Many traditional researches focus on feature space improvements for better recognition accuracy and classifier/meter quantity reduction. But practically, cost of each classifier/meter will influence the optimal NILM/SILM solution. Saving construction cost of load monitoring system is one of the most attracting advantages from NILM and SILM. Various researches on NILM or SILM are trying to decrease the cost by decreasing numbers of metering points [69].

The big data in Smart Grid is generated from various sources, such as power utilization habits of users; phasor measurement data for situational awareness; energy consumption data measured by the widespread smart meters; energy market pricing and bidding data collected by automated revenue metering (ARM) systems; management, control and maintenance data for devices and equipment in the power generation, transmission and distribution networks acquired by intelligent electronic devices; operational data for running utilities, such as financial data; and very large data sets, not directly obtained through the grid measurement but widely used in decision-making, such as weather data, data from the National Lightning Detection Network (NLDN), and Geographic Information System (GIS) data [70].

With the huge increase in data size introduced by an increasing number of physical devices, the large-scale advent of incomplete and uncertain data cannot be avoided, such as data from smart grids [70]. For sparse data, the number of data points is inadequate for making a reliable judgment. In machine learning and data mining applications, redundant data can seriously deteriorate the reliability of models trained from the data. Data uncertainty is a phenomenon in which each data point is not deterministic but subject to some error distributions and randomness. This is introduced by noise and can be attributed to inaccurate data readings and collections [71].

Current methods for NILM problems assume that the number of appliances in the target location is known, however, this may not be realistic. In real-world situations, the initial setup of the site can be known but new appliances may be added by users after a period of time, especially in a household or nonrestrictive scenarios. In this sense, current methods without detecting new appliances may not accurately monitor loads of different appliances and scenarios. In this work, a novel new appliance detection method is proposed for NILM with imbalance classification for appliances switching on or off. The prediction of an appliances being switched on or off is an important step in load monitoring and it is inherently heavily imbalanced, e.g. air conditioning are rarely switched off while some appliances, e.g. coffee machine are rarely switched on that is, the switching on frequencies for coffee machine and air conditioning in a household are different, making the problem imbalanced [72].

NILM only requires data from a smart meter to disaggregate appliance-level data. The NILM is cost-effective and friendly to the new installation and replacement of appliances. The disaggregation problem is usually solved by machine learning methods, e.g. sparse coding and Hidden Markov Model. The NILM problem can be transformed into a multi-label classification problem, such that the ON/OFF state of each appliance is classified simultaneously at each time step. When the NILM problem is treated as a multi-label classification problem, it is inherently a class imbalance problem because some appliances are frequently used (e.g. refrigerators) while others may only be occasionally used (e.g. coffee machines). Class imbalance is a common issue in many real-world applications, such as diagnosis of rare diseases, faulty diagnosis, and anomaly detection. Class imbalance problems occur when one class severely out-represents another [71], i.e. a class consists of much more samples (i.e. majority class) than other classes (i.e. minority class).

Another important issue is that current NILM models are built based on the assumption that the number of appliances during training and testing is fixed. However, in a real-world setting, this assumption seldom holds true because users may add new appliances after a period of time, especially in households and nonrestrictive locations. In this sense, existing algorithms may not be able to accurately monitor loads in the target location. With the introduction of appliances in the loads, it may imply that that the resident has gradually changed the resident behavior due to personal issue or external factors. For example, driven by the price differences, residents may prefer to purchase and use more energy-saving appliances, or users choose to use more electrical appliances during the valley load period than during the peak load period. Better capturing of these patterns may help to infer residents' potential interests and more personalized electricity plan and energy saving appliances could be recommended.

NILM only requires data from a smart meter to disaggregate appliance-level data. Most researchers of the NILM focus on monitoring switching events on a single appliance while monitoring a set of the same type of appliances may be more meaningful. Although an appliance does not consume much energy, there may be many such type of appliances (such as lights) in the house and users may switch them all ON or OFF together. In addition to smart meter data, appliance usage characteristic is another important information for load monitoring because the usage period (being turned ON) of some appliances may be relatively fixed (e.g. coffee machine in the morning but rarely during mid-night). Usually, the obtained data contains noise, which may affect the classification results. Therefore, a noise filter-based method is applied in this work to better handle the noisy imbalance problem in the NILM system.

In standard NILM problems, a classifier or an ensemble of classifier is trained using a given dataset to learn the multi-label classification of the ON/OFF states of a set of given appliances in a house. In real-world situation, new appliances may be added while existing appliances may be removed from the house. The multi-label state classification problem is a steaming problem in which a chunk of data (readings of the smart meter) arrives in every time step. In each time step, the new appliance detection and training (NADT) in the proposed method detects deployment of any new appliance and learns the behavior of this new appliance for multi-label state classification.

Current NILM researches focus on the classification of the ON/OFF states of an appliance or a group of appliances. However, all of them assume that the number and types of appliances are previously fixed prior to the NILM. This is unrealistic because people always buy new appliances and plug them to the power network. In this common scenario, current methods will fail because of the unexpected addition of new appliances. Therefore, the detection and adaptation of addition of new appliance is an important new challenge to the NILM researches. Better identification of the introduction of new appliances would improve the overall classification performance of the multi-label learning machine.

When NILM is considered as a multi-label classification problem, the target is to classify if an appliance is switched ON (1) or OFF (0) at a given time step. Some appliances (e.g. air conditioning) are rarely switched off while some appliances (e.g. coffee machine) are rarely switched on. Therefore, class imbalance is unavoidable in NILM problem. Proper techniques should be employed to improve the robustness and effectiveness of these systems. Resampling is effective in handling the class imbalance problems and is one of the key elements for successful operation of many complex systems such as smart grids [73].

With the consent of the householder, data from their smart meter could be used to help them, for example, if there were no signs of electrical usage or heating in the house of an elderly person, a text alert could be sent out to a carer or trusted relative suggesting that they check up on them. As there are millions of smart meters available, it is possible to create a platform to support future services at large-scale and at good value. It is foreseen that energy data can be analyzed to recognize behavioral patterns and assist with monitoring particular health conditions. The potential is there to help shorten the length of stay of people in hospital, and even prevent people from going into hospital to fully make use of resources.

The installation of smart meters at home sets up a platform from which innovative new products and services can be built upon. Reference [74] reported what this might look like in the rapidly advancing world of digital health and care, with potentially huge benefits for vulnerable groups in the population. While many of the ideas in this report are at an early stage of development, it is clear from the pace and direction of innovation in the health sector that the use of energy data will have a significant role in the United Kingdom's future.

Smart meters are being rolled out for the benefits they can bring to energy consumers and the energy system in the United Kingdom, but the data produced may also be useful in health and care applications. Reference [74] examines what research, innovation, and commercial activity has been conducted so far in this domain, and what the opportunities and challenges for its development could be. The report gives an overview on the use of smart meter data in health and care applications. A small number of research projects have presented evidence of the ability to use digital energy data to recognize activities or usage patterns that could be associated with a variety of health conditions. A number of companies also integrate such data into their health monitoring service offerings alongside other technology. As yet there is no clinical trial evidence of the effectiveness of using digital energy data to improve health outcomes. Potentially recognizable health-relevant features include inactivity, such as through falls, sleep disturbance, memory problems, changes in activity patterns, low activity levels, occupancy and unhealthy living conditions. Much of the small amount of research in this area so far has been applied to Alzheimer's disease and dementia. Other key targets are a range of mental illnesses, such as depression and care of people who are vulnerable in some respect. Proposed applications include issuing alerts to carers when unusual activity patterns are recognized, and monitoring of things like the progress of conditions to inform treatment needs or of living conditions such as use of heating or showers. The report also raises the possibility of using digital energy data to inform diagnosis and public health, drawing parallels with initiatives in other areas, but there is no evidence at the moment that this will be possible or practical. But the future potential could be huge.

There are significant challenges. Assurances around privacy may be important. Sharing smart meter data with public services may be controversial. Advances also need to be made in more accurately recognising specific electrical appliances from dwelling-level data, which would increase the usefulness of smart meter-based systems in identifying activities and reduce false alarms. This could likely be improved if the sampling rate of energy data could be pushed beyond the current 10 second limit. Even if better activity recognition can be achieved, there is much more work to be done in reliably and usefully connecting observable energy use patterns with health conditions. The manner in which smart meter data is stored and shared is tightly governed by regulation. However, using this data in health contexts is likely to involve taking it out of the regulated smart meter infrastructure to share it with third parties. Given the sensitive use to which such data will be put, ensuring good data security and privacy after data has left the currently regulated system should be an important focus for regulators considering its use in health contexts. The level of failure tolerance for health critical uses is also likely to be lower than for standard energy metering applications, with potential implications for how the system is regulated. Questions will also need to be considered about where responsibility lies when systems fail with potential health consequences.

There are many user acceptance issues around smart meter data in health contexts shared in common with other telehealthcare approaches, there should be an early focus on understanding what questions users for both practitioners and patients might have. For smart meter health applications to be used at large scale, they will be part of much larger smart metering and health data infrastructures. Early attention should be given to what system-level issues might be expected to arise and how they can be addressed.

17.3 Case Study

There are many applications in the use of new techniques to healthcare. Few of them will be illustrated in this chapter.

17.3.1 Continuous Glucose Monitoring

Diabetes provides a good reason for the research and development of smart devices, since the condition affects roughly 10% of the adult population, and requires regular monitoring and administration of treatment. Pricking your finger and collecting a drop of blood to check glucose levels is an invasive but integral part of daily diabetes management. Using conventional diabetes management system for checking your blood is often painful and unpleasant, such as using the lancets, lancing device, and measurement device presented in [75].

In general, lancing devices are very much the same. They are small tubes that launch an inserted lancet (small needle) into the skin to make a puncture that produces a drop of blood. All lancing devices tend to use the same type of spring-loaded mechanism to push the lancet needle into the skin. Most lancing devices have an adjustable setting to determine the depth of the jab. The depth is measured by the number of millimeters the jab penetrates the skin. All of the manufacturers promise less or no pain when using their device. These claims are based on some combination of the following product characteristics such as using a thinner lancet needle; coating the lancet needle with a material that reduces friction with the skin and adjusting the depth of the jab. Figures 17.2 and 17.3 show conventional blood glucose measuring components and a conventional blood glucose meter respectively. There are other factors to keep in mind as well when purchasing a glucose testing kit. The cost of a test strip, the quality of the lancing device, how much blood the meter requires for an accurate reading and whether it delivers a fast result.

Noninvasive methods have been developed to give real-time measurements of blood glucose. The device continuous glucose monitor (CGM) helps monitoring blood glucose on a continual basis by

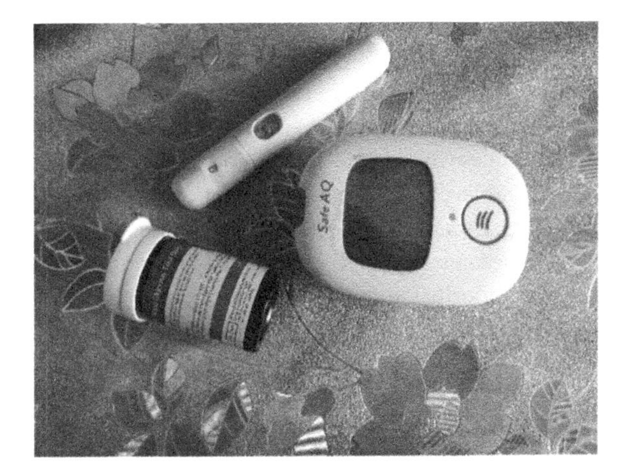

Figure 17.2 Conventional blood glucose measuring components.

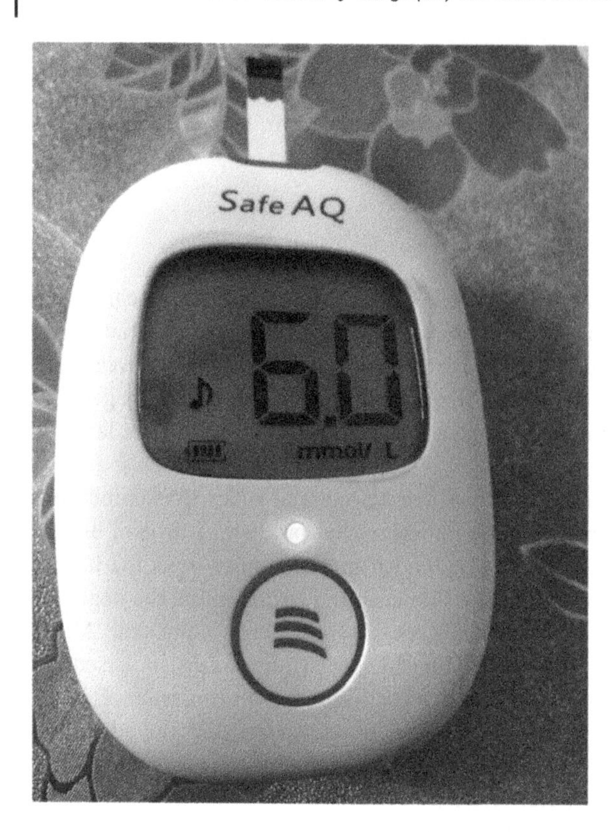

Figure 17.3 Conventional blood glucose meter.

insulin-requiring people with diabetes, e.g. people with type I, type II diabetes, or other types of diabetes for several days at a time, by taking readings at regular intervals. The first CGM system was approved by the US Food and Drug Administration (FDA) in 1999, and in recent years, a number of smart CGMs have hit the market. The CGM that has a sensor that lasts up to 180 days. It can give glucose readings and trend arrows for every five minutes and can help you manage your diabetes before going too high or too low. It can send you pop-up messages on your phone, sound an alert or vibrate on your arm to alert you to highs and lows, based on your customized settings [76].

Advantages of using a CGM include tracking sugar levels all through the day and night; checking levels at times when one would not normally test, for example, during the night; monitoring trends when sugar levels are starting to rise or drop, so patients can take action earlier; reducing the number of finger-prick checks and hypoglycemic episodes as they can see a downward trend before they actually become hypoglycemic [77]. When you have diabetes, keeping your blood sugar in check is a huge priority. No one enjoys pricking their fingers all day and testing with traditional blood glucose monitors, which is why a CGM can be a more convenient solution [78]. However, there are disadvantages of using a CGM as well such as irritation due to wearing the sensor and data overloading.

Smart CGMs send data on blood glucose levels to an app on iPhone, or Apple Watch, allowing the wearer to easily check their information and detect trends. The app also allows for remote monitoring by caregivers, which could include the parents of diabetic children or the relatives of elderly patients. For people without diabetes, the normal range for the hemoglobin A1c level is between

4 and 5.6%. Hemoglobin A1c levels between 5.7 and 6.4% mean you have prediabetes and a higher chance of getting diabetes. Levels of 6.5% or higher mean you have diabetes. You can improve hemoglobin A1c (HbA1c) level as you can tailor your insulin doses more carefully.

Continuous blood sugar meters are worn for long periods of time on your body and offer you continuous monitoring, without constantly having to stop and prick your finger. CGMs offer more intensive monitoring of a diabetic patient's sugar levels. Therefore, CGM are particularly well suited to patients who have to check multiple times a day, or to patients who want to get more frequent feedback during the day.

Continuous blood glucose meters can also give you and your doctor more information than a standard meter, such as if your blood sugar begins to drop too low, the device can warn you, this can be a very helpful feature in helping patients avoid hypoglycemia. Doctors may also recommend a continuous monitor for other personal reasons that have to do with your own health circumstances and lifestyle. Not everyone is able to get one since certain conditions may prevent you from using one, which is another reason why you should ask your doctor if a CGM is right for you.

There are also options for CGM that come with some high-tech features like Bluetooth capability so you can send the results from your blood sample straight to your phone. There is a continuous glucose monitor (CGM) with a thin sensor that you insert under the skin and wear all day. You then scan the sensor with the device to get your glucose reading. Continuously monitors glucose levels in your body from a small sensor inserted in the skin, and the reading is sent wirelessly to a handheld device, or you can work with your smartphone or Apple Watch via an app. The blood sugar monitor is simple to use and provides a 5-second reading, plus it connects to the Contour Diabetes app via Bluetooth to keep a log of your readings to share with the doctor. Figure 17.4 shows a CGM system.

Regarding the automated insulin delivery, it is one of the most fascinating areas in IoT medicine in which there is an Open Artificial Pancreas System (OpenAPS). OpenAPS is a type of closed-loop insulin delivery system, which differs from a CGM in that as well as gauging the amount of glucose in a patient's bloodstream, it also delivers insulin so to close the loop. The artificial pancreas tracks blood glucose levels using a CGM and automatically delivers the hormone insulin when needed using an insulin pump.

Automating insulin delivery offers a number of benefits that can change the lives of diabetics. This can keep blood glucose within a safe range, preventing patients from hyperglycemia, in which the blood glucose is greater than 125 mg/dL (milligrams per deciliter) while fasting for at least eight hours. In general, a person with a fasting blood glucose greater than 125 mg/dL has diabetes. Similarly, it can also prevent patients from hypoglycemia, that is, there is a deficiency of glucose in the bloodstream.

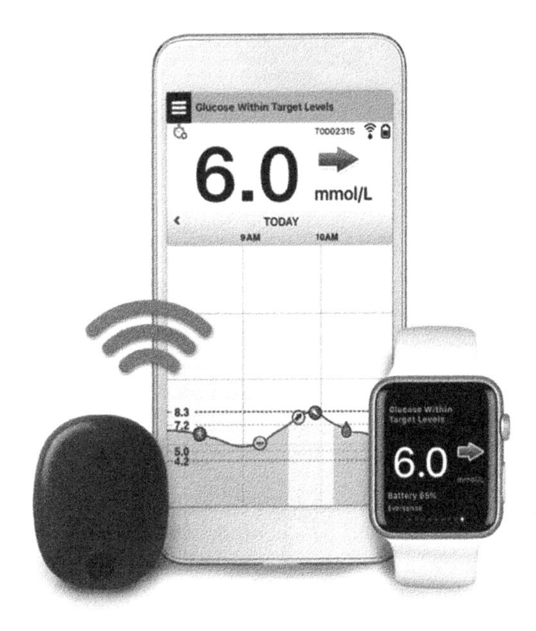

Figure 17.4 A continuous blood glucose meters [76].

17.3.2 Smart Pet

While the main initiatives in smart cities are related to areas, for example, smart energy, smart transport and smart buildings. In [79],

Figure 17.5 The pug – A lovely pet or an ugly dog? *Source:* Niroz Shrestha/ Adobe Stock.

Figure 17.6 A collar-wearing pet cat, living with her owner.

it mentioned that smart cities are defined along six axes or dimensions, namely, economy, mobility, environment, people, and living and governance. Where to include the role of smart services for pets in a smart city is not yet clear. Some urgent efforts would be needed to develop animal bylaws under the present environment in areas such as pet monitoring in smart cities. Some people consider their pets as members of their family and a trend is growing where more pet devices are becoming available to meet the needs of pet owners and their pets. Smart pet collars for cats and dogs have become available and can track and analyze vitals and activity, and alert owners when something is wrong. As is the case for humans, it is good to monitor the health condition of your pets, such as whether they get enough exercise or have a healthy diet. It would be useful to notice any change in behavior of your dogs or cats as this can signal ill health. Smart health monitoring collars are noninvasive wireless devices that continuously collect vital signs and behavior patterns [80].

Figure 17.5 shows a pug which is a breed of dog with physically distinctive features of a wrinkly, short-muzzled face, and curled tail. The breed has a fine, glossy coat that comes in a variety of colors, most often light brown or black, and a compact, square body with well-developed muscles. The thing with pugs is that you will either absolutely love their smooshed faces or cannot stand their aesthetic. Figure 17.6 shows a beautiful pet cat with a collar living with her owner. There is a great opportunity for IoT to improve pet liveability in smart cities [81]. Figures 17.7 and 17.8 show that people take pets seriously as part of the family members.

There are 3.3 billion people living in cities today, a number that will be doubled by 2050. The average number of pets per household varies in different countries. There is no documentation of the average number of pets per household in the world. However, in the United Kingdom 59% of households own pets [82] whereas for the United States the average figure is 67% [83]. As human population increases, the pet population will increase too.

In August 2013, as many as 50 000 starving stray dogs swarmed Detroit when people left the poor city, leaving their pets behind. Whether you are a passionate activist fighting for animal rights or you are a person who does not like dogs or cats, governing bodies must recognize the problem, collect information and develop infrastructure and tools to formulate bylaws that foster a safe, and sustainable environment for people and animals simultaneously. For example, municipal councils should have the power to anticipate and help solve these problems through animal bylaws [84].

Figure 17.7 Pets are dressed as family members.

Monitoring of pet animal in smart city is a big challenge for authorities concerned. Fortunately, technologies that are emerging can help smart cities, their shelters, and pet owners to improve livability for city residents. There is social media, facial recognition software, interactive website, licensing apps, and wearable GPS app [85]. With pictures of dogs and cats, there is a possibility to attract people and find homes for animal needing adoption. Snapshots are used to match the faces of lost dogs with those that have been found or admitted to shelters, and it allows pet owners, shelters, or anyone to look for the animals with a smartphone or computer. Through interactive website, pets can be linked to people looking to adopt one. It is also noted that 90% of lost pets that are microchipped are found within 48 hours. To help owners easily track their pets or to identify or predict aggressive behavior and avoid developing huge problems. There is a wearable GPS app to help owners track their pets. This gives peace of mind to pet owners that they will never lose their best friend. Many smart pet-devices will be developed to create new apps and solutions to make the integration of pets and people in sharing a common environment.

Figure 17.8 People live with pets as family members.

Animal biometrics-based recognition systems are considered a good alternative for the health management, tracking, identification, and security of pet animals. Reference [86] reported a pet recognition system for the monitoring of pets in the smart cities. The system recognizes the individual dogs based on their biometric facial feature. The system uses the one-shot similarity and distance metric-based learning methods for matching of facial features of the pet. It was claimed that the efficacy of proposed pet animal recognition system is 96.87% recognition rate.

Indeed, feeding your pet is a delicate balance. Some pets with medical conditions need regular meals to keep blood sugar or hormone levels stable, but it can be difficult to provide round-the-clock feeding. There is a need to create a product that would help solve all of these problems, making life easier for people and happier for pets. A smart pet feeder with multiple settings to allow for ease of use was developed as reported in [87]. This feeder connects to your phone via an app and allows you to set up profile for your pet. This profile includes weight, age, activity level, and breed, and then matches you with suggested feeding amounts and schedules. If your pet needs a different feeding schedule than the suggested schedule, you have complete control over that and can manually set a feeding schedule that works for your pet and your home.

Also taking pets away when enjoying a break is an increasing trend. For instance, in the United Kingdom, nearly two million dog owners and around one million cat owners take their pets away when enjoying a break away from home. Many people who do not wish to be parted from their pets when visiting a city for pleasure would need services that meet their and their pet's particular needs. IoT can play a crucial role in assisting tourists traveling with pets during their trip and allow city managers to coordinate pet's tourism services and monitor the status of their use at all times to guarantee quality pet public services. It is possible to have the ability of computation, communication, and control technologies to improve human interaction with pets, and with an activity tracker that logs the pet's exercise, body temperature, and exposure to sunlight, providing the top of the fitness goals for the pet.

17.3.3 Smart Meters for Healthcare

Alerting relatives, carers or health practitioners to events which may require a response, such as when someone has been incapacitated by a fall. This functionality has already been the subject of research as described in [88]. Technological advancements in the field of electrical energy distribution and utilization are revolutionizing the way consumers and utility providers interact. In addition to allowing utility companies to monitor the status of their network independently in autonomous fashion, data collected by smart meters as part of the wider advanced metering infrastructure, can be valuable for third parties, such as government authorities. The availability of the information, the granularity of the data, and the real-time nature of the smart meter, means that predictive analytics can be employed to profile consumers with high accuracy and approximate, for example, the number of individuals living in a house, the type of appliances being used, or the duration of occupancy, to name but a few applications. As utilities transition to dynamic customer engagement strategies they can deliver personalized alerts, recommendations, and other communications that generate greater customer energy and cost savings, inspire behavior change.

Before the availability of smart meter data, utilities relied upon substation data or customer participation in home audits. They might have also leveraged local government records to see what appliances might be installed in a home. But the data was limited and it was not real-time or seasonally specific. As a result, most utility marketing was blanketed to entire service territories [89]. The problem with blanket marketing was two-fold. First, without insight into a household's unique energy habits, preferences and usage patterns, utilities communications were frequently not relevant or timely to individual customers. Second, when it comes to make better energy decisions, there is no universal motivator or set of rules that applies to all customers, makes it impossible for any blanket call to action to have a meaningful impact.

Within the United Kingdom, energy suppliers and the government are funding the cost of the smart meter roll out and ongoing maintenance. It is foreseen that by investigating advanced machine learning and load disaggregation techniques of this highly accurate sensing network, detailed habits of an individual's interactions with electrical devices can be assessed; smart meter electricity readings can be used to support social care that meets a person's individual needs, maximizes independence, and promotes a sense of security for those living alone [90].

Existing monitoring services, such as motion sensors, cameras, fall detectors and communication hubs, wearable body networks, are intrusive, expensive and are met with patient resistance. Additionally, current technical solutions are tailored to a specific application and do not meet the ongoing changing requirements of a patient; whereas building on the smart meter infrastructure does not require user interaction. The technology creates a personalized profile of the user's behavior at home.

A short overview on smart health with smart meters was reported in [91] and a review was given in [92]. Therefore, a structural health-monitoring system is needed to come out from the problem associated due to the rapidly growing population of elderly and the healthcare demand. Reference [93] discussed the consumer's electricity usage data, from the smart meter, how to support the healthcare sector by load profiling the normal or abnormal energy consumption. The measured dataset is taken from 12 households and collected by the smart meter with an interval of an hour for one month. The dataset is grouped according to the features pattern, reduced by matrix-based analysis and classified with K-Means algorithm data mining clustering method. It was shown how the clustering result of the sum square error (SSE) has connection trend to indicate normal or abnormal behavior of electricity usage and leads to determine the assumption of the consumer's health status.

EPSRC supported a study which involves testing a new remote patient monitoring system for people with self-limiting conditions such as Dementia (causes problems with memory, thinking, and behavior). The system requires no direct interaction from the patient. It enables patients to live more safely at home and maintains their independence for as long as possible. It aims to develop novel technology to assess an individual's personal physical and mental-health by monitoring their electricity usage at home. This is achieved by processing data collected from smart meters, which captures detailed habits of an individual's interactions with electrical devices. The technology identifies any anomalies in a person's routine, which is the result of a health-related condition. For example, an Alzheimer's patient leaving an oven on or person suffering with depression remaining awake at night. This is achieved by employing advanced data analytics, known as machine learning, to understand trends in electricity usage. The system can identify when an individual gets up, goes to bed, eats, their location within the home and a bad night's sleep. Essentially, the technology creates a personalised profile of the user's behaviour at home. Existing monitoring services such as motion sensors, cameras, fall detectors and communication hubs, and wearable body networks are intrusive, expensive and are met with patient resistance. Additionally, current technical solutions are tailored to a specific application and do not meet the ongoing changing requirements of a patient; whereas our approach would require minimal installation, and builds on the smart meter infrastructure, without the need for user interaction.

As mentioned, smart meters have allowed the monitoring of consumers' energy usage with a high degree of granularity. Detailed electricity usage patterns and trends can be identified to help understand daily consumer habits and routines. There is an opportunity to exploit the usage patterns and recognise when sudden changes in behaviour occur. This would allow detailed, round the clock monitoring of a person's wellbeing and would be particularly useful for tracking individuals suffering from self-limiting conditions such as, Alzheimer's, Parkinson's disease, and clinical depression. Reference [94] explored this idea further and presented a new approach for unobtrusively monitoring people in their homes to support independent living. The posited system uses data classification techniques to detect anomalies in behavior through personal energy usage patterns in the home. Some results show that it was possible to obtain good accuracy with a Neural Network classifier. However, level of confidence must be assessed, and the application is still some ways to go.

Analytics are tailored to an individual's health condition for monitoring, early intervention, detection and prediction of self-limiting conditions. If abnormal behavior is detected, an alert could then be sent to a carer or family member. It is required to support and enable a larger number of people to remain independent whilst living with long-term health conditions, such as Alzheimer's. For example, in the United Kingdom, around one in five adults are registered disabled and more than one million of those currently live alone. These conditions place significant demands on healthcare services globally. Specifically, research is needed to devise a system that can detect when an Alzheimer's patient has left an oven on or remained awake at night.

Healthcare and social care providers will rely heavily on the patient monitoring system to ensure patient safety and welfare in the future, particularly for individuals living alone with Alzheimer's disease. Smart meter reading may directly benefit various policymakers, who are setting guidelines for the prioritization of the smart meter rollout. Vulnerable people must be given priority and seamless access to smart meters to ensure that state-of-the-art assistive healthcare monitoring services are implemented quickly. Therefore, appropriate policies and guidelines must be established between energy regulators, energy providers, charities, and health and social care providers. This is important to ensure that a framework is developed between the various parties. The technology creates a personalized profile of the user's behavior at home. The system is a technological solution within telehealth/telemedicine. There is no requirement for the deployment of extra sensors around the home to provide peace of mind and remote patient care, compared to current technologies available on the market today.

The UK NHS has conducted a trial to access how data from smart meters could help monitor at-home patients. The work was to see how information on energy usage can be interpreted by AI algorithms and intelligent machines to check on patients [95]. The technology could be used to notify a patient's carer whenever unusual behavior is detected. The Department for Business, Energy and Industrial Strategy (BEIS) mentioned being able to monitor a patient at home shows how innovative technologies enabled by smart meters can improve many aspects of patent's live, not just the energy use. This use of smart meter to smart health has the potential to change someone's quality of life, and their families' lives, for the better by helping patients with long-term conditions stay at home and remain independent for longer. By keeping a close eye on the energy use of the elderly and vulnerable, carers may be able to monitor when things go wrong, and find out more quickly when help is needed.

Reference [92] focuses on the potential they could have to contribute to one of the most important domains to both individuals and the provision of health and care. With an aging population and medical advances, health and care providers are increasingly looking to digital solutions to help provide care in a more personalized, responsive, and cost-effective way. The analysis of digital energy data from smart meters can give insights into activities within the home. To achieve these purposes, a systematized review was conducted and consulted with experts in the use of digital energy data in health contexts, NILM, ambient sensing and digital health. It is this electrical load monitoring capacity that is the key to health and care applications.

NILM is a method for identifying individual electrical appliance use from analysis of the current and voltage at the home electricity meter through analysis of electricity data sampled at frequencies around once every 10 seconds, up to hundreds of thousands of times per second. While NILM was first developed in the early 1990s, academic interest in the field has increased rapidly since 2010, and commercial interest since 2014, both driven by an increased focus on energy demand in buildings combined with rapid reductions in the costs of sensing technology, and equally rapid improvements in the machine learning algorithms.

The capacity to identify use patterns of individual appliances allows for a greater understanding to be developed of the behavioral patterns of occupants. Studying the use patterns and changes in use patterns of individual appliances offers the potential to detect abnormal patterns of behavior linked to various health conditions. For example, unusual energy use overnight may be evidence that an occupant is experiencing sleep disturbances. Furthermore, analysis of the combination of uses of appliances, and variations in these, offers the potential to infer different forms of household practices which can then be linked back to a more understanding of the social purposes of that sequence of appliance use. Such combinations can be used to tell whether somebody is simply getting up in the night to go to the toilet, or whether they are getting up in the night to eat, make a cup of tea, and watch television for a period.

It is in interpretation of these activities of daily living that the strongest potential of using smart energy data to support health and care lies. However, such applications would be just a small part of the much wider domain of digital health. Digital health includes concepts such as the digitization of health records, health data analytics, mHealth and wearables, and tele-healthcare. This involves the provision of health and/ or care services remotely, such as telemedicine for remote consultations, often using audio/visual equipment; tele-health, for remotely sharing of clinical data and tele-care.

By using NILM to recognize patterns and abnormalities in people's appliance use and activities of daily living, it may be possible to provide tele-healthcare services without needing to rely on dedicated tools and sensors which may be costly and obtrusive. Indeed, since the vast majority of homes are expected to be equipped with smart meters by the end of the decade, such services could potentially be introduced with minimal expense and disruption to people's daily lives. However, there are also significant technical, social, and regulatory barriers which would need to be overcome.

Presently, there is no established way to classify the behaviors and pathologies focused by smart meter health studies. It is suggested to take a pragmatic approach to track the diminishing range of daily tasks, such as cooking and washing that a patient can perform as self-limiting conditions worsen. In this way, occupant health conditions might be detected through changes in appliance use by observing smart meter data. It is believed that patterns of eating and sleeping as well as changes in behavior, routine, and activity levels could be determined. Each of these changes may be related to memory problems, the decline of social relationships, the deterioration of personal hygiene and hyperactivity or inactivity. In turn, such changes may indicate Alzheimer's disease and other dementias, as well as other mental health problems mentioned.

Depression is mentioned in [96] as a condition which their outline energy use monitoring architecture may be able to detect at its early stages, thus allowing for more effective treatment. Reference [97] described the positive health benefits that detecting and reducing sedentary behaviors may have by reducing abnormal blood glucose levels. However, they also highlight that linking specific behaviors to conditions is difficult in practice. A sleep disturbance detection protocol is described in [98]. References [99, 100] outlined the structure of an ambient-assisted living monitoring system and the progress of early lab testing, with the hope that the scheme will be extended a practical trial. Sleep disturbance as a characteristic of Alzheimer's is also considered in [101]. Wider infrastructural architectures required to integrate smart meters with home area networks, consumer access devices (CAD) and cloud computing facilities was reported in Reference [102].

Reference [94] reported a study with a random sample of energy use data of people aged over 70 drawn from a large Australian smart meter dataset. In order to detect the activities of occupants, and in turn understand if they change over time, the energy use data is typically disaggregated into specific appliance use patterns. Higher energy use appliances are usually easier to detect. References [88, 97, 103] reported that appliances that have been focused on include kettles, cookers, and microwaves. The alarm system monitors kettle activity and sends a notification when the kettle has not been used as expected. Reference [104] used an energy monitor at 1–10 second measurement frequency to train a device identifier, extracting a unique energy use profile for each device. The data were categorized into abnormal and normal usage pattern. Such processes are common in NILM and as the sophistication of NILM increases it is likely that a wider range of devices will be able to be detected with greater accuracy. In another use case, Reference [101] gave social workers the progression of health, such as night-time activity related to the progression of Alzheimer's disease. Remote activity monitoring on electricity use record provided an extra tool for the social worker to aid their understanding of the development of conditions.

By using AI, techniques that model patterns in electricity usage, and a person's day-to-day routines at home could be achieved. The proposed solution is able to identify when individual appliances are used in the home and model both normal and abnormal behavior. The current system can identify microwave, cooker, and washing machine usage. Interactions with these devices are used to help detect significant variations in activities of daily living, and can be used to safeguard the patients.

Reference [25] reported an initial six-month clinical trial has been completed. Energy readings were monitored every 10 seconds and used to identify interactions with appliances in multiple homes. By detecting when appliances are used (how often and when), routine behavior was established. This provided the possibility to identify any abnormal behavior when the interaction pattern changed and can be used to raise alerts when required. The results demonstrate that the system can monitor and support patients in an unobtrusive and personalized manner.

It is now planned to evaluate the technology using 50 patients with mild to moderate dementia, living in their own homes over a 2.5-year longitudinal study. If the study proves successful, a much larger case control clinical trial will be undertaken to determine to effectiveness of the technology as a clinical decision support system.

There are conflicting views on the impact of smart meters on health. On the worst case, only the negative impact due to smart meters will be discussed here. Because radiofrequency (RF) radiation is a possible carcinogen, and smart meters give off RF radiation, it is possible that smart meters could increase cancer risk. Common Smart Meter Radiation Symptoms such as headaches, dizziness, short-term memory loss, a fuzzy head, irritability, itchiness, sleep disturbance, digestive problems, heart-rate changes, stress, and irritability, skin problems such as rashes and facial flushing, arthritis and body pain, heart problems, eye problems include burning or painful eyes, reduced visual acuity, floaters, pressure within or behind the eyes, cataracts, and seizures. Epilepsy is a central nervous system disorder in which brain activity becomes abnormal, causing seizures and sometimes loss of awareness [105, 106]. People's exposure to wireless Smart Meter radiation varies considerably between properties. Therefore, the number of people affected by short-term symptoms will also vary according to individual conditions. Surveys suggest that about 30% of the population are slightly allergic to radio exposure, usually without knowing it, 3% moderately, and under 1% severely.

Electromagnetic radiation can affect the autonomic nervous system, protein expression and the thyroid. Mechanisms include calcium efflux at ion cyclotron resonance on cell membranes, reduced melatonin, mast cell degranulation, free radicals, DNA effects, biogenic magnetite, cryptochromes, and metal implants. Some genetic variants are more sensitive to EM radiation. Long-term or high-level exposure to similar radiation is linked with cancers and neurological diseases. UK wireless smart meter trials began in 2009 so there are no long-term studies yet.

Regarding protection due to smart meter “radiation,” it is difficult and expensive to shield a whole house, and almost impossible to shield a garden, from a neighbor’s wireless smart meter. Mobile phone radiation can remain strong for over 1 km in semi-rural areas and it goes through walls. Unless you can build from scratch, it is usually better to start with a single room, using special carbon paint. Some people have tried lining walls with high quality kitchen aluminum foil. These need to be earthed [107]. The closer your sleeping or sitting area to a wireless smart meter, the stronger the signal and the more likely you are to suffer ill health. Take especial care if a wireless smart meter is on the other side of a wall from a room which people regularly use, or where the radiation will pass through the wall and room to reach the mast directly. It is especially important to ensure that a wireless smart meter is not located close to a child’s bedroom or to where an elderly person spends much time. This can include a room close to a neighbour’s house if the neighbour’s wireless smart meter is nearby. Carefully check the location of wireless smart meters in a block of flats. If all the wireless Smart Meters are banked on the exterior wall of one room, that room could have high levels of radiation. There are some smart meter protection strategies to reduce radiation exposure such as investing in a smart meter cover; and painting the wall behind the smart meter with electromagnetic field (EMF) radiation blocking paint.

17.3.4 Other Case Studies

17.3.4.1 Cancer Treatment

The trial used a bluetooth-enabled weight scale and blood pressure cuff, together with a symptom-tracking app, to send updates to patients’ physicians on symptoms and responses to treatment for head and neck cancer every weekday. The patients who used this smart monitoring system experienced less severe symptoms related to both the cancer and its treatment when compared to a control group of patients who carried on with regular weekly physician visits with no additional monitoring. It is foreseen that the smart technology can help simplifying care for both patients and their care providers by enabling emerging side effects to be identified and addressed quickly and efficiently to ease the burden of treatment [108]. The study demonstrates the potential benefits of smart technology when it comes to improving patient contact with physicians, and monitoring of patients’ conditions, in a way that causes minimal interference with their daily lives. With Google developing machine learning algorithm to identify cancerous tumors on mammograms, the application of machine learning in the healthcare systems has opened up new avenues in the smart healthcare market [109].

Looking forward to future applications, there are some promising signs that the model could potentially increase the accuracy and efficiency of screening programs, as well as reduce wait times and stress for breast cancer patients. But getting practical, it will require continued research, prospective clinical studies and regulatory approval to understand and prove how software systems could improve patient care. This work looks into detection and diagnosis of breast cancer, not just within the scope of radiology, but also pathology. A deep learning algorithm has been developed to help doctors spot breast cancer more quickly and accurately in pathology slides. With shared decision-making and smart electronic health records, machine learning enables major stakeholders in the healthcare industry to gain advantage and speed up the care delivery curve.

17.3.4.2 Connected Inhalers

Asthma is a condition that impacts the lives of hundreds of millions of people across the world. Connected inhalers is beginning to give them increased insight into and control over their symptoms and treatment. One type of connected inhalers is a sensor that attaches to an inhaler or bluetooth spirometer. It connects up to an app and helps people with asthma and COPD (chronic obstructive pulmonary disease) to understand what might be causing their symptoms, track uses of rescue medication, and also provides allergen forecasts.

One of the benefits of using a connected inhaler is improved adherence, in other words, medication is taken more consistently and more often. The sensor generates reports on inhaler use that can be shared with a doctor, and show whether they are using it as often as is prescribed. For patients, this provides motivation and also clarity, showing how the use of their inhaler is directly improving their condition.

17.3.4.3 Ingestible Sensors

When you ingest something, you swallow it or otherwise consume it. If you do not ingest enough iron, you will feel tired and weak and you will look pale. Trees ingest carbon dioxide, and humans ingest the oxygen that trees in turn produce. We also ingest a lot of other things, such as fry rice, ice cream, and burger. Ingestible sensors are an example of how smart medicine can monitor adherence. According to a study by the WHO in 2003, 50% of medicines are not taken as directed [110]. Pills have been created to dissolve in the stomach and produce a small signal that is picked up by a sensor worn on the body. The data are then relayed to a smartphone app, confirming that the patient has taken their medication as directed.

Ingestible sensors can help to track and improve how regularly patients take their medication, as well as allowing them to have a more informed dialogue with their physician about treatment. While the idea of taking pills with a sensor might seem invasive, the system is opt-in on the part of patients, and they can discontinue sharing some types of information, or opt out of the program altogether, at any time. As more of these devices are brought to market and even become available as prescription medication, digital healthcare will start to become the rule rather than the exception.

17.3.4.4 Elderly People

5G has made the smart technology and IoT a reality. Some examples on how IoT-connected smart devices are facilitating medical treatment, preventing illness and aiding healthcare are reported in [111]. One of them is about elderly people. When elderly people live alone without a helper nearby, there are serious risks to their health if they have a problem and are not able to get assistance, or if they forget to carry out tasks such as taking medicines. IoT offers a number of potential solutions to this problem that allow the elderly to maintain their independence while still getting the assistance they might need in an emergency.

The available technology ranges from wearable IoT devices that can detect a fall to more extensive and sophisticated home monitoring systems that learn an individual's movements and habits, and are thus able to alert a caregiver if there is a major change or if no movement is detected for an abnormal amount of time. Apple has incorporated fall detection into the Apple Watch to detect a sudden, sharp fall and will sound an alert and display a message prompting the user to confirm whether they need assistance. If no movement is detected for more than a minute, the device will proceed to alert the emergency services and send a message to emergency contacts with the user's location. Some remote monitoring systems are also able to interface with other tele-health devices such as blood pressure monitors.

17.4 Conclusions

Advances in science and technology are changing medical practice. But the benefits have also added complexity, making it increasingly difficult for healthcare professionals to be confident that they base their decisions on the latest information. Clinical decision support tools can help. The growing availability, standardization, and integration of data from disparate sources across systems and institutions raises new challenges. Successful digitalization goes beyond implementing new technologies and tools, but has an impact on the broader healthcare infrastructure.

Connected IoT-based healthcare apps can improve transparency of processes for decision-makers. Innovating new products and technologies and integrating them in the healthcare industry is the reason behind the rapid transformation. With digital solutions, we are moving toward a completely new era of patient treatment, monitoring, and managing health. Some practical and potential case examples have been given in the chapter.

Through the mHealth technologies developed using IoT, Industries can reduce the cost of treatment with effective use of asset and resources and hence increasing the productivity. With the help of sensors and the internet, the industry can track the devices/objects which will generate an analytics and real-time insights and hence help in making smarter decisions for the industry. The progress and convergence of data, processes, and things on the internet would make such connections more relevant and important, creating more opportunities for people, businesses, and industries.

The image grade therapy system allows the doctor to observe the patient remotely and assess the real need for the operation. Through the app, the doctor keeps himself updated remotely on the patient's condition. Another example of technology linked to smart healthcare is smart health which includes home tele-monitoring and remote diagnosis to allow a significant improvement in the quality of life. Smart healthcare will move data and not people. There is a huge potential offered by the interaction between technology and medicine.

The technical challenges of the smart health in a smart city include (i) reducing cost for a balanced service to achieve the financial sustainability, (ii) gaining acceptability of the patients and clients to obtain the social sustainability. These challenges can be subdivided into (i) hacking and unauthorized use of internet of health technology, (ii) lack of standards and communication protocols, (iii) errors in patient-data handling, (iv) data integration, (v) managing device diversity and interoperability, and (vi) scalability, data volume and performance. The sustainability can be obtained by addressing the issues of (vii) physician compliance, (viii) data overload, and (ix) privacy and security policy. These challenges should be addressed to provide a real-time, robust, and acceptable healthcare facility for citizens.

Technological evolution in the health sector leads to quality, results, and value. Patients deserve effectiveness and personalization of services and for this reason it is important to invest in this direction. Digital healthcare is a trend that involves the use of new technologies, precisely to improve the level of assistance while keeping the costs, where it is possible. Smart healthcare not only means adopting new products and technologies for diagnosis and treatment, but it includes a greater exchange of information among the parties, a more active role of patients during treatment and, finally, a better management of clinical data.

Advances in tele-health will be well positioned to better manage everything from dangerous battlefield medical care to providing state-of-the-art care to economically disadvantaged people and those living in remote geographical areas. Gigabit LTE networks are expected to provide enormous improvement to the tireless efforts of emergency management personnel and first responders.

An explanation of the causal chain underlying smart meter data for health monitoring is from a change in health will lead to a change in behavior, and so the use of appliance will be changed and results in the change in energy use patterns. Such ongoing monitoring is already an established part of tele-healthcare, however, there is currently only indicative evidence of the potential use. This kind of resource should be fully looked into carefully. Developing software for medical devices or software as a medical device compliant with standards is not a trivial thing. It is required to develop software in line with its intended use and compliant with many standards for example, ISO 13485, ISO 14971, and IEC 62304 standards for the certification of quality management systems and software development.

References

1 World Population Ageing (2019). https://www.un.org/en/development/desa/population/publications/pdf/ageing/WorldPopulationAgeing2019-Report.pdf (accessed 19 September 2021).

2 Bommer, C., Sagalova, V., Heeseemann, E. et al. (2018). Global Economic Burden of Diabetes in Adults: Projections From 2015 to 2030. *Diabetes Care* 41 (5): 963–970.

3 IPPOCRATE AS (2020). Smart healthcare and technologies in the healthcare sector. 20 January0. https://www.ippocrateas.eu/smart-healthcare-and-technologies-in-the-healthcare-sector/ (accessed 25 September 2021).

4 Kyriakoulis, K.G., Kollias, A., Anagnostopoulos, I. et al. (2019). Diagnostic accuracy of a novel cuffless self-blood pressure monitor for atrial fibrillation screening in the elderly. *J Clin Hypertens.* 21: 1797–1802.

5 Watanabe, N., Bando, Y.K., Kawachi, T. et al. (2017). Development and validation of a novel cuff-less blood pressure monitoring device. *JACC: Basic to Translational Science* 2 (6): 631–642.

6 Fortin, J., Rogge, D.E., Fellner, C. et al. (2021). A novel art of continuous noninvasive blood pressure measurement. *Nature Communications* 12: 1387.

7 Alcinesio, A., Meacock, O.J., Allan, R.G. et al. (2020). Controlled packing and single-droplet resolution of 3D-printed functional synthetic tissues. *Nature Communications* 11: 2105.

8 Fan, C., Davison, P.A., Habgood, R. et al. (2020). Chromosome-free bacterial cells are safe and programmable platforms for synthetic biology. *PNAS* 117 (12): 6752–6761.

9 Kim, J. and Franco, E. (2020). RNA nanotechnology in synthetic biology. *Current Opinion in Biotechnology* 63: 135–141.

10 Farren, J.I. (2006). RNA synthetic biology. *Nature Biotechnology* 24: 545–554.

11 Dolgin, E. (2021). The tangled history of mRNA vaccines. NEWS FEATURE, Nature (14 September).

12 Meghan Bradway, Carme Carrion, Bárbara Vallespin, Omid Saadatfard, Elisa Puigdomènech, Mireia Espallargues, and Anna Kotzeva (2017). mHealth assessment: conceptualization of a global framework. May. https://www.ncbi.nlm.nih.gov/pmc/articles/PMC5434253/ (accessed 25 September 2021).

13 World Health Organization (2011). mHealth -New horizons for health through mobile technologies. World Health Organization. https://www.who.int/goe/publications/goe_mhealth_web.pdf (accessed 18 September 2021).

14 Global Health Care Outlook (2021). https://www2.deloitte.com/global/en/pages/life-sciences-and-healthcare/articles/global-health-care-sector-outlook.html (accessed 19 September 2021).

15 Rialle, V., Duchene, F., Noury, N. et al. (2004). Health "Smart" home: information technology for patients at home. *Telemedicine Journal and e-Health* 8 (4): https://doi.org/10.1089/153056202260507530.

16 Suzuki, T., Tanaka, H., Minami, S. et al. (2013). Wearable wireless vital monitoring technology for smart health care," 7th International Symposium on Medical Information and Communication Technology (ISMICT), Tokyo, Japan (6–8 March 2013). IEEE.

17 Econsultancy (2018). Embracing Technology and Innovation in Digital Marketing – Pharma and Healthcare. November. https://econsultancy.com/reports/embracing-technology-and-innovation-in-digital-marketing-pharma-and-healthcare/ (accessed 10 September 2021).

18 NCS Pte Ltd (2021). Make extraordinary happen - smarter healthcare for a healthier future. https://www.ncs.co/extraordinary?gclid=EAIaIQobChMIovWhkKL38gIVLdxMAh1YXwQtEAAYASAAEgIKq_D_BwE&gclsrc=aw.ds#healthcare?utm_source=sa360&utm_medium=paid-search&utm_campaign=dentsu-brand-awareness-ncsmakeextraordinaryhappen-healthcare-transport-applications&utm_term=searchads (accessed 10 September 2021).

19 Kalidasan, V., Yang, X., Xiong, Z. et al. (2021). Wirelessly operated bioelectronic sutures for the monitoring of deep surgical wounds. *Nature Biomedical Engineering* 5: 1217–1227.

20 Hina Sattar; Imran Sarwar Bajwa; Riaz Ul Amin; Nadeem Sarwar; Noreen Jamil; M. G. Abbas Malik; Aqsa Mahmood; Umar Shafi, "An IoT-based intelligent wound monitoring system," IEEE Access, Volume: 7, 2019, pp. 144500 – 144515, DOI: https://doi.org/10.1109/ACCESS.2019.2940622

21 Technavio Blog (2018). Top 5 healthcare technologies changing the hlobal smart healthcare market in 2018. https://blog.technavio.com/blog/top-5-healthcare-technologies-changing-global-smart-healthcare-market (accessed 25 September 2021).

22 WHO (2011). Global health and aging. https://www.who.int/ageing/publications/global_health.pdf (accessed 19 September 2021).

23 World Economic Forum (2015). The untapped potential of the elderly. https://www.weforum.org/agenda/2015/06/the-untapped-potential-of-the-elderly/ (accessed 1 January 2022).

24 TechNavio (2018). Global Big Data Spending Market in Healthcare Sector 2018-2022. https://www.technavio.com/report/global-big-data-spending-in-healthcare-market-analysis-share-2018 (accessed 19 September 2021).

25 Sikdar, S., Parker, P., Chalmers, C., and Fergus, P. (2019). Utilising smart meters in assisted living. National Health Executive (NHE). https://www.nationalhealthexecutive.com/Comment/utilising-smart-meters-in-assisted-living- (accessed 2 October 2021).

26 Huawei (n.d.). 5G+ Smart Healthcare Is Now Booming. https://carrier.huawei.com/en/success-stories/Industries-5G/Medical/5G-Smart-healthcare-development (accessed 19 September 2021).

27 Zhang, W., Wen, Y., Lai, L.L. et al. (2021). Cost optimal data center servers: a voltage scaling approach. *IEEE Transactions on Cloud Computing* 9 (1): 118–130.

28 Zhang, W., Wen, Y., Lai, L.L. et al. (2020). Electricity cost minimization for interruptible workload in data center servers. *IEEE Transactions on Services Computing* 13 (6): 1059–1071.

29 TechNavio (2021). Healthcare Cloud Computing Market by Product and Geography - Forecast and Analysis 2021-2025. https://www.technavio.com/talk-to-us?report=IRTNTR41148&type=sample (accessed 24 September 2021).

30 Marketsandmarkets (2021). Wearable Healthcare Devices Market by Type (Diagnostic (ECG, Heart, Pulse, BP, Sleep), Therapeutic (Pain, Insulin)), Application (Fitness, RPM), Product (Smartwatch, Patch), Grade (Consumer, Clinical), Channel (Pharmacy, Online) - Global Forecast to 2025. https://www.marketsandmarkets.com/Market-Reports/wearable-medical-device-market-81753973.html?gclid=EAIaIQobChMI3L37kfqX8wIVmE5gCh3NkgKXEAAYASAAEgKTZ_D_BwE (accessed 24 September 2021).

31 Haim, T. (2020). How Smart Technology Can Lead To Healthier Lives. https://www.healthtechzone.com/topics/healthcare/articles/2020/05/05/445317-how-smart-technology-lead-healthier-lives.htm (accessed 25 September 2021).
32 Davies, R.J.H. (1995). Ultraviolet radiation damage in DNA. *Biochemical Society Transactions* 23 (2): 407–418.
33 Kiefer, J. (2007). Effects of ultraviolet radiation on DNA. In: *Chromosomal Alterations* (ed. G. Obe and Vijayalaxmi), 39–53. Springer.
34 Chatterjee, N. and Walker, G.C. (2017). Mechanisms of DNA damage, repair, and mutagenesis. *Environmental and Molecular Mutagenesis* 58 (5): 235–263.
35 Lai, Q.H. (2021). Role of the Polymerase Associated Factor I Complex (PAF1C) in Transcription Following DNA Damage. Transfer of Status Report, Sir William Dunn School of Pathology, University of Oxford, United Kingdom. (For internal communication).
36 Lai, Q.H. (2019). The Cause and Consequences of Transcriptional Readthrough. Sir William Dunn School of Pathology, University of Oxford, United Kingdom. (For internal communication).
37 Hanawalt, P.C. and Spivak, G. (2008). Transcription-coupled DNA repair: two decades of progress and surprises. *Nature Reviews Molecular Cell Biology* 9 (12): 958–970.
38 Brueckner, F., Hennecke, U., Carell, T., and Cramer, P. (2007). CPD damage recognition by transcribing RNA polymerase II. *Science* 315 (5813): 859–862.
39 Lans, H., Hoeijmakers, J.H.J., Vermeulen, W., and Marteijn, J.A. (2019). The DNA damage response to transcription stress. *Nature reviews Molecular Cell Biology* 20 (12): 766–784.
40 Wellener, P., Zale, J., and Ashton, H. (2019). Pathways to faster innovation. https://www2.deloitte.com/us/en/insights/topics/innovation/faster-innovation-patents-exponential-technologies.html (accessed 19 September 2021).
41 Wells, S. (2020). Engineers reveal a prosthetic leg with a mind of its own. https://www.inverse.com/innovation/ai-prosthetic-leg (accessed 22 September 2021).
42 Gilliland, N. (2018). How AI is transforming healthcare. https://econsultancy.com/how-ai-is-transforming-healthcare/ (accessed 25 September 2021).
43 Madison, D. (2018). The future of augmented reality in healthcare. https://healthmanagement.org/c/healthmanagement/issuearticle/the-future-of-augmented-reality-in-healthcare (accessed 22 September 2021).
44 Alsop, T. (2022). Augmented (AR), virtual reality (VR), and mixed reality (MR) market size 2021–2028. https://www.statista.com/statistics/591181/global-augmented-virtual-reality-market-size/ (accessed 22 September 2021).
45 Number of Internet of Things (IoT) connected devices worldwide from 2019 to 2021, with forecasts from 2022 to 2030. https://www.statista.com/statistics/1183457/iot-connected-devices-worldwide/ (accessed 1 January 2022).
46 Vailshery, L.S. (2022). IoT connected devices worldwide 2030. https://www.statista.com/statistics/802690/worldwide-connected-devices-by-access-technology/ (accessed 15 September 2021).
47 Statista Research Department (2016). Internet of Things (IoT) connected devices installed base worldwide from 2015 to 2025. https://www.statista.com/statistics/471264/iot-number-of-connected-devices-worldwide/ (accessed 24 September 2021).
48 A guide to the Internet of Things. https://www.intel.com/content/dam/www/public/us/en/images/iot/guide-to-iot-infographic.png (accessed 24 September 2021).
49 Telit (2020). World Class Healthcare for All: The Rise of Gigabit LTE Networks on the Path to 5G. https://www.iotforall.com/gigabit-lte-networks-5g-path (accessed 22 September 2020).
50 Slawomir Wolf (2020). 5 Key Considerations for Large-scale IoT Deployments. https://www.iotforall.com/5-key-considerations-large-scale-iot-deployments (accessed 22 September 2021).

51 IEEE 1451. https://en.wikipedia.org/wiki/IEEE_1451 (accessed 24 September 2021).

52 Working Group Chair: Tsang, K.F. (n.d.). IEEE P2668: Standard for Maturity Index of Internet-of-things: Evaluation, Grading and Ranking. https://standards.ieee.org/project/2668.html (accessed 24 September 2021).

53 Pandza, J. (2021). Consumer IoT EN 303 645: The European Standard on connected device security. https://www.enisa.europa.eu/events/cybersecurity_standardisation_2021/presentations/04-03-pandza (accessed 24 September 2021).

54 Consumer IoT security. https://www.etsi.org/technologies/consumer-iot-security (accessed 24 September 2021).

55 ISO 27000 Series of Standards. https://www.itgovernance.co.uk/iso27000-family (accessed 24 September 2021).

56 IEC 80001-1:2021. https://www.iso.org/standard/44863.html (accessed 24 September 2021).

57 IT applications inhealth care technology. https://www.iso.org/ics/35.240.80/x/ (accessed 24 September 2021).

58 Health informatics, ISO/IEEE 11073-10419:2019. (accessed 24 September 2021).

59 Glucose meter, ISO/IEEE 11073-10417:2017. (accessed 24 September 2021).

60 Blood pressure monitor, ISO/IEEE 11073-10407:2010. (accessed 24 September 2021).

61 Basic electrocardiograph (ECG), ISO/IEEE 11073-10406:2012. (accessed 24 September 2021).

62 Hao Ran Chi, K. F. Tsang, K. T. Chui, Henry Chung, Bingo Wing Kuen Ling and Loi Lei Lai, "Interference-mitigated ZigBee based advanced metering infrastructure," *IEEE Transactions on Industrial Informatics*, Vol. 12, No. 2, 2016, pp. 672-684.

63 Zobaa, A., Vaccaro, A., and Lai, L.L. (2016). Guest editorial: enabling technologies and methodologies for knowledge discovery and data mining in smart grids. *IEEE Transactions on Industrial Informatics* 12 (2): 820–823.

64 Xu, Z., Lai, L.L., Wong, K.P. et al. (2017). Guest editorial special section on emerging informatics for risk hedging and decision making in smart grids. *IEEE Transactions on Industrial Informatics* 13 (5): 2507–2510.

65 Ferguson, M. (2012). Architecting a big data platform for analytics. IBM Data Magazine.

66 Einav, L. and Levin, J. (2014). Economics in the age of big data. *Science* 346 (6210): 1243089. https://doi.org/10.1126/science.1243089.

67 Lynch, C. (2008). How do your data grow? *Nature* 455: 28–29.

68 Vaccaro, A., Pisica, I., Lai, L.L., and Zobaa, A.F. (2019). A review of enabling methodologies for information processing in smart grids. *International Journal of Electrical Power and Energy Systems, Elsevier* 107: 516–522.

69 Xu, F., Huang, B., Cun, X. et al. (2018). Classifier economics of semi-intrusive load monitoring. *International Journal on Electrical Power and Energy Systems, Elsevier* 103: 224–232.

70 Lai, C.S. and Lai, L.L. (2015). Application of big data in smart grid. *Systems, Man, and Cybernetics (SMC), 2015 IEEE International Conference on, IEEE*, Hong Kong (9–12 October 2015), 665–670.

71 Lai, C.S., Tao, Y., Xu, F. et al. (2019). A robust correlation analysis framework for imbalanced and dichotomous data with uncertainty. *Information Sciences* 470: 58–77.

72 Zhang, J., Chen, X., Ng, W.W.Y. et al. (2019). New appliance detection for non-intrusive load monitoring. *IEEE Transactions on Industrial Informatics* 15 (8): 4819–4829.

73 Ng, W.W.Y., Zhang, J., Lai, C.S. et al. (March 2019). Cost-sensitive weighting and imbalance-reversed bagging for streaming imbalance and concept drifting in electricity pricing classification. *IEEE Transactions on Industrial Informatics.* 15 (3): 1588–1597.

74 Fell, M., Kennard, H., Huebner, G. et al. (2017). Energising health: a review of the health and care applications of smart meter data. Smart Energy GB. https://www.smartenergygb.org/media/qbvkvf3a/energising-health-final-report.pdf (accessed 1 October 2021).

75 Cornejo, C. (2021). 10 Top Diabetes Lancing Devices. https://www.healthline.com/diabetesmine/ten-top-diabetes-lancing-devices?slot_pos=article_4&utm_source=Sailthru%20Email&utm_medium=Email&utm_campaign=diabetes&utm_content=2021-09-14&apid=34865723&rvid=8d344b4a4a376347b5922699bdebcc29268ed88375053c70e1b967d265c7f341 (accessed 15 September 2021).

76 eversense (n.d.). A CGM System. https://global.eversensediabetes.com/ (accessed 20 September 2021).

77 Diabetics. https://www.diabetes.org.uk/guide-to-diabetes/managing-your-diabetes/testing/continuous-glucose-monitoring-cgm (accessed 15 September 2021).

78 Livingston, M. (2022). Best continuous glucose monitors for 2021. https://www.cnet.com/health/best-continuous-glucose-monitors-for-2021/ (accessed 15 September 2021).

79 Manville, C., Cochrane, G., Cave, J., et al. (2014). Mapping Smart Cities in the EU. European Union. https://www.europarl.europa.eu/RegData/etudes/etudes/join/2014/507480/IPOL-ITRE_ET(2014)507480_EN.pdf (accessed 21 September 2021).

80 PetPace (n.d.). The Petpace smart collar health monitoring solution for cats and dogs. https://petpace.com/ (accessed 21 September 2021).

81 Maroto, P. (n.d.). Smart cities, pets and regulations. https://pacomaroto.wordpress.com/smart-cities-series/is-there-room-for-pets-in-smart-cities/ (accessed 21 September 2021).

82 The Pet Food Manufacturers' Association (n.d.). Pet Population 2021. https://www.pfma.org.uk/pet-population-2021 (accessed 30 October 2021).

83 American Veterinary Medical Foundation (n.d.). U.S. pet ownership statistics. https://www.avma.org/resources-tools/reports-statistics/us-pet-ownership-statistics (accessed 30 October 2021).

84 Tandon, G.H. (2018). Animals in Smart Cities. https://www.slideshare.net/gauravhtandon1/animals-in-smart-cities (accessed 21 September 2021).

85 Enbysk, A. (2014). 5 smart technologies to help cities cope with soaring pet populations. https://smartcitiescouncil.com/article/5-smart-technologies-help-cities-cope-soaring-pet-populations (accessed 21 September 2021).

86 Kumar, S. and Singh, S. (2018). Monitoring of pet animal in smart cities using animal biometrics. *Future Generation Computer Systems* 83: 553–563.

87 Cosgrove, N. (2022). Petnet – One of the first smart pet feeders. https://petkeen.com/petnet-one-of-the-first-smart-pet-feeders/ (accessed 21 September 2021).

88 Alcalá, J., Parson, O., and Rogers, A. (2015). Detecting anomalies in Activities of Daily Living of elderly residents via energy disaggregation and cox processes. *Presented at the BuildSys 2015 – Proceedings of the 2nd ACM International Conference on Embedded Systems for Energy-Efficient Built*, Seoul, Republic of Korea (4 November 2015), 225–234.

89 Gupta, A. (2021). Portland General Electric Turns to Data to Achieve "Net Zero by 2040" Goal. https://www.bidgely.com/blog/portland-general-electric-turns-to-data-to-achieve-net-zero-by-2040-goal/ (accessed 2 October 2021).

90 EPSRC (n.d.). Data Analytics for Health-Care Profiling using Smart Meters. https://gow.epsrc.ukri.org/NGBOViewGrant.aspx?GrantRef=EP/R020922/1 (accessed 5 October 2021).

91 Lai, Q.H. (2021). Smart health with smart meters – a short overview. *IEEE Smart Cities Newsletters.* https://smartcities.ieee.org/newsletter/december-2021/smart-health-with-smart-meters-a-short-overview (accessed 22 December 2021).

92 UCL Energy Institute (2017). UCL-Energy report highlights potential for smart meter data to support healthcare. https://www.ucl.ac.uk/bartlett/energy/news/2017/may/ucl-energy-report-highlights-potential-smart-meter-data-support-healthcare (accessed 25 December 2021).

93 Kelati, A., Plosila, J., and Tenhunen, H. (2019). Smart Meter Load Profiling for e-Health Monitoring System. *2019 IEEE 7th International Conference on Smart Energy Grid Engineering (SEGE)*, Oshawa, ON, Canada (12–14 August 2019).

94 Chalmers, C., Hurst, W., Mackay, M., Fergus, P. (2015). Smart Meter Profiling for Health Applications. *2015 International Joint Conference on Neural Networks (IJCNN)*, Killarney, Ireland (12–17 July 2015). https://doi.org/10.1109/IJCNN.2015.7280836

95 Holmleigh (2021). NHS to trail the use of smart meter data for mental health patients. https://www.holmleigh-care.co.uk/nhs-to-trial-the-health-patients/ (accessed 5 October 2021).

96 Ghassemian, M., Auckburaully, S.F., Pretorius, M., and Jai-Persad, D. (2011). Remote elderly assisted living system - A preliminary research, development and evaluation. Presented at the IEEE International Symposium on Personal, Indoor and Mobile Radio Communications, PIMRC, Toronto, ON, Canada (11–14 September 2011), 2219–2223.

97 Kalogridis, G. and Dave, S. (2014). Privacy and eHealth-enabled smart meter informatics. Presented at the 2014 IEEE 16th International Conference on e-Health Networking, Applications and Services, Healthcom, Natal, Brazil (15–18 October 2014), 116–121.

98 Chiriac, S., Saurer, B.R., Stummer, G., and Kunze, C. (2011). Introducing a low-cost ambient monitoring system for activity recognition. Presented at the 2011 5th International Conference on Pervasive Computing Technologies for Healthcare (PervasiveHealth) and Workshops, Dublin, Ireland (23–26 May 2011), 340–345.

99 Chiriac, S. and Rosales, B. (2012). An ambient assisted living monitoring system for activity recognition – results from the first evaluation stages. In: *Ambient Assisted Living, Advanced Technologies and Societal Change* (ed. R. Wichert and B. Eberhardt), 15–28. Berlin Heidelberg: Springer.

100 Chiriac, S., Röll, N., Parada, J., and Rosales, B. (2012). Towards combining validation concepts for short and long-term ambient health monitoring. Presented at the 2012 6th International Conference on Pervasive Computing Technologies for Healthcare (PervasiveHealth) and Workshops, San Diego, CA, USA (21–24 May 2012), 268–274.

101 Noury, N., Berenguer, M., Teyssier, H. et al. (2011). Building an index of activity of inhabitants from their activity on the residential electrical power line. *IEEE Transactions on Information Technology in Biomedicine* 15: 758–766.

102 Chalmers, C., Hurst, W., Mackay, M., and Fergus, P. (2015). Smart health monitoring using the advanced metering infrastructure. Presented at the Proceedings – 15th IEEE International Conference on Computer and Information Technology, CIT, Liverpool, UK (26–28 October 2015), 2297–2302.

103 Song, H., Kalogridis, G., and Fan, Z. (2014). Short paper: Time-dependent power load disaggregation with applications to daily activity monitoring. 2014 IEEE World Forum on Internet of Things, WF-IoT, Seoul, Korea (South) (6–8 March 2014), 183–184.

104 Chalmers, C., Hurst, W., MacKay, M., and Fergus, P. (2016). Smart monitoring: an intelligent system to facilitate health care across an ageing population. EMERGING 2016: The Eighth International Conference on Emerging Networks and Systems Intelligence. Presented at the Eighth International Conference on Emerging Networks and Systems Intelligence, IARIA XPS Press, Venice, Italy (9–13 October 2016), 34–39.

105 ElectroSensitivity UK (ES-UK) (2012). My ill health from wireless smart meters. http://www.es-uk.info/wp-content/uploads/2018/05/13%20-%20Ill%20Health%20from%20wireless%20Smart%20Meters.pdf (accessed 1 October 2021).

106 https://beatemf.com/smart-meter-radiation-symptoms/ (accessed 2 October 2021).

107 Mitchell, J. (2019). 7 Smart Meter Protection Strategies. 12 April. https://beatemf.com/smart-meter-protection/ (accessed 1 October 2021).

108 News Medical Life Sciences (2018). Smart technology helps improve outcomes for patients with head and neck cancer, ASCO Perspective. https://www.news-medical.net/news/20180517/Smart-technology-helps-improve-outcomes-for-patients-with-head-and-neck-cancer.aspx (accessed 14 September 2021).

109 Shetty, S. and Tse, D. (2020). Using AI to improve breast cancer screening. https://blog.google/technology/health/improving-breast-cancer-screening/ (accessed 19 September 2021).

110 Med, A.S. (2003). *Adherence to Long-Term Therapies - Evidence for Action*. World Health Organization, ISBN 92 4 154599 2.

111 Rebecca Sentance (2021). 7 examples of how the internet of things is facilitating healthcare. https://econsultancy.com/internet-of-things-healthcare/ (accessed 25 September 2021).

18

Criteria Decision Analysis Based Cardiovascular Diseases Classifier for Drunk Driver Detection

18.1 Introduction

The Global status report on road safety 2018, launched by World Health Organization (WHO) highlights that the number of annual road traffic fatalities has reached 1.35 million. Road traffic injuries are now the leading killer of people aged 5–29 years. Road traffic crashes continue to place on society both in terms of human suffering and average costs to countries of 3–5% of their annual gross domestic product, which makes road safety an urgent development priority [1–6]. In spite of this massive – and largely preventable – human and economic toll, action to combat this global challenge has been insufficient. Drastic action is needed to put these measures in place to meet any future global target that might be set and save lives.

Ref. [4] reports that ministers and other representatives of States and Governments, assembled at the United Nations on 30 June and 1 July 2022, for a high-level meeting with a dedicated focus for the first time on improving global road safety. It stresses that road traffic deaths and injuries are a major public health problem for all countries. An integrated approach is promoted for road safety not only to save lives and stop injuries, but also to positively impact the achievement of the Sustainable Development Goals.

Naturally, impaired driving is dangerous. It's the cause of more than half of all car crashes; it means operating a motor vehicle while you are intoxicated by alcohol or drugs. Road safety is a shared responsibility. It is essential to promote good practices and adopt comprehensive, effective, and science-based legislation on key risk factors. The merits of the adoption and enforcement of comprehensive legislation on road crashes risk factors, including medical conditions and medicines that affect safe driving, and low visibility, as well as the implementation of proven measures to mitigate such risks and traffic law enforcement actions, supported by intelligent risk monitoring practices must be promoted.

Zagrebelna [7] proposes simple, effective, and low-cost solutions to prevent drinking and driving that can be implemented on a national or local level. It targets governments, non-governmental organizations, and road safety practitioners, particularly those in low and middle-income countries. Ref. [8] reports that drink-driving is a major road safety problem in many countries, although the extent of the problem is often unclear – especially in low and middle-income countries. Even in quite modest amounts, alcohol impairs the functioning of several processes required for safe road use, including vision and motor skills. Alcohol impairment increases the chance that all road user groups, including drivers, riders, and pedestrians, will be involved in a crash.

The WHO has recently updated its fact sheet on Road Traffic Injuries. The 2030 Agenda for Sustainable Development has set an ambitious target of halving the global number of deaths and injuries from road traffic crashes by 2020 [5, 9]. Impaired Driving remains the most frequent factor

Smart Energy for Transportation and Health in a Smart City, First Edition. Chun Sing Lai, Loi Lei Lai and Qi Hong Lai.

contributing to fatal crashes. Among the 1640 traffic fatalities that occurred from 2016–2018, 57.8% involved drivers, walkers, or bicyclists impaired by alcohol or drugs. Impaired driving contributed to 46.3% of fatalities. Impaired walking or biking was involved in 9.6% of fatalities. An additional 1.9% involved with both an impaired driver and walker or bicyclist. Impaired walking and biking involvement in fatal crashes has increased 87% over the previous three-year period [10].

Every day, about 32 people in the United States die in drunk-driving crashes – that's one person every 45 minutes. In 2020, 11 654 people died in alcohol-impaired driving traffic deaths – a 14% increase from 2019. These deaths were all preventable [11]. Alcohol is a substance that reduces the function of the brain, impairing thinking, reasoning, and muscle coordination. All these abilities are essential to operate a vehicle properly and safely. As alcohol levels rise, the negative effects on the central nervous system increase. Alcohol is absorbed directly through the walls of the stomach and small intestine. Then it passes into the bloodstream where it accumulates until it is metabolized by the liver. The alcohol level is measured by the weight of the alcohol in a certain volume of blood. This is called Blood Alcohol Concentration, or BAC. At a BAC of 0.08 g of alcohol per deciliter (g/dL) of blood, crash risk increases exponentially.

However, even a small amount of alcohol can affect driving ability. In 2020, there were 2041 people killed in alcohol-related crashes where a driver had a BAC of 0.01–0.07 g/dL. BAC is measured with a breathalyser, a device that measures the amount of alcohol in the breath of a driver, or by a blood test.

About 30% of all traffic crash fatalities in the United States involve drunk drivers (with BACs of 0.08 g/dL or higher). On average, over the 10-year period from 2011–2020, about 10 500 people died every year in drunk-driving crashes. Car crashes are a leading cause of death for teens, and about a quarter of fatal crashes involve an underage drinking driver. In 2020, 29% of young drivers 15–20 years old who were killed in crashes had BACs of 0.01 g/dL or higher. The highest percentage of drunk drivers (with BACs of 0.08 g/dL or higher) were the 21–24-year old age group and 25–34-year old age groups. 5268 people operating a motorcycle were killed in traffic crashes. Of those motorcycle riders, 1436 (27%) were drunk (BAC of 0.08 g/dL or higher). Among children (14 and younger) killed in motor vehicle crashes, over one-fifth (21%) were killed in drunk-driving crashes. Of those deaths, more than half the time (57%) the child killed was in the vehicle driven by the drunk driver.

Therefore, it is worth developing drunk driver detection (DDD) to reduce the losses from drunk-related traffic accidents. Basically, there are three types of DDD approaches, namely, direct approach, vehicle-based approach [12, 13], and bio signal-based approach [14]. Direct approaches are widely adopted. The approaches require collecting drivers' breath, blood, or urine and then detect the drivers' BACs from the collected samples. The vehicle-based approaches mainly detect the differences of the drivers' behavior between normal cases and drunk cases. If large variation is detected, the driver will be classified as drunk. However, these two approaches cannot hardly provide real-time, automatic detection and early warning at the same time. However, the bio-signal based approaches can achieve this. The plethysmogram signal was used to detect the variations in organ volume and the corresponding status of the drivers [14]. But the drawback is that the method requires long processing time. Among the bio-signals, electrocardiogram (ECG) and electroencephalography (EEG) have been proven to provide status of the human in a timely way. As compared to EEG, ECG is easier for wearable applications implementation.

There are five main waves that can be provided from the ECG signals, namely, P, Q, R, S, and T waves and these waves relate to the dedicated electrical activities of the human heart. In the future, more and more wearable ECG sensors will be available in the market and this will facilitate the development of ECG-based DDD.

18.2 Cardiovascular Diseases Classifier

Electrocardiogram (ECG) signals, are important information for cardiovascular disease diagnosis conducted by cardiologists. Such a diagnosis requires the development of a cardiovascular diseases classifier (CDC). Generally, a CDC mainly comprises feature vectors extraction and machine learning algorithms like an Artificial Neural Network or Support Vector Machine. Features can be divided into three categories, that is non-fiducial features, fiducial features, and hybrid features. Non-fiducial features normally refer to features that do not characterize the ECG signals using P waves, Q waves, S waves, QRS complexes, and T waves [15–19], and vice versa for fiducial features [20, 21]. Hybrid features refer to feature vectors constructed by both non-fiducial and fiducial features [22–24]. In this investigation, a Support Vector Machine (SVM) is used to construct the CDC for the four most common types of cardiovascular diseases, namely bundle branch block, myocardial infarction, heart failure, and dysrhythmia. Seven criteria, including overall accuracy (OA), sensitivity (S_e), specificity (S_p), area under the curve (AUC), training time (T_r), testing time (T_e), and number of features (N_f), which are features to indicate the speed and accuracy of detection, are used as the essential parameters to compute the analytic hierarchy process (AHP) score to aid the multiple criteria decision analysis (MCDA) for the evaluation of the optimal CDC.

Since decisions involve many intangible assets that need to be weighed. They must be measured along with tangible assets, and the measures of those tangible assets must also assess how well they serve the goals of decision-makers. AHP is a theory of measurement through pairwise comparisons, relying on expert judgment to derive priority scales that measure intangible assets relatively. Comparisons are made using a scale of absolute judgment, which represent the degree to which one element dominates another with respect to a given property. Judgments may be inconsistent, and how to measure inconsistency and improve judgments, where possible to achieve better consistency, is the consideration of AHP [25].

Turning to multicriteria decision analysis (MCDA), in the last decade or so, MCDA has been increasingly used to support decision-making and policy setting in healthcare [26–29]. Decision-making in healthcare is complex, as it usually involves confronting trade-offs between multiple objectives, requires the involvement of many stakeholders, under various constraints. MCDA can quantify benefits, risks, and uncertainties arising in decision-making, by considering a set of criteria and their relative importance under a practical and transparent process, while incorporating a wide range of views from different stakeholders to express a more societal aspect [30].

Moreno-Calderon et al. [31] analysed the use of multicriteria software in health priority settings and found that only a few studies used MCDA software in healthcare decision-making. Adunlin et al. [32] showed that MCDA has been applied to a broad range of areas in healthcare, with the use of a variety of methodological approaches. Glaize et al. [33] reported a scoping review by assessing 70 case studies about the application of MCDA from three areas, that is, health services, type of interventions, and healthcare.

Traditional work usually aims at the highest overall accuracy and/or lowest testing time. In reality, every end user has to specify the weights between criteria. It is not uncommon to find a ratio setting by quick understanding or simply a direct 1 : 1 assignment is adopted. It is noted that the needs of volunteers are neglected or not required. In the new method, assignments of criteria are devised for AHP analysis. The incorporation of AHP analysis in the classifier enables the consideration of the need of volunteers.

18.2.1 Design of the Optimal CDC

Figure 18.1 summarizes the block diagram of the new method. After the retrieval of ECG data, feature vectors are extracted. The SVM classifiers are then designed based on the feature combinations. Therefore, N configurations can be obtained. The best model is selected among configuration f_1 to configuration f_N based on seven criteria, namely overall accuracy, sensitivity, specificity, area under the curve, training time, testing time, and number of features, with the aid of MCDA via AHP. The details of the new method are illustrated in the following figure.

18.2.2 Data Pre-Processing and Features Construction

The data is obtained from an online and open access database [34, 35]. A group of healthy candidates as well as candidates with the four most common types of cardiovascular diseases are selected. They are 52 candidates from health control, 15 bundle branch block candidates, 148 myocardial infarction candidates, 18 heart failure candidates and 14 dysrhythmia candidates. The unequal sample size in each class will lead to a bias of the SVM classifier [36]. The ECG signal is further partitioned into 30 seconds sub-signals to obtain 500 samples of healthy candidates and 125 samples of unhealthy candidates for each type of cardiovascular disease. This process aims at equalizing the number of samples in each healthy/unhealthy class. Before the introduction of these four diseases, the notations are briefed. Denote RR-interval to be the consecutive R points between consecutive ECG signals, QRS complex is the time between Q wave and S wave where point R is between

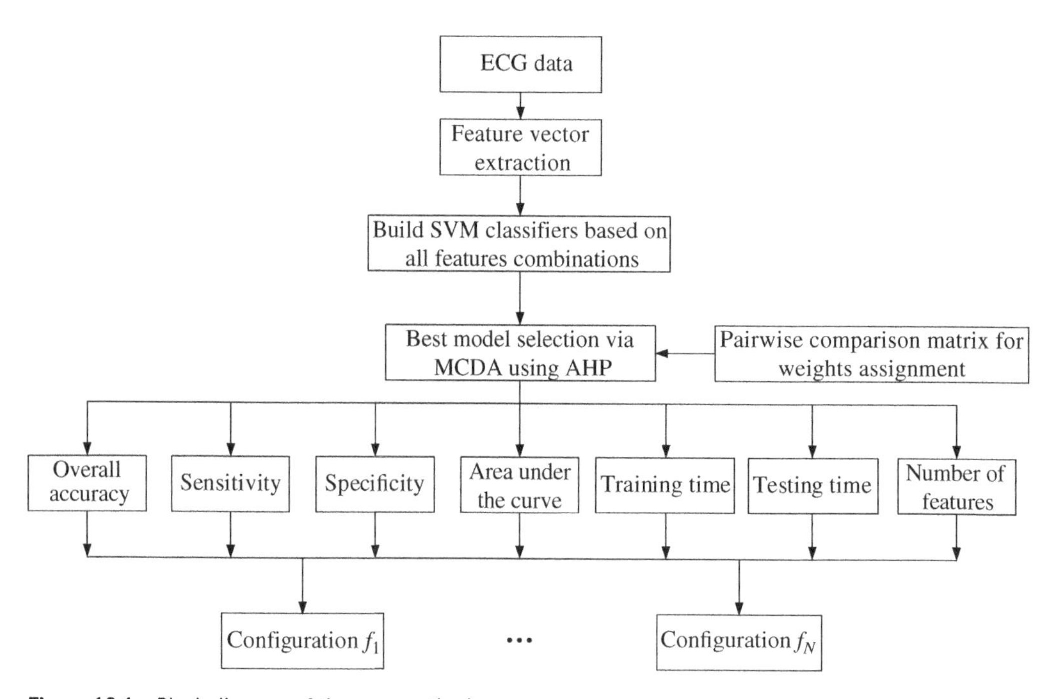

Figure 18.1 Block diagram of the new method.

Q wave and S wave. Similarly, QT interval refers to the time between point Q wave and T wave. The background of these four diseases is presented as follows:

i) Myocardial Infarction: Irregular heartbeat and thus irregular RR-interval may occur in the ECG signal of the patients [37];
ii) Bundle Branch Block: Patients have QRS complex with value exceeding 0.12 ms [38];
iii) Dysrhythmia: The heartbeat can be more than 100 beats per minute or less than 60 beats per minute. Thus, RR-interval is different from the normal ECG signal. Also, the QT interval may increase if the type of cardiovascular disease is ventricular arrhythmias [39];
iv) Heart Failure: A finding of prolonged QT interval in the ECG signals of the patients [40].

As a result, Q wave, R wave and S wave, QRS complex, and RR-interval are representative features to identify between healthy persons and cardiovascular patients. The feature vector consists of 10 features using the average and standard deviation of these five parameters. Before detecting and computing the features, the ECG signals will undergo data preprocessing [41]. The maximum frequency of an ECG signal is typically less than 60 Hz, thus a bandpass filter with cutoff frequencies at 1 and 60 Hz is implemented. A derivative filter is then applied to sharpen the Q, R, and S wave. Finally, signal squaring and sliding window integration are utilized for the location of Q, R, and S wave.

18.2.3 Cardiovascular Diseases Classifier Construction

The CDC is constructed by employing SVM with a 10-dimensional feature vector. This algorithm uses a Lagrange Multiplier with a set of support vectors, a set of weighting and an offset bias [42, 43]. This section focuses on the design of CDC. The performance of CDC is dictated by OA, S_e, S_p, AUC, T_r, T_e, and N_f. It directly classifies the ECG signal into healthy (negative response) candidates and unhealthy (positive response) candidates. OA, S_e, S_p, and AUC are related to the accuracy of CDC. T_r is the time required to train the CDC and T_e is the time needed to detect the ECG signal. In this investigation, CDC will be trained up and validated with the ECG datasets. For the analysis of positive response – Class 0, 500 healthy patients are used. For the analysis of positive response – Class 1, 125 bundle branch block patients, 125 myocardial infarction patients, 125 heart failure patients and 125 dysrhthmia patients are retrieved from the database. Table 18.1 lists the datasets for CDC with binary classifier.

The CDC will use a 10-fold cross validation for performance evaluation [44] and the polynomial kernel function (third order) is used for SVM analysis. There is a total of 1023 combinations $\left(\sum_{n=1}^{10} {}_{10}C_n\right)$, thus 1023 configurations can be formulated from a selection (from 1 to 10) of the

Table 18.1 Database specification of ECG data for CDC.

Class 0 (Healthy/Negative response)	Number of samples	Class 1 (Unhealthy/Positive response)	Number of samples
PTB diagnostic (Healthy)	500	Bundle branch block	125
		Myocardial infarction	125
		Heart failure	125
		Dysrhthmia	125

Table 18.2 CDC of each configuration.

f_j	OA	S_e	S_p	AUC	T_r (s)	T_e (s)	N_f
f_1	0.324	0.350	0.298	0.321	3.5	2.3	1
f_2	0.310	0.324	0.296	0.303	3.4	2.5	1
f_3	0.298	0.288	0.308	0.287	3.6	2.4	1
...	...	...	...	...	...	...	...
f_{1021}	0.986	0.988	0.984	0.972	4.9	3.4	10
f_{1022}	0.964	0.970	0.958	0.946	5.1	3.4	10
f_{1023}	0.970	0.974	0.966	0.949	4.3	3.5	10

10 features. For the jth configuration where $j = 1,...,1023$, namely f_j, its corresponding criteria, OA, S_e, S_p, AUC, T_r, T_e, and N_f are recorded. The main settings of SVM are summarized as follows, in general, the default setting is adopted in the MATLAB toolbox:

i) Number of classes: Two;
ii) Class 0: 500 Healthy candidates;
    Class 1: 125 bundle branch block candidates, 125 myocardial infarction candidates, 125 heart failure candidates, and 125 dysrhthmia candidates;
iii) Feature vector: The maximum dimensionality is 10, which consists of Q wave average, Q wave standard deviation, R wave average, R wave standard deviation, S wave average, S wave standard deviation, QRS complex average, QRS complex standard deviation, RR-interval mean, and RR-interval standard deviation;
iv) Kernel function: Third order polynomial;
v) Fold of cross validation: Ten-fold 1023 classifiers are constructed in 1023 configurations; the results are tabulated in Table 18.2.

18.3 Multiple Criteria Decision Analysis of the Optimal CDC

In Table 18.2, seven criteria, namely OA, S_e, S_p, AUC, T_r, T_e, and N_f, are employed for performance evaluation of the 1023 scenarios. Multiple criteria decision-making (MCDM) has been utilized in many areas. It entails using the particular characteristics of cardiovascular diseases. By allocating appropriate weightings, the AHP is adopted to evaluate and analyse the best scenarios among the 1023 scenarios investigated. The allocation of weightings confronts the feedback from an AHP analysis of 200 volunteers from which a pairwise comparison 7×7 matrix A_m $(m = 1,...,200)$ is formulated. It is intuitively understood that T_e should be as low as possible and that the accuracy should be kept to an acceptable level. Since the speed of detection is essential, the analysis on MCDA reveals that high weightings should be assigned to OA, S_e, S_p, AUC, T_e. These five parameters are referred as the primary parameters. While N_f is typically preferred to be small for speedy detection, it is noted that T_r will not affect the detection time. Hence N_f and T_r are classified as the secondary parameters.

The volunteers are required to fill in the $a_{m,ij}$, where i and j are between 1 and 7, in Table 18.3. The AHP based MCDA CDC is referred as the new classifier (NC). Traditional classifiers (TC) in [17, 21, 22] are also evaluated. Both the NC and the TC are applied to the three feature groups, namely, non-fiducially features, fiducially features and hybrid features in [17, 21, 22]. The performance comparison between the NC and the TC is tabulated in Table 18.4. Based on

Table 18.3 Pairwise comparison 7 × 7 matrix A_m.

	OA	S_e	S_p	AUC	T_r	T_e	N_f
OA	1	$a_{m,12}$	$a_{m,13}$	$a_{m,14}$	$a_{m,15}$	$a_{m,16}$	$a_{m,17}$
S_e	$a_{m,21}$	1	$a_{m,23}$	$a_{m,24}$	$a_{m,25}$	$a_{m,26}$	$a_{m,27}$
S_p	$a_{m,31}$	$a_{m,32}$	1	$a_{m,34}$	$a_{m,35}$	$a_{m,36}$	$a_{m,37}$
AUC	$a_{m,41}$	$a_{m,42}$	$a_{m,43}$	1	$a_{m,45}$	$a_{m,46}$	$a_{m,47}$
T_r	$a_{m,51}$	$a_{m,52}$	$a_{m,53}$	$a_{m,54}$	1	$a_{m,56}$	$a_{m,57}$
T_e	$a_{m,61}$	$a_{m,62}$	$a_{m,63}$	$a_{m,64}$	$a_{m,65}$	1	$a_{m,67}$
N_f	$a_{m,71}$	$a_{m,72}$	$a_{m,73}$	$a_{m,74}$	$a_{m,75}$	$a_{m,76}$	1

Table 18.4 Performance of NC versus TC.

Method	Datasets (number of samples)	Features	Results (related work TC)	Results (new work NC)
Two-layered hidden Markov model [3]	MIT-BIH database (34 799 samples from 16 Arrhythmia candidates)	PR interval, QRS complex interval and T sub-wave interval	OA = 0.992	OA = 0.987
			$S_e = 0.993$	$S_e = 0.99$
			$S_p = 0.992$	$S_p = 0.984$
			AUC = 0.971	AUC = 0.966
			$T_r = 3.7$ s	$T_r = 3.4$ s
			$T_e = 2.7$ s	$T_e = 1.9$ s
			$N_f = 3$	$N_f = 2$
Cross wavelet transform with a threshold based classifier [14]	The PTB Diagnostic ECG database (18 489 samples from 52 healthy control candidates and 148 myocardial infarction candidates)	Total sum of wavelet cross spectrum value and total sum of wavelet coherence	OA = 0.976	OA = 0.966
			$S_e = 0.973$	$S_e = 0.978$
			$S_p = 0.988$	$S_p = 0.958$
			AUC = 0.949	AUC = 0.933
			$T_r = 6.2$ s	$T_r = 5.6$ s
			$T_e = 4.1$ s	$T_e = 2.8$ s
			$N_f = 6$	$N_f = 4$
SVM [15]	CU database, VF database, and AHA database (40 956 samples from 67 Ventricular fibrillation and rapid ventricular tachycardia candidates)	Leakage, count 1, count 2, count 3, A1, A2, A3, time delay, FSMN, cover bin, frequency bin, kurtosis, and complexity	OA = 0.952	OA = 0.947
			$S_e = 0.951$	$S_e = 0.952$
			$S_p = 0.951$	$S_p = 0.942$
			AUC = 0.943	AUC = 0.937
			$T_r = 4.8$ s	$T_r = 4.5$ s
			$T_e = 2.7$ s	$T_e = 1.6$ s
			$N_f = 13$	$N_f = 10$

the discussion for AHP formulation, the assignment of values of $a_{m,ij}$ are based on the following guidelines:

i) Write 1 if there is equal importance of i and j;
ii) Write 3 if i is slightly more important than j;
iii) Write 5 if i is more important than j;
iv) Write 7 if i is strongly more important than j;
v) Write 9 if i is absolutely more important than j.

The pairwise comparison 7 × 7 matrix A_m is then normalized, and an $Anorm_m$ can be obtained by modifying the matrix entries $a_{m,ij}$ in A_m into matrix entries $anorm_{m,ij}$ in $Anorm_m$:

$$anorm_{m,ij} = \frac{a_{m,ij}}{\sum_{l-1}^{7} a_{m,lj}} \tag{18.1}$$

By averaging each row of Equation (18.1), the corresponding 7 × 1 priority matrix w_m with entries $w_{m,k}$ for $k = 1,...,7$ is given by:

$$w_{m,k} = \frac{1}{7}\sum_{l=1}^{7} anorm_{m,kl} \tag{18.2}$$

Let $C_{p,q}$, ($p = 1,...,7$ and $q = 1,...,1023$) be the pth criteria, and q^{th} scenario of CDC. $C_{p,q}$ is normalized to become $C_{p,q,norm}$. The final score for each scenario, AHP_q, is evaluated by:

$$AHP_q = \sum_{l=1}^{7} C_{p,q,norm}\left(\frac{1}{200}\sum_{m=1}^{200} w_{m,l}\right) \tag{18.3}$$

To avoid inconsistency in the construction of pairwise comparison matrices, the optimal CDC is concluded from the highest value of AHP_q [45]. It is evaluated that the optimal CDC is obtained from scenario f_{652}, with feature vector composes of average of Q, standard deviation of Q, standard deviation of S, average of QRS mean, standard deviation of QRS, average of RR-interval, and standard deviation of RR-interval, with AHP_{652} as follows: OA = 0.988, S_e = 0.992, S_p = 0.985, AUC = 0.982, T_r = 4.5 seconds, T_e = 2.8 seconds, N_f = 7.

18.4 Analytic Hierarchy Process Scores and Analysis

The performance scores between the NC and the TC [17, 21, 22] are evaluated and tabulated in Table 18.4. In this investigation, the algorithms in related work have been evaluated, with the addition of MCDA using AHP to obtain the best scenario by assigning weights to the seven criteria. As the new work and related works are in the same application area, the classification of cardiovascular diseases, the weight assignment can be reused to facilitate performance comparisons. From Table 18.4, the percentage changes are evaluated as follows:

1) Percentage change compared with AHP scores from [17]: OA = −0.504%, S_e = −0.302%, S_p = −0.807%, AUC = −0.515%, T_r = −8.109%, T_e = −29.630%, and N_f = −33.333%. It is concluded that there is an improvement of 30% in speed of detection of cardiovascular diseases @~99.5% accuracy.

2) Percentage change compared with AHP scores from [21]: OA = −1.025%, S_e = 0.514%, S_p = −3.036%, AUC = −1.686%, T_r = −9.677%, T_e = −31.707%, and N_f = −33.333%. It is concluded that there is an improvement of 30% in speed of detection of cardiovascular diseases @~99% accuracy.
3) Percentage change compared with AHP scores from [22]: OA = −0.525%, S_e = 0.105%, S_p = −0.946%, AUC = −0.636%, T_r = −6.250%, T_e = −40.741%, and N_f = −23.077%. It is concluded that there is an improvement of 40% in speed of detection of cardiovascular diseases @~99.5% accuracy.

The analysis reveals that in the NC, the speed of detection has been increased by 30–40% while the accuracy is retained at ~99–99.5% of the TC. It is seen that there the reduction of OA, S_e, and S_p are less than 1%. Thus, the AHP based MCDA CDC is a reliable and speedy detection scheme for cardiovascular diseases.

To collect ECG data, the ECG sensors will be implemented to convert the raw data into meaningful representation. There are four stages in the development of ECG wearable sensors, which are pre-amplification stage, filtering stage, tertiary amplification and DC high-pass filtering, and voltage level shifter. Then, support vector machine (SVM) will be used to classify two classes, normal situation and drunk situation. The ECG samples of normal situation and drunk situation will be first collected and then pre-processed. After that, feature extraction will be performed and the extracted features are then utilized to construct the kernel function for the classifier. The kernel function responses to transform the data into high dimensional space.

The result demonstrates that the accuracy of proposed DDD achieves a satisfied accuracy compared to those conventional methods. Besides, using ECG-based detection can realize early detection and fully automated detection.

18.5 Development of EDG-Based Drunk Driver Detection

To collect ECG data, the ECG sensors will be implemented to convert the raw data into meaningful representation. There are four stages in the development of ECG wearable sensors, which are pre-amplification stage, filtering stage, tertiary amplification and DC high-pass filtering, and voltage level shifter. Then, support vector machine (SVM) will be used to classify two classes, normal situation and drunk situation. The ECG samples of normal situation and drunk situation will be first collected and then pre-processed. After that, feature extraction will be performed and the extracted features are then utilized to construct the kernel function for the classifier. The kernel function responses to transform the data into high dimensional space.

The result demonstrates that the accuracy of proposed DDD achieves a satisfied accuracy compared to those conventional methods. Besides, using ECG-based detection can realize early detection and fully automated detection.

The ECG sensor front-end consists of four stages, preamplification, filtering, tertiary amplification & DC-offset high pass filtering, and voltage level shifter. The front-end is responsible to convert the raw ECG data into meaningful representation and attenuate noises and interferences. After ECG preprocessing, the classifier will be developed using the collected data. Feature extraction will then be performed and utilized in the kernel functions for building classifiers. To test the DDD classifiers' performance, a widely adopted method, *K* fold cross validation, is performed [46]. The overall accuracy is recorded.

18.5.1 ECG Sensors Implementations

Instrumentation amplifier is necessary for the ECG sensor design. The instrumentation amplifier is the combination of the differential amplifier with two non-inverting amplifiers as buffering input. Therefore, the input impedance matching can be neglected. The instrumentation amplifier has high common mode rejection ratio which means that the whole amplifier will just amplify the difference of the input without amplifying the input noise. This is an important feature especially for the bio-signal with tiny amplitude such as ECG signal. Also, the low DC input and voltage noise can be achieved by comparing with the typical operational amplifier. There are four stages [47] that can be found in the front-end of ECG sensors. The operation will be explained below:

Stage 1: Pre-Amplification Stage

The gain of the pre-amplify stage is chosen to be 50 which is based on the consideration of making the suitable range of the input ECG signal. The input should be within the power supply range of +/− 9V without any collapsing and allows for the next stage filtering.

Stage 2: Filtering Stage

The relative energy spectra of ECG signal after Fourier transform are summarized in Table 18.5 [46].

There are variety of components with various frequency range coexisted in the ECG signals. Therefore, it is required to determine the cutoff frequencies of the band-pass filter in order to filter out the unnecessary frequency components. The frequency range was chosen from 0.6 to 15.9 Hz which contains P, Q, R, S, T waves of the ECG signals. This frequency range can exclude most of the unwanted noises in the ECG signals. Then, the first order passive type low-pass filter will be cascaded with the first order active type low pass filter to form the second order filter. The higher order of filter can become more likely to be formed by cascading another passive filter since the active filter is constructed from the non-inverting operational amplifier with the defined input impedance.

Stage 3: Tertiary Amplification and DC-Offset High Pass Filtering

The gain is designed to be controllable in order to adapt various situations and provide convenient measurement. Based on the dedicated frequency range selection from the ECG signal power spectrum evaluation, the first order high-pass filter will be constructed in passive way with the controllable gain and behind the cutoff frequency 0.7 Hz.

Stage 4: Voltage Level Shifter The level shifter is to make the positive feedback to the voltage level of the positive input and shift the voltage level up when the DC-offset is found at the negative input. The design of the voltage level shifter should be adapted to the variable input of the DC-offset input and so that suitable DC-offset can be implied to shift the entire ECG signal level up.

Table 18.5 Energy spectral of ECG signal.

Useful components	Noises and interferences
P-wave: 0.6–5 Hz	Muscle noise: 5–50 Hz
T-wave: 1–7 Hz	Respiratory noise: 0.12–0.5 Hz
QRS complex: 10–15 Hz	Line frequency: 50 or 60 Hz
	Human DC offset: 0.01–0.4 Hz

18.5.2 Drunk Driving Detection Algorithm

To classify whether the driver is drunk or not, two sets of data, normal ECG signals and drunk ECG signals, are collected using the ECG sensor. The collected data will be used to develop the drunk driving classifier using support vector machine (SVM). SVM is a famous learning machine for data analysis and classification. A high dimensional space can be obtained after training a set of data. In this case, there are two sets of data. Hence, after training, these two sets of data will be separated in the resultant space as far as possible. They are usually separated by a hyperplane used to classify which class the input data belongs to. For low-dimensional input information, the kernel function is used to convert the input information to high-dimensional for classification. Sampling will be performed on the ECG signals. The sampling data points will be the feature and they will be utilized to customize the kernel function and classifier. Large margin between the two classes (normal and drunk) would be required. SVM is a kind of machine learning algorithm to recognize pattern and classify the unknown input data to the appropriate category. Therefore, Lagrangian dual optimization is considered for maximizing the margin distance using SVM to solve the maximum margin problem [47]:

$$\tilde{L}(\alpha) = \arg\max_{\alpha} \left\{ \sum_{i=1}^{N} \alpha_i - \frac{1}{2} \sum_{i=1}^{N} \sum_{j=1}^{N} \alpha_i \alpha_j S_i S_j K\left(\chi_i, \chi_j\right) \right\} \tag{18.4}$$

Subject to

$$\alpha_i \geq 0, \; i = 1,\ldots,N$$

$$\sum_{i=1}^{N} \alpha_i s_i = 0$$

where α is defined as the Lagrange multiplier, s belongs to {1,−1} which is the class label of input ECG signals, $K(x_i, x_j)$ denotes the kernel function which is used to transform the input data to the desired high-dimensional feature dimension.

18.6 ECG-Based Drunk Driver Detection Scheme Design

Figure 18.2 shows a conventional drunk driving test. With the advancement in computing power, electronic devices and communication technology, Figure 18.3 reviews some current devices for electrocardiogram (ECG) measurement.

The proposed ECG-based scheme consists of four stages including (i) signal preprocessing for ECG data, (ii) feature extraction and building classifier, (iii) multiple criteria decision-making (MCDM), and (iv) *K*-fold validation. The similarities of ECG samples of different status of the human are extracted as feature vector. Then, the feature vector is weighted with respect to the importance of data points. After that, since the dimensionality of feature vector affects detection performance, MCDM is applied to select the best classifier by creating a number of scenarios with various feature dimensions.

Figure 18.4 shows the flow of a proposed ECG-based drunk driver detection scheme. Two human conditions are considered, namely Class 0: Normal, Class 1: Drunk. Four stages are categorized in the development of the ECG DDD scheme. The first stage is signal preprocessing for ECG samples. At this stage, raw ECG signals used in training consist of noise, interference, and offset and so they cannot be directly used in the training. Signal preprocessing is carried out for noise suppression and ECG sample segmentation. The second stage includes feature extraction and building classifiers for

Figure 18.2 Conventional drunk driving test.

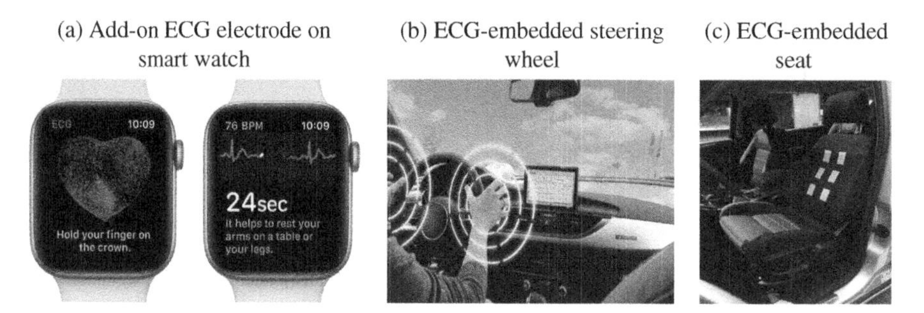

Figure 18.3 Devices for electrocardiogram (ECG) measurement.

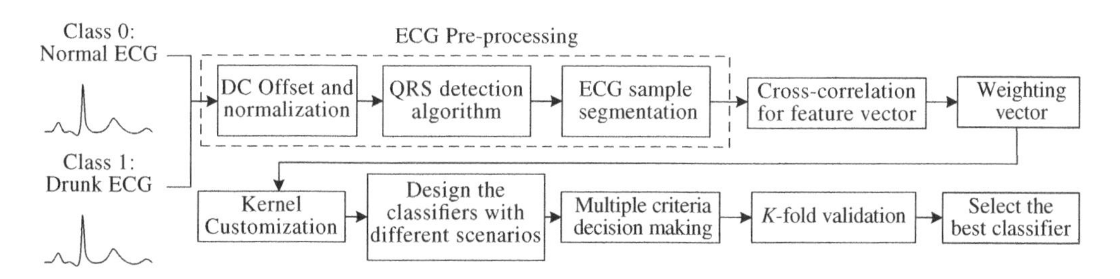

Figure 18.4 Flowchart of proposed ECG-based drunk driver detection scheme.

the ECG DDD scheme. The features are extracted from the segmented ECG samples at Stage 1 and then transformed to high-dimensional feature vector for classification. To improve detection accuracy, independent weighting factors are assigned to all features during building classifiers. The assignment of features is based on multiple criteria decision-making (MCDM) and it is considered as Stage 3. K-fold cross validation is a widely adopted method for training and evaluating the performance of a classifier [48]. At initialization phase, all available data are divided into K sets randomly. At the first fold of validation, $K - 1$ data sets, considered as training sets, are picked up for training classifier, and the remaining data set, considered as validation set, is used to validate a classifier. At the next fold, the prior validation set will become training set and it will not be selected for validating classifier for the rest of the folds. One of the previous training sets is selected to be validation set. The procedure repeats until all K folds are completed. The resultant performance of a classifier is an average result of all folds. Therefore, at the fourth stage, K-fold cross validations are made and the performance of the classifiers are evaluated. The classifier with the best overall performance is selected for the ECG DDD scheme.

The training ECG signals collected from body sensors are usually interrupted with noises and interferences. Therefore, signal preprocessing is necessary to suppress the noises and interferences, and divides the whole ECG signals into multiple ECG samples in one heartbeat duration. As a result, feature extraction can be extracted from ECG samples at Stage 2. It is worth pointing out that ECG contains five main peaks generated by P, Q, R, S, and T waves. Each wave represents an electrical signal transmitted to various heart muscles. In most cases, heart related condition and diseases can be revealed by examining those ECG peaks. So, they are regarded as peak-of-interest. There are three steps of ECG preprocessing as explained below:

Step 1: Bandpass filter

Normally, the largest peak of ECG signal is located within QRS complex. Locating QRS complex facilitates ECG segmentation. Theoretically, QRS complex is usually found from 5 to 15 Hz of ECG frequency spectrum. Based on this ECG characteristic, a low pass filter with high cutoff frequency of 11 Hz and a high pass filter with low cutoff frequency of 5 Hz are cascaded to form the bandpass filter. The frequency components outside the range-of-interest such as muscle noise, cable noise, and wave interferences could be filtered.

Step 2: Derivative filter

An effective way to determine the peaks of interest is to search the turning points and this can be achieved by using differentiation. As such, five-point derivative filter is used to determine the change of slope on ECG signal with short time interval. The change of slope represents turning points in QRS complex and so the locations of Q, R, and S peaks could be determined.

Step 3: Squaring and moving window integration

Change of slope is not sufficient to determine QRS complex as the slope can be varied due to several factors such as heart condition and signal noises. To improve the determination of QRS complex, squaring and moving window integration is used. First, all data points are turned into positive values by squaring. Then, moving window integration is carried out to sort out more parameters, such as the interval between two waves, for determining QRS complex. Finally, multiple QRS complexes can be determined using sorted parameters and the change of slopes.

ECG samples with one heartbeat duration are sorted out using bandpass filter, derivative filter, and moving window integration. By applying different thresholds with respect to the typical values of ECG waves, the amplitudes, durations, and intervals of P, Q, R, S, and T waves can be found. Figure 18.5 shows the pre-processing of ECG signal.

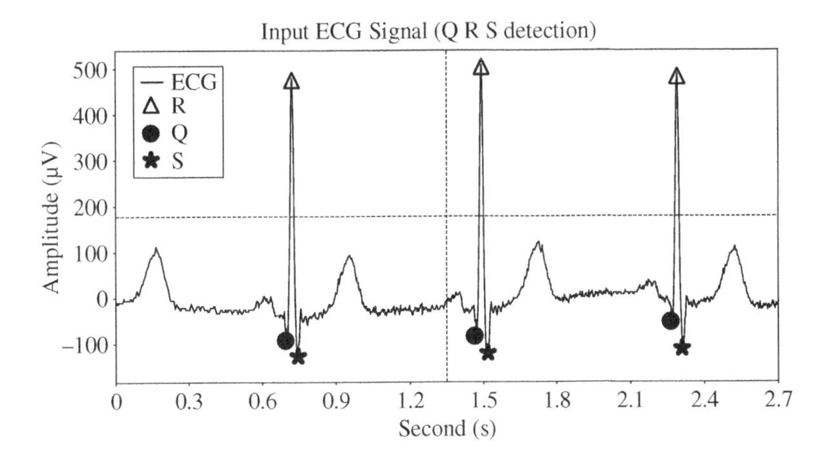

Figure 18.5 Preprocessing of ECG signal.

18.7 Result Comparisons

The performance of the ECG-based DDD classifier was evaluated by using the K-fold validation. First, all ECG samples will be divided into K groups. During a fold, one group will be selected to evaluate the classifier and the rest of the groups will be used to train the classifier. The process will repeat for $K - 1$ times and so all samples will be evaluated. Also, the accuracy of the classifier will be calculated at each fold and the overall accuracy will be obtained by averaging them. Normally, the accuracy can be calculated by taking average of sensitivity and specificity where sensitivity depends on true positive ratio and specificity is determined by true negative ratio [49].

The comparisons between the proposed work and other methods are summarized in Figure 18.6 below.

18.8 Conclusions

In real-life applications, early detection and fully automated detection are the important considerations. For the direct methods, they cannot meet the requirement of real-time protection and fully automated detection as they involve manpower to monitor the data collection process. For vehicle-based method, it cannot provide early detection since it is determined by the changes of the vehicle motions. As such, only bio-signal-based method can fulfil the requirement of DDD since it can achieve simultaneous monitoring on the drivers and so early detection and fully automated detection can be provided.

Acknowledgements

The permission granted in using materials from the following papers is very much appreciated.

Lee, W.C., Hung, F.H., Tsang, K.F. et al. (2015). A speedy cardiovascular diseases classifier using multiple criteria decision analysis. *Sensors, MDPI* 15: 1312–1320.

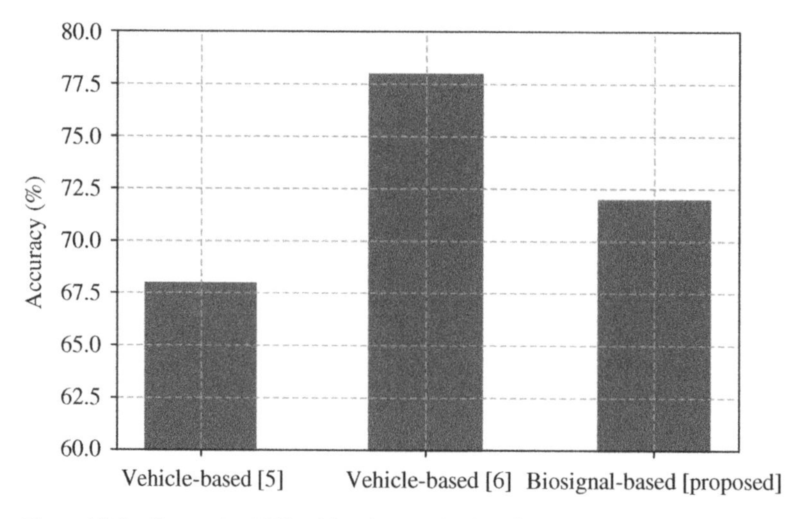

Figure 18.6 Comparing DDD with other methods [50].

References

1 World Health Organization (2018). Global status report on road safety 2018. Geneva: World Health Organization; 2018. Licence: CC BYNC-SA3.0 IGO.

2 Ibero-American Federation of Victims' Associations Against Road Violence (FICVI) (2019). Report on the Decade of Action for Road Safety. Progress and Challenges from the Perspective of Ibero-American Victims' Associations, Ibero-American Federation of Victims'Associations Against Road Violence (FICVI) Fundación MAPFRE, ISBN: 978-84-9844-749-1.

3 Dalal, K., Lin, Z., Gifford, M., and Svanström, L. (2013). Economics of global burden of road traffic injuries and their relationship with health system variables. *International Journal of Preventive Medicine* 4: 1442–1450.

4 Political Declaration of the High-Level Meeting on Improving Global Road Safety (2022).The 2030 horizon for road safety: securing a decade of action and delivery Final draft as of 23 May. https://www.un.org/pga/76/wp-content/uploads/sites/101/2022/05/Final-draft-PD-on-road-safety-23-May-2022.pdf (accessed 21 June 2022).

5 (n.d.). WHO updates fact sheet on Road Traffic Injuries. https://communitymedicine4all.com/2020/02/25/who-updates-fact-sheet-on-road-traffic-injuries-2/ (accessed 22 April 2022).

6 (2021). United Nations Sustainable Transport Conference. Beijing China. https://www.un.org/sites/un2.un.org/files/media_gstc/FACT_SHEET_Road_safety.pdf (accessed 7 February 2022).

7 Zagrebelna, A. (2018). Drinking and driving: a road safety manual for decision-makers and practitioners. https://collaboration.worldbank.org/content/sites/collaboration-for-development/en/groups/eastern-partnership-transport-panel/groups/road-safety/documents.entry.html/2018/10/15/drinking_and_driving-51OJ.html (accessed 21 October 2021).

8 Global Road Safety Partnership (2007). Drinking and driving: a road safety manual for decision-makers and practitioners. Geneva, Global Road Safety Partnership. https://collaboration.worldbank.org/content/usergenerated/asi/cloud/attachments/sites/collaboration-for-development/en/groups/eastern-partnership-transport-panel/groups/road-safety/documents/jcr:content/content/primary/blog/drinking_and_driving-51OJ/Drinking%20Driving_English.pdf (accessed 13 December 2021).

9 UN General Assembly (2015). Transforming our world: the 2030 Agenda for Sustainable Development. (21 October). https://sustainabledevelopment.un.org/content/documents/21252030%20Agenda%20for%20Sustainable%20Development%20web.pdf (accessed 26 July 2021).

10 Washington Traffic Safety Commission (2019). WASHINGTON 2019 TRAFFIC SAFETY ANNUAL REPORT. http://wtsc.wa.gov/wp-content/uploads/dlm_uploads/2014/12/FFY2019AnnualReport-REVISED.pdf (accessed 26 December 2021).

11 NHTSA (n.d.). Drunk Driving, Department of Transportation (US), National Highway Traffic Safety Administration (NHTSA). https://www.nhtsa.gov/risky-driving/drunk-driving (accessed 28 June 2022).

12 Li, Z., Jin, X., and Zhao, X. (2015). Drunk driving detection based on classification of multivariate time series. *Journal of Safety Research* 54: 61–67.

13 Dai, J., Teng, J., Bai, X. et al. (2010). Mobile phone based drunk driving detection. *2010 IEEE 4th International Conference on Pervasive Computing Technologies for Healthcare (PervasiveHealth)*, Munich, Germany (22–25 March 2010), 1–8.

14 Murata, K., Fujita, E., Kojima, S. et al. (2011). Noninvasive biological sensor system for detection of drunk driving. *IEEE Transactions on Information Technology in Biomedicine* 15: 19–25.

15 Lee, Y.S. and Chung, W.Y. (2012). Visual sensor based abnormal event detection with moving shadow removal in home healthcare applications. *Sensors* 12: 573–584.

16 Noh, Y.H. and Jeong, D.U. (2014). Implementation of a data packet generator using pattern matching for wearable ECG monitoring systems. *Sensors* 14: 12623–12639.

17 Liang, W., Zhang, Y., Tan, J., and Li, Y. (2014). A novel approach to ECG classification based upon two layered HMMs in body sensor networks. *Sensors* 14: 5994–6011.

18 Staa, T.-P.V., Gulliford, M., Ng, E.S.-W. et al. (2014). Prediction of cardiovascular risk using Framingham, ASSIGN and QRISK2: how well do they predict individual rather than population risk? *PLoS One* 9: 1–10.

19 Sanz, J.A., Galar, M., Jurio, A. et al. (2014). Medical diagnosis of cardiovascular diseases using an interval-valued fuzzy rule-based classification system. *Applied Soft Computing* 20: 103–111.

20 Tseng, K.-K., He, X., Kung, W.-M. et al. (2014). Wavelet-based watermarking and compression for ECG signals with verification evaluation. *Sensors* 14: 3721–3736.

21 Banerjee, S. and Mitra, M. (2014). Application of cross wavelet transform for ECG pattern analysis and classification. *IEEE Transactions on Instrumentation and Measurement* 63: 326–333.

22 Li, Q., Rajagopalan, C., and Clifford, G.D. (2014). Ventricular fibrillation and tachycardia classification using machine learning approach. *IEEE Transactions on Biomedical Engineering* 61: 1607–1613.

23 Sun, L., Lu, Y., Yang, K., and Li, S. (2012). ECG analysis using multiple instance learning for myocardial infarction detection. *IEEE Transactions on Biomedical Engineering* 59: 3348–3356.

24 Xie, B. and Minn, H. (2012). Real-time sleep apnea detection by classifier combination. *IEEE Transactions on Information Technology in Biomedicine* 16: 469–477.

25 Saaty, T.L. (2008). Decision making with the analytic hierarchy process. *Journal of Service Science* 1 (1): 83–98.

26 Dai, Z., Xu, S., Wu, X. et al. (2022). Knowledge mapping of multicriteria decision analysis in healthcare: a bibliometric analysis. *Frontiers in Public Health* https://doi.org/10.3389/fpubh.2022.895552.

27 Hamid, R.A., Albahri, A.S., Albahri, O.S., and Zaidan, A.A. (2021). Dempster-Shafer theory for classification and hybridised models of multi criteria decision analysis for prioritisation: a telemedicine framework for patients with heart diseases. *Journal of Ambient Intelligence and Humanized Computing* 1–35.

28 Diaby, V., Campbell, K., and Goeree, R. (2013). Multi-criteria decision analysis (MCDA) in healthcare: a bibliometric analysis. *Operations Research for Health Care* 2: 20–24. https://doi.org/10.1016/j.orhc.2013.03.00110.

29 Marsh, K., Goetghebeur, M., Thokala, P., and Baltussen, R. (2017). *Multi-Criteria Decision Analysis to Support Healthcare Decisions*. Cham: Springer https://doi.org/10.1007/978-3-319-47540-0.

30 Angelis, A. and Kanavos, P. (2017). Multiple criteria decision analysis (MCDA) for evaluating new medicines in health technology assessment and beyond: the advance value framework. *Social Science & Medicine* 188: 137–156. https://doi.org/10.1016/j.socscimed.2017.06.024.

31 Moreno-Calderon, A., Tong, T.S., and Thokala, P. (2020). Multi-criteria decision analysis software in healthcare priority setting: a systematic review. *PharmacoEconomics* 38: 269–283. https://doi.org/10.1007/s40273-019-00863-9.

32 Adunlin, G., Diaby, V., and Xiao, H. (2015). Application of multi criteria decision analysis in healthcare: a systematic review and bibliometric analysis. *Health Expectations* 18: 1894–1905. https://doi.org/10.1111/hex.12287.

33 Glaize, A., Duenas, A., DiMartinelly, C., and Fagnot, I. (2019). Healthcare decision-making applications using multi criteria decision analysis: a scoping review. *Journal of Multi-Criteria Decision Analysis* 26: 62–83. https://doi.org/10.1002/mcda.1659.

34 Bousseljot, R.-D. (2004). PTB Diagnostic ECG Database. Version: 1.0.0 (25 September). https://www.physionet.org/content/ptbdb/1.0.0/ (accessed 21 June 2022).
35 Goldberger, A.L., Amaral, L.A.N., Class, L. et al. (2000). PhysioBank, PhysioToolkit, and PhysioNet: components of a new research resource for complex physiologic signals. *Circulation* 101: e215–e220.
36 Vapnik, V.N. (1995). *The Nature of Statistical Learning*. Berlin, Germany: Springer.
37 Kuchar, D.L., Thorburn, C.W., and Sammel, N.L. (1987). Prediction of serious arrhythmic events after myocardial infarction: Signal-averaged electrocardiogram, holter monitoring and radionuclide ventriculography. *Journal of American College of Cardiology* 9: 531–538.
38 Rotman, M. and Triebwasser, J.H. (1975). A clinical and follow-up study of right and left bundle branch block. *Circulation* 51: 477–484.
39 Krowka, M.J., Pairolero, P.C., Trastek, V.F. et al. (1987). Cardiac dysrhythmia following pneumonectomy. Clinical correlates and prognostic significance. *Chest* 91: 490–495.
40 Gottdiener, J.S., Arnold, A.M., Aurigemma, G.P. et al. (2000). Predictors of congestive heart failure in the elderly: the cardiovascular health study. *Journal of American College of Cardiology* 35: 1628–1637.
41 Tompkins, W.J. (2000). *Biomedical Digital Signal Processing C-Language Examples and Laboratory Experiments for the IBMPC*, 236–264. Upper Saddle River, NJ, USA: Prentice Hall.
42 Bishop, C.M. (2006). *Pattern Recognition and Machine Learning*, 325–343. Singapore: Springer.
43 Cortes, C. and Vapnik, V. (1995). *Support-Vector Networks*, Machine Learning, vol. 20, 273–297. Kluwer Academic Publishers.
44 McLachlan, G.J., Do, K.A., and Ambroise, C. (2004). *Analyzing Microarray Gene Expression Data*, Supervised Classification of Tissue Samples, 221–251. New York, NY, USA: John Wiley & Sons.
45 Ozdemir, M.S. (2005). Validity and inconsistency in the analytic hierarchy process. *Applied Mathematics and Computation* 161: 707–720.
46 Lee, W.C., Hung, F.H., Tsang, K.F. et al. (2015). A speedy cardiovascular diseases classifier using multiple criteria decision analysis. *Sensors* 15: 1312–1320.
47 Tompkins, W.J. (2000). *Biomedical Digital Signal Processing C-Language Examples and Laboratory Experiments for the IBM®PC*, 245–264. New Jersey: Prentice Hall.
48 Wu, C.K., Tsang, K.F., Chi, H.R., and Hung, F.H. (2016). A precise drunk driving detection using weighted kernel based on electrocardiogram. *Sensors* 16: 659–667.
49 Zhu, W., Zeng, N., and Wang, N. (2010). Sensitivity, specificity, accuracy, associated confidence interval and ROC analysis with practical SAS® implementations. *NESUG Proceedings: Health Care and Life Sciences* 1–9.
50 Wu, C.K., Tsang, K.F., and Chi, H.R. (n.d.). A wearable drunk detection scheme for healthcare applications. *2016 IEEE 14th International Conference on Industrial Informatics (INDIN)*, Poitiers, France (19–21 July 2016).

19

Bioinformatics and Telemedicine for Healthcare

19.1 Introduction

Bioinformatics is a subject in life sciences that combines biology and information technology study. Its application includes the investigation of molecular sequences and genomics data. Being a combination of different branches of life sciences, the objective of bioinformatics is to develop methodologies and tools to study large volumes of biological data in order to organize, store, systematize, visualize, explain, query, understand, and interpret those data. Bioinformatics utilizes modern computer science that includes cloud computing, statistics, mathematics and even pattern recognition, reconstruction, machine learning, simulation, and molecular modelling. In brief, bioinformatics involves the application of computer technology to manage large volumes of biological information.

The introduction of remote technology for accessing data or helping patient/doctor contact has been a huge step in the healthcare service development. Telemedicine includes remote delivery of healthcare which encompass many areas of engineering including wireless access, remote patient monitoring, distributed databases, and mobile computing systems. Telemedicine aims to bring health services closer to patients. This technology can be applied to several situations for example, it allows the provision of health services at home. Telemedicine attends to different patient needs to provide healthcare services by offering inexpensive medical services in underserved and remote areas [1, 2]. Remote health monitoring is an important issue in telemedicine. Hence, various network technologies and wireless communications have been developed to provide healthcare services [3, 4]. Remote patients outside hospital can utilize telemedicine applications [5]. The severity of remote patients is often addressed by classifying them into different triage levels.

The triage process aims to quickly examines patients who are taken to the hospital to determine the most seriously ill. The integration of technology and decision theory has the potential to help nurses in getting priorities by collecting data on the changing clinical information of patients and analyzing it. This study reported in Claudio et al. [6] investigated the potential to develop a dynamic decision support system (DSS) for patient treatment prioritization. It has the potential benefits of combining technology with decision theory to assist medical workers in prioritizing patients.

Regarding the importance of remote patient classification, it is important to categorize patients' situations and determining the order by which help is needed according to the disease severity and urgency. Automation of this method may effectively improve the provision of quality healthcare services and hospitalization, thereby improving living quality and saving lives. The population

Smart Energy for Transportation and Health in a Smart City, First Edition. Chun Sing Lai, Loi Lei Lai and Qi Hong Lai.

of older persons is increasing both in numbers and as a share of the total. For example, a recent figure shows that the share of the global population aged 65 years or above is projected to rise from 10% in 2022 to 16% in 2050 [7]. The growth of an ageing population and an increase in the frequency of natural disasters can overload the capacity of healthcare systems and the number of already insufficient doctors. Prioritization is essential to provide prompt healthcare services. Turning to the prioritization of remote patients, it is important to handle large amounts of data and deliver patient records quickly and reliably. Improper classification and patient prioritization can lead to incorrect strategic decisions which can endanger patient lives [3, 4].

Digitized healthcare presents numerous opportunities for reducing human errors, improving clinical outcomes, and tracking data over time. Artificial intelligence (AI) methods from machine learning to deep learning assume a crucial function in numerous well-being-related domains, including improving new clinical systems, patient information and records, and treating various illnesses [8, 9]. The AI techniques are also most efficient in identifying the diagnosis of different types of diseases. AI provides new occasions to recuperate patient and clinical groups, and decrease costs. The models used are not only limited to give patients and medical service experts with data creation, recommendations, predication, and sharing; it can also help in recognizing the statistical data related to the population and particular groups or the elements such as climate and odor that form components of the environment in which disease or high-risk behaviors are frequently present.

Traditional pathology approaches have played an integral role in the delivery of diagnosis, semi-quantitative or qualitative assessment of protein expression, and classification of disease. Technological advances and the increased focus on precision medicine have recently paved the way for the development of digital pathology-based approaches for quantitative pathologic assessments, namely whole slide imaging and AI-based solutions, allowing us to explore and extract information beyond human visual perception [10].

Data are from wearables; bio data from in vitro diagnostics (IVD) which are tests done on samples such as blood or tissue that have been taken from the human body. IVD can detect diseases or other conditions, and can be used to monitor a person's overall health to help cure, treat, or prevent diseases. IVD may also be used in precision medicine to identify patients who are likely to benefit from specific treatments. IVD can include next generation sequencing tests, which scan a person's DNA to detect genomic variations. By combining massive medical data with AI, this will generate insights to assist doctors in health alerts, diagnosis, treatment, monitoring, and long-term care. Lifestyle and behavioral modifications may help slow down the biological decline. AI may help generate strategies for longevity to extend lifespan.

Dementia is a chronic disease that span over decades of life. It causes cognitive impairment, affect mood and behavior, leading to loss of ability to live independently. Because of the subtle and insidious changes in memory and other functioning, dementia frequently escapes attention. Early identification of symptoms of dementia may not totally cure the disease, but will help in treating other health conditions that may make the degenerative process worse. Drugs and no drug interventions may also help optimizing the brain's ability to cope with degeneration.

Being healthy is a basic requirement for life with quality. However, it is frequently missed that psychological health is equally important as physical health. Without good psychological health, it is extremely hard to enjoy a state with subjective well-being. Also, ample research suggest that people with chronic physical health problems are at higher risks of having mental illnesses. Depression and anxiety are common in older people. They may aggravate cognitive problems and pose additional risks to the person. For example, depression is associated with poor prognosis after cardiac attacks or strokes. As mental disorders are highly treatable conditions, early recognition will have significant positive impact to quality of life and functioning.

As health is the most important element of living. Various funding sources are available for promoting and advancement of health research, development, and services. In the Industrial Strategy Challenge Fund, these challenges aim to put the United Kingdom at the forefront of AI and data revolution. AI is transforming many industries, including healthcare [11]. The Industrial Strategy Challenge Fund's ageing society challenges aim to support innovation in technologies and services for an ageing population, and improved therapies and treatments to keep them healthy in later life. Therefore, the building of a national resource of volunteers to support the development of new diagnostic tools and technologies would be needed [12].

At present the healthcare systems across the developed world are facing an increasing burden of managing chronic diseases and cancers seen in ageing populations. Many of these conditions are currently only detected at an advanced stage when they are much more difficult to treat effectively. The program aims to combine health and other data in conjunction with AI to accelerate how diagnosis can be made earlier and more accurately. It will support preventative strategies and earlier and more effective treatments to be developed. To achieve this, volunteers will donate biological samples, typically blood for analysis and digital data to support research on early disease detection. The investment aims to improve national health service (NHS)-industry-academic collaboration and innovation; improve risk prediction, early detection, and early intervention; develop a unique research and development environment; and improve data sharing, digital connectivity and access to data; develop products and services that help people remain independent and active into older age [13].

In summary, the challenge is to address businesses, including social enterprises, to develop and deliver products, services and business models that will be adopted at large-scale which support people as they age. This will allow people to remain active, productive, independent, and socially connected across generations. The challenge is funding social, behavioral, and design research, drawing on a wide range of academic disciplines, to provide market insight and evidence that will enable businesses to maximize their commercial opportunities and address key challenges in the field of healthy ageing. The fund supports large-scale whole genome sequencing for precision medicine. It supports a network of digital pathology, radiology, diagnostics, and AI research centers, to increase value of radiology and medical imaging for early diagnostics. It aims to address challenges in the development of digital health innovations and to grow the digital health area. The investment in funding to UK businesses and academia aims to accelerate the use of research and health data, as presently there is insufficient data to train AI applications.

Turning to Asia, there is also funding in Hong Kong. Population ageing is expected to rise and has become a major public health concern in the region. An update on prevalence estimates of dementia and late life mental disorders is required for good healthcare planning. To understand the needs of mental health services of older people, there is Hong Kong Mental Morbidity Survey for older people (HKMMSOP) which is a study funded by the Food and Health Bureau of The Government of Hong Kong SAR. From 2019 to 2022, the research team from the Department of Psychiatry of the Chinese University of Hong Kong invites 6000 older adults randomly recruited from different districts for a free assessment on cognition and psychological health [14].

19.2 Bioinformatics

As mentioned in the Introduction, bioinformatics is an interdisciplinary field that develops methods and software tools to understand biological data. Bioinformatics is an important core of computational biology which includes the development and application of data-analytical and theoretical methods, mathematical modelling and computational simulation techniques to the study of biological, behavioral, and social systems. In short, bioinformatics is about applying statistical and computational techniques to the analysis of biological systems. This is of particular interest in relation to the genome, where a growing number of academics and organizations are interested in the applications of genetic analysis in the fields of clinical care, digital health, and personalized medicine. Bioinformatics is the combination of health information, data and knowledge. It uses computational techniques and tools to analyze the enormous biological databases.

The diseases such as metabolic disorders, urea cycle disorders, inborn errors, and path-aligner can be identified at the early stage using various bioinformatics computational tools. These tools are used to process genetics and proteomics data and compare with healthcare data. Healthcare data consists of physiological data from various muscles or organs obtained by using various electronic components, cost reports, bill claims and surveys related to patient satisfaction. Healthcare analysts are helpful to provide best healthcare at the lowest price which is good for family economic system. This chapter briefly explains what is bioinformatics. The brief information about various kinds of computational tools and databases with respect to the applications are mentioned. The current research trends in the bioinformatics are discussed.

Bioinformatics is the mix of different fields such as, software engineering, computer science, statistics, informatics, and engineering which assess and outline biological and genomic information. Since the completion of the Human Genome Project in 2003, a number of projects have sought to sequence the genomes of large numbers of individuals in order to better understand the variation in genetic information across populations, and how it is linked to the different characteristics of individuals. For example, in 2012 the UK government announced the 100 000 Genomes project, administered by Genomics England, which aimed to sequence the genomes of 100 000 NHS patients. The project focussed in particular on individuals with rare diseases or conditions, and was completed in 2018.

Although around 99.9% of human DNA is identical between individuals, analyzing the possible interactions between variations in the remaining 0.1% still represents a difficult problem. Analyzing these data is the focus of the rapidly growing field of genomics, a subfield of bioinformatics specifically concerned with the analysis of whole genomes. Falling costs of genome sequencing, and huge increases in computing power over recent years, have made it possible to analyze genetic data on a scale not previously foreseen. Sophisticated statistical tools are used to analyze correlations between variations in the DNA of an individual, and their physical characteristics. The results can provide powerful insights into how these variations affect risk of disease or response to particular stimuli or drugs.

These insights are of particular interest to pharmaceutical industry and healthcare providers. The response of an individual to a particular drug or treatment is often difficult to predict, and a large amount of time and money is wasted in finding which treatment option is most effective for a given individual. Genomics has the potential to much more accurately predict an individual's response based on their genetics. This could revolutionize care by allowing health service providers to give patients highly personalized and more effective treatment. And there is plenty of work going on in this field right here in the United Kingdom. For instance, the world's largest health-focused big data institute opened in Oxford a few years back. This Big Data Institute (BDI), brings together researchers in genomics, statistics, health data, and other related fields. The BDI is already working in partnership with Novartis, for example, to bring forward various genomics-related programs.

A hard problem in the genomics field, though, is the accurate and reliable identification of correlations between genetics and physical characteristics. The interactions between different regions of the genome are complex, and the effects of any individual genetic variation may be small – this means you may only detect a physical difference by looking across a range of small genetic variations. This type of problem lends itself to sophisticated statistical algorithms, and computational analysis. A lot of work is being invested in developing new genomic algorithms and techniques. For protection of the tools and algorithms that companies developed, it is very important to allow them to commercialize their ideas, encourage investment, and continue developing new and improved techniques. There is a common misconception that this type of invention, typically implemented in software, is not eligible for patent protection. In fact, the European Patent Office, and other offices around the world, routinely grant applications for new technical software inventions. These can provide valuable protection for companies working in this sector [15].

For the applications of bioinformatics in medicine, bioinformatics has proven useful as the complete sequencing of the human genome has helped to unlock the genetic contribution for many diseases. Its applications include drug discovery, personalized medicine, preventative medicine, and gene therapy. Regarding personalized medicine, it is a model of healthcare that is tailor-made to each person's unique genetic make-up to analyze data from genome sequencing or microarray gene expression analysis in search of mutations or gene variants that could affect the response of a patient to a particular drug or modify the disease prognosis. The genetic profile of a patent can assist the doctor to predict the lack of ability to resist to certain diseases, provide proper medication and with the proper dose to reduce side-effects. It is applied in the treatment of personalized cancer medicine, diabetes-related disease, and HIV.

Turning to preventive medicine, it focuses on the health of individuals, communities, and defined populations. It uses various research methods, including biostatistics, and epidemiology, to understand the patterns and the causes of health and disease, and to transform such information into programs designed to prevent disease, disability, and death. An example of preventive medicine is the screening of newborns immediately after birth for health disorders, such genetic diseases or metabolic disorders, that are treatable but not clinically evident in the newborn period. To develop such screening tests to identify the disease at an early stage, researchers use bioinformatic tools to analyze genomics, proteomics, and metabolomics data for possible disease biomarkers.

Bioinformatics aims to study and analyze the large biological databases to understand the disease and such predictions are useful to relate with the healthcare data [16]. Suresh et al. [17] delivers various bioinformatics computational methods to offer solutions for a more personalized approach to healthcare. Applying deep learning techniques for data-driven solutions in health information allows automated analysis in supporting the problems arising from health-related matters. Raza and Dey [18] gives an overview of the main principles of bioinformatics, health informatics, and real-case applications of translational bioinformatics in healthcare. Bioinformatics computational tools are used to process genetics data and compare with healthcare data which consist of physiological data from various muscles or organs [19]. Shi et al. [20] gave an explanation on the importance of AI in gene selection tools to classify the microarray cancer data. Serra et al. [21] provided a comprehensive review on machine learning algorithms used in bioinformatics for dimensionality reduction of large-scale and complex data, feature selection for assigning biomarker includes things from blood pressure and heart rate to basic metabolic studies and X-ray findings to complex histologic and genetic tests of blood and other tissues in raw data usage. Single nucleotide polymorphism is high dimension and can be characterized by different temporal resolutions with analysis and classification using machine learning. Drug repositioning and classification of patient from neuroimaging data are some applications of AI in bioinformatics [21].

Gene data plus medical data and AI could improve the accuracy of diagnosis and is an innovative digital solution that changed the diagnosis and treatment paradigm. AI techniques such as machine learning to deep learning are applied to healthcare for disease diagnosis, and patient risk identification. Numerous medical data sources such as ultrasound, magnetic resonance imaging, mammography, genomics, and computed tomography scan are required to better diagnose diseases with the techniques. Furthermore, AI can help patients to continue their rehabilitation at home, that is, care that can help the patient get back, or improve abilities needed for daily life. These abilities may be physical, mental, and/or cognitive, for example, thinking and learning [22].

19.3 Top-Level Design for Integration of Bioinformatics to Smart Health

Smart meters not only measure energy utilization, but also provide a communication gateway function that gives information required by the utility from the network gateway to the consumer appliances. In addition to this, a data concentrator could be used over an Automated Metering Infrastructure (AMI) wireless network and communicated to the utility over a wide area network (WAN) [23]. Table 19.1 shows the number of electricity, gas, and water smart meters worldwide from 2014 to 2020 [24]. A typical challenge for setting up smart meters is how to include them within the infrastructure and create custom-tailored smart metering use cases. One approach to realize these goals is to use an IoT platform that provides out-of-the-box solutions for smart metering as reported in [25].

Regarding the impact of smart meters to health, Public Health England has conducted an extensive study to determine exposures from smart meters as the technology deployed in the United Kingdom. It concludes that exposure to radio waves from smart meters is well below the guidelines given by the International Commission on Non-Ionizing Radiation Protection. The results were published in Bioelectromagnetics [26]. The research conclusion also suggested that exposure to the radio waves produced by smart meters is likely to be much lower than that from other everyday devices such as smart phones. For a quantitative comparison, the readers could refer to Fong et al. [27].

Paxman et al. [28] identified several barriers to adopting smart meter data at a large-scale level. There is the unavailability of funding and a lack of collaboration between computer science, governments, engineering, energy and healthcare – therefore studies remain only theoretical, with little evidence of real-world validity and demonstration projects. The penetration of smart meters remains slow in the majority of countries. For example, in the United Kingdom, despite rapid

Table 19.1 Number of electricity, gas, and water smart meters between 2014 and 2020.

Year	Number in millions
2014	575
2015	730
2016	840
2017	1000
2018	1190
2019	1400
2020	1645

adaptation, smart meters remain the lesser-used metering technology by households due to connectivity issues. In addition, there are other challenges identified for using smart meters. First, the distributed nature of smart metering systems makes their implementation costly. Second, the installation of smart meters raises data privacy concerns. Third, compliance with industry protocols and guidelines to ensure various standards is needed. Fourth, smart meters communicate with mobile technology and a weak signal can disrupt communication. However, it is believed that opportunities are more than challenges, and the use of smart meters in telecare should be a reality. More financial support is needed to carry out wider clinical trials to formulate policies and business models.

Based on nonintrusive load monitoring, it can identify the type of appliances using the aggregated load profile measured with a smart meter. The load disaggregation can be solved with machine learning. One possible implementation is to get all the aggregated load patterns from the smart meter transmitted with wireless communications technology to a data center or cloud and then analyze them with an intelligent algorithm to obtain various individual loads of each appliance by a service provider. By comparing historical information with many well-developed techniques such as deep neural networks, it can make recommendations with a DSS. In summary, a top-level logic flowchart of the smart health based on a smart meter system is summarized in Figures 19.1 and 19.2 gives the component and module diagram for the system.

This system could promote bioinformatics which is the integration of different subjects such as data science, biology, and statistics to capture and analyze biological information. Combined with the reducing cost of high-throughput DNA and RNA- sequencing, bioinformatics has made genomic and proteomic data analysis more efficient and effective, but has also created new challenges concerning data storage and security.

Bioinformatics presents a powerful tool to model and monitor biomedical and healthcare data with massive potential to improve the understanding of how diseases develop and/or spread. For example, the probability of developing metabolic disorders increases with certain genetic mutations, and these can be identified by comparing individual- to population-level genetic data using

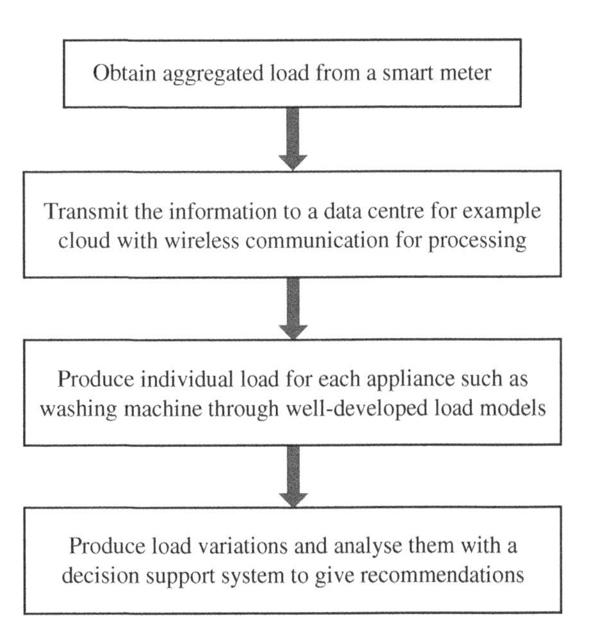

Figure 19.1 A top-level smart health with smart meter system design [29].

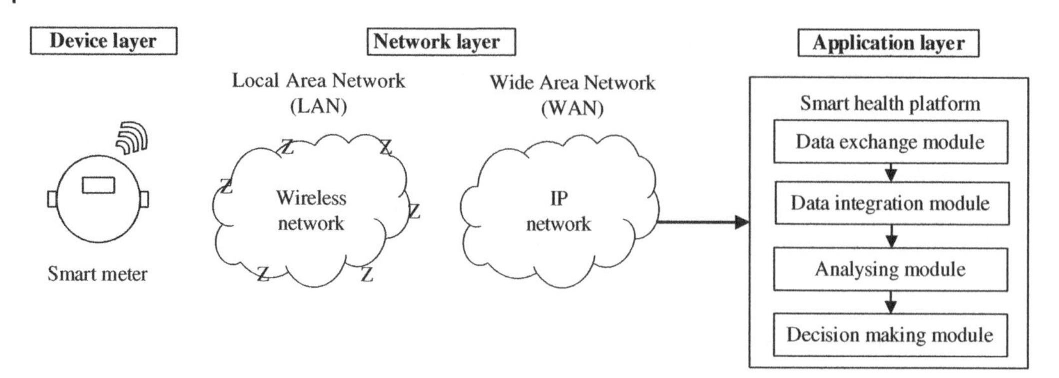

Figure 19.2 A component and module diagram for smart health with a smart meter system. Wireless network - NB-IoT, Lora, etc.; Internet Protocol (IP) Network – gateway, data concentrator, etc. Data exchange module consists of historical information database, weather information, energy usage data. Data integration module is for data checking, data prepossessing. Analyzing module produces energy usage patterns with AI methods such as deep learning. Decision-making module produces a list of recommendations based on risk assessment techniques for healthcare workers and doctors for action [29].

bioinformatics tools. Combined with the use of electronic health records, i.e. patient history and physiological data, such as electroencephalogram and electrocardiogram obtained from various electronic components, a personalized treatment plan can be derived.

The present proposed system integrates the use of smart meter data and could be used to develop predictive models to give more accurate, relevant, and timely health warnings to patients and their care providers. Integrating smart meter data with existing translational bioinformatics resources will accelerate the implementation of data-driven medicine in the clinical setting, and will play an imperative role in a smart city environment to improve disease diagnosis, prevention, and intervention.

A deep learning approach to integrate various datasets is proposed to extract complementary knowledge from multiple domains of sources [30]. By integrating multiple forms of numerical data including time series and non-time series data to extract supporting features from the multimodal data, clinical, and multimodal imaging data can be used to predict disease progression. Figure 19.3 shows the schematic diagram for neural network training using multimodal data to predict health status of the individual with the feature importance and selection approach to infer causal relationships and recommendations.

19.4 Artificial Intelligence Roadmap

AI transforms research, development, and innovation. First, it is necessary to continue to strengthen basic research on AI itself. Researchers working across disciplines need to address some fundamental challenges that remain, such as establishing a common language to develop interoperability between data science projects or datasets, driving intelligent collaboration. AI systems often rely on a narrow disciplinary base that needs to be expanded to cover a range of fields in a smart way.

The second is that AI technology needs to be better promoted to fields without structured data, and effectively play a guiding role. AI is transforming research fields such as bioinformatics. The

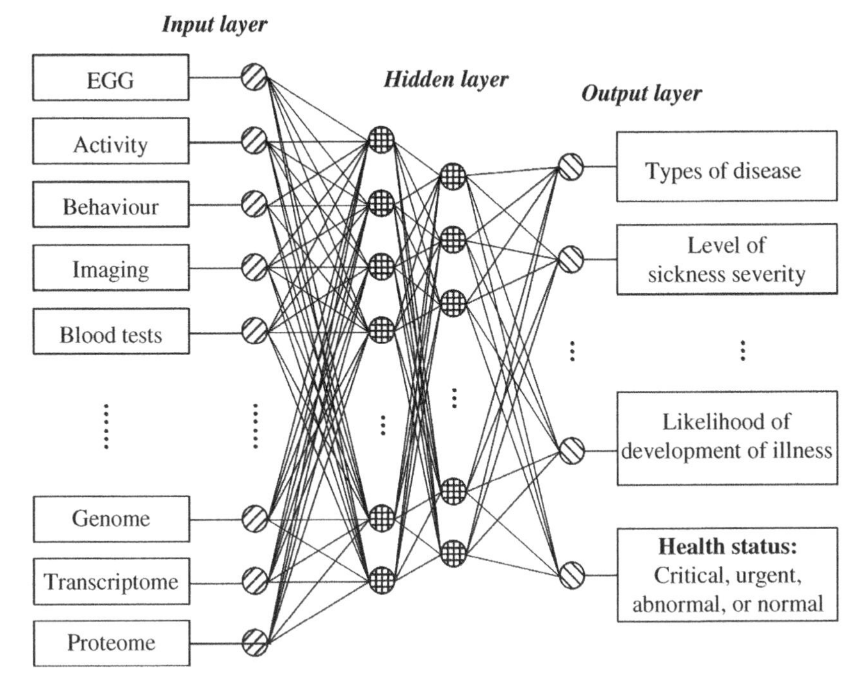

Figure 19.3 A deep learning approach in decision-making for health status.

potentially disruptive impact of AI and other data sciences has begun to be incorporated into all forms of knowledge creation, including the social sciences and humanities. Given the availability of massive datasets and computer power, this disruption will not only transform the current inefficient routine tasks of searching text or large observational datasets, but will also cross disciplinary boundaries, uncover new patterns, and raise new questions and insights.

Despite all the possible benefits, there are challenges from skills and diversity. As such, the AI roadmap should ensure the provision of first-class postgraduate courses with global appeal and to safeguard a prestigious research base. To value the diversity of people working with AI, the AI community needs to reflect diverse populations. Benchmarking and forensic tracking of levels of diversity and inclusion in AI is recommended, data-led decisions on where to invest, and ensuring that underrepresented groups should be included in any funding programs.

For AI education, students leaving school should gain basic AI knowledge, not just to understand the basics or mathematical concepts of programming or quantitative reasoning, and ethics; but to be a conscious and confident user of AI-related products. They should know what questions to ask, what risks to look for, what societal implications might arise, and what opportunities AI might offer. Comprehensive AI curriculum resources must be available to provide incentives and opportunities for teachers to learn and develop AI knowledge. A lifelong learning habit must be developed and encouraged.

For data, infrastructure, and public trust, consolidating and accelerating the required infrastructure to increase the data accessible to AI technologies are important. There are requirements to invest in organizations that connect general principles to specific applications and take steps to inspire innovation and enable valuable and secure data sharing. For example, the United Kingdom wants to be the best place in the world to access and use secure and high-quality data, to develop new applications, and business models; to be the best place for the public to scrutinize and input automated decision making, to ensure the public can trust AI. The infrastructure required for this is not only physical facilities such as broadband networks and high-performance computing capabilities, but also virtual and social infrastructures such as procedures, people, standards, governance, good practice, recommendations, and practices.

AI thrives on data and places new demands on data collection, management, and use. Governments should focus on developing plans to provide more public sector data safely and securely. For the private sector, in addition to reviewing compliance with personal data protection, regulators should help companies deploy appropriate privacy enhancements and create conditions for data to be used for AI. Translating the idea of a data-sharing agreement into a regulatable effort should be accelerated to establish a legal framework around different data-sharing structures such as data trusts, cooperatives, and contracts. Access to data is fundamental to the development of AI, to develop or revise appropriate standards such as data interoperability and shaping the future data governance framework are at a high priority.

The AI roadmap must also include the steps to drive the adoption of AI technologies in key sectors such as healthcare, in both the commercial and public sector. The COVID-19 pandemic has accelerated the adoption of digital systems such as e-commerce platforms and marketplaces, as well as tools to support remote work. The government should ensure that civil servants have access to information on all AI-based tools and projects being conducted across sectors. AI should be used in outcome-based projects, and only when absolutely appropriate, and departments should quickly learn from each other what AI is working and what needs to be developed. The focus should then be on linking public sector data with private sector data, with more sophisticated adoption of AI through various datasets. Governments should fund exploratory projects across industries, sectors, and disciplines as a means of demonstrating the potential of AI to enhance public service capabilities.

Improvements in disease interventions through prioritization and personalization based on large amounts of data can maximize benefits in the health sector. The UK's NHS is at the forefront of adopting AI technologies. The United Kingdom has three particular areas of strength: first, to come up with a health data strategy to obtain cleansed, coded, and real-time data through a strong industry data governance framework; second, to build partnerships for small- and medium-sized enterprises and the NHS and new models of incentives; third, AI has the potential to conduct large-scale national trials more efficiently, building on the UK's world-leading life sciences industry and the increasingly mature digitalization of the UK's National Health Service, making the UK ideal for biomedical research.

In order to ensure an advantage in the new round of scientific and technological revolution and industrial transformation, many developed countries have regarded AI as the focus of technology and industrial layout, from strengthening national strategy formulation, establishing national research institutions, organizing, and implementing major projects, and strengthening personnel training. and the introduction of other aspects to actively promote the development of AI. It is necessary to make long-term key investments and increase talent training in order to take the initiative in the long-term AI competition.

The roadmap must involve strengthening the construction of public service platforms such as databases and standards. Currently, in terms of databases and standards, companies are mainly based on decentralized construction. The existing problems are: first, there is a hidden danger of data monopoly, which is not good to the growth of small and medium-sized enterprises and the healthy development of the overall industry; second, the massive data formed by a large number of public institutions has not been fully exploited and utilized by the society. Based on the practice of the NHS in the United Kingdom to organize and open electronic medical record data, support companies to develop novel diagnostic methods and treatments; advancing and accelerating the application of AI technology in the medical field is an essential and positive direction. The recent formation of the Global Partnership on Artificial Intelligence aims to guide the responsible and human-centric development and use of AI. It is clear that to maximize the benefit from AI, stronger cooperation globally would be needed.

The integration of AI and pathology can reduce the shortage of pathologists. Pathology diagnostics is hailed as the gold standard of diagnosis by the healthcare community. Pathology field is facing a huge deficit in supply of pathologists. Current pathology reports' timeline and accuracy do not meet the oversized demands from patients. Integrating AI technology to improve the efficiency pathology diagnostics is the present direction. The use of AI in pathology is gaining momentum. This technology allows pathologists and radiologists to process images and slides more quickly and accurately, which subsequently leads to higher volume and faster turnaround time for both private practice groups and hospitals. The AI automatically classifies cells into different cell groups and identifies patterns and cell conditions using deep learning to diagnose the patients. The developers are training AI programs with image analysis algorithms to diagnose problems using deep learning and big data from healthcare industry. AI is expected to change pathology with faster yet accurate diagnosis and quality control and reduction in human error.

Machine learning (ML) and deep learning (DL) are being utilized for brain tumor detection, cervical cancer detection, breast cancer detection, COVID-19 detection, physical activity recognition, thermal sensation detection, and cognitive health assessment of dementia individuals. AI recognition products used in cervical cancer cell screening have successfully been implemented [31]. Khamparia et al. [32] reported the use of deep convolutional and variational autoencoder network for cervical cancer prediction and classification. Abbas et al. [33] applied whale optimization based efficient features and extremely randomized tree algorithm to breast cancer detection. Iwendi et al. [34] described the use of boosted random forest algorithm for COVID-19 patient health prediction. Rehman et al. [35] accounted a personalized thermal comfort model to predict thermal sensation votes for smart building residents. Javed et al. [36] explored a smartphone sensors-based personalized human activity recognition system for sustainable smart cities. Aslam et al. [37] reported Blockchain and ANFIS empowered IoMT application for privacy preserved contact tracing in COVID-19 pandemic. Iwendi et al. [38] demonstrated the classification of COVID-19 individuals using adaptive neuro-fuzzy inference system and Tripathy et al. [39] used classification, regression model for predicting various diseases.

Although there are many opportunities from AI deployment, there are challenges as well to AI healthcare start-ups and the major reasons are technology commercialization occurred far slower than originally planned. Cash was also spent faster than could be fundraised. In addition to this, most hospitals were unwilling to pay. There is slow product development too, for example, a combination of slow to respond to local market such as the US-based developers are unable to catch up with Chinese clinicians' demand and cross-cultural team issues.

Regarding bottlenecks with AI and healthcare, there are at last two of them. The first one is that the existing data is not sufficient for AI model training due to that fact that data to train AI needs to be large sets; structured; well labelled, and closed loop. The second bottleneck is that healthcare value chain is not ready for AI implementation. Healthcare industry directly associates with human lives therefore requiring more prudence in the design of AI applications considering ethics and regulations. Healthcare industry's values chain is long, with complex stakeholders including buyers, distributors, decisions makers, users and consumers. Therefore, AI application providers have to respect and confirm with existing industry practices and standards.

19.5 Intelligence Techniques for Data Analysis Examples

With different types of data, it is required to have advanced intelligence techniques for data analysis. Currently, many techniques are available for data analysis applications. A possible framework is given in this section.

Due to the limitation of a stand-alone machine learning method, the hybrid model combining multiple machine learning methods was conducted to improve forecasting accuracy [40]. The Multi-view Deep Forecasting (MvDF) framework is proposed for forecasting the behavior, activities of patients. A forecasting error correction model for MvDF, the Radial Basis Function Neural Network (RBFNN) trained via minimization of Localized Generalization Error (L-GEM), is proposed to compensate the forecasting error of the MvDF. The MvDF and the RBFNN constitute the Multi-view Deep Forecasting method with error correction is shown in Figure 19.4.

Three representative deep neural networks for RBFNN prediction are introduced and discussed which should generate different representations from the same input sequences. Then, the proposed MvDF is explained and the robust RBFNN for error correction of MvDF is illustrated below as well.

For multi-view construction, three types of deep learning models for time series prediction are adopted to provide three different views [41]. They are Temporal Convolutional Neural Network (TCN), Bi-directional Long Short-Term Memory Neural Network with Temporal Attention (BLSTMattn), and Convolutional Gated Recurrent Unit Neural Network (C_GRU). All these three deep neural networks capture the temporal features in different ways. Each layer of TCN learns the short-term relationship of the input sequences. The long-term useful temporal representations for patent's behavior and activities forecasting could be established by layer-by-layer stacking the convolutional layers. BLSTMattn learns the context information through BLSTM and focuses on key temporal information for time step activities forecasting by the temporal attention mechanism. C_GRU utilizes CNN as the filter for extracting higher-level local temporal patterns which are fed into GRU to learn the representations for time step activities forecasting.

However, these three models have their own limitations. As for TCN, in the case of its local kernel and layer-by-layer stacking characteristics, the information of those closer to the predicted time may not be well focused while this information is important for time step activities forecasting. On the contrary, BLSTMattn can more easily remember the recent information of the input

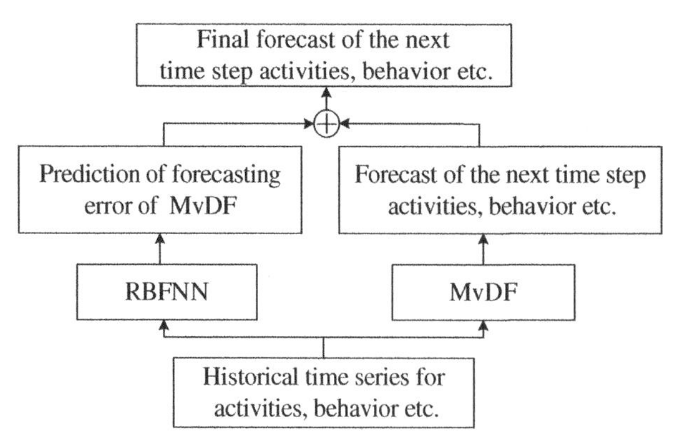

Figure 19.4 Multi-view deep forecasting (MvDF) framework.

sequences, but it cannot solve the vanishing gradients problem completely. Although the hybrid model C_GRU can solve the drawbacks of the two above mentioned models to some extent, it easily gets affected by different random weight initializations due to the more complex model architecture and hence behaves in an unstable way. Therefore, the multi-view learning framework could be used to utilize the knowledge from these three models (views) to establish a more effective model for time step activities forecasting. Brief introductions of these three deep neural networks are given as follows:

TCN adopts the 1D dilated causal convolution in each layer to capture the local temporal patterns. The 1D dilated causal convolution is implemented with the left zero padding such that the output of each layer at time t is convolved only with features from time t and earlier in the previous layer. Besides, it supports an exponential expansion of the receptive field without loss of coverage, which enables a longer time dependency to be captured. Suppose x^{l-1} is an input of the l^{th}convolutional network layer consisting of multiple dilated causal convolutions, and the convolutional kernel of lth layer is $C = \{f_1, f_2 ... f_w\}$, then the output of the lth convolutional layer at time step t is defined as:

$$C(t) = \sum_{i=1}^{k} f_i x_{t-d(k-i)}^{l-1} \tag{19.1}$$

where d is the dilation factor and k is the kernel size. By stacking the convolutional layers, more past information can be fused in the deeper layer and thus more high-level temporal features can be learned. In addition, the identity mapping is also used in TCN to stabilize a deeper and larger network [42]. A residual block illustration is given in Figure 19.5 [43]:

BLSTMattn applies the Bi-directional Long Short-term Neural Network to capture the temporal relation of the input sequences. In addition, the temporal attention mechanism is used in BLSTMattn to enhance the importance of different hidden states at different time. The overall architecture of BLSTMattn is shown in Figure 19.6.

BLSTM consists of two unidirectional LSTMs with opposite directions so that it can capture the context information. The LSTM consists of four basic components, that is, the input gate i_t, the forget gate f_t, the output gate o_t, and the cell state c_t. The forget gate f_t takes current input x_t and previous hidden state h_{t-1} as inputs to determine how much information of the previous cell state c_{t-1}

Figure 19.5 A residual block of TCN with the black lines as kernels and the blue line for the identity mapping.

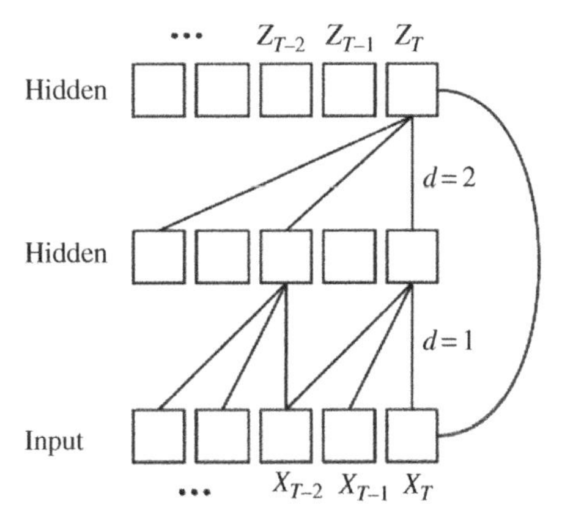

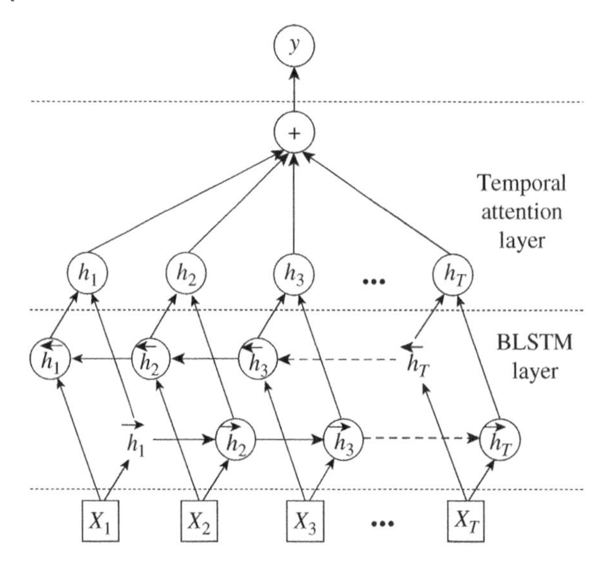

Figure 19.6 BLSTMattn architecture [43].

will be forgotten using the sigmoid function $\sigma(\cdot)$. x_t and h_{t-1} are also used to calculate a new candidate cell state $\tilde{c}_t$. The input gate i_t determines which information of $\tilde{c}_t$ should be updated into the stored cell state c_t. The output gate o_t controls how much information c_t is given and thus results in the hidden state h_t as given by the following equations:

$$f_t = \sigma\left(W_f \cdot [h_{t-1},\ x_t] + b_f\right) \tag{19.2}$$

$$\tilde{c}_t = \tanh\left(W_{\tilde{c}} \cdot [h_{t-1},\ x_t] + b_{\tilde{c}}\right) \tag{19.3}$$

$$i_t = \sigma(W_i \cdot [h_{t-1},\ x_t] + b_i) \tag{19.4}$$

$$c_t = f_t * c_{t-1} + i_t * \tilde{c}_t \tag{19.5}$$

$$o_t = \sigma(W_o \cdot [h_{t-1},\ x_t] + b_o) \tag{19.6}$$

$$h_t = o_t * \tanh(c_t) \tag{19.7}$$

where, denotes the dot product and $*$ denotes the element-wise product. W_j and $b_j (\forall j \in \{i, f, o, \tilde{c}\})$ are learnable parameters.

BLSTM concatenates two hidden states of opposite directions at the same time t as the final hidden state:

$$h_t = \left[\overrightarrow{h}_t, \overleftarrow{h}_t\right] \tag{19.8}$$

where $\overrightarrow{h}_t$ and $\overleftarrow{h}_t$ denote the tth hidden states of two LSTMs with forward and past-ward direction, respectively.

The temporal attention mechanism is applied to the hidden states to denote the importance of each h_t by the corresponding attention weight α_t. The weighted sum of all hidden states is denoted as r_H which could be used for time step activities:

$$u_t = \tanh(h_t) \tag{19.9}$$

$$\alpha_t = \frac{\exp(u_t \cdot u_w)}{\sum_t \exp(u_t \cdot u_w)} \tag{19.10}$$

$$r_H = \sum_t \alpha_t * h_t \tag{19.11}$$

where u_w is a learnable parameter vector.

For C_GRU, as shown in Figure 19.7, it adopts convolutional layers to learn the local temporal features of the input sequences while the gated recurrent unit (GRU) neural network uses the local temporal patterns to learn the long-term dependencies [44].

GRU consists of the reset gate, update gate, and the hidden state. The reset gate r_t is used to decide whether to ignore the previous hidden state h_{t-1}. The update gate z_t is used to determine how much memories should be updated by a new candidate hidden state $\tilde{h}_t$, as given in the following equations:

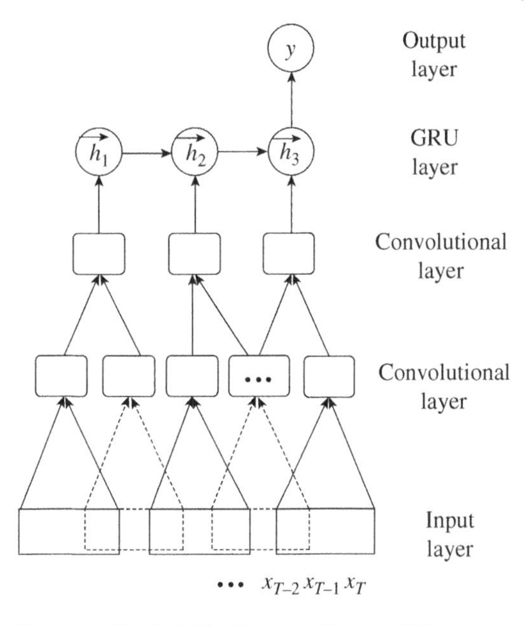

Figure 19.7 C_GRU schematic diagram [43].

$$r_t = \sigma(W_r \cdot [h_{t-1}, x_t] + b_r) \tag{19.12}$$

$$\tilde{h}_t = \tanh\left(W_{\tilde{h}} \cdot [r_t{}^* h_{t-1}, x_t] + b_{\tilde{h}}\right) \tag{19.13}$$

$$z_t = \sigma(W_z \cdot [h_{t-1}, x_t] + b_z) \tag{19.14}$$

$$h_t = (1 - z_t) * h_{t-1} + z_t{}^* \tilde{h}_t \tag{19.15}$$

Considering a single deep neural network has its own characteristic and limitation to capture the temporal information of the input sequences, three deep neural networks (TCN, BLSTMattn, and C_GRU) could be adopted to consider the temporal characteristics of the same input sequences in different perspectives to describe the data more comprehensively and accurately. As such MvDF is proposed for data analysis. The multi-view learning paradigm is utilized to make full use of the information provided by each deep neural network. The success of multi-view learning strongly depends on complementary and consensus principles.

The complementary principle states that utilizing different representations of the data can describe the data more accurately since each representation may contain some information that others do not know. Three different deep neural networks could forecast the next time step activities by using extracted representations of their own. The view attention is utilized to combine the forecasting outputs of three deep neural networks appropriately under the complementary principle. Let O denote the matrix consisting of outputs of three networks. That is $O=[o_1,o_2,...o_V]$ and the number of views $V=3$. Let $W_v = [w_1,w_2,...,w_V]$ where w_v $(v=1,2,...,V)$ denotes the weight of the output of the vth deep neural network. The following equations show how to calculate the view weight matrix W_v:

$$M = \tanh(W_w \cdot O) \tag{19.16}$$

$$W_v = softmax(u_v \cdot M) \quad (19.17)$$

where W_w and u_v are learnable parameters. Therefore, the complementary loss is established as shown in the following:

$$complementary\ loss = \sum_{n=1}^{N} \left\| \left(\sum_{v=1}^{V} w_v f_v(s_n) \right) - y_n \right\|_2^2 \quad (19.18)$$

where $f_v(\cdot)$, s_n, y_n, and N represent the vth deep neural network, the nth sample, the label of the corresponding n^{th} sample, and the total number of samples, respectively.

The consensus principle aims at minimizing the disagreement on multiple distinct views. Dasgupta et al. [45] reported that the probability of the disagreement of two independent hypotheses f_1 and f_2 on unseen samples is the upper bound of the error rate of each hypothesis, as shown in the following:

$$P(f_1 \neq f_2) \geq \max\left(P_{error}(f_1), P_{error}(f_2)\right) \quad (19.19)$$

Thus, by minimizing the disagreement probability of two hypotheses on unseen samples, the forecasting accuracy of each hypothesis can be improved. The consensus loss function could also be established to minimize the disagreement of three deep neural networks' forecasting outputs on the perturbed input samples $\tilde{s}_n$ which are generated by adding small Gaussian noises to the original samples s_n:

$$consensus\ loss = \sum_{n=1}^{N} \sum_{i=1, i<j}^{V} \left\| f_i(\tilde{s}_n) - f_j(\tilde{s}_n) \right\|_2^2 \quad (19.20)$$

Therefore, the final objective function for training the proposed MvDF model is given by minimizing both the complementary loss and consensus loss:

$$\min \sum_{n=1}^{N} \left\| \left(\sum_{v=1}^{V} w_v f_v(s_n) \right) - y_n \right\|_2^2 + \lambda \sum_{n=1}^{N} \sum_{i=1, i<j}^{V} \left\| f_i(\tilde{s}_n) - f_j(\tilde{s}_n) \right\|_2^2 \quad (19.21)$$

where λ is the hyper-parameter to balance the complementary and consensus principles.

It should be noted that MvDF could not be trained from scratch. This is because a deep neural network with randomly initialized weights may fail to accomplish the difficult next time step activities forecasting task and thus each network cannot provide useful information to each other to improve the forecasting accuracy. Therefore, each deep neural network is pre-trained with the labeled data first, and as shown in Figure 19.8, they are jointly fine-tuned according to Eq. (19.21). The blue color indicates the data flow for calculating the complementary loss while the red color indicates the data flow for calculating the consensus loss. The circles represent the outputs of the deep networks. The output of the View Attention is the view weight matrix W_v.

The purpose of RBFNN training is to find a network structure and connection weight to minimize the generalization error. In fact, once the number of hidden neurons is determined, centers and widths of hidden neurons can be obtained by K-means clustering [46]. After fixing both centers and widths, connection weights can be calculated by a pseudo-inverse technique. Therefore, the

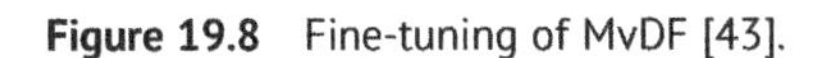

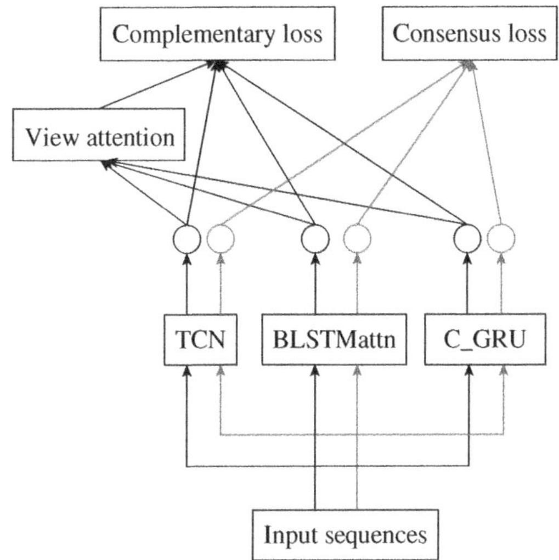

Figure 19.8 Fine-tuning of MvDF [43].

objective of RBFNN training can be simplified to the finding of the optimal number of hidden neurons which minimizes the generalization error.

Since the proposed MvDF may still suffer from systematic errors in prediction [47], RBFNN trained via minimizing the Localized Generalization Error (L-GEM) is used to further enhance the next time step activities forecasting accuracy. More specifically, RBFNN takes the same input as MvDF but tries to predict the next time step activities forecasting error generated from MvDF rather than the next time step activities forecasting.

19.6 Decision Support System

Many real-life decision making cases are subject to bounded rationality in which the technical and economic evaluation of all solution choices is restricted by the consideration of dominant subjective constraints. Traditionally, DSS was introduced as a system that intended to support decision makers in semi-structured problems that could not be fully assisted by algorithms. DSSs could act as an aid for planners to expand their capabilities for searching a better solution. DSSs could provide the means to complement decision makers by quantitatively supporting decisions that could otherwise be based on personal skill, ideas, and experience. In addition to the traditional DSS characteristics, i.e. data and models, the inclusion of an intelligent knowledge base would be required to quantify the impacts of both hard such as technical, economical and soft such as subjective constraints.

As a matter of fact, nowadays, decision-making is more complicated than it was in the past for two main reasons. Firstly, growing technology and communication systems have introduced a greater number of feasible solution alternatives from which a decision maker can select. Secondly, the increased level of problem complexity can result in a chain reaction that leads to an unacceptable cost if a poor solution is chosen.

DSS is an information system, which is based on interactive computer to support decision-making in planning, management, operations for industry, business, and organizations. The rapid changes on decision-making, which cannot be easily identified, can be assisted by communication technologies and computer-based system compiling information gathered from a wide range of resources like raw data, documents, experience and knowledge of experts, and business models [48]. Originally, decision support concept came from the theoretical studies of decision-making for organizations by Carnegie Institute of Technology and technical practice on interactive computer systems by Massachusetts Institute of Technology in the 60s [48]. In the earlier 90s, DSS applications are spread into different areas through data warehousing and on-line analytical processes. Since mid-90s, DSS starts to apply web-based analytical process [49]. Recently, with the development of cloud computing technology, DSS based on cloud computing technology has been proposed [50–52]. Throughout the development of DSS in the last 40 years, there are more than 20 methods to implement the DSS. It is difficult to distinguish the best principle to solve decision-making problems from others since proposed DSS systems are usually project-oriented. DSS has been involved in a plenty of areas which include healthcare, electricity, transportation, resource dispatch, and so on.

A DSS involves a number of scientific areas such as computer science, simulation technology, software programming, cognitive science, and so on. Basically, there are three types of problems for decision-making, namely structured, unstructured, and semi-structured. Structured problems can be solved by standard solution techniques with clearly specified procedures to make a decision. Whereas the procedures of unstructured problems are unspecified in advance, and most of the decision procedures are followed only once. In semi-structured problems, procedures for decision-making can be specified but the optimal decision-making cannot be verified. For different levels in organizations and business companies, the objectives of the DSSs are not the same. There are three different levels in companies and organizations. Firstly, strategic planning, including long-term policies planning, is used for governing resource acquisition, utilization, and disposition. Secondly, management control ensures that the resources can be obtained and used effectively and efficiently to achieve the organization objectives. Finally, operation control ensures effective progress. For problem solving, there are Knowledge base and Database which are transmitted to an area for information processing through a communication system. The problem solution may also be influenced by other information systems. During the processing, there is a need to have optimization tools, online analytical processing (OLAP) tools, data mining, and Knowledge management system. Decision output and recommendation are given to the user through the Man-machine interface. This forms a simple but practical DSS architecture as shown in Figure 19.9.

In general, there are five types of DSSs to achieve a decision-making, namely, communication-driven DSS, data-driven DSS, document-driven DSS, knowledge-driven DSS, and model-driven DSS [53]. No matter what kind of DSS is applied to a certain project, there are some generic

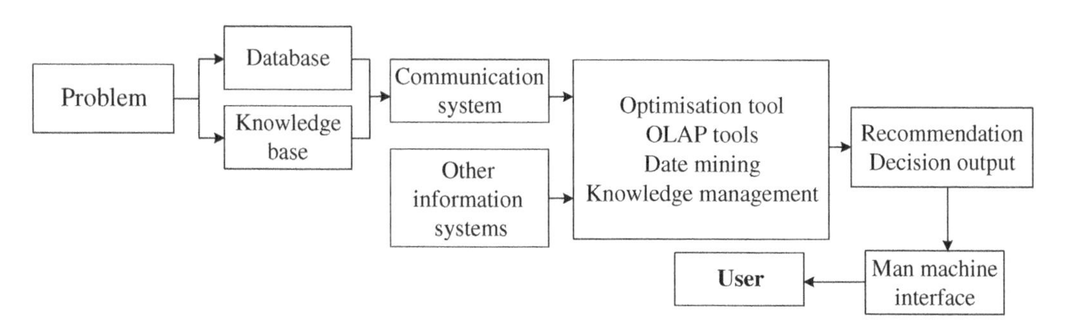

Figure 19.9 DSS architecture.

requirements such as an ability to work in both standalone and web-based environment. Many areas have been involved with DSS. Campanella and Ribeiro [54] described a small helicopter landing case to present a DSS framework for dynamic multiple-criteria decision making. As an important trend for communication development, cloud computing technologies will be more involved for decision-making [50, 55]. Li et al. [51] and Miah and Ahamed [52] gave examples of cloud-based DSS for traffic control and driver safety applications.

Clinical decision support systems (CDSSs), represent a paradigm shift in healthcare and are used to help clinicians in the complex decision-making processes. CDSS are applications that are based on computers which collect and analyze data from electronic health records (EHRs) and offer cues and reminders to assist healthcare practitioners in following evidence-based clinical standards at the point of treatment, reminding them to screen for cardiovascular disease (CVD) risk factors, flagging cases of hypertension or hyperlipidemia, providing information on treatment protocols, prompting questions about medication adherence, and providing tailored recommendations for health behavior changes. Since their first use in the 1980s, CDSS have a rapid evolution. They are now commonly administered through electronic medical records and other computerized clinical workflows. Despite the advancement, there remain unknowns regarding the effect of CDSS on the patient outcomes and costs. There have been numerous published examples in CDSS success stories, but CDSSs are not without risks. Sutton et al. [56] gave a state-of-the-art overview on the use of CDSSs in medicine, including the different types, use cases, common pitfalls, and potential harms.

A CDSS is intended to improve healthcare delivery by enhancing medical decisions with targeted clinical knowledge, patient information, and health information [57]. A traditional CDSS is comprised of software designed to be a direct aid to clinical-decision making, in which the characteristics of an individual patient are matched to a computerized clinical knowledge base and patient-specific assessments or recommendations are then presented to the clinician for a decision [58].

Computer-based CDSSs can be traced to the 1970s. At that time, they had poor system integration, were time intensive, and often limited to academic activities [59, 60]. There were also ethical and legal issues raised with the use of computers when using system recommendations which are lack of scientific reasons [61]. Presently, CDSS often makes use of web-applications or integration with EHRs and computerized provider order entry (CPOE) systems. They can be administered through desktop, tablet, smartphone, but also other devices such as biometric monitoring and wearable health technology. These devices could produce outputs directly on the device or be linked into EHR database [62].

CDSSs have been divided into various categories and types, including intervention timing [57, 63]. CDSSs are frequently classified as knowledge-based or non-knowledge based. In knowledge-based systems, rules such as IF-THEN statements are generated [64], with the system retrieving data to evaluate the rule, and producing an action or output [63]. Rules can be developed using literature-based, practice-based, or patient-directed evidence [58]. CDSS that are non-knowledge based still require a data source, but the decision leverages AI, machine learning (ML), or statistical pattern recognition, rather than being programmed to follow expert medical knowledge [63]. The systems are composed of the rules that are programmed into the knowledge-based system, namely expert system. The inference engine uses the programmed or AI-determined rules, and data structures, and applies them to the patient's clinical data to produce an output, which is presented to the end user for example, medical doctors through the communication mechanism such as the website, or EHR frontend interface [65]. Cloud-based clinical surveillance systems can connect data across multiple facilities, providing flexible, real-time insights into patient outcomes and the impacts of clinical decisions. To understand the potential that advanced clinical decision support holds for health systems, there are case studies for shortening response times to sepsis and reducing mortality rates that exemplify its successes and setbacks [66]. In a separate study published in the Journal of the American College of Surgeons, researchers established that implementing EHRs alone did not improve clinical outcomes. To minimize the problem, they installed a real-time bedside clinical visualization system. As a result, hospital mortality fell by a third, from a 10.8% baseline to 7.2% in intervention data [67]. If real-time data is constantly available, and CDSS is always in action, clinicians can more easily identify the most critical patients.

The cloud is a key factor in enabling clinical surveillance and clinical decision support to become scalable and also capable of drawing in all relevant data sources. While implementing cloud-based clinical surveillance, which offers multiple benefits, it is critical to get ahead of potential difficulties. Health systems are heavy users of Internet of Medical Things (IoMT) devices, including sensor-based technologies, stand-alone devices, and wearables that support patient monitoring and clinical surveillance. These devices generate huge amount of data, but virtual systems must be able to aggregate and synchronize that data and present it in an easily-scanned visual format for it to be useful to physicians with patients located remotely. Common challenges include customizing a system for a hospital, integrating surveillance technology with existing systems, and training providers to use clinical surveillance and the data it yields effectively. Use of cloud-based solutions can also help health systems become more adaptable, just as they are facing massive shifts in how healthcare is delivered during and after a pandemic. These shifts to virtual care, along with patient and provider expectations driving the push toward decentralized care models, mean that the cloud will enable the future of healthcare.

The healthcare industry is under the challenge of data integrity, confidentiality, and large operational data cost. Therefore, the cloud-based clinical decision support system (CCDSS) is adopted to maintain the EHRs, where patients and healthcare professional can easily have access to their legitimate records irrespective of the geographical location and time. Oyenuga et al. [68] reported the building of a DSS based on a private Cloud-based System Architecture to ensure the patient data security, protection of data, and the regulatory compliance policies in the health organizations. It explored the pros and cons, challenges that could be faced during implementing a CCDSS in healthcare for optimization of the resources with time and money. Interoperability is another common concern with clinical surveillance. Therefore, integration with EHRs is a priority. Most importantly, new technologies can give stress to providers without proper testing and training. Health systems must ensure that technology makes the lives of providers easier.

Despite ongoing development for the last 30 years, CDSS and even EHRs suffer from the interoperability concern. Many CDSS exist as stand-alone systems, or exist in a system that cannot communicate effectively with other systems. However, interoperability standards are continuously being developed and improved, such as Health Level 7 (HL7) and Fast Healthcare Interoperability Resources (FHIR). These have already been adopted in commercial EHR vendors [69]. Several government agencies, medical organizations and informatics bodies are actively supporting and some even mandating the use of these interoperability standards in healthcare systems [70, 71]. The cloud also offers a potential solution to interoperability [72]. Cloud EHRs have open architecture, newer standards, and more flexible connectivity to other systems [73]. Web-based EHRs are required to store data in high-level storage centers with advanced encryption and other safeguards. They must comply with national data security standards including for example, the Health Insurance Portability and Accountability Act (HIPAA) in the United States, Personal Information Protection and Electronic Documents Act (PIPEDA) in Canada, or the Data Protection Directive and General Data Protection Regulation (GDPR) in Europe. Efforts are around worldwide to promote the widespread implementation at present. The demand is expected to rise further as a result of government recognition and activities to stimulate the deployment of CDSS and EHR systems. The Global Clinical Decision Support Systems' Market size is expected to reach $8.2 billion by 2027, recording a marked growth of 10.7% CAGR during the forecast period [74].

19.7 Conclusions

An overview of smart health with wireless communication and smart meters is given. Advances in science and technology are changing medical practice. However, the benefits have also added complexity, making it increasingly difficult for healthcare professionals to be confident that they base their decisions on the latest information. Clinical decision support tools, particularly those based on AI, can help. The growing availability, standardization, and integration of data from disparate sources across systems and institutions raise new challenges. Successful digitalization goes beyond implementing new technologies and tools but has an impact on the broader healthcare infrastructure.

The scientific reason underlying smart meter data for health monitoring is that a change in health will lead to a change in behavior, and so the use of appliances will change and result in a change in energy use patterns. Such ongoing monitoring is already an established part of telehealthcare, however, there is currently only indicative evidence of the potential use. This kind of resource should be fully looked at, carefully. Developing software for medical devices or software as a medical device, compliant with standards is not a trivial thing but an essential requirement for the certification of quality management systems and software development.

CDSSs have been shown to augment healthcare providers in a variety of decisions and patient care tasks, and today they actively and ubiquitously support delivery of quality care. Some applications of CDSS have more evidence behind them, especially those based on CPOE. Support for CDSS continues to mount in the age of the electronic medical record, and there are still more advances to be made including interoperability, speed and ease of deployment, and affordability. At the same time, it is essential to stay vigilant for potential downfalls of CDSS, which range from simply not working and wasting resources, to fatiguing providers and compromising quality of patient care. Extra precautions and conscientious design must be taken when building, implementing, and maintaining CDSS. A portion of these considerations were covered in this chapter, but further review will be required in practice, especially as CDSS continues to evolve in complexity through advances in AI, interoperability, and new sources of data.

As the healthcare industry continues to evolve, medical professionals need to stay informed on new technologies and tools that can improve patient care. AI has already been implemented into several healthcare fields including radiology, and many more. With the potential to save lives and make diagnoses faster, it will be a crucial part of healthcare in the years to come. Bioinformatics can certainly add great value. Through bioinformatics, healthcare system and biomedical industry could target various diagnostic kits, analysis programs, and with the help of AI to predict and classify different types of data for disease identification and also to find the genetical cause for a particular disease.

References

1 Mohammed, R.T., Zaidan, A., Yaakob, R. et al. (2021). Determining importance of many-objective optimisation competitive algorithms evaluation criteria based on a novel fuzzy-weighted zero-inconsistency method. *Int J Information Technology & Decision Making* 21 (1): 195–241. https://doi.org/10.1142/S0219622021500140.

2 Mohammed, T.J., Albahri, A., Zaidan, A. et al. (2021). Convalescent-plasma-transfusion intelligent framework for rescuing COVID-19 patients across centralised/decentralised telemedicine hospitals based on AHP-group TOPSIS and matching component. *Applied Intelligence* 51: 2956–2987.
3 Kalid, N., Zaidan, A., Bahaa, B. et al. (2018). Based real time remote health monitoring systems: a review on patients prioritization and related 'Big Data' using body sensors information and communication technology. *Journal of Medical Systems* 42 (2): 30. https://doi.org/10.1007/s10916-017-0883-4.
4 Kalid, N., Zaidan, A., Zaidan, B. et al. (2018). Based on real time remote health monitoring systems: a new approach for prioritization 'Large Scales Data' patients with chronic heart diseases using body sensors and communication technology. *Journal of Medical Systems* 42 (4): 69. https://doi.org/10.1007/s10916-018-0916-7.
5 Okura, T., Enomoto, D., Miyoshi, K.-I. et al. (2016). The importance of walking for control of blood pressure: proof using a telemedicine system. *Telemedicine and e-Health* 22 (12): 1019–1023. https://doi.org/10.1089/tmj.2016.0008.
6 Claudio, D., Kremer, G., Bravo-Llerena, W., and Freivalds, A. (2014). A dynamic multi-attribute utility theory–based decision support system for patient prioritization in the emergency department. *IIE Transactions on Healthcare Systems Engineering* 4 (1): 1–15. https://doi.org/10.1080/19488300.2013.879356.
7 United Nations Department of Economic and Social Affairs, Population Division (2022). World Population Prospects 2022: Summary of Results. UN DESA/POP/2022/TR/NO. 3.
8 Uysal, G. and Ozturk, M. (2020). Hippocampal atrophy based Alzheimer's disease diagnosis via machine learning methods. *Journal of Neuroscience Methods* 337: 1–9. https://doi.org/10.1016/j.jneumeth.2020.108669.
9 Zebene, A., Årsand, E., Walderhaug, S. et al. (2019). Data-driven modeling and prediction of blood glucose dynamics: machine learning applications in type 1 diabetes. *Artificial Intelligence in Medicine* 98: 109–134. https://doi.org/10.1016/j.artmed.2019.07.007.
10 Baxi, V., Edwards, R., Montalto, M., and Saha, S. (2022). Digital pathology and artificial intelligence in translational medicine and clinical practice. *Modern Pathology* 35: 23–32. https://doi.org/10.1038/s41379-021-00919-2.
11 Artificial intelligence and data economy. https://www.ukri.org/what-we-offer/our-main-funds/industrial-strategy-challenge-fund/artificial-intelligence-and-data-economy/
12 Ageing society. https://www.ukri.org/what-we-offer/our-main-funds/industrial-strategy-challenge-fund/ageing-society/
13 Accelerating detection of disease challenge. https://www.ukri.org/what-we-offer/our-main-funds/industrial-strategy-challenge-fund/ageing-society/accelerating-detection-of-disease-challenge/
14 Chinese University of Hong Kong (n.d.). The Hong Kong Mental Morbidity Survey for Older People. http://hkmmsop.org
15 Morris, R. (2021). Bioinformatics: Revolutionising Healthcare? 14 January. https://jakemp.com/en/knowledge-centre/insights/bioinformatics-revolutionising-healthcare (accessed 18 July 2022).
16 Chen, R. (2015). On bioinformatic resources. *Genomics, Proteomics & Bioinformatics* 13: 1–3.
17 Suresh, A., Vimal, S., Robinson, Y.H. et al. (ed.) (2022). *Bioinformatics and Medical Applications: Big Data Using Deep Learning Algorithms*, 1e. Wiley.
18 Raza, K. and Dey, N. (2021). *Translational Bioinformatics in Healthcare and Medicine*, 1e. Elsevier.
19 Majhi, V., Paul, S., and Jain, R. (2019). Bioinformatics for healthcare applications. *2019 Amity International Conference on Artificial Intelligence (AICAI)*, Dubai, United Arab Emirates (4–6 February 2019). https://doi.org/10.1109/AICAI.2019.8701277

20 Shi, T.W., Kah, W.S., Mohamad, M.S. et al. (2017). A review of gene selection tools in classifying cancer microarray data. *Current Bioinformatics* 12: 202–212.

21 Serra, A., Galdi, P., and Tagliaferri, R. (2018). Machine learning for bioinformatics and neuroimaging. In: *Data Mining and Knowledge Discovery (ed. WIREs Authors)*, e1248. *Wiley* Periodicals 33 pages. https://doi.org/10.1002/widm.1248.

22 Kumar, Y., Koul, A., Singla, R., and Ijaz, M.F. (2022). Artificial intelligence in disease diagnosis: a systematic literature review, synthesizing framework and future research agenda. *Journal of Ambient Intelligence and Humanized Computing* 1–28. https://doi.org/10.1007/s12652-021-03612-z.

23 Lee, R.P.K., Lai, L.L., and Lai, C.S. (2014). Design and application of smart metering system for micro grid. *Proceedings of the International Conference on Systems, Man and Cybernetics*, Manchester, UK (13–16 October 2013).

24 Number of smart meters (electricity, gas & water) worldwide from 2014 to 2020 (in millions) [Online]. https://www.statista.com/statistics/625890/worldwide-smart-meter-deployment/ (accessed 1 January 2022).

25 (n.d.). IoT smart metering solutions and smart meter data visualization with Things Board [Online]. https://thingsboard.io/smart-metering/ (accessed 7 July 2022).

26 UK Health Security Agency (2020). Smart meters: radio waves and health. 23 April. [Online]. https://www.gov.uk/government/publications/smart-meters-radio-waves-and-health/smart-meters-radio-waves-and-health (accessed 8 July 2022).

27 Fong, B., Fong, A.C.M., and Li, C.K. (2020). *Telemedicine Technologies: Information Technologies for Medicine and Digital Health*, 2e. UK: Wiley.

28 Paxman, J., Matt James, Enrico Costanza and Julia Manning (2020). Examining possibilities for smart meter data in health and care support. [Online]. https://2020health.org/publication/smart-future-of-healthcare/ (accessed 4 July 2022).

29 Lai, Q.H. and Lai, C.S. (2022). Healthcare with wireless communication and smart meters. *IEEE Consumer Electronics Magazine* 10 pages, https://doi.org/10.1109/MCE.2022.3181438.

30 Lee, G., Kang, B., Nho, K. et al. (2019). MildInt: deep learning-based multimodal longitudinal data integration framework. *Frontiers in Genetics* 10: https://doi.org/10.3389/fgene.2019.00617.

31 Mehmood, M. et al. (2021). Machine learning assisted cervical cancer detection. *Frontiers in Public Health* 23: https://doi.org/10.3389/fpubh.2021.788376.

32 Khamparia, A., Gupta, D., Rodrigues, J.J., and de Albuquerque, V.H.C. (2021). DCAVN: cervical cancer prediction and classification using deep convolutional and variational autoencoder network. *Multimedia Tools Applications* 80: 30399–30415. https://doi.org/10.1007/s11042-020-09607-w.

33 Abbas, S., Jalil, Z., Javed, A.R. et al. (2021). BCD-WERT: a novel approach for breast cancer detection using whale optimization based efficient features and extremely randomized tree algorithm. *PeerJ Comput. Sci.* 7: e390. https://doi.org/10.7717/peerjcs.390.

34 Iwendi, C., Bashir, A.K., Peshkar, A. et al. (2020). COVID-19 patient health prediction using boosted random forest algorithm. *Front. Public Health* 8: 357. https://doi.org/10.3389/fpubh.2020.00357.

35 Rehman, S.U., Javed, A.R., Khan, M.U. et al. (2020). PersonalisedComfort: a personalised thermal comfort model to predict thermal sensation votes for smart building residents. *Enterprise Inform Syst.* 1–23. https://doi.org/10.1080/17517575.2020.1852316.

36 Javed, A.R., Faheem, R., Asim, M. et al. (2021). A smartphone sensors-based personalized human activity recognition system for sustainable smart cities. *Sustain. Cities Soc.* 71: 102970. https://doi.org/10.1016/j.scs.2021.102970.

37 Aslam, B., Javed, A.R., Chakraborty, C. et al. (2021). Blockchain and ANFIS empowered IoMT application for privacy preserved contact tracing in COVID-19 pandemic. *Pers. Ubiquit. Comput.* 1–17. https://doi.org/10.1007/s00779-021-01596-3.

38 Iwendi, C., Mahboob, K., Khalid, Z. et al. (2021). Classification of COVID-19 individuals using adaptive neuro-fuzzy inference system. *Multimedia Systems* 1–15. https://doi.org/10.1007/s00530-021-00774-w.

39 Tripathy, B., Parimala, M., and Reddy, G.T. (2021). Innovative classification, regression model for predicting various diseases. In: *Data Analytics in Biomedical Engineering and Healthcare*, 179–203. Elsevier.

40 Sinha, N. Loi Lei Lai, Palash Kumar Ghosh, Yingnan Ma (n.d.). Wavelet-GA-ANN based hybrid model for accurate prediction of short-term load forecast. *Proceedings of the International Conference on Intelligent Systems Applications to Power Systems*, Kaohsiung, Taiwan (5–8 November 2007). IEEE.

41 Han, Z., Zhao, J., Leung, H. et al. (2019). A review of deep learning models for time series prediction. *IEEE Sensors Journal.*

42 Li, D. and Wang, Z. (2017). Video superresolution via motion compensation and deep residual learning. *IEEE Transactions on Computational Imaging* 3 (4): 749–762.

43 Zhong, C., Lai, C.S., Ng, W.W.Y. et al. (2021). Multi-view deep forecasting for hourly solar irradiance with error correction. *Solar Energy* 228: 308–316.

44 Zhao, R., Wang, D., Yan, R. et al. (2018). Machine health monitoring using local feature-based gated recurrent unit networks. *IEEE Transactions on Industrial Electronics* 65 (2): 1539–1548.

45 Dasgupta, S., Littman, M., and McAllester, D. (2002). PAC generalization bounds for co-training. *Advances in Neural Information Processing Systems* 375–382.

46 Lai, C.S., Jia, Y., and McCulloch Zhao Xu, M.D. (2017). Daily clearness index profiles cluster analysis for photovoltaic system. *IEEE Transactions on Industrial Informatics* 13: 2322–2332.

47 Liu, H. and Chen, C. (2019). Data processing strategies in wind energy forecasting models and applications: a comprehensive review. *Applied Energy* 249: 392–408.

48 Decision support system. http://en.wikipedia.org/wiki/Decision_support_system (accessed 16 March 2012).

49 A Brief History of Decision Support Systems by D. J. Power. http://dssresources.com/history/dsshistory.html (accessed 16 March 2012).

50 Bhat, M.A., Shah, R.M., Ahmad, B., and Bhat, I.R. (2010). Cloud computing: a solution to information support systems (ISS). *International Journal of Computer Applications* 11 (5): 5–9.

51 Li, Z., Chen, C., and Wang, K. (2011). Cloud computing for agent-based urban transportation systems. *IEEE Intelligent Transportation Systems* 26: 73–79.

52 Miah, S.J. and Ahamed, R. (2011). A cloud-based DSS model for driver safety and monitoring on Australian roads. *Int. J. Emerg. Sci.* 1 (4): 634–648.

53 Shiry, S. (n.d.). Decision Support Systems. Introduction http://ceit.aut.ac.ir/~shiry/lecture/DSS/Introduction.ppt (accessed 3 March 2012).

54 Campanella, G. and Ribeiro, R.A. (2011, Elsevier Science Publisher B. V., North-Holland). A framework for dynamic multiple-criteria decision making. *Decision Support Systems* 52: 52–60.

55 Zhang, H.T., Yang, Q., Lai, C.S., and Lai, L.L. (2012). New trends for decision support systems. *Proceedings of the 2012 IEEE International Conference on Systems, Man and Cybernetics*, Seoul, Korea (South) (14–17 October 2012).

56 Sutton, R.T., Pincock, D., Baumgart, D.C. et al. (2020). An overview of clinical decision support systems: benefits, risks, and strategies for success. *npj Digital Medicine* 3: 17. https://doi.org/10.1038/s41746-020-0221-y.

57 Osheroff, J., Teich, J.M., Levick, D. et al. (2012). *Improving Outcomes with Clinical Decision Support: An Implementer's Guide*, 2e. HIMSS https://doi.org/10.4324/9781498757461.

58 Sim, I., Gorman, P., Greenes, R.A. et al. (2001). Clinical decision support systems for the practice of evidence-based medicine. *J. Am. Med Inf. Assoc. Jamia.* 8: 527–534. https://doi.org/10.1136/jamia.2001.0080527.

59 De Dombal, F. (1992). Computers, diagnoses and patients with acute abdominal pain. *Archives of Emergency Medicine* 9: 267–270. https://doi.org/10.1136/emj.9.3.267.

60 Shortliffe, E.H. and Buchanan, B.G. (1975). A model of inexact reasoning in medicine. *Mathematical Biosciences* 23 (3–4): 351–379. https://doi.org/10.1016/0025-5564(75)90047-4.

61 Middleton, B., Sittig, D.F., and Wright, A. (2016). Clinical decision support: a 25 year retrospective and a 25 year vision. *Yearbook of Medical Informatics* 25 (S 01): S103–S116. https://doi.org/10.15265/IYS-2016-s034.

62 Dias, D. (2018). Wearable health devices – vital sign monitoring, systems and technologies. *Sensors* 18 (8): 2414. https://doi.org/10.3390/s18082414.

63 Berner, E.S. (ed.) (2016). *Clinical Decision Support Systems: Theory and Practice (Health Informatics)*, 3e. Springer.

64 Lai, L.L. (1995). Computer assisted learning in power system relaying. *IEEE Transactions on Education* 38 (3): 222–228.

65 Deo, R.C. (2015). Machine learning in medicine. *Circulation* 132: 1920–1930.

66 Morgan, C.K., Amspoker, A.B., Howard, C. et al. (2021). Continuous cloud-based early warning score surveillance to improve the safety of acutely ill hospitalized patients. *Journal for Healthcare Quality* 43 (1): 59–66. https://doi.org/10.1097/JHQ.0000000000000272.

67 Berger, D.H., Howard, C., Holcomb, J.B., and Herlihy, J.P. (2018). Improving Survival in Critically Ill Patients after Implementing a Visual Decision Support System. *Journal of the American College of Surgeons* 227 (4): e1–e2. https://doi.org/10.1016/j.jamcollsurg.2018.09.013.

68 Oyenuga, S.O., Garg, L., Bhardwaj, A.K., and Shrivastava, D.P. Cloud-based clinical decision support system. In: *Conference Proceedings of ICDLAIR2019*, Lecture Notes in Networks and Systems, vol. 175 (ed. M. Tripathi and S. Upadhyaya). Cham: Springer https://doi.org/10.1007/978-3-030-67187-7_24.

69 (2022). FHIR standard for health care data exchange. Index - FHIR v4.3.0. https://www.hl7.org/fhir/index.html (accessed 21 June 2022).

70 HER Intelligence (2015). 5 Ways States Mandate Health Information Exchange. https://ehrintelligence.com/news/5-ways-states-mandate-health-information-exchange (accessed 22 April 2022).

71 EU Recommendation (2019). European Electronic Health Record Exchange Format. cepPolicyBrief No. 15/2019.

72 Bresnick, J. (2015). Interoperability, Low Costs Make Cloud-Based EHRs a Favorite. Health IT Analytics. https://healthitanalytics.com/news/interoperability-low-costs-make-cloud-based-ehrs-a-favorite (accessed 17 July 2022).

73 Fernández-Cardeñosa, G., de la Torre-Díez, I., López-Coronado, M., and Rodrigues, J.J.P.C. (2012). Analysis of cloud-based solutions on EHRs systems in different scenarios. *Journal of Medical Systems* 36: 3777–3782.

74 (2022). Global Clinical Decision Support Systems, Industry Analysis Report and Forecast, 2021–2027. https://www.reportlinker.com/p06240978/?utm_source=GNW (accessed 4 July 2022).

20

Concluding Remark and the Future

20.1 The Relationship

One of the most important and dynamic aspects of smart cities is the evolution of the roles and relationships between the key participants involved in envisioning and creating them; namely, the government bodies aiming to transform the lives, wellbeing and safety of their citizens; and private sector players helping to realize these aspirations, by building, managing, and funding digital urban infrastructure and services.

It is foreseen that in the future, in the society, many of the new jobs being created are highly skilled. There is a need to reallocate the workforce from low-skill domains to high-skill ones, and education is going to play a significant role in helping people transition between the jobs.

As well as investing in education and re-training, governments may have to look at expanding housing supplies in and around cities, as this is where the job booms fuelled by new technology tend to take place. However, traditional roles such as kindergarten teachers or nurses will continue to be in demand as these roles require a multitude of interpersonal skills, such as social intelligence, perception, and manipulation. We can probably automate some basic teaching because artificial intelligence (AI) can fulfill the role of an interactive tutor, up to a certain point. But teaching children in smaller groups, writing essays, holding debates and discussions to foster social skills and creative thinking are the domains where humans still hold the comparative advantage in the foreseeable future.

It is clear that smart education is key for the future. Smart education is a model of learning adapted to new generations of digital society. In comparison to traditional classroom teaching models, smart education is an interactive, collaborative, and visual model, designed to increase student engagement and enable teachers to adapt to students' skills, and interests. People could complete their degrees at different speeds, as it suits them.

Use of smart learning system increases the interaction of teachers and students, as they follow each other in the process and it becomes easy for teachers to track on students learning power. Also, the use of smart classes and modern technology eases the learning process for all students. In general, the advantages of smart classes include improvement of student–teacher interaction and communication; real-time blended teaching and learning with interactive lessons, game-based activities, collaboration tools and formative assessments; giving students a better understanding of concepts; introducing students and instructors to education technology; and improving visualization and creativity. However, this approach would be much more costly, need skilled faculty, and maintenance for electronic gadgets.

Smart Energy for Transportation and Health in a Smart City, First Edition. Chun Sing Lai, Loi Lei Lai and Qi Hong Lai.

The progress in technology, industry, agriculture, and public health has been accompanied by unprecedented environmental challenges from factors such as climate change, biodiversity loss, soil degradation, and water scarcity. To address this urgent need to safeguard the health of our future generations. A new, multi-disciplinary approach that brings together scientific knowledge of both human and ecosystem health to know about economic trends, market behavior, and policy making is important to be applied to the society practically and to solve complex environmental and development challenges [1].

World leaders and experts from business, government, international organizations, civil society, and academia must work together to bridge knowledge gaps on the links between economic development, natural systems, and human health to compel collaboration across disciplines and coordinated action to address the future complex challenges.

The future goal is to advance economies that expand opportunities for shared prosperity and building resilience by helping people and communities to prepare and withstand shocks and stresses. To achieve this goal, it is necessary to address the root causes of emerging challenges and create systemic change; to catalyze and scale transformative innovations, create unlikely partnerships that span sectors, and take risks others cannot or will not.

World leaders must give support to address global challenges. Investments in research to tackle a wide range of issues such as climate change, disease and inequality. The support includes novel, high risk, and multi-disciplinary projects that may break boundaries and produce results that could dramatically improve the quality of life of the future generations. Underpinning all the research is the need to translate academic excellence into impact from innovations in science, medicine, and technology, to provide expert advice and policy recommendations.

There is a need to have research on subjects as diverse as the relationship between diet, human health and the environment, avoiding future pandemics, and developing more efficient models of drug discovery. Whether working with machine learning experts to understand vast, complex biomedical datasets or ethicists to develop a new understanding of collective moral responsibility in the face of infectious disease, to bring together researchers from different disciplines to advance human health and improve quality of life.

There is a need to compare the potential scale and cost of different low-carbon utilization pathways to operate at large-scale and at low cost, so that it could be big business in the future.

The evolution of smart cities needs more than technology. It requires good relationships between key stakeholders. Government and the private sector must partner to turn their vision of connected and efficient services into reality.

20.2 Roadmap

According to [2], devising cost-effective national and regional net zero roadmaps demands cooperation among all parts of government that integrates energy into every country's policy making on finance, labor, taxation, transport, and industry. Energy or environment ministries alone cannot carry out the policy actions needed to reach net zero by 2050.

In many countries today, taxes on diesel, gasoline, and other fossil fuel consumption are an important source of public revenues, providing as much as 10% in some cases. In the net zero pathway, tax revenue from oil and gas retail sales falls by about 40% between 2020 and 2030. Managing this decline will require long-term fiscal planning and budget reforms.

The net zero pathway relies on international cooperation among governments, especially on innovation and investment. This is not simply a matter of all governments seeking to bring their national emissions to net zero, it means tackling global challenges through coordinated actions and accelerating the energy transition from fossil fuel to renewables worldwide.

Governments must work together in an effective and mutually beneficial manner to implement coherent measures that cross borders. This includes carefully managing domestic job creation and local commercial advantages with the collective global need for clean energy technology deployment. Accelerating innovation, developing international standards and coordinating to scale up clean technologies needs to be done in a way that links national markets. Cooperation must recognize differences in the stages of development of different countries and the varying situations of different parts of society. For many rich countries, achieving net zero emissions will be more difficult and costly without international co-operation. For many developing countries, the pathway to net zero without international assistance is not clear. Technical and financial support is needed to ensure deployment of key technologies and infrastructure. It is essential to have greater international cooperation to achieve global CO_2 emissions falling to net zero by 2050.

20.3 The Future

20.3.1 Smart Energy

There is no question that the world must move toward a more sustainable energy future. It is a global challenge to meet increasing global energy demand, reduce environmental impact and the carbon footprint. At the same time, societal development needs to be addressed for societal resilience. Modern cities are facing the challenge of combining competitiveness on a global city scale and sustainable urban development to become smart cities.

As smart cities strive toward using green and renewable energy, sustainable energy, plays a critical role. A smart city has its energy generation, supply, transmission, distribution, and management components to be modular, automated, reliable, safe, and controllable. A smart city energy system must provide reliability, efficiency, robustness against attacks and a high level of supply availability and redundancy. All these are key indicators for a practical and useful energy system.

Smart cities are emerging as highly complex technological endeavors that combine technology from many engineering disciplines including energy, transportation, building management, cybersecurity, and others that constitute only a small part of technologies that are fundamental to the development of smart cities. Computing tools and artificial intelligence approaches have emerged to address most of the important technologies involved in smart cities. Challenges exist in many directions, including technological, economic, and social aspects.

The huge concern in increased greenhouse gas emissions, and evolving laws and regulations leads to a focus in sustainable and clean energy to reduce energy demand, and improve energy efficiency. A radical restructuring of energy supply is underway, and it is needed to ensure sustainable prosperity, and quite possibly the survival of the human species. This transformation includes the introduction of new elements at all levels in the chain of production, delivery and use, new network configurations, new design and operational procedures, new incentives and business models, new social structures, and new policies.

Sustainable energy is the basic need that will enable a true transformation of our cities through which one can have safe, sustainable, connected, and smart environments for the majority of the world's population as the energy development and utilization get smarter. There are various supporting technologies and requirements of sustainable energy for smart cities. When equipped with smart home systems, data-driven energy modeling and simulation techniques can significantly transfer the world of smart cities. However, significant technical and commercial challenges must be overcome.

The energy sector is the source of around three-quarters of greenhouse gas emissions today and holds the key to averting the worst effects of climate change, perhaps the greatest challenge mankind has faced [2]. The global pathway to net-zero emissions by 2050 detailed in this report requires all governments to significantly strengthen and then successfully implement their energy and climate policies. In any case, the power and energy systems must be stable and reliable to serve mankind. Due to the increase in the high penetration of renewables in the future, to achieve net zero in 2050, the use energy storage is an option. However, for smart energy point of view, it is essential to have an acceptable, economical, and technical solution.

A recent enquiry on the role of batteries and fuel cells in achieving net zero was conducted by the UK Parliament. A written evidence was considered [3].

There were no policy and incentive schemes to promote deployment of energy storage prior to the introduction of electricity market reform. For instance, energy storage was not defined within the legislation for the Renewable Obligation or feed-in tariff scheme.

From the commercialization perspective, the current energy policies in supporting energy storage (including batteries) for power grid applications are not well implemented.

The impacts of energy policies for supporting low-carbon infrastructure on the economic and financial performance of battery and thermal energy storage in the United Kingdom when coupled with a wind farm were explored by Lai and Locatelli [4]. As the net present value reduces with increased energy storage capacity (when coupled with generation), low-carbon incentives are, unintentionally, barriers to the development of energy storage due to: (i) current generator incentives give a favorable return on investment and energy storage would diminish it; (ii) energy storage cannot participate in generator only incentives. Specifically, policy mechanisms (such as contract for difference, CfD) designed to support low-carbon technologies could affect the energy storage adoption. There is a need for energy policy schemes to support and protect the energy storage market.

CfD for low-carbon power generators can harm the energy storage market if the strike price is set to be above the maximum wholesale price. The strike price also needs to be above the average wholesale price for CfD to be useful. Renewable energy technologies, especially onshore windfarm and solar photovoltaic, have plummeted in cost. However, energy storage systems, in particular batteries, are still expensive, and there are no long-term policy mechanisms in place to promote energy storage growth.

Some of the following recommendations to supportgrid-scale battery commercialization could be considered.

Price floor mechanism for energy storage: A price floor is a regulatory policy, with the government to enforce price limit or control on how low a price can be charged for a good, service, product, or commodity [5]. Price floor mechanisms have been implemented for the energy sector, including for the carbon price [6]. In April 2013, the UK government introduced a carbon price floor to reduce carbon market price uncertainty and worked well with the emissions trading scheme [6]. The carbon price floor sets a minimum market price for carbon and was developed to deal with the low-carbon prices in the EU emissions trading scheme, as a consequence by the oversupply of permits and the economic recession. The carbon price has been effective to reduce greenhouse gas emissions by increasing the economic viability of low-carbon technologies.

Generators and energy storage systems influence the wholesale electricity market [5]. With greater uncertain power demand and generation due to larger penetration of intermittent renewables and reliance on electricity, wholesale market price volatility is a major challenge to be dealt with. Indeed, negative wholesale prices have appeared in recent years [7].

With a price floor mechanism, the Government may pay the system operator a certain value (£/MWh) to store energy when the wholesale market reaches a level of the low market price. Consider that many system operators may store the energy driven by the price floor mechanism, the reduction in power generation will increase the wholesale price. The energy stored will be sold during a period of high prices, but given this availability of energy, there will be less shortage of electricity; therefore, the "peak period" will be reduced. In summary, creating a price floor would make battery economically viable and, at the same time, contribute to reducing the volatility of the electricity market.

Upfront subsidy to meet energy storage cost: The UK government does not provide direct subsidies for the deployment of large-scale or behind-the-meter energy storage systems [8]. However, upfront subsidies can promote the development of certain technologies. For instance, the Green Deal was a UK government policy initiative that let the domestic sector to pay for energy-efficient home improvements, including solar panels and heat pumps through the savings on their energy bills [9]. Batteries could benefit from comparable initiatives due to the relatively high cost for energy storage and the balance of plant. Similarly, the research question of "How much upfront subsidy is required to promote batteries deployment?" is relevant to be addressed, considering the different energy storage technologies type and system conditions.

The following was considered as well: What are the costs and benefits of using battery and fuel cell technologies in their various applications, including when integrated into the wider energy system?

For energy storage including batteries, capital costs are the upfront cost consisting of both "hard costs" (e.g. materials costs including electrolyte) and "soft costs" (e.g. licensing fees and the engineering, procurement, and construction costs) [10, 11]. Operation and maintenance costs occur during the system life cycle and include labor, repair, regular servicing, and electricity purchasing (energy storage charging cost) [10]. The Balance of System cost is associated with the auxiliary equipment (e.g. power converters) for a power system [11, 12].

Grid-scale energy storage including batteries have been built with solar farms and wind parks to minimize electricity curtailment [13]. Consider the economic benefits, the developed sources of revenue streams for system operators with grid energy storage are.

Electricity markets: Spot price/ wholesale market: Electricity is a commodity that can be traded in the wholesale market from various energy technologies. The wholesale price increases with electricity demand. Nord Pool AS is a European power exchange and is responsible for delivering power trading across Europe [7].

National Grid (grid services).

Firm Frequency Response (FFR): FFR complements other categories of frequency response (e.g. primary response) and provides firm availability. The service can be either dynamic (i.e. continuously provided service for managing the usual second-by-second system changes) or static (i.e. usually a discrete service triggered at a set frequency deviation). The minimum power capacity to provide the FFR service is 1 MW [14]. FFR can be from generators, energy storage, and aggregated demand response.

Short-term Operating Reserve (STOR): The provider gives a contracted level of power when called by National Grid, UK to achieve energy reserve requirement [15]. The STOR provider must provide a minimum of 3 MW of steady demand reduction or generation for two hours minimum. STOR technology requirement is the same as FFR.

Fast Reserve (FR): FR provides rapid active power by reducing the demand or increasing the generation, as requested by the National Grid Electricity System Operator [16], to participate in controlling frequency variations. FR needs all units to be able to begin service delivery within two minutes following an order, at a rate of 25 MW/min or more and deliver a minimum of 50 MW. The FR provider needs to give power consistently for a minimum of 15 minutes [16]. FR technology requirement is the same as FFR.

Distribution network operators: Super Red Credits (SRCs): Distribution network operators provide SRC payments to non-intermittent generators for providing energy during peak demand times (i.e. super red periods). These generators allow the distribution network to defer the reinforcement or grid upgrade. To receive these credits, generators must be connected to the extra high voltage grid. Participation in SRC payments is possible for renewable sources with batteries [17].

Electricity Market Reform: Capacity Market (CM) aims to create enough reliable capacity (both supply and demand-side) for secure electricity supplies, in particular during critical periods for the system (e.g. poor weather conditions) [18, 19]. CM allows the market to determine a price for competitive capacity. Capacity agreements are given to providers of current and new capacity, from one year (T-1) to four years (T-4) ahead. This gives investors certainty and confidence about future revenues, under intermittent generation and uncertain market conditions. The CM provides revenue in monthly capacity payments. Capacity payments are paid monthly during the delivery years to capacity providers [18].

From a long-term (i.e. decades) perspective, there is a need to examine the technical and economic impacts on the wider energy system with prolific installation of large-scale batteries. Long-term electrical power system models support decarbonization studies and energy technology assessments, including batteries and fuel cells [20]. There is no comprehensive study on the aforementioned revenue sources with regard to the technical and economic impacts to the UK energy system.

To gain insight into how the industry and companies are achieving sustainable energy goals in a complex, challenging situation that is undergoing increasing and evolving environmental issues, regulations, and expectations, it is important to drive innovation and technology deployments that will ultimately enable vital data collection and analysis needed to assess, monitor, and maintain sustainable energy across smart communities, and smart cities as a whole.

Future efforts are required for stakeholders to collaborate in a multi-disciplinary way to converge toward a unique, complex, adaptable, marketable, and resilient global energy solution. It will also provide a practical aid for technology developers, planners, policymakers and technical, economic and policy advisers to push for low-carbon economy, supported by scientific and technological contributions, which make sustainable energy for smart cities feasible.

The following future topics must be considered:

1) Microgrid design, optimization, operation, and control involving renewable energy generation characteristics.
2) Virtual power plant design, optimization, control involving renewable energy plants such as solar and wind.
3) Design and optimization of power converter control with consideration of network.
4) Renewable energy development and feasibility study project involving new modeling and methodology of design optimization.
5) Sharing operating experience of biomass and other community energy development project with technical rigor and methodology.
6) Reliability, security, privacy, and trust.
7) Applications, business, and social issues.
8) Standards development.
9) Big data collection and analysis for sustainable energy connected to power networks.

10) Artificial intelligence-based future sustainable energy generation modeling and simulation.
11) Sustainable energy systems interfaced with industrial internet of things.
12) Energy forecasting and optimization methods to improve renewable energy resources.
13) Advances in energy storage technology and grid integration methods for smart cities.
14) Sustainable energy system financing and delivery mechanisms.
15) Policy formulation toward successful smart sustainable energy cities implementation.
16) Behavioral changes and monitoring: challenges, strategies, and lessons learned.
17) Demonstration projects.
18) ICT diffusion.

20.3.2 Healthcare

One essential future area in healthcare is to consider the latest technological developments in electronics that blend into everyone's health and safety in smart cities. Smart wearables, such as watches and wristbands, have been widely used in health monitoring in recent years. These connected consumer electronics devices not only provide indicative information for consumers' health management, but also play a vital role in communicable disease management and prevention. In the context of consumer technology as part of a smart city infrastructure, wireless communication and sensing technologies have evolved substantially over the past decade and have seen an accelerating trend of technical advances, particularly in the consumer healthcare sector. Optimization techniques have been developed significantly and smart devices have become more efficiently designed and implemented for the benefit of both individual consumers and health service providers.

The benefit offered by the present 5G and later on 6G is much more than the increased maximum throughput. It has many current technical advances including industrial Internet of Things (IIoT). As the IIoT integrates many heterogeneous networks, such as wearable healthcare systems, wireless sensor networks, wireless local area networks, and mobile communication networks, it is critical to design self-organizing and smart protocols for heterogeneous ad hoc networks in various internet of things (IoT) applications, such as cyber-physical-social systems, cloud computing for heterogeneous ad hoc networks, large-scale sensor networks, data acquisition from distributed smart devices, green communication and applications, trading platform, environmental monitoring and control. In particular, IoT plays an important role to connect everything together and to the Internet through specific protocols for information exchange and communications, achieving intelligent recognition, location, tracking, monitoring, and management.

According to the United Nations, World Health Organization on 11 November 2013 warned of the serious implications for billions worldwide due to the shortage of healthcare workers, which is estimated to grow to 12.9 million by 2035 from the current deficit of 7.2 million [21]. This is due to an ageing health workforce with staff retiring or leaving for better paying jobs without being replaced, coupled with not enough young people entering the profession or being adequately trained.

Hence, there is a need to develop wearable healthcare systems to perform self-health monitoring. In general, wearable healthcare systems demand low power consumption and high measurement accuracy. Smart technologies including green electronics, green radios, deep learning approaches, and intelligent signal processing techniques play important roles in the developments of the wearable healthcare systems. It is important to have the ability to design and implement consumer healthcare devices when keeping up with the rapidly changing consumer healthcare market demands up-to-date technology information acquired through gaining practical knowledge for creating reliable and comfortable wearables that protects the internal electronic modules and systems for health monitoring.

There are no questions that healthcare touches our lives and is essential to our communities. Together with governments and healthcare organizations, modern technology is used to improve care today and help shape a healthier future. New standards of care with robots to assist health workers in their day-to-day tasks are developed for a smarter healthcare for a healthier future. For example, [22] reported the use of Robotic Nurse Assistant to provide valuable assistance to nurses by carrying out essential but repetitive tasks like measuring patients' vital signs, synchronizing patients' data for real-time follow-up and delivering medicines.

The trend of pet humanization coupled with rising populations of small animals will drive the market revenue. Pet owners are considering their animals an integral part of their family and increasing spending on their health and wellbeing. Safety remains one of the top priorities of pet owners, as evidenced by the popularity of advanced pet tech products such as GPS trackers, smart doors, and smart collars. According to a study by Consumer Technology Association (CTA), 46% of pet owners use an app that provides health information directly to their vets, and 40% use pet tech for philanthropic purposes such as finding shelters for rescued animals after natural disasters. Figures 20.1 and 20.2 show two examples.

As of 2021, the global pet tech market (including smart collars, harnesses and feeders) is worth over USD 5.5 billion in 2020. This is projected to grow by 22% between 2021 and 2027 to over USD 20 billion, with most of the growth predicted to be in the European market (Global Market Insights) [24].

The European pet wearable market size captured around 20% revenue share in 2020 and is expected to gain traction due to the growing trend of pet humanization in countries such as the United Kingdom, France, and Spain. The United Kingdom ranks first in the number of pet population with 26% of adults in the country owning a pet dog. The regional industry size from pet healthcare applications is projected to witness around 18% CAGR from 2021 to 2027 due to the incessant need to prevent pet obesity and chronic diseases.

Pets, like humans, experience complex emotions and suffer from physical ailments, but unlike humans, pets cannot tell us their pain. As a result, by the time we notice a change in their behavior, they have suffered from more pain than needed, and the disease has advanced further rendering it more difficult and expensive to treat. This is why euthanasia is one of the most commonly

Figure 20.1 Pet owner with her dog.

Figure 20.2 Owner exercising with the pet. *Source:* [23] Magdalena Smolnicka/Unsplash.

performed veterinary procedures, to put them out of misery. Monitoring vital parameters is one way to catch and prevent diseases, giving us the opportunity to take care of our pets before it is too late. Ideally, regular vet visits will provide the most beneficial and timely help but is not practical as it is too expensive for most citizens. As mentioned, pet owners are considering their animals a part of their family and the rising prevalence of preventable diseases encourages owners to invest in innovative pet tech devices to monitor the health.

In addition, at every economic downturn there is a surge in pet abandonment. This time it is exacerbated by the loneliness of the pandemic which heightened pet adoption; resumption of normal busy life results in a larger than usual dump of pandemic puppies. Not only are stray pets potentially hazardous, but they also suffer from unimaginable emotional disarray that follows them for life. Even when rehomed in a loving family, their emotional trauma renders them difficult to care for. The ongoing COVID-19 pandemic has resulted in a decline in sales of pet tech products in 2020 due to the closure of manufacturing facilities, supply chain disruptions and reduced consumer spending ability due to economic instability and increased unemployment rates. However, there has been a spike in the number of pet ownership and coupled with improving economic conditions and easing of restrictions, a boost in market growth and a rise in consumer spending is expected. It would be helpful to have wireless pet healthcare systems developed.

To tackle the challenges in keeping up with the rapid healthcare technology development, future research should pay attention to some of the following topics:

- Emerging health and/or pathogen sensing technologies for smart cities
- Smart consumer electronic devices and systems
- Smart and assistive healthcare technology in smart cities
- Highly integrated front-end circuits for smart health
- Integrated circuits for body area networks and/or infrastructural communications
- Novel multimedia processing for health monitoring
- Self-powered monitoring devices
- Privacy and/or security enhancing technologies for healthcare provisioning in smart cities
- Health applications such as baby monitoring, elderly care, and tracking of pets in smart cities

20.3.3 Smart Transportation

Researchers need to examine the effectiveness of various technical solutions in cutting the greenhouse gases emitted by cars and other road transport. The aim is to understand what is needed for the implementation of practical low-carbon transportation.

The overall transportation system will be much improved in smart cities. Public transportation options will be more convenient and efficient. People who drive will be able to find parking more easily, without having to fuss with parking meters and other devices.

A city should have excellent public transportation, which makes it possible to enjoy the city without having to worry about parking at all. Smart parking and smart transportation could give tourists and residents a better experience.

Clean and efficient transportation is essential. In the hope of optimizing mobility, many cities are turning to smart technologies to ease traffic congestion and provide users with real-time updates.

Driven by fast-moving technology, new business models and changing societal expectations, the pandemic has kept millions at home and disrupted every aspect of the transportation sector, and the transportation of the future will be more accessible than it is today. Sustainable, fairer, more efficient, and more convenient are needed, even if the precise timing and nature of this transition are uncertain [25].

Ground transportation is being shifted toward more efficient and green transportation. Wireless networking, sensing, computing, and control advances have significantly changed the way the city works. Smart transportation systems are getting smarter and have become highly multi-disciplinary. They require an ever-increasing integration of electrical/electronic, control, information, and mechanical disciplines. Emerging innovative technologies, such as automated driving and electrified vehicles, are also profoundly promoting connection, automation, and electrification of the current transportation sector.

Intelligent electrified and automated vehicles are featured by increasingly strict requirements to ensure safe, sustainable, smart, and smooth operations of vehicular mechatronic systems, under highly changing driving conditions and road situations.

Intelligent vehicles encompass complex interactions among human driver, traffic environment, cognition, sensing, and automatic control systems. Intelligent driving has been gaining increasing attention due to the great potential for enhancing vehicular safety and performance as well as overall traffic efficiency.

Data is at the heart of the future of mobility, requiring the private and public sectors to agree on standards, legal frameworks, and financial terms for secure, robust data exchange. At the same time, issues about personal privacy and cybersecurity are growing and remain a priority.

In cities, agencies might use policy and regulatory tools (including fees and pricing) to encourage greater overall throughput and reduced congestion, for example, or to incentivize or mandate access to underserved communities. In a freight supply network, companies or organizations might limit delivery speed, transparency, and flexibility to better match loads to vehicles across the network.

Regarding market structures for new services and technologies, should governments proactively create policy, legislation, and regulation that set guidelines within which the private sector must act, or allow businesses a more open market-based approach to drive the pace of innovation and let regulation follow? Governments can set the agenda and articulate a vision but let the private sector look for the best way to realize that vision using a light-touch regulatory approach.

It is required to develop methods which are robust to background clutter and produce good categorization performance. For example, ensemble methods function with more information for the object classes. Greater improvement could be gained through the use of different clustering algorithms to capture various structures [26]. With different types of image presentations, a classification ensemble could be learned based on the different expression data sets from the same training image set. The use of a classification ensemble to categorize new images can lead to improved performance. Ensemble approaches must be further improved to have better resistance to variations in view, lighting, and occlusion.

It is foreseen that it is not possible to have a smart city without smart transportation. And in densely populated places such as a city, the importance of ensuring millions of people can move around smoothly and efficiently is very clear.

Technology holds the key to driving smarter and more seamless transport infrastructure and traffic management, enabling the future of the city to grow with transport that works for all. The future of urban mobility depends on better understanding places and the movement of people to shape strategies that are built around how we live. It is suggested that building a satellite connected transport network for smarter journeys would be essential [27].

There are many future tasks such as research on controllers to address the obstacle avoidance for an intelligent vehicle; an object classification method based on deep learning for autonomous vehicles; integration of electric vehicles, smart loads, and distributed renewables into smart grids; real-time energy management strategies sudden change in system configuration; coordination of transportation and power networks; standards development; and business modeling.

20.3.4 Smart Buildings

The huge concern in increased greenhouse gas emissions, and evolving laws and regulations leads to a focus in smart building to reduce energy demand and improve energy efficiency.

Smart building is the basic building block that will enable a true transformation of our cities through which one can have safe, sustainable, connected, and smart environments for the majority of the world's population as the buildings get smarter. There are various supporting technologies and requirements of smart buildings for smart cities. When equipped with smart home systems, data-driven energy modeling and simulation techniques can significantly transfer the world of smart cities.

It is essential to consider not only the smart building environmental impact, but also its overall social performance and cost reduction over the building's lifecycle. For buildings, one of the future directions is to enhance further connectivity across applications and data to transform buildings into dynamic and efficient high performance facilities. Building owners, developers, real estate managers, and investors can support connected solutions to gather information from structural integrity to occupant behavior and implement value-adding improvements.

According to [28], the global market for smart buildings is expected to grow at a compound annual growth rate (CAGR) of 32%, reaching $43 billion by 2022. To gain insight into how the industry and companies are achieving smart building sustainability goals in a complex, challenging situation that is undergoing increasing and evolving environmental issues, regulations, and expectations, it is important to drive innovation and technology deployments that will ultimately enable vital data collection and analysis needed to assess, monitor, and maintain sustainability in a smart building and across campus environments, smart communities, and smart cities as a whole.

Smart buildings technology deployment is not an easy exercise in plug-and-play, a systematic approach would be helpful. IoT-based solutions require integrating thousands of connected devices and networks to traditional building hardware with data analytics, machine learning, and mobile applications.

Due to this complexity, for the future, the following must be considered:

- Define the main objectives of smart building enhancement efforts.
- Collaborate with technology teams to ensure design plans reflect business goals,
- Plan, design, and choose the technology that can maximize the investment and minimize any potential risk.

Future effort must support and encourage researchers, engineers, managers, academics and practitioners who collaborate in a multidisciplinary way to converge toward a unique, complex, adaptable, marketable, and resilient global solution. A practical aid should also be developed for decision-makers to push for low-carbon economy, supported by scientific contributions, which make smart buildings for smart cities feasible.

Policy and business models must be formulated based on multidisciplinary research for novel, scientific, technological insights, principles, algorithms, and experiences on technologies, case studies, novel approaches, and visionary ideas related to data-driven innovative solutions and big data-powered applications to cope with the real world challenges for smart buildings.

Topics from now and beyond must include the following:

- Big data collection and analysis for smart and connected buildings.
- Social computing big data and networks for smart buildings.
- Big data security and privacy for smart buildings.
- Practical deployment and case studies, and industry applications.
- Net zero energy buildings.
- Architectures and protocols for smart buildings.
- Building management system.
- Power electronic for management of energy flows in buildings.
- Sensors networks.
- Prosumer side management.
- Smart homes .
- Blockchain .
- Cyber-physical systems and society.
- Reliability, security, privacy, and trust.
- Applications, business, and social issues.
- Data fusion strategies for energy efficiency for individual buildings.
- Building certificate.
- Standards development.
- Policy formulation.
- Data-driven simulation for energy-consumption prediction in smart buildings.
- Artificial intelligence-based future smart building modeling and simulation.

References

1 (n.d.). The Rockefeller Foundation Economic Council on Planetary Health. The ROCKEFELLER FOUNPATION. [Online]. https://www.planetaryhealth.ox.ac.uk/ (accessed 22 May 2021).

2 (2021). Net Zero by 2050: A Roadmap for the Global Energy Sector. IEA Report, International Energy Agency. https://www.iea.org/reports/net-zero-by-2050 (accessed 22 May 2021).

3 Lai, C.S., Taylor, G., Darwish, M., and Locatelli, G. (2021). Written evidence - Role of batteries and fuel cells in achieving net zero. UK Parliament (15 April). https://committees.parliament.uk/writtenevidence/25232/pdf/ (accessed 5 September 2021).
4 Lai, C.S. and Locatelli, G. (2021). Are energy policies for supporting low-carbon power generation killing energy storage? *Journal of Cleaner Production* 280: 124626, https://doi.org/10.1016/j.jclepro.2020.124626.
5 Gissey, G.C., Subkhankulova, D., Dodds, P.E., and Barrett, M. (2019). Value of energy storage aggregation to the electricity system. *Energy Policy* 128: 685–696.
6 Anuta, O.H., Taylor, P., Jones, D. et al. (2014). An international review of the implications of regulatory and electricity market structures on the emergence of grid scale electricity storage. *Renewable and Sustainable Energy Reviews* 38: 489–508, https://doi.org/10.1016/j.rser.2014.06.006.
7 Nord Pool AS (n.d.). N2EX day ahead auction prices: February 2009–March 2009. NORDPOOL. [Online]. https://www.nordpoolgroup.com/Market-data1/GB/Auction-prices/UK/Hourly/?view=table (accessed 13 March 2021).
8 Potau, X., Leistner, S., and Morrison, G. (2018). Battery promoting policies in selected member states. N° ENER C2/2015-410, European Commission. [Online]. https://ec.europa.eu/energy/sites/ener/files/policy_analysis_-_battery_promoting_policies_in_selected_member_states.pdf (accessed 3 May 2021).
9 (n.d.). Green Deal: Energy saving for your home. Gov.UK. [Online]. https://www.gov.uk/green-deal-energy-saving-measures (accessed 6 July 2021).
10 Li, X., Chalvatzis, K., and Stephanides, P. (2018). Innovative energy islands: life-cycle cost-benefit analysis for battery energy storage. *Sustainability* 10: 3371.
11 International Renewable Energy Agency (2017). Electricity storage and renewables: Costs and markets to 2030. International Renewable Energy Agency. [Online]. https://www.irena.org/-/media/Files/IRENA/Agency/Publication/2017/Oct/IRENA_Electricity_Storage_Costs_2017.pdf (accessed 3 August 2021).
12 Office of Energy Saver (n.d.). Balance-of-system equipment required for renewable energy systems. Energy Saver, energy.gov [Online]. https://www.energy.gov/energysaver/balance-system-equipment-required-renewable-energy-systems (accessed 4 July 2021).
13 Lai, C.S., Jia, Y., Lai, L.L. et al. (2017). A comprehensive review on large-scale photovoltaic system with applications of electrical energy storage. *Renewable and Sustainable Energy Reviews* 78: 439–451.
14 National Grid (n.d.). Firm frequency response (FFR). National Grid ESO. [Online]. https://www.nationalgrideso.com/balancing-services/frequency-response-services (accessed 4 November 2021).
15 National Grid (n.d.). Short term operating reserve (STOR). National Grid ESO. [Online]. https://www.nationalgrideso.com/balancing-services/reserve-services/short-term-operating-reserve-stor (accessed 19 November 2021).
16 National Grid (n.d.). Fast reserve. National GridESO. [Online]. https://www.nationalgrideso.com/balancing-services/reserve-services/fast-reserve (accessed 12 August 2021).
17 Office of Gas and Electricity Markets (2012). Distribution connection and use of system agreement (DCUSA) DCP108- Availability of the non-intermittent generator tariff. Office of Gas and Electricity Markets. [Online]. https://www.ofgem.gov.uk/ofgem-publications/62523/dcp108d-pdf (accessed 19 January 2021).
18 Electricity Settlement Company (2018). G17 - Capacity provider payments EMRS guidance. EMR Settlement Limited, Electricity Settlements Company. [Online]. https://www.emrsettlement.co.uk/documentstore/guidance/g17-capacity-provider-payments.pdf (accessed 1 April 2021).
19 Lai, L.L. (ed.) (2001). *Power System Restructuring and Deregulation: Trading, Performance and Information Technology*. UK: John Wiley & Sons.

20 Lai, C.S., Locatelli, G., Pimm, A. et al. (2021). A review on long-term electrical power system modeling with energy storage. *Journal of Cleaner Production* 280: 124298, https://doi.org/10.1016/j.jclepro.2020.124298.

21 UN News (2013). Global shortage of health workers expected to keep grow agency warns, United Nations. [Online]. https://news.un.org/en/story/2013/11/455122-global-shortage-health-workers-expected-keep-growing-un-agency-warns (accessed 5 September 2021).

22 NCS Pte Ltd (2021). Make extraordinary happen - smarter healthcare for a healthier future. https://www.ncs.co/extraordinary?gclid=EAIaIQobChMIovWhkKL38gIVLdxMAh1YXwQtEAAYASAAEgIKq_D_BwE&gclsrc=aw.ds#healthcare?utm_source=sa360&utm_medium=paid-search&utm_campaign=dentsu-brand-awareness-ncsmakeextraordinaryhappen-healthcare-transport-applications&utm_term=searchads (accessed 10 September 2021).

23 Dong owner. https://unsplash.com/s/photos/dog-owner (accessed 10 Sepetember 2021).

24 Pet Tech Market (2021). Global Market Insights. https://www.gminsights.com/industry-analysis/pet-tech-market (accessed 7 September 2021).

25 Corwin, S., Zarif, R., Berdichevskiy, A., and Pankratz, D. (2020). The futures of mobility after COVID-19 Scenarios for transportation in a post coronavirus world. 22 May. https://www2.deloitte.com/us/en/insights/economy/covid-19/future-of-mobility-after-covid-19-transportation-scenarios.html (accessed 6 June 2021).

26 Luo, H.-L., Wei, H., and Lai, L.L. (2011). Creating efficient visual codebook ensembles for object categorization. *IEEE Transactions on Systems, Man and Cybernetics: Part A: Systems and Humans* 41 (2): 238–253.

27 NCS Pte Ltd (2021). Make extraordinary happen - taking smart traffic in a new direction. https://www.ncs.co/extraordinary?gclid=EAIaIQobChMIovWhkKL38gIVLdxMAh1YXwQtEAAYASAAEgIKq_D_BwE&gclsrc=aw.ds#healthcare?utm_source=sa360&utm_medium=paid-search&utm_campaign=dentsu-brand-awareness-ncsmakeextraordinaryhappen-healthcare-transport-applications&utm_term=searchads (accessed 10 September 2021).

28 PwC (n.d.). The future of the smart building: deploying technology to enhance operations and occupant experience. [Online]. https://www.pwc.com/us/en/industries/capital-projects-infrastructure/library/smart-building-tech-deployment.html (accessed 5 September 2021).

Index

Smart Energy for Transportation and Health in a Smart City, First Edition. Chun Sing Lai, Loi Lei Lai and Qi Hong Lai.

c

d

j

k

l

q

t

IEEE Press Series on Power and Energy Systems

Series Editor: Ganesh Kumar Venayagamoorthy, Clemson University, Clemson, South Carolina, USA.

The mission of the IEEE Press Series on Power and Energy Systems is to publish leading-edge books that cover a broad spectrum of current and forward-looking technologies in the fast-moving area of power and energy systems including smart grid, renewable energy systems, electric vehicles and related areas. Our target audience includes power and energy systems professionals from academia, industry and government who are interested in enhancing their knowledge and perspectives in their areas of interest.

1. *Electric Power Systems: Design and Analysis, Revised Printing*
   Mohamed E. El-Hawary

2. *Power System Stability*
   Edward W. Kimbark

3. *Analysis of Faulted Power Systems*
   Paul M. Anderson

4. *Inspection of Large Synchronous Machines: Checklists, Failure Identification, and Troubleshooting*
   Isidor Kerszenbaum

5. *Electric Power Applications of Fuzzy Systems*
   Mohamed E. El-Hawary

6. *Power System Protection*
   Paul M. Anderson

7. *Subsynchronous Resonance in Power Systems*
   Paul M. Anderson, B.L. Agrawal, and J.E. Van Ness

8. *Understanding Power Quality Problems: Voltage Sags and Interruptions*
   Math H. Bollen

9. *Analysis of Electric Machinery*
   Paul C. Krause, Oleg Wasynczuk, and S.D. Sudhoff

10. *Power System Control and Stability, Revised Printing*
    Paul M. Anderson and A.A. Fouad

11. *Principles of Electric Machines with Power Electronic Applications*, Second Edition
    Mohamed E. El-Hawary

12. *Pulse Width Modulation for Power Converters: Principles and Practice*
    D. Grahame Holmes and Thomas Lipo

13. *Analysis of Electric Machinery and Drive Systems*, Second Edition
    Paul C. Krause, Oleg Wasynczuk, and S.D. Sudhoff

14. *Risk Assessment for Power Systems: Models, Methods, and Applications*
    Wenyuan Li

15. *Optimization Principles: Practical Applications to the Operations of Markets of the Electric Power Industry*
    Narayan S. Rau

16. *Electric Economics: Regulation and Deregulation*
    Geoffrey Rothwell and Tomas Gomez

17. *Electric Power Systems: Analysis and Control*
    Fabio Saccomanno

18. *Electrical Insulation for Rotating Machines: Design, Evaluation, Aging, Testing, and Repair*
    Greg C. Stone, Edward A. Boulter, Ian Culbert, and Hussein Dhirani

19. *Signal Processing of Power Quality Disturbances*
    Math H. J. Bollen and Irene Y. H. Gu

20. *Instantaneous Power Theory and Applications to Power Conditioning*
    Hirofumi Akagi, Edson H. Watanabe, and Mauricio Aredes

21. *Maintaining Mission Critical Systems in a 24/7 Environment*
    Peter M. Curtis

22. *Elements of Tidal-Electric Engineering*
    Robert H. Clark

23. *Handbook of Large Turbo-Generator Operation and Maintenance,* Second Edition
    Geoff Klempner and Isidor Kerszenbaum

24. *Introduction to Electrical Power Systems*
    Mohamed E. El-Hawary

25. *Modeling and Control of Fuel Cells: Distributed Generation Applications*
    M. Hashem Nehrir and Caisheng Wang

26. *Power Distribution System Reliability: Practical Methods and Applications*
    Ali A. Chowdhury and Don O. Koval

27. *Introduction to FACTS Controllers: Theory, Modeling, and Applications*
    Kalyan K. Sen and Mey Ling Sen

28. *Economic Market Design and Planning for Electric Power Systems*
    James Momoh and Lamine Mili

29. *Operation and Control of Electric Energy Processing Systems*
    James Momoh and Lamine Mili

30. *Restructured Electric Power Systems: Analysis of Electricity Markets with Equilibrium Models*
    Xiao-Ping Zhang

31. *An Introduction to Wavelet Modulated Inverters*
    S.A. Saleh and M.A. Rahman

32. *Control of Electric Machine Drive Systems*
    Seung-Ki Sul

33. *Probabilistic Transmission System Planning*
    Wenyuan Li

34. *Electricity Power Generation: The Changing Dimensions*
Digambar M. Tagare

35. *Electric Distribution Systems*
Abdelhay A. Sallam and Om P. Malik

36. *Practical Lighting Design with LEDs*
Ron Lenk and Carol Lenk

37. *High Voltage and Electrical Insulation Engineering*
Ravindra Arora and Wolfgang Mosch

38. *Maintaining Mission Critical Systems in a 24/7 Environment*, Second Edition
Peter Curtis

39. *Power Conversion and Control of Wind Energy Systems*
Bin Wu, Yongqiang Lang, Navid Zargari, and Samir Kouro

40. *Integration of Distributed Generation in the Power System*
Math H. Bollen and Fainan Hassan

41. *Doubly Fed Induction Machine: Modeling and Control for Wind Energy Generation Applications*
Gonzalo Abad, Jesús López, Miguel Rodrigues, Luis Marroyo, and Grzegorz Iwanski

42. *High Voltage Protection for Telecommunications*
Steven W. Blume

43. *Smart Grid: Fundamentals of Design and Analysis*
James Momoh

44. *Electromechanical Motion Devices*, Second Edition
Paul Krause, Oleg Wasynczuk, Steven Pekarek

45. *Electrical Energy Conversion and Transport: An Interactive Computer-Based Approach*, Second Edition
George G. Karady and Keith E. Holbert

46. *ARC Flash Hazard and Analysis and Mitigation*
J.C. Das

47. *Handbook of Electrical Power System Dynamics: Modeling, Stability, and Control*
Mircea Eremia and Mohammad Shahidehpour

48. *Analysis of Electric Machinery and Drive Systems, Third Edition*
Paul C. Krause, Oleg Wasynczuk, S.D. Sudhoff, and Steven D. Pekarek

49. *Extruded Cables for High-Voltage Direct-Current Transmission: Advances in Research and Development*
Giovanni Mazzanti and Massimo Marzinotto

50. *Power Magnetic Devices: A Multi-Objective Design Approach*
S.D. Sudhoff

51. *Risk Assessment of Power Systems: Models, Methods, and Applications*, Second Edition
Wenyuan Li

52. *Practical Power System Operation*
Ebrahim Vaahedi

53. *The Selection Process of Biomass Materials for the Production of Bio-Fuels and Co-Firing*
Najib Altawell

54. *Electrical Insulation for Rotating Machines: Design, Evaluation, Aging, Testing, and Repair*, Second Edition
Greg C. Stone, Ian Culbert, Edward A. Boulter, and Hussein Dhirani

55. *Principles of Electrical Safety*
Peter E. Sutherland

56. *Advanced Power Electronics Converters: PWM Converters Processing AC Voltages*
Euzeli Cipriano dos Santos Jr. and Edison Roberto Cabral da Silva

57. *Optimization of Power System Operation*, Second Edition
Jizhong Zhu

58. *Power System Harmonics and Passive Filter Designs*
J.C. Das

59. *Digital Control of High-Frequency Switched-Mode Power Converters*
Luca Corradini, Dragan Maksimoviæ, Paolo Mattavelli, and Regan Zane

60. *Industrial Power Distribution*, Second Edition
Ralph E. Fehr, III

61. *HVDC Grids: For Offshore and Supergrid of the Future*
Dirk Van Hertem, Oriol Gomis-Bellmunt, and Jun Liang

62. *Advanced Solutions in Power Systems: HVDC, FACTS, and Artificial Intelligence*
Mircea Eremia, Chen-Ching Liu, and Abdel-Aty Edris

63. *Operation and Maintenance of Large Turbo-Generators*
Geoff Klempner and Isidor Kerszenbaum

64. *Electrical Energy Conversion and Transport: An Interactive Computer-Based Approach*
George G. Karady and Keith E. Holbert

65. *Modeling and High-Performance Control of Electric Machines*
John Chiasson

66. *Rating of Electric Power Cables in Unfavorable Thermal Environment*
George J. Anders

67. *Electric Power System Basics for the Nonelectrical Professional*
Steven W. Blume

68. *Modern Heuristic Optimization Techniques: Theory and Applications to Power Systems*
Kwang Y. Lee and Mohamed A. El-Sharkawi

69. *Real-Time Stability Assessment in Modern Power System Control Centers*
Savu C. Savulescu

70. *Optimization of Power System Operation*
Jizhong Zhu

71. *Insulators for Icing and Polluted Environments*
Masoud Farzaneh and William A. Chisholm

72. *PID and Predictive Control of Electric Devices and Power Converters Using MATLAB®/Simulink®*
Liuping Wang, Shan Chai, Dae Yoo, and Lu Gan, Ki Ng

73. *Power Grid Operation in a Market Environment: Economic Efficiency and Risk Mitigation*
Hong Chen

74. *Electric Power System Basics for Nonelectrical Professional*, Second Edition
Steven W. Blume

75. *Energy Production Systems Engineering*
Thomas Howard Blair

76. *Model Predictive Control of Wind Energy Conversion Systems*
Venkata Yaramasu and Bin Wu

77. *Understanding Symmetrical Components for Power System Modeling*
J.C. Das

78. *High-Power Converters and AC Drives*, Second Edition
Bin Wu and Mehdi Narimani

79. *Current Signature Analysis for Condition Monitoring of Cage Induction Motors: Industrial Application and Case Histories*
William T. Thomson and Ian Culbert

80. *Introduction to Electric Power and Drive Systems*
Paul Krause, Oleg Wasynczuk, Timothy O'Connell, and Maher Hasan

81. *Instantaneous Power Theory and Applications to Power Conditioning*, Second Edition
Hirofumi, Edson Hirokazu Watanabe, and Mauricio Aredes

82. *Practical Lighting Design with LEDs*, Second Edition
Ron Lenk and Carol Lenk

83. *Introduction to AC Machine Design*
Thomas A. Lipo

84. *Advances in Electric Power and Energy Systems: Load and Price Forecasting*
Mohamed E. El-Hawary

85. *Electricity Markets: Theories and Applications*
Jeremy Lin and Jernando H. Magnago

86. *Multiphysics Simulation by Design for Electrical Machines, Power Electronics and Drives*
Marius Rosu, Ping Zhou, Dingsheng Lin, Dan M. Ionel, Mircea Popescu, Frede Blaabjerg, Vandana Rallabandi, and David Staton

87. *Modular Multilevel Converters: Analysis, Control, and Applications*
Sixing Du, Apparao Dekka, Bin Wu, and Navid Zargari

88. *Electrical Railway Transportation Systems*
Morris Brenna, Federica Foiadelli, and Dario Zaninelli

89. *Energy Processing and Smart Grid*
James A. Momoh

90. *Handbook of Large Turbo-Generator Operation and Maintenance*, 3rd Edition
Geoff Klempner and Isidor Kerszenbaum

91. *Advanced Control of Doubly Fed Induction Generator for Wind Power Systems*
Dehong Xu, Dr. Frede Blaabjerg, Wenjie Chen, and Nan Zhu

92. *Electric Distribution Systems*, 2nd Edition
Abdelhay A. Sallam and Om P. Malik

93. *Power Electronics in Renewable Energy Systems and Smart Grid: Technology and Applications*
Bimal K. Bose

94. *Distributed Fiber Optic Sensing and Dynamic Rating of Power Cables*
Sudhakar Cherukupalli and George J Anders

95. *Power System and Control and Stability,* Third Edition
Vijay Vittal, James D. McCalley, Paul M. Anderson, and A. A. Fouad

96. *Electromechanical Motion Devices: Rotating Magnetic Field-Based Analysis and Online Animations,* Third Edition
Paul Krause, Oleg Wasynczuk, Steven D. Pekarek, and Timothy O'Connell

97. *Applications of Modern Heuristic Optimization Methods in Power and Energy Systems*
Kwang Y. Lee and Zita A. Vale

98. *Handbook of Large Hydro Generators: Operation and Maintenance*
Glenn Mottershead, Stefano Bomben, Isidor Kerszenbaum, and Geoff Klempner

99. *Advances in Electric Power and Energy: Static State Estimation*
Mohamed E. El-hawary

100. *Arc Flash Hazard Analysis and Mitigation*, Second Edition
J.C. Das

101. *Maintaining Mission Critical Systems in a 24/7 Environment,* Third Edition
Peter M. Curtis

102. *Real-Time Electromagnetic Transient Simulation of AC-DC Networks*
Venkata Dinavahi and Ning Lin

103. *Probabilistic Power System Expansion Planning with Renewable Energy Resources and Energy Storage Systems*
Jaeseok Choi and Kwang Y. Lee

104. *Power Magnetic Devices: A Multi-Objective Design Approach*, Second Edition
Scott D. Sudhoff

105. *Optimal Coordination of Power Protective Devices with Illustrative Examples*
Ali R. Al-Roomi

106. *Resilient Control Architectures and Power Systems*
Craig Rieger, Ronald Boring, Brian Johnson, and Timothy McJunkin

107. *Alternative Liquid Dielectrics for High Voltage Transformer Insulation Systems: Performance Analysis and Applications*
Edited by U. Mohan Rao, I. Fofana, and R. Sarathi

108. *Introduction to the Analysis of Electromechanical Systems*
Paul C. Krause, OlegWasynczuk, and Timothy O'Connell

109. *Power Flow Control Solutions for a Modern Grid using SMART Power Flow Controllers*
Kalyan K. Sen and Mey Ling Sen

110. *Power System Protection: Fundamentals and Applications*
John Ciufo and Aaron Cooperberg

111. *Soft-Switching Technology for Three-phase Power Electronics Converters*
Dehong Xu, Rui Li, Ning He, Jinyi Deng, and YuyingWu

112. *Power System Protection*, Second Edition
Paul M. Anderson, Charles Henville, Rasheek Rifaat, Brian Johnson, and Sakis Meliopoulos

113. *High Voltage and Electrical Insulation Engineering*, Second Edition
Ravindra Arora and Wolfgang Mosch

114. *Modeling and Control of Modern Electrical Energy Systems*
Masoud Karimi-Ghartemani

115. *Wireless Power Transfer*
Zhen Zhang and Hongliang Pang

116. *Coordinated Operation and Planning of Modern Heat and Electricity Incorporated Networks*
Mohammadreza Daneshvar, Behnam Mohammadi-Ivatloo, and Kazem Zare

117. *Smart Energy for Transportation and Health in a Smart City*
Chun Sing Lai, Loi Lei Lai, and Qi Hong Lai

Printed and bound by CPI Group (UK) Ltd, Croydon, CR0 4YY

16/04/2025

14658352-0005